Verticillium Wilts

This monograph is dedicated to the memories of I. Isaac,
W.G. Keyworth, P.W. Talboys and I.W. Selman.

VERTICILLIUM WILTS

G.F. Pegg
Professor Emeritus
School of Plant Sciences
University of Reading, UK

and

B.L. Brady
Formerly of the
International Mycological Institute
UK

CABI *Publishing*

CABI *Publishing* is a division of CAB *International*

CABI Publishing
CAB International
Wallingford
Oxon OX10 8DE
UK

Tel: +44 (0)1491 832111
Fax: +44 (0)1491 833508
Email: cabi@cabi.org
Web site: www.cabi-publishing.org

CABI Publishing
10 E 40th Street
Suite 3203
New York, NY 10016
USA

Tel: +1 212 481 7018
Fax: +1 212 686 7993
Email: cabi-nao@cabi.org

A catalogue record for this book is available from the British Library, London, UK.

Library of Congress Cataloging-in-Publication Data

Pegg, G. F. (George Frederick), 1930-
Verticillium wilts / G.F. Pegg and B.L. Brady.
p. cm.
Includes bibliographical references (p.).
ISBN 0-85199-529-2 (alk. paper)
1. Verticillium. 2. Wilt diseases. I. Brady, B. L. (Beryl Ledsom) II. Title.

SB741.V45 P44 2002
632'.45--dc21

2001037313

ISBN 0 85199 529 2

Typeset in Photina by Columns Design Ltd, Reading
Printed and bound in the UK by Cromwell Press, Trowbridge

Contents

Acknowledgements

The authors wish to thank the Directors of the International Mycological Institute (now CABI *Bioscience*) and the Royal Botanic Gardens, Kew for permission to use their libraries. B.L. Brady is particularly indebted to Dr D. Minter of CABI *Bioscience*, whose expertise with computers and friendly indulgence have enabled her to carry on with the compilation of the bibliography. The authors express their gratitude to Rosalind and Alistair Feakes for painstaking attention to detail in the typing of the manuscript. G.F. Pegg expresses his special indebtedness to Mary Pegg and to Anne Burgess for their constant encouragement and help during the progress of the book.

Introduction 1

The genus *Verticillium* Nees represents one of the world's major pathogens, affecting crop plants mostly in the cool and warm temperate regions, but has also been reported from subtropical and tropical areas. There are some seven major pathogenic species affecting trees, herbaceous plants, plantation crops and mushrooms: *V. dahliae* Kleb., *V. albo-atrum* Reinke et Berth., *V. nigrescens* Pethybr., *V. nubilum* Pethybr., *V. tricorpus* Isaac., *V. theobromae* (Turc.) Mas. & Hughes and *V. fungicola* (Preuss) Hassebrauk (= *V. malthousei* Ware). Of these the polyphagous wilt pathogens, *V. dahliae* and *V. albo-atrum*, stand out in importance both agriculturally and in coverage in the scientific literature. *V. nigrescens, V. nubilum* and *V. tricorpus* are also wilt pathogens but, in general, are of less major importance. *V. theobromae* causes a fruit-rot of banana, and *V. fungicola* a devastating sporophore infection of the cultivated mushroom.

In the 185 years that have elapsed since Nees von Esenbeck erected the genus, no comprehensive review of the pathogenic verticillia has appeared. This is surprising in view of the great volume of published work that has followed Reinke and Berthold's description of the first wilt pathogen, *V. albo-atrum* in 1879. Rudolph (1931) and Englehard (1957) published extensive host lists, and Panton (1964) and Pegg (1974) short reviews. The five wilt-inducing species have been described by Hawksworth and Talboys (1970) and these, together with a number of soil saprophytic species, but omitting *V. tricorpus*, were also described by Domsch *et al.* (1981a,b).

In 1969, G.F. Pegg (Wye College, University of London) and P.W. Talboys (East Malling Research Station, UK) planned the first international meeting on *Verticillium* following proposals made at the International Botanical Congress, London 1968, for a *Verticillium* workshop. I. Isaac and W.G. Keyworth were invited to join in the formation of an *ad hoc* committee, the work of which led to the First International *Verticillium* Symposium at Wye College in 1971, with scientists from 18 countries. An International Standing Committee subsequently was established to organize future symposia and to serve as a forum for the dissemination of information and research collaboration, including the exchange of scientific workers. The proceedings of the first meeting were included in the review *Verticillium* diseases, by Pegg (1974). Subsequent Symposia were held in Berkeley, USA (Wilhelm, 1976); Bari, Italy, 1981 (proceedings and extended reviews of this meeting were published in Cirulli (1984), a special publication of the Mediterranean Phytopathological Union); Guelph, Canada, 1986; Leningrad, 1990; Dead Sea, Israel 1994 (abstracts published in *Phytoparasitica* (1995)) and the 7th Silver Jubilee Symposium at Cape Sounion, Greece, 1997. Expanded abstracts and posters from the 25th Jubilee Symposium were published by the American Phytopathological Society (Tjamos *et al.*, 2000). With the exception of the first, third and seventh symposia, there were no published proceedings in full, and symposial abstracts are available only in limited circulation. The 8th International *Verticillium* Symposium was held in Cordoba, Spain in 2001. (See note added in proof p. 539.)

For the remainder, accounts of *Verticillium* research have been incorporated into general reviews or books on vascular wilts. There have been a number of these. Immediately preceding the 1st *Verticillium* Symposium, an International Meeting on Pathological Wilting of Plants was held in Madras, India (1971). This was devoted largely to *Fusarium* but included *inter alia* studies on *Verticillium* (Sadasivan *et al.*, 1978). A NATO meeting held in Greece, 1989, followed a similar pattern but included prokaryote wilt-inducing pathogens and, for the first time in a wilt symposium, molecular genetics aspects of the organisms. One-third of the papers were devoted to *Verticillium* (Tjamos and Beckman, 1989).

Specialist conference proceedings on cotton dealing with all aspects of *Verticillium* wilt can be found in Proceedings of the Beltwide Cotton Research Conferences: e.g. Hot Springs, Arkansas (1968); Atlanta, Georgia (1971); and Lubbock, Texas (1973). General reviews dealing with different aspects of *Fusarium* and *Verticillium* pathogenesis have been written by Dimond and Waggoner (1953a,b); Waggoner and Dimond (1954); Talboys (1964, 1968); Sadasivan (1961); Dimond (1955, 1970 and 1972); and Pegg (1985). The most comprehensive treatment of fungal wilt diseases and their pathogens including *Verticillium* is *Fungal Wilt Diseases of Plants* (Mace *et al.*, 1981). The chapters in this deal extensively with 'Life cycle and epidemiology'; 'Genetics and biochemistry of the pathogen'; 'Biochemistry and physiology of pathogenesis'; 'Water relations'; 'Sources and genetics of host resistance in field, fruit and vegetable crops and shade trees'; 'Biochemistry and physiology of resistance'; 'Anatomy of resistance'; and 'Biological and chemical control'. While *Verticillium* receives much attention, the literature reviewed is far from comprehensive.

Verticillium wilt pathogens are also dealt with by Beckman (1987) in the book *The Nature of Fungal Wilt Diseases of Plants*. This text, which originally was intended as a revision of J.C. Walkers' 1971 monograph on *Fusarium* wilt of tomato, is largely biased towards this genus. The book deals analytically with infection, determinative phases (establishment), expression of disease (pathogenesis), genetic variation of host and pathogen, and the role of environmental factors in disease development and control. There are 600 references on wilt pathogens, but no index. Selective reference to *Verticillium* diseases of tropical crops is made by Holliday (1980) and a condensed account of *Verticillium* wilts in *The Dictionary of Plant Pathology* (Holliday, 1989). *Verticillium* spp. as plant pathogens have also been described in numerous publications as minor references (Phillips and Burdekin, 1983), or in occasional papers dealing with specific hosts (Pethybridge, 1916).

In this monograph, the pathogenic wilt-inducing *Verticillium* spp. are considered under the following headings: 'Taxonomy', 'Morphogenesis and Morphology', 'Cytology and Genetics', 'Aetiology', 'Ecology', 'Physiology and Metabolism', 'Pathogenesis', 'Resistance', 'Control' (physical, chemical, biological, integrated, legislation and quarantine, and resistance breeding), 'Hosts' and 'Techniques and Methodology'. Many host responses are common to susceptible and resistant hosts alike, making impossible a clear demarcation between sections on resistance and pathogenicity. Similarly, with other sections, there is overlap which has necessitated duplication and cross-referencing. For many years, North American scientists failed to recognize *V. dahliae* as a valid species, referring to Ms and Dm strains of *V. albo-atrum*. This led to much confusion in subsequent citations. Following the 1976 *Verticillium* Symposium, there was general international agreement to recognize the five species, including *V. dahliae* as described by Isaac (1949, 1953b). Where the original author clearly indicated the microsclerotial form, it has been referred to as *V. dahliae sensu* Kleb. Quanjer (1916) introduced the term tracheomycosis, and Pethybridge (1916) hadromycosis for a fungus confined to the xylem (hadrome). These terms are still in occasional use today and are cited where appropriate. Literature cited in the text is complete to December 2000. Studies in which *Verticillium* features only as a minor topic or as one of several test organisms have not been included.

The output of publications on *Verticillium* over the last 50 years has been exponential. While the monograph has been extensive in cover of the literature, it is by no means exhaustive. The numbers of publications from different countries are frequently proportional to the importance of particular susceptible crops to their national economies. They also reflect the evolution of scientific research in developing and (some developed) countries. At the start of the Third Millennium (2001), it is still apparent in worldwide literature that the pressure to achieve publication targets takes precedence over the creation of original and innovative research. It is thus sad to record in 2000 (and earlier) the publication of papers repeating and confirming results achieved 20 and 30 years before.

It is perhaps a vain hope that future reference to a history of *Verticillium* research would avoid such repetitive studies. Indeed, a principal objective of the monograph was to produce in one volume a discursive compendium of information on *Verticillium* to enable young (and older) research workers to see what has already been achieved and to identify the many new areas of research in which *original* contributions could be made to future our understanding and control of this most important pathogen and its diseases.

Taxonomy 2

Throughout the monograph, it has been difficult to avoid overlap in subject matter between different chapters. This is particularly a problem with multi-subject papers. Aspects of taxonomy may also be found in Chapters 3, 4 and 7 dealing with Morphogenesis and Morphology, Cytology and Genetics, and Physiology and Metabolism, and others, especially in the separation of *V. dahliae* and *V. albo-atrum*. Where a biochemical study is designed specifically with taxonomic objectives, it is considered under taxonomy.

Verticillium Nees 1817

V. albo-atrum Reinke & Berthold (1879)
  = *V. albo-atrum* var. *caespitosum* Wollenweber (1929)
  = *V. albo-atrum* var. *caespitosum* f. *pallens* Wollenweber (1929)
  = *V. albo-atrum* var. *tuberosum* Rudolph (1931)
V. dahliae Klebahn (1913)
  = *V. dahliae* var. *longisporum* Stark (1961)
  = *V. albo-atrum* var. *medium* Wollenweber (1929)
  = *V. albo-atrum* auct. *pro parte*
  = *V. ovatum* Berkeley & Jackson (1926)
V. nigrescens Pethybridge (1919)
V. nubilum Pethybridge (1919)
V. theobromae (Turconi) Mason & Hughes in Hughes (1951)
V. tricorpus Isaac (1953)
V. intertextum Isaac & Davies (1955)

V. longisporum Karapapa Stark (1997)
= *V. dahliae* var. *longisporum* Stark (1961)

The following taxa, either invalidly or inadequately described, are referred to in the literature and probably belong in *V. dahliae*:

V. traceiphyllum Curzi (1925)
V. armoricae Klebahn (1937)
V. albo-atrum var. *dahliae* Nelson (1950)
V. albo-atrum var. *menthae* Nelson (1950)
V. dahliae forma *cerebriforme* van Beyma (1940)
V. dahliae forma *restrictus* van Beyma (1940)
V. dahliae forma *zonatum* van Beyma (1940)

V. fumosum Seman (1968), isolated from cotton, is not readily recognizable from the illustrations, but is also quoted in the literature (e.g. Kuznetzov, 1979; Muromtsev and Strunnikova, 1981; Strunnikova and Muromtsev, 1984, 1987).

Verticillium Nees is a genus of the Deuteromycotina characterized by conidiophores which, when branched, bear these branches in whorls, and where the conidiogenous cells are themselves disposed several at one level forming whorls which frequently are reduced to single or paired cells. A few species have been shown to have ascomycete teleomorphs. More than 50 species have been described and include groups of species parasitizing insects, nematodes and other fungi and, in particular, dicotyledonous plants, where they are among those fungi causing wilt diseases (Schippers and Gams, 1979).

Gams and van Zaayen (1982) proposed the section Nigrescentia for those species of the genus with dark resting structures, either dark inflated hyphae, dark conidiophores or microsclerotia, thus comprising the six species listed above. Only *V. albo-atrum*, *V. dahliae*, *V. tricorpus* and, to a lesser extent, *V. nigrescens* cause wilt diseases and are examined in detail here, although *V. nubilum*, which is associated with 'coiled sprout' disorder of potato, is so often included in surveys of this group of fungi that it will be mentioned frequently. *V. theobromae*, similarly a non-wilt pathogen, is responsible for a rot of banana fruit described as 'cigar end', and is not treated in detail here.

Nomenclature of the *Verticillium* Wilt Fungi

Much confusion has surrounded the identity and naming of the wilt-inducing species of *Verticillium* involving a controversy which is almost but not entirely resolved today. Isaac (1949, 1967) gives detailed accounts of the history of the argument which is summarized briefly here.

Reinke and Berthold (1879) described *V. albo-atrum*, the fungus causing potato wilt, as having dark brown to black resting mycelium which, by the pressing together of contiguous hyphae formed cellular masses he described as 'Dauermycelien', 'Sklerotien' or 'Zellhauf'.

In 1913, Klebahn isolated a fungus from wilted dahlia which differed from *V. albo-atrum* by forming true sclerotia by irregular septation in both transverse and longitudinal directions in three planes with budding cells which then darkened. Ever since the description of *V. dahliae*, there has been controversy as to whether one large variable species, one species with two or more varieties, or two distinct species are involved (see Hansen, 1938).

The dispute hinged on whether or not the fungus described by Reinke and Berthold formed true sclerotia. Klebahn (1913), Pethybridge (1919), Van der Meer (1925, see Chapter 11), Berkeley *et al.* (1931), Ludbrook (1933), Van Beyma (1940), Isaac (1949, 1967), Robinson *et al.* (1957), Smith (1965), Skadow (1969b) and Schnathorst (1973) (see also Schnathorst in Schnathorst *et al.* (1973)) agreed that it did not, and thus two species, *V. albo-atrum* and *V. dahliae*, were involved, while Wollenweber (1929, see Chapter 11), Rudolph (1931, see Chapter 11), Presley (1941), Wilhelm and Taylor (1965, see Chapter 10), Van den Ende (1958) and Brandt (1964a, see Chapter 4) considered that the fungi forming sclerotia and those forming dark resting mycelium were conspecific as *V. albo-atrum*. Wollenweber (1929) maintained that the sclerotial fungus was the usual form of *V. albo-atrum* and the fungus forming dark resting mycelium should be considered as a variety *caespitosum* of that species; Rudolph (1931, see Chapter 11) named a similar fungus var. *tuberosum*. Both of these varieties Isaac (1949) considered merely to be the original *V. albo-atrum* Reinke & Berth., while the sclerotial fungus belonged in *V. dahliae* Kleb.

Notwithstanding the absence of Reinke and Berthold's definitive type material for reference, the identity of *V. albo-atrum* and *V. dahliae* as separate species each with distinct characteristics of their own gradually became accepted.

The temperature difference for growth and survival for the microsclerotial (Ms) and dark resting mycelial (Dm) types of *Verticillium* constitutes the single most important character for the separation of *V. albo-atrum* and *V. dahliae* as biologically distinct species. The worldwide geographic distribution of the two species is based on this character. However, the practice with some authors, notably from the USA, former USSR states (now CIS; Commonwealth of independent states) and some developing countries, of grouping all such wilt pathogens collectively as *V. albo-atrum* still persists. It is always advisable, therefore, when referring to the literature to establish whether microsclerotia were recorded. If so, the fungus should be considered as *V. dahliae* regardless of the designation of '*V. albo-atrum*'. This rule has been followed here, and if any reference is made by the author of a chapter to microsclerotia the fungus is referred to here as *V. dahliae* even if '*V. albo-atrum*' appears in the title of the communication.

In a number of publications, the author, while designating the pathogen *V. albo-atrum*, has omitted description of diagnostic characters to permit a true judgement to be made. In such circumstances, a crude 'rule of thumb' guide is that if field studies are based on locations with a summer average ambient temperature of 25°C or greater, the pathogen in question is likely to be *V. dahliae* Kleb. The exception to this is the high temperature lucerne strain of *V. albo-atrum* R et

B (see Basu and Butler, 1991), but on lucerne the pathogen has always been identified correctly.

Verticillium nigrescens and *V. nubilum* were described by Pethybridge (1919) as forming chlamydospores only as dark resting structures, and *V. tricorpus* (Isaac, 1953b) forms resting mycelium, microsclerotia and chlamydospores (see also Isaac, 1953c). A number of other characters have been used, with greater or lesser success, in the classification of these fungi, and the literature is summarized below. Most studies have been concerned with the *V. albo-atrum* versus *V. dahliae* question, and information on the other three species is relatively limited.

Wollenweber (1929) isolated a pathogen from wilting carnation (*Dianthus caryophyllus*) with conidiophores which could be regarded as verticillate. On this basis, he erected the species *Verticillium cinerescens*. Several authorities, including Isaac (1949) and Garrett (1956), subsequently recognized this species. Van Beyma (1939, 1940) considered Wollenweber's fungus was not verticillate and belonged in the genus *Phialophora* Medlar. *P. cinerescens* (Wollen.) Van Beyma is now the recognized pathogen on carnation and *V. cinerescens* is invalid.

Various studies have been conducted using *Verticillium* antigens as inducers of blood serum antibodies. Antibodies have been coupled to fluorescent dyes or chemical markers to give a quantitative measure of specificity as in the enzyme-linked immunosorbent assay (ELISA). The degree of sharing of common antigens has been used as an indication of the relatedness of species or strains. In all the early studies, the problem has been compounded by the use of non-specific antibodies. Whitney *et al.* (1968), Hall (1969), Milton *et al.* (1971), Pelletier and Hall (1971) and Selvaraj and Meyer (1974) examined simple protein patterns of *V. dahliae* and *V. albo-atrum* by gel electrophoresis to resolve the then question of species separation. Greater dissimilarity was found in protein patterns between isolates of the two fungi than between isolates of either species. This was regarded as a basis for separating the two species. Teranisihi *et al.* (1973), however, found no serological resemblance between *V. tricorpus* and either *V. albo-atrum* or *V. dahliae*. Fitzell *et al.* (1980b) in similar gel diffusion studies showed very close affinities for *V. albo-atrum* and *V. dahliae*, while *V. nigrescens* and, to a lesser extent, *V. tricorpus* showed very little antigenic conformity with these species. The antisera in this work were derived from mycelial preparations which were considered less specific than those from conidia. Guseva (1972) considered the water soluble mycelial proteins as taxonomic indicators, Guseinov and Runov (1971) examined nucleic acids of various species. As subsequent research confirmed, neither of these substances could be used to separate species.

Studying the serological relationships of cotton verticillia and other species, Strunnikova and Muromtsev (1984) found common antigens a, b and c in all species of *Verticillium* except *V. nigrescens* from cotton. Antigens d, e, f, g and k were found in *V. dahliae*, *V. albo-atrum* and the Russian described species *V. fumosum*. *V. dahliae*, *V. albo-atrum* and *V. fumosum* were antigenically identical. *V. tricorpus*, *V. nubilum* and, to an even lesser extent, *V. lateritium* and *V. chlamydosporium* from soil showed antigenic identity. In a subsequent study

(Strunnikova and Muromtsev, 1987) using binary cross-immunoelectrophoresis on the same species with isolates from aubergine, pepper and cotton, considerable heterogeneity was found in antigenic composition of *V. albo-atrum* and *V. dahliae*. *V. fumosum* had few reacting antigens, and other species showed no cross-reactions with *V. dahliae* and *V. albo-atrum*. On this and previous evidence, the authors argued in favour of considering them a single species.

Intraspecific or strain identification in the laboratory presents an even greater challenge. This is particularly a problem with perennial and woody hosts in relation to plant breeding or quarantine and legislation, as in the hop where field testing takes 1 year to establish a strain or pathotype identity reliably.

Using immunoelectrophoresis, Wyllie and DeVay (1970b) compared the defoliating (P_1, formerly T_9) and non-defoliating (P_2, formerly SS_4) cotton strains of *V. dahliae* with the mildly pathogenic *V. nigrescens*. The two species differed, as did the two strains, but the P_2 strain appeared to be more closely related serologically to *V. nigrescens* than to the P_1 strain of *V. dahliae*. Charudattan and DeVay (1972) reported common precipitin bands between cotton, *V. albo-atrum* (*V. dahliae*), *V. nigrescens* and some *Fusarium* spp.

Nachmias *et al.* (1982a) prepared an antiserum to an extracellular protein–lipopolysaccharide (PLP) antigen from culture fluids of a potato strain of *V. dahliae*. This antiserum detected the antigen in extracts of the tubers, stems and leaves of potato plants infected by *V. dahliae*, but not in healthy plants or in those infected by other pathogens, nor did it react with potato isolates of *V. tricorpus*, *V. nigrescens* or *V. nubilum*. The authors claim that this PLP antigen is likely to be pathogen-specific and a useful tool in diagnosis.

Using ten 'progressive' (V_2) and ten 'fluctuating' (V_1) strains of *V. albo-atrum* from hops, Mohan and Ride (1982) found that strains could be divided into three antigen groups dependent on the presence in high or low concentration or absence of antigen 21. Hopes that antigen 21 concentration might be associated with virulence were disappointed by further work (Mohan and Ride 1983, 1984) where the apparent association was shown to have been fortuitous and none of the serological characters could be correlated with virulence to hop. Protein and enzyme patterns in strains of *Verticillium* were described by Webb *et al.* (1972).

The preliminary use of ELISA antiserum prepared against *V. albo-atrum* hop strains for the rapid diagnosis of hop wilt strains was reported by Swinburne *et al.* (1985). Lazarovits *et al.* (1987) found the technique sensitive for detection of *V. dahliae* antigen and discusses its possible use for diagnosis. Polyclonal antibodies (PAbs) raised against *V. dahliae* isolate 373 from rape was tested for specificity against total soluble proteins from 17 fungal species (Fortnagel and Schlosser, 1995). Biotinylated PAbs in combination with streptavidin–horseradish peroxidase were used for the double monoclonal antibody sandwich (DAS)-ELISA. With *V. dahliae* reacting positively, 16 fungi were negative, except *B. cinerea* which gave a non-specific response due to lectin.

In a different approach to strain identification, Kuznetsov *et al.* (1977) described the intracellular lipids of *V. dahliae* including cardiolipin, monoglycerides, sterols, free fatty acids and triglycerides. Non-pathogenic strains contained more triglycerides than pathogenic ones, and free fatty acids were 9–10 times more abundant in the pathogenic than in the non-pathogenic strains.

Zhao *et al.* (1997), working in China, claimed that esterase isozymes from 14 isolates of *V. dahliae* from cotton were sufficiently distinct to provide a reliable test for both species identification and pathogenicity. Using polyacrylamide gel plate electrophoresis, total esterases from strains from different hosts were determined after 7–12 h. Bands E_3 and E_6 were associated with pathogenicity and pathotype. Shang *et al.* (1998) in a similar study successfully distinguished between *V. dahliae*, *V. nigrescens*, *V. nubilum* and *V. tricorpus* on esterase isozyme patterns. The authors were equivocal on the possibility of positively separating *V. albo-atrum* from *V. dahliae* on esterase isozymes, but suggested that the method was satisfactory to identify *V. albo-atrum* from lucerne. Accounts of enzyme activities in different strains and species with their limited value as taxonomic discriminators are also presented in Chapter 7.

Morphogenesis and Morphology 3

Cell Wall Composition

In a classic paper, Wang and Bartnicki-Garcia (1970) placed *Verticillium* in their group V category. Conidial walls of *V. dahliae* were shown to possess an outer granular, alkali-soluble surface consisting of a heteropolysaccharide–protein complex containing: mannose, galactose, glucose, glucuronic acid, glucosamine and amino acids. The alkali-insoluble inner wall consisted of a microfibrillar network of 1,4-β-glucan and chitin lipid (2.9–3.4%), and traces of phosphate were also present. Wang and Bartnicki-Garcia (1970) suggested that lysine and histidine – the only amino acids remaining after prolonged alkali–acid–alkali digestions – formed the linkage between chitin and protein in the wall. Benhamou (1989), using a mollusc gonad lectin–gold conjugate, found galacturonic acids in the inner cell wall. Using a gold-tagged *exo* glucanase purified from *Trichoderma harzianum* cellulase, Benhamou *et al.* (1990) found 1,4-β-glucan in conidial but not in hyphal walls of *V. albo-atrum*. Failure to obtain binding following cellulase digestion suggested a cellulosic molecule reinforcing the conidium wall architecture. Konnova *et al.* (1995) described the monosaccharide composition of an alkaline hydrolysate of cell walls of a cotton strain of *V. dahliae* following polysaccharide separation on Sephadex G-50 and Acrilex P-4 gels.

Cytoplasm

The predominant L-amino acids and amides in *V. albo-atrum* cell cytoplasm are: threonine (14%); proline (13.2%); glycine (10%); serine (8.6%); glutamine

(8.5%); valine (8.4%); asparagine (8.2%); and alanine (7.9%) (Laskin and Lechevalier, 1973). Extractable lipids in *V. dahliae* were 8% in the mycelium and 14% in conidia on a dry matter basis. Walker and Thorneberry (1971) described the lipid content of *V. albo-atrum*. The main fatty acids were palmitic (32%); stearic (35%); oleic (21%); and linoleic (7%). Benhamou (1989) found galacturonic acids in the plasma membrane of *V. albo-atrum*. The addition of photodynamic dyes to the culture medium of *V. dahliae* by Ageeva (1999) led to the quantitative reorganization of cell membrane lipids: phospholipids, sterols and fatty acids. Lipids comprised, phospholipids + monoglycerides, sterols, triglycerides, sterol ethers and free fatty acids. Bengal pink was the most effective dye. It was concluded that the single oxygen effect on fungal cells led to membrane permeability changes from quantitative changes to lipids which promoted membrane complex stabilization.

Hyphae

Hyphae and conidia of *Verticillium* spp. are mostly haploid (Tolmsoff, 1973). Most cells are monokaryon but hyphal tips may be multinucleate in *V. albo-atrum* and other species (MacGarvie and Isaac, 1966) and in *V. dahliae* (Tolmsoff, 1973). Hyphal septa are perforate but nuclei have not been reported traversing the pore (Typas and Heale, 1976a). Brandt (1964a) and Brandt and Reese (1964) claimed that while extension is directly proportional to the availability of growth requirements, diffusible morphogenic factors (DMFs) exist in *V. dahliae* which inhibit hyphal elongation and induce lateral branching. Light prevents the formation of DMFs in culture (Brandt, 1967). Lateral branches may contribute to the growing front of a colony or may take part in conidiogenesis or may anastomose with other hyphae. Anastomoses are usually confined to mature areas of the mycelium (Puhalla and Mayfield, 1974), but may occur between hyphal tips or conidial germ tubes (Tolmsoff, 1973). The frequency of anastomoses falls rapidly with higher incubation temperatures (Puhalla and Mayfield, 1974)

Loss of melanin pigmentation characteristic of *V. dahliae* and the dark sectors in *V. albo-atrum* cultures giving hyaline colonies have been described by Presley (1941), Pegg (1957), Robinson *et al.* (1957), Brandt and Roth (1965) and Boisson and Lahlou (1980). These white, often fluffy variants of either species showed no loss of virulence in pathogenicity tests correlated with loss of pigmentation or resting structures.

Conidiophores

Conidia, phialospores, are formed in clusters in a mucilaginous slime on elongated conidiogenous cells called phialides. These phialides are borne in whorls on branched aerial hyphae (see Hawksworth and Talboys, 1970). *V. nigrescens* and *V. nubilum*, while possessing verticillate conidiophores, have simpler branches and fewer whorls than *V. albo-atrum*, *V. dahliae* or *V. tricorpus*. Typical structures are found on fungi growing on natural substrates or on selected media. In culture, however, variations in conidiophore morphology of virulent isolates of *V. albo-atrum* can range from forms resembling *Acremonium* to *Cephalosporium*. *V. intertextum* may form synnemata with or without carotinoid pigmentation (Isaac and Davies, 1955). Similar structures have been described by Pegg (1957) for a hyaline variant of *V. albo-atrum*. Valadon and Heale (1965) describe several carotenoid pigments in a UV mutant of *V. albo-atrum*. Reinke and Berthold (1879) illustrated a darkened base of the conidiophore in their description of *V. albo-atrum*, and Klebahn (1913) made the absence of such pigmentation one of the distinctive characters of *V. dahliae*. Van der Meer (1925, see Chapter11), Berkeley *et al.* (1931), Isaac (1949, 1967) and Smith (1965) all described larger conidiophores with dark pigmented bases of the Dm type (*V. albo-atrum*), when compared with the smaller, completely hyaline conidiophores of the Ms type (*V. dahliae*). This difference is especially noticeable on the host and in strains which recently have been brought into artificial culture; the character may be lost in *V. albo-atrum* on prolonged culture, however. The non-wilt banana pathogen *V. theobromae* also forms remarkably dark conidiophores, but among wilt fungi the character is unique to *V. albo-atrum*.

When mycelium grows from infected debris into the soil, the first-formed conidiophores are verticillate (Sewell, 1959). Further penetration of the soil leads to the development of simpler conidiophores and finally to single conidiogenous cells. In stirred aqueous culture, with the suppression of the mycelial phase, individual conidia may function as conidiogenous cells.

Conidium Ontology and Morphology

In all species, the first-formed conidium is holoblastic, each successive conidium forming enteroblastically (Hawksworth *et al.*, 1983, Figure 6D) the form of development earlier described as phialidic. Puhalla and Bell (1981) report a general tendency among wilt fungi to reduce to a 'yeast phase' when present in vascular fluids or liquid media, a condition often described as 'dimorphism'. Garber and Houston (1966), describing the presence of *V. dahliae* conidia in cotton plant vessels, write 'it is difficult to see how the conidia are formed, however the process appeared to be one of budding from either the tips or sides of the mycelium', which infers a holoblastic ontogeny. Buckley *et al.* (1969), describing germination of conidia of the same species under similar conditions, observe it as extrusion and growth of a second conidium from the first, and state that no budding process was observed, thus indicating that conidiation is enteroblastic. Keen *et al.* (1971) showed that *V. dahliae* in liquid culture continues to grow as conidia where initial conidium concentration is increased from 10^4 to

10^8 conidia ml^{-1} and is depressed by compounds such as semicarbazide, phenylhydrazine, deoxyadenosine, gossypol or 5-fluorodeoxyuridine.

Shevtsova and Zummer (1988, see Chapter 4) suggested that the *myd* gene controlling dimorphism in mutants and wilt-type *V. dahliae* is extrachromosomal. Brisson *et al.* (1978) describe two forms of conidium in scanning electron microscope studies of *V. dahliae* in chrysanthemum petioles: normal or α-conidia and sickle-shaped or β-forms. Since these studies did not involve pure cultures, it is possible that the second type of conidium originated from a second fungus.

In an early paper, Van der Meer (1925, see Chapter 11) claimed that microsclerotial (Ms) and dauermycelien = dark or resting mycelial (Dm) strains had conidia of different sizes. Isaac (1949) failed to distinguish spore size differences and recorded sizes for both species in the range 3.5–10 × 2–4 μm. Smith (1965), in a key paper, showed that first-formed conidia produced by *V. albo-atrum* on host or agar substrates are larger and usually more abundant than those produced by *V. dahliae* under similar conditions. Conidia of many strains of both species were shown to be frequently, but not always, somewhat longer and slightly wider in *V. albo-atrum* than in *V. dahliae*. The occasional 1-septate conidia of *V. albo-atrum* are also larger, as are the 1-septate conidia of *V. tricorpus* in which conidial size is similar to that in *V. albo-atrum*. Of nine strains of *V. albo-atrum*, conidial measurements range from 3.5–7 × 1.8–2.6 to 5–13 × 1.8–2.5 μm, with 1-septate conidia measuring 10–12 × 3–3.5 μm in one strain and 8 × 3 μm in another. In 11 strains of *V. dahliae* conidial measurements range from 3–5 × 1.3–2 to 4–6 × 1.8–2 μm. Conidia in two strains of *V. tricorpus* measured 4–11 × 2–3 (1-septate conidia 11 × 3) and 2–11 × 1–3 μm (1-septate conidia 8–15 × 3–4 μm). Smith compared his measurements with those in the literature, including measurements made from the drawings of the type of *V. albo-atrum* published by Reinke and Berthold.

Devaux and Sackston (1966), measuring conidia of three strains each of *V. albo-atrum*, *V. dahliae* and *V. nigrescens* in lactophenol, found no statistical significance in size between those of the first two species. Conidia of *V. nigrescens* were significantly longer but did not differ in width from those of the first two species.

Pelletier and Aubé (1970) showed that conidial size differed in various culture media, at different temperatures and after different periods of growth, and considered that conidial size alone was not a reliable character to use in species determination in *Verticillium*. Conidia for the most part are single-celled and haploid (Tolmsoff and Wheeler, 1974). One-septate conidia of *V. albo-atrum* have been recorded (Pegg, 1957), but with no record of the ploidy. Larger diploid conidia occur regularly as a small proportion of a normal haploid colony (Ingram, 1968; Tolmsoff, 1972; Tolmsoff and Wheeler, 1974).

V. dahliae var. *longisporum* Stark (1961), described with consistently larger conidia, is now recognized (Typas and Heale, 1980; Puhalla and Bell, 1981; see Chapter 4) as a homozygous diploid and described as a new species, *V. longisporum*, by Karapapa *et al.* (1997b,c). Support for this distinction was presented by

Subbarao *et al.* (1998) based on conidial DNA and polymorphisms in the intergenic spacer region of the nuclear DNA.

Resting Structures

In *V. albo-atrum*, hyphal sections differentiate into thick-walled melanized cells, the 'dauermyzel' of Reinke and Berthold (1879) (Isaac, 1949; Schnathorst, 1965; Devaux and Sackston, 1966; Tolmsoff, 1973). *V. dahliae* forms clusters of thick-walled heavily melanized cells which separate as discrete bodies from the parent mycelium. These are the 'microsklerotien' of Klebahn (1913). Bell *et al.* (1976a) showed that the number of microsclerotia was directly proportional to the number of hyphal fusions. Catechol stimulates the production of dark mycelium and microsclerotia in *V. albo-atrum* and *V. dahliae* (Robinson *et al.*, 1957; Bell *et al.*, 1976a). Presley (1950), Brandt (1964) and Brandt and Reese (1964) have described unidentified 'diffusible morphogenic factors' produced by *V. dahliae* affecting microsclerotial production. A few hyaline cells remain in the cell mass; Gordee and Porter (1961) and Schnathorst (1965) claim that these are the only cells that can germinate. Schreiber and Green (1963) and Isaac and MacGarvie (1966), however, maintain that lightly melanized cells also germinate.

Isaac (1949), in a comprehensive description of the pathogenic isolates of *Verticillium*, described in detail as seen under the light microscope, details of all the resting structures of the *Verticillium* species. In *V. dahliae*, septation and swelling occur in contiguous hyphae which continue to bud until globular, almost spherical cell masses form. These later become melanized as the typical microsclerotia. Nadakavukaren (1963) in the first transmission electron microscopy (TEM) study, described in *V. dahliae* microsclerotia, thick-walled cells containing mitochondria, cytoplasmic inclusions and large vacuoles, and thin-walled, empty cells with the exception of possible nuclei. Thin-walled cells were observed germinating while thick-walled ones were thought to contain food reserves. Griffiths (1970), using TEM, confirmed Isaac's observations that all cells were identical and thin walled before differentiation. Some cells autolyse, while membrane-bound autophagic vesicles accumulate, resulting in living and dead cells in the microsclerotium. Fibrillar material is secreted between cells and subsequently melanizing particles are extruded from living cells into the surrounding fibrillar material. Outermost cells have the thickest deposit. Brown and Wyllie (1970) using scanning electron microscopy (SEM) and TEM described early degeneration of the peripheral microsclerotial cells, leaving non-functional hyaline cells embedded in a pigmented mucilaginous matrix among heavily pigmented functional cells. Pigmented cells are connected by septal pores, each retaining an organized cytoplasm and nucleus. A similar study on *V. albo-atrum* resting mycelium by Griffiths and Campbell (1971) showed a development similar to *V. dahliae* but with the absence of budding.

Using wild-type and hyaline mutants of *V. dahliae*, Wheeler *et al.* (1976) confirmed the findings of Griffiths (1970) and demonstrated that pigmentation could be induced in hyaline microsclerotia in the mutant by the addition of the essential precursor (+)-scytalone. A detailed TEM study of the chlamydospores of *V. nigrescens* and *V. nubilum* was carried out by Griffiths (1982). Endospores described only by Aubé and Pelletier (1968) in *V. albo-atrum* may be the extensions described by Brown and Wyllie (1970). An alternative possibility is endoparasitism by fungus or protozoan (see Chapter 10).

Early stages of the formation of *V. dahliae* microsclerotia *in planta* are described by Wright and Abrahamson (1970), and the nutritional regulation of microsclerotia by Hall and Ly (1972b). Other microscopic descriptions of pathogens *in planta* are described in Chapter 8.

Cultural and morphological variability of species of *Verticillium* grown in the presence of antibiotics was described by Litvinov and Babushkina (1978).

Cytology and Genetics 4

In the absence of sexual reproduction and a known teleomorph for any of the six vascular pathogen species of *Verticillium*, progress in genetic research has been slow (cf. *Neurospora crassa*) and has had to await the discovery of new analytical techniques developed for other fungi or organisms. The literature on *Verticillium* genetics therefore falls very roughly into three phases of development, not all, alas, separated in an orderly, chronological sequence.

Prior to the early 1960s, observed differences in morphological or pathological behaviour of species or strains were dealt with on a descriptive basis, and much of the genetic interpretation was speculative. Following work on *Aspergillus nidulans* in the 1950s, great progress was made on conidial anamorph studies in the 1960s and subsequently the derivation of nutritional (and other) mutants and a recognition of the significance of heterokaryosis and mitotic recombination. This has continued up to the present. What might be called the third phase in *Verticillium* genetics, that involving DNA manipulations and the development of molecular gene probes, started in the late 1980s; the number of publications in this field to date is still in the late 'lag phase'. This chapter is concerned solely with the genetics of the fungus. Other aspects of genetics involving host plants are dealt with in Chapter 10. The use of molecular genetics to attempt to distinguish between species, strains and host forms of *Verticillium* is covered in this chapter rather than in Chapter 2. Other reviews specifically on *Verticillium* species are given by Hastie and Heale (1984), Heale (1988, 1989) and a general review of wilt pathogens by Puhalla and Bell (1981), Bell (1992b) and Rowe (1995).

Nuclear State

All mycelial segments are uninucleate except the apical cell which is binucleate or, rarely, multinucleate (Typas and Heale, 1977). The conidia of all the species are predominantly uninucleate (Hastie, 1962, 1964; Roth and Brandt, 1964b; Heale *et al.*, 1968; Tolmsoff, 1973; Puhalla and Mayfield, 1974). MacGarvie and Isaac (1966) claimed that 1% of conidia of *V. nubilum* were binucleate. Pegg (1957) described bicellular conidia in *V. albo-atrum*, each cell of which was uninucleate. Each of the individual cells of the microsclerotia of *V. dahliae* were shown by MacGarvie and Isaac (1966) to be uninucleate; this was confirmed by Typas and Heale (1980), who found the same condition in cells of the resting mycelium of *V. albo-atrum*.

Ploidy

Hastie and Heale (1984) claimed that wild-type strains of all species, with very few exceptions, are all haploid. Buxton and Hastie (1962) found a straight-line relationship between UV dose and lethality in *V. albo-atrum* conidia. This 'one-hit curve' is typical for haploid organisms. Recessive mutants only able to grow on a complete but not minimal medium are called auxotrophs. Auxotrophs are not directly detectable if diploid cultures or nuclei are treated, since the expression of the mutant allele would be prevented by the remaining non-mutant dominant allele. Hastie and Heale (1984) thus argued that auxotrophs derived from wild-type strains must be from haploid cultures. The segregation of recessive genes affecting drug resistance markers from a suspected diploid has been used as evidence for heterozygosity and hence diploidy (Fordyce and Green, 1964; Ingram, 1968; Typas and Heale, 1976b). Decreased radiation sensitivity and mutability has also been used by Ingram (1968), Hastie (1970), Puhalla and Mayfield (1974) and Molchanova *et al.* (1978) to distinguish the less sensitive diploids.

While wild-type strains of *Verticillium* are predominantly haploid, some stable diploids do occur in *V. albo-atrum* and *V. dahliae*. The first naturally-occurring stable diploid was isolated by Stark (1961) – *V. dahliae* var. *longisporum* Stark (= *V. longisporum* Karapapa) from diseased horseradish. Subsequently, Puhalla and Hummel (1984) isolated two others from sugarbeet and rape. When treated with *p*-fluorophenylalanine, they haploidized to small-spored stable strains. In the wild, however, these diploids are stable. Homozygous diploids could arise by failure of mitosis or somatic nuclear fusion in a homokaryotic cell (Hastie and Heale, 1984). Typas and Heale (1980) estimated the incidence of homozygous diploidy in wild-type haploids as 1 in 10^3–10^4. The large size of the conidia and the way UV-derived haploid auxotrophs from the wild-type paired to produce typical *V. dahliae* var. *longisporum* colonies was considered by Ingram (1968) as evidence for diploid status. Conidial size differences alone are unreliable, since

Smith (1965) reported that first-formed conidia, often on single phialides, on host substrate or agar are larger than the later secondary conidia. Chaudhuri (1923) first described this effect where primary conidia were larger on host tissue than on agar. The conidial size difference between *V. dahliae-longisporum* and authentic isolates of *V. dahliae* and *V. albo-atrum* was confirmed by Typas and Heale (1977) analysing large populations of conidia from each species and artificially-induced diploids of *V. dahliae* var. *longisporum* using a micro-particle counter. A subsequent paper (Typas and Heale, 1980) showed a twofold increase in DNA in suspected diploid compared with haploid spores (see later in this chapter). Tolmsoff and Wheeler (1974) found that nuclear DNA levels in haploid nuclei of *V. dahliae* and *V. albo-atrum* were comparable to the haploid states of other fungi. Tolmsoff and Bell (1971) considered that continuous changes in ploidy occurred in *V. dahliae*, *V. albo-atrum*, *V. nubilum*, *V. tricorpus* and *V. nigrescens* throughout their life cycles. Homozygous diploids of each species with varying degrees of stability were induced from haploid cultures grown on a minimal medium containing NH_4 ions. Cultures of *V. dahliae* gave 49 and 20% diploid conidia 13 and 37 days after plating, respectively. *V. tricorpus* cultures similarly produced 74 and 26% diploid spores after 14 and 25 days, respectively. The authors claimed that microsclerotia were produced from both haploid and diploid cells. Diploid cells all resulted in the formation of microsclerotia within homozygous haploid colonies. Haploids occurred from diploid hyphae at the site of microsclerotial formation. More especially, haploids which failed to form diploids lost the ability to produce microsclerotia. While the implications of this paper were far-reaching, the experiments were not part of a study on parasexual recombination and diploidy was assumed from the shape and size of conidia which were twice the length of haploids. Roth and Brandt (1964b) reported a large-spored variant of *V. dahliae* ('*V. albo-atrum*' *sic*); many of the conidia showed two or more nuclei and mycelium derived from these conidia had nuclei in groups of three, four or six; Hastie (1970) considered that this was possibly a diploid. Two further wild-type strains of *V. dahliae* from rape and sugarbeet from Sweden were suspected to be stable diploids by Puhalla and Hummel (1983) and were proved to be so by Jackson and Heale (1985) using the criteria of conidial volume and relative DNA content. These authors considered them to belong in *V. dahliae* var. *longisporum*, which was corroborated by subsequent work (Karapapa *et al.*, 1997a,b,c). Typas and Heale (1980) found that cells in the young (6–8 days) microsclerotia of *V. dahliae* and resting mycelium (9–12 days) of *V. albo-atrum* were uninucleate and haploid, while Tolmsoff (1972, 1973) maintained that there were considerable numbers of diploid cells in ageing microsclerotia.

Only a limited number of studies on chromosome cytology have been carried out. MacGarvie and Isaac (1966) observed two rows of three granules staining positively with aqueous Azure-A during conidial mitosis in *V. dahliae*. The nature of these structures (0.2 μm diameter) was not resolved. Mitosis in *Verticillium* spp. appears unusual, and conflicting reports have appeared.

Brushaber *et al.* (1967) observing HCl Giemsa-stained nuclei of *V. albo-atrum*, claimed that the nuclear membrane remained intact, chromosomes condensed and a spindle and centrioles were visible. Chromosomes were attached to the centriole by spindle microfibrils, which were closely associated with the nuclear membrane. The axis of division was perpendicular to the long axis of the hypha. Anaphase was unilateral and uncoordinated. Heale *et al.* (1968) using Giemsa, Feulgen and acid fuchsin staining, found $n = 4$ for *V. albo-atrum*, and claimed that chromosomes were joined on a thread-like structure. Following an anaphase-like stage, the nuclear membrane constricted between two sets of chromosomes, resulting in two daughter nuclei. In some nuclei, four large and one small paired set of Giemsa-stained bodies suggested $n = 5$. This was supported subsequently by genetic studies (Typas and Heale, 1978) indicating four large and one small linkage group. Tolmsoff (1973), based on a modal value for several fungi of 50 fg of DNA per cell and $n = 10$ (excluding mitochondrial DNA) calculated the condensed length of an average *V. dahliae* chromosome as approximately 0.29 μm. Typas and Heale (1980), however, found 28 fg of DNA per cell for *V. albo-atrum*, corresponding to approximately 2.8×10^7 nucleotide pairs per haploid genome.

Tolmsoff (1972, 1973) using TEM and time-lapse photography described an unusual situation in *V. dahliae*. Eight DNA-containing subunits connected in tandem to form a chain were found in haploid cells. Each subunit appeared tadpole-shaped with a rounded head and a narrow tail. The subunits were attached head to tail, with one end of the chain bearing a free head, the other a free tail; the chain was referred to as a 'chromosome' and the subunits as 'chromomeres'. Diploid nuclei were described as being formed by end-to-end connection between two haploid chromosomes. Tolmsoff maintained that microsclerotia became polychromosomal during ageing, and by temporary disconnection between chromomeres followed by reconnection, great opportunity for genetic recombination within the microsclerotium was possible. Tolmsoff (1980, 1983) further considered that heteroploidy was a mechanism of variability in *Verticillium*. Typas and Heale (1980), using microdensitometry of Feulgen-stained nuclei in conidia, hyphae and resting structures of both *V. albo-atrum* and *V. dahliae*, found no major differences in the amount of genetic material of the nuclei of resting structures and maintained that they were dormant haploid phases in the life cycle and not centres of changes of ploidy and genetic recombination. This represents the current view of most workers in the field.

The cytology of *V. dahliae* was described by Safiyazov *et al.* (1972). Reports by a group of CIS workers (Abdukarimov *et al.*, 1990; Ibragimov and Khodzhibayeva, 1990; Safiyazov *et al.*, 1990) for the existence of a virus or plasmid-like source of DNA in *V. dahliae* mycelium from cotton have not been supported by comparable observations in other countries, although Barbara *et al.* described double-stranded RNA in *V. albo-atrum*. Reports claimed the existence of 'virus' particles 30–40 nm in diameter and DNA electrophoretically distinct from the fungal genome at 89.1 kb. Claims for an association between such particles and fungal virulence (Safiyazov *et al.*, 1990) or, more remarkably, of a combination of this DNA with the cotton genome using *V. dahliae* as a vector, are wholly without substantiation and must be discounted.

Mutants and Mutagenesis

Verticillium is unique among wilt fungi in forming spontaneous natural auxotrophs. Milton and Isaac (1967) found a natural biotin-requiring isolate. Puhalla (1977) reported a high frequency of nicotinamide-requiring *V. dahliae* auxotrophs. Natural variants requiring methionine, arginine, adenine and pyridoxine were recovered at low frequency. Earlier, Puhalla and Mayfield (1974) showed that from the mutability of a particular gene, 16% of auxotrophs recovered from the T9 cotton strain of *V. dahliae* were nicotinamide requiring.

Roth and Brandt (1964c) found the highest frequency of morphological mutants of *V. dahliae* in cultures grown at temperatures >28°C. Tolmsoff (1972) found a higher frequency and range (9.1%) of morphological variants from microsclerotia compared with 0.5% from conidia.

Since spontaneous nuclear markers and readily available stable phenotypes are scarce in nature, much effort has gone into the production of induced mutants using various means. A variety of mutagenic agents has been used; one of the earliest was UV radiation (Robinson *et al.*, 1957; Buxton and Hastie, 1962; Fordyce and Green, 1964; Heale, 1966; Hastie, 1973; Puhalla, 1973a; Tolmsoff, 1973; Ingle and Hastie, 1974; Typas and Heale, 1976b; Typas, 1981). Hastie and Gadd (1981) induced UV mutants in *V. albo-atrum* in which conidia germinated while still attached to the conidiogenous cell. This *isg* (*in situ* germination) character appeared to be metabolically linked to melanin production in the resting mycelium. Several workers (Robinson *et al.*, 1957; Puhalla, 1973a; Typas and Heale, 1976a) noted an increased sensitivity to UV of *V. albo-atrum* compared with *V. dahliae*. It is generally assumed that *V. albo-atrum* has a less effective repair mechanism than *V. dahliae*. The data of Buxton and Hastie (1962) showing that 0.5% of UV-induced auxotrophs were produced at the 3% survival level were confirmed subsequently by Heale (1966), Puhalla and Mayfield (1974) and Typas and Heale (1976a). The reduced yield of auxotrophs resulting from the incubation in light of UV-irradiated conidia confirmed the existence of a photo-repair system (Puhalla, 1976).

Fordyce and Green (1964), Hastie (1973), Ingle and Hastie (1974) and Clarkson and Heale (1985a) used *N*-methyl-*N'*-nitro-*N*-nitrosoguanidine (NTG). The last authors found that 0.5% of NTG-treated conidia at the 5.8% survival level were auxotrophic. Several workers in the CIS (formerly USSR) have obtained γ-radiation mutants of cotton isolates of *V. dahliae* using ^{60}Co as a γ source (Kasyanenko and Portenko, 1978b; Shevtsova, 1978; Portenko and Kasyanenko, 1987). These authors also used *N*-nitroso-*N*-methyl urea. Herbicides and insecticides have also been shown to act as mutagens on

Verticillium (Hubbeling and Basu Chaudhary, 1970; Galperina, 1990). Mutants resistant to toxic substances have also been studied. Hastie (1962) selected a spontaneous mutant resistant to acriflavine by plating out large numbers of *V. albo-atrum* conidia on a complete medium containing 100 μg ml^{-1} acriflavine. The gene influencing acriflavine-tolerance has been studied in natural populations of both *V. albo-atrum* and *V. dahliae* (Typas and Heale, 1976a). The isolate of *V. albo-atrum* from tomato was diploid and heterozygous for the gene for acriflavine resistance. Acriflavine was used in both species to induce hyaline mutants (*hyl*$^-$) from (*hyl*$^+$) melanin-forming wild-types. More recently, Typas (1981) used acridine orange and ethidium bromide to induce mutation especially in mitochondrial DNA.

The auxotrophic mutants of *Verticillium* spp. consist predominantly of those with amino acid requirements, mostly adenine, argenine and methionine, also histidine, isoleucine, lysine, serine and tryptophan (Fordyce and Green, 1964; Ingle and Hastie, 1974; Typas and Heale, 1976b; Kasyanenko and Portenko, 1978b). Pirozhenko and Shevtsova (1988), studying adenine-dependent mutants of *V. dahliae*, showed five complementary groups segregating in the progeny of heterozygous diploids which they deduced to correspond to five different genes controlling adenine biosynthesis. Requirements for vitamins (aneurin, *p*-amino benzoic acid, inositol, nicotinic acid and pyridoxine) are cited by Hastie (1978).

Puhalla (1976) used an ingenious glycerol technique for auxotroph selection. A mutagen-treated conidial suspension of *V. dahliae* was incubated on a minimal medium prohibiting auxotroph growth but permitting growth of phototrophs, which were then killed by the glycerol. Auxotrophs were recovered by overlaying the medium with a complete medium. However, this technique did not work with *V. albo-atrum* (McGeary, 1980). In addition to acriflavine resistance, Typas (1981) found mutants resistant to antimycin A and cyanide. Benomyl-resistant mutants were also obtained by Kasyanenko and Portenko (1978b). This character, shown to confer cross-resistance in *V. dahliae* to methylthiophanate, was due to a single, dominant nuclear gene (Koroleva *et al.*, 1978). Talboys and Davies (1976a,b) had earlier shown that *V. dahliae* could increase tolerance to benomyl gradually up to 12 p.p.m. (*hyl*$^-$) variants of *V. dahliae* were consistently more tolerant to benzimidazoles than Ms types. Typas (1984) showed that nuclear mutations were also responsible for antimycin A, azide and cyanide resistance; whereas amytal and chloramphenicol resistance were due to cytoplasmically inherited factors. Resistance to antimycin A (Typas, 1984) was reflected earlier in the recovery of coloured mutants from *V. dahliae* by Ezrukh and Babushkina (1978) following treatment with metabolites from an unknown actinomycete. Typas (1981) studying mitochondrial DNA preparations from *Verticillium* spp. demonstrated linkage between *hyl*$^-$ (hyaline) mutants and *amy*$^\alpha$ (amytal resistance). Using reciprocal micromanipulation of DNA preparations carrying nuclear and cytoplasmic markers, Typas obtained rare mitochondrial DNA recombinants between *hyl*$^-$ and *amy*$^\alpha$.

Mutants affecting colour and resting structures

The study of melanin biosynthesis, which has relied heavily on the use of selected mutants, is considered in Chapter 7 (Brandt, 1964b; Heale and Isaac, 1964; Gafoor and Heale, 1971a,b; Bell *et al.*, 1976; Stipanovic and Bell, 1976; Kasyanenko and Portenko, 1978b; Shevtsova, 1978; Typas and Heale, 1978).

Puhalla (1975, 1979) and Puhalla and Hummel (1981) used mutant techniques on a worldwide range of isolates from many hosts to study strain evolution and isolation. UV and γ irradiation mutants of *V. tricorpus* yielded types with only resting mycelium, or chlamydospores or microsclerotia. Others produced only two of these three features found in wild-types. Tolmsoff (1973) and Molchanova *et al.* (1978) suggest that an orange carotenoid pigment seen in newly isolated *V. tricorpus* is indicative of diploidy (cf. Valadon and Heale, 1965). Hastie (1968) described a sooty (*so*) mutant of *V. albo-atrum*. The *so* mutants develop rapid melanogenesis throughout the culture within 4–5 days compared with resting mycelium formed in approximately 10–14 days in older (central) parts of wild-type cultures, depending on the medium. Hastie (1968) found *so* and *arg9* linked on the same chromosome arm and was shown by Typas and Heale (1978) to be in the smallest (5th) linkage group. Li *et al.* (1998a) derived *nit* mutants of *V. dahliae* lacking microsclerotia ms^- and carbendazim-tolerant mutants. Progeny of heterokaryons of $ms^- \times ms^+$ and $ms^- \times ms^-$ pairings derived from single conidia showed that microsclerotial character was unstable and was reduced and lost after repeated subcultures. Virulence of these isolates on cotton was intermediate between strongly and weakly virulent parents. Progeny of heterokaryons of microsclerotial-forming strains and non-microsclerotial mutants using a *nit* phenotype as marker were unstable and scattered (see Tian *et al.*, 1997; Tian *et al.*, 1998b). The authors, in the absence of firm evidence, suggested that cytoplasmic control of microsclerotial production could migrate from cell to cell in anastomoses (see Li *et al.*, 1997). A naturally-occurring variant of *V. dahliae* with orange-brown microsclerotia was described by Seman (1970).

Effects of hosts on mutagenesis and of mutants on pathogenesis

Robinson *et al.* (1957) found no alteration in virulence of *Verticillium* isolates following repeated passage through potato. In contrast, Fordyce and Green (1964) found two of ten isolates of *V. dahliae* from peppermint that became virulent to tomato following two successive inoculations in that host. Wild-type isolates from peppermint were invasive to tomato but non-pathogenic. The changed isolates *ex* tomato subsequently were non-pathogenic to peppermint and failed to develop microsclerotia. There are many reports in the literature of apparent changes in virulence following inoculation, but few with detailed documentary evidence.

The role of mutants in pathogenesis (see Chapter 8) has centred exclusively on pectolytic enzymes. The reasons for this are: (i) that such mutants are readily obtained by UV and easily detected by plate test; and (ii) pectic enzymes are attributed to be one of the few reported single cause and effect mechanisms for *Verticillium* disease induction. The evidence for enzymes and disease is discussed fully in Chapter 8. The technical difficulties in the use of mutants may be summarized as:

1. Mutants rarely exhibit zero enzyme activity.
2. Multiple isoenzymes are often involved.
3. One enzyme, e.g. pectin lyase (PL), may substitute for a deficiency in another, e.g. polygalacturonase (PG).
4. The activity of enzymes *in vitro* is often very different from their *in vivo* activity.

Only a limited number of the many reports on pectolytic enzymes has described the use of mutants with one or more constitutive enzymes deleted. While it is possible for the host to act as an inducing substrate restoring activity lacking in the mutant, the reisolation of the mutant and confirmation of the continued loss of a specific activity is taken as strong evidence for the role (or lack of) of pathogenic enzymes *in vivo*. The evidence, however, has been conflicting. Puhalla and Howell (1975), working with single enzyme-deficient mutants of *V. dahliae* from cotton, found a reduction in symptoms associated with loss of pathogen enzyme capacity. In a subsequent paper, Howell (1976) derived UV mutants of the T9 strain deficient in PL, PME and endo-polygalacturonase. Such mutants comprised 0.025–0.5% of the 5–10% irradiated survivors. Stem inoculation by these mutants led to normal wild-type disease symptoms. A general criticism of single mutants was the duplication of their role by non-deleted enzymes. To counter this, Howell (1976) derived surviving mutants deficient for PL and PG by repeated mutagenesis. These produced normal wilt symptoms and remained deficient for these enzymes on reisolation. A more recent comprehensive study by Durrands and Cooper on *V. albo-atrum* in tomato (Durrands and Cooper, 1988a,b,c; Cooper and Durrands, 1989) described some six PL and 25 PG isozymes in wild-type virulent pathogens. Three main mutants were selected from 10% survivors using the alkylating agent, ethylmethane sulphonate (EMS) at 1% as mutagen. Mutants had variously reduced PL and PG activities. Isolate C23 had 3 and 9% of the wild-type PL and PG, respectively, and included all the isozymes. One mutant, 34i with 9% PL and 7% PG, had only a single basic PG isozyme and could not utilize galacturonides. All mutants had some pectolytic activity and other hydrolases such as cellulase, β-glucosidase and β-galactosidase and leucine arylamidase. Unlike Howell (1976), Durrands and Cooper tested root infectivity by root inoculation.

Symptoms of epinasty, chlorosis and wilting were absent, reduced or delayed in plants inoculated with the mutants. C23-inoculated plants remained generally healthy, with the exception of slight chlorosis and mild wilting in 50% of the plants. Pathogenicity (infectivity) proceeded in the relative absence of PL. Mutant C23 produced higher levels of PG and PL than another mutant (34i) but was less virulent. These studies are the first seriously to implicate PL in pathogenicity. The presence of other enzymes and the complexity of induction in the host, however, leaves many questions unanswered and suggests a fruitful field of investigation with genetically modified isolates independent of mutagens.

The Parasexual Cycle

Genetic recombination in the absence of sexual reproduction, first described by Pontecorvo *et al.* (1953) and Pontecorvo (1954) in *Aspergillus niger*, was established in *V. albo-atrum* in a signal contribution by Hastie (1962), a former student of Pontecorvo. The topic has been reviewed variously by Pegg (1974), Puhalla and Bell (1981), Hastie (1981), Hastie and Heale (1984) and Heale (1988). Much of the work is based on Hastie (1962, 1964, 1967, 1968), Puhalla and Mayfield (1974) and Typas and Heale (1978).

A prerequisite for parasexuality is hyphal anastomosis and the formation of heterokaryons. The fusion of haploid homokaryotic hyphae with limited and restricted nuclear migrations results in the establishment of a haploid heterokaryotic mycelium. Isolated somatic nuclear fusion may occur between haploid heterokaryons to form diploid nuclei, heterozygous at the complementary gene loci of the original homokaryon. Based on their selective advantage, the heterozygous diploids multiply by mitosis. Identical daughter genotypes are replicated by normal mitosis but, in addition, in irregular mitoses, genetic recombination occurs and non-disjunction resulting in unstable novel diploid segregants. The progressive loss of chromosomes (a process called 'haploidization') until the haploid status is regained results in a mycelium containing novel recombinant haploid, the parental homokaryon types, and heterozygous aneuploid and diploid nuclei.

Heterokaryosis

Anastomosis occurs commonly in *Verticillium* spp. between conidial germ tubes, first described in *V. albo-atrum* by Reinke and Berthold (1879). Using UV, Hastie (1962) produced diauxotrophic mutants of *V. albo-atrum* from hop which were forced on a minimal medium. Fordyce and Green (1964) used similar auxotrophs of *V. albo-atrum* and *V. dahliae* to form prototrophic diploids from interspecific anastomoses with recombinant characters. Anastomosis has been found between conidial germ tubes and hyphae as well as between adjacent germ tubes (Schreiber and Green, 1966). Puhalla and Mayfield (1974) showed that *V. dahliae* heterokaryons consisted mainly of uninucleate cells while binucleate cells were confined to a 1–2 mm zone behind the colony front where limited migration occurred. Heale (1966) demonstrated nuclear migration between anastomozing conidia of a lucerne *V. albo-atrum* auxotroph. Anastomosed cells in *V. dahliae* heterokaryons provide the auxotrophic requirements for large homokaryotic areas including the colony edge. Heterokaryon mycelium is a mosaic, the margins of which are unstable with imbalanced nuclear ratios (Heale, 1966, 1988). Hastie (1973) described a mosaic of resting mycelium and microsclerotia in interspecific heterokaryons of auxotrophic mutants of *V. dahliae* and *V. albo-atrum*. Puhalla (1973b) showed that complementary auxotrophs of *V. dahliae* T9 formed heterokaryons that were stable at 21°C. The results agreed with those of Hastie (1973) that complementation was due to a mosaic of heterokaryotic and homokaryotic regions with some hyphal tips growing syntrophically. Various techniques have been described for the production of heterokaryons (Hastie, 1962, 1973; Heale, 1966; Ingle and Hastie, 1974; Typas and Heale, 1976a, 1979). Typas and Heale (1979) produced heterokaryons by microinjection, yielding 21% from 80% of injected survivors. Typas (1983) also formed heterokaryons by protoplast fusion. Five complementary groups from eight adenine-dependent *V. dahliae* mutants were found by Pirozhenko and Shevtsova (1988). Heterozygotic diploids were found in a number of combinations.

Heterozygous diploids

Direct evidence for somatic nuclear fusion in a *V. dahliae* heterokaryon (a rare natural event) was provided by Puhalla and Mayfield (1974) in phase contrast photographs showing single large nuclei in some cells and two small, presumed haploid nuclei, in other cells. Forced heterokaryons at 30°C which have ceased growing frequently produce prototrophic diploid sectors. Heterokaryons in *V. albo-atrum* are more unstable. While prototrophic diploid conidia are recovered routinely from *V. albo-atrum*, they are seldom found in *V. dahliae* heterokaryons. This difference between the two species may reflect their different temperature tolerances and requirements (Puhalla and Bell, 1981). Ingle and Hastie (1971, 1974) indeed showed that the frequency of prototrophic diploid conidia from *V. albo-atrum* was stimulated at temperatures above 22°C. Hastie (1962, 1964) and Typas and Heale (1976a) obtained heterozygotes by plating dense heterokaryon conidial suspensions on a minimal medium. Hastie (1973) and Ingle and Hastie (1971, 1974) incubated mixed inocula of *V. albo-atrum* on a glucose, nitrate minimal medium which favours the growth of heterozygous diploids which emerge as relatively fast growing sectors in a background of slower growing homo- and heterokaryotic mycelium. Puhalla (1973b) used a high incubation temperature of 30°C to select *V. dahliae* heterozygous diploids. These conditions according to Ingle and Hastie (1974) promote nuclear cycle synchrony which enhances the rate of nuclear fusion. The rates of mitotic recombination have been estimated in conidiophore phialides as 0.2 per nuclear division. The frequency of diploid formation from haploids is given as 10^{-6} per nuclear division and haploidization as 10^{-2} per nuclear division (Hastie and Heale, 1984).

The conditions favouring intraspecific heterozygous diploids also favour interspecific crosses. Fordyce and Green (1964), Hastie (1973) and Typas and Heale (1976a) all obtained *V. albo-atrum* × *V. dahliae* hybrids, but no strong evidence for their natural occurrence. Selection pressure favours haploid prototrophic recombinants more than heterozygous diploids, since the latter have a much slower growth rate than the wild-type and reduced sporulation (Heale, 1988). Since heterozygous diploids obtained from crosses of *V. albo-atrum* and *V. dahliae* showed infrequent haploidization and restricted recombination, Hastie (1973) suggested that this represented a non-homology of the genomes and strong evidence for specific distinction. Schnathorst (1973) reported prototrophic growth of auxotrophic isolates of paired cultures of *V. albo-atrum* × *V. dahliae*, *V. dahliae* × *V. nubilum* and *V. albo-atrum* × *V. nubilum*. In Hastie's (1971) experiments, diauxotrophic mutants each of wild-type *V. albo-atrum* and *V. dahliae* were obtained by UV treatment and cultured on a minimal medium. Heterozygous diploids were obtained from hybrid crosses. Homozygous diploids were also obtained from selfed crosses of each species. Selfed diploids formed aneuploids and haploid segregants after culturing. After 3 weeks, conidial analysis gave 4, 7 and 89% for diploid, aneuploid and haploid, respectively. With hybrid diploids, the same values were 73, 27 and 0.2%. This illustrated that haploidization is abortive and therefore non-effective in nature.

In a later study (Hastie, 1978), interspecific diploids between *V. albo-atrum* × *V. tricorpus* and *V. dahliae* × *V. tricorpus* similarly gave a very low frequency and variety of viable recombinants. If such hybrids occurred in nature, only a restricted gene flow between the two species populations would be possible. Interspecific diploids of *V. tricorpus* appeared as bright orange sectors, confirming the findings of Tolmsoff (1973) and Molchanova *et al.* (1978) that diploidy was associated with the occurrence of this pigment. Sporulation occurs more frequently in haploid than diploid mycelium. Colonies of *Verticillium* spp. derived from single heterozygous diploid conidia revert to haploid status after 4 weeks (Hastie, 1978). McGeary and Hastie (1982), however, recovered a more stable diploid from paired diauxotroph cultures obtained from tomato and lucerne isolates. Heale (1988) reported the synthesis of a semi-stable diploid of *V. albo-atrum* derived from auxotrophs from hop isolates. Antirrhinum plants inoculated with complementary auxotrophs yielded diploids with moderate pathogenicity to hop which remained stable for 6 weeks. Hastie (1970) considered that the variable stability of diploids could be attributed to heterozygosity for chromosome aberrations caused by the mutagen. McGeary and Hastie (1982) found stable and unstable diploids from two diauxotroph crosses which supported this argument.

Heterokaryon compatibility (compatibility grouping)

In general terms, paired auxotrophic isolates of either *V. albo-atrum* or *V. dahliae* from the same host have a greater propensity to form heterokaryons than intraspecific crosses of auxotrophs from different hosts (Heale, 1966). Typas and Heale (1976a), using a more comprehensive range of isolates, confirmed that in some intraspecific crosses heterokaryon formation was low, as was the result in all interspecific crosses. Interpretation was made difficult, however, by the pleiotropic effects of heterokaryon markers. In a signal contribution, Puhalla (1979) and Puhalla and Hummel (1983, 1984) presented the first real evidence for specific compatibility groupings and the existence of incompatibility barriers based on a study of 94 worldwide isolates of *V. dahliae* from different hosts. UV-derived mutants produced hyaline microsclerotia without allomelanin (alm) and brown microsclerotia (brm). Compatibility was shown between paired microsclerotial pigment-deficient mutants by a line of black microsclerotia. Incompatibility was illustrated by confluent growth of the colonies. Secretory cross-feeding effects in the absence of hyphal fusion were eliminated by the careful choice of mutants which did not secrete melanin precursors. All 94 isolates were placed in 16 compatibility (het-c) groups. Nine severe defoliating cotton isolates were grouped in het-c group P_1; seven of nine tomato isolates were assigned to P_2. Four of six pepper isolates – a very host-specific isolate – were in het-c P_5. Aubergine isolates, a universal susceptible host, occurred in all het-c groups. Typas (1983), using protoplast fusions of *V. dahliae* and *V. albo-atrum* to circumvent possible hyphal wall anastomosis barriers, found the yield of heterozygous diploids increased from 1 in 10^7 to 1 in 10^5.

Typas and Heale (1976a), working with *V. albo-atrum* and *V. dahliae*, and Clarkson and Heale (Heale, 1988) studying mild and progressive hop isolates of *V. albo-atrum*, found no clear evidence of incompatibility groups. Whereas Puhalla used unforced trials of compatibility, Heale *et al.* used intensive selection pressure on diauxotrophs to form heterokaryons and heterozygous diploids. Using auxotrophs derived by UV or NTG mutagenesis, O'Garro and Clarkson (1988b) explored the possibility of heterokaryon compatibility between race 1 and race 2 isolates of North American, European and Australian isolates of *V. dahliae* from tomato. North American and Australian isolates of race 1 and race 2 were each 100% compatible within each geographical group, but crosses between country isolates were wholly incompatible. Two out of three European isolates formed heterokaryons with both US and Australian isolates. Thirteen of 30 crosses producing heterokaryons formed prototrophic diploid conidia. Diploidy was greatest in crosses showing 100% compatibility. Such heterozygous diploids derived from race 1 and race 2 crosses highlight the potential for field variability arising through parasexuality. Ivanova and Kasyanenko (1990) using auxotrophic mutants reported hybridization between *V. dahliae* and *V. tricorpus* (see Schnathorst, 1973). Interspecific heterokaryons were produced in 70.7% of the crosses and in 80% of intraspecific matings. Heterozygous diploids were produced in almost equal numbers in both inter- (6.5%) and intraspecific crosses (6.5%). Correll *et al.* (1988) used vegetative compatibility grouping (VCG) to look for genetic affinities between a wide range of strains of *V. albo-atrum* on ten different host species, comparing geographical origin, host specificity and virulence. Nitrate non-utilizing (*nit*) mutants were obtained on a minimal medium supplemented with 1.5% potassium chlorate. Chlorate-resistant sectors were cultured on a minimal medium containing nitrate as a sole nitrogen source. Nitrate-resistant (*nit* mutants) sectors grew as thin resupinate colonies with no aerial growth. Compatibility was demonstrated in complementation tests when paired *nit* mutants produced dense aerial growth (Wilhelm, 1954) indicative of prototrophic heterokaryon formation. *Nit* mutants produced typical wild-type growth on a complete medium. Two phenotypically distinct mutants, *nit l*, unable to utilize nitrate but able to utilize hypoxanthine, and *nit M*, unable to utilize either nitrate or hypoxanthine, were found in each strain of *V. albo-atrum*. *Nit 3*, mutants for the structural locus of nitrite reductase and major nitrogen regulatory locus, were not identified in this study. *Nit l* and *nit M* testers were paired to assign host forms and strains to a particular VCG. Fifteen strains from lucerne from worldwide sources were compatible with each other but incompatible with all other strains from different hosts. The lucerne strain was regarded as a genetically homologous population and assigned to VCG1. This was confirmed for Polish lucerne isolates by Furgal-Wegrzycka (1997) who also found five self-incompatible isolates which were also incompatible with non-pathogenic isolates of lucerne. Strains from diverse hosts (*Pelargonium*, hop, potato, cucumber and *Ceanothus*) were all compatible and placed in VCG2. Four of six hop strains of progressive and fluctuating types from the UK were self-incompatible with both *nit* mutants and wild-type hyphal anastamosis (cf. Puhalla and Hummel's (1983) non-reacting strains). The authors caution the validity of forced auxotrophs as indicators of intrinsic compatibility (cf. Clarkson and Heale, 1985a,b). In a valuable reassessment of VCGs in *V. dahliae*, Joaquim and Rowe (1990) examined the 15 VCGs erected by Puhalla and Hummel (1983) using *nit* mutants instead of microsclerotial colour mutants (strains based on Ms colour mutants and considered to be incompatible in VCG tests were compatible when *nit* mutants were used). The 22 strains originally assigned to 15 groups only fell into four VCGs on the basis of *nit* complementation tests. One strain PU was heterokaryon self-incompatible. A subsequent study (Joaquim and Rowe, 1991) based on 187 wild-type 'strains' [*sic*] of *V. dahliae* from 22 potato fields in Ohio demonstrates the complexity of a VCG-based taxonomy. Using chlorate-derived *nit* mutants, two strains were assigned to VCG1, 53 to VCG2 and 128 to VCG4. Four strains failed to produce *nit* mutants. An additional 47 strains from nine US states were placed as two in VCG2 and 45 in VCG4, which was subdivided into VCG4A and VCG4B. Isolates from VCG4A were weakly compatible with VCG3, but VCG4B strains were wholly incompatible with VCG3. The use of the term 'strain' in this and other studies as an apparent substitute for 'isolate' leads to much misunderstanding, especially for example in the context of the P_1 and P_2 strains where a more profound taxonomic distinction exists. Pathogenicity and virulence of isolates and VCGs were determined by computing areas under foliar senescence progress curves (AUSPC) for weekly foliar ratings, during a 14–56 day period using an integrating formula (Campbell and Madden, 1990). Most potato isolates in Ohio and in other US states came under VCG4, but the most virulent were in VCG4A compared with VCG2 and VCG4B. Corsini *et al.* (1985) earlier found that potato isolates in VCG4 were more virulent to potato than potato isolates in VCG3 (using Puhalla and Hummel's (1983) mutant technique) on a cotton isolate in VCG1. On this evidence, Joaquim and Rowe (1991) inferred the existence of two potato pathotypes (races), but the pathogenicity tests employed only a single cultivar, cv. Superior. Strausbaugh (1993) found essentially the same pattern in Idaho but with an additional group (VCG4A/B). Using *nit* mutants, Strausbaugh *et al.* (1992) re-examined the 26 strains placed by Puhalla and Hummel (1983) in 16 VCGs. These authors placed three isolates in VCG1, 13 in VCG2, seven in VCG4 and one to a newly assigned VCG5. The isolates placed by Joaquim and Rowe (1990) in VCG3 were assigned to VCG4 by Strausbaugh *et al.* (1990). Potato isolates are normally found in VCG4, but Strausbaugh *et al.* (1990) found isolates from one site in California which all fitted in VCG1. The authors claimed that VCG isolates of *V. dahliae* were very stable and never mutated to another group. Additional isolates tested in the Joaquim and Rowe (1991) study were from cotton (assigned to VCG1); pepper and pistachio (VCG2) and tomato (VCG3). Following this work, Rowe *et al.* (1997) extended the study to potatoes in Washington, Oregon and eastern Canadian provinces using *nit 1*, *nit M* or *nit 3* mutants, paired against known tester strains; all isolates were VCG4. Western US isolates were 78% VCG4A, 15% VCG4B and 7% VCG4AB. From 400 western and eastern North American and Canadian tubers, 25 and 21%, respectively, were carrying *V. dahliae*.

Nagao *et al.* (1994) failed to establish VCGs in Japanese isolates using melanin synthesis-deficient mutants alone. Subsequently using *nit* mutants, Nagao *et al.* (1994, 1995) demonstrated substantial VCG diversity in Japanese isolates of *V. dahliae*. *V. dahliae* isolates were placed into six pathotypes based on the response of five hosts. *Nit* mutants were induced on a minimal agar medium with 3% $KClO_3$. Two complementary mutants, *nit*-I and *nit*-M, were paired with all combinations on the minimal medium for 20 days. Three main VCGs were found: VCGJ1 (pepper pathotype), VCGJ2 (tomato) and JCGJ3 (aubergine). VCGJ1 was compatible with J2 and J3, but J2 and J3 were incompatible with each other. An isolate pathogenic to both tomato and pepper was compatible with J1 and J3, but surprisingly was incompatible with J2. Ebihara *et al.* (1997) attempted a comparison of VCGs of Japanese isolates with the now established standard ones of Rowe and co-workers. Only tomato and pepper were cited as hosts. Thirty-two isolates designated VCGJ2 and VCGJ3 corresponded to Joaquim and Rowe's (1991) VCG2A and 2B, respectively. The position of VCGJ1 is questionable since this reacted with VCG2B. Adding further confusion, a tomato isolate of race 2 was apparently compatible with VCG2B and VCG4. Seven of 42 isolates would not produce *nit* mutants (see also Wakatabe *et al.*, 1997). A valuable attempt to correlate Dutch *nit 1* and *nit M* complementary mutant isolates with European (UK and Greece) and American isolates was carried out by Rataj-Guranowska and Hiemstra (1997). However, the arbitrary allocation of group numbers based on a limited selection of isolates from different host genera adds confusion to a complex situation. Thus, Hiemstra and Rajaj-Guranowska (1997) designated isolates pathogenic to ash, maple, potato, strawberry, phlox and from soil as VCG NL1, and those from *Forsythia, Syringa Rubus, Ribes, Rosa* and *Chrysanthemum* to VCG NL2. Group NL1 corresponded to the UK β, and the Greek VCG1, and to two US groups VCG3 and VCG4A and 4B. The Dutch group NL2 corresponded to UK α, Greek VCG11 and the US groups VCG1 and VCG2 (Rataj-Guranowska and Hiemstra, 1997). The Dutch findings illustrate clearly the need for full international cooperation and the exchange of universally recognized and designated testers, as with the OARDC reference strains distributed by Rowe and co-workers from Ohio USA.

Genetic relationships in populations of cotton strains and isolates of *V. dahliae* have received much attention. In the USA, a survey of 100 New Mexico isolates from cotton and *Capsicum annuum* (chilli pepper) grouped according to plant or soil source origin was conducted by Riggs and Graham (1995). Using *nit* mutant testers, all cotton isolates were of VCG4A and those of pepper were VCG3. A similar survey of 27 *V. dahalie* strains from Africa, Asia, Europe and the USA was based on approximately 500 *nit* mutants (Daayf *et al.*, 1995). The P_1 strain and race 3 on cotton both fell into VCG1 and were non-pathogenic on tomato. Non-defoliating (P_2 strain) types and races from tomato were included in VCG2 and VCG4. Hyal mutants derived from wild-type isolates always came into the parental VCG. The authors indicated that subpopulations (VCGs) of *V. dahliae* might not be completely genetically isolated.

The cotton-growing republics of the CIS, Kyrgyzstan, Uzbekistan, Kazakhstan, Turkmenistan, Tajikistan and Azerbaijan, have been the centres of much research on the genetics of pathogenicity of *V. dahliae*. Akimov (1997) in a limited study on 28 strains from Tajikistan and 10 from Uzbekistan mostly from cotton but including some from soil, okra, tomato and cucumber, found that all were readily self-compatible and compatible with a single (undesignated) *nit* tester strain. The findings of Akimov and Portenko (1996) and partially of Akimov (1997) were refuted by Portenko and Akimov (1997). In their later study, the dominant strain in Middle Asia (previously assigned to VCG1) using OARDC testers 115 and T-9 was VCG B. VCG1 (P_1 strain) was a minor strain, apparently undetected by Akimov (1997).

In Greece, 23 cotton isolates were VCG2 (= P_2), which also included isolates from tomato and watermelon. Two tomato isolates only were assigned to VCG4. Nine isolates failed to complement any of the testers (Elena, 1997). A subsequent study (Elena and Paplomatas, 1998) employing 44 isolates of *V. dahliae* from various diseased hosts identified three groups VCG2A or B (17 isolates),

VCG3 (two) and VCG4A or B (eight). Seventeen isolates could not be correlated with known VCGs. In a study of 71 Greek cotton isolates, Elena (1999) found 46 were VCG2, two were VCG4 and VCG1 (a first report for Greece), while 22 strains [*sic*] (isolates) were non-compatible with any tester. In a later study (Elena, 2000), all 17 Greek watermelon isolates of *V. dahliae* corresponded to VCG2. Gennari *et al.* (1997), in a study not correlated with testers from other laboratories, examined 79 monospore cultures of *V. dahliae* from tomato, pepper and melon using *nit* mutants. *Nit M* was the dominant mutant. Three groups were recognized: VCGA, including isolates from all hosts; VCGB, confined to some tomato isolates; and VCGC, to some pepper isolates. Of the isolates tested, 42% were self-compatible and hence did not belong to a VCG group. *No correlation could be found between VCG group and pathogenicity, a finding in common with other reports.* Details of *nit* mutant derivation and the production of random DNA probes were described by Paplomatas and Elena (1995). Tian *et al.* (1998a) found that 5-tricyclazole-tolerant, 5-carbendazim-tolerant and *nit* mutants all lost tolerance in culture to a greater or lesser extent, reverting to wild-type. Mutant phenotypes of *nit* and carbendazim-tolerant mutants, however, became stable after inoculation on cotton.

The complexity and diversity of VCG studies is well illustrated by a comprehensive countrywide survey in Israel by Korolev *et al.* (1997, 1999). Several hundred isolates were assigned on the basis of *nit* mutant complementation using OARDC tester strains as follows: VCG2A (26 isolates) occurring 8% in northern Israel and 3% in southern regions; VCG2B (128 isolates) all from the north; VCG4B (375 isolates) all from the south. There was no correlation with host origin. Most crops in the north (cotton, aubergine, weeds and chrysanthemum) were infected with VCG2B and the remainder with VCG2A. All southern crops (cotton, potato, aubergine, tomato, groundnut and weeds) were infected with VCG4B and seldom with VCG2A. The distribution of pathogenicity was similarly diverse: VCG2A and most VCG2B, irrespective of crop origin, induced weak symptoms on cotton and severe symptoms on aubergine (the universal suscept). Two cotton isolates of VCG4B induced severe symptoms on cv. Acala SJ-2 cotton and cv. Black Beauty aubergine, whereas all cotton isolates of VCG2B induced severe symptoms and death in cotton but only moderate symptoms in aubergine. VCG2B isolates from other hosts were more severe on aubergine (the more usual response) than on cotton (see also Bao *et al.* (1998) for the non-correlation of host isolate and VCG). In a comprehensive survey, involving 565 isolates from 13 host species at 47 geographical sites, Korolev *et al.* (2000) confirmed that 92% of 158 isolates of VCG2B were found in northern Israel, 90% of 378 isolates of VCG4B in the south while 28 isolates of VCG2A were geographically scattered. The authors concluded that all isolates in VCG2A and 86% of isolates in VCG4B, irrespective of location, induced weak to moderate symptoms on cotton, corresponding to strain P_2, while inducing severe symptoms on aubergine, whereas all isolates in VCG2B corresponded to the defoliating strain P_1 on cotton but caused only mild symptoms on aubergine. In Israel, at least, there appears to be some geographical divergence of genotype, but the complexity is such that the significance of this in fungal genetics is difficult to interpret at present. The situation on cotton in Israel was compared with Spain where cotton is now an important crop (Korolev *et al.* 1997). VCG2A and VCG4B, both of the P_2 non-defoliating strain with different infection frequencies, were distributed throughout cotton fields in both Mediterranean countries, whereas VCG1 corresponding to the P_1 strain was exclusive to Spain and VCG2B, described as 'defoliating-like', was restricted to Israel. In China, Xia *et al.* (1998) placed 102 isolates of *V. dahliae* from cotton and aubergine in VCG2, with variable compatibility and variable non-defoliating disease severity on cotton. Eleven isolates designated VCG1 confirmed the presence of the P_1 strain in China. One isolate was self-incompatible. It is by no means clear how these results correlate with those of Ma *et al.* (1998) or Zhou *et al.* (2000).

One technical difficulty in reaching conclusions with many VCG studies is the relatively limited range of hosts and isolates employed and the often restricted range of sampling. Similarly, the significance of the term 'subgroup' without clarification, viz-à-viz repeated and confirmed results, and the confusion of laboratory culture coding with a widely-accepted terminology is often misleading (Nagao *et al.*, 1998). A study of 22 isolates of *V. dahliae* from green soybean, udo (*Aralia cordata*), horseradish (*Amoracia rusticata*), sweet pea, or *Chenopodium album* by Ebihara (1999a) found that all were of weak pathotype E. These were divided into soybean pathotype and isolates non-pathogenic to soybean. *Nit M* and *nit 1* mutants of each isolate were paired with VCG testers. Fourteen isolates from soybean and udo were assigned to J3, and U108 to J2. An isolate from horseradish was not compatible with any VCGJ tester. The soybean pathotype E could not be differentiated from other isolates of E by VCG. The current picture in Japan based on a limited number of isolates appears to be of three VCGs: VCGJ1 (eight isolates), VCGJ2 (seven) and VCGJ3. J1 was compatible with J2 and J3, but J2 and J3 were only weakly reacting. Ebihara *et al.* (1999b) described the provisional assignment of 56 Japanese isolates to four subgroups corresponding to: J1 = VCG2A/B; J2 = VCG2A; J3 = VCG2B; and J4 = VCG2A/B and 4A. The picture was complicated by cross-compatibility of 'bridging strains'. No Japanses isolates correponded to VCG1 and VCG3. The apparently simple distribution of tomato and pepper pathotypes into single VCGs does not accord with the picture in other countries. A signal study on tomato isolates in Ontario, Canada by Dobinson *et al.* (1998) illustrates this point well. This research is particularly valuable in that it represents one of only few attempts to analyse VCGs using DNA manipulation. Fourteen tomato isolates analysed by restriction fragment length polymorphisms (RFLPs), random amplified polymorphic DNAs (RAPDs) and DNA fingerprints detected by hybridization to a dispersed, repetitive genomic probe were classified into five DNA types. These were type I (two non-pathogenic isolates); type II (four race-2 and three non-pathogenic); type III (one race 2); type IV (one race 2 and two race 1) and type V (one race 1). Isolates of the same DNA type were compatible, as were types II and III = VCG4B; type IV = VCG2A; type V = VCG2B. This result belies a simple interpretation of the significance of VCGs and points to multiple origins of the race 2 tomato pathotype, at least in Ontario. Thirty-one vegetable isolates of *V. dahliae* in Italy were placed into three unclassified VCG groups based on *nit* mutant complementation. Pathogenicity was variable and, in common with most studies, was not correlated with VCG (Gennari *et al.*, 1997).

Rowe and Botseas (1995) describe an interesting synergism between *V. dahliae* and the potato root nematode *Pratylenchus penetrans* in the potato early dying syndrome. No difference in virulence on potato was found between isolates from VCG4A and 4B in eelworm-free soil. In *P. penetrans*-infested soil, however, VCG4A isolates showed higher disease severity accompanied by lower tuber yields than VCG4B isolates. The authors speculate on the existence of several VCGs in field soil and the difficulty in interpreting field data based on the interaction, or not, of one or more VCG groups' isolates with soil nematodes.

Chen (1994) found that 42 isolates of *V. dahliae* from a diverse range of woody ornamental hosts could all be assigned to three VCG groups, 30 to VCG1, two to VCG2 and four to VCG4. The weakly pathogenic soilborne species *V. nigrescens* and *V. tricorpus* exhibited a greater genetical diversity than *V. dahliae* (Korolev *et al.*, 1997). Mutants were mostly *nit 1* and a small percentage which were unable to utilize hypoxanthine were designated *nit M*; *nit 3* mutants were only recovered from *V. nigrescens*. Biochemical complementation occurred between different mutant phenotypes derived from the same parent strain viz: *nit 1* × *nit 3*; *nit 1* × *Nit M* and *nit 3* × *Nit M*. There was no correlation between host or geographical location and VCG. For *V. tricorpus* VCGs, one, two, four, seven and eight were from potato; three and four from weeds; and one, three, five and six from soil. In *V. nigrescens* VCGs, one, four, five and six from potato; four and five from weeds; two from cotton and aubergine; and three, seven and eight from soil. Isolates of both species failed to induce symptoms in cotton and aubergine but colonized root and hypocotyls.

At the present state of our knowledge, it is still not clear whether *V. dahliae* should be regarded as a single interbreeding population, or *sensu* Puhalla and Hummel (1983) a series of genetically independent subpopulations each one (a V-C group) capable of hyphal anastomosis resulting in heterozygous diploids and parasexual genetic recombination (Hastie, 1981). Moreover, in view of the differential findings of Puhalla and Hummel (1983, 1984), Typas and Heale (1976a) and Clarkson and Heale (1985a), it is not clear whether the patterns and mechanism of genetic variability in *V. albo-atrum* are the same as in *V. dahliae* or merely represent differences in the techniques used to derive auxotrophs.

In nature, however, reports of diploid wild-type strains in either species are rare, and the role of parasexuality (rather than the possibility of its occurrence) in the field, therefore, remains to be demonstrated. It is none the less clear, however, that strains of *V. dahliae* (Puhalla and Hummel, 1983) and *V. albo-atrum* (Hastie, 1981) exist which are incapable of self-anastomosis or heterokaryon formation.

Phialide Analysis

The conidiogenous cells and conidia of *Verticillium* species are uninucleate. Hastie (1967, 1968) developed the technique of studying the formation of new genotypes by the micromanipulation of each successive conidium as it seceded from the conidiogenous cell. Each conidium was grown to discover its nutritional requirements, from which results the nuclear division of the conidiogenous cell in which recombination had occurred could be deduced. The collective descendants of each conidiogenous cell were referred to as a 'phialide family'. Conidia are either diploid, haploid or aneuploid. (Hastie (1981) estimated the frequency of aneuploid formation from diploid nuclei as 0.035 per nuclear division, with mitotic recombination and haploidization occurring in single phialide populations.) Hastie (1968), using phialide analysis to map centromeres, placed eight auxotrophic marker genes in four linkage groups and also linked *so-1* to *arg-9*. Typas and Heale (1978), using recombination frequencies in the absence of phialide analysis in *V. albo-atrum* and *V. dahliae*, placed 33 marker loci for both species in three large and two small linkage groups corresponding to five chromosomes. The existing imprecision with mapping techniques to date, however, precluded the construction of a meaningful chromosome map (Heale, 1988).

Cytoplasmic Inheritance

The instability of isolates of *V. dahliae* and *V. albo-atrum*, whereby wild-type pigmented cultures lose the ability to form microsclerotia or resting mycelium and revert to a hyaline (hyl^-) state, is well known (Isaac, 1949; Pegg, 1957; Robinson *et al.*, 1957). Most commonly it is a laboratory cultural phenomenon, but also occurs *in planta* when only hyl^- cultures can be reisolated from hyl^+ wild-type inoculations (Pegg, 1957). Various mechanisms have been postulated to account for hyalinity, such as mutation (Pethybridge, 1916), genetic recombination (Robinson *et al.*, 1957), ploidy changes (Tolmsoff, 1972, 1973) and cytoplasmic factors (Hastie, 1962). Hastie (1962) eliminated a chromosomal basis for hyalinity by demonstrating that heterokaryons of auxotrophs of hyaline and dark cultures derived from hop isolates of *V. albo-atrum* failed to segregate at conidiogenesis unlike nutritional markers carried on nuclear genes. Heale (1966) using lucerne and tomato strains of *V. albo-atrum* obtained similar results, which were attributed to the failure of a cytoplasmic particle to pass from the phialide to the conidium. Hastie and Heale (1984) designated wild-type dark and hyaline cultures as *hyl*$^+$ and *hyl*$^-$, respectively. Heale (1966) reported *hyl*$^-$ variants which reverted spontaneously to *hyl*$^+$. Those cultures with a greatly reduced ability to produce normal melanin were sensitive to medium and environment and were termed partial hyaline (*hyl*$^+\rho$) by Typas and Heale (1976a) and may reflect the number of *hyl*$^+$ factors per cell. The sooty (*so*) mutant was shown to be a nuclear character (Hastie, 1968) requiring *hyl*$^+$ cytoplasm for its expression (Typas and Heale, 1978).

Stable hyaline variants produce more aerial mycelium, have a higher linear growth rate, but form fewer conidia than the dark pigmented wild-type. KCN 10^{-3} M inhibited O_2 uptake in *hyl*$^-$ hyphae and the converse in *hyl*$^+$, suggesting a mitochondrion origin of *hyl*$^+$ (Pilkington and Heale, 1969). Confirmation of the mitochondrion as the source of *hyl*$^+$ was provided by Typas and Heale (1979), who restored melaninogenesis by micro-injection of *hyl*$^+$ mitochondria into *hyl*$^-$ mycelium. Typas (1984) found that amytal (*amy*$^\alpha$) resistance, a mitochondrial marker, was linked to *hyl*$^+$. Acriflavine was used by Typas and Heale to induce mitochondrial mutations. Bell (1992a), described three distinct hyaline or albino mutants in *V. dahliae*: (i) true albino (*alm*) mutants which make normal numbers of albino microsclerotia; (ii) largely hyaline colonies (*rms* or *hyal* mutants) (these form dark microsclerotia only if treated with catechol and are similar to Typas and Heale's (1976a) *hyl*$^+\rho$ mutant); and (iii) hyaline *rms* or *hyl*$^-$ mutants due to mitochondrial mutations and which do not respond to catechol treatment. The *alm* and *rms* variations are due to mutations or changes in nuclear DNA. Both *alm* and *rms* (*hyal*) mutants lack aerial growth and have faster growth rates than the wild-type. Based on a conidial analysis of heterokaryons of UV-derived auxotrophs from *V. dahliae* from cotton, Shevtsova and Zummer (1988) considered that the character for a mycelial–yeast dimorphism (*myd*) mutant was cytoplasmically carried and could be reversed by cytoplasmic transfer in heterokaryon bridges. Each mycelial cell contained 20–30 mitochondria. Successive passage of *V. dahliae* through resistant hosts led to the loss of microsclerotia and development of a hyaline mycelium controlled by a mitochondrial gene *hyl*. Microsclerotia were thought to be controlled by nuclear genes influenced by *hyl*$^+$ mitochondrial factors (Shevtsova *et al.*, 1982). Evidence on pathogenicity and survival of *hyl*$^-$ variants in nature is conjectory and contradictory. Pegg (1974) and Tjamos (1981) reported no loss of pathogenicity with *hyl*$^-$ cultures, while Isaac (1949), Smith (1965) and McHugh and Schreiber (1984) reported to the contrary. Daayf *et al.* (1998) obtained hyaline subclones from a P_1 strain wild-type *V. dahliae* clone from cotton. One subclone (V7-2) exhibited weak pathogenicity to cotton, unlike the other (V7-7) which had wild-type (V-7) defoliating virulence. V7-7 grew better than V7-2 over a wider range of temperature, and while both subclones grew on NH_4 on an N_2 source, V7-7 grew more vigorously on NO_3-N_2 and exclusively on NO_2. Both subclones belonged to the same VCG, illustrating that the above differences were not related to heterokaryon compatibility. Root-dip and stem inoculation would minimize differences in field virulence of *hyl*$^-$ and *hyl*$^+$ (Heale, 1988). Mohan and Ride (1984) reported that *hyl*$^+$ (serotype 1) isolates of *V. albo-atrum* from hop with high sporulation and low PG production changed to *hyl*$^-$ (serotype 2) with aerial mycelium, low sporulation and high PG. Neither the loss of pigmentation of sporulation nor the PG-producing potential *in vitro* were correlated with pathogenicity. While loss of pigmentation theoretically would threaten survival outside the host from lysis and UV mutation (Heale, 1988), experimental evidence is lacking and hyaline soil fungi are commonplace (but see Bell, 1992b).

Genetics of Pathogenicity – Strains, Races and Virulence

Since there is still little information regarding the origin of pathogenic variation in the field and an inadequate understanding of the basis of host resistance, our state of knowledge of the genetics of host–pathogen relationships must at present be regarded as rudimentary. Until evidence suggests otherwise, each species must be regarded as independent and considered under particular host groups. However, work on VCGs and DNA polymorphism comparisons suggests the existence of populations of subspecific strains with stable affinities forming complex groups of pathotypes, with notable exceptions crossing host boundaries (see also Chapter 10).

Cotton

Following the findings of Schnathorst and Mathré (1966a) and Schnathorst (1971) of the existence of populations of *V. dahliae* varying in virulence, two strains have emerged. The most virulent, a defoliation type (comparable with progressive isolates of *V. albo-atrum* in hop) formerly designated the T_1 or T_9 types, was shown by Puhalla and Hummel (1983) and by Joaquim and Rowe (1990) to be a genetically isolated self-compatible group. Using Joaquim and Rowe's (1990) terminology, all defoliating isolates are designated P_1. P_2 isolates represent the second strain, less virulent than P_1 and non-defoliating, which formerly was termed SS_4. The P_1 strain is indigenous to most of the cotton-growing areas of the USA and with the P_2 strain in California. The P_1 strain may have been distributed via seed transmission to other countries (Bell, 1992a) and has been found in Peru (with the P_2 strain) (Schnathorst, 1969), Spain (Blanco-Lopez *et al.*, 1989), China (Oingii and Chiyi, 1990) and Mexico (Schnathorst, 1971). In the CIS, an extensive cotton-growing area, five races (0, 1, 2, 3 and 4) have been recognized; these are identified in terms of virulence on cultivars 108F and Tashkent 1 and a *Gossypium hirsutum* subspecies *neglectum* line 0144 of *G. arboreum* L. (Popov *et al.*, 1972; Portenko and Kas'yanenko 1978, 1987). Races 2 and 3 predominate in Tajikistan, defoliate Tashkent cultivars which are similar in reaction to *G. hirsutum* cv. Acala 4–42. CIS races 2 and 3 are equivalent to the P_1 strain, while races 1 and 4 non-defoliating types approximate to P_2. The P_1 strain is also differentiated on *G. hirsutum* cv. Deltapine 15 and *G. barbadense* cv. Tanguis 2885. Much confusion surrounds the meaning and the weighting of the terms 'strain', 'race' and 'pathotype', especially in CIS studies. The genetics of cotton species and cultivars is more complex than that of tomato (*Lycopersicon*), where single-gene host-specific reactions have led to the recognition of clearly defined races. The work of Schnathorst and Mathré (1966a), Schnathorst (1969, 1971) and, more recently, Puhalla and Hummel (1983) and Joaquim and Rowe (1990) has condensed all the cotton strains to two pathotypes (collective strains). Schnathorst and Evans (1971) compared the relative virulence of US and Australian isolates of *V. albo-atrum* [*sic*] (*V. dahliae*) and Schnathorst and Sibbett (1971) compared the virulence of Californian isolates on cotton and olive. Bell (1973), on the basis of screening a world collection of cotton *V. dahliae* isolates on cotton cultivars from all continents and ten additional hosts, concluded that a gradient of virulence existed in the different isolates which was strongly affected by environment, such that a separation into specific races was not justified. Bell (1973) claimed that environment could change a P_1 reaction to a P_2 type. It is abundantly apparent from the literature that the terms strains, races and occasionally pathotype are used loosely, frequently without correlation with other studies. Such terms are cited here in their original context.

The selection of strains of *V. dahliae* in the field and their role on the useful life of cotton cultivars was discussed by Ashworth *et al.* (1984). Kas'yanenko (1987) claimed that virulence of *V. dahliae* was controlled by at least three nuclear genes located on different chromosomes. Six races were identified in Tajikistan, race 3 arising either by recombination of races 1 and 2 or by a mutation of an *av* gene to virulence in one of these races. Subsequently, Portenko and Kas'yanenko (1987) recognized races 0, 1, 2, 3 and 4. Race 0 was avirulent on all cotton testing cultivars. A UV mutant of race 1 showed all the characteristics of race 3, appearing to involve a single virulence gene. In a study of the virulence of cotton isolates in soil with a previous mixed cropping history, Koroleva and Kas'yanenko (1987) found that cv. Tashkent 1 race 1 was nearly totally eliminated and replaced by a dominant race 2. After 3 years rotation with lucerne, race 2 predominated together with less aggressive strains of race 2 together with R1, R3 and R4. Line 108F and Tashkent 1 were strongly attacked but not line 0144 of *G. arboreum*. New races and biotypes were attributed to genetic recombination following anastomosis (Khokhryakov, 1976). UV-derived mutants of *V. dahliae*, showing increased virulence over the wild-type, had higher levels of ploidy (Safiyazov and Cherkasova, 1979). The incorporation of marker genes into biotypes of races 2 and 3 did not alter their virulence on *G. hirsutum* cultivars 108F and Tashkent 1 or *G. arboreum* cv. 0144. The reproduction of race 2 on cv. 108F, however, was lower than race 1 following mixed inoculation (Kas'yanenko, 1980). Kas'yanenko and Portenko (1978a) proposed that somatic recombination could be the basis for the source of new races of

V. dahliae in cotton. Molchanova and Kas'yanenko (1978) described the production of induced mutants of *V. tricorpus* used in cotton wilt studies.

Portenko (1990), based on parasexual studies, claimed that virulence to cotton was controlled by recessive alleles of nuclear genes with cytoplasmic (mitochondrial?) factors affecting aggressiveness. Ryabova (1990) described race 2 as dominant in Tadzhikistan and Uzbekistan, but in the presence of races 1, 3 and 4. Races 1, 3 and 4 are morphologically identical.

Kas'yanenko and Shevtsova (1981), on the basis of hybrid analysis of conidial progeny of heterozygous diploids, postulated that the *V. dahliae–Gossypium* host–parasite system involves three virulence genes, $V1_{v1}$, $V2_{v2}$ and $V3_{v3}$, and three resistance genes, $R1_{r1}$, $R2_{r2}$ and $R3_{r3}$, with disease development possible only with the combination of respective virulence and resistant genes, all gene pairs being independent. Each virulence gene is considered to have a series of alleles controlling aggressiveness (Kas'yanenko, 1990). Ibragimov and Ismailov (1976) claimed that an avirulent strain 26 (race 0) of *V. dahliae* had a lower rate of aminoacyl-tRNA synthesis for lysine, phenylalanine and aspartic acid than race 2 or the cotton cv. 450-555. Ismailov and Rysbayeva (1990) found that aggressive mutants of *V. dahliae* obtained chemically and by UV irradiation possessed more isozymes of oxidases, oxidoreductases and hydrolases than less aggressive mutants. Baryshnikova (1990), studying heterokaryon diploids in a genetic analysis of 13 adenine-independent *V. dahliae* mutants, found five complementation groups corresponding to five loci controlling adenine biosynthesis.

In the most recent study, Portenko *et al.* (1995) conducted a comprehensive survey of vegetative compatibility based on *nit* mutants, of 24 cotton strains of *V. dahliae* from Middle Asian cotton and soil from races 0, 1, 2, 3 and 4 on cotton cultivars. These strains were also compared with 33 strains from cotton, tomato, aubergine, pepper, strawberry, cucumber and okra, variously from the USA, France, Spain, the UK, Morocco, Moldavia, Israel, Russia and Tajikistan. All Middle Asian isolates belonged to a single VCG corresponding to the P_1 group of Joaquim and Rowe (1990). Of the 33 strains, 31 were in VCG1 and a non-defoliating cotton strain was in VCG1 (P_2). A UK strawberry isolate (strain 345) was confined to a separate VCG. No self-incompatible strains were found. Five banding patterns of non-specific esterases as seen in polyacrylamide gel electrophoresis were found, three for P_1, one for P_2 and one for strain 345. The confusion existing around the identification of 'strain', 'collective strain' or 'isolate' extends to recent reports from China, where the cotton crop and *Verticillium* wilt are important. Wu *et al.* (1995) describe 16 strains [*sic*] of *V. dahliae* isolated from the Shandong province. Most of these, following testing in cultivars Lumian, Sumian and Zhong at 1.3×10^7 conidia ml^{-1} inoculum concentration, corresponded to P_2 (non-defoliating). Symptoms were confined to foliar streak, chlorotic lesion and wilting. Three Chinese isolates [strains *sic*] VD8 and two from Shandong SD5 and SD13 were defoliating, P_1 types. In compatibility tests, these strains were all in VCG1. However, the authors claim that strain SD6, a P_2 type based on symptoms, was compatible with VCG1 isolates. A study by Song *et al.* (1995) reduces Chinese (Anyang) strains to three types: filamentous, nucleate and intermediate in apparent order of pathogenicity, inferring that nucleate and filamentous types all corresponded to P_1 strains. In the Hebei cotton-growing region, Ma *et al.* (1998) using cluster analysis of disease indices on six cultivars, but on three *Gossypium* spp., classified 40 of 60 'strains' [*sic*] (= isolates?) from 37 major cotton counties into three virulence groups (VGs). VG1 was virulent on Sea Island and upland cultivars but arvirulent on *G. aboreum*. 'Strains' of VGII were moderately, and VG3 weakly virulent. Zhou *et al.* (2000) working in the Jiangsu province describe three virulence pathotypes. Using a range of *G. hirsutum* cultivars, R01–R14, the P_1 (T_9) strain from the USA was virulent on cultivars R02, R04, R05, R06, R08, R09, R11 and R14. The Chinese strain of P_1 (isolate VD8 [*sic*]) was non-pathogenic to R05, R06, R09, R11 and R14. Cv. R01 was tolerant to the US race of P_1 used in their experiment and to all Chinese isolates tested. The authors, with little or no corroborative evidence, tentatively proposed the adoption of seven races of the Chinese non-defoliating P_2 strain.

These papers emphasize most clearly the need for the adoption of a universal system of identifying and cataloguing strains or pathotypes and clearly separating these from mere isolates. Where the term 'strain' or 'race' is used, it must be widely recognized and accepted in that country, based very clearly on differential host/cultivar responses. This need for a commonly agreed international nomenclature – a challenge for joint laboratory and fieldworker collaboration – will require a comparison of molecular studies and an exchange of specific host germplasm, not only between countries, but also between different groups working in the same country.

Tomato

By comparison with hop and cotton, the host pathogen genetics in tomato are much simpler. A single gene *Ve* from the Peruvian wild species *Lycopersicon pimpinellifolium* was incorporated into commercial cultivars of *L. esculentum* by Blood in 1925 (Bryan, 1925; Schaible *et al.*, 1951; Goth and Webb, 1981) and confers resistance to race 1 of *V. dahliae* and to strains (host pathotypes from tomato and hop) of *V. albo-atrum* (Pegg and Dixon, 1969). Race 2 pathogenic to race 1 resistance appeared in 1962 (Alexander, 1962) and subsequently in France (Laterrot and Pécaut, 1966), Italy (Cirulli, 1969) and Morocco (Besri *et al.*, 1984; Besri, 1990). The origin of race 2 is purely conjectural and is assumed to have arisen by mutation or parasexual recombination. Grogan *et al.* (1979) suggested that race 2 already in existence with race 1 increased under the selection pressure from an increased planting of Ve lines. Race 2 isolates showed a continuum of virulence suggestive of genetic variation at loci other than the *Ve* gene.

An interesting report by Besri (1990) suggests that increased susceptibility to colonization in race 1 and race 2 resistant cultivars could affect the selection pressure on the emergence of new races. If mutations occurred in host plants predisposed to colonization, the possibility exists of new virulent pathotypes becoming naturalized in erstwhile resistant cultivars. To date, however, there is little or no field evidence to support this hypothesis.

The genetics of resistance and hence the interpretation of pathogen behaviour are not entirely straightforward. In addition to the single dominant gene labelled 1965 (Shapolov and Lesley, 1940), some cultivars released prior to 1940 had multigenic resistance, and many modern 'isogenic' cultivars possess different gene modifiers. Okie and Gardner (1982a) claim that F_1 hybrids heterozygous for the *Ve* gene were less resistant to race 1 than homozygous F_1 hybrids, suggestive of incomplete dominance. Bender and Shoemaker (1984) considered that the expression of virulence in race 1 and race 2 isolates may differ in relation to the nature of toxic metabolites produced *in vivo*. This could also reflect the reaction of different cultivars to *v* and *av* pathotypes where colonization was similar in R and S cultivars while S plants remained symptomless (Blackhurst and Wood, 1963b). Field resistance to *V. dahliae* race 2 of a polygenic nature was reported by Hubbeling *et al.* (1971) for cv. Heinz 1350 but was incomplete and depended on soil type and pH. This is reminiscent of the findings of Bell (1973) for cotton and Sewell and Wilson (1984) for hop. Hubbeling and Basu Chaudhary (1969), earlier working on *V. albo-atrum*, demonstrated the effect of environment on virulence, or resistance. Susceptible plants grown in soil with a high pH and an excess of Ca^{2+} ions escaped infection or showed only weak atypical symptoms. A comparative pathogenicity trial using four isolates of race 1 and three of race 2 showed that these could only be distinguished on appropriate cultivars (GCR-26 Ve/Ve and GCR-218 ve/ve). Only one isolate (R1) caused foliar symptoms on tobacco; all isolates (R1 and R2) induced mild foliar symptoms on cabbage and none on French bean. All isolates (R1 and R2) caused foliar symptom stunting and loss of dry rot in aubergine, but stunting and dry weight loss were variable symptoms in other hosts, as was recovery of the fungus in cabbage, tobacco and pepper (Mingochi and Clarkson, 1994).

A similar study from Greece (Vloutoglu *et al.*, 1997) involving 29 isolates of *V. dahliae* from cotton, 19 from tomato and 17 from watermelon showed that all isolates were highly virulent on watermelon cv. Sugar Baby 67. Cotton isolates were mildly virulent or avirulent on tomato, whereas tomato and watermelon isolates were highly virulent on a race 1 susceptible tomato cv. Early Pak No 7. Five tomato isolates (presumably race 2) and six watermelon isolates caused moderate symptoms on the race 1 resistant tomato cv. Ace 55 VF. Most tomato isolates were moderately virulent on cotton. A long-term study by Goverova and Govorov (1997a) involving 2000 isolates of *Verticillium* (mainly *V. dahliae*) from 20 cultivated and wild species from southern regions of Russia and the CIS, revealed six 'physiological races' on tomato, three on pepper, three on aubergine and six on strawberry. Since this region covers one of Vavilov's 'centres of plant diversity', a comparison *inter alia* of these tomato races with the recognized tomato races of the rest of the world is required to establish what is regarded as a discriminating genotype.

Hop

The first record of hop wilt caused by *V. albo-atrum* was by Harris (1925b) on *Humulus lupulus* cultivars Fuggle and Tolhurst. Since the 1940s, the pathogen has spread to all hop-growing areas in the UK and is now endemic in Germany, the USA, Poland, Bulgaria and New Zealand (Keyworth, 1944b; Wilhelm, 1981). Isaac and Keyworth (1948) recognized two types of isolate – a weakly aggressive one causing mild wilt designated the fluctuating strain (indicative of an intermittent symptom pattern) and a virulent one called the progressive strain. The original virulent strain, subsequently called race PV_1, was pathogenic to cv. Fuggle – a universal suscept. Keyworth (1947) released a series of cultivars with increased resistance ('tolerance') to PV_1, including cultivars Keyworth's Early, Whitbread's Golding, Defender Density and Janus, all derived from a Manitoba wild hop. Keyworth introduced the term 'tolerance', since the new clones were freely invaded by the fungus and showed incomplete (multigenic) resistance in the form of diminished symptoms. In 1972, Neve (1976) introduced cv. Wye Target with high resistance to PV_2, a progressive race with increased virulence to which all existing cultivars were susceptible, including cv. Wye Challenger resistant to PV_1. Within a few years of the introduction of Wye Target, a new (superprogressive) race PV_3 appeared pathogenic on all cultivars in current production (Sewell and Wilson, 1978, 1984). The interpretation of the genetics of pathogenicity is complicated by a very limited knowledge of *H. lupulus* genetics. Attempts have been made to identify mild and progressive races *ex planta* (Connell and Heale, 1985; Swinburne *et al.*, 1985) but no individual or sets of characters has separated the two types absolutely. Sewell and Wilson (1984) considered that the host sampling of the population of hop *Verticillium* was too limited to obtain an accurate picture. In their view, races PV_1, PV_2, PV_3 and mild represented an artificial separation while in reality the field population represented a continuum of pathogenicity, the final outcome being determined by environment (cf. Bell, 1973). In an important contribution, Clarkson and Heale (1985b) paired complementary auxotrophic mutants of three fluctuating (M18, M33 and M50) and three progressive (PV_1, PV_2 and PV_3) isolates to seek heterokaryon compatibility. Compatibility was based on prototrophic growth on a glucose minimal medium and diploid conidia determined by Feulgen microdensitometry. Prototrophic diploidy was greater between pairings of PV isolates and between PV_1 and the three M isolates. Dual inoculation of hop with various auxotrophic mutants failed to yield prototrophic colonies on reisolation (although both original auxotrophs could be recovered). Similar dual inoculation of *Antirrhinum* (a universal suscept) with the auxotrophs M18 *nic 4 cob 26* and PV_3 *arg 8 pyr 2* yielded a recombinant haploid prototroph with pathogenicity to hop cultivars intermediate between the two parental isolates. On cv. Wye Northdown, the mean wilt scores for M18, PV_3 and M18/PV_3 were 2.2, 7.1 and 8.0, respectively, and for cv. Wye Challenger, 1.8, 7.2 and 2.0. Large-spored diploid isolates were obtained from *Antirrhinum* following inoculation with M18/PV_3 heterozygous diploid conidia. One diploid was highly pathogenic to hop and was recovered from hop as a stable diploid. Notwithstanding the constraints surrounding forced heterokaryons (Joaquim and Rowe, 1990, 1992) and the absence of conclusive proof of parasexual recombination in nature, the work of Clarkson and Heale (1985c) demonstrates that parasexuality is possible and is an effective source of new pathotypes. The further possibility exists that weed non-hosts of *Verticillium*, widespread in most field situations, could function like *Antirrhinum* as a vehicle for heterokaryon formation and a passive source of new pathotypes with no elimination of the host.

Rataj-Guranowska *et al.* (1995) confirmed earlier studies that 16 isolates from different pathogenicity groups (based on virulence on hop cultivars) formed one RFLP group (NL). Two lucerne isolates were from RFLP group L. Of 17 *nit* mutants derived from these isolates, nine were self-incompatible, five were self-compatible, but four reverted to prototrophy. Two isolates from the same pathogenicity group were placed in VCG1, one from another group in VCG2 and a lucerne strain in VCG3. Unlike all other reports, these workers claimed that one hop isolate (527) was vegetatively compatible with a lucerne isolate in VCG3. If substantiated, this is the first evidence that *V. albo-atrum* from lucerne does not represent a unique genotype, and is in marked contrast to the findings of Correll *et al.* (1988), who found that 15 isolates from lucerne all comprised a single VCG, while 17 other isolates from diverse hosts all formed another single VCG.

Lucerne

Wilt of *Medicago sativa* L. caused by *V. albo-atrum* was first described in Sweden in 1918 (Hedlund, 1923). Afterwards it spread to Germany (Richter and Klinkowski, 1938), Great Britain (Noble *et al.*, 1953; Isaac, 1957c) and Europe and the Soviet Union (Kreitlow, 1962). In 1964, lucerne wilt was reported in Canada (Aubé and Sackston, 1964) and in 1976 in the Pacific Northwest of the USA (Graham *et al.*, 1977, 1979; Christen and Peaden, 1981). Although lucerne culture was widespread in central Europe and Asia and records in Greece and China go back to 500 BC and 200 BC, respectively (see Pegg, 1984), distribution of *Verticillium* wilt in Europe and North America from the first report of the disease took a mere 58 years. Since the lucerne pathotype is very host specific (Isaac and Lloyd, 1959; Heale and Isaac, 1963; Delwiche *et al.*, 1981; Christen and French, 1982), its origin, so late in the cultivated history of the crop, and its subsequent widespread occurrence is difficult to explain. Isaac

(1957c) demonstrated that seed transmission and within-crop spread can occur by means of conidial spread on mowing equipment under wet conditions (Pegg, 1984). In one of very few international comparative experiments (Christen *et al.*, 1983), four North American and three European isolates of *V. albo-atrum* were inoculated on to cultivars Agate and Apalachee from the USA and on to the European cultivars Kabul, Sabilt, Europe and Vertus in experiments conducted in the USA and the UK. Notwithstanding that inoculation techniques were different, the results failed to show differences between the European and North American isolates previously suggested (Müller, 1969; Ranella *et al.*, 1969; Gondran, 1976, 1977, 1984; Flood *et al.*, 1978a). Isaac (1957c) and Christen and French (1982) earlier had found no difference in virulence of regional or geographical isolates. It is now generally agreed the North American outbreak was due to the introduction (most probably by contaminated seed) of a European strain. The European epidemic since 1918 was most likely due to the distribution of infected seed possibly from an original Scandinavian centre of origin. Sources of resistance to the pathogen have been limited. Zaleski (1957) found that all UK and French cultivars in the late 1950s were susceptible. Panton (1965) identified highly resistant field-grown plants in crops of susceptible cultivars. The resistance was heritable and under polygenic control. Crosses between parents with high resistance and intermediate resistance, or high and low resistance gave progeny with transgressive segregation for resistance, with the accumulation of resistance in later generations (Panton, 1967a,b,c,d).

The cv. Vertus selected from inoculated healthy plants which none the less were colonized (Lundin and Jonsson, 1975) was released as a tolerant cv. Vertus is no longer reliable in the UK as a resistant (tolerant) cultivar (Pegg, 1984). How much this reflects environmental effects on disease expression rather than the emergence of a more virulent strain is open to question. Basu and Butler (1994) in a test on 32 Canadian isolates of *V. albo-atrum* from lucerne found evidence only for a single strain. Panton (1965) tested four species of *Medicago*, including *M. hemicycla*, and found that all were susceptible. Rogers (1976), however, incorporated resistance from this wild species into a commercial variety to give the resistant Maris Kabul. Latunde-Dada and Lucas (1983) found that clonal variants derived from mesophyll protoplasts of the susceptible Europe were highly tolerant, a character associated with higher ploidy levels.

V. dahliae also affects *Medicago*, but to date is only a minor pathogen. The pathogenicity of 12 isolates of *V. dahliae* from nine host species was examined on the weed milk vetch (*Astragalus adsurgens*) and *inter alia* lucerne by Wei and Shang (1998). Milk vetch was infected by isolates from mung bean (*Vigna radiata*), watermelon, aubergine, cotton, sunflower, tomato, potato, radish and milk vetch, inducing typical symptoms. Mung bean and watermelon isolates induced mild symptoms in lucerne, which was a symptomless host to all other host isolates (see also Strunnikova *et al.*, 1994). Specific races of *Verticillium* in addition to *V. albo-atrum* on lucerne have been described for *V. dahliae* on mint (Fordyce and Green, 1960) and Brussels sprout (Isaac, 1957a). In general and unlike formae of *Fusarium oxysporum*, *V. albo-atrum* and *V. dahliae* attack a wide range of hosts but invariably strains are most virulent on the common dominant hosts. Serial passage of less virulent isolates usually results in increased virulence. Where selection pressure is on pathogen growth correlated with symptom severity, this would be expected. However, McGeary and Hastie (1982) found diauxotrophic isolates of *V. albo-atrum* from potato, tomato and lucerne which showed a pleiotrophic reduction in virulence. Some auxotrophs colonized hosts in the absence of symptoms. Hastie and Heale (1984) similarly reported extensive colonization of the tolerant lucerne cv. Vertus with minimal wilt symptoms. The origin and specificity of host-specific strains of *Verticillium* remain unanswered; nuclear gene mutation of existing soil or pathogenic strains appears to be the most feasible explanation in the absence of another compatible heterokaryon group.

Shevtsova (1990) proposed a structured developmental sequence for *V. dahliae*, *V. tricorpus* and *V. nigrescens*. Critical phases of development termed 'phenocrises' were envisaged, four each in *V. dahliae* and *V. nigrescens* and five in *V. tricorpus*. The last 'phenocrisis' in each species – melanogenesis – was controlled by six nuclear genes, with a special gene regulator in *V. dahliae* and *V. nigrescens*. Sorbicillinogenesis (mycelial orange pigment biosynthesis), the third stage of *V. tricorpus*, was controlled by four nuclear genes and three extrachromosomal genes. Shevtsova postulated that the triggering of the different phases was by extrachromosomal genes to the selective advantage of the species in a particular ecological niche. The hypothesis, which is highly speculative, relies on the existence of a 'single conversion pathway of morphogenes'.

Other host species

A screen of five *Verticillium* species, *V. albo-atrum*, *V. dahliae*, *V. tricorpus*, *V. nigrescens* and *V. nubilum*, *inter alia* with nematophagus and entomophagus species for subtilisin activity (a serine-type protease associated with a wide range of archaebacteria, eubacteria, fungi and higher eukaryotes) by Segers *et al.* (1999) found only trace activity in the four common wilt-inducing species. Substantial levels were present in *V. nubilum* comparable with those found in entomophagus and nematophagus species. There was no cross-reactivity of the *V. nubilum* protease with antisera against subtilisins from source fungi, neither did its *Eco* R-restricted DNA hybridize with probes from these. Since the other wilt pathogens readily produce broad-spectrum trypsin proteases, the association of subtilisin with *V. nubilum* is seen as a reflection of its low virulence and high saprotrophicity. A statement by Bidochka *et al.* (1999) that insect, mushroom and nematode *Verticillia* differ from plant pathogenic species in their ability to produce chitinase does not accord with the published record (Pegg and Young, 1982). The virulence of ten isolates of *V. dahliae* from eight different hosts to an assortment of rapid cycling rape genotypes showed that only rape isolates (three) induced severe symptoms on rape, whereas six from other hosts induced only mild symptoms and had no effect on yield. Curiously, a potato isolate induced very weak symptoms in plants which subsequently gave significantly higher yields than uninoculated controls (Zeise, 1995). Resende *et al.* (1994) compared the effects of four Brazilian and one Ugandan isolate of *V. dahliae* from cacao (*Theobroma cacao*) and isolates from other hosts on cacao, aubergine, tomato, cotton and pepper. In general, isolates were more aggressive when inoculated on to the original host. Cacao isolates induced severe symptoms on aubergine but were only mildly pathogenic to tomato and avirulent to pepper, but, in both hosts, systemic colonization occurred. The Ugandan isolate was significantly more virulent on cotton and pepper than the Brazilian. A Brazilian isolate readily colonized weed species from Brazil some of which were symptomless. Carder (1989) examined the cellulose isozyme patterns of five wilt-causing *Verticillium* spp. which were used to distinguish all the species but were inadequate for pathotype identification.

Molecular Genetics

Following pioneering studies on *F. oxysporum* in the late 1980s (see Heale, 1989), the first reports of molecular genetic studies on *Verticillium* spp. appeared in the 5th International *Verticillium* Symposium held in Leningrad in 1990. The main thrust of current research in conjunction with vegetative compatibility studies has been on the relationship between *V. albo-atrum* and *V. dahliae* and other authentic species, and their uniqueness. A further major objective together with VCGs has been an exploration of subspecific affinities or groupings, involving strains, pathotypes and host-adapted types or groups.

Techniques currently available use restriction endonucleases to cleave total genomic DNA of *Verticillium* cultures or selected specific DNA fractions. Specific nucleases cut DNA at particular nucleotide sequences. DNA fragments thus obtained are separated on the basis of molecular size by gel electrophoresis. These DNA fragments may be hybridized with synthesized radiolabelled probes following electroblotting from gel to nitrocellulose membrane (Nazar *et al.*, 1991). Genetic differences in isolates resulting from differences in restriction sites will be seen as polymorphic patterns (RFLPs). Thus the closer the homology between two RFLPs from different isolates, the closer the genetic affinity. Where only small fragments of DNA of known or unknown sequence exist which are difficult to isolate and identify, their amplification by polymeric growth (polymerase chain reaction; PCR) has proved invaluable. Short lengths of single-stranded DNA (primers) which bind only to the ends of target DNA sequences are used. If the target nucleotide sequence is known, specific primers may be synthesized to bind to key diagnostic sites. The selected target DNA is amplified in the presence of a polymerase in repeated cycles of temperature change to denature, anneal (bind) and elongate the molecule (Nazar *et al.*, 1991). Using appropriate primers, unpurified target DNA in plants or systems can be amplified and identified. More recently, RAPD analysis, a variation of PCR, has been used in comparative studies where details of the target DNA are not known. The method employs short random sequence primers which amplify a number of unidentified target sequences to produce a 'fingerprint' of DNA following gel electrophoresis. Comparisons of fingerprints are used to seek out genetic homology (Koike *et al.*, 1995b). Polymorphisms have been described in mitochondrial and ribosomal DNA (Nazar *et al.*, 1991; Typas *et al.*, 1992) and in ribosomal RNA (Typas *et al.*, 1992).

Identification of *V. albo-atrum* and *V. dahliae*

Notwithstanding the established morphological and physiological distinction between the two fungi, much effort based on DNA sequencing has gone into the genetic identities of Reinke and Berthold's (1879) and Klebahn's (1913) original species. Robb *et al.* (1990) examined nucleotide variability in non-conserved intervening transcribed spacer regions (ITSs) of rRNA genes. RNA probes identified two bands, a 1.9-kb fragment including ITS1 and a 2.6-kb fragment including ITS2. These ITSs separated three mature RNA sequences in both *V. albo-atrum* and *V. dahliae* (18S, 5.85S and 25S rRNAs). Differential oligonucleotide probes were synthesized based on two divergent sequences each in ITS1 and ITS2 and amplified by PCR. Consistent differences were reported for several isolates for each of the two species from different hosts. Carder and Barbara (1991) used RFLP analysis to probe Southern blots with random genomic clones from *V. albo-atrum*. All isolates of *V. dahliae* were differentiated from all isolates of *V. albo-atrum*. All isolates from lucerne were clearly different from all other isolates of *V. albo-atrum*. Much interspecific variation was found in *V. dahliae*. The authors claimed in this paper that the wide nucleotide diversity between *V. albo-atrum* and *V. dahliae* justified their separation as distinct species.

In a subsequent paper, Okoli *et al.* (1993) probed Southern blots derived from 17 isolates of *V. dahliae* with 71 random genomic clones from *V. dahliae*. Fifteen isolates fitted clearly into two RFLP groups designated A and B, which also correlated with isozyme patterns. *V. albo-atrum* isolates similarly separated into two groups, L and NL. A further 13 isolates of *V. dahliae* separated nine in group A and four in group B. Unlike *V. albo-atrum* isolates, two *V. dahliae* isolates showed combinations of the polymorphisms distinguishing both A and B. Groups A and B showed no correlation with host plant or geographic origin. Cellulase isozyme patterns correlated with species, and in *V. dahliae* with isolate group. All but two of the *V. albo-atrum* isolates of both groups gave identical patterns, but two out of four L isolates gave a different pattern. Kening *et al.* (1995) found PCR primers that amplified a 700-base pair (bp) region of the mitochondrial rRNA gene of *Verticillium*. These primers bound uniformly with conserved sites in ten species of *Verticillium* and 60 isolates of *V. dahliae*. Although priming sites were conserved, the region amplified provided probes specific for *V. albo-atrum*, *V. dahliae* and *V. tricorpus*. A 600-bp DNA fragment obtained by RAPD was unique for *V. dahliae*. RAPD data from 32 isolates of *V. dahliae* established that VCG1 was unique but other VCGs were similar to one another (see Joaquim and Rowe, 1990, 1991); subgroups identified in VCGs 2, 3 and 4 did not match VCG groupings. In another study, RAPD analysis of genomic DNA from 15 isolates of *V. albo-atrum*, 10 isolates of *V. dahliae* and 19 isolates of *V. lecanii* compared amplification patterns by unweighted paired-group matching analysis of the clusters. *V. albo-atrum* and *V. dahliae* showed a high degree of similarity, while *V. lecanii* isolates were widely dissimilar from the other species (Roberts *et al.*, 1995). In contrast, Barbara *et al.* (1995) found by RFLP that *V. albo-atrum* and *V. dahliae* isolates from the UK divided into clear groups with little variation between groups, unlike reports for variation reported outside the UK. Intergenic long sequence repeats were found in both species. Subrepeat structures divided isolates into three groups: (i) haploid isolates of *V. dahliae*; (ii) diploid *V. albo-atrum* isolates; and (iii) diploid isolates of *V. dahliae*. Significant differences between isolates in the number of subrepeats were found only in haploid isolates of *V. dahliae* and were not correlated with RFLP group. Sequence differences of ITSs for rRNA suggested a closer affinity between diploid isolates of *V. dahliae* and *V. albo-atrum* than with haploid *V. dahliae* isolates. These results are at variance with earlier reports and present a more complex picture (see Robb *et al.*, 1990; Carder and Barbara, 1991; Okoli *et al.*, 1994).

The complexity of the genetic picture in *V. albo-atrum* and haploid and diploid *V. dahliae* isolates was illustrated by Morton *et al.* (1995). These authors attempted to use subrepeat sequences (uncommon in fungi) in intergenic regions of rRNA obtained from PCR products as a basis for the subspecific identification of isolates. *V. albo-atrum* isolates separated as previously into L (Lucerne) and NL groups, but no variation in the sequencing of subrepeat groups could be found between isolates. Two diploid isolates of *V. dahliae* from sugarbeet and rape similarly showed no variation in PCR product or in AluI restrictase digestion of sequenced products between isolates. A total of 67 haploid *V. dahliae* isolates from different hosts, including hop, strawberry, tomato, pepper, potato, sugarbeet, maple, aubergine, *Chrysanthemum*, Chinese cabbage and mint, showed considerable variation in PCR product from 290 to 610 bp. This variation in haploid *V. dahliae* subrepeat sequences, however, did not correlate with RFLP or VCG group, or any other obvious characteristic, and was of no value in subspecific discrimination. The authors conclude on the basis of subrepeat variation (or lack of it) that 'pathogenic *Verticillium* species' [*sic*] can be regarded as three distinct species, i.e. *V. albo-atrum*, haploid *V. dahliae* and stable diploid *V. dahliae*. A further distinction for *V. albo-atrum* into group I and group II was proposed by Carder and Barbara (1999). Based on molecular evidence, group I consists of, pathogenic (i) lucerne isolates and (ii) all other isolates, while group II contains two non-pathogenic species which appear, on molecular characters, to be closely related to *V. psalliotae* and *V. fungicola* (*syn. V. malthousei*). Messner *et al.* (1996) derived phenograms of 79 distinct fragments from the use of four primers of 34 isolates of *V. dahliae* from nine dicot host genera. These it was claimed identified two groups only; one made up entirely of rape isolates which agreed with *V. longisporum* (Karapapa *et al.*, 1997b) and the other consisting of all other isolates from a wide range of hosts. The authors claimed that the calculated phylogenetic tree derived from the sequenced gene for the 18S rRNA associated *V. dahliae* with the sexual system of the ascomycetes. The complete sequence of the nuclear rRNA gene complex from a cotton isolate of *V. dahliae* was described by Pramateftaki and Typas (1997). A comprehensive study of multiple isolates of *V. dahliae* from various hosts and countries of origin by K.N. Li *et al.* (1993, 1999) demonstrated the uniqueness of an RAPD fragment to identify *V. dahliae* specifically against all other *Verticillium* species and other soil-borne fungal genera. RAPD primer E20 (AACGGTGACC) yielded a 567-bp band shared only by other *V. dahliae* isolates. Southern blot analysis showed that the PCR product specifically hybridized to *V. dahliae* genomic DNA. Preliminary attempts at quantitation using a cloned PCR fragment as target detected 50–500 copies ≡ 0.01–0.1 ng of DNA.

Identification of other species

Mukhamedov *et al.* (1990) found differences between *V. dahliae*, *V. tricorpus* and *V. nigrescens* based on RFLP differences in the ITS coding and DNA in the ribosome. Later work (Moukhamedov *et al.*, 1994) showed that while 5.8S rRNA and DNA sequences for all thee fungi were conserved, the internal transcribed spacer regions for the 18–28S rRNAand DNA of *V. tricorpus* were unique. These were used to synthesize a specific primer set for *V. tricorpus* identification (see later). Genomic DNA digested with the restriction endonucleases *Eco*RI or *Hae*III and hybridized with a *V. albo-atrum* homologous rRNA gene probe found RFLP polymorphisms which distinguished *V. lateritium*, *V. lecanii*, *V. nigrescens*, *V. nubilum* and *V. tricorpus* (Typas *et al.*, 1992). *Eco*RI digestions failed to provide RFLPs to distinguish *V. albo-atrum* from *V. dahliae*. However, digestion of genomic and *mit* DNA with *Hae*III showed distinctive patterns for both species. Kening *et al.* (1995) and Li *et al.* (1994) identified PCR primers that amplify a 700-bp region of the mitochondrial rRNA gene of *Verticillium*. Tests on nine species of *Verticillium* and 60 isolates of *V. dahliae* showed that primers bound to sites well conserved across the genus. The region amplified, however, was variable between species. By cloning and sequencing these regions, specific probes for *V. albo-atrum*, *V. dahliae* and *V. tricorpus* were identified. Using RAPD, a 600-bp DNA fragment unique to *V. dahliae* was found. RAPD data on 32 isolates of *V. dahliae* showed that VCG1 appeared to be a separate and unique group. VCGs 2, 3 and 4 were not distinguished by RAPD, but some subgroups, not confined to any one VCG, were found.

Examination of strains (pathotypes) and host-adapted forms in *V. albo-atrum* and *V. dahliae*

V. albo-atrum

There is now much evidence from pathogenicity tests, and classical and molecular genetics that the lucerne strain of *V. albo-atrum* is unique from all other strains (Carder and Barbara, 1990, 1991; Typas *et al.*, 1992; Griffen *et al.*, 1994). RAPD studies by Koike *et al.* (1996a) separated 15 isolates of *V. albo-atrum* into two sub-clusters. One, RAPD type IV included one lucerne pathotype of group 2 (Vaa2), three potato pathotypes and three undetermined isolates of limited provenance. Details of the lucerne isolates in RAPD type V are insufficient to establish whether the apparent molecular uniqueness of the lucerne pathotype (RAPD type IV) was compromised. Griffen *et al.* (1994) cloned DNA coding for the rRNA gene complex (rDNA) from a PV_1 (progressive) hop isolate. rDNA was mapped using endonucleases; functional units of the intergenic spacer regions 18S, 5.8S and 25S were located by hybridization to specific rDNA probes from *Aspergillus nidulans*. The rDNA repeat, 7.6 kb in length, was used to probe the repeat size in 18 hop isolates, including PV_1, PV_2 and PV_3 types, and a lucerne isolate. All hop isolates had the same 7.6-bp repeat, while the lucerne isolate was quite distinct where the rDNA complex was 8.4 kb. This isolate appeared to be atypical and, though originally pathogenic on lucerne cv. Du Puits in the field, was avirulent in their tests. This emphasizes the great need in this work for studies on multiple isolates newly obtained from the field. A comprehensive analysis of DNA polymorphisms on 35 isolates of *V. albo-atrum* from hop, seven from lucerne and five from potato, tomato, *Chrysanthemum* sp. or *Antirrhinum* sp. based on mtDNA using Southern hybridization, was conducted by Griffen *et al.* (1997). Amplified polymorphic DNA was analysed using primers based on primers from intergenic spacer and 25S regions. rDNA and RFLP delineated group I with 44 isolates, group II with two atypical hop isolates and group III with a single av lucerne isolate. MtDNA RFLPs separated DNA group I into a subgroup of 38 isolates and another containing all virulent lucerne isolates. Analysis of amplified polymorphic DNA (APD) separated 16 phenotypes, 12 of which contained most hop isolates but with no correlation of origin, hop cultivar, pathogenicity or date of isolation. One APD phenotype neatly contained all virulent lucerne isolates. Two further APD phenotypes equated to rDNA group II. Barasubiye *et al.* (1995a,b) in a comparative study of potato and lucerne isolates, found that lucerne strains were more pathogenic on lucerne than potato strains. Lucerne was avirulent on potato. Using RFLP on mtDNA, little DNA polymorphism was found between the two strains. In contrast, four RAPD markers present in lucerne isolates but absent in potato isolates were found (see Carder and Barbara, 1990). The clonal origin of the lucerne strain and its genetic and pathogenic uniqueness suggest that it could be elevated to specific ranking (*Verticillium medicaginis*) following additional confirmatory evidence from different countries.

V. dahliae

Host strains and pathogenicity of Japanese isolates of *V. dahliae* appear to show better correlation and consistency than has been described for other strains and countries. Horiuchi *et al.* (1990) and Koike *et al.* (1995b,c, 1996a) recognized five pathogenicity groups: (A) aubergine strain, pathogenic to aubergine and turnip; (B) tomato strain, pathogenic to tomato, aubergine and turnip; (C) sweet pepper strain, pathogenic sweet pepper, aubergine and turnip; (D) crucifer strain, solely pathogenic to sweet pepper and turnip; and (E) a weakly virulent strain (see also Li *et al.*, 1998). Initially Koike *et al.* (1995b,c) identified the five groups based on DNA polymorphisms detected in 12 of 20 primers used in RAPD. These were ABC/DE, AC/B/DE, A,BC/DE, A/B/C/DE and AC/B/D,E . Later Koike *et al.* (1996b) recognized four pathogenicity groups in three RAPD sub-clusters. One sub-cluster, RAPD type I, contained two pathotypes, group A (tomato) and group C (sweet pepper). A second sub-cluster, RAPD type II, contained group B (tomato), and a third, RAPD type III, contained four diploid isolates of group D, *Brassica* pathotype, and one haploid isolate. M. Koike *et al.* (1997) used the RAPD analysis to study phytogenetic relationships of the three *V. dahliae* and two *V. albo-atrum* molecular groups and 70 mostly Canadian isolates of haploid and diploid *V. dahliae*, *V. albo-atrum* and *V. albo-atrum* group 2 (Vaa2) and *V. tricorpus*. From these, three phylogenetically distinct groups emerged: haploid *V. dahliae*; diploid *V. dahliae*; and *V. albo-atrum* potato and lucerne pathotypes, Vaa2 and *V. tricorpus*. Of the haploid *V. dahliae*, Canadian tomato or potato were closely related to the Japanese tomato. Both were well separated from non-tomato isolates. Similarly, diploid *V. dahliae* were distant from haploid *V. dahliae* and *V. albo-atrum*. A third group included four subgroups: (i) potato pathotype of *V. albo-atrum*; (ii) lucerne type of *V. albo-atrum*; (iii) Vaa2; and (iv) *V. tricorpus*; (iii) and (iv) were closer to *V. tricorpus* than *V. albo-atrum*. A phylogenetic tree constructed from 143 restriction sites from PCR-RFLP analyses of histone-4 and β-tubulin genes illustrates the enormous complexity of *Verticillium* genetics. In this tree, diploid *V. dahliae* (*Brassica* pathotype) was closely allied to non-tomato pathotype haploid *V. dahliae*, while the tomato pathotype haploid *V. dahliae* isolates were far removed from *V. albo-atrum* and other diploid and non-tomato pathotype isolates. While differences in repetitive sequences were different for *V. dahliae* from other species, no differences could be found in cotton strains (races [*sic*]) of *V. dahliae* (Mukhamedov *et al.*, 1990).

Pérez-Lara *et al.* (1995) described the spread of cotton defoliating and non-defoliating strains of *V. dahliae* and the spread of the defoliating strain as a highly virulent pathotype of olive. Using RAPD, it was possible to identify cotton isolates. Harris and Yang (1995) studying 43 isolates of *V. dahliae* from British host plants found a good correlation between *nit* mutant VCG groups (two) and two corresponding RFLP groups. Using a strawberry hybrid (*Fragaria vesca* × *F. chiloensis* × *F. virginiana*) as a test plant, the mean pathogenicity for each RFLP/VCG group was different, but the range overlapped. There was no correlation for host adaptation and RFLP group.

Using random genomic clones from isolates of *V. dahliae* to probe Southern blots, Okoli *et al.* (1994) tested isolates of *V. dahliae* from peppermint (host-adapted haploid pathotype) and *V. dahliae* from crucifers (stable diploid, host-adapted). The mint isolate formed a distinct subspecific RFLP group M, based more on host specificity than on geographical origin, as did the crucifer isolate (group D). The mint group M is equivalent to the subspecific groups A and B (Okoli *et al.*, 1993), defined among non-host adapted types. The authors consider that the diploid cruciferous isolate could be considered a separate isolate. The same view is expressed by Karapapa *et al.* (1994). *V. dahliae* var. *longisporum* is a serious pathogen of oilseed rape in Sweden, Germany, France and Poland, but not yet in the UK. All isolates were shown by Feulgen DNA microdensitometry to be diploid. Based on the pioneering studies by Stark (1961) and Hastie and co-workers, and the uniqueness in RAPD, PFGE (pulse-field gel electrophoresis) characteristics, stable diploidy, large conidia, irregular, elongate microsclerotia, absence of PPO activity on tannic acid substrate and pathogenicity on oilseed rape, *Arabidopsis*, Chinese cabbage and Japanese radish, Okoli *et al.* (1993) proposed a new species for this type – *Verticillium cruciferarum* sp. nov. (see also Subbarao *et al.*, 1995). A subsequent detailed examination of the strain by Karapapa *et al.* (1997b,c) found three oligonucleotide primers with RAPD band profiles clearly different from *V. albo-atrum* and *V. dahliae*. Based on this and conidial, phialide, microsclerotial characters, conidial nuclear diameter (4′,6-diamidino-2-phenylindole (DAPI) fluorescence), near diploid DNA microdensitometry and pathogenicity tests, the authors proposed the name *Verticillium longisporum* comb. nov. A more controversial proposal was that the species may have evolved by parasexual hybridization between strains of *V. albo-atrum* and *V. dahliae*.

Cotton has become an important crop in Australia, and *V. dahliae* is one of the major problems. Work by Ramsay *et al.* (1996) on rDNA sequences following amplification sequencing and restriction digestion proved to be specific only for genus and species but not for pathotype. RAPD-PCR analysis using 13 decamer primers showed differences between isolates, but the authors were equivocal about the identification by molecular analysis of virulence *in planta*. An analysis of 99 isolates of *V. dahliae* from eight major cotton-growing regions of South-Eastern Australia was conducted by Zhu *et al.* (1999b) using RAPD and PCR. From a total of RAPD bands amplified using ten random decamer primers, 54.4% were found to be polymorphic. Cluster analysis revealed 15 RAPD groups, 10 of which were subdivided into three clades which correlated with region of fungal isolation. Five, however, showed no such correlation. Similar work from Spain and Israel has produced more positive results. Pérez-Lara *et al.* (1995) using RAPD and 10-mer commercial oligonucleotides as random primers identified 15 cotton non-defoliating isolates of *V. dahliae* (P_2 strain) and 11 defoliating isolates (P_1 strain) from cotton. The defoliating pathotype was also lethal on olive. A comparative study of cotton pathotypes from Spain and Israel (Perez-Artes *et al.*, 1997) used three primers capable of amplifying specific DNA bands for the identification of P_1 and P_2 strains (D = defoliating and ND = non-defoliating in their terminology). Spanish cotton isolates were identified as P_1 (22 isolates) and P_2 (26 isolates). The situation in Israel was less clear, with no isolates showing the polymorphism specific of P_1; all were characterized as P_2 (ND). Using PCR and gel electrophoresis, and UPGMA cluster analysis, the Spanish pathotypes fell into two distinct clusters. The Israel isolates similarly separated into two groups both of the ND, P_2 type, but one which had 95% similarity to the Spanish P_2 group showed some isolates to be of a 'D-like' character with some stunting and partial defoliation after typing on cultivars Coker 310 and Deltapine Acala 40. This result indicates that virulence of cotton pathotypes is not a simple qualitative distinction, as work from the Asian republics (CIS) using other criteria has indicated.

The main thrust of genetic research in China is a resolution mainly by RAPD-PCR studies of the *V. dahliae* strain, race or pathotype complex in cotton. The preliminary results available from different cotton regions are variable and lacking correlation. Zhou *et al.* (1999) compiled eight RAPD cluster groups from 26 isolates; unlike a similar study on Australian isolates, the author found a correlation with virulence but not geographical location. A similar analysis of 26 cotton strains [*sic*] (isolates) from 12 cotton regions by X.T. Liu *et al.* (1999b) claimed *inter alia* that genomic variation was found in single spores [*sic*] (cultures) from one isolate. The identification of *V. albo-atrum* [*sic*] RAPD cluster A compared with B in other strains [*sic*] (*V. dahliae*) from cotton, however, does not inspire confidence in the report. A similar study by Wang and Shi (1999) analysing *V. dahliae* isolates from Hebe, Henan and Shandong provinces similarly confuses strains with isolates. Two specific RAPD bands were found in defoliating strains/isolates. P_1 type isolates from northern Chinese regions were closer to the US 9 strain than those from Jiangsu province.

RAPD-PCR analysis of 38 olive isolates from southern, central and northern Morroco generated 66 polymorphic DNA fragments from 10 of the 40 primers used (Cherrab *et al.*, 2000). The authors successfully correlated RAPD groups with the regional origin of the isolate. With one exception, no DNA correlation with colony morphology was found. The genetic diversity and complexity of *Verticillium* is well illustrated by the work of Dobinson *et al.* (1998) based on RFLP and RAPD studies on race 2 isolates of *V. dahliae* from tomato in southwestern Ontario. Using DNA polymorphisms and DNA fingerprints obtained by hybridization to a dispersed repetitive genomic DNA probe, five DNA types were identified: type I (two non-pathogenic isolates); type II (four race 2 and three non-pathogenic isolates); type III (one race 1); type IV (one race 2 and two race 1 isolates); and type V (one race 1). Isolates of the same DNA type were compatible, as were type II and III isolates (= VCG4B); type IV (= VCG2A); andtype V (= VCG2B). These salutory findings suggest not least a multiple origin for race 2 in Ontario first reported by Dobinson *et al.* (1996). (See X.T. Liu *et al.* (1999a) for improved extraction methods of genomic DNA and RAPD analysis of *V. dahliae*.)

The use of molecular genetics for the identification and quantitation of *Verticillium in planta*

The identification and quantitative measurement of *Verticillium* spp. *in planta* presents an intrinsically difficult problem. Relatively crude visual estimates of fungal colonization have been made by Waggoner and Dimond (1954), Talboys (1958c) and Pegg and Dixon (1969), and by culturing colonies from tissue sections (Keyworth, 1964). Ride and Drysdale (1971, 1972), Wu and Stahmann (1975) and Toppan *et al.* (1976) have estimated fungal biomass from the chemical detection of fungal chitin. This method, however, is time consuming, influenced by host chemicals and incapable of distinguishing living from dead mycelium, and hence the true pathogenic potential of the fungus. Various authors have attempted to overcome this limitation by plating out fungus–host tissue comminutes on nutritive media with or without the incorporation of antibiotics (Matta and Dimond, 1963; Busch and Schooley, 1970; Busch and Hall, 1971; Pegg, 1978; Pegg and Jonglaekha, 1981; Pegg and Street, 1984). The problems associated with the host comminute plating method, including seed-borne transfer, conidiation and the effect of host substances on colony establishment, have been summarized by Pegg (1978).

In many *Verticillium*-infected host plants, colonization is by a single species and often a single strain, forming a pure colony in the host. The main problem here is how to measure (or estimate from samples) the total pathogen biomass and its distribution in the host. *Verticillium* infection of potato, however, presents a different challenge. Where potato is grown in temperate regions of the USA and Israel, the eponymous potato early dying syndrome (PED) is associated with multiple infection by combinations of *V. dahliae*, *V. albo-atrum*, *V. tricorpus* and the eelworm (*Pratylenchus penetrans*). Potato cultivars vary in resistance to *Verticillium* spp. (Platt, 1986), and environment and/or cultivar changes can shift the occurrence or dominance of a particular species (Celetti and Platt, 1987b). Against this background, Robb *et al.* (1994) developed a PCR assay for the quantitative detection of species. Based on Nazar *et al.* (1991) and the non-conserved ITS region of rRNA approximately 295 bp long, five nucleotide differences were found between *V. dahliae* and *V. albo-atrum*, 17 between *V. dahliae* and *V. tricorpus* (Robb *et al.*, 1993) and 12 between *V. albo-atrum* and *V. tricorpus* (Moukhamedov *et al.*, 1994). These differences were used to develop differential primer sets for the three species. Using PCR assays based on genomic DNA of fungus and host potato from petioles and stems, tubers and leaves, quantitative detection of the three species was made. A straight line relationship was obtained for fungal DNA from 10^{-4} to 10^{-2} ng of DNA and the PCR product ratio. Two subgroups of *V. albo-atrum* were discovered. Isolates from *V. albo-atrum* 2 showed close affinity to *V. tricorpus* and were found in Canada, the UK and The Netherlands.

The PCR assay has the advantage over other methods in identifying the species or subgroup and detecting lower quantities of fungus than might be detected using other techniques. The method, however, is not without problems, since Hu *et al.* (1993) found endogenous plant inhibitors of the PCR. Control reactions using an internal control are necessary to avoid false negatives where fungal quantities are small. *V. albo-atrum* was readily detected from lucerne stems (strong reaction) and roots in plants just developing wilt symptoms. The signal strength from primer-treated tissue was proportional to the level of colonization based on plating experiments. No signal could be detected from plants inoculated with a non-pathogenic *V. dahliae* isolate. Preliminary experiments on sunflower and cotton, inoculated with *V. dahliae* isolates, and potato and tomato inoculated with *V. albo-atrum* gave encouraging results (see also Robb *et al.*, 1994). Carder *et al.* (1994) described a method for identifying and quantitatively assessing some isolates of *V. dahliae* and *V. albo-atrum* based on four main primers, only three of which could be used for *in planta* assays. Their claim that RFLPs could 'divide both species into major subspecific groups with little genetic variation' belies reports to the contrary from several sources, e.g. Griffen *et al.* (1994). The application of their methods for the successful quantitation of the pathogen *in planta* seems fraught with practical difficulties and overambitious, where a wide range of pathotypes might be encountered as colonists of infected plants. Thus while some primers are claimed to detect 200 pg of DNA from infected plants, the complications and combinations of primers required make the scheme impracticable. A more realistic assay for the identification and quantitation of *V. dahliae*, *V. albo-atrum*, Vaa2 and *V. tricorpus* in a single host (potato) by Robb *et al.* (1994), describing in detail the *in planta* assay and incorporating a recovery stage, inspires more confidence in the method. A specific primer based on 18–28S rDNA of the internal transcribed spacer sequences of *V. tricorpus*, described earlier (Moukhamedov *et al.*, 1994), was also used successfully for both identification and quantitative assessment in potato. A PCR-based assay for the *in planta* and soil detection of *Verticillium* spp. and strains was described by Mahuku *et al.* (1999) for potato pathogens. The assay, using specific primers, was compared with plating assays on selective media. The PCR assay was completed in 2 days compared with 4 weeks for other methods. *V. albo-atrum* aggressive strain (VA1) was detected in soil and stems; however, neither technique could detect a weakly pathogenic strain (VA2) although the PCR method detected both strains in soil and *V. tricorpus* from previously inoculated plots. The method was recommended for routine detection and epidemiological studies. Zhu *et al.* (1999a) described two 26-bp PCR primers designed according to the ITS sequence of rRNA from *V. dahliae*: P1, 5′CATCAGTCTCTCTGTTTATACCAACG, and P2, 3′CGATGCGAGCTGTAACTACTACGCAA. Assays with the primers amplified a 324-bp rRNA gene fragment from genomic DNA of *V. dahliae* and diseased plant tissue – presumably cotton. Details of limits and/or difficulties with the detection method were not given.

The techniques of molecular genetics represent the most sophisticated methods of analysing the population complex of *Verticillium* spp. at the subspecific level into strains (pathotypes) and also quantitatively as biomass *in planta*.

The results to date are tantalizing in that while some groups of isolates may be clearly identified, there is no uniform comparability with RFLP or RAPD group, and VCGs, or host specificity. While the lucerne strain of *V. albo-atrum* and *V. dahliae* var. *longisporum* Stark (Ingram, 1968) appear to justify specific ranking, for reasons discussed earlier, individual isolates of erstwhile uniform strains appear to have similar uniqueness. In the light of the most recent research, and the still unfolding genetic complexity of this remarkable genus, it is hardly surprising that early investigators using unsophisticated methodology were confused and frustrated by particular isolates in their preliminary attempts to identify and classify field populations and pathotypes.

Aetiology 5

Survival

Survival of *Verticillium* spp. may be regarded as from susceptible crop to crop or from season to season. In soil, perennation is associated with torulose or dark thick-walled mycelium in *V. albo-atrum*, microsclerotia in *V. dahliae*, chlamydospores as seen in *V. nigrescens* and *V. nubilum*, or all three types in *V. tricorpus*. In the short term, hyaline mycelium and conidia may exist in soil, while in the living plant all types of fungal structure can be found, and in dead residue microsclerotia and chlamydospores predominate, together with dark mycelium (Powelson, 1970; Pegg, 1974, 1985; Schnathorst, 1981; DeVay and Pullman, 1984). Heale and Isaac (1963) found resting mycelium of *V. albo-atrum* in dead lucerne stems in soil could survive from 9 to 10 months; DeVay and Pullman (1984) cite a longevity in fallow soils of 2–3 years, and Luck (1953) at least 4 years; microsclerotia, however, are much more durable (Schnathorst, 1965; Schnathorst and Mathré, 1966b; Schnathorst and Fogle, 1973; DeVay *et al.*, 1974). Wilhelm (1955a) resuscitated 13-year-old dried agar cultures of *V. dahliae* maintained at room temperature and, in the field, recorded *V. dahliae* in test tomato plants after 14 years in the absence of a tomato crop, though at a very low incidence. No infection in vetch, wheat, lucerne or barley cropped in the same soil at that time could be detected. Earlier, Wilhelm (1950a) had established the presence of *V. albo-atrum* [*sic*] (*V. dahliae*) in immune, (non-host) crops, but had found no evidence for saprophytic growth in the soil (Wilhelm, 1951b; see also Rao, 1959). Schreiber and Green (1962, 1963) found that hyaline hyphae and conidia survived for only 3–4 weeks. In contrast to Wilhelm's (1955a) results, Green (1980) working with soil at Ψ0.001 bar to air dry and a temperature of 28°C found a rapid decline in microsclerotial viability surviving after 3 years in sand and 4.5 years in loam. *V. dahliae* in potato in south-western Ontario survives in soil for several years, but *V. albo-atrum* in the same host rarely overwinters (McKeen and Thorpe, 1981). Keinath and Millar (1986) confirmed Heale and Isaacs' (1963) results with *V. albo-atrum* in lucerne, in which only 3% of the pathogen survived in stems after 11 months.

The highest number of microsclerotia occur in the top 10 cm of soil, decreasing almost to zero at 40 cm (Ben-Yephet and Szmulewich, 1985). No microsclerotia survived in soil transferred to the laboratory after 5 years, but 4% of the field population survived after a 7-year crop rotation. In relation to inoculum potential, Green (1969) determined that 50×10^3 *V. dahliae* conidia g^{-1} of soil were required to give 100% infection of tomato compared with only 100 microsclerotia g^{-1} of soil. While conidial infection declined from 100% to 0% over 3 weeks, no reduction occurred with microsclerotia over 7 weeks. This mirrored the relative recoveries of conidia and microsclerotia from soil. Evans *et al.* (1967) recorded a seasonal decline of microsclerotia in cotton field soils in the upper 10 cm during spring and summer, with a replenishment at harvest when infested plant debris was released to the soil. The highest levels of 335–417 microsclerotia g^{-1} of soil occurred at planting time. Rotating cotton with barley for one season resulted in a spectacular decline in inoculum and a corresponding increase in yield in the succeeding cotton crop. The results of a 6-year field experiment on potato showed no reduction of soil-borne inoculum of *V. dahliae* following 5 years of weed-free fallow or continuous cropping with maize. While microsclerotia and resting mycelia appear to have no inherent dormancy or a requirement for exogenous nutrients *per se* (Green, 1971), fluctuations in temperature, irrigation and the addition of organic amendments, such as sucrose or glutamic acid, all lead to decreased survival. The inference from Green's (1971) work was that root exudates and other chemical and physical changes in the soil contrived to stimulate or repress soil fungistasis with a concomitant effect on microsclerotial numbers. Evans (1971b) confirmed the importance of rhizosphere and root exudates. Microsclerotia of *V. dahliae* isolated from field soils exhibit a bacterium-determined dormancy which may be reversed by surface sterilization or by air drying the soil to 30–40% relative humidity for 6 weeks to inactivate the bacteria (see also Steiner and Lockwood, 1970; DeVay *et al.*, 1974; Butterfield and DeVay, 1975a). The exudates from young cotton roots overcome the bacterial inhibition (Butterfield and DeVay, 1975b). In a subsequent paper, Green (1976) studied the effect of soil type, soil water capacity (saturation capacity, field capacity or air dried), constant or fluctuating, and temperature. Survival was poorest in a silt loam at soil capacity and 28°C. No viable microsclerotia remained under this treatment after 8 months. In all other treatments, microsclerotia survived (from <1% to >30%) after 48 months. Survival of *V. dahliae* in the cotton crop in the CIS was discussed by Sidorova (1990). Using immunofluorescent staining of *V. dahliae* on membrane filters in soil, Vishnevskaya *et al.* (1990) described the production of soil conidia in several cycles during the year. A 6-year field experiment on *V. dahliae* in Idaho potato fields showed no evidence for a reduction in soil-borne microsclerotia following 5 years of either weed-free fallow or of continuous cropping with maize. The absence of weeds during fallowing suggested a loss of soil suppressiveness, since levels of microsclerotia in these plots rose rapidly after cropping with susceptible cv. Russet Burbank. The return of *V. dahliae*-infested potato haulm was 2.5 t ha^{-1} for the susceptible cultivar compared with 3.6 t ha^{-1} for a resistant clone. Notwithstanding this level of inoculum, soil microsclerotial populations did not reach a maximum until after the 4th year. Similar results were obtained with *V. tricorpus*, emphasizing the microbial suppression of microsclerotia (Davis *et al.*, 1990). The effect of temperature on survival is discussed in Chapter 6. Keyworth (1942) working on hops showed that cultural practices effective in controlling *V. albo-atrum* were ineffective in controlling *V. dahliae*, which was capable of much greater survival in soil. While dormant propagules of *Verticillium* spp. may be capable of germinating in distilled water, in field soils fungistasis and microbial antagonism appear to play an important role in maintaining dormancy (Sewell and Wilson, 1966). Isaac and MacGarvie (1962, 1966) reported that resting structures of *V. albo-atrum*, *V. dahliae* and *V. nubilum* all exhibited dormancy, but those of *V. tricorpus* did not. Dormancy was not broken by several techniques, including freezing, thawing, heat shock, high O_2 levels, root exudates, alternate wetting and drying, vitamins, enzymes, detergents, soil extracts and root exudates. In the absence of soil microorganisms, Isaac and MacGarvie (1966) induced germination by soaking the different propagules in distilled water for 12 h and plating on nutrient agar. They suggested that dormancy resulted from inhibitors which needed to be removed by one or more agencies prior to germination. *Verticillium* appears to have little ability to survive for prolonged periods under anaerobic conditions. When cotton gin waste contaminated with *V. dahliae* was composted, only the pathogens present in the outermost layer survived (Staffeldt, 1959; Sterne *et al.*, 1979). McKeen (1976) reported that while *V. albo-atrum* caused earlier and more severe symptoms on Kennebec and Irish Cobbler potato cultivars than *V. dahliae*, survival in Ontario soils was poor. Recovery from soil after harvest declined rapidly, with little fungus surviving after winter. McKeen claimed that environmental conditions obtaining during the saprophytic phase ('saprogenesis') [*sic*] were possibly more important in limiting occurrence and distribution than those of pathogenesis.

Since in some experiments on survival, conidiation in soil occurs before viability of resting structures is affected, the distinction between inoculum production and survival is not always clear. Ioannou *et al.* (1976) showed that low O_2 and high CO_2 inhibited both microsclerotial formation and survival. While flooding soil for periods of 10, 20 and 40 days with concomitant low O_2 and high CO_2 levels inhibited microsclerotial *production*, experiments on existing microsclerotial populations showed no effect of any of the flooding treatments on survival.

The survival of *V. albo-atrum* was studied in artificially inoculated lucerne seeds (Huang *et al.*, 1995). Survival was less than 10% when stored for 10 weeks at 20 and 30°C, but over 90% when stored at −10 and −20°C. Similarly, results were low (10%) in soil incubated for 3 months at 20°C and high (90%) in soil kept for 10 months at −5°C. The incidence of seed-induced wilt decreased rapidly with the duration of seed storage. After 1 month's storage, wilt incidence was 40% and zero after 12 or 18 months' storage, respectively. No viable fungus was found in seed stored at 30°C for 6 months. Somewhat in contrast to earlier reports on anaerobic behaviour, the survival of *V. dahliae* on seed buried at 10 cm was greater than seeds buried at 2 cm. Microsclerotia of *V. dahliae* vary in diameter between <53 μm and 106 μm. Larger microsclerotia in the range 75–106 μm survived long-term storage at 24°C better than those in the <53–74 μm range. Melanized microsclerotia survived better after 35 weeks storage at 24°C than non- or partly melanized sclerotia. Albino microsclerotia, obtained either by the use of tricylazole, a reductase inhibitor in the melanin pentaketide pathway, or from an *alm* (albino) mutant, showed 50% reduced germination after 2 h UV irradiation of 254 nm and were killed after 6 h. Non-melanized microsclerotia showed poor survival when buried in non-sterile soil for 3–4 weeks. The resultant colony growth was a good indicator of lethal and sublethal trends (Hawke and Lazarovits, 1995a).

Germination and Infection

Under field conditions, propagules of *Verticillium* spp. often fail to germinate for a variety of physical, chemical and biological reasons. Conidia and resting structures have no innate or constitutive dormancy, and the observed inhibition in soil is due to mycostasis (Pegg, 1985). Soil fungistasis may be reversed by either increasing the available nutrient or decreasing (often diluting) the inhibitory agent (Hora *et al.*, 1977; Griffin and Roth, 1979). In one of the earliest papers on *Verticillium* mycostasis, Schreiber and Green (1963) found that germination of conidia and microsclerotia of *V. dahliae* was reduced by 80% in soil compared with sterile deionized water. This inhibition was reversed when microsclerotia were in contact with roots of a host (tomato) or a non-host (wheat) and, to a lesser extent, with extracts of these plants. The average percentage germination for water, wheat root and tomato root were 6.4, 36.1 and 89.7%, respectively. The greater stimulation by the susceptible plant was illustrated by the root extracts giving 5.8, 10 and 75.2%, respectively. Since the greatest activity was in a basic fraction, Schreiber and Green postulated that basic amino acids or nitrogen compounds were the active principal. Schreiber and Green (1963) also suggested that the reversing principal in host and non-host root extracts could be the same, with differences between different plants merely reflecting different concentrations. Van Wyk and Baard (1971) found that conidia of *V. dahliae* germinated on soil sterilized by steam or propylene oxide but not on unsterilized soils. Five per cent peptone, but not glucose lignin or wheat straw, reversed fungistasis. The authors described a complex picture of fungistasis and its reversal, suggesting that nutrient deficiency could be implicated. Studies on *V. dahliae* microsclerotia in Uzbekistan cotton soils using membrane filters by Strunnikova and Vishnevskaya (1995) claimed that edaphic factors, while not affecting germination, influenced morphological development and turnover rate. Root observation boxes for the effect of roots on microsclerotia were described by Mol and Van Riessen (1995). Using such boxes, Mol (1995a) found that roots of all crop plants stimulated germination, but potato cvs Element and Astarte had a much greater effect than roots of cv. Ostara, pea, flax, sugarbeet or onion. The author concluded that the use of a 'stimulant' crop to reduce microsclerotial inoculum would be inefficient due to the very low level of actual root contact with the pathogen. The reports of the benefit of grass and other non-host crops grown in a rotation with *Verticillium*-susceptible hosts (Easton and Nagle, 1986) may result from a stimulation of dormant propagules with their subsequent destruction by soil microorganisms. Various organic substances have been shown to be effective reversers of mycostasis: glucose, sucrose and, to a lesser extent, $NaNO_3$ (Green and Papavizas, 1968); glucose, galactose, sucrose, alanine and glutamic acid (Emmatty and Green, 1969); soils amended with sugar, amino acid mixtures or wheat root exudates (Fitzell *et al.*, 1980a); B-group vitamins and glucose (Zilbertsem *et al.*, 1983); and mixtures of methionine and riboflavin (DeVay *et al.*, 1987).

Griffin *et al.* (1975) using agar discs over a variety of soils in plates, showed that conidial germination could be stimulated by volatile substances. Ethylene, a potent soil gas, however, found at enhanced levels (6.5 p.p.m.) by Ioannou *et al.* (1976), had no effect on growth, sporulation or microsclerotial production. Increasing CO_2 concentration from normal to 1% at 21% O_2 enhanced microsclerotial germination. Increasing CO_2 concentration and decreasing O_2 concentration to 13% decreased germination. Nevertheless, some germination occurred at O_2 levels down to 1% (Zilbertsem *et al.*, 1983). The competition for root exudate nutrients between *V. dahliae* and bacteria was demonstrated by Olsson *et al.* (1987) measuring the population density of fungus stained with fluoroscein diacetate and rhodamine-B isothiocyanate, and bacteria stained with acridine orange on rape seedling roots. Mycostasis of *V. dahliae* microsclerotia in cotton field soil following prolonged autumn rain was attributed by Ashworth *et al.* (1976) to the release of cupric ions. The toxicity of aluminium ions to *V. dahliae* reported by Orellana *et al.* (1975) could be induced by soil acidification (Baard and Pauer, 1982). Mol (1995a,c) found that roots of host crops had a stronger stimulatory effect on microsclerotial germination than those of non-hosts. Potato cultivars differing in resistance similarly had a differential effect on germination. The total elimination of soil microsclerotia by break crops may be low since germination can occur twice. The role of melanin in the survival and germination of microsclerotia was studied by Hawke and Lazarovits (1995b) using tricyclazole, a reductase inhibitor in the pentaketide–melanin pathway, or using the *alm* albino mutant. Microsclerotial size in both melanized and non-melanized structures was an important determinant of germination. Microsclerotia in the 75–106 μm range germinated faster and more synchronously than those in the 53–75 or <53 μm range. Colony growth rate showed a similar relationship. Germination of microsclerotia lacking or low in melanin was only slightly retarded following short (<24 h) periods of 254 nm UV irradiation, but the resultant colony growth was greatly reduced, compared with non-irradiated controls. Microsclerotia in which melanin synthesis was increasingly blocked showed correspondingly reduced germination when incubated in non-sterile soil amended with organic nitrogen sources or in sterile soil inoculated with *Verticillium* antagonists.

Little critical work has been carried out on the factors limiting germination in the recovery (quantitative estimation) of microsclerotia from soil or plant debris (see Harris *et al.*, 1993). Thus, laboratories obtain different results ostensibly using the same methods. Termorshuizen (1995) has proposed the use of a recovery factor based on the percentage germination from plant tissue added to soil or direct to a soil suspension plate, as a fraction of the percentage germination of sterile-water-washed microsclerotia germinated on soil suspension-free agar plates. The possibilities of variation in studies on viability, dormancy and inoculum potentials are considerable, not the least of which is the genetic status and provenance of the microsclerotia. Thus the standard 'control' of water-washed microsclerotia will be of one VCG, but those in soil samples will be unknown regardless of the dominant pathotype.

Infection

Root systems of higher plants growing in the abrasive medium of soil and in the presence of a plant-eating microfauna may suffer extensive damage to root hairs and the piliferous layer in an erstwhile healthy root system. Microsites of dead cells resulting from adverse physical or physiological conditions also offer a non-living infection court to *Verticillium* spp. (Pegg, 1985). Bewley (1922) first suggested that *V. albo-atrum* penetrated intact root hairs of tomato. This was confirmed by Selman and Buckley (1959). Isaac (1946) had earlier demonstrated the direct passage of *V. dahliae* through root hairs of sainfoin (*Onobrychis sativa*) as well as through ruptures at the emergence sites of lateral roots. Perry and Evert (1983) also showed that *V. albo-atrum* entered tomato root epidermal cells and not through lateral root emergence sites. The infection of cotton roots by *V. dahliae* was shown by Garber and Houston (1966) to occur on the root cap and in the zone of elongation. Penetration was both inter- and intracellular, with impedence of the invading hyphae at the endodermis. Fitzell *et al.* (1980a) using fluorescein isothiocyanate (FITC) staining of germinating microsclerotial hyphae found penetration 2–3 cm behind the root tip, suggestive of local stimulation. Hyphae enter the cortex directly without forming appressoria. Changes in isolated cotton root cells following inoculation with *V. dahliae* conidia were reported in detail by Vlasova (1986). For infection, roots need to be in close proximity to the pathogen, since Sewell (1959) reported that germinating hyphae were never more than 2 mm distant from infection sites. Penetration and the establishment of infection take place in most hosts over 4 days. Nelson (1950) described penetration of peppermint roots by *V. dahliae* after 6 h. In cotton, the elongation region of young roots is invaded within 24 h, but the xylem is not breached until after 72 h (Garber, 1973); in peppermint, this is 96 h (Nelson, 1950). Garber claimed that while root hairs of cotton could be penetrated directly, few such infections actually reached the xylem. Gerik and Huisman (1988), also using FITC staining, showed that the density of *V. dahliae* on cotton roots increased with distance from the tip, reaching a maximum 1 cm behind the apex. Hyphae were present in the cortex growing towards the stele and concentrating at its surface. Selman and Buckley (1959) also showed a radial orientation of invading hyphae across cortex and pericycle, indicating that the xylem is a preferred rather than accidental habitat (see Pegg, 1985). Mycelial density appears to be a function both of inoculum concentration and available carbohydrate (Selman and Buckley, 1959). Griffiths and Isaac (1966a) and Griffiths (1971a) studied the colonization of cellophane membranes by conidia of *V. dahliae* and described attenuation of the hyphae at the point of entry; the penetration rate was also dependent on external sources of carbohydrate. The relationship between pathogenicity and penetration has been the subject of much interest. A comparison of species and strains of *Verticillium* in the penetration of tomato roots showed a positive correlation between rapid entry and the more highly pathogenic strains (Griffiths and Isaac, 1966b). This result appeared unrelated to the number of infection sites, since Lacy and Horner (1966) found that the number of invasion sites on *V. dahliae* susceptible and resistant mint cultivars were identical, but there was greater invasion of the susceptible cultivars. Identical findings were made for cotton by Evans and Gleeson (1973). In lucerne seedlings at least, infection by *V. albo-atrum* resulted in reduced mitotic activity and greater chromosome aberration in susceptible cultivars compared with healthy ones. The resistant (tolerant) cv. Vertus showed less chromosome damage and greater mitotic activity than susceptible cultivars (Czaplinska and Cebrat, 1975). 'Focal micro infection' of new cotton cultivars by *V. dahliae* and *F. oxysporum* f.sp. *vasinfectum* was reported by Guseva *et al.* (1974), but while it was attributed to synergism the condition was not defined. Xylem colonization rather than infection *per se* appears to distinguish the normal susceptible host from the non-host. Huisman (1988a) showed that *V. tricorpus* penetrates cotton and tomato with equal facility, but symptoms only develop in tomato in which the stele is colonized. Root damage at transplanting, or by deliberate wounding, resulted in rapid invasion of chrysanthemum cuttings by *V. dahliae* (Alexander and Hall, 1974). Root-dip inoculation methods result typically in a massive uptake of bud cell or conidial inoculum through severed roots. The ultrastructural details of root invasion were described by Bishop and Cooper (1983b). Selman and Buckley (1959) described the formation of callosities at cortical cellular penetration sites, but which failed to arrest the pathogen. The growth rate and density of the invading fungus were enhanced in the presence of sucrose. A similar result was reported by Regnani and Matta (1986) in which direct penetration of pepper seedling roots only occurred with the addition of sucrose to the conidial inoculum. Regnani and Matta (1986) claimed that abundant callosities retarded but did not prevent fungal passage. Wounds favoured infection and especially broken roots in spore suspensions. While it is assumed that most plant infections originate from multiple root infection courts, Bejarano-Alcazar *et al.* (1999) successfully infected aubergine cv. Black Beauty at a single site below the main root tip of young seedlings. The minimum number of microsclerotia (26 per site) produced up to 65% of successful infections. After 6–7 weeks, the number of infection sites declined from the 65–100%, 2–4 weeks after inoculation, to 10–29%. Apparently, infections which failed to become systemic died out.

While roots provide the main channel of entry into the plant several novel methods of infection have been described. Morehart *et al.* (1985) successfully inoculated susceptible tomato, aubergine and tulip tree seedlings (*Liriodendron tulipifera*) with naked protoplasts of *V. albo-atrum* prepared by using a lytic enzyme from *Streptomyces* sp. Normal mycelium was isolated from infected plants. Cirulli (1974) infected tomato leaves with a conidial suspension of *V. albo-atrum*. Infection was only possible by abrading leaves. Only local lesions occurred, and no systemic symptoms. Hung and Whitney (1976) successfully infected resistant and susceptible potato leaves with *V. albo-atrum*. Penetration occurred directly in both cultivars, through epidermis, leaf hairs and stomata. In resistant cultivars, infection hyphae were confined to a subenticular locus resulting in a hypersensitive reaction giving pin-point necrotic spots. Spreading lesions in susceptible cultivars resulted from widespread colonization of epidermal and palisade tissue which led to leaf death after 5 days. Natural leaf infection of *V. dahliae* in olive was reported by Tjamos and Despina (1986) based on successful plating experiments of surface-sterilized laminae. Whether foliar infection as described by Hung and Whitney (1976) occurrred is not clear, since mycelial growth from colonized minor veins (which is commonplace) would be possible. In a subsequent paper (Tjamos and Tsongriani, 1990), this question is not clarified. The formation of microsclerotia in at least 10% of leaves from identified infected branches emphasized the importance of leaves as a source of inoculum capable of widespread dispersal. Novel artificial infection of lucerne pollen by *V. albo-atrum* conidia was achieved by Huang and Kokko (1985). Penetration of the exine wall was largely but not exclusively through germination pores, leading to the complete degeneration of the pollen cell contents which were replaced by a cluster of hyphal cells. This technique suggests a possible rapid method (~48 h) for screening a range of host germplasm, assuming that resistance might be expressed or reflected in the pollen grain.

Colonization

The special feature of wilt pathogens is their containment in the host plants' xylem vessels. The question, after a century of research, of whether this represents a preferred or a constrained habitat remains largely unanswered (Pegg, 1985).

Following entry into the plant, either by hyphal growth or by uptake of phialoconidia or blastoconidia through broken roots or stem puncture, colonization proceeds either gradually, radially and basipetally, or rapidly in xylem elements and basipetally by spores and periodic sporulation. Reinke and Berthold (1879) must be credited with the first description of vascular conidia in their original description of *V. albo-atrum* in potato. The significance of spores in effecting a rapid invasion of the host was emphasized by Banfield (1941) with *V. dahliae* in elm trees. Rapid conidial movement was recorded in hop (*V. albo-atrum*) by Christie (1956), Talboys (1962) and Sewell and Wilson (1964a); in tomato (*V. albo-atrum*), including bud cells, by Pegg and Dixon (1969); in cotton (*V. dahliae*) by Presley (1966); in tobacco (*V. dahliae*) by Wright (1969); and in lucerne (*V. albo-atrum*) by Pennypacker and Leath (1983). In cotton (and other hosts), phialoconidia may be formed on simple conidiophores or modified single-celled phialides (Buckley *et al.*, 1969); alternatively, yeast-like blastoconidia may form (Schnathorst, 1981). The physiological (and/or other) significance of these different spores, if any, has yet to be determined.

Leyendecker (1950) failed to observe spores in cotton roots and stems; Garber (1973), however, recorded 50–100 conidia in individual vessels, more in assays of tracheal fluid from susceptible than tolerant cultivars (Schnathorst *et al.*, 1967). Two groups of intravascular conidia of *V. dahliae*, one group 2 × 6 μm found on hyaline mycelium measuring 1.3 μm in diameter, and a group of larger spores 5 × 10 μm derived from larger (4.0 μm diameter) darker hyphae. Rapid colonization resulting in isolation of the pathogen from leaves after 4 days was described by Garber and Houston (1966, 1967). Stem colonization was complete after 48 h, with more spores in susceptible than resistant cotton cultivars. Presley *et al.* (1966) found free-floating conidia at the top of tall (115 cm) cotton plants 24 h after conidial inoculation. Large numbers of conidia were found in hop (6870 *V. albo-atrum* conidia ml^{-1} of xylem fluid) (Sewell and Wilson, 1964a). Similar numbers were found in cotton xylem fluid by Schnathorst *et al.* (1967). The concentration of stem conidia of *V. dahliae* in susceptible mint (*Mentha piperita*) was 100–15,000 conidia mm^{-1} of stem, whereas in the resistant species (*M. crispa*) the number was reduced to 10–57 conidia mm^{-1} of stem (Brandt *et al.*, 1984). Talboys (1962) reported conidia in a number of host species. In cotton infected with *V. dahliae* (Garber, 1973) and potato infected with *V. albo-atrum* (Perry and Evert, 1983), conidia were shown to accumulate in vessel end plates with germ tubes penetrating the adjacent vessel or moving from vessel to vessel by penetrating pit membranes. Hyphae have also been reported in the lumena of paravascular parenchyma which stained darker than normal. Knoll (1972) maintained that the age of lucerne plants was critical for the rapidity of internal spread. Vascular colonization may thus proceed at speed by the trapping of conidia on end plates, hyphal growth to the next vessel, followed by conidiation and a repeat of the process at the next obstacle.

Colonization by hyphae, a slower process than by conidia, has been shown to be, in part, a function of inoculum concentration. At low inoculum concentrations, colonization of cotton by *V. dahliae* rarely extends beyond cortical (pericyclic) layers of the hypocotyl. With higher levels, invasion progresses as a mycelium mat to the stele and vascular elements (Garber, 1973). Savov (1978a), working on the same system, showed a slower progression in the root than in the stem, but the most rapid spread in leaves. Hall and Busch (1971) claimed that chrysanthemum plants colonized by *V. dahliae* only developed symptoms after foliar colonization. A technique based on electrical resistance to detect stem colonization in the absence of aerial symptoms developed by Tattar (1976) showed that resistance to a pulsed electrical current was significantly lower in symptomless, infected *Acer saccharinum*, *A. platanoides* and *A. rubrum* than in corresponding healthy plants (Newbanks and Tattar, 1982).

Talboys (1958b) found that *V. albo-atrum* entered hop plants through intact or damaged roots. Two phases of colonization were postulated (Talboys, 1964), the extravascular and the vascular. The intensity and speed of extravascular colonization was claimed to determine the quantity of pathogen entering the xylem and hence the severity of aerial symptoms (Harris, 1958; Talboys, 1972). Keyworth (1953a,b) using a series of semi-resistant and susceptible root and stem–graft combinations showed that dense stem colonization was a prerequisite for severe wilt development. He suggested that stems and not roots were the determinants of wilt severity. Sinha and Wood (1967a) found that colonization of tomato with *V. albo-atrum* reached a peak in the stem and then declined, paradoxically, while symptoms became more severe. Resistant plants which initially had only sparse colonization eventually lost all invading hyphae. Dixon and Pegg (1969) showed a 70% reduction in the number of tomato stem vessels containing *V. albo-atrum*. Pegg and Vessey (1973) and Pegg (1976b) demonstrated that extensive colonization was correlated with high chitinase and 1,3-β-glucanase activities. *V. dahliae* in olive and apricot showed a similar but seasonal decline in biomass interpreted as mycelial quiescence over the summer to winter period (Ragazzi *et al.*, 1987) or complete elimination followed by seasonal reinfection (Wilhelm and Taylor, 1965; Taylor and Flentje, 1968).

In comparative studies of *V. dahliae* and *V. tricorpus* on field cotton and other plants, colonies were obtained throughout the growing season by plating washed roots on a growth-restrictive medium, either from the rhizoplane or the cortex. Colonization by *V. dahliae* was unaffected by changes in temperature or soil moisture, whereas that by *V. tricorpus* was greater at 20–23°C than at 28–31°C. To mid-July, colonies formed at random, but thereafter occurred as clusters coincident with increases in soil inoculum density. Colonization was not related to pathogenicity since *V. tricorpus* readily colonized cotton and tomato but only invaded the vascular system of tomato (Huisman, 1988a,b). Strunnikova *et al.* (1994) reported on the infection of lucerne by *V. albo-atrum* [*sic*] grown in rotation with cotton in Kyrgystan, Uzbekistan, Kazakhstan, Turkmenistan, Tajikistan and Azerbaijan. Infection and colonization of roots occurred in the absence of symptoms, providing a carry over for the succeeding crop. Since *V. albo-atrum* is not a pathogen of cotton in these republics, the authors were presumably referring to *V. dahliae*.

Leaf and petiolar colonization have been shown to be important determinants of symptom severity in chrysanthemum with *V. dahliae* (Hall and Busch, 1971; Robb *et al.*, 1975), in tomato with *V. albo-atrum* (Dixon and Pegg, 1969) and in hop with the same species (Talboys, 1958b). Talboys (1958b) considered that the stelar boundary layer – the endodermis – represented a determining barrier to vascular colonization and hence symptom severity. Suberin or callose deposition on cortical radial walls (Talboys, 1958b,c; Selman and Buckley, 1959), and as an endodermal casparian band (Talboys, 1958c) were envisaged as lateral barriers to pathogen ingress. In young plants, however, casparian strips exist as only a light deposit, but passage cells could still serve as entry ports to the stele. It is difficult to interpret wall deposits as absolute determinants of colonization *sensu* Talboys, since unless a comparable barrier (correlated with resistant cultivars) existed in the stele, hyphal growth over a short period would restore stelar concentrations of the pathogen equal to those on the outside of the endodermis.

Much has been written on the effect of tyloses on the rate and extent of vascular colonization (Talboys, 1958b) (see Chapter 9). Dixon and Pegg (1969) showed in tomato, at least, that colonization preceded formation of tyloses and their presence, correlating with symptom intensity, appeared to contribute to xylem (and water) occlusion. A study involving the infection of tomato cultivars with hop and tomato strains of *V. albo-atrum* found that tylosis only occurred with particular host cultivar–strains combination but was not associated with a resistant reaction, or an avirulent response. A similar but longer term reduction in colonization was noted by Ciccarese *et al.* (1990) in peach trees in southern Italy infected by *V. dahliae*. Following severe infection, trees recovered over 3 years, with 41, 14 and 0.8% of plants showing infection, respectively. On each of three successive years, positive isolations were made from 72, 18 and 1.7% of the diseased trees. Blanco-Lopez *et al.* (1990) reported that while 30% of commercial olive orchards of trees younger than 15 years in southern Spain were infected with *V. dahliae*, trees rarely died. Recovery of *V. dahliae* infection in young ash (*Fraxinus excelsior*) is common, with approximately 50% of infected trees becoming symptomless after 1 year and decreasing further. This was attributed to the production of new fungus-free xylem. Symptoms rarely occurred in recovered trees. Initial infection, however, stunted trees and, subsequently, competition by healthy adjacent trees often resulted in death (Hiemstra, 1995). Tosi and Zazzerini (1998) reported discontinuous infection of olive in Perugia, Italy apparently related to temperature fluctuations; only during a mild winter (1995) could *V. dahliae* be reisolated continuously.

Transmission and Dispersal

Root contact

Isaac (1953b) considered that root to root contact would be the most common and rapid means of spreading *Verticillium* spp. The propensity for sporulation in soil, natural root damage (Pegg, 1985) and the infection of intact roots (Bewley, 1922; Nelson, 1950) support this view.

Air dispersal

Air-borne conidia of *V. albo-atrum* have been detected using a slit sampler at several sites above lucerne fields in the UK (Davies and Isaac, 1958). The authors also found conidia in the atmosphere of a London garden and in house dust in Edinburgh. Lindeman *et al.* (1982) confirmed Davies and Isaac's lucerne findings using an Anderson sampler. Of 102 isolates, 50% were derived from 7 μm or >7 μm particles; the remainder were from 2–4 μm particles. Under laboratory conditions, Termorshuizen (1997b) induced sporulation on wilting leaves and stems of *Arabidopsis thaliana* root-dip inoculated with *V. dahliae*. No sporulation was reported on heavily-infested potato plants in the field. While approximately 10% infection of *Arabidopsis* was achieved from soil watered with 1×10^6 conidia ml^{-1}, air-dispersed conidia were not found and determination of their role in establishing field infection is problematic. The rapid spread of *V. albo-atrum* in lucerne crops in the UK and North America in recent years has stimulated much interest in the dispersal mechanisms in land previously uncropped with lucerne. The possibility of successful aerial infection was discounted by Jimenez-Diaz and Millar (1988) based on the low ratio of *Verticillium* to other spores they found over the crop and their previous finding (Jimenez-Diaz and Millar, 1986) that *V. albo-atrum* conidial inoculum would not infect uninjured leaves. Bonifacio (1974) and Bonifacio and Parrini (1975) postulated in the absence of firm evidence that *V. dahliae* infection of olive in Tuscany resulted from (*V. albo-atrum* [*sic*] *V. dahliae*) spores. The inference from the work of Tjamos and Despina (1986) and others who found microsclerotial inoculum in detached olive leaves is that such debris could be wind-borne to provide distant soil-borne infection. This was not demonstrated. The likelihood of wind-borne inoculum in olive and other field crops was demonstrated elegantly by Easton *et al.* (1969). Using the ethanol–streptomycin selective medium of Nadakavukaren and Horner (1959), they isolated *V. dahliae* colonies from loess dust 20 feet above a potato crop in Washington State, USA, equivalent to 100 propagules g^{-1} dust.

Water transmission

In a valuable contribution, Easton *et al.* (1969) found *V. dahliae* microsclerotia in irrigation channels supplying a potato field. Microsclerotia were absent at the source of the channels, but at the ends numbers were 31,599 l^{-1} in June, rising to 250,873 l^{-1} in July. An irrigation settling pond, 1 mile from a diseased potato crop, contained 4388–15,410 l^{-1} microsclerotia. The importance of these findings may be considered in relation to the possible spread of disease from recycled irrigation water.

Seed transmission

Seed transmission of resting mycelium or microsclerotia is now well established. The epidemiology and control of seed-borne *Verticillium* was reviewed by Sackston (1983b). Isaac and Heale (1961) showed that resting mycelium of *V. albo-atrum* remained viable on lucerne seed for 13 months. Christen (1982b) demonstrated *V. albo-atrum* between the cells of the outer integument, the highest infection being in small seeds. This was confirmed by Gilbert and Peaden (1988). Seed transmission in Mexican lucerne was found only in some cultivars (Lomeli Osuna *et al.*, 1991). Inoculation of the stigmatic surface led to latent infection in the style during seed development, which in humid conditions became active, invading the pod and seed coat (Huang *et al.*, 1985). The authors (see also Insect transmission) suggest that this process could be insect initiated. In a novel experiment, Huang and Kokko (1985) infected the extine layer of lucerne pollen with *V. albo-atrum*. The fate of this or the implications were not explored. Sheppard and Needham (1980) claimed that *conidia* could remain viable on seed for 8 months, leading to disease on virgin land from imported seed. In a more detailed investigation, Huang *et al.* (1994) found that the survival of *V. albo-atrum* on artificially inoculated seed in air and buried in soil was low at temperatures above zero to 30°C. At subzero temperatures, 90% of the pathogen survived after 10 months in air and soil. It was claimed that seed storage at 30°C for 1 year would eliminate the fungus. Following conidial stem inoculation of lupin species with *V. albo-atrum*, Parnis and Sackston (1979) found internal seed infection in *Lupinus luteus* but only superficially in *L. albus*. Urosevic (1983, published 1987) similarly recorded *V. albo-atrum* on the surface of *Quercus robur* acorns. The production of inoculum (resting mycelium or microsclerotia) on plant organs capable of detachment from the host may be an important aspect of fungal transmission and dispersal. Leaves of olive were shown to contain viable microsclerotia of *V. dahliae* (Tjamos and Tsongriani, 1990). Similarly, petioles of ash (*Fraxinus excelsior*) from wilted trees were shown to be effective in transmitting microsclerotia (Rijkers *et al.*, 1992). Petioles from diseased Norway maple (*Acer platanoides*) could also disperse *V. dahliae*, either as mycelium or subsequently as sources of microsclerotial inoculum (Hiemstra, 1997). While *V. dahliae* was found in the vascular tissue of the lower trusses of several tomato cultivars, no evidence was found for seed-borne transmission (Besri, 1978).

Seed transmission of *V. dahliae* has been recorded in a variety of hosts (Van der Spek, 1973): spinach (Snyder and Wilhelm, 1962) and with up to 80% infection from plants in infested soil (Van der Spek, 1972); and chickpea (Maden, 1987). In groundnut, Melouk and Wadsworth (1990) found varietal susceptibility to seed infection; a Spanish type gave 79% infection from infected plants. Seed transmission in safflower (*Carthamus tinctoria*) is well documented (Schuster and Nuland, 1960; Zimmer, 1962; Klisiewicz, 1974a,b, 1975). Microsclerotia were found externally and on the testa and pericarp. Torulose and hyaline hyphae were also present in the sclerenchyma of the pericarp. Schnathorst and Mathré (1966b) showed that the P_1 cotton strain of *V. dahliae* caused severe wilt in safflower with implications for the seed dissemination of the cotton pathogen. In *Capsicum*, seed transmission of *V. dahliae* has been described in Korea (Park and Kim, 1986), Albania on 22 cultivars (Ibrahimllari, 1987) and Spain (Palazon and Palazon, 1989). Infection of the hull and testa of sunflower seed was described by Sackston and Martens (1959). Sackston (1980, 1983a) considered that seed of resistant cultivars could transmit the pathogen. Seed of 11 samples naturally infected with *V. dahliae* collected from different production areas in Pakistan by Bhutta *et al.* (1997) all showed impaired germination. *V. dahliae inter alia* was also found on sunflower seed from Northern Nigeria (Ataga and Akueshi, 1996). In cotton, microsclerotia are found on the fuzzy seed coat, deposited during harvest and ginning; occasionally internal infestation has been found (Evans *et al.*, 1966b; Bell, 1992a). Minton (1978) and Shen (1985) described treatments and strategies for reducing cotton seed transmission of *V. dahliae*. In China, seed infestation with *V. dahliae* (hyphae and microsclerotia) was found in cotton seed from 24 counties of eight cotton-producing provinces. Only one region, Xinjang Uygur, was found free from seed infestation (Sun *et al.*, 1998a). In Turkey, Karaca *et al.* (1973) reported a 0.7% seed contamination. Gaibullaev (1972) suggested that grading cotton seed to a specific weight excluded seed (presumably lower weight) containing the pathogen. The rapid spread of *V. dahliae* in oilseed rape in Europe, closely mirroring the large expansion in its cultivation, has led to a search for seed-borne transmission (Cappelli *et al.*, 1998). While Heppner and Heitefuss (1995) induced an approximately 4% seed infestation in glasshouse-infected *Brassica napus* L. cv. *oleifera* Metzger, a screen of rape cv. Ceres collected from 74 sites in Germany found seed contamination in only one sample (0.5%).

Significantly, *V. dahliae* has been recorded in the seed of weed hosts, *Xanthium pungens* (Evans, 1968) and *Senecio vulgaris* (Schippers and Schermer, 1966). In the latter host, *V. dahliae* is seed-borne but not seed transmitted. In this case, the seed is small and the presumption is that the pathogen is suppressed by soil antagonists. Other weed hosts capable of seed transmission are *Xanthium spinosum* and *Carthamus lanatus* (Bell, 1992a,b).

Vegetative transmission and spread

The use of the potato tuber (*Solanum tuberosum*) as a convenient food and antiscorbutant by explorers, pioneers, sailors and troops has undoubtedly provided the major vehicle for the dispersal of *Verticillium* spp. within and between continents (Pegg, 1985). Transmission of *V. albo-atrum* in tubers was reported by McKay (1921, 1926). Easton *et al.* (1972b) showed that 13–51% of certified seed potatoes from the USA and Canada were infected with *V. dahliae* (*V. albo-atrum* R and B microsclerotial type [*sic*]). Platt (1986) confirmed that both *V. albo-atrum* and *V. dahliae* were tuber-borne. Wilt incidence from planting tubers from infected plants was greatest in cv. Irish Cobbler, with <3% in cv. Russet Burbank; the wilt incidence for a range of other cultivars was from 5 to 20%. A mean wilt incidence of 50–80% resulted in yield losses of 25–40%. Beckman *et al.* (1969) found both species in certified seed tubers. Christen (1982b, 1983) has demonstrated seed-borne infection of lucerne seed and external contamination, both able to establish outbreaks in hitherto *Verticillium*-free soils. *V. dahliae* has been determined *in situ* by FITC immunostaining (Nachmias and Krikun, 1984a,b); 36–34% in Israeli tubers (Tsror *et al.*, 1999). In Quebec province, Tartier and Devaux (1977) found that most tuber cultivars used and most potato-growing areas were wilt infested. Notwithstanding the implications of tuber-borne inoculum for disease carry over and the initiation of new epidemics, Robinson and Ayers (1961) in Canada found (unlike Platt, 1986) in all cultivars tested, that internally infected tubers consistently resulted in a lower wilt incidence than healthy tubers infested with surface-borne inoculum. Moreover, a preponderance of tubers produced by wilted plants gave rise to wilt-free plants. The significance of even isolated outbreaks from infected tubers lies in the massive production of mycelial or microsclerotial inoculum on buried haulm. In the UK, MacGarvie and Hide (1966) found that 79% of 225 certified seed potato stock was contaminated with *Verticillium* spp. but not *V. albo-atrum* or *V. dahliae*; 72% of the infected stock contained *V. tricorpus*, 10.2% *V. nubilum* and 8.5% *V. nigrescens*, with most isolations made from young tuber sprouts. Between 1963 and 1976, samples of commercial seed stock of four cultivars yielded similar results: *V. tricorpus*, 78%, *V. nigrescens*, 9% and *V. nubilum*, 3% (Hide, 1981). The role of the potato in spreading *V. dahliae* into previously wilt-free virgin land in Israel was described by Krikun and Susnoski (1971;

see Pegg, 1974). From its introduction in tubers in the reclaimed irrigated pathogen-free Negev, *V. dahliae* spread via infested haulm to tomato, aubergine, squash, cucumber, pepper, onion, melon, mango, apricot, groundnut, watermelon, olive, avocado and cotton. In Ontario, potato was infected more or less equally with *V. albo-atrum* and *V. dahliae*, with some *V. nigrescens* (Busch, 1967); however, whereas *V. dahliae* could be isolated throughout the winter using aubergine as a very susceptible host, *V. albo-atrum* could not be isolated after November. It was concluded that *V. albo-atrum* overwinters in tubers and from these is reintroduced annually into Ontarian soils (McKeen, 1970; Beckman, 1973). The survival of *Verticillium* spp. in many weed and other non-host plants has been well documented (Martinson, 1964; Lacey and Horner, 1966; Harrison and Isaac, 1969; McKeen, 1969; Evans, 1971a; Isaac and Levy, 1971; Evans and Gleeson, 1973; Benson and Ashworth, 1976; Schnathorst and Davis, 1978; Thanassoulopoulos *et al.*, 1984). Evans and Gleeson (1973) included examples from the *Chenopodiaceae*, *Labiatae*, *Compositae*, *Papilionaceae*, *Convolvulaceae*, *Malvaceae*, *Gramineae* and *Solanaceae*. While perennating non-hosts may serve as a source of inoculum between crops where mycelium does not persist in the ground, little is known of their possible role in transmission. *Verticillium* spp. can be disseminated readily by man in vegetatively propagated infected plant material; one example of this is *V. dahliae* in cuttings of *Chrysanthemum* (Alexander and Hall, 1974). Potato propagated by stem cuttings in Tasmania was also infected with *V. albo-atrum* (Sampson, 1977).

Insect transmission

Popushoi and Kulik (1976) surmised that *V. dahliae* wilt of apricot in Moldavia is transmitted by the fruit bark beetle *Scolytus rugulosus*. The involvement of the potato leaf hopper in the spread of potato wilt in the USA was described by Johnson *et al.* (1987). Although not an insect, earthworms (*Lumbricus terrestris*) have been shown to gather and pull underground stems and leaves of peppermint (*Mentha piperita*) infected with *V. dahliae*. Microsclerotia from worm casts were pathogenic to peppermint (Melouk and Horner, 1976). Price (1976) found viable conidia and fragments of microsclerotia in faeces of the bulb mite *Rhizoglyphus echinops* fed on cultures of *V. dahliae*. Most other accounts of insect transmission relate to lucerne and are largely from the work of Huang and co-workers in Canada. Huang and Richards (1983, 1987) detected *V. albo-atrum* on leaf fragments used in cell construction by the leaf cutter bee (*Megachile rotundata*). This species is exploited commercially in Canada for pollination. Some 30% of foraging bees had mouths and bodies contaminated with conidia, which were also found in styles and sigmata (Huang and Richards, 1983). There was 2% infection of pods from symptomless plants (Huang *et al.*, 1986b). Huang *et al.* (1983) also showed that the pea aphid *Acyrthosiphon pisum* is a vector of *V. albo-atrum*: 19–90% of aphids, alate and apterous were shown to be carrying conidia on legs and antennae. In greenhouse experiments, Kalb and Millar (1984, 1986) demonstrated disease spread as fungus gnats (*Bradysia* spp.) flew from diseased to healthy plants. A similar experiment by Harper and Huang (1984) successfully transferred the pathogen on the lucerne weevil (*Hypera postica*) – and also on predators of lucerne insect pests – to healthy lucerne plants. Leath and Hower (1988) reported *V. albo-atrum* colonization of *Sitona hispidulus* larval feeding sites on lucerne roots, but transmission was not confirmed.

Huang and Harper (1985) and Huang *et al.* (1987) found that *V. albo-atrum* fed experimentally to grasshoppers (*Melanopus sanguipes* and *M. bivittatus*), lucerne weevils (*H. postica*) and woolly bears (*Apantesis blakei*) survive in the digestive tract of each, appearing in faeces 1 day later, for 3 days. Grasshopper faeces containing *V. albo-atrum*, when buried near healthy lucerne seedling roots, developed wilt 6 weeks later. *V. albo-atrum* in aphid (*Acyrthosiphon pisum*) bodies, or grasshopper (*M. sanguipes*) faeces survived well at −40°C for 24 months (Harper *et al.*, 1988). At 15°C, the pathogen in aphids survived for 5 months and in grasshopper faeces for 21 months. This result illustrates the effectiveness of insects in survival and transmission of lucerne *V. albo-atrum*. Huang (1989) implicated leaf vein-feeding insects in the transmission of lucerne wilt, based on local isolated infected leaf lesions. Leath and Hower (1988) found *V. albo-atrum* colonizing *Sitona hispidulus* larval feeding sites, but there was no evidence for insect transmission. Bedlan (1987) described *V. albo-atrum* transmission in pepper and aubergine. The nutritive value of wilt-infected lucerne to insects is lower than that of healthy plants. Larvae of the Lepidopteran *Spodoptera eridania* increased the consumption rate on *V. albo-atrum*-infected plants to equal the growth rates on healthy plants. Larval growth rates were the same on both types of plant (Kingsley *et al.*, 1983).

Husbandry practices

When infected lucerne stems were buried in a manure pile, the pathogen remained viable after 6 weeks. It was also recovered from sheep faeces 2 days after feeding with wilt-infected hay (Huang *et al.*, 1986b). Lopez-Escudero and Blanco-Lopez (1999) also isolated *V. dahliae* from sheep manure spread in Andalucian olive plantations. No mention was made of the pathogenicity of this isolate to olive or how it compared with that isolated from plantation soil. Baudin and Lepoivre (1984) demonstrated the transfer of *V. dahliae* to tomatoes via microsclerotial transmission in sewage sludge. By far the most wholesale spread (and propagation) of wilt is by man's cultural and harvesting practices. The multiplication of resting mycelium and microsclerotia on crop residues and the distribution of this by accidental transport, wind and soil cultivation is a major source of new infection (Howard, 1985). This is discussed elsewhere. The distribution of infected plant material in hop *V. albo-atrum* was spread both locally and nationally in the UK by poor hygiene, transport of infected propagation material and by mechanical cultivation and harvesting methods (Pegg, 1985). The spread of infective hop wilt material by workers was described by Keyworth (1942), and the influence of tillage and non-cultivation on hop *V. albo-atrum* movement by Sewell and Wilson (1974). The effect of mechanical techniques on spread is illustrated in hop picking machines (Sewell and Wilson, 1961), and the transmission of *V. albo-atrum* in lucerne by cutting equipment (Isaac, 1957c; Gondran, 1977; Haas and Martin, 1977). Similarly, *V. dahliae* in groundnut wilt is spread mechanically (Purss, 1961). Christen and Peaden (1982) investigated various methods of spreading lucerne wilt: (i) plants were irrigated with conidia-contaminated water; (ii) ground prior to sowing had infected hay incorporated; (iii) the cutter bar of the mower was contaminated with conidia; (iv) seed prior to sowing was coated with conidia; and (v) resting mycelium and infected plant material was scattered over ground immediately after cutting. Of all these methods, only the last two were effective in inducing significant yield loss. Nevertheless, it is the primary establishment of the pathogen in virgin soil or plant material regardless of the *initial* severity which is of crucial significance. The converse of the foregoing control by husbandry practices, was described by Green (1958) who recommended, in the absence of alternative methods, deep ploughing to give partial control of *Verticillium* in muck soils.

The role of infected nursery stock in proliferating and spreading inoculum is a major world problem. Thanassoupopoulos (1993) attributed the relatively recent widespread disease in newly planted olive groves to infected Greek nursery stock especially in the Kassandra peninsular. One problem (which was resolved in the UK hop industry) was the location of nurseries in areas of heavy infection. *Verticillium*-infected olive nursery stock was also implicated in field disease spread combined with leaf and weed inoculum (Naser and Al-Raddad Al-Momany, 1998). An obvious but often overlooked source of transmission in ornamental and landscape horticulture is the use of mulches around trees and shrubs derived from composted infected plants (Hoitink and Krause, 1999).

Inoculum Density and Assessment of Inoculum

Research on inoculum has been concerned with: conditions favourable for microsclerotial production, minimal levels capable of causing disease, the relationship between disease severity and propagule numbers and attempts to assess inoculum potential (Powelson, 1970). The important question of the optimal timing and location of microsclerotial production in relation to harvest time and plant part was examined by Mol and Scholte (1995a,b). A comparison was made of leaf blade, aerial stem, subterranean stem, stolon and root, harvested while the canopy was green and at maturity. After 4 weeks incubation and air drying, maximum microsclerotial production was found in haulm harvested at maturity and more in cv. Element than cv. Mirka. Petiole and aerial stem yielded most inoculum, while roots yielded more than stolons. In a subsequent experiment (Mol, 1995b), unfortunately confined to pot conditions in a glasshouse, a comparison of microsclerotial formation on four potato cultivars and the intercrops, pea, sugarbeet, onion, flax, spring barley, field beans, spring wheat and spring rape, found that highest numbers were on potato, flax and barley following root-dip, rather than inoculation from infested soil. Of the intercrop plants, flax bore the most microsclerotia. Curiously, two potato cultivars Ostara and Mirka produced higher microsclerotial numbers following infection from contaminated soil, this however may merely reflect variability in the experiment. As Mol *et al.* (1996b) later showed, the results are not surprisingly conditional on the isolate of *V. dahliae* used, with potato isolates showing more virulence to potato than bean and vice versa; hence, the previous year's crop of the same cultivar was more important than a higher level of residual inoculum of a different pathotype.

Green (1969) working on tomato found that the minimal level of inoculum for 100% infection was 50×10^3 *V. dahliae* conidia g^{-1} of soil or 100 microsclerotia g^{-1} of soil. However, the *potential* infectiveness of this inoculum over time was very different; whereas conidial viability decreased to zero after 3 weeks, microsclerotial numbers were not diminished after 7 weeks. In field experiments, Ashworth *et al.* (1979a) showed that the rate of tomato plant infection increased with the number of microsclerotia between 0.1 and 27 microsclerotia g^{-1} of soil. At the end of the season, all plants were infected regardless of inoculum density (ID). These inoculum values are purely relative, depending *inter alia* on inoculum viability, host origin, soil type and environmental factors, which have led authors to consider the concept of *inoculum potential sensu* Garrett (1960) (see Dimond and Horsfall, 1965). Illustrating this concept, different inoculum requirements have been cited for cotton infection (Bell, 1992b). The threshold for cotton plant infection under Californian field conditions was 0.03 microsclerotia g^{-1} of soil. At the end of the growing season, with 0.3–1.0 microsclerotia g^{-1} of soil, infection was between 20 and 50%. With 3.5 microsclerotia g^{-1} of soil or greater, infection was 100% (Ashworth *et al.*, 1972a). The disease intensity of susceptible cotton increases progressively, with 100% infection achieved by 10 microsclerotia g^{-1} of soil (Butterfield, 1975; Ashworth, 1983; DeVay and Pullman, 1984). The rate of percentage infection (calculated from progress curves) increases with increased ID from 0 to 50 microsclerotia g^{-1} of soil. Greater IDs are required for the P_2 strain than for the P_1 to cause the same degree of damage (Bell, 1992b). The relationship between disease severity in cotton and inoculum levels of *V. dahliae* from non-cotton hosts was described by Schnathorst (in Dimond and Horsfall, 1965) (Evans *et al.*, 1966a). Disease severity of a cotton isolate showed a straight line relationship to $\log_{10}$ spore concentration. An identical slope was obtained for several isolates plotted against log spore viability × turbidity of the inoculum solution. From this, disease severity S = K logI where I = VN, where V = the percentage of viable conidia and N = conidial numbers ml^{-1} of suspension. Disease severities (termed 'inoculum potential' by Schnathorst) thus calculated for different isolates tested on *Gossypium hirsutum* cv. Delta Pine 15 were: P_1 cotton isolate, 13,000; almond, 5000; honeydew melon, 1600; watermelon, 1000; cherry, 300; and flax 75. Based on the work of Schnathorst and Mathré (1966a), a threshold of 100 propagules g^{-1} of soil of the P_1 strain was required before visible symptoms were seen, whereas the less virulent P_2 strain required 1000 g^{-1} of soil. In these glasshouse experiments, resting mycelium and microsclerotia constituted the propagules. Similar results were obtained with conidial inoculum (Schnathorst, 1963). This author gives the minimum ID of 100 in areas where the disease is epidemic. The numerical relationship of microsclerotium or hyphal fragment and a successful infected root is not clear and would not be the same with different strains and conditions (Schnathorst, 1981). In soil assay plates, each propagule averaged ten infective hyphae (Schnathorst and Fogle, 1973). Pullman and DeVay (1982a) in a 7-year study found that the ID in May was related to the mid-September (post-38°C ambient temperature period) foliar symptoms, with a linear slope at ID <40 microsclerotia g^{-1} of soil. Mol *et al.* (1996a) proposed a mathematical model based on correlations to describe IDs of *V. dahliae* microsclerotia in soil over a long time period. The model took into account systemic root infections in relation to soil ID. Microsclerotial production in debris and reduced crop growth was related to systemic infections. The equation also incorporated microsclerotial release into soil and mortality. The function fitted experimental data on potato (four cultivars), flax, pea, barley, sugarbeet, onion and field bean. The highest ID occurred at 2.3 thermal time units (TTU) of 3600° days (base 0°C). Ten per cent of initial inoculum was still present after 4.5 TTU. Bejarano-Alcazar *et al.* (1995b, 1997) found no correlation between the ID of *V. dahliae* determined at sowing time in Spanish cotton field soils with vascular discolouration, incidence of foliar symptoms or disease intensity index at harvest time. Linear regressions were analysed on data transformed to several mathematical models. The P_1 strain induced earlier and more severe symptoms than P_2 from similar IDs. During 1986 and 1987, soil inoculum of P_1 rose to a threshold of 24–44 and 44–75 c.f.u. g^{-1} of dry soil, respectively. The authors claim that the severity of wilt epidemics in cotton were described more accurately by the descriptive parameters of the disease intensity index over physiological time than by the corresponding increase in the incidence of foliar symptoms. Koroleva *et al.* (1986) found a seasonal variation in microsclerotia in Tadzhikistan (CIS) cotton soils, with a summer (August) low value of 76 g^{-1} of soil increasing to 272 in October, 246 in February and 417 in May at sowing time. While the microsclerotial content in >500 soil samples varied from 0 to >800 g^{-1} of soil (average 15–90), there was no overall increase over a 6-year period. Grishechkina (1990) showed that different cotton organs contributed variously to the winter soil reservoir. In dry leaves the microsclerotia varied from 82×10^3 to 7×10^6 microsclerotia g^{-1} of dry weight leaf; stems, 2×10^3–827×10^3 and roots 172 to 300×10^3. As would be expected, the wilt-susceptible *G. hirsutum* cultivars Tashkent and Andizhan 5 produced more inoculum than the resistant *G. barbadense* cv. Termez 7. In Spain, the P_1 (defoliating) strain of *V. dahliae* in *G. hirsutum* cv. Coker 310 produced soil inoculum of 3.8–59.8 microsclerotia g^{-1} of soil over 2 years, while the P_2 strain yielded 9–29.1 g^{-1} of soil with proportional field infection and disease severity (Jimenez-Diaz *et al.*, 1990).

The yield of the moderately tolerant cv. Acala SJ-2 in California was reduced 20–25% where soil microsclerotia totalled 21–23 g^{-1} of soil. The more tolerant Acala SJ-4 outyielded SJ-2 with >9–13 microsclerotia g^{-1} of soil but at lower ID; SJ-2 gave the best yield (Ashworth and Huisman, 1980). One infected cotton stalk distributed in soil 0.8 m^{-2} gave 100% infection (Leyendeker, 1950). The incorporation of gin trash on land at 25 t ha^{-1} for 3 years in Texas increased wilt incidence from 32 to 81% and 53 to 86% (Waker and Onken, 1969).

DeVay *et al.* (1974) showed that the viable microsclerotia from air-dried cotton field soils ranged in size from 11 to 225 μm. After 6 h at 24–26°C on cellophane overlying potato dextrose agar (PDA), each microsclerotium germinated with up to 36 germ tubes, each of which formed several verticillate conidiophores bearing many conidia. There was, however, no direct correlation between the proportion of diseased plants and the soil microsclerotial content (Bejarano-Alcazar and Jimenez-Diaz, 1997; cf. Bejarano-Alcazar *et al.*, 1995b). Little was known of the precise infection process. Support for DeVay *et al.* (1974) findings were provided by Strunnikova *et al.* (1997) who found no differences in microsclerotial density regardless of the virulence of the pathogen or cotton grown in continuous cultivation or in rotation. They found mycelium from germinated microsclerotia in greatest abundance in the soil and on cotton roots following soil amendment with hydrolysed lignin, but this was claimed to be saprophytic since disease levels were low. Up to 90% of viable microsclerotia in air-dried soil were bound to plant debris (Ashworth *et al.*, 1974b). These authors claimed that the free microsclerotia in soil represented the potentially effective inoculum. The frequency distribution of *V. dahliae* propagules in Oregon potato fields was described by Johnson *et al.* (1988). Menzies and Griebel (1967) showed by plate counts that the population of *V. dahliae* in soil increased three- to fivefold during the first 10 days of sampling and then subsided to a relatively constant level. The early increase was due to conidiation and germ tube formation, both of which were short lived, leaving microsclerotia which had mostly germinated. The importance of microsclerotial conidia (if any) has yet to be demonstrated. Farley *et al.* (1971), however, showed that soil artificially inoculated with microsclerotia, air dried and remoistened over nine successive cycles was reinfested, largely with conidia after each treatment. The evidence largely suggests that mycelium and conidia are short lived, requiring infection within a few weeks of production, unlike microsclerotia (Schreiber and Green, 1963). The *V. dahliae* ID in Tadzhikistan (CIS) cotton fields averaged 19.6 ± 3.9 c.f.u. g^{-1} of dry soil (Molchanova, 1990). The relative abundance of strains (Soviet classification) after 3 years of lucerne following cotton, was race 2, 52.4%; race 3, 32%; race 1, 10%; and race 0, 5.4%. The effect of the lucerne rotation was a threefold reduction in inoculum and a relative increase in races 1, 3 and 0 in the absence of a complementary cotton cultivar for race 2. Khokhlacheva (1990) described the reduction in inoculum in soil by microbial activity when cotton was grown on ridges. The susceptible cotton cv. Tashkent 1 had maximum infection after 4 years' continuous cultivation but was halved after 3 years' cultivation of cultivars 108-F and S-6030. This was explained in terms of reduced levels of infested cotton debris returning to the soil. No specific resistance to race 2 was encountered (Tribunskii *et al.*, 1976). A commendable attempt to analyse the spatial distribution of microsclerotia of *V. dahliae* of cauliflower in Californian field soil was attempted by Xiao *et al.* (1997). At each site, an 8 × 8 grid divided into 64 2 × 2 m plots was sampled to a depth of 15 cm with a 2.5 cm diameter probe. Plots were sampled simultaneously for disease incidences and microsclerotia using an Anderson sampler. Field aggregation of microsclerotia was assessed by Lloyd's index of patchiness (LIP). Spatial autocorrelation and geostatistical analyses were also used. The LIP <1 indicated aggregation but the degree of this at most sites was low. An aggregated pattern of wilt was detected at five of 12 sites. High disease incidence of 77–98% at 11 of 12 sites was due to a high ID. A further model of *V. dahliae* development, in potato, claimed to be effective by the authors Termorshuizen and Rouse (1993), coupled a *Verticillium* subroutine to an existing potato crop growth model. The model described disease development in a single plant with a stochastic variable to calculate stem base infection. Simulations of root infection and the incidence of stem colonization agreed with field data. The success of this model depends on adequate field data and details of fungal dynamics. Optimal runs were at pathozones of 30 μm, suggesting that infection only occurs when a microsclerotium is in close proximity to a root. The relationship between *V. dahliae* inoculum density and wilt incidence and severity in cauliflower was studied in 1.2 m^2 microplots in which known numbers of microsclerotia were added to fumigated soil (Xiao and Subbarao, 1998). Inoculum levels were checked by Anderson sampling, planting on a Na polypectate agar. Four microsclerotia g^{-1} of soil led to a 16% wilt incidence and 10 microsclerotia g^{-1} of soil to 50% wilt. Inoculum levels above 20 microsclerotia g^{-1} gave no significant increase in wilt incidence or severity. In general, there was a direct relationship between increased inoculum and the speed of symptom appearance. The authors maintain that wilt management methods should be directed to a reduced microsclerotial level of fewer than 4 microsclerotia g^{-1} of field soil. In contrast to the foregoing, Khan *et al.* (2000) failed to find a simple relationship between *V. dahliae* inoculum density and brown root discoloration in horseradish and hence a disease-forecasting system based on soil microsclerotial numbers. Knowing field soil inoculum levels could prevent growers planting in highly infested fields, but from the data of Khan *et al.* (2000) field soils with low inoculum levels would not preclude a disease-free crop. Filatova *et al.* (1987) constructed a mathematical model to evaluate host *V. dahliae* interactions.

A detailed spacial distribution of *V. dahliae* inoculum in Chinese field plots and aubergine wilt sampling was conducted by Guo and Yan (1999). Horizontal sampling was at a depth of 15 cm from five locations in 64 grids in 64 m^2 plots in each of nine fields planted with the susceptible cv. Luoyang Zaoqingque; and vertical at depths of 0–20, 20–30, 30–50 and 50–80 cm. Microsclerotia ranged from 8.63 g^{-1} to 39.93 g^{-1} of soil. Microsclerotia were shown to have a congregated distribution. *Verticillium* wilt had a congregated distribution when the incidence was <75% and was uniform at >75% incidence. No data for vertical inoculum distribution were presented. (See also Slattery and Eide, 1980; Slattery, 1983, on inoculum potential in potato soils.)

Assessment methods

One of the pioneering attempts quantitatively to measure *V. dahliae* propagules in field soil was by Harrison and Livingston (1966) who used a modified Andersen air sampler. This technique was used subsequently by Butterfield and DeVay (1975b) and Schnathorst and Fogle (1973). Wet-sieving of soil for microsclerotia in cotton fields was used by Ashworth *et al.* (1972a) with 37, 53 and 74 μm sieves, plating residues on a cellophane film overlying a low-sugar Czapeks' agar with streptomycin sulphate. This was also used for cotton soils by Huisman and Ashworth (1974a,b) and for *V. dahliae* in potato field soil by Smith and Rowe (1984). A comparative study by Butterfield and DeVay (1977) found that a modification of the Andersen sampler technique yielded a 2.8-fold increase in c.f.u. compared with the Huisman and Ashworth (1974a,b) wet-sieving technique. The lowest mesh size used by Ashworth *et al.* (1972a) would not recover conidia, small mycelial fragments or the 11 μm diameter microsclerotia reported by DeVay *et al.* (1974). Anomalous counts of soil propagules by the fragmentation of microsclerotia by milling processes after prolonged storage at 40% relative humidity were described by Ashworth *et al.* (1974b). In particular, 2 mm sieving was especially damaging. After an examination of several techniques, Koroleva (1985) concluded that wet sieving and the use of an Andersen sampler were the most successful. Nicot and Rouse (1983, 1987a) compared three methods of assessing ID. Wet sieving appeared to be the most precise but also the most time-consuming and with bias. Andersen sampling was considered to be unbiased but the least precise, with dilution plating least time consuming, least biased and, in their opinion, giving 100% recovery (see Termorshuizen, 1995). The statistical evaluation of sampling variation was presented by Evans and Gleeson (1980) for *V. dahliae* soil populations. Evans *et al.* (1967) isolated microsclerotia from soil using a simple density separation in a water column. Caesium chloride, which has high specific gravity, low viscosity and low toxicity to *V. dahliae*, was used for the gravity separation of low levels of microsclerotia from soil (Ben-Yephet and Pinkas, 1976). Using 5 g of air-dried soil shaken with 20 ml of 1:1 (w/v) caesium chloride for 10 s, microsclerotia separated in the upper layer with a 55% recovery. Differences in detection rates by various workers may be interpreted in part by the antimicrobial effects of co-isolated microorganisms. This was recognized by Evans *et al.* (1972), Popov and Stepanova (1973), Harris *et al.* (1993) and others, and in the various antibiotics and nutrients employed. Stepanova (1975) and Grishechkina and Sidorova (1984b) used a modification of Isaac *et al.* (1971) medium with streptomycin, biomycin and levomycin supplementing a phosphate–pectinate agar. Using Czapek's agar supplemented with (PCNB), ethanol, chloramphenicol, polyoxin AL, blasticidin S and biotin, Itoh *et al.* (1989) found that *V. dahliae* colony numbers on Japanese radish were linearly related to ID when plotted on a log–log scale. When PCNB was added to soil, increased infection occurred with no increase in ID. Rice straw incorporated decreased ID and reduced the number of root lesions. In fields of Chinese cabbage, the summer soil microsclerotial content was one-tenth of the autumn level. Microsclerotial numbers ranged from 6 to 2000 g^{-1} of soil. A linear relationship was found between the log value of the half-life of the number of microsclerotia and temperature over the range 5–31°C. The original alcohol–agar medium of Nadakavukaren and Horner (1959) was modified by Easton using 15 ml of solidified ethanol–streptomycin agar with a supplementation of 50 p.p.m. penicillin G. Taylor (1969) obtained good recovery of *V. tricorpus* from soil using cellulose and biotin as the sole nutrient sources. Camporota and Rouxel (1977) augmented Isaac *et al.*'s (1971) medium by increasing PCNB to 100 p.p.m. and adding 1% streptomycin, chlortetracycline and chloramphenicol, plating out dry sieved soil. An adaptation of Komada's selective medium for *Fusarium oxysporum* combining L-sorbose as carbon source with L-asparagine salts and streptomycin sulphate by Christen (1982a) gave good recovery of *V. albo-atrum* from hop and lucerne soils. Although not directly concerned with inoculum density, Ausher *et al.* (1975) described the value of an ethanol–streptomycin medium augmented with PCNB for the successful isolation of *V. dahliae* from senescent tomato tissue contaminated with *Fusarium* and saprophytes. Lu *et al.* (1990) employed a PDA, streptomycin, PCNB medium similar to that of Evans *et al.* (1972) . The authors found that field disease development in cotton was proportional to numbers of propagules in the soil. The top 0–40 cm of soil contained 89% of the pathogen. Only 11% of the soil pathogen population was deeper than 40 cm. No differences in microsclerotia were found in resistant or susceptible host rhizospheres. Termorshuizen (1995), in attempting to rationalize results of different workers, employed a correction factor based on the ratio of the germination percentage of microsclerotia collected from field-grown potato stems added to a soil suspension or added directly to a soil dilution plate compared with the germination rate of sterile, water-washed microsclerotia plated in the absence of a soil suspension. This factor is multiplied to the value obtained from a normal soil plating. Variable results were obtained; in general the soil suspension had an inhibitory effect on germination (see Henis, 1971; Harris *et al.*, 1993).

Sewell (1959) reported a technique for direct observation of soil inoculum using glass-walled boxes. Growth from infected plant debris seldom exceeded 2 mm penetration into soil. Keinath and Millar (1986) similarly found hyphal growth no more than 5 mm through the soil. One of the simplest methods of measuring ID, but one difficult to quantitate, is the seeding of soil with tomato and subsequent sectioning and plating of tissue (Schreiber and Green, 1962). Tsror and Nachmias (1990) similarly used a crude 'presence or absence' screen for arable and plantation soil crops. Ten soil samples per hectare were transferred to a glasshouse and planted with aubergine (a universal suscept for *V. dahliae*) for 6–8 weeks. Some 10×10^3 ha^{-1} in Israel are mapped annually for *V. dahliae* in this way. Harris and Yang (1990) and Harris (1990b) used a wet sieving technique and semi-selective media to determine the pre-planting risk of strawberry to *V. dahliae*. The method which could detect 1 c.f.u. 10 g^{-1} of soil relied on its success by using particles <20 μm, drying the soil sample and applying an optimum amount per plate. *V. dahliae* at 0.5–1 c.f.u. g^{-1} of soil corresponded to a 5% wilt incidence. A close correlation between crop infection and soil microsclerotial numbers was shown by Stepanova *et al.* (1977). Barley, millet, maize, pea and tomato were shown to be symptomless carriers of *V. dahliae*, but symptom-developing cotton increased soil inoculum 1.5-fold compared with the other plants.

A sieving technique for determining inoculum levels in cotton leaves was refined by Tsai and Erwin (1975); fractions containing microsclerotia were counted in a Hawkley eelworm counting chamber after a critical blending period. *V. albo-atrum* in potato stems buried for 7 months survived better (47–69%) than *V. dahliae* (23–61%) (Slattery, 1981). Cultivars Red Pontiac and Superior contained most *V. albo-atrum* inoculum, and cultivars Russet Burbank, Kennebec and Irish Cobbler most *V. dahliae*. Many descriptions of the use of conidial inoculum have appeared as minor features of research contributions. Erwin *et al.* (1965) standardized relative spore density to %T in a colorimeter (nephelometry); 50% of cotton plants stem injected with 50×10^4 *V. dahliae* conidia ml^{-1} developed epinasty after 4 days. Bugbee and Presley (1967) found that field observations of resistance in cotton corresponded to glasshouse experiments in which 4 to 7-week-old plants were stem injected with $2–3 \times 10^6$ conidia ml^{-1} of spore suspensions. In sunflower, Moser and Sackston (1973) found a linear relationship between symptoms, including stunting, and inoculum density following stem injection. Some susceptible cultivars became infected at one *V. dahliae* conidium per plant, but an average of 10 spores per plant gave uniform infection, an inoculum level orders of magnitude lower than those used by most experimenters.

Sivaprakasam and Rajagopalan (1974), injecting aubergine stems with *V. dahliae* conidia, found no symptoms at 1×10^3 ml^{-1}, mild chlorosis at 1×10^4–5×10^5 conidia ml^{-1} and severe symptoms at concentrations above this to 1×10^7 spores ml^{-1}. Nipoti (1982a,b) used conidial concentrations of 102–1010 spores ml^{-1} on aubergine. In all these studies, the important criteria for a strict comparison of crops and spore levels is the virulence of the pathogen strain, the *actual* concentration of spores entering the plant and the number of vascular strands pierced. The inoculum concentration is only a relative indicator of these, unless plants are injected to excess of the minimal requirement.

Ecology 6

Interaction with Other Organisms

Nematodes

For many years, simple field observations recognized a synergistic relationship between nematodes and vascular wilt diseases in a wide range of crops. More precise information of the interaction has only been forthcoming during the last 30 years. In a pioneering study, Mountain and McKeen (1960) confirmed an increase in the incidence of *V. dahliae* wilt in aubergine in the presence of *Pratylenchus penetrans* (although the association of nematodes and *Fusarium* wilt had been reported 20 years earlier). In subsequent studies (McKeen and Mountain, 1960; Mountain and McKeen, 1962a), a similar relationship was demonstrated for *P. penetrans* and *V. albo-atrum* and *V. dahliae* on aubergine and tomato. Mountain and McKeen (1962b) and Faulkner and Skotland (1965) showed that *P. penetrans* on aubergine and *P. minyus* on peppermint, respectively, exhibited increased reproduction in the presence of *V. dahliae*. The presence of *V. albo-atrum* or two other fungi on tomato roots increased the ratio of male and female *Heterodera rostochiensis* (potato root eelworm) (Ketudat, 1969).

Mint species

Work by Faulkner and Skotland (1965) on wilt of peppermint showed that when cuttings were grown in the glasshouse with *V. dahliae* microsclerotia alone or with *Pratylenchus minyus*, the presence of the nematode increased both the incidence of wilt and the severity of its symptoms. The authors suggested that the nematodes increase entry points for the fungus and liberate plant materials which enhance fungal germination. It was also proposed that the fungus may release growth-promoting substances for the nematodes. Faulkner *et al.* (1970), with the same two parasites of peppermint, developed a technique for assessing whether the influence of the nematode was merely the provision of infection routes for the fungal hyphae. Detached stems were rooted at two locations to form two separate root systems each of which was accommodated in a separate plastic pot. It was found that the nematode influenced both the period of incubation and the incidence and severity of wilt even when each organism parasitized a separate root system of the same plant.

In a series of split-root experiments on peppermint plants with *P. minyus* and *V. dahliae* var. *menthae* with pre- and post-infection with both organisms, singly and in reciprocal combination, Faulkner *et al.* (1970) showed that the nematode increased the severity and incidence of wilt even though infection was confined to an isolated part of the root. The authors concluded that physiological changes occur concomitantly on infection by the nematode, which increases plant susceptibility independently of any wound infection caught. A different effect on peppermint (*M. piperita*) and spearmint (*M. cardiaca*) by the northern root-knot nematode *Meloidogyne hapla* and *V. dahliae* was reported by Eshtiaghi (1975). Peppermint was more susceptible to the nematode than spearmint. Significantly fewer galls were produced by *M. hapla* in the presence of *V. dahliae*. Similarly, the nematode reduced the severity of *V. dahliae* symptoms in peppermint.

Tomato

Conroy *et al.* (1972) studied the influence of *P. penetrans* on infection levels by *V. dahliae* at controlled inoculum densities of both organisms and under strictly controlled environmental conditions, a feature lacking in earlier experiments. Infection of cv. Bonny Best was 100% at 200 microsclerotia g^{-1} of soil and became progressively lower at 100, 75, 50 and 25 microsclerotia g^{-1} soil. Consistent increases in infection occurred at all inoculum levels in the presence of the nematode. No stimulation of nematode reproduction was shown in this experiment. Conroy and Green (1974) described a similar interaction between *Meloidogyne incognita* and *Trichodorus christiei* and *V. dahliae* on tomato. Jones and Overman (1976a) described a synergistic stimulation of *V. dahliae* infection by *Belonolaimus longicaudatus* (the stunt ectoparasite) and *Trichodorus christiei*.

Similar effects were described on tomato with the ring nematode *Criconemoides* sp., *M. incognita* and *Belonolaimus longicaudatus* at 28°C; at 21°C, the effects were lessened (Overman and Jones, 1977). The addition of *Meloidogyne javanica* to R and S tomato cultivars inoculated with *V. dahliae* enhanced wilt incidence and severity in the susceptible cultivar (*Orion and Krikun*, 1976). Conversely, Shoemaker and Barker (1979) could demonstrate no synergism between *M. incognita* and *V. dahliae*. A similar effect was described by

Price *et al.* (1980) using *M. incognita, M. javanica, M. hapla* and *V. dahliae* on four hybrid tomato cultivars, resistant to all pathogens and a susceptible cultivar, Rouge de Marmande. Unlike *F. oxysporum*, the nematodes had no effect on *V. dahliae* infection. Using very young tomato seedlings at high humidity, Janowicz and Mazurkiewicz (1982) showed that the nematode *M. hapla* enhanced *V. dahliae* wilt and shortened its incubation period. Miller (1975) described the same effect with *Heterodera tabacum* and *V. albo-atrum* on tomato. The majority of reports confirm a synergistic enhancement of *Verticillium* infection by different nematode genera and species. There is less evidence for the breaking of genetic resistance to *Verticillium* wilt as has been confirmed for nematode predisposition to *Fusarium* attack. *V. dahliae* reduced the resistance of three cultivars with varying resistance to *M. incognita*. Egg production was highest when seedlings were inoculated 2 weeks prior to nematode inoculation (Hasan and Khan, 1985). Pessoa *et al.* (1988) produced a tomato cultivar Nemadaro in Brazil resistant to *M. incognita, M. javanica, M. arenaria* and *V. dahliae*.

Potato

The interactions between nematodes, especially *P. penetrans* and *Heterodera (Globodera) rostochiensis*, and potato wilt caused by *V. dahliae* and *V. albo-atrum* are well documented. *Verticillium* wilt of potato in the USA was later to be identified as a fungal, nematode, environmental disease complex eponymously called 'potato early dying syndrome' (PED) and has been well documented (Morsink and Rich, 1968; Burpee and Bloom, 1974, 1978; Jacobsen, 1974; Jensen, 1978; Jacobson *et al.*, 1979; Martin *et al.*, 1982; Riedel *et al.*, 1985; Francl *et al.*, 1987, 1988; Powelson and Rowe, 1993; Botseas and Rowe, 1994).

Relatively few experiments have examined the infection interface of root with fungus and nematode. Bowers *et al.* (1996) in factorial greenhouse experiments examined the interaction of *V. dahliae* with *P. penetrans* or *P. crenatus* in PED. Soil was infested with known densities of the organisms in combination and plants were harvested destructively. After 5 weeks, potato roots in the presence of *V. dahliae* alone had only 1.2% infected root tips, while *V. dahliae–P. penetrans* and *V. dahliae–P. crenatus* combinations led to 2.3 and 2.5% infection, respectively. Similarly, colonization by *V. dahliae* was greater (0.13 cm colonization m^{-1} of root in the presence of *P. penetrans* or 0.02 cm colonization m^{-1} of root with *P. crenatus*). Subsequent stem colonization was 58% after 5 weeks and 100% after 7 weeks. Since root penetration by the pathogen was not observed at nematode feeding sites, the synergistic response of the two organisms was considered to be biochemical rather than due to mechanical damage.

Morsink and Rich (1968) described the interaction of potato cv. Katahdin, *P. penetrans* and *V. albo-atrum* [*sic*] (*V. dahliae*) in which the fungus appeared primarily to affect shoot growth and aerial wilt symptoms while the nematode was largely responsible for tuber weight. Shoot dry weights were higher in fungus–nematode combinations than with *V. dahliae* only. Nitrogen increased tuber yields in nematode only and nematode–fungus combinations. Hide and Corbett (1971), Corbett and Hide (1971), Hide and Corbett (1974) and Hide *et al.* (1984) found that symptoms appeared 2 weeks earlier when potatoes were inoculated with *V. dahliae* (3.5×10^4 microsclerotia ml^{-1} of compost) and *H. rostochiensis* UK pathotype A at 10 eggs g^{-1} of compost. No evidence was found for the nematode providing infection sites for the fungus, but colonization was seen after 8 weeks in giant cells in epidermis, cortex, pericycle, phloem parenchyma and around young females in ruptured root cortex. Similar results were found for the same species by Harrison (1971) and for *V. albo-atrum* [*sic*] (*V. dahliae*) and *P. penetrans* by Burpee and Bloom (1974) and Gould (1974), who also found that timothy, rye and buckwheat used as cover crops all served as reservoirs of the nematode, and for *M. hapla* and *V. dahliae* by Jacobsen (1974).

In a detailed screening of 218 clones of *Solanum tuberosum* to *M. hapla* in naturally infested *Verticillium* and nematode soil, Hoyman (1974a,b) found no interaction between them. MacDonald (1976) found that the development of *V. albo-atrum* wilt symptoms in Norland potatoes was directly proportional to the level of *P. penetrans* in the soil. Observed populations of 1, 12, 19 and 86 nematodes 100 ml^{-1} of soil were associated respectively with yield losses of 1.8, 32.3, 44.3 and 52.1%. In controlled laboratory experiments with *Globodera* (*H.*) *rostochiensis* on tomato and potato, Harig (1976) found that in the presence of the nematode, infection by *V. dahliae* increased but had little effect on yield; in contrast, *V. albo-atrum* killed most tomatoes but infection of potato was slight.

In Israel, Krikun and Orion (1977, 1979) found that high levels of *P. thornei* alone had little effect on potato yields, but *V. dahliae* in the presence of the nematode reduced the yield of *Verticillium*-tolerant cultivars Blanka and Desirée by 30–40%. Rotations with wheat led to high nematode levels. Similar results were described by Siti *et al.* (1979). In a repeat of their 1974 work, Burpee and Bloom (1978) confirmed an initial (2–3 weeks) increase in disease severity in nematode–fungus combinations. Subsequently, no effect on symptom severity or yield in either susceptible or resistant cultivars could be demonstrated. *V. dahliae* appeared to suppress nematode populations.

G. rostochiensis was tolerated by resistant potato cultivars Maris Piper and Maris Anchor better than by the susceptible cvs Maris Peer and Pentland Crown. The tolerance of cv. Maris Anchor was broken in the presence of *V. dahliae* which behaved like cv. Maris Peer (Evans, 1982). Continuous cropping for 5 years with *V. dahliae*-resistant clones A66107-51 and A68113-4 compared with susceptible cv. Russet Burbank showed significantly reduced soil populations of *V. dahliae* and disease severity when land was cropped with cv. Russet Burbank. A similar reduction in *Pratylenchus neglectus* in soil and roots of Russet Burbank followed 5 years cropping with cv. Butte (Davis *et al.*, 1983b). In subsequent field and glasshouse studies (Davis *et al.*, 1986), cv. Butte was found to be highly resistant to *P. penetrans* and *P. neglectus* and led to a reduction in soil and root populations in 2 months. Since cv. Butte is very susceptible to *V. dahliae*, it was suggested that yield increases and a reduction in the incidence of *V. dahliae* were unrelated to nematode control. Lacey *et al.* (1984), using UK cultivars, found little evidence of an interaction between *V. dahliae* and *Globodera* spp. The authors infer that some nematicide treatments may control *Verticillium*, giving misleading results.

Under Dutch conditions, Scholte and S'Jacob (1989) found that single infestations of *Meloidogyne* spp. or *V. dahliae*, but not *P. neglectus*, reduced tuber yields. Both nematodes, however, reacted synergistically with *V. dahliae* to suppress yield. Scholte (1989a) found that *V. dahliae* in soil heavily contaminated with *Streptomyces* spp. causing netted scab reduced haulm growth before wilt symptoms were evident, and tuber yield, but not number. This author claimed that treatment with the nematicide oxamyl delayed *V. dahliae* infection by controlling mainly *P. thornei*.

A detailed ecological study of nematode spp. in Maine by Huettel *et al.* (1991) may throw some light on the variable results reported. *P. penetrans* and *P. crenatus* were encountered most frequently; the highest levels of detection (19%) occurring in May–June rather than September–November. All field rotations supported *Pratylenchus* spp. *M. hapla* was encountered in 14% of early samples and 19% of the late. Oats–potato, potato–potato and clover–potato successions supported the highest populations. Lower frequencies of *Pratylenchus*, *Criconemella*, *Helicotylenchus* and *Tylenchorhynchus* spp. were recorded. Two plant samples from 27 fields yielded multiple associations of *V. dahliae* and *P. penetrans* (seven fields), *V. albo-atrum* and *P. penetrans* (one field), and *V. dahliae*, *V. albo-atrum* and *P. penetrans* (two fields). A similar field survey in New Brunswick, Canada, found *Pratylenchus* spp. in 43 of 46 fields in 1990 and in 37 of 43 fields in 1991, with respective populations of 910 and 410 nematodes g^{-1} of dry root and 1030 and 720 nematodes kg^{-1} of dry soil. *V. dahliae* was detected in all potato fields sampled. Previous crops had no effect on nematode populations, except in one instance in 1991 where root nematode numbers were higher where potato followed cereal compared with peas. In general, *P. crenatus* was more prevalent than *P. penetrans*, while *M. hapla* was detected only at low levels at a few sites. The authors (Kimpinski *et al.*, 1998) found no significant correlation between nematode and fungal populations in the soil. Johnston *et al.* (1994) in Prince Edward Island reported a reduced incidence of wilt in potato following a field sowing mixture of *Trifolium hybridum* and *T. pratense* than following a previous crop of cereal. In this study at least, no correlation was found between *P. penetrans* populations and disease severity. Saeed *et al.* (1998) used growth chambers and a *V. dahliae* inoculum density of 5.4 propagules g^{-1} of soil to investigate the effect of initial *P. penetrans* population on growth and wilt symptoms in cv. Russet Burbank potato. In the absence of *V. dahliae*, only the highest concentrations of nematode, 2.1–8.8 and 7.5–32.4 nematodes cm^{-3} of soil, affected shoot growth and tuber yield. In the presence of the fungus, however, 0.8 and 7.5 nematodes cm^{-3} of soil reduced yield by 15 and 45%. Their data showed that the nematode–fungus pathogenic interaction occurs at <1 nematode cm^{-3} of soil but levels need to be substantially higher before additional disease develops.

Many reports have dealt specifically with the fungus–nematode interaction, i.e. PED (Rowe, 1985). It is not clear whether all references to *Verticillium* wilt complexes in potato refer specifically to PED. Martin *et al.* (1982) using field microplots in Ohio showed that *V. dahliae* and *P. penetrans* in combination, but not individually, induced PED. Using split-root inoculations on glasshouse plants, Kotcon and Rouse (1984) showed that root deterioration in PED was associated with premature senescence of foliar tissue and was independent of root colonization by pathogens. A more complex interaction involving *V. dahliae*, *P. penetrans*, *Colletotrichum coccoides* and *Rhizoctonia solani* was described by Kotcon *et al.* (1985). Studies on three populations of *P. penetrans* and two of *V. dahliae* in field microplots by Rowe and Riedel (1984) and Riedel and Rowe (1985) found that PED occurred in the presence of both organisms at populations which individually had no effect. *P. scribneri* reacted slightly and *P. crenatus* not at all with *V. dahliae*. Orlowsky *et al.* (1988) found no association between *M. chitwoodi* and *V. dahliae* in a fine sand loam in central Oregon. Increased irrigation interacted with *V. dahliae* to give more severe symptoms and lower yield. No reference was made in this study to soil temperature or the evaporative cooling effect of irrigation water. Studies by MacGuidwin and Rouse (1990a,b) illustrate the variable effect of fungal pathogen and nematode species.

In a microplot study involving non-yield-reducing levels of *V. dahliae* and *M. hapla*, nematode but not fungus populations increased over 3 years. PED symptoms were induced by *V. dahliae* alone, but only *M. hapla* caused loss of tuber yield, which was up to 70% compared with controls. No synergism for symptoms or tuber loss were found in plots inoculated with both organisms (MacGuidwin and Rouse, 1990a). In a subsequent study, MacGuidwin and Rouse (1990b) used the same cultivar, Russet Burbank, with *V. dahliae* and *P. penetrans*. Foliar symptoms of PED were caused by *V. dahliae* alone but were more severe when nematodes were present. The dry weight of tubers (but not tuber number) was reduced by *V. dahliae* alone but not by *P. penetrans*. In combination, tuber yields were reduced by up to 36% – a result at variance with the findings of Morsink and Rich (1968). MacGuidwin and Rouse (1990b) found the results variable in different years, as were the population dynamics of *P. penetrans*, which either decreased or remained unaffected in the presence of *V. dahliae*. A multifactorial experiment with *V. dahliae*, *P. penetrans* and *Verticillium*-susceptible cultivars Superior and Kennebec and resistant cultivars Tobique, Russette and Reddale, was conducted by Rowe and Riedel (1990). The resistant cultivars showed neither aerial symptoms nor yield loss to *V. dahliae* alone. In combination with *P. penetrans*, however, the resistance was reduced for all cultivars. Microplot experiments on PED in Ohio by Wheeler and Riedel (1994) found a significant yield loss in each of 3 years (1986–1988) from *V. dahliae* and a significant interaction of *V. dahliae* and *P. penetrans* in two of 3 years. *P. penetrans* alone caused yield losses in 2 years and *P. scribneri* alone caused losses in one of 3 years. *V. dahliae* had a negative effect on nematode populations of each nematode species. Wheeler *et al.* (1994b) conducted a detailed statistical analysis of linear transect samples of *V. dahliae* and nematode populations and potato yields in commercial Ohio potato fields. Yields were negatively correlated with pre-plant densities of *V. dahliae* and *P. penetrans* and their interaction in three of seven fields and with *M. hapla* in three of ten fields. Fields were also negatively correlated with *V. dahliae* and *P. penetrans* individually, and positively with interactions between *M. hapla* and *V. dahliae*.

An additional overlay to the interaction of organisms in the soil environment is the complexity of genetic relationships. The work of Rowe and Botseas (1995) indicates that failure to understand the genetics of a system may lead to misinterpretation of results. Using vegetative compatibility groupings (VCGs) based on nitrate-non-utilizing *nit* mutants (see Chapter 4), these authors showed that isolates of *V. dahliae* from VCG4A may interact synergistically with *P. penetrans*, but isolates from VCG4B do not interact with the nematode or do so only weakly. The possibility of many such genetically compatible or incompatible groups for different host plants existing in the soil is a pitfall for superficial interpretations of data based on a limited knowledge of the organisms.

An additional important factor in PED which had been largely ignored in earlier studies, with the exception of Orlowsky *et al.* (1988), is soil water capacity. Powelson *et al.* (1995) have shown (in the absence of fungus and nematode measurements) that PED is more severe if soil water content is excessive. Two levels of irrigation were applied, presented as percentages of estimated consumptive use (ECU), during a critical period between emergence and tuber initiation. Plants were grown at 50% of water requirement (deficit) and 150% ECU (excess) and then returned to 100% ECU. Aerial senescence was variously 9.1 and 29.5% lower in the deficit treatment, while final yield was 17% higher than in the high irrigation level treatment. The implications of this work in relation to fungus activity at reduced (evaporative) temperatures, fungus and nematode populations and different nematode species, notwithstanding other unidentified factors, are enormous and will provide much scope for the difficult multifactorial and labour-intensive experiments required. PED in Israel is attributed to an interaction between *V. dahliae* and *P. mediterraneus* against which methyl bromide fumigation has been effective (Maharshak *et al.*, 1995b). Other species of nematode studied in association with *V. dahliae* are *G. pallida* (Franco and Bendezu, 1985; Storey and Evans, 1987), and *G. pallida* and *G. rostochiensis* (Evans, 1987).

Cotton

A general review of cotton pest management systems including nematodes is presented by DeVay *et al.* (1989). An early experiment by Bugbee and Sappenfield (1972) on 3- to 6-week-old cotton plants showed that symptoms of *V. albo-atrum* [*sic*] (*V. dahliae*) were more severe in the presence of *M. incognita*. Khoury and Alcorn (1973) confirmed this synergistic effect of *M. incognita*.

Dominant nematode species associated with cotton were described by Kostyuk (1973). Experimental results were somewhat inconclusive; large numbers of nematodes in the rhizosphere were thought to reduce wilt resistance. No evidence was found for nematode transmission of *V. dahliae* to either agar culture or cotton plant. An Indian cultivar of *Gossypium hirsutum* developed more severe *V. dahliae* wilt symptoms in the presence of *Haplolaimus seinhorstii*. Results agreed with field observations where *H. seinhorstii* was the most prevalant of three species in 'wilt-infested' soil (Shanmugam *et al.*, 1977). *Rotylenchus reniformis* was shown to be pathogenic to cotton, reducing shoot fresh weight and dry weight at high inoculum concentration. In combination with *V. dahliae*, the effect on stunting and wilting was multiplicative. *V. dahliae* also appeared to stimulate *R. reniformis* populations, but this would be expected if root tissue became necrotic. Sudakova (1978) attributed increased *Verticillium* infection to nematode-stimulated amino acid and nitrate exudation from cotton roots. Nematode populations were doubled in the presence of *V. dahliae*. Four nematodes were shown to enhance fungal pathogenesis: *Chiloplacus quintastriatus*, *Paraphelenchoides capsuloplanus*, *Ditylenchus* sp. and *Pratylenchus brachyurus*. Garber *et al.* (1983) described the selection from Acala cotton of two *M. incognita*-tolerant strains that were also tolerant to *V. dahliae*. Nematode resistance was associated with fewer root galls, reduced wilt, better growth and higher yields (24–69%) compared with other cultivars (Hyer *et al.*, 1983). (See also Erturk *et al.*, 1975 on the interaction of plant parasitic nematodes and *V. dahliae* infection in relation to control methods.)

Mukhina *et al.* (1990) found that a 1.5- to 3-fold increase in unspecified nematodes was associated with a 27–37% increase in infection, with losses in the range of 24–41%. Cultivating cotton on ridges led to a 67% fall in nematodes. A winter rye rotation similarly resulted in a 52% decrease. Resistance to *V. dahliae* in six Chinese cotton cultivars was negatively correlated with rhizosphere nematode populations but positively correlated with fungi and actinomycetes (H.-L. Li *et al.*, 1998).

Strawberry

Müller (1973a,b, 1977) used *Impatiens balsaminea* as a test plant for inoculation experiments using a strawberry isolate of *V. dahliae* and *P. penetrans*, concluding that the apparent synergistic effect was due merely to wounds caused by the nematode facilitating entry. However, when other *Pratylenchus* species were tried, although all penetrated young seedling roots, only combinations of *V. dahliae* with *P. penetrans* and *P. vulnus* induced wilt symptoms, while combinations with *P. crenatus*, *P. fallax* and *P. thornei* did not. Müller (1973c) first reported a synergism between *V. dahliae* in strawberry and *P. penetrans*. Field and plot experiments confirmed that *P. penetrans* hastened the onset and increased the severity of *V. dahliae*-induced wilt in strawberry cultivars (Grosse *et al.*, 1975, 1976). McKinley and Talboys (1979), in pot and 'microplot' experiments on *V. dahliae* wilt of strawberries, using isolates of *P. penetrans* and *P. fallax*, found that the presence of *P. penetrans* increased the rate of wilt development except when the concentration of microsclerotia in the soil was very low. The nematode appeared to cause local changes in the root cortex which aided penetration of adjacent tissues by the fungus. Szczygiel (1983) found that the tolerance limit to *P. penetrans* was low – 50 nematodes 100 g^{-1} of soil; this was reduced to 25 nematodes 100 g^{-1} of soil when *V. dahliae* was included in the inoculum.

Other host species

Ditylenchus dipsaci enhanced the appearance and severity of *V. albo-atrum* symptoms in lucerne. *P. penetrans* acted similarly in the field, but not in glasshouse tests (Vrain, 1987). Gubis (1982) obtained lucerne selections resistant to the clover root nematode *Ditylenchus dipsaci*, *Corynebacterium insidiosum* and *V. albo-atrum*. Insunza and Ericksson (1981) considered that preliminary infection of oilseed rape by the beet cyst nematode *Globodera (Heterodera) schachtii* predisposed plants to *V. dahliae* infection. Bergeson (1972) and Green (1981) described the effect of nematodes in increasing root exudation which reversed fungistasis. Exudates were also considered to suppress the growth of rhizosphere antagonists to *Verticillium* spp. A synergistic role for *P. penetrans* in wilt of flax by *V. dahliae* and *V. albo-atrum* has been demonstrated under certain conditions (Coosemans, 1977). A minimal population of nematode and fungus was required, otherwise one organism masked the other. While peach and cherry are both susceptible to *V. dahliae* and *V. albo-atrum*, field wilt in New York State was caused wholly by *V. dahliae* associated with intercropping with tomato or legumes. The effect of *V. dahliae* on cherry was greater when either *M. hapla*, *P. penetrans* or, to a lesser extent, *Tylenchorynchus claytoni* was present (Ndubizu, 1977). No resistance to *M. incognita* was found in 300 cultivars of *Solanum melongena*, but *S. sisymbrifolium* showed resistance to the nematode *V. albo-atrum* [*sic*] (*V. dahliae*), offering the possibility of interspecific crosses (Fassuliotis and Dukes, 1972).

Pratylenchus vulnus but not *M. hapla* caused a severe wilt disease of Manetti rose rootstock. Following fungus–nematode inoculations, *V. dahliae* could not be implicated (Santo, 1974). A similar situation was described for the ornamental *Codiaem variegatum* by Maas and Klaare (1973). While *V. albo-atrum* or *Rotylenchulus reniformis* were able to kill plants within 6 months, no interaction between them was found, and in practice *V. albo-atrum* was the principal pathogen (Tchatchova and Sikora, 1983). Dwinell and Sinclair (1967) claimed that *P. penetrans* increased the invasive ability of elm seedlings following phosphorus treatment but not when potassium was added. Nitrogen levels in elm and peach were positively correlated with nematode numbers. *H. humuli* inhibited young hop plants in pots when the population exceeded 500 eggs g^{-1} of soil. *H. humuli* and *V. albo-atrum* in combination synergistically enhanced disease in a wilt-susceptible cultivar (unnamed) but not in cvs Northdown or Target (the pathogenic strain was unidentified). *V. albo-atrum* had no effect on nematode reproduction in greenhouse or field, but hatching of juveniles was delayed by 2 weeks (von Mende and McNamara, 1995).

Fungi

A distinction is made between the use of antagonistic fungi and other organisms used in field experiments to control or ameliorate wilt symptoms and laboratory or greenhouse studies reporting on the phenomenon or mechanism of antagonism, or surveys of antagonistic organisms. The former is discussed in Chapter 10. Similarly, the interaction of species and strains of *Verticillium* in inducing cross-protection is dealt with elsewhere.

In an early study, Kerr (1961) reported on tomato root-surface organisms antagonistic to *V. albo-atrum*. Askarova and Mamadaliev (1974) reported more fungi antagonistic to *V. dahliae* than antagonistic bacteria and actinomycetes in the rhizosphere of cotton. This was confirmed for Chinese cotton by H.-L. Li *et al.* (1999) who found *Gliocladium*, *Myrothecium*, *Aspergillus*, *Penicillium*, *Cephalosporium*, *Fusarium* and *Doratomyces*. *Gliocladium* and *Myrothecium* were dominant in the rhizosphere of susceptible cotton. Species antagonistic to *V. dahliae* were present in greater numbers around the roots of tolerant cultivars, a difference more apparent in seedlings than in mature plants. Species of *Trichoderma* commonly occur in soil and in the plant rhizosphere (Domsch *et al.*, 1980b) where they are both antagonistic to some fungi and active parasites of others. *T. viride* is reported as antagonistic to *V. albo-atrum* (Aubé, 1967) and to *V. dahliae* (Catani and Peterson, 1966). Tarunina (1980) showed that under laboratory conditions, activity of *T. viride* against *V. dahliae* decreases as the temperature is increased from 24 to 27°C. Shadmanova *et al.* (1981) examined the effectiveness of *T. viride* against both *V. dahliae* and *F. oxysporum* f. sp. *vasinfectum*. Czaplinska (1973), studying the mycoflora of roots of 1-, 2- and 3-year-old lucerne plants in Poland, showed that *T. viride* and species of *Fusarium*, *Phytophthora* and *Pythium* inhibit *V. albo-atrum in vitro*, and suggest that agricultural methods favourable to these antagonists should be adopted to discourage *Verticillium* infection.

Subbarao and Bailey (1961), studying rhizosphere fungi associated with tomato plants, found *Fusarium* to be dominant in that of susceptible varieties, and *T. viride* dominant in that of resistant ones. When roots of susceptible varieties were injected with both *Verticillium* and *Trichoderma*, disease intensity was reduced compared with injection with *Verticillium* alone. Sportelli *et al.* (1983) also reduced the severity of wilt symptoms using *T. viride*. Most of the Russian reports only consider *T. viride* (syn. *T. lignorum*). D'Ercole *et al.* (1984) list as active *in vitro* against *V. dahliae* and isolated from the soil the following species: *T. viride* (29 strains), *T. harzianum* (14), *T. koningii* (six), *T. hamatum* (two), *T. poly-*

sporum (two) and *T. pseudokoningii* (one). D'Ercole and Niputi (1986) reported success in the control of tomato wilt in glasshouse experiments using the first three species. Population analysis of *Trichoderma harzianum* in California soils showed that densities of *Verticillium nubilum* were negatively correlated with increases in *T. harzianum*. *T. harzianum* populations were highest during wet winter months, with a second peak in early summer (Eastburn and Butler, 1988).

A general survey of fungi antagonistic to *Verticillium* spp. in Uzbekistan cotton-growing soils by Babushkina (1971, 1974) found that numbers were greatest against *V. dahliae*, with fewer against *V. albo-atrum* and least against *V. tricorpus*, no doubt reflecting the frequency of occurrence of these pathogens. Rhizospheres of healthy cotton plants were richer in fungal antagonists than those of *V. dahliae*-infected plants. Babushkina (1982) found the following fungi to be highly antagonistic to one cotton wilt strain of *V. dahliae* and moderately so to another: *Aspergillus flavus*, *A. fumigatus*, *A. lutescens*, *A. nidulans*, *A. terreus*, *Fusarium equiseti*, *Penicillium cyclopium*, *P. claviforme*, *P. notatum*, *R. roquefortii*, *P. rubrum* and *T. viride*. Sezgin *et al.* (1982) list the following as forming inhibition zones in laboratory experiments with *V. dahliae*: *Aspergillus ochraceus*, *A. sulphureus*, *A. terreus*, *Gliocladium roseum*, *G. virens*, *Myrothecium roridum*, *M. verrucaria*, *Penicillium patulum*, *Trichoderma harzianum* and *T. viride*. Henni (1987a) found that both *T. harzianum* and *P. griseo-fulvum* inhibited germination of microsclerotia of a *V. dahliae* strain causing wilt of aubergines (see Napier *et al.*, 1956). Turhan and Grossmann (1988) assayed *Neocosmospora vasinfecta* var. *africana* for antagonism against 15 soil fungi and found *V. dahliae* to be very resistant compared with the other fungi tested.

Gliocladium sp. was shown by Muromtsev (1980) to inhibit the *V. dahliae* strain of cotton wilt in laboratory and field experiments. *G. roseum* isolated from the cotton rhizosphere was more effective as a *V. dahliae* antagonist than *Trichoderma* (Globus and Muromtsev, 1990). Marois *et al.* (1982) in glasshouse experiments tested 34 fungi isolated from soil for their ability to control wilt of aubergine caused by *V. dahliae*. They found that *Aspergillus alutaceus*, *Gliocladium virens*, *Paecilomyces lilacinus*, *Talaromyces flavus*, *T. harzianum* and *T. viride* when present in the soil reduced the incidence of wilt to 0–20% compared with 90% in controls, but when applied in field tests only *T. flavus* reduced aubergine wilt leading to increased yield. Wilderspin and Heale (1984) obtained a 50% reduction of infection in antirrhinum by *V. albo-atrum* using *T. flavus* in laboratory experiments. Fravel *et al.* (1987) showed *T. flavus* to produce a metabolite that retards radial growth and kills microsclerotia of *V. dahliae in vitro*. Kim *et al.* (1988) subsequently identified this metabolite as glucose oxidase. The literature on *Trichoderma* and *Gliocladium* as potential biological control agents was reviewed by Papavizas (1985). Henis and Fahima (1990) described the effectiveness of *T. flavus* as an antagonist of *V. dahliae* (see Chapter 10) which has been shown to colonize the rhizospheres of cotton, solanaceous hosts, artichoke and cotton (Tjamos and Fravel, 1995b). Studies on roots of tomato, potato and aubergine showed that *T. flavus* preferentially colonized root tips more effectively than normal rhizosphere or non-rhizosphere organisms. The authors infer that the 'antagonism' to *V. dahliae* measured in terms of disease control may result from a competition for root sites, which would be infection courts for *V. dahliae*. An additional or alternative explanation was proposed by Saito *et al.* (1995). A wild-type strain IG and a mutant T4 both inhibited growth of *V. dahliae* by 73%. In non-sterile soils infested with *V. dahliae*, wilt was reduced by both strains by 44%, whereas in autoclaved soil containing the pathogen the T4 strain both reduced the disease and stimulated aubergine growth. An ethyl acetate-soluble inhibitor extracted from *T. flavus* culture filtrates was strongly inhibitory to *V. dahliae* and other fungi. Control measures which superficially offer an obvious explanation may operate quite differently, employing alternative mechanisms. Tjamos and Paplomatos (1987a) showed that solarization of *Verticillium*-infested soil led to an increase in specific suppression of *V. dahliae* by *T. flavus* rather than thermal killing of microsclerotia exclusively. Similarly, Easton *et al.* (1975) found that soil fumigation of *V. dahliae*-infested potato fields with 1,3-dichloropropene (Telone) reduced wilt in potatoes without reducing the pathogen population. The Dutch potato cv. Amethyst is unaffected by *Colletotrichum coccodes*; in combination with *V. dahliae* however, yield was reduced by 19%. Yield loss in this cultivar from *V. dahliae* alone was 10% (Scholte *et al.*, 1985). See Otazu *et al.* (1978) for the interaction of *C. atramentarium* and *V. dahliae* on potato wilt in North Dakota. *Rhizoctonia solani* increased the susceptibility of cotton to *V. dahliae*, especially in young plants (Khoury, 1970; Takacs and Khoury, 1970). Other fungi reported to enhance wilt development include *Alternaria*, *Curvularia*, *Macrophomina*, *Penicillium*, *Trichoderma* and *Volutella* (Mostafa, 1967; Gazikhodzhaeva and Becker, 1968). These authors also claim that the following genera all reduce wilt incidence: *Aspergillus*, *Botryodiplodia*, *Cephalosporium*, *Chaetomium*, *Colletotrichum*, *Gliocladium* and *Gonatorrhodiella*. In laboratory experiments, a biotrophic contact mycoparasite, *Melanospora zamiae*, was shown to obtain all necessary nutrients from parasitizing washed living mycelia of *V. dahliae* and *V. albo-atrum* in addition to other fungi (Jordan and Barnett, 1978).

Keinath *et al.* (1990) screened 25 fungi, 25 actinomycetes and seven bacteria for antagonism to *V. dahliae* microsclerotia buried in nylon mesh (15–20 per mesh square) in soil to which individual putative antagonists had been added. After 2–7 weeks, only selected isolates of *G. roseum* were effective in killing all microsclerotia in soil microplots. Some *G. roseum* isolates were ineffective. There was no correlation between the virulence of an isolate and its ability to produce antibiotics on Weindling's medium. Results varied with type of amendment and inoculum density; in general, the inoculum (colonized vermiculite) at 1% (w/w) soil was high. Bonfante *et al.* (1972) reported that growth of *V. albo-atrum* in culture was inhibited by *Tuber melanosporum*. All the foregoing reports refer to ascomycetes or ascomycete anamorphs. Tolmsoff and Wheeler (1977), however, isolated a member of the Mastigomycotina,

Hypochytrium catenoides, parasitizing individual cells of *V. dahliae* microsclerotia. *Zygorhynchus moelleri* (Mucorales), too, was reported to lyse hyphal walls of *V. albo-atrum*. See also Kuter (1984) on hyphal interactions between *V. dahliae* and *Rhizoctonia solani*.

Limited but conflicting information is available on the effect of mycorrhizas on *Verticillium* infection. Davis *et al.* (1979) claimed that *V. dahliae* wilt severity increased in cotton previously inoculated with the vesicular-arbuscular species *Glomus fasciculatus* compared with non-inoculated plants. De Melo *et al.* (1985) similarly found that aubergine in mycorrhizal association with *G. leptotrichum* or *G. macrocarpus* sustained increased wilt. Conversely, plants in association with *G. heterogama* and *G. marginata* had reduced infection. It is not clear how the specificity of *Glomus* spp. operated positively or negatively in wilt infection, and no additional work has been forthcoming in this difficult experimental field. Baath and Hayman (1983) in glasshouse experiments on *V. albo-atrum* wilt of tomato pre-inoculated with *G. mosseae* and *G. caledonium* found no evidence for vesicular–arbuscular (VA) mycorrhizas on wilt severity. None the less, Harrison *et al.* (1985) showed that 95% of Scottish potato crops had VA mycorrhizas from which the most commonly isolated *Verticillium* was *V. tricorpus*. A reinvestigation of the little understood condition in potato called 'pink-eye' by Goth *et al.* (1993) found significant positive correlations among 12 clonal advanced breeding lines for *Verticillium* resistance following inoculation with *V. albo-atrum* and *V. dahliae* individually and in combination. Pink-eye was enhanced by *Verticillium*, but the presence of the pathogen was not essential for pink-eye development.

Rhizospere populations of *V. dahliae* in (*Bacillus thuringiensis*) (*B.t.*) var. *tenebrionis* transgenic potato cv. Russet Burbank showed no difference from appropriate transgenic controls. Since *Pythium* spp. and *Fusarium* spp. numbers were also the same, it is assumed that populations of recognized antagonists of *V. dahliae* (not studied) were similarly identical.

Bacteria

Klingner *et al.* (1971) observed that *Erwinia carotovora* was present in the rhizosphere of cotton plants which escaped *Verticillium* wilt, and demonstrated that the strain isolated exhibited *in vitro* antagonism against *V. dahliae*. Dipping cotton seeds in a bacterial suspension or amending the soil with a bacterium–carrot mixture before planting reduced the incidence of wilt (see Rahimian and Mitchell, 1984). Kerr *et al.* (1978) reported the inhibition of *V. dahliae* by a *Rhizobium* isolate and by 15 strains of *Agrobacterium tumefaciens*, and Ezrukh (1978) showed metabolites of actinomycetes from a cotton rhizosphere to interact with *V. dahliae*. Chi (1963) reported fungistatic and fungicidal effects of *Streptomyces rimosus* on both *V. albo-atrum* and *V. dahliae* (see also Napier *et al.*, 1956).

Hall *et al.* (1986) described how several isolates of *Bacillus subtilis*, taken from healthy maple stem tissue, when introduced into wounds in the stems of both Norway and silver maple seedlings, brought about a reduction of wilt symptoms on subsequent inoculation of the seedlings with a conidial suspension of *V. dahliae*. Azad *et al.* (1987) recorded a tendency for bacteria, notably *Bacillus* spp., which show antagonism towards *V. dahliae in vitro*, to be distributed differentially in the rhizosphere of wilt-resistant, rather than susceptible potato cultivars. Other bacteria antagonistic to *V. dahliae* were species of *Pseudomonas*, *Gluconobacter*, *Flavobacterium* and *Streptomyces*. Of 15 bacteria screened against *V. dahliae*, *V. albo-atrum*, *V. nigrescens* and *V. nubilum*, *Bacillus luteus* and *B. mesentericus* strongly inhibited growth of *V. dahliae* while *B. vitreus* was highly antagonistic to *V. nigrescens*. Antagonism resulted in lysis and necrosis of chlamydospores, microsclerotia and resting mycelium (Ezrukh and Strelkova, 1979).

Several instances have been cited of *Verticillium* and bacteria combining synergistically in pathogenesis. A bacterium resembling *Pseudomonas marginalis* pv *marginalis* with pectolytic activity, increased in number in association with *V. dahliae* in a root deterioration of horseradish (Mueller *et al.*, 1982). While *V. dahliae* appeared to be the primary initiator, the bacterium combined to allow the egress of other organisms, leading to a rapid breakdown of tissues. The synergistic interaction of *V. dahliae* with *Erwinia* spp. was described by Kritzman *et al.* (1987). Tsror *et al.* (1991) reported a similar interaction with *V. dahliae*, and *Escherichia coli* transformed with cloned *pelB* or *pelE* pectate lyase genes from *Erwinia chrysanthemi* in potato. In this case, however, *E. coli* alone transformed with either gene induced symptoms similar to *Erwinia carotovora*. While *V. dahliae* is a major cause of PED in fields new to potato or those fumigated and with a concomitantly low population of the fungus, *E. carotovora* subsp. *carotovora* or *E. carotovora* subsp. *atroseptica* are able to induce PED (see also, Zink and Secor, 1982; Powelson, 1985).

Other organisms

Alabouvette *et al.* (1979) isolated and characterized a mycophagous amoeba, *Thecamoeba granifera* subsp. *minor*, which fed on *V. dahliae*, penetrating hyphal and conidial walls. Amoeba numbers increased in soil to which fungal cultures were added. The nematode *Aphelenchus avenae* is capable of feeding and developing on pure cultures of *V. albo-atrum*. Soil insects, too, have been implicated with *Verticillium* (Curl and Gudauskas, 1985); two species of springtail (Collembola), *Proisotoma minuta* and *Onychiurus encarpatus*, reduced the effective germination of microsclerotia by feeding on the emergent germ tubes. Price (1976) described the passage of *V. albo-atrum* [*sic*] (*V. dahliae*) conidia and microsclerotial fragments through the alimentary canal of the bulb mite (*Rhizoglyphus echinops*), a common scavenger on cotton debris. Both propagules were viable, conidia in pellets gave the highest percentage recovery.

Barbara *et al.* (1987) claimed to have found double-stranded RNA – a suspected mycovirus in hyphae of *V. albo-atrum*. A similar claim was made by Safiyazov *et al.* (1990) for the presence of globular virus particles 30–40 nm diameter responsible for the lysogeny of fungal hyphae. A completely unsubstantiated claim inferred a relationship between virulence and the presence of a 'virus'. Karban *et al.* (1987) described glasshouse experiments in which cotton seedlings previously exposed to spider mites (*Tetranychus urticae*) were less likely to develop wilt symptoms after *V. dahliae* inoculation compared with mite-free controls. Conversely, *T. urticae* grew less rapidly on fungus-inoculated plants – an unexplored interaction. The interrelationship between *V. dahliae* and two other potato pathogens, the virus PVX and *Colletotrichum atramentarium* was described by Goodell *et al.* (1982). Similarly Johnson *et al.* (1986) examined the effect of interactions of populations of *Aternaria solani*, *V. dahliae* and the potato leaf hopper (*Empoasca fabae*).

Effect of Amendments on Survival and Disease

A substantial literature exists on the effect of mineral nutrition and other soil amendments on wilt diseases. For the most part, results are presented as either diminishing or enhancing wilt with little or no knowledge of the mechanism involved. The two main features, however, are likely to be either an increase or decrease in field inoculum based on enhanced sporulation or decreased inoculum; or affecting the expression of virulence of the pathogen or resistance of the host (Pennypacker, 1989).

Nitrogen

The generalized understanding is that low nitrogen regimes regardless of N-type are associated with decreased wilt severity and incidence. The literature, however, is conflicting, reflecting the very different conditions (mostly unquantitated) of the experimental systems described and the likely possibility that single elements studied in isolation give an incomplete or confused picture, often ignoring soil type, pH, cation exchange capacity, soil moisture and the status of other elements. Low levels of nitrogen have been related to decreased susceptibility in *Antirrhinum* (*V. albo-atrum*; Isaac, 1957b), aubergine (*V. dahliae*; Sivaprakasam and Rajagopalan, 1974) and hop (*V. albo-atrum*; Keyworth and Hewitt, 1948; Sewell and Wilson, 1967; Talboys, 1987). Conversely, Davis and Everson (1986) found that susceptibility of potato to *V. dahliae* increased when N levels were low, but both ammonium and nitrate ions were employed. High planes of N not unexpectedly increased susceptibilty in lucerne (Isaac, 1957c), *Antirrhinum* (Isaac, 1957b), aubergine (Sivaprakasam and Rajagopalan, 1974), hop (Keyworth and Hewitt, 1948) and tomato (Roberts, 1943; Jones and

Overman, 1986). High planes of nitrogen nutrition were found to reduce susceptibility in *Antirrhinum* (Isaac, 1956; Dutta and Isaac, 1979a), potato (Huber and Watson, 1970; Huber, 1980; Davis, 1985; Davis and Everson, 1986), tomato (Wilhelm, 1951a) and cotton (Bell, 1973).

In cotton, the effects of N are often equivocal. While heavy dressings of N regardless of type increase the severity of wilt, increases in yield (presumably in healthy plants) proportional to nitrogen dressing compensate for wilt losses (Makhmudov *et al.*, 1978).

Such a situation was described in aubergine by Elmer and Ferrandino (1994). A 3-year study using high and low levels of *V. dahliae* showed that fertilizing with $(NH_4)_2SO_4$ resulted in more root and foliar non-structural carbohydrate compared with a $Ca(NO_3)_2$ treatment. There was no effect of either N_2 source on total rhizobacteria or fluorescent pseudomonads. After anthesis and symptom appearance, neither fertilizer affected symptoms, colonization or yield. The $(NH_4)_2SO_4$ led to increased root and foliar N, P and Mn. In soils with low inoculum, $(NH_4)_2SO_4$ was superior to $Ca(NO_3)_2$, giving larger leaves and a 33–44% increased marketable yield (see Mirpulatova *et al.*, 1977, 1978). Similar results were recorded much earlier on hop by Marocke *et al.* (1977). $Ca(NO_3)_2$ at 175 units ha^{-1} enhanced wilt from *V. albo-atrum* compared with $(NH_4)_2SO_4$ or the least wilt-favouring N_2 source, urea. High nitrogen often in K-deficient situations has been reported to increase wilt severity by Presley (1950), Presley and Dick (1951), Young *et al.* (1959), Longenecker and Hefner (1961) and Chernyayeva *et al.* (1984). Low or medium levels of N have been shown to have little material effect on wilt (Hinckle and Staten, 1941; Baard and Pauer, 1981; El-Zik, 1984). Split applications of N through the growing season may reduce wilt and increase yield (El-Zik, 1985, 1986) or as a supplementary dressing halfway through the season (Neal and Sinclair, 1960). Differences in wilt susceptibility have been noted with the use of different N sources. By comparison with the NO_3 radical, NH_4-N increased resistance in *Antirrhinum* (Isaac, 1957b; Dutta and Isaac, 1979a) and potato (Huber and Watson, 1970). Wilhelm (1951a) reduced the inoculum potential of *V. albo-atrum* in tomato soil by the addition of $(NH_4)_2SO_4$, bone or fish meal. Isaac (1959) recorded delayed symptom expression in lucerne wilt after using NH_4-N. In hop, however, high nitrogen of either source was shown by Sewell and Wilson (1967) to enhance wilt symptoms; conversely, low regimes of either NO_3-N or NH_4-N led to amelioration. In the USA, *V. dahliae* wilt of cotton is most prevalent in neutral to alkaline loam and clay soils (Bell, 1989). In sand–nutrient culture NH_4-N, NO_3-N and urea at low or high concentrations all benefit cotton wilt resistance (Ranney, 1962). Urea applied as a 2% foliar spray or as a prill fertilizer 2 weeks before planting, suppressed field wilt. Chernyayeva *et al.* (1984) reported the same effect in cotton fields using NH_3-N. Carns *et al.* (1964) suggested that wilt reduction following urea application was due to the deleterious effect on the pathogen of NH_4^+-derived toxins. Turakulov *et al.* (1977) applied N as $(NH_4)_2SO_4$ at 300 kg ha^{-1} to heavily *V.*

dahliae-infected Uzbekistan cotton fields together with 220 kg ha^{-1} P_2O_5 and 150 g ha^{-1} K_2O. Compared with a comparable treatment substituting N as NH_4NO_3, wilt infection was reduced and cotton yield was significantly increased by 20%. This result was confirmed in a model experiment by Muromtsev and Chernyaeva (1979) in which different types of N fertilizer (applied at 0.5g kg^{-1} soil) were added to serozemic and derno-podzolic soils inoculated with *V. dahliae*. Infective propagules were increased following N applied as NO_3-N and decreased with the NH_4-N treatment. Soya flour (5 g kg^{-1} of soil) and lucerne meal and starch (20g kg^{-1} of soil) similarly reduced propagules. As will be seen in sections on P, K and trace elements, it is not possible to consider the effect of N in isolation. NH_4-N may inhibit cation uptake resulting in deficiencies in Ca and Mg uptake. Rhizosphere pH is also altered by NH_4-N application by proton efflux from the cytoplasm to balance the charge. NO_3-N application conversely causes an influx of cations with a corresponding increase in rhizosphere pH caused by an efflux of OH^- or HCO_3^- (Pennypacker, 1989). Nitrate reduction in the plant and the incorporation of NH_4-N into amino acids and amides both require a source of root carbohydrates. However, when soil NO_3-N is high, NO_3-N is xylem transported for reduction in the shoots, thus maintaining a higher level of root carbohydrates than in an NO_3-N-deficient soil when NO_3-N reduction occurs in the root, there depleting the soluble carbohydrates. In this context, the experiment of Roberts (1944) who induced resistance to *V. albo-atrum* in tomato by reducing root carbohydrates is of especial interest and illustrates the importance of a knowledge of the soil or plant nutrient status.

Direct effects of N ions on *Verticillium* have been recorded. $NH_4H_2PO_4$ (Chernyayeva *et al.*, 1984; Stapelton *et al.*, 1987), NH_4NO_3 (Baard and Pauer, 1981) and urea (Chernyayeva *et al.*, 1984; Grishechkina and Siderova, 1984a,b) all have been reported to reduce propagule numbers of cotton *V. dahliae*. Stapleton *et al.* (1987) also show that NH_4-N fertilizers enhance pathogen killing in soil solarization. NH_4-N fertilizers stimulate more biological activity in the soil, which could include *V. dahliae* antagonists, whereas NO_3-N fertilizers increase microsclerotial numbers (Chernyayeva *et al.*, 1984). Kaufman and Williams (1965) claimed that N (unspecified) actually increased total soil organisms but greatly reduced antagonists to *V. dahliae*. Chan and Wilhelm (1984) presented another facet of the effect of N source on resistance; NH_4-N fertilizers specifically were shown to induce massive production of stelar terpenoid phytoalexins in cotton.

Phosphate

For the most part, P appears to exert little influence on *Verticillium* wilt. Davis *et al.* (1988) noted a decrease in *V. dahliae* colonization of potato stem when N and P jointly were optimal. No differences were seen when N and P separately were at optimal level. In a subsequent study on continuously cropped potato cv. Russet Burbank (Davis *et al.*, 1994b), controlled levels of N and P were applied to an Idaho calcareous silt loam with initial levels of N as NO_3 at 0.9 mg kg^{-1} and P at 3.5 mg kg^{-1}. After 2 years following annual dressings of 0, 120 and 240 kg P ha^{-1} and split applications of N at 300 kg N ha^{-1}, soil P levels were 7, 25 and 50 mg kg^{-1}, respectively. Under these treatments, *V. dahliae* wilt was reduced by up to 95% and the initial level of inoculum (9 c.f.u. g^{-1} of soil) greatly reduced. Subsequent control was effected with an annual treatment of split N at 300 kg ha^{-1} and 240 kg P ha^{-1}. The most severe *Verticillium* wilt was observed in N- and P-deficient plants. When lucerne was grown at high levels of superphosphate, Isaac (1957c) recorded increased wilt with the weak pathogen *V. dahliae* but not with *V. albo-atrum*. Similarly, no effect was seen in tomato by Roberts (1943) or in *Cacao* by Emechebe (1980) under altered levels of P. In cotton in field conditions, superphosphate application alone had no effect on wilt (Hinkle and Staten, 1941; Young *et al.*, 1959). The effect of P appears to depend on its abundance in soil and its interaction with other elements, thus, when P was combined with high N, *V. dahliae* severity increased (Longenecker and Hefner, 1961). When K was deficient, P application with low N increased wilt (Presley and Dick, 1951). When K levels were adequate, no effect of P was found (Baard and Pauer, 1981). In glasshouse experiments with a P-deficient soil, Davis *et al.* (1979) found that the addition of superphosphate at 20 μg of CaH_2PO_4 g^{-1} of soil increased P_1 strain wilt severity in plants inoculated with the mycorrhizal fungus *Glomus fasciculatus* compared with non-mycorrhizal plants. VA mycorrhizas are known to facilitate P uptake from deficient soils. Plants fertilized with a higher level 300 μg of P g^{-1} of soil showed equally severe wilt in mycorrhizal and non-mycorrhizal plants. High levels of phosphate inhibited mycorrhizal formation. In trials in Uzbekistan cotton, Yarovenko *et al.* (1975) found that N (source not described) and P as P_2O_5 in the ratio of 1:0.7 or 1:1 at 150 or 250 kg N ha^{-1} decreased *V. dahliae* infection and gave the highest cotton yields compared with other N:P ratios. The effectiveness of N decreased in the absence of P.

Potassium

Potassium nutrition has been known for many years to affect wilt development and is critical for the resistance of cotton (Bell, 1989). Many reports describe the effect of K fertilization on reduced *V. dahliae* wilt severity (Presley, 1950; Presley and Dick, 1951; Young *et al.*, 1959; Isaev, 1963; Abdel-Raheem and Bird, 1967; Hafez *et al.*, 1975; Ashworth *et al.*, 1982, 1985; Song *et al.*, 1995). Presley and Dick (1951) showed that infected plants responded to K only in K-deficient soils. In California, heavy soils tend to make K less available to plants (Schnathorst, 1981). Schnathorst (1975) also showed that cotton cultivars sensitive to marginal K levels show lower field tolerance to wilt. Young *et al.* (1959) noted that soils in south-western USA where K levels are non-limiting are unresponsive to K additions to ameliorate wilt.

An interesting feature of K metabolism is the association of a K-deficiency disease complex involving (but not exclusively) *V. dahliae*. Approximately 15–20% of Californian cotton fields are affected, in which plants develop a foliar bronzing and metallic shield frequently associated with a marginal chlorosis and necrosis (Hafez *et al.*, 1975; Ashworth *et al.*, 1982; Weir *et al.*, 1987, 1988). Weir *et al.* (1989) reported the reversal of K deficiency by soil solarization. K deficiency thus appears to be a function of pathogenesis rather than a cause of root infection. Smith and Richards (1962) showed that the foliar symptoms were due to accumulation of the polyamines putrescine (1,4 diaminobutane), from the decarboxylation of ornithine, and agmatine (4(aminobutyl)guanidine) from the decarboxylation of arginine. They may substitute to some extent for cellular K^+ and Mg^{2+} but their relationship to K deficiency is not known. Ashworth *et al.* (1982) have shown that soils in which apparent K-deficient plants are growing have adequate K levels. K fertilization increases boll yield but does not alleviate foliar symptoms. Presley and Dick (1951) and Young *et al.* (1959) showed that soil solarization reduced wilt incidence, eliminated K deficiency symptoms and increased yield substantially, but only marginally increased soil K. Soil fumigation (Hafez *et al.*, 1975) achieves the same result. These findings suggest that the 'potassium' effect may result from disordered host amino acid metabolism by *V. dahliae* or its products. This, however, does not exclude the possibility of a soil K deficiency – particularly in K-binding soils which may affect *Verticillium* infection.

As in the USA, soils in 47% of the 217,500 ha of cotton land in Andizhan province of Uzbekistan are low in K. In these regions, application of 50 and 100 kg K_2O ha^{-1} decreased *V. dahliae* wilt from 23 to 20 and 17%, respectively with a corresponding increase in yield (Satlarov, 1976). The author recommended an N:P:K ratio of 1:0.6:0.6 as most effective. A similar result on Azerbaijan cotton was reported by Novruzov (1976). Minton and Ebelhar (1991) described a complex interaction on two cotton cultivars DES 1190 and Stoneville 825 in the sandy loam soils of the Mississippi Delta showing K deficiency. Application of 112 kg K ha^{-1} reduced *V. dahliae* wilt K deficiency symptoms from 12 to 7%, but did not affect lint yield. The *Meloidogyne incognita* index was reduced from 2.7 to 2.5 and from 2.7 to 2.4 by K and aldicarb-quintozene treatments, respectively. Fibre strength was increased by the addition of K. Fibre quality and yield were a function of the different cultivars. A somewhat confused account by Alimov (1980) claimed that treatment of cotton soil with the herbicides dipropetryn and fluometron increased soil P and K levels. Increased wilt susceptibility from fluometron application was attributed to a higher plant N content, but no evidence was presented to substantiate this. Application of foliar sprays of KNO_3 to six *G. hirsutum* cultivars, with controls treated with N to compensate for NO_3-N, showed reduced *V. dahliae* incidence and that fruit constituted the major K sink (Janes *et al.*, 1993). This was confirmed by DeVay *et al.* (1997) who demonstrated that leaf K levels from inoculated plants only became deficient as plants neared peak boll load. Notwithstanding that the authors postulate that infection impairs uptake and translocation of K, the possible effect of pathogen products on putrescine metabolism is well worthy of investigation.

Ashworth *et al.* (1985, 1986) found that pistachio trees which were deficient in K had a higher incidence of *V. dahliae* infection than optimal level controls. Application of 1.5 kg of K salt per tree reduced infection by 35%. This effect was only observed at low (0.02–0.2 microsclerotia g^{-1} of soil) inoculum and was ineffective at 5 microsclerotia g^{-1} of soil. The authors postulated that inadequate K levels reduced the velocity of root growth and extended the contact time of root to microsclerotia. There is little real evidence to support this contention which may be seen to be oversimplified in the fullness of time when a more complex biochemical explanation may unfold. Kichanova *et al.* (1990), using cell sap mineral analyses of cotton in Tashkent, reported that a deficiency in nitrate-N, P or K resulted in increased disease incidence and yield loss. The latter would be expected regardless of disease severity. In contrast to the foregoing, Isaac (1957c) found that increased levels of K_2SO_4 shortened the incubation period to symptoms in *V. albo-atrum* infection of lucerne but did not affect disease incidence. Song *et al.* (1995) reported that application of 12 kg K_2O mu^{-1} (mu = 0.067 ha) to Chinese cotton soils led to increased plant fresh–dry weight leaf areas and phenolic contents while decreasing infection and disease severity, mostly in cv. Handan. In tomato infected with *V. albo-atrum*, Pegg (1985) described an increase in divalent and polyvalent (but not monovalent) cations in leaves compared with controls. The increase commenced prior to the appearance of foliar symptoms.

Krikun *et al.* (1990) reported K deficiency in the *V. dahliae* hosts potato, tomato, aubergine, groundnut and rapeseed. While Alimukhamedov *et al.* (1977) recommend the use of mineral fertilizers against cotton wilt, Savov (1978b) maintained that increased nutrition, especially in combination with irrigation, favours infection. K applied as K_2O to *V. dahliae*-infected aubergines at rates equivalent to 150, 300 and 600 kg ha^{-1} lengthened the incubation period and reduced its intensity. The highest rate was phytotoxic (Sivaprakasam and Rajagopalan, 1971). Plants showing resistance with applied K had decreased foliar alanine, methionine and proline. Alanine and methionine injections into 45 day infected plants increased their susceptibility. Healthy plants injected with proline developed symptoms identical to those with the pathogen alone (Sivaprakasam *et al.*, 1972).

Trace elements

Most studies on the beneficial role of microelements on the alleviation of wilt disease have come from the former USSR and on *V. dahliae* and cotton (Askarova *et al.*, 1973). Bell (1989) speculated that this could be due to the generally low nutrition thresholds in soils of cotton-growing republics where measurable response to microelement amendments would be expected (Mirpulatova, 1961; Isaev, 1963, 1966; Babaev and Bagirov, 1967; Mekhmanov, 1967; Askarova *et al.*, 1970; Shatrova and Buimistru, 1976; Sultanov *et al.*, 1977; Uronov and Sultanov, 1978; Mustakimova and Muratmukhamedov, 1982; Miller and Bekker, 1983; Savov, 1986b). The most common trace elements to diminish the effect of *V. dahliae* are Co and Mn (Askarova *et al.*, 1970), Co, Cu and Zn (Sultanov *et al.*, 1977), and Co, Zn, Cu and Mo (Miller and Bekker, 1983). Most treatments have been applied as sprays from bud formation to fruiting, but soil applications before planting and from seed treatment have decreased wilt (Isaev, 1963; Babaev and Bagirov, 1967). In the USA, results have been more variable. Microelement applications had no effect in Arkansas (Young *et al.*, 1959) or Texas (Longenecker and Hefner, 1961).

Mamadaliev *et al.* (1976) and Mamadaliev and Altunina (1976) found that application of 190 kg ha^{-1} ammonium polyphosphate containing 0.4% Co to Uzbekistan cotton fields decreased wilt and increased seed cotton yields more effectively than the N, P treatment alone. A similar effect was recorded by Gazizov *et al.* (1975) where cotton treated with ammonium polyphosphate supplemented with 1% Cu in a 150 kg N + 120 kg P_2O_5 + 40 kg K_2O + 3 kg Cu ha^{-1} of mixture resulted in a decrease in severely infected plants and a slight increase in yield compared with a non-copper mixture. The result was confirmed by Artykbaev and Urunov (1978) who found that the Cu addition stimulated host protein and nucleic acid synthesis. Several authors have reported on the effectiveness of glauconite or powdered glauconitic rock in suppressing *V. dahliae* or its effects. This mineral has long been known to be an important component of some wilt-suppressive soils. The application of 675 kg ha^{-1} pre- and postharvesting of cotton led to a slight reduction in wilt but a substantial increase in yield (Urunov *et al.*, 1976). Abduraupov and Urunov (1976) described a synergistic effect of glauconite when applied with different N, P, K treatments in which the yield increase was greater than the sum of the individual fertilizer or glauconite components. In a subsequent paper, Abduraupov and Kichanova (1977) found that cotton yields and the percentage of wilting plants increased with increasing N up to 300 kg N ha^{-1} and 35% wilt. The addition of glauconite at 500 kg ha^{-1} reduced wilt to approximately 20% the level of non-fertilized control plots.

In a series of glasshouse experiments, Ashworth *et al.* (1982) found that sprays of $CuSO_4$ prior to inoculation decreased wilt severity. Similarly sprays of B and Mn had a beneficial effect (Desai and Wiles, 1976). When transferred to the Californian and Mississippi (respectively) fields, however, no effect was observed. When high concentrations of Zn (Joham, 1971), Al (Orellana *et al.*, 1975), Mn (Shao and Foy, 1982) or Cu (Bell, 1973, 1989) were used in sand–nutrient or water culture, marked reductions in wilt severity were recorded, but with phytotoxic results. Mirpulatova (1961) achieved field control combining Zn in sprays of 2% urea. Positive effects have been recorded on other hosts. Shatrova and Buimistru (1976) obtained increased resistance to *V. albo-atrum* and increased yield of aubergine following a combined treatment of Mn and B with urea and KCl. Dutta and Bremner (1981) also found that dipping tomato seedlings in a solution containing Cu, B and Mn prior to *V. albo-atrum* inoculation significantly reduced infection. Kibishauri and Tsiklauri (1979) claimed to reduce *V. dahliae* infection of apricot trees following applications of $FeSO_4$, $MnSO_4$ and Mo salts, together with a chelating agent. Pre-infection liming of *Verticillium*-resistant and susceptible lucerne cultivars increased the number of resistant plants and delayed or reduced aerial symptoms of susceptible plants (Kudela and Pirkl, 1978). Calcium has been assumed to have a role in disease resistance based on the formation of enzyme-resistant cell walls with calcium pectate skeletons. Selvaraj (1974a), indeed, claimed that by enhancing calcium levels, tomato and aubergine developed an increased resistance to *V. dahliae*. A further claim that supplementation of a *V. dahliae* culture medium with $CaCl_2$ up to 0.05 M increased endoPG production did not support the resistance claim and was also at variance with the results of Cooper *et al.* (1978) who showed with careful experimentation that the addition of 0.01 M Ca to a host cell wall endoPG reaction mixture suppressed enzyme activity by 20%.

Salinity

Natural soil salinity or salinity induced by the recycling of salt-rich irrigation water affects the distribution and severity of wilt disease. Christensen *et al.* (1954) found in Arizona cotton fields that when conductance of soil extracts exceeded 5 mΩ cm^{-1}, the wilt incidence was very low. Above 10 mΩ cm^{-1}, hardly any wilt occurred. The foliar Na content from plants in these areas was fourfold that in plants badly wilted in lower saline soils. These observations by Christensen *et al.* (1954) are at variance with the work of Holmes (1976) on sugar maple and the more recent findings in Israel where salinity, especially in the Negev Desert growing region, is a limiting factor to crop production. Pot experiments on young sugar maple found that high NaCl levels killed all trees. Low salt concentrations alone induced foliar scorch, but combined with *V. albo-atrum* inoculation resulted in a total killing of plants (Holmes, 1976). Livescu *et al.* (1990) and Nachmias *et al.* (1993, 1995a) simulated certain field conditions in potato crops in Israel by adding NaCl and $CaCl_2$ 4:1 in irrigation water to 5 mΩ cm^{-1}. Drought stress was also induced by reducing irrigation to 25%. These treatments led to an increase in symptom severity, increased pathogen growth in the potato vascular system and reduced yield of tubers. The authors postulated a common mechanism causing increased susceptibility resulting variously from high salinity, drought, short photoperiods and nematode infection. Similar results were obtained in potato by Orenstein *et al.* (1995) using a salt concentration equivalent to 10.2 mΩ cm^{-1} and a drought stress induced by 10% polyethylene glycol (PEG). Potato growth was restricted under these conditions while the c.f.u. count of *V. dahliae* increased compared with controls. Abscisic acid (ABA)-treated plants (simulating ABA accumulation from drought stress) led to increased c.f.u. production. Besri (1990) claimed that high salt levels in irrigation water in Morocco led to a 100% breakdown of resistance of tomato cultivars resistant to race 1 of *V. dahliae*. It was also claimed that susceptibility to race 2 resistant cultivars also increased with increasing soil salinity. Work in California on two lucerne cultivars (Howell *et al.*, 1994), cv. NK-89786, more resistant to *V. albo-atrum* than cv. Moapa-69, showed that the former was more susceptible to salinity (simulated by irrigation with NaCl or $CaCl_2$) than cv. Moapa-69, with yield reduction thresholds of 0.8 and 1.6 α S m^{-1}, respectively. *V. albo-atrum* inoculation alone had little effect on foliage yield. However, when infected cv. NK-89786 was treated with salt at 3.0, 5.0 or 7.6 α S m^{-1}, yield was significantly reduced compared with non-inoculated controls. A comparable yield loss in infected Moapa-69 was only achieved at a saline conductivity of 7.5 α S m^{-1}. *In vitro* studies on *V. dahliae ex* potato cultured on aqueous and solid media containing 0.05–0.35 M NaCl showed that while growth was significantly retarded, conidial differentiation and sporulation density were enhanced (Danti and Broggio, 1997).

pH

V. dahliae on cotton normally occurs on neutral to alkaline soils, pH 6–9. At pH 5.5 and below, growth, microsclerotial production and survival are all inhibited. Bell (1992b) surmised that this might result from the accumulation of Mn and Al ions as evidenced by the increased concentration of these ions in cotton leaves from plants grown in acid soils compared with neutral. Rudolph (1931) reported severe *V. dahliae* wilt in cotton in sandy loam, clay soils and soils rich in organic matter. The incidence of wilt in sandy soils, however, cannot be dissociated easily with irrigation treatments, making sandy soils more comparable with clay. Brinkerhoff (1973) described cotton wilt in neutral to acid clay soils in Oklahoma and in the Mississippi Delta. Where acidic soils are limed to increase the pH, wilt severity invariably is increased, for example with tomatoes (Jones *et al.*, 1971) and cotton (Shao and Foy, 1982), where P_1 and P_2 strains were used. Foy *et al.* (1981) showed that liming reduced the Mn content of cotton; Mn in an acid (pH 4–5) environment inhibits mycelial growth at 120 mg l^{-1} and microsclerotial production at 30 mg l^{-1} (Shao and Foy, 1982).

Aluminium which also accumulates in plants at low pH, suppresses mycelial growth and almost totally inhibits microsclerotia formation at 8.0 mg l^{-1} (Orellana *et al.*, 1975). When soil pH is reduced to 4.5 with $Al_2(SO_4)_3$, viable propagules are greatly reduced (Wilhelm, 1950b; Miller *et al.*, 1967; Baard and Pauer, 1981).

Orellana *et al.* (1975) showed that liming Al-toxic soil from pH 4.4 to 5.4 increased *V. dahliae* wilt in sunflower. This treatment, as with Mn, decreased the Al content of cotton plants (Soileau *et al.*, 1969). In a grosser experiment, application of N, P, K (6, 8, 12) at 600 lb acre^{-1} in an Arkansas cotton field decreased wilt from 32.5 to 24.4% over 3 years, while adding ground limestone at 2000 lb acre^{-1} increased the incidence of wilt to 38.2% (Young *et al.*, 1959). Babaev and Bagirov (1967) showed that, in general, mixtures of microelements that reduce wilt in the field also inhibit *V. dahliae* in culture. *V. albo-atrum* (and bacterial wilt) infection of lucerne leads to pH changes in the plant (Chlumská and Krátká, 1984).

Plant residues

The principal problem associated with the burial of plant residues is the danger of microsclerotial propagation. This is particularly true with the incorporation of infected crop residues, e.g. potato haulm, in the soil. The benefit of added organic matter is attributed to the augmentation of antagonists. McKeen (1943) showed that addition of green plant residues to a soil containing *V. dahliae* microsclerotia resulted in a suppression of wilt. A similar effect was observed by Dutta and Isaac (1979b) following the addition of chitin and green manure to soil in which infected *Antirrhinum* plants were growing. This result was attributed to an increase in the actinomycete population. Muromtsev *et al.* (1979) and Marshunova and Federova (1980) claimed that the saprophytic microflora, increasing *V. dahliae* antagonist populations, was greatly stimulated by ploughing in rye, pea and mustard, resulting in a decreased wilt incidence. Grishechkina (1990) reported that autumn plantings of mustard and rape as green manure have reduced inoculum potential of cotton strains of *V. dahliae* by 80–95%. Indeed Melouk *et al.* (1995) established that volatile compounds (mustard oils) from rapeseed meal were inhibitory to *V. dahliae in vitro*. Similarly, Solarska (1994b) also recommended rye, *Phacelia*, *Vicia* and mustard as intercrop green manures to alleviate hop wilt. Subsequently, Solarska *et al.* (1996) found rye or, secondarily, rye–vetch as the best intercrop green manure to limit symptoms caused by *V. albo-atrum*. Wilderspin and Heale (1984, 1985) incorporated 1% chitin or dried *Laminaria* (a complex (mostly 1,3-β-)glucan) in hop strain *V. albo-atrum*-infested soil. Resting mycelium was stimulated to form short-lived mycelium and conidia. *Antirrhinum* test plants, susceptible to hop strains, subsequently showed a 50% reduction in infection compared with non-amended soil controls. Grishechkina (1988) described similar effects on cotton *V. dahliae* strains following the addition of ground residues of barley, wheat, maize, lucerne, clover, tomato, cucumber, mustard, pepper or aubergine to soil in pot experiments. Harrison (1976) found a reduction of microsclerotial inoculum in *V. dahliae*-infested potato fields following the addition of 0.4–1.6% by weight of chopped barley straw to the soil. The addition of N as NO_3-N or NH_4-N had no effect on the barley straw treatment. As reported elsewhere in the text, field soil-incorporated residues of broccoli (a wilt-resistant plant) have provided effective control of *V. dahliae* wilt of cauliflower. Under ninefold experimental conditions, the effect of dry (1% w/w) and fresh (8.7% w/w) broccoli incorporation at a range of temperatures was analysed by Subbarao and Hubbard (1996) using the Anderson sampler for subsequent microsclerotial numbers. At all temperatures, both dry and fresh broccoli reduced microsclerotia. At 30°C or below, fresh broccoli was more repressive than dry. At the highest temperature (45°C), microsclerotial numbers were reduced in control (unamended) soil. However, in the presence of dry or fresh broccoli residues at this temperature, all microsclerotia were eliminated. Regardless of temperature, maximum reduction of inoculum occurred within 15 days.

The results of a factorial experiment in Salinas California (Xiao *et al.*, 1998), evaluating broccoli residues or crops on *V. dahliae* infection in cauliflower and irrigation methods and regimes, showed a 94% reduction in propagules after two broccoli crops in contrast to a fivefold increase after two cauliflower crops in the absence of broccoli residues. Disease incidence and severity were reduced by broccoli treatments. Irrigation in this experiment applied in various ways increased disease incidence and severity compared with a water deficit.

Shetty *et al.* (2000) using immunohistochemical staining with *V. dahliae*-specific monoclonal antibodies (mAbs) showed, in the presence of high soil populations of *V. dahliae*–soil broccoli amendments, markedly reduced root colonization of both cauliflower and broccoli; the latter, although resistant, is normally root colonized. In addition to suppressing microsclerotial viability in soil, broccoli residues may inhibit the root-colonizing potential of surviving propagules. The addition of lignin as an organic amendment supplementing mineral fertilization in newly reclaimed serozem sandy soils in Russia increased raw cotton production while reducing the incidence of wilt (Batirov, 1986). Davis *et al.* (1996) confirmed a widely held view in regions of the USA that green manure derived from an intercrop of *Sorghum vulgare* var. *sudanense* (sudan grass) was *inter alia* the most successful plant residue soil amendment to reduce wilt in potato. This effect, though successful in Idaho, could not be repeated in the Columbia basin of Washington. In a comparative experiment carried out at Oregon in microplots containing Idaho and Washington soil treatments, Idaho sudan grass residues led to a reduced disease severity of 9% as measured by the reduction in area under the senescence progress curve (AUSPC); in contrast, Washington sudan grass residues induced a 12% *increase* in disease. Significant population changes were recorded (Idaho treatments gave an initial population of *V. dahliae* propagules of 33% lower than the fallow control, an intermediate period population of 26–43 *higher* than fallow followed by a 3% reduction over control numbers at the end of the season). The corresponding figures for the Washington sudan grass treatment were 25% lower than controls at the beginning of the season, followed by mid-season levels 25–29% higher, ending the season with 57% lower than fallow. With these curious and variable results, Parks and Powelson (1997) using the same treatment could not attribute regional differences to inoculum levels. At present, no satisfactory explanation has been given to explain this phenomenon. In another report, Huisman *et al.* (1995) found that one crop of Sudan grass followed by 1 year of barley was equivalent to 2 years of Sudan grass in aleviating potato wilt. The net effect was to increase soil suppressiveness to disease with a concomitant increase in yield, but with no effect on the density of *V. dahliae* in the soil.

La Mondia *et al.* (1999) and Gent *et al.* (1999), confirming the interaction between *V. dahliae* and the nematode *P. penetrans* in disease severity and loss of tuber yield, found that incorporation of spent mushroom compost or straw mulch into soil alleviated symptoms in part. Photosynthesis, as measured by CO_2 exchange, and transpiration were reduced in plants infected by both pathogens and reversed in part by the compost treatment. This result, however, is almost certainly due to reduced levels of infection, rather than any intrinsic effect of organic matter on host metabolism. Huisman and Davis (1997), using sudan grass residues, found that soil fungal populations, especially *Fusaria*, increased from several hundred to 2000–10,000 c.f.u. g^{-1} of soil. *V. dahliae* colony densities on roots were reduced to 46% of those in non-amended soils. Oat seed-treated soils led to a reduction of 10% of the controls. Comparable root colonization rates for non-amended versus amended treatments were 0.19 and 0.42 mm per colony. An important feature of this paper was that a bioassay of root colonies based on planting on selective media detected only 10–35% of colonies identified using a *V. dahliae* mAb-based immunoassay.

In a more experimental extension of the field/glasshouse experiments, Gilbert and Griebel (1969) eliminated *V. dahliae* microsclerotia from soil which had been exposed for 24 h to a distillate from lucerne. Volatile products from decomposing debris can profoundly affect spore germination. Berestetskii *et al.* (1982) found that methanol from pea and maize tissue, butanol from pea and acetaldehyde from lupin residues all completely suppressed spore germination. Using Ontario potato field soil under laboratory conditions, Lazarovits (1997) reported that 7–10 days following the incorporation of meat (unspecified) or bonemeal and incubating at 24°C and 50% water holding capacity, *all* microsclerotia of *V. dahliae* were killed. Using the same amendments, microsclerotia were killed in the headspace of the incubation vessel by a volatile product suggested to be NH_3. These results were accompanied by a 10,000-fold increase in bacterial and fungal populations. In a field extension of this experiment, Lazarovits *et al.* (1999) incorporated soybean meal (SM) and meat and bonemeal (MBM) into the top 15 cm of soil at 37 t ha^{-1} in two Ontario commercial potato fields in spring 1996 and monitored infection until 1998. Wilt was reduced to zero in 1996 and remained reduced in 1997; by 1998, however, disease levels were equal to, or higher than, controls. Soil bacteria increased 100-fold immediately after application but returned to control levels after one season. Similar ephemeral increases in soil pH from 6.0 to 8.5, and increasing ammonia, nitrite and nitrate levels also occurred. In the first season, SM and MBM reduced microsclerotial viability by 72–100% ($r = 0.9$, $P < 0.0001$). Field plot experiments incorporating chicken manure (66 t ha^{-1})

15 cm deep reduced wilt (*V. dahliae*) and nematodcs to near zero levels. Microsclerotia in known concentrations buried in nylon bags showed similarly reduced viability. Pig (55 hl slurry ha^{-1}) and cattle manure (100 t ha^{-1}) had equivocal effects on disease control and microsclerotia (Conn and Lazarovits, 1999). Olive oil water, a waste product from oil-pressing (considered to be a potential stimulant for microsclerotial development in Crete), when used for olive irrigation was shown by Fodale *et al.* (1999), to the contrary, to have an antifungal role. Complete inhibition of *V. dahliae in vitro* occurred with incorporation of >300 ml l^{-1} of PDA. This study requires field application with microsclerotial sampling and disease assessment.

Herbicides and insecticides

Comparisons of triazine herbicides showed that sazol leached less than aminotriazole on clay and sorbed more; it was toxic to *V. albo-atrum in vitro* (Gomaa and Matolcsy, 1974). The Israeli group at Rehovot have shown that dinitroaniline herbicides nitralin and trifluralin, added to soil at 1 mg g^{-1} of soil, reduced *V. dahliae* wilt in tomato and aubergine by 97%. *In vitro* inhibition of the pathogen by trifluralin was at much higher concentrations than the 1 mg g^{-1} in the soil or the 0.84 mg g^{-1} in the hypocotyl detected in the aubergine. The resistance occurred in spite of phytotoxicity, which could also be a causal effect (Grinstein *et al.*, 1976).

Tolkachev (1978) showed that the triazine herbicide (prometryne) at 1.5 kg ha^{-1} and the nitroaniline (trifluralin) applied at 2 kg ha^{-1} in association with the culture of Tashkent 1 wilt-resistant cotton had no effect on heterotrophic microbiological soil processes or on soil mycolytic activity. The reduction in the weed population led to a general decrease in saprophytic bacteria but a large increase in cotton rhizoplane and rhizosphere bacteria. Application of Cotoran (fluorometuron) at 1.1 kg a.i. ha^{-1} increased the soluble N content while decreasing the protein, organic P, sugar contents and oxidative enzymes of cotton cultivars. These changes were correlated with a decreased resistance to *V. dahliae* (Alimov, 1980). In contrast to the foregoing, Solyanova (1978) and Tolkachev (1978) applied fluorometuron and prometryne at 1 and 1.5 kg ha^{-1} at cotton sowing and trifluralin at 1 kg ha^{-1} at sowing over a period of 5 years. No effect on the virulence or incidence of *V. dahliae* wilt was observed at any time. Grinstein *et al.* (1981) showed that trifluralin and other dinitroaniline herbicides were not directly toxic to *V. dahliae*, but suggested that pre-inoculation treatment of tomatoes stimulated the synthesis of a water-soluble phytoalexin, such that susceptible, herbicide-treated cultivars behaved like resistant ones. In glasshouse experiments on cotton, Mohamed (1982) claimed that fluorometuron, but not trifluralin, reduced the symptoms of wilt. Compounds such as oxadiazon, pendimethalin, fluridone, alachon and trifluralin, while inhibiting *V. dahliae* growth *in vitro* appeared to stimulate fungal growth in treated cotton plants.

Macedo and Chiba (1985) described the fungistatic nature of a number of herbicides *in vitro* to *V. albo-atrum*. Jordan *et al.* (1987) noted that the yield of *V. albo-atrum*-infected strawberry plants treated with herbicides was greater than uninfected plants. In the non-herbicide category, catechol added to soil reduced the incidence of *V. dahliae* wilt in tomato (Chet *et al.*, 1978). The resilience of *Verticillium* as a soil pathogen is well illustrated by its tolerance to phosphinothrin tested to 50 mM. Phosphinothrin (glufosinate), a microbial toxin synthesized as a selective weed killer in the presence of transgenic phosphinothrin-resistant plants, substantially suppressed the soil microfora in treated areas; fungi (excluding *V. albo-atrum*) by 20% and bacteria by 40% (Ahmad and Malloch, 1995). Minor *in vitro* and *in vivo* effects of Bion R [benzo (123) thiadiazole-7-carbothioic acid-*S*-methyl ester] (leaf and root) and the herbicide oryzalin (roots only) in partially suppressing *V. dahliae* in aubergine were reported by Hijwegen and Termoshuizen (2000). Treatment of *Verticillium*-infected cotton with the herbicide DPC (mepiquat) increased root exudates and ion concentration (Dong *et al.*, 2000). The claim that mepiquat induced resistance in cotton was not substantiated.

Insecticides may also affect wilt diseases; *O,S*-dimethyl acetylphosphoamidothioate (acephate) applied to cut potato tubers at 1 and 2 oz a.i. cwt^{-1} delayed the appearance of *V. dahliae* and *V. albo-atrum* wilt symptoms but did not affect the progress of the diseases.

Effects of Crop Rotation

Many reports have noted reductions in inoculum levels and infectivity on test plants following the rotation of susceptible with resistant or non-host plants. The efficacy of crop rotation as a control measure, however, is limited for two reasons: (i) resting structures, especially microsclerotia, remain viable in soil for many years and will infect the re-introduced susceptible plant (Huisman and Ashworth, 1976); and (ii) *Verticillium* spp. can perennate in non-host root systems, particularly weed plants, providing 'carry over' reservoirs of infection (Pegg, 1974). This was clearly demonstrated in cotton by Brown and Wiles (1970).

Green (1967) found that rotation of *V. dahliae*-infested peppermint with the non-hosts, maize or grass, reduced subsequent disease incidence. Potatoes or soybeans had little effect. Minton (1972) found the incidence of cotton wilt increased following 4 years of sorghum (a non-host) cultivation when the grain crop was infested with the weed *Aramanthus retroflexus*. Huisman and Ashworth (1976) found that populations of microsclerotia had not fallen to ineffective levels 6 years after the last cotton harvest. Monocultures of cotton cv. Acala SJ-2 in California led to increases in propagule concentration of 13–15 microsclerotia g^{-1} $year^{-1}$ up to a plateau level of 60–70 or up to 300 in soils with either the P_1 and P_2 or P_1 strain only. Only 10 microsclerotia g^{-1} of soil were required for 100% wilt. Various break crops have been suggested to reduce the incidence of wilt in cotton. Lucerne in conjunction with systemic fungicides (Tellyaev, 1978) or alone for 3 years or annual break crops of maize (Kostenko and Nesterov, 1971) have proved effective. Egamov (1979) used irrigation to stimulate microsclerotial germination, followed by 3 years of lucerne which resulted in a 68% reduction in wilt in a succeeding cotton crop. Alimukhamedov *et al.* (1977) similarly obtained a 15–20% reduction following a rotation with lucerne and maize. Plants that reduce inoculum potential as well as wilt incidence in cotton are grasses, legumes and crucifers. The most effective crops (in order) reducing *V. dahliae* inoculum were pea, clover, barley, rye and maize. No root penetration was seen in these wilt-resistant crops (Sidorova, 1974). Ranney (1973) in a comprehensive study on seven cotton cultivars grown on land severely infested with *V. albo-atrum* [*sic*] (*V. dahliae*) reported that cv. Coker 201 produced the highest yield of seed cotton, which increased by 18% after 1 year's fallow. Deltapine 16 gave the highest yield on moderately infested soil, but this was outyielded by Coker 201 after 1 year's fallow. In a further test of 17 cultivars in irrigated and non-irrigated land, the lowest wilt incidence was in irrigated cv. Stoneville 7 and cv. Auburn 56 under non-irrigation. Cv. Rex gave the highest lint yields under both conditions. In Uzbekistan (CIS), winter rye, barley, winter peas, vetch, winter rape and white mustard grown as green manure all decreased the incidence of cotton wilt and increased cotton yields. White mustard was the most effective crop (Mannapova, 1976). Similar results were reported in laboratory tests by Marshunova and Muromtsev (1975). A summer break crop of maize followed by a winter crop of mustard (*Brassica juncea*) in a Uzbekistan monoculture of cv. 108-F cotton decreased *V. dahliae* infection (wilt?) by 31% and increased seed cotton by 0.79 t ha^{-1} (Sobirov *et al.*, 1980). Similar results were reported by Marupov (1990). An earlier trial (Egamov, 1976) reported larger yield increases following a rye or sorghum break, but here the monoculture wilt incidence was lower. Solarska (1994b, 1995) achieved a similar effect of rye intercropped with hop. This led to a big increase in the *V. albo-atrum*-antagonistic fluorescent *Pseudomonas* spp., more so than with mustard. Butterfield *et al.* (1978) and Pullman and DeVay (1981) found that a single season rice crop decreased *V. dahliae* propagules to undetectable levels; flooding was less effective. Zhdarkina *et al.* (1981) found the effect of rotation with maize, rape, rye or vetch led to a 1.8-fold decrease in disease incidence, while rotation with rice gave a fourfold decrease. Grishechkina (1990) reported a 80–95% reduction in inoculum potential following autumn sowing of mustard and rape for green manure. In an early study in Arkansas, Hinkle and Fulton (1963) found that soybean followed by barley or sorghum and cotton in the third year substantially reduced wilt. The use of a 6-year rotation with wheat, pea and vetch reduced inoculum by 96% but did not eradicate the pathogen (Ben-Yephet *et al.*, 1989).

The avoidance of cotton hosts susceptible to a particular race of *V. dahliae* can act as a palliative in suppressing levels of that race. Sidorova (1978) described a reduction in the percentage of infection in *G. hirsutum* cv. Tashkent

1 (susceptible to R_2 and resistant to R_1) following cropping alternation with cv. 108F and *G. barbarense* cv. S6030, 'both' [*sic*] susceptible to R_1 and R_2. The presence in soil of R_1 equal in quantity to R_2 inhibited wilt in Tashkent 1 by R_2. Somewhat different results with the same cultivars were described by Matveev and Tribunskii (1979). Culture of cv. 108F followed by Tashkent 1 had no effect on wilt suppression, but wilt in Tashkent 1 grown as a monoculture at 59.3% fell to between 6.2 and 25% following cv. S 6030, described as fairly resistant to both races. Bejarano-Alcazar *et al.* (1995a) rotated cotton with sunflower, maize or wheat. Cotton infected with the defoliating strain built up rapid levels of microsclerotia in soil and reached higher levels than that infected with the non-defoliating strain.

Rape is regarded as unsuitable as an intercrop in intervals of less than 4 years due to its susceptibility to *V. dahliae* and other soil-borne diseases (Daebeler *et al.*, 1987; Ohlsson, 1988). Alternation of red clover with winter rape is also discouraged since red clover is also susceptible to *V. dahliae* (Luth and Pfeffer, 1989). Shetty *et al.* (1999) advocated rotating strawberry with vegetable crops as an environmentally acceptable but less efficient control of *V. dahliae* than soil fumigation with methyl bromide. The incidence of *V. dahliae* in tomato, pepper and aubergine in Bulgaria was lower following previous cropping with rice (Milev and Nechev, 1973; cf. Butterfield *et al.*, 1978). Linseed (*Linum usitatissimum*), grown increasingly in the UK, has been shown to be a host of *V. dahliae*. The significance of this in cropping programmes with rape, potato, soft fruit and ornamentals was stressed by Harris and Fitt (1995).

Verticillium wilt of potato occurs in all areas of potato production in the world, with ambient temperature selecting either *V. dahliae* or *V. albo-atrum* as the principal pathogen. In some temperate areas, both organisms may be isolated. *V. dahliae* is associated with a warm climate and light soils, as found in central USA, Canada and Israel, and *V. albo-atrum* in cooler, wetter conditions in heavier soils more typical of northern Europe, eastern Canada and the USA. *V. tricorpus* has also been associated with potato wilt (Platt *et al.*, 1995). The PED associated with American potato culture is a complex phenomenon associated with potato nematodes (Botseas and Rowe, 1994), irrigation (Powelson *et al.*, 1995) and soil type (Francl *et al.*, 1988) (see also Powelson and Rowe, 1993).

Emmond and Ledingham (1972) obtained highest yields of potatoes in Saskatchewan from a single potato crop in a 3-year rotation with sweet clover and fallow. Two crops in a 6-year rotation reduced yield substantially. The main pathogen was *V. albo-atrum*. Mol *et al.* (1995) confirmed the fact, established some 30 years earlier, that removal of post-cropping foliar debris from sequential field crops, in their case, potato, field bean and barley, reduced disease incidence in the succeeding crop and that infection was proportional to soil inoculum density (microsclerotia). An extensive study of two fields in potato monoculture and 15 fields in a potato–wheat rotation was conducted by Joaquim *et al.* (1988). During 4 years, microsclerotia from fields in rotation varied from 9.6 to 19.6 per 10 g of oil. Following potato, a general increase in microsclerotia occurred but there was much variation in fields. The importance of weeds as reservoirs of microsclerotia in potato crop rotations in Ontario was highlighted by Busch *et al.* (1978). The soil population of *V. dahliae* peaked in July or August regardless of continuous potato or potato–wheat cropping (Joaquim *et al.*, 1988). Scholte (1989b) confirmed most of the earlier work in that stem infection correlated with wilt incidence and that early season plant vigour enhanced *V. dahliae* infection and increased with cropping frequency. Scholte also showed that stem infection decreased with nematicide application but this had no effect on microsclerotial levels in the soil. Two valuable reports by Davis *et al.* (1997a,b) illustrate the value of sudan grass as a break crop with potato. Results showed that more than a single factor could be attributed to the grass in the suppression of *V. dahliae* disease. The presence of *Fusarium equiseti* in soil and on potato roots was the most positive indicator of *V. dahliae* suppression. Mn, mineralizable N and *F equiseti* were all positively correlated with potato yield: 74% of field variability for wilt incidence could be accounted for by the equation Y = –12.18 + 0.5143 *V. dahliae* (c.f.u. g^{-1} of soil) + 3.4997 Mn (initial pre-plant level) – 0.01421 *F. equiseti* (c.f.u. g^{-1} of soil). Similarly, 85% of field yield variability could be determined by Y (US# tubers 280 g) = 82.6 – 0.7980 (*V. dahliae*) + 2.269 (mineralizable N) + 0.02025 *F. equiseti* (Davis *et al.*, 1997b). Davis *et al.* (1997a) showed by using controlled levels of soil inoculum that the least colonized crops used successfully in rotation with potato were sudan grass, maize and lucerne. Cereals used as break crops were all colonized by potato isolates of *V. dahliae*, with oats the most severely affected.

The effect of 10–11 years of individual continuous crops of onions, maize, flax, kidney bean, field bean and peas on 2 years of potato culture was studied on new polder soil in Holland. In both potato years following field bean, *V. dahliae* ratings (based on stem plating and microsclerotial numbers on stems) reached a maximum of 84%; only 10% yield reduction could be attributed to *V. dahliae*. It was concluded that a 44 and 36% rating following maize and onion was due to poor soil structure and reduced yield following flax due to *V. albo-atrum* (Lamers, 1995). Following an extensive survey of UK potato field soils, Locke and Buck (1997) established a mean propagule count of 6.6 propagules g^{-1} of soil. Counts from 36 sites where low susceptibility crops such as cereals, maize, grass, onion, leek, carrot, parsnip, cane fruit and brassica had been grown showed a propagule level of 0.5 g^{-1} of soil. Where linseed was grown as an alternative crop, microsclerotia rose to 14.4 propagules g^{-1} of soil (based on 69 field samples) compared with 3.8 in 185 non-linseed cultured soils. The danger to the potato from such a rotation is all too evident since Fitt *et al.* (1992) had already established *V. dahliae* as a contemporary pathogen of linseed in the UK and Germany. Sunflower, a prime host for *V. dahliae*, benefits from a crop rotation. Kernasyuk and Radzievskii (1981) recommended a 7–8 year rotation with maize and sugarbeet. The use of crop rotation was proposed as an alternative to methyl bromide fumigation against *V. dahliae* in strawberry (Profic-Alwasiak, 1982; Shetty *et al.*, 1999).

Temperature

Temperature is perhaps the single most important environmental factor, governing not only the germinating of *Verticillium* spp. and disease development, but also the geographic distribution of *V. albo-atrum* and *V. dahliae*. World maps of major crop disease caused by the two species (Pegg, 1984) show a distribution of each in cool to warm temperate regions to 60°N and 50°S of the equator. *V. albo-atrum*, however, is completely excluded in a belt between 20° North and 20° South of the equator.

Isaac (1949) in a definitive study to eliminate confusion surrounding earlier reports (Edson and Shapovalov, 1920; Chaudhuri, 1923) based often on incorrect nomenclature, found that *V. dahliae*, *V. albo-atrum* and *V. nigrescens* all showed optimum growth at 22.5°C, but at 30°C *V. albo-atrum* was reduced to a yeast-like budding while the other species maintained moderate growth. These findings were broadly confirmed by Robinson *et al.* (1957), Smith (1965), Devaux and Sackston (1966) and Skadow (1969b). Similarly, Zattler and Chrometska (1960) confirmed Isaac's temperature for *V. albo-atrum* and Wilhelm (1948) that for *V. dahliae*.

The lucerne strain of *V. albo-atrum* has a higher optimum temperature than those from other hosts, probably reflecting its ability to withstand North American summer temperatures (Heale, 1985). A range of 38 isolates from western Canada and north-western USA had an optimal temperature in culture of 25°C (Christen and French, 1982). Growth occurred at a higher temperature on osmotically adjusted agar. At −25 to −60 bars, growth was tenfold greater at 30°C than at 0 to −10 bars. Optimal growth at −19 to −90 bars was in the range 20–27°C. Comparative infectivity experiments on four North American and three European lucerne strains of *V. albo-atrum* in London and Washington State, using temperatures of 17–30°C and 12–21°C, showed that disease severity was greater at the higher temperature in each country (Christen *et al.*, 1983). The evidence was used to support a claim that the US outbreak originated from a European strain of the pathogen. Minimum temperatures at which growth occurs in all species varies between 4.5 and 6°C. Schnathorst (1973) and Bell (1973) have reviewed the temperature relationships of *Verticillium* species in culture with special reference to cotton strains of *V. dahliae*. Wyllie and DeVay (1970a) and Schnathorst *et al.* (1975) found that the cotton-defoliating strain had an optimal temperature of 27°C and could germinate at 33°C. The non-defoliating strain, however, had an optimum of 24°C and could not germinate at the higher temperature. Devaux and Sackston (1966) reported that resting mycelium of *V. albo-atrum* formed only between 14 and 26°C, and while growth of *V. dahliae* was possible at 33°C, no microsclerotia formed above 30°C. These authors described the optimal temperature for *V. cinerescens* as 28°C, with minimal and maximal limits of 6 and 35°C, respectively.

Effect of temperature on symptom expression

A classic and comprehensive study by Ludbrook (1933) on aubergine infected with *V. albo-atrum* concluded that air rather than soil temperature was the most important environmental factor influencing symptoms. In comparison with plants grown at 24.5°C, plants grown at 30°C remained disease free but were abnormally spindly. Ludbrook was one of the first people to suggest a separation of Ms and Dm types of *Verticillium* into distinct species based on their temperature responses. Wilting occurred when the temperature was lowered to 26°C. Severe wilting occurred when plants were grown with a soil temperature of 20–24°C and with an air temperature of 23–26°C. The shortest incubation of 17 days was found when soil and air temperatures were at 20°C. McKeen (1943), using Wisconsin soil tanks, found that infection of potato by *V. albo-atrum* [*sic*] (*V. dahliae*) occurred between 12 and 32°C, but symptom expression was greatest between 20 and 28°C. From observations of *V. dahliae* in culture and in *Chrysanthemum*, McKeen (1943) concluded that the optimal temperature of 22–24°C for symptom development represented the optimal temperature for vegetative growth in the xylem. Working with *V. albo-atrum* in tomato, Bewley (1922) found glasshouse temperatures of 17.4 and 21.3°C (optimal) to be favourable for rapid wilt, which did not appear at 25°C. Bewley (1922) recommended elevated temperatures as a means of wilt control. Williams (1962) found that shading tomato plants by day when night temperatures lay between 10 and 15°C, decreased symptoms caused by *V. albo-atrum*, but not *V. dahliae*. Using *Arabidopsis thaliana* as an experimental short-cycle plant, Soesanto and Termorshuizen (1997) found that microsclerotia from three isolates were produced over the pre- and post-infectional temperature range of 5–25°C, although slowly at 5°C. Maximum production with roots > shoots was over the broad range 15–20°C. *V. dahliae* on rape in Poland showed microsclerotial germination from 6 to 34°C, but optimally between 15 and 28°C. There is no indication of the genetic homogeneity of the fungi used in these experiments to explain these rather broad temperature optima.

Early studies on temperature relationships failed to distinguish between the effect on the infection (penetration) process and vegetative growth or conidiation in the xylem. Edgington and Walker (1957) not surprisingly found that soil temperature in part determined tomato stem temperature and hence *V. dahliae* wilt incidence. In this context, the inoculation method (soil incorporation of inoculum or stem puncture) is all important. Isaac (1949) comparing wilt production in tomato by *V. albo-atrum* grown at 21.5, 25, 27 and 29°C showed that plants wound-inoculated just below soil level only wilted at 21.5°C. The pathogen was still recovered from the plant at the three higher temperatures, however. Griffiths and Isaac (1963) found similarly with lupin and sunflower that symptom appearance was more rapid at lower temperatures. No confirmation has been achieved for the inference by Ludbrook (1933) that temperature could predispose plants to infection (but see D'Ercole and Nipoti, 1980).

Variations in temperature response have been reported for different host cultivars and (but not identified) for different pathogen strains or races. Nelson (1950) recorded maximum symptoms in tomato at temperatures of 24 and 28°C (*V. dahliae*). *Mentha piperita*, *M. spicata*, *M. crispa* and *M. piperita* × *M. crispa* hybrids showed greatest disease severity to *V. dahliae* at 25°C early in the experiment, but later became more severe at 20°C. Most clones later recovered at 25°C. Host and pathogen were inhibited at 10°C and the fungus at 30°C (Berry and Thomas, 1961). Symptoms were maximal in pepper at 25–30°C (Kendrick and Middleton, 1959). Sewell and Wilson (1976) claimed that hop plants inoculated with mild strains (M) expressed symptoms at low but not high temperatures. Symptoms caused by virulent strains (V_1, V_2 or V_3), however, were unaffected in symptom type by temperature. Sewell and Wilson (1977) later demonstrated that the distinction of a particular symptom type per isolate by temperature was conditional on root infection from soil-incorporated inoculum; when stem spore inoculation by puncture was used, the temperature effect was lost.

Much work has been done on the effect of temperature and the geographic distribution of *V. dahliae*-free (or mildly wilted) cotton. Disease severity is greatly reduced when air and soil temperatures are high (Schnathorst, 1981). Brinkerhoff *et al.* (1967) obtained severe symptoms on *V. dahliae*-inoculated cotton grown at 28°C day and 18°C night, typical of the defoliating strain. At 36°C day and 18°C night, symptoms failed to appear. The 3°C higher optimum temperature for the virulent strain compared with the non-defoliating strain may account for its successful spread to become the dominant pathotype. Halisky *et al.* (1959) and Erwin (1977b) have both implicated high summer temperatures in specific regions in California and Arizona for the relative unimportance of *Verticillium* wilt. Temperature may also change the symptom pattern in susceptible or resistant cultivars. At 28°C and above, the severe symptoms seen in a susceptible cotton cultivar infected by the defoliating strain appear to be caused by a non-defoliating type. Similarly, a susceptible cultivar infected by a non-defoliating strain at that temperature appears like a tolerant cultivar (Garber and Presley, 1971). Similar results were reported for cotton in the CIS (formerly USSR) by Ikramov and Mirdadaev (1974), Ter-Avanesyan (1976), Ikramov *et al.* (1976, 1979) and Ikramov (1979); in Turkey by Esentepe (1974); in the USA by Minton and Gipson (1978); and in Venezuela by Malaguti *et al.* (1973). Minton *et al.* (1979) reported on the effects of controlled night temperatures. Besri (1980a) found that soil temperatures in Morocco in spring and autumn favoured infection of tomato by *V. dahliae*. Air temperature during the same period encouraged pathogen growth in the stem, but not symptom development. In plastic tunnels, temperature is the limiting factor for tomato wilt, 90% occurring between November and April (Besri and Zrouri, 1983). In other crops, Sewell *et al.* (in Crosse *et al.*, 1977) found that eight isolates of *V. albo-atrum* were less severe in three hop cultivars at 62°F than at 52°F, at which temperature two V_3 strains induced severe wilt in the erstwhile resistant cv. Wye Target. *V. dahliae* was most commonly isolated from red clover when grown at the low temperature of 10°C (Nan *et al.*, 1991). Francl *et al.* (1990) studying the effect of environment on PED found that high temperatures had a negative effect on yield of *V. dahliae*- and *V. dahliae* + *Pratylenchus penetrans*-infected potato plants, an aggravated stress effect. These influences were modelled by Johnson (1988) in relation to potato yield.

Sackston (1971) and Wilanowski-Guzdziol and Sackston (1972) have explored the effect of fluctuating temperatures on *V. dahliae in vitro* and *in planta*. Temperature fluctuations 3 and 6°C above and below means of 22, 25, 28 and 31°C were investigated with one or four cycles per 24 h. Growth on V-8 agar was greater with fluctuations around means above optimum, compared with a constant optimal mean. Conversely, fluctuations around means below optimum gave lower growth than at a constant suboptimal mean temperature. Contrary to generally accepted findings at constant temperatures, experiments on *V. dahliae*-infected sunflowers with fluctuations around means showed that at 22°C no effect on symptoms was apparent, while at fluctuations around 25°C, symptoms were *reduced*, and at 28°C plants showed *increased* symptom severity. No explanations have been advanced to explain these unusual results.

Pullman and DeVay (1982b) used physiological time, degree days – an accumulation of daily temperature above 11.9°C, the cotton developmental threshold – to measure disease progress. Evidence of a temperature predisposition in strawberry to *V. dahliae* was shown by D'Ercole and Nipoti (1980) after plantlets were stored at low temperature. Micropropagated plants developed earlier and showed more severe symptoms than those derived from runners. An unusual temperature study by Huang *et al.* (1986a) showed that mycelium from the lucerne strain of *V. albo-atrum* survived for 30 months at –40°C in grasshopper (*Melanoplus sanguinipes*) faeces, or on pea aphid (*Acyrthosiphon pisum*) bodies. At 15°C, however, survival on aphid was 5 months and in faeces 21 months. The results are interpreted to suggest survival of the pathogen on vectors over a Canadian winter.

Howell and Erwin (1990) reported one of the lowest geographical recordings of *V. albo-atrum* at approximately 34°C in irrigated lucerne in southern Californian desert regions. Exceptionally, this strain (which justifies species rank) grew optimally on PDA at 23–25°C and grew at up to 30°C. Summer field temperatures rose to 40–44°C. Although canopy temperatures were 42°C, evaporative cooling in soil gave root temperatures of 25°C. Notwithstanding this, the pathogen was recovered after 6 months from lucerne straw baled at 40°C. Erwin and Howell (1998) consistently found *V. albo-atrum* surviving in lucerne in 1989 and 1990 in the Californian Mojave at air temperatures above the supposed maximum of 30°C for growth and sporulation. While lucerne isolates of *V. albo-atrum* from Europe, northern USA and California had similar isozymes, US isolates were more severe on US cultivars than UK isolates. There was clear evidence that the Californian strain had evolved or adapted to a higher temperature, as was confirmed in culture against the northern US and UK isolates. The Californian pathogen was recovered from lucerne stem buried in sterile field soil for up to 16 months stored at 30°C; for up to 3 months stored at 35°C and for up to 16 months in lucerne stem kept at 41°C but not in soil (Howell and Erwin, 1995).

Thermal death points in soil for strains of both species were calculated by Bollen (1969). *V. albo-atrum* failed to recover after exposure at 52.5°C for 30 min and *V. dahliae* at 57.5°C for the same period. Host recovery from infection at high temperature has been reported for avocado (Zentmeyer, 1949), apricot (Taylor, 1963), olive (Wilhelm and Taylor, 1965) and cotton (Bell and Presley, 1969b).

Effect of Soil Type

V. albo-atrum and *V. dahliae* characteristically are non-aggressive soil residents which rarely venture more than a few millimetres from their propagule base (DeVay and Pullman, 1984). Rudolph (1931) considered that cotton wilt in the USA was generally severe on sandy loam, clay loam soils and those high in organic matter. Heavy soils are cooler and hence present a more optimal temperature in summer. On sandy soils, it is difficult to exclude the deleterious effect of irrigation on the disease, since irrigation frequency is usually higher than on heavy loams. *Verticillium* wilt is a problem of neutral or alkaline soils, rarely acid ones (Bell, 1989; see Jones *et al.*, 1971). Schnathorst (1981) has pointed out that while the incidence of *V. dahliae* in cotton in virgin, semi-desert acidic soils is very low and may exist in native plants, populations can increase to give disease in epidemic proportions in 5 years following irrigation, amendments and planting with susceptible cultivars.

Dutta and Isaac (1979c) investigated the effect of soil pH, temperature and organic matter content on fungistasis on *V. albo-atrum* and *V. dahliae* in topsoil, subsoil and dune soil. No relationship was found between seasonal variation in organic matter, water content, pH and fungistasis in top soil. Lower fungistatic activity in subsoil and dune soil was attributed to lower soil moisture and nutrients. Increased soil temperature enhanced fungistasis equally against both species. Some cultivars and breeding lines of groundnut (*Arachis hypogea*) found to be resistant or highly tolerant of *V. dahliae* in alluvial clay were susceptible when grown on a calcareous loess. The addition of an iron chelate restored plant vigour and yield comparable with uninfected controls. The role of iron in resistance was not explained or explored further (Krikun and Frank, 1975). The importance of the soil microflora in fungistatic regulation of *V. dahliae* in natural cotton soils was emphasized by Sidorova (1990). Mukhamedzhanov and Zubenko (1990) claim that improving cotton soil structure by ploughing to 60 cm, incorporating 30 t ha^{-1} of manure and inter-row cropping with lucerne and rye so improves the antagonistic microbial flora that wilt is reduced by 41% compared with resistant cultivars grown under a normal cultural schedule. Wittman (1972) discussed *inter alia* the effect of soil type on the incidence of *V. albo-atrum* and *V. dahliae* in hop.

Effect of Irrigation on Disease

The effect of water on the progress of *Verticillium* disease depends very much on the crop, soil type and ambient climate. Under UK field conditions, Talboys (1970) found no effect of rainfall on hop wilt over an extended period. In a pioneering study on the latex-yielding, guayule (*Parthenium argentatum*) in semi-arid desert conditions, Schneider (1948) found that *V. dahliae*-induced wilt increased with the frequency of watering. Plants watered every 4 weeks had 42% infection, plants watered every 2 weeks, 62% infection and those watered weekly 77%. The non-irrigated plants had only 8% infection. Essentially similar results were recorded in cotton by Leyendecker (1950). Schnathorst (1981) reported that drought-resistant hosts are generally unaffected by wilt where irrigation is limited to natural rainfall, and problems with these crops only arise when artificial irrigation is introduced. It is clear from the work of Savov (1978b) and Babaev *et al.* (1979) that much of the severity of cotton wilt in the CIS is associated with the application of semi-continuous irrigation. Two possible explanations of the relationship of water to wilt severity and incidence relate to the survival and germination of microsclerotia. In a signal study, Karaca *et al.* (1971) showed that the soil temperatures in Anatolian cotton fields in July, prior to irrigation, at depths of 5 and 15 cm were 34.7 and 33.3°C, respectively, temperatures at which no infection is likely and survival of the pathogen is threatened. As a result of evaporative cooling following irrigation, the temperature fell to 28.0 and 29.4°C, respectively, permitting infection to proceed. It has been claimed that microsclerotial germination is particularly stimulated by periodic wetting and drying (Farley *et al.*, 1971). Microsclerotia germinated after nine cycles of air drying, followed by wetting. Their evidence suggested that microsclerotia could germinate and sporulate several times in the proximity of organic substrates or non-host rhizospheres, given suitable physical conditions. Egamov (1979) also stated that periodic irrigation was a stimulant of germination. Completely opposite results were obtained by Baard and Pauer (1981) who found no effect of drying and rewetting soil on microsclerotial numbers, or on infection of cotton or *Datura stramonium*. Curiously, under their conditions, N and P fertilizers had no effect on disease incidence, but reduced soil propagule numbers. Non-cotton hosts respond similarly to irrigation.

Symptom severity in irrigated lucerne was greater than in non-irrigated plants (Palatinus, 1983). A survey of the south-western Saskatchewan grain belt found that cutting frequency and overhead irrigation of lucerne led to rapid and severe wilt (Gossen and Jesperson, 1990). The incidence of *V. dahliae* infection in olive groves was similarly enhanced by irrigation, as was the reduction in incubation time in maple seedlings (Shufelt and Linderman, 1987). A similar finding was made by Al-Ahmad and Mosli (1993) for Syrian, irrigated, new olive groves where ploughing frequency was also correlated with increased disease. Irrigation in the presence of *V. dahliae* also predisposes aubergine to wilt in plastic tunnels – especially under Thessaloniki (Greece) summer temperatures (Bletsos *et al.*, 1999). Fruit quality and colour was enhanced by irrigation, but the combined effect of irrigation and wilt reduced early yield and total production, and fruit quality.

Powelson and colleagues in the USA have shown a close correlation between PED and irrigation. Field studies in the Columbia basin of the Pacific North West established that the severity of foliar symptoms in potato is determined by the level of irrigation in the period (~ 4 weeks) between emergence and tuber initiation (Powelson *et al.*, 1995). Pre-tuber-initiation irrigation at 150% estimated consumptive use (ECU) led to a 22% increase in AUSPC compared with a deficit at 50% ECU. Symptoms of PED were most apparent 850 degree days after planting (Cappaert *et al.*, 1994). Potato cv. Russet Burbank grown under excessive soil moisture (–0.01 MPa) in the presence of *V. dahliae* resulted in reduced aerial biomass and increased root:shoot ratio. In contrast, root colonization was suppressed at –0.01 MPa compared with –0.15 MPa with a comcomitant reduction in propagule production. The authors (Gaudreault *et al.*, 1995) postulate that high soil water pressures may enhance PED by slowing root growth and/or indirectly increasing the rate of microsclerotial germination. A valuable contribution to irrigation studies by Arbogast and Powelson (1997) may explain variations reported in the literature: cultivars under the same treatment differ in the degree of foliar senescence and, importantly, in their response to plant available water (PAW) (a function of ECU). Senescence was most severe in cv. Russet Burbank and least in cv. Katahdin which of six cultivars was the only one tolerant to water stress. When this cultivar was irrigated at 100% PAW, the AUSPC was 28% larger than at 75 or 50% PAW. Cv. Katahdin differed from other cultivars in having lower recoverable populations of *V. dahliae* propagules in stem apices (see Gaudreault *et al.*, 1995). Irrigation is also closely interactive with plant density and fertilization. Excess water is especially damaging in conjunction with high N_2 (Bell, 1992a). In the Texas High Plains cotton region, losses from wilt declined by 9% over 20 years, paralleling a concomitant reduction in N_2 and irrigation (Wanjura and Barker, 1987). Similarly, a plant density of 19×10^4 plants ha^{-1} with three irrigation cycles yielded 40% more than 5×10^4 plants ha^{-1} with four cycles. A density of 15×10^4 with one cycle yielded as much as a conventional control treatment (Palomo Gil and Quirarte, 1976). The picture is similar in Arkansas. Nitrogen and irrigation generally increase cotton lint yields except in the presence of *V. dahliae*, or when the crop is late sown or if the season is excessively wet. High irrigation frequencies and high N rates (90–150 lb split dose urea acre^{-1}) predispose cotton to *Verticillium* wilt (McConnell *et al.*, 1998, 1999) (see also Verhalen, 1967; Garber *et al.*, 1976).

Soil flooding for 6 weeks killed microsclerotia of *V. dahliae*, as did anaerobic conditions in the laboratory by keeping soil under N_2 for 3 weeks (Nadakavukaren, 1960; Menzies, 1962). Pullman and DeVay (1981) found that flooding of cotton soil or rotation with paddy rice were equally effective in eradicating *V. dahliae*. Quite the opposite results were reported for *V. albo-atrum* infection of *Liriodendron tulipifera* by Morehart and Melchior (1982). *V. albo-atrum* was significantly more aggressive in trees stressed by periods of low water, compared with trees flooded intermittently. In the latter, symptoms were atypical and the pathogen could not be isolated from above ground parts. In contrast, plants grown at optimal or reduced water levels showed extensive colonization of the petioles. The effects of salinity on wilt severity have not been clearly separated from the effect of water and the frequency of irrigation cycles (see Effect of amendments on survival and disease). Besri (1981) and Kaufman *et al.* (1990) reported that saline irrigation enhanced *V. dahliae* infection and induced premature senescence in the absence of the pathogen in tomato and potato, respectively. Manandhar and Bruehl (1973) claimed that the optimal water potential of the cotton strain of *V. albo-atrum* [*sic*] (*V. dahliae*) was −10 to −30 bars at 25°C in soil or on agar. Growth at −100 bars was near zero. Besri (1980b), however, established the optimal growth of *V. dahliae* at −41.79 bars. While growth was possible at −100 bars, a reduction in water potential from −1.5 to −13.79 bars increased growth. Raising the temperature to 35°C inhibited growth, but this was restored on the addition of salt. Manandhar and Bruehl (1973) found a highly significant temperature–water potential interaction which has serious implications for subtropical, saline, field conditions. No growth of *V. dahliae* occurred on cornmeal agar at 35°C but, when the water potential was reduced to −30 to −40 bars with KCl, considerable growth occurred. This effect was confirmed by Besri (1980b).

In a 3-year experiment on cotton cv. SJ-2 in the San Joaquin valley, California (Huisman and Grimes, 1989), the effect of *V. dahliae* at 40–80 ms g^{-1} of soil was studied on plants grown at a density of 25,000–200,000 ha^{-1} and subjected to different irrigation periods (1–5) and timings (June or August). Disease incidence was only linear at low levels of inoculum, agreeing with the results of Ashworth *et al.* (1979b). Disease incidence was inversely related to plant density (see Minton *et al.*, 1972). With a low frequency of irrigation (0–1), severe symptoms (defoliation) decreased with increased plant density. With an increased number of irrigations (3–5), however, the percentage defoliation increased in the higher plant densities. The timing of irrigation was more important than the total number. Disease severity had a linear, inverse relationship with the time of the first post-plant irrigation (44% for June and 7% for August).

As would be expected, delays in irrigation led to reduced yield due to water stress (Grimes and Yamada, 1982), making control a delicate balance of timing in relation to disease and yield potential.

Effect of Plant Density on Disease

Most studies on the effect of plant density on the incidence and severity of wilt have been carried out on cotton. The relationship is complex, depending on both irrigation and edaphic factors. In general, reports confirm the early findings of Leyendecker (1950) and Blank *et al.* (1953) that the percentage infection (incidence) of *V. dahliae* wilt is inversely proportional to the seed sowing rate per hectare. Increasing plant density from 50–60 10^3 ha^{-1}, the normal seeding rate, to 10–15 × 10^4 ha^{-1} reduced wilt and increased yield (Leyendecker, 1950). Blank *et al.* (1953), also working in New Mexico, found that increasing plant populations from 14 × 10^3 ha^{-1} to 45 × 10^3 increased yield by 43%; when stands were increased to 70 × 10^3 ha^{-1}, yields increased by 69%. These results were dependent on irrigation, without which no effect was seen.

Planting two rows on a 15 inch raised bed (Leyendecker *et al.*, 1952), or narrowing the width between rows, or increasing the density per row (Minton *et al.*, 1972), all reduced disease incidence and increased yield. Wilhelm *et al.* (1970) obtained similar results. Ray (1976) described the selection of cotton cultivars suitable for narrow row cultivation. Minton (1980) conducted a comprehensive study with two cultivars, Rilcot 90 (susceptible) and Paymaster 266 (resistant) grown in one, two or four rows per bed, 100 cm apart. Foliar symptoms were greater in cv. Rilcot than cv. Paymaster, but the percentage infection declined with close row spacing. The number of diseased plants of each cultivar was higher with closer row spacing but did not increase directly with plant populations. Huisman and Grimes (1989) found a linear relationship between mean number of infections versus number of plants per hectare and concluded that the expected disease incidence would be proportional to the size of the plant root system, based on the assumption that root and canopy densities on an area basis are only minimally influenced by plant density for most of the season. This rationale is based on a constant infection rate per unit root density; thus a linear relationship between disease incidence and the inverse of plant density would be expteced. Brody *et al.* (1990) challenged this concept of the declining probability of infection with increasing plant density. In their spacing experiments, roots were severed, in spite of which the denser stands showed fewer infections. This result was attributed to a host-determined mechanism in which the denser plants developed an increased resistance.

The relationship between symptom severity and plant density was closely related to irrigation (Huisman and Grimes, 1989). Under a low (1) post-planting water regime, defoliation decreased with increasing plant density. With increased (3–5) irrigations, the highest percentage of defoliated plants occurred in the high density plots. Virulence (% defoliation in infected plants) was also greatest in the higher plant densities.

Field results in the CIS (former USSR), particularly under Uzbekistan conditions, are similar to those in the USA. Abrarov and Gubanov (1973), using early and mid-season cotton cultivars, increased plant density to 25×10^4 ha^{-1} from 10×10^4 ha^{-1} and obtained 0.8–1.0 t lint ha^{-1} overall increased yield with reduced *V. dahliae* infection. The initial picking, however, was 30% lower.

Abrarov and Ibragimov (1976) increased yield per hectare under disease conditions by increasing plant density to 50×10^4 ha^{-1} with only 2–3 bolls per plant. Abrarov and Ibragimov (1976) reported that at a plant density of 90×10^3 ha^{-1}, 43% of the plants' dry matter is in the 'generative organs' (flowers); at denser sowings, this percentage is reduced to 35%, but with increased yields and increased *Verticillium* resistance. Trushkina *et al.* (1974), Ibragimov *et al.* (1977) and Savov (1978a,b) also report increased yield and decreased wilt when planting density was increased. These effects are not confined to cotton. Williams (1975) and Shatrova (1977) found that *V. dahliae* or *V. albo-atrum* infection was reduced concomitant on a reduction in plant spacing in strawberry and aubergine, respectively. A multi-hybrid experiment studying intraspecific competition and *V. dahliae* infection in sunflowers (Sadras *et al.*, 2000) found that the percentage of symptomatic leaves accounted for 28–32% of control plot yields, while response to competition accounted for 60–75% of the residual variation. A negative relationship was shown between yield and response to competition in crops with *Verticillium*.

Physiology and Metabolism 7

Nutrition of the Pathogens

Many different general media have been used for diagnostic purposes and rapid production of inoculum (Gour and Dube, 1982; Gubanov *et al.*, 1985). Talboys (1960) used a prune extract, lactose, yeast (PLY) agar to enhance rapid melanin formation in the identification of *V. dahliae* and *V. albo-atrum*. Potato dextrose (PDA) and cornmeal agars (CA) induced typical dark resting mycelium of *V. albo-atrum*, whereas on Czapek-Dox agar, chains of discrete spherical segments gave rise to mycelial knots superficially resembling *V. dahliae* microsclerotia (Campbell and Griffiths, 1974). This effect on Czapek's has not been reported by other workers.

Carbon Sources

Species of *Verticillium* are capable of metabolizing a wide range of carbon sources, including glucose, fructose, arabinose, galactose, mannose, rhamnose, sucrose, maltose, cellobiose, ribose and xylose (Isaac, 1949; Malca *et al.*, 1966; Cooper and Wood, 1975). As a sole carbon source, arabinose and ribose delay growth of *V. dahliae*. Similarly, a period of adaptation is required for galactose (Malca *et al.*, 1966). *V. dahliae* and *V. albo-atrum* grew well on the fungal sugar trehalose and glycerol. *V. albo-atrum* grew better on the other polyols, D-mannitol, D-glucitol and ribitol; both species grew poorly on galactitol, D-xylitol, D- and L-arabitol and i-inositol. *V. dahliae* produced few or no microsclerotia on D-mannitol, D-glucitol or erythritol (Vega and Le Tourneau, 1971). Whitney *et al.*

(1969) showed that carboxymethyl cellulose could be metabolized as a sole source of carbon. Starch (Villalobos, 1972) and ethanol are weakly utilized by *V. albo-atrum*. Other utilizable sole sources of carbon include arbutin by *V. albo-atrum* (Demetriades and Emmanouil, 1971); glucose, glycerol, acetate and succinate by the P_1 strain of *V. dahliae* (Hartman *et al.*, 1972); D-galacturonic acid or D-gluconic acids by *V. albo-atrum* (Rattigan and Ayres, 1975); and starch by *V. albo-atrum*, *V. dahliae* and *V. nigrescens* (Ezruch and Babushkina, 1973). Simple ortho- but not metaphenols can be metabolized by *V. dahliae* (Le Tourneau *et al.*, 1976). Le Tourneau (1974) also demonstrated the production of alcohol-soluble sugars and sugar alcohols by *Verticillium* spp. from glucose (see Pfyffer *et al.*, 1990). Organic acids generally are unsuitable substrates, but *V. dahliae* grows well on succinate and acetate (Hartman *et al.*, 1972).

The discovery by Hartman *et al.* (1972) that *V. albo-atrum* [*sic*] (*V. dahliae*) could fix CO_2 is of considerable interest in relation to the presumed low carbohydrate, high CO_2 environment of the xylem vessel. Labelled carbon from $^{14}CO_2$ was incorporated into fungal protein (42%), nucleic acid (34%), lipids (3%), low molecular weight components (17%) and extracellular metabolites (<2%). The enzymes involved in anaplerotic CO_2 fixation, i.e. pyruvate carboxylase and phosphoenolpyruvate carboxykinase, were described by Hartman and Keen (1973, 1974a,b). The enzymes were found in the cytosol and the mitochondria. Oxalacetate was the primary fixation product. Hartman *et al.* (1972) showed that CO_2 was required for growth when glucose or glycerol were used as sole carbon sources.

Nitrogen Sources

Species of *Verticillium* can grow on a wide range of inorganic and organic nitrogen sources but to varying effect depending on type and concentration and whether supplementing other nitrogen compounds. Isaac (1949) showed that L-asparagine and peptone gave best growth of *Verticillium* isolates at 0.1% concentration; at 1.0% concentration, ammonium salts were better. In a subsequent paper (Isaac, 1957b), *V. albo-atrum*, *V. dahliae* and *V. nigrescens* were shown to give best growth on Czapek-Dox medium with supplemented nitrate. Working on *V. albo-atrum* Reinke et Berth [*sic*] (*V. dahliae*) grown at a phosphate-buffered pH of 5.9–6.3, Malca *et al.* (1966) showed that the ammonium ion was better utilized than nitrate. L-Arginine, L-alanine, L-serine, L-γ-aminobutyric acid (GABA), proline, L-lysine and the amides L-asparagine and L-glutamine, however, were better N sources than inorganic nitrogen. L-α-alanine gave a 71% increase in growth after 9 days compared with ammonium nitrate. L-Glycine, L-histidine and L-leucine gave poorer growth than ammonium N. Babushkina (1979) described a range of variants of *V. dahliae* from cotton and *V. tricorpus* from tomato induced respectively by *Aspergillus terreus* and *Fusarium sporotrichiella*. Lactose was the best carbon source and $NaNO_2$ [*sic*], asparagine and peptone generally were the best nitrogen sources. Growth of *V. dahliae* variants was less than the wild-type, and growth of *V. tricorpus* greater than the wild-type. Deshkova *et al.* (1977) working with *V. dahliae*, *V. albo-atrum* and *V. nigrescens* found, in agreement with Isaac (1949), that $NaNO_3$ and asparagine were the best N_2 sources, but found that NH_4NO_3 was unsuitable for growth (see also Mishina, 1974). A somewhat conflicting result for *V. dahliae* was shown by Duncan and Himelick (1986). Czapek-Dox amended with $(NH_4)_2SO_4$ resulted in reduced growth and sporulation. Unlike Malca *et al.* (1966), Duncan and Himelick (1987) using amino acids as *sole* nitrogen sources found that mycelial dry weight was greatest on L-leucine and, in descending order, L-valine, L-proline, L-serine and L-alanine gave similar results, followed by a similar group of basic amino acids, L-phenylalanine, L-histidine and L-arginine, and, finally, threonine, asparagine and glutamine. Conidia were produced best, in descending order, on asparagine, glutamine, arginine, alanine, proline, serine, histidine, leucine, valine, phenylalanine and threonine. Mussell (1973) found that protein could be hydrolysed to support *V. dahliae*, reflecting Malca *et al.*'s 1966 results for casein hydrolysate. Glucosamine, a normal plant constituent, inhibits radial growth of *V. albo-atrum* (White and Gadd, 1983).

A limited number of studies has been made on the ratio of carbon to nitrogen compounds. Interpretation of this ratio is difficult in relation to the sources of C and N and the influence of other culture factors. Selvaraj (1974a) compared growth of a single strain each of *V. albo-atrum* and *V. dahliae* under a range of C/N ratios. Maximum microsclerotial production of *V. dahliae* occurred at 30 mg ml^{-1} glucose and a C/N ratio of 50. Under soil conditions, the addition of glucose, sucrose and sodium nitrate 1% (w/w) led to increased production of viable microsclerotia and conidia. Addition of ribose led to a decline (Green and Papavizas, 1968). Costache *et al.* (1981) studied the effect of nitrogen sources on growth and sporulation of vascular pathogens *inter alia V. dahliae*. Ammonia-linked respiration in conidia of *V. albo-atrum* [*sic*] was reported by Tolmsoff (1969).

Other Compounds and Elements

In general, *Verticillium* species have no absolute requirements for vitamins, although Kaiser (1964a), Roth and Brandt (1964a) and Milton and Isaac (1967) found that isolates of *V. dahliae* growing on a glucose, nitrate, salts medium showed stimulated melanin synthesis and microsclerotia formation following the addition of biotin, thiamine, pyridoxine and inositol. Certain isolates were deficient for biotin for growth sporulation and microsclerotial production (Kaiser, 1964a; Milton and Isaac, 1967). The latter is a common requirement for UV-induced mutants which cannot grow on a minimal medium.

While *Verticillium* species may have a requirement for trace elements, the concentration required is usually provided by impurities in laboratory 'analar' grade chemicals and many sources of 'deionized' water. Brandt (1962) recorded a stimulation of melanin synthesis and microsclerotial numbers by thc addition of manganese. Similarly, Thorneberry (1973) reported stimulated growth of *V. dahliae* in aqueous Czapek-Dox medium with 0.15–0.2 p.p.m. zinc. Sodium fluoride at 1×10^{-3} M was claimed to stimulate *V. albo-atrum* growth in culture; at 2.5×10^{-3} M, growth was inhibited (Treschow, 1965).

Two references show an interesting relationship between *Verticillium* and phosphate. Zhivopistseva and Voronkov (1974) claimed that up to 12% of the total polyphosphates in the cotton cell could be accumulated in the mycelium of *V. dahliae*. Most of the polymers had more than 60 phosphate groups in the molecule. The content and form of the polyphosphates depended on age and pathogenicity of the isolate. Singh *et al.* (1982) reported that *V. albo-atrum* [*sic*], probably *V. dahliae*, was very active in solubilizing P from rock phosphate. In relation to the enhanced severity of wilt under saline conditions, Danti and Broggio (1997) showed that *in vitro*, NaCl from 0.05–0.35 M incorporated into solid or aqueous media, while reducing colony growth increased conidial differentiation and sporulation.

Atmospheric Gases

Early but simplistic experiments by Kessler (1966) showed that cultures of *V. albo-atrum* [*sic*] (*V. dahliae*) grown in limiting O_2 concentrations of less than 1–6% O_2 had reduced sporulation. At the highest concentration (6%), only occasional microsclerotia appeared and none at lower concentrations. In an atmosphere of 15% CO_2, no microsclerotia formed but sporulation was good. In aerated aqueous cultures sporulation was reduced sixfold compared with non-aerated cultures. Ioannu *et al.* (1977a,b) showed that radial growth of *V. dahliae* was unaffected at subatmospheric O_2 concentrations. Concentrations down to 0.5% O_2 gave a maximum growth rate but reduced microsclerotial numbers. Variants which had lost the capacity to form microsclerotia grew better in hypobaric oxygen conditions. Increased CO_2 concentration suppressed microsclerotia but enhanced radial growth, confirming Kessler's earlier results.

Pilkington and Heale (1969) earlier showed that O_2 uptake by *V. albo-atrum* was greater during the formation of the dark resting mycelium, a fact borne out by the lower oxygen demand per unit dry weight of hyaline variants. Ioannu *et al.* (1977a,b) similarly interpret increased radial growth of *V. dahliae* under low O_2 conditions to the suppression of microsclerotia. The above results are at variance with those of Luck (1954) who claimed that microsclerotia from *V. dahliae* from peppermint did not form at concentrations of CO_2 less than the 0.03% of normal air. Ioannu *et al.*'s results, however, showed that while O_2 concentrations were held at 17–18%, microsclerotia formed up to 12% CO_2. Conidia and radial growth were still 20 and 70% of maximum at 20% CO_2. Microsclerotia viability was unaffected by 3 months' exposure to low oxygen and high CO_2 (Ioannu *et al.*, 1977a,b). The results of

Kessler and Ioannu are consistent with the growth and limited sporulation observed in the xylem vessel lumen, an environment which is assumed to be oxygen-deficient.

Water Potential

Mozumder and Caroselli (1966) reported a linear decrease in radial growth of *V. dahliae* with decreasing water potential (Ψ). Conidial germination similarly fell with decreasing osmotic potential (Ψs) down to Ψs −45 bars (Mozumder *et al.*, 1970). Subsequently, Manandhar and Bruehl (1973) claimed that the fungus could grow at −100 bars controlled either osmotically or matrically (Ψm). These potentials are much lower than the host plant can tolerate (usually higher than Ψs or Ψm −15 bars). Congly and Hall (1976) showed that growth of *V. dahliae* actually increases with decreasing Ψs, reaching a maximum at −10 to −30 bars. The minimum Ψs permitting conidial germination, mycelial growth and sporulation was −100 to −120 bars (Ioannou *et al.*, 1977c). Radial growth on agar was maximal at Ψs values of −10 to −20 bars, agreeing in general with the results of Congly and Hall (1976). Growth as dry matter in aqueous culture (more comparable with the pathogenic environment), however, decreased with decreasing Ψs of the medium from −2 to −120 bars. Conidial production *in aqua* increased logarithmically with falling Ψs from −2 to approximately −50 bars, maximal at −50 bars. No or few microsclerotia were produced between −70 and −80 bars, even though radial growth and dry weight were reduced by only 60% at these values. In non-sterile soil experiments, maximum microsclerotia were produced in infected tomato stems at Ψ −32 bars and a temperature of 24°C.

In saturated soil or at Ψ −100 bars, microsclerotia were greatly inhibited. These results have relevance for flooding as a control measure to reduce inoculum (see Chapter 10).

pH

Conidial germination of *Verticillium* species appears optimal in the pH range 5.0–7.0. Below pH 3.0, little or no growth occurs (Puhalla and Bell, 1981). Maintenance of a constant pH was the single most important factor in maintaining good growth of *V. dahliae* (Malca *et al.*, 1966). Isaac (1949) obtained only poor growth of *Verticillium* spp. on unbuffered media containing ammonium salts or organic nitrogen sources; growth was restored on the addition of $CaCO_3$. pH values up to 10.0 can be tolerated by *V. albo-atrum* and *V. dahliae* (Isaac, 1949; Robinson *et al.*, 1957; Malca *et al.*, 1966; Selvaraj, 1971). With nitrate as a sole N_2 source in unbuffered media, the pH rises from an initial 6.0 to 9.0. Using Dox, phosphate-buffered Dox and PDA media, Isaac (1949) found that the initial pH optimum for *V. albo-atrum* was 5.3–7.2 and for *V. dahliae* 8.0–8.6. Robinson *et al.* (1957) claimed that these differences disappear on an adequately buffered medium. The use of strong phosphate buffers to counter pH drifts during culture growth may inhibit *V. dahliae* and other species (Bell *et al.*, 1971). Optimal growth of *V. dahliae* in aqueous media was between pH 5.9 and 6.3 (Malca *et al.*, 1966).

Temperature (see also Chapter 2)

V. albo-atrum grows better than other species at temperatures in the range 5–20°C, but growth ceases above 30°C (Edson and Shapovalov, 1920; Isaac, 1949; Devaux and Sackston, 1966). *V. dahliae* has an optimal temperature of 22–27°C, but limited growth may occur above 32°C. Kaiser (1964c) reported growth of *V. albo-atrum* [*sic*] (*V. dahliae*) up to 36°C. *V. nigrescens* may tolerate even higher temperatures. Devaux and Sackston (1966) found minimum, maximum and optimal temperatures for strains of *V. nigrescens* of 6, >35 and 28°C, respectively. Resting mycelium of *V. albo-atrum* was found only between 14 and 26°C, but with *V. dahliae* no microsclerotia were found above 30°C, although some growth occurred at 33°C (Devaux and Sackston, 1966). Wilhelm (1948) earlier reported few microsclerotia developing between 25 and 31°C. Bell *et al.* (1976a) showed that microsclerotia numbers were directly proportional to the number of hyphal fusions. At 30°C, hyphal fusions and microsclerotia in *V. dahliae* are dramatically reduced (Puhalla and Mayfield, 1974). Brinkerhoff (1969) examined *inter alia* the influence of temperature on microsclerotial development in abscinded *V. dahliae*-infested cotton leaves. Abundant microsclerotia formed in 2–5 days between 18 and 30°C and after 30 days at 5°C. No microsclerotia formed at 32°C. Optimal conidial germination temperatures depend upon species, strain and cultural factors. Wyllie and DeVay (1970a) reported that the conidia of the T9 strain of *V. dahliae* from cotton (P_1 strain of Joaquim and Rowe, 1990) germinated at 33°C; the non-defoliating SS4 (P_2) strain did not. Conidia germinate well between 22 and 27°C, those of *V. albo-atrum* alone, however, will not germinate at 30°C. *V. dahliae* conidia germinate optimally at 30°C when water availability is low, but this is reduced to 25°C (Mozumder *et al.*, 1970) when water is abundant. The thermal death points in soil for *V. albo-atrum* and *V. dahliae* after an exposure of 30 min were 52.5 and 57.5°C, respectively (Bollen, 1969).

Light

Experiments on the quantitative and qualitative effects of light on growth, conidiation and resting structure formation are difficult to interpret in the absence in many cases, of precise spectral wavelengths used, photoperiods, the relative energies involved and other cultural factors. Pethybridge (1916) originally observed that resting mycelium of *V. albo-atrum* was produced sooner in total darkness than in daylight. Heale and Isaac (1965) found a temperature interaction in that little resting mycelium was found in continuous light at 24°C and none at 15°C. Under alternating 48 h cycles of light and dark, banding of pigmentation occurred. *V. dahliae* formed microsclerotia evenly under continuous darkness or light. A report by Chaudhuri (1923) that zoning (dark pigmentation) of *V. albo-atrum* was independent of light or dark may have referred to *V. dahliae*.

V. tricorpus, similarly to *V. albo-atrum*, produced fewer resting structures at 24°C in continuous light than in darkness. Yellow pigmentation independent of light failed to develop at 15°C. In a poorly described experiment, Caroselli *et al.* (1964) claimed that 'green light' inhibited *V. albo-atrum* [*sic*] (*V. dahliae*) growth. More microsclerotia were formed in 'red light' and darkness. Kaiser (1962, 1964b) reported that conidial production of *V. albo-atrum* [*sic*] (*V. dahliae*) was greatly enhanced, but microsclerotia production completely inhibited by blue (478.8 μm) light. Similarly, continuous light stimulated conidial numbers but totally inhibited microsclerotia in single-spore isolates, a result at variance with the findings of Heale and Isaac (1965). Under blue light, an 'orange pigment' (probably carotene; see Isaac and Davies, 1955; Pegg, 1957; Valadon and Heale, 1964, 1965) formed. Biotin stimulated conidial production under all wavelengths but did not reverse blue light inhibition of microsclerotia. Alternating 18 h dark, 6 h light overcame the blue-light effect. The radiant energy at all wavelengths was 7000 ergs cm^{-2} s^{-1} ≡ 130 fc white fluorescent light. This result agrees in part with an earlier report by McClellan *et al.* (1955a) who found abundant sporulation in *V. albo-atrum* [*sic*] (*V. dahliae*) under blue or fluorescent white light but not under red or far red (>7000 Å). This wavelength was greater than Kaiser's (1964), cited as 631.6 μm, under which sporulation occurred. Marsh *et al.* (1959) in a general review stated that sporulation occurs more freely in light, and Leach (1962) that it is good in near UV.

Kaiser (1964b,c) found vegetative growth inhibited slightly at 478.8 μm. At a lower wavelength in the near UV spectrum 3200–4000 Å and under lower radiant energy, colony diameter of *V. albo-atrum* [*sic*] (*V. dahliae*) was greater than in complete darkness (Brandt, 1967). The effect was not a stimulation of growth but rather a reversal of dark-induced inhibition leading to a lag phase after 12 days. Brandt and Reese (1964) earlier claimed the existence of an uncharacterized diffusible morphogenetic factor (DMF) which stimulates melanin formation and inhibits hyphal growth and sporulation. Inhibition of melanin synthesis by near UV is thought to be by a suppression of DMF and hence increased growth (Brandt, 1964b, 1967). The original report that near UV (320–400 μm) completely inhibited microsclerotia was by Leach (1962). It is the last of Isaac's (1949) five developmental stages of microsclerotium formation which is inhibited by light or near UV.

Strain differences to UV sensitivity arc wcll known. Robinson *et al.* (1957) found a strain of *V. albo-atrum* was more sensitive to UV than one of *V. dahliae*. The survival rates of 15 isolates each of *V. albo-atrum* and *V. dahliae* to 30-s exposure to UV light was 0.02–0.24% and 8–46%, respectively, illustrating the UV screening effect of melanin. A lucerne isolate of *V. albo-atrum* was intermediate in sensitivity, indicating the somewhat different nature of the lucerne pathogen. Cultures of *V. dahliae* irradiated with red light (600–680 μm) showed reduced colonization of cotton following inoculation, compared with non-radiated cultures (Akhmedzhanov *et al.*, 1990). Red-light irradiation reduced the level of protein–lipopolysaccharide (PLP) toxin in aqueous culture. The level of the antifungal isohemigossypol in infected cotton, thought to be induced by PLP, was also greatly reduced.

Enzymes

The role of enzymes in the biology of *Verticillium* spp. is so extensive that the subject cannot be confined to a single chapter. In this chapter, enzyme references relate for the most part to *in vitro* studies on the organism; many papers, however, include reference to *in vitro* and *in planta* studies or as a subordinate part of a different study. For these reasons, enzymes are also discussed in the chapters on Pathogenesis, Resistance, Morphology, Taxonomy and elsewhere.

The extensive literature on pectolytic and cellulolytic enzymes has been considered mostly under Pathogenesis (Chapter 8). Detailed reviews have also been presented by Dimond (1955, 1970), Pegg (1981a, 1985), Beckman (1987), Heale (1988) and Bell (1992b). Several authors writing in the 1980s and 1990s on enzymes appear unaware of the importance of substrate composition and concentration, and hence the problem of enzyme induction and substrate and product repression, considered in the seminal paper by Cooper and Wood (1975). Many papers on pectolytic enzymes lack novelty and are repetitions of much earlier studies.

Sagdieva *et al.* (1974) showed that gossypol added to the culture medium suppressed pectin-lyase (PL) activity of *V. dahliae*. Spore concentrations of 10^5 ml^{-1} of medium in shake cultures led to mycelial growth, but not at 10^6 ml^{-1} of medium and above (Sagdieva, 1974). A series of papers from India and the CIS describe variously PL, PG and cellulase production by *V. dahliae*, mostly from cotton, and *V. albo-atrum*, referring to the effect of substrate and other factors (Dube and Mathur, 1975, 1977; Vasil'eva and Gladkikh, 1976; Mathur and Dube, 1978; Gladkikh *et al.*, 1979; Dube and Mathur, 1982). Gupta (1973) described the stimulation of endoPG in *V. albo-atrum* [*sic*]. Catechol, catechin, gallic acid, orcinol, phloroglucinol and ruffianic acid had neither stimulatory nor inhibitory effects on pectolytic activity from *V. dahliae* and *V. albo-atrum*. Oxidation products of the phenolic compounds reduced enzyme activity 10–30% (Mathur and Dube, 1982). Multiple forms of pectolytic enzymes secreted by *V. albo-atrum* were described by Gupta (1977). Three exo-PL components were produced on a polypectate medium and four on pectin. Three components of PME also found were in the polypectate medium and four in the pectin, all separated on Sephadex G100. Vasil'eva *et al.* (1984), using ultra thin-layer immunoelectrophoresis on polyacrylamide gels, similarly described several isozymes of PG from a cotton isolate of *V. dahliae*. Endo-PG from *V. albo-atrum* [*sic*] (*V. dahliae*) was purified and characterized by Wang and Keen (1970).

Several attempts have been made to correlate pectolytic activity and isolate virulence (Kostroma, 1973; see also Chapter 8). Constitutive levels of endo-PG and PL were higher in a virulent cotton strain (511) of *V. dahliae* than in a weak strain (1211). Phenolic extracts from resistant cotton cv. Tashkent 1 inhibited PL more than comparable extracts from a susceptible line cv. 108F. The results were confirmed subsequently in two races and 11 strains of *V. dahliae*, where it was found that endoPG activity of avirulent strains, unlike virulent ones, were unaffected by substrate composition. The hypothesis was advanced that variations in varietal resistance of cotton might reflect differences in endogenous pectin content. Howell (1970) examined differential enzyme synthesis by haploid and diploid forms of *V. albo-atrum*.

Enzymes concerned with the degradation of substrates other than pectin substances and cellulose include variously galactanase, xylanase, arabinase, glucosidase, glucanase, chitinase and protease. Keen *et al.* (1970) described the induction and repression of D-galactosidase from cultures of *V. albo-atrum*. *V. albo-atrum* possesses at least two galactanases, an exo-xylanase and an endo-arabinase (Cooper and Wood, 1975). Mussell and Strouse (1971) showed that *V. dahliae* secreted extracellular proteases capable of hydrolysing protein when used as a sole carbon source. Vessey and Pegg (1973) and Young and Pegg (1982) described endo-chitinase, 1,4-β-acetylglucosaminidase, 1,4-β-glucosidase and exo-1,3-β-glucanase in *V. albo-atrum*. Hydrolysis of purified *V. albo-atrum* cell wall preparations was complex, involving host (tomato) and fungal enzymes in combination (Young and Pegg, 1982). Adylbekov *et al.* (1991) characterized a 1,3-β-glucanase from *V. dahliae* present in culture filtrates and mycelium. It was claimed that the purified enzyme could hydrolyse cell wall polysaccharides of *V. dahliae*. Bahkali (1991) confirmed Cooper and Wood's (1975) xylanase results and showed that glucose and xylose repressed xylanase synthesis, while cellulose, cellobiose and carboxymethylcellulose induced some enzyme stimulation. In a comparison of *V. albo-atrum*, *V. dahliae*, *V. tricorpus*, *V. nubilum* and *V. nigrescens*, *V. dahliae* had the highest levels of cellulolytic enzymes and the highest pathogenicity to tomato (Bakhali, 1989a). Bakhali (1989b) described the induction of endo-PG, exo-PG and PL in cultures of *V. tricorpus* isolated from wilted potato plants. Bakhali (1987) described a similar result on isolated date palm (*Phoenix dactylifera*) cell walls. The production of PL only was proportional to mycelial growth rate.

Low concentrations of enzyme substrate enhance enzyme synthesis (Cooper and Wood, 1975). Conversely, high concentrations of product or substrate repress synthesis (Talboys and Busch, 1970). Glucose, a common carbon source of the medium, inhibits most polysaccharases at its nutrient concentration (Talboys, 1958a; Keen and Erwin, 1971; Selvaraj, 1974). Sucrose, lactose and galactose act as similar repressors (Talboys, 1958; Bell *et al.*, 1971; Cooper and Wood, 1975). Other factors influencing enzyme activity in *Verticillium* species are: host tissue (Bell *et al.*, 1971; Cooper and Wood, 1975), nitrogen source, e.g. peptone (Talboys and Busch, 1970), Ca^{2+} and PO_3^{3+} ions (Talboys and Busch, 1970; Bell *et al.*, 1971; Mussell and Strouse, 1972; Selvaraj, 1974), pH (Cooper and Wood, 1975) and the agitation of aqueous media (Keen and Erwin, 1971). Whereas most enzymes concerned with basic metabolism are constitutive, many extracellular (secretory) enzymes are induced by specific substrates. The use of pectin-deficient mutants and the dependence on specific substrates has been studied in detail by Durrands and Cooper (1988a,b,c).

Hartman and Keen (1973, 1974a,b) described the incorporation of ^{14}C from $Na_2{}^{14}CO_3$ into aspartate or citrate by pyruvate carboxylase and phosphoenolpyruvate carboxykinase in cell-free extracts from *V. dahliae*. The presence of arginine decarboxylase from *V. dahliae* (Khan and Minocha, 1989) established a putrescine biosynthetic pathway formerly known only in higher plants and bacteria. Kandrov *et al.* (1990a,b) and Vasiliev *et al.* (1991) purified a casein kinase II (CK_2) considered to be important in the transition of intercellular signals. The enzyme consisted of three peptide chains of 38, 41 and 53 kDa. Mycelium of *V. dahliae* contained more CK_2 than bud cells. The enzyme phosphorylated serine and threonine residues from casein using ATP and GTP. *V. dahliae* grown in aqueous culture with a protein source produced an extracellular alkaline serine protease. Dobinson *et al.* (1997) identified by N-terminal amino acid analysis, a 30 kDa protein co-purifying with the serine protease in an 80-fold purification, as a trypsin-like protein. The purified enzyme hydrolysed Nα-benzoyl-DL arginine *p*-nitroanilide hydrochloride (BAPNA), a reaction inhibited by leupeptin, confirming the trypsin-like nature of the enzyme.

Many references report correlations between isolate virulence and enzyme activities. When cotton leaves of line S 4727 were added to substrates containing homogenates of av and v mycelial [*sic*] strains and a virulent microsclerotial strain all showing acid phosphatase activity, all showed an increase in bound acid phosphatase. In the virulent strains only, free acid phosphatase increased (Tukeeva and Tukeyeva, 1977; see also Zhivopistseva and Voronkov, 1974). A pathogenic isolate of *V. dahliae* from cotton had higher G6 phosphate dehydrogenase and malate dehydrogenase than a less pathogenic isolate (Miryakubova, 1977). In tests with four isolates of the 'Bukhara' strain of *V. dahliae*, catalase and invertase activities declined in the mycelium with age but increased in the culture medium. Addition of P or S salts to the medium and the substitution of NH_4NO_3 or $CO(NH_2)_2$ for $NaNO_3$ greatly reduced catalase but slightly increased invertase activities. Guzhova *et al.* (1976) reported that mitochondria of a pathogenic form of *V. dahliae* were more highly phosphorylated than those of non-pathogenic forms. Huang and Mahoney (1999) purified to homogeneity an endo-PG from *V. albo-atrum* culture filtrates. The basic protein had a molecular weight of 37 kDa, a pI of 8.6 and contained 1.7% carbohydrate. The enzyme lost activity above 30°C in the absence of substrate and was sensitive to changes of ionic strength at low salt levels. Maximum activity was at pH 4.6–5.0 and 46°C, and was stable in the pH range 3–11 at 30°C.

Melanin Biosynthesis

Melanin occurs as electron-dense granules in the outer cell wall of hyphae and microsclerotia and protects against parasitism, dehydration and toxic chemicals (Bell and Wheeler, 1986) and presumably against UV radiation. Early studies (MacMillan and Brandt, 1960; Heale and Isaac, 1964) showed that 3,4-dihydroxyphenylalanine (DOPA) and catechol, induced a dark pigmentation and agar staining in cultures of *V. dahliae* and *V. albo-atrum*. Dihydroxyphenols, catechol and DOPA, but not tyrosine, reversed the UV-induced inhibition of melanin (Brandt, 1964b; Heale and Isaac, 1964). Bell and Puhalla (1974) described naturally-occurring cyclic DOPA which they then considered to be the precursor of melanin. Gafoor and Heale (1971a) suggested that catechol was a precursor of *Verticillium* melanin, assuming a shikimate→DOPA→indole 5,6-quinone→5,6-dihydroxyindole pathway. Bell *et al.* (1976a,b), however, showed that the melanin was an oxidized polymer of tetrahydroxy binaphthalene. Using UV 254 nM-derived mutants of *V. dahliae* with normal microsclerotia but abnormal pigmentation, four types were obtained – albino (*alm*-1–4) and brown microsclerotia (*brm*-1–4). Mutants *brm*-1 and *brm*-3 accumulated a metabolite shown by Bell *et al.* (1976a) to be (+)scytalone (3,4-di-hydro-3,6,8-trihydroxy-1(2H) naphthalenone), which restored normal melanin synthesis. The full sequence of melanin biosynthesis was shown to be: condensation of five acetate residues to tetrahydroxy naphthalene, which after dehydrogenase activity was converted to (+)scytalone. Butler *et al.* (1988) described in *V. albo-atrum* and *V. dahliae* a dehydratase which converted scytalone to trihydroxynaphthalene (THN). A second dehydrogenase converts THN to (–)vermelone (dihydro-dihydroxy 1(2H) naphthalenone). (–)Vermelone under the action of a further dehydratase is converted to dihydroxy naphthalene, two molecules of which condense to tetrahydroxy binaphthalene (Bell *et al.*, 1976a,b; Stipanovic and Bell, 1976, 1977; Wheeler *et al.*, 1976, 1978). Bell *et al.* (1976a,b) and Wheeler *et al.* (1978) found, like Heale and Isaac (1964), that *alm-1* converted catechol and DOPA to a black pigment, but significantly this did not resemble normal *V. dahliae* melanin. Bell *et al.* (1976b) suggested that an oxidase, possibly a laccase, oxidized THN to melanin. Melanin derived from a bi-naphthalene quinone would be structurally very different from the indole quinone–dihydroxyindole polymer usually associated with fungi.

Confirmation of the pentaketide pathway and a dihydroxynaphthalene origin for melanin in *V. dahliae* was provided by Tokousbalides and Sisler (1979) and Lazarovits and Stoessl (1988). The fungicide tricyclazole, a pentaketide inhibitor, led to the formation of melanin shunt products, 2-hydroxyjuglone and flaviolin. Increasing tricyclazole led to reduced 2-hydroxyjuglone and increased flaviolin, excluding the possibility of shikimate-derived melanin. The fungicide carpropamid (2-dichloro-*N*-[1-(4-chlorophenyl) ethyl]-1-ethyl-3-methyl (cyclopropanecarboxamide)) strongly inhibited pigmentation of, *inter alia*, *V. dahliae*, depositing white crystals of scytalone and smaller amounts of vermelone in the culture (Kurahashi *et al.*, 1998). Carbendazim and tricyclazole (TCZ) at 0.1 μg ml^{-1} inhibited microsclerotial production and induced albino 'variants'. TCZ at 0.5 μg ml^{-1} initially inhibited melanin biosynthesis resulting in a reddish pigmentation. Melanin variants were reversible after transfer to PDA but albino variants remained stable after many transfers. Albino variants were also obtained from fungicide-treated *V. dahliae*-infected cotton. Hawke and Lazarovits (1995b) confirmed the role of TCZ in inhibiting the reductase of the pentaketide melanin pathway. Non-melanized (TCZ-treated) microsclerotia were smaller and showed delayed germination; when exposed to 254 nm UV, all were killed after 6 h whereas radiated melanized microsclerotia were unaffected after a 48 h exposure. Work in the CIS by Tyshchenko *et al.* (1990) confirmed the origin of melanin by the dehydrogenation of tetraoxy-1-1-binaphthalene. These authors postulated the existence of stable, free, pigment bi-radicals which deposit on hydrophobic membranes following a decrease in cellular pH. The prime site of tricyclazole action was inhibition of the reductive conversion of THN to vermelone. Brandt (1962) found that manganese stimulated melanin synthesis in microsclerotia, possibly as an oxidase co-factor. Shevtsova *et al.* (1982, 1983) reported that stages in melanin biosynthesis in *V. dahliae* and *V. tricorpus* were under the control of homologous systems of nuclear genes.

Other pigments associated with *Verticillium* are carotenoids (Valadon and Heale, 1964, 1965). In *V. albo-atrum*, lycopene is converted sequentially to γ-carotene, torulene and neurosporaxanthin. γ-Carotene may also give rise to β-carotene. Carotenoid pigmentation is usually associated with variants of wild-type strains (Pegg, 1957; Valadon and Heale, 1964, 1965) and exposure to UV or white light. Isaac and Davies (1955), however, reported orange pigmentation in a hyaline species, *V. intertextum*, closely similar to the hyaline variant from *V. albo-atrum* described by Pegg (1957).

Metabolism

Primary metabolism

Under aerobic conditions, glucose is metabolized in *Verticillium* spp. via the EMP or the hexose monophosphate pathways (Brandt and Wang, 1960; Malca *et al.*,

1968). Respiration involves the tricarboxylic acid (TCA) cycle during glucose catabolism (Brandt and Wang, 1960; Tokunaga *et al.*, 1969; Hartman and Keen, 1973). The glyoxalate shunt operates whether acetate or citrate is metabolized but is repressed by glucose (Tokunaga *et al.*, 1969; Hartman *et al.*, 1972). Many enzymes have been described for the TCA and glyoxalate cycles, and cytochrome enzymes have been found in mitochondria and microsomes.

Pilkington and Heale (1969) found that both immature (hyaline) and mature dark hyphae of a wild-type dark strain of *V. albo-atrum* had a higher O_2 uptake than immature or mature hyphae from a hyaline variant. Dark cultures were cyanide insensitive compared with hyaline. The range of carbon substrates utilized by *Verticillium* and the enzymes involved have been described earlier; see also Chapter 8. Bechtol and Thorneberry (1966) found that *V. albo-atrum* [*sic*] (*V. dahliae*) from cotton used in addition to glucose, glucose-6-phosphate, xylose, acetate, alanine, glutamate and leucine as respiratory substrates. Heale and Gupta (1970) demonstrated that cellobiose could be used as a sole carbon source by *V. albo-atrum* and suggested that cellobiose phosphorylase found in mycelial extracts was instrumental in its use as a respiratory substrate.

Nitrogen metabolism

In *Verticillium* spp., nitrate is reduced to nitrite, hydroxylamine and ammonia prior to assimilation (Puhalla and Bell, 1981). Ammonia is converted to the amides glutamine and asparagine and some is fixed with α-ketoglutaric acid to form glutamic acid (Malca *et al.*, 1968) which is transaminated further (Hartman and Keen, 1973). Mussell (1973) found inducible proteolytic enzymes in 32 isolates of *V. dahliae*, stimulated in culture by gelatin, galacturonic acid or low concentrations of glucose. Kasymova *et al.* (1990) reported the presence of an 18–20 kDa protease inhibitor in seed of resistant cotton which reduced *V. dahliae* protease activity following infection. *V. dahliae* from sunflower grown with sunflower cell walls as a sole N_2 source secreted three proteases: PI, PII and PIII. PI and PII in lesser amounts were considered to be autolysis products of PIII. This enzyme was an endopeptidase with a mol. weight of 12,750 Da and optimum activity between pH 7.5 and 9.5. The active site of the enzyme probably contained free SH groups (Lambert and Pujarniscle, 1984). Adenine and N_6-benzyladenine or kinetin at 10 mM added to Czapek-Dox broth led to a fourfold stimulation of growth. Protein kinase activity was greatly enhanced and also responded to cAMP – a minimal response in non-supplemented medium (Atmar *et al.*, 1976).

Few studies have been carried out on DNA base ratios. Shmotina *et al.* (1971) found very similar GT levels in *V. albo-atrum* (59.5%) and *V. dahliae* (58.7%). Small differences in DNA structure were found in three isolates of *V. dahliae* by Guseinov and Runov (1974). Guseinov *et al.* (1974c) found an AT ratio of 1.19 in isolates from susceptible cotton and of 1.7 in isolates from resistant.

Secondary metabolism

Lipids

Intercellular lipids of *V. dahliae* include cardiolipin, monoglycerides, sterols, free fatty acids and triglycerides (Walker and Thorneberry, 1971). Non-pathogenic isolates contained more triglycerides than pathogenic ones, whereas pathogenic ones had nine- to ten-fold increased fatty acids than non-pathogenic isolates (Kuznetsov *et al.*, 1977). Correlations of metabolites and virulence based on relatively few isolates in the absence of hybridization studies may be of little value as markers in identifying absolute differences between particular strains or isolates. Safiyazov (1977) apparently studying quantitative and qualitative differences in lipids between a weakly virulent *V. albo-atrum* and a highly virulent *V. dahliae* (mycelial and microsclerotial strains of *V. dahliae* [*sic*]) found no differences in unsaturated fatty acids, but more linolenic acid in *V. albo-atrum*. *V. dahliae* contained laurinic, pentodecanic and cerotinic acids which were absent in *V. albo-atrum*. This example illustrates the need for replication of isolates of precisely identified species. Lipids and polyketides were identified in *V. dahliae*, *V. tricorpus* and *V. nigrescens*. Lipid metabolism is seen as an important prerequisite for the synthesis of the protein–lipopolysaccharide toxin (Stepanichenko *et al.*, 1990). Razakov *et al.* (1985), using high and low resolution mass spectroscopy identified β-sitosterol, stigmasterol and cholesterol from *V. dahliae*. The neutral lipids and pentaketides of *V. dahliae* were described by Ten *et al.* (1977b) and a phytotoxic naphthoquinone pigment [PKZh1] (Ten *et al.*, 1977a). In a comparative study of *V. albo-atrum*, *V. dahliae* and *V. tricorpus* metabolites by thin-layer chromatography (TLC), 13 unspecified metabolites not characteristic of any individual species were produced. Ergosterol was formed by most isolates of the three species and (+)scytaline by a single isolate of *V. dahliae* (Williams *et al.*, 1992).

Proteins and peptides

Balandina and Shvetsova (1977), studying several cotton strains (isolates?) of *V. dahliae*, claimed that increased virulence was correlated with increases in total protein and in components in the globulin, albumin and glutenin groups. Highly virulent clones were distinguished by proteins with a higher thial content. In a wholly unsubstantiated paper by Rubin *et al.* (1977), the claim was made that proteins with malate dehydrogenase from *V. dahliae* and a compatible host (susceptible cotton line 4727) following extraction, 'hybridized'. Such hybrid proteins only showed enzymic activity when host and pathogen were compatible. In the case of incompatible isolate and cultivar, newly formed [*sic*] proteins had no enzyme activity. A claim by Ibragimov (1990) that 'plasmid DNA from *V. dahliae* combined with the infected cotton nuclear genome' [*sic*] and regulated cotton genes at the translational level has received no confirmation or support in the scientific literature. In a cytofluorometric study of DNA and proteins in cotton *V. dahliae* isolates differing in pathogenicity and biochemistry, Safiyazov and Shaabdurakhmanova (1974) found correlations between cell fluorescence virulence and nuclear polyploidy.

Ten *et al.* (1981, 1990) described a range of peptidic phytotoxins from cultures of *V. dahliae* from cotton; a cyclic hexapeptide, verticilline (a peptide) and a 27 amino acid polypeptide (PKG1). Insufficient structural and chemical details of these, and the peptides described by Nachmias *et al.* (1985, 1987) and Buchner *et al.* (1982) exist to be able to make a meaningful comparison or condensation of the claims from the different laboratories. Andreev and Serebryakova (1974) carried out an electrophoretic analysis of proteins from several species of *Verticillium*.

Ten *et al.* (1990) claim that the different peptides when introduced into cotton affect photophosphorylation and ion transport. Lipid and naphthoquinone-type compounds were involved in peroxide oxidation. The authors suggest a chemical inter-relationship between toxin formation and pentaketide synthesis and melanin formation. The behaviour of isolates of the same pathogen but from different hosts in relation to *in vitro* and *in vivo* metabolite production has yet to be clarified. Studies from different laboratories are rarely made on identical media or involve organisms grown under precisely the same cultural conditions. Moreover, there is little attempt on the part of many workers to seek a common identity with a previously described compound. Thus 'unique' differences *per se* claimed for a particular molecule may merely reflect differences in the cultural conditions or inadequacies in the chemical isolation and identification.

Growth-regulating substances

Indolylacetic acid (IAA) at 0.04 $\mu g\ ml^{-1}$ medium was produced by a tomato isolate of *V. albo-atrum* with sodium nitrate as the sole N_2 source in Czapek-Dox medium (Pegg and Selman, 1959). Wiese (1968) and Bhaskaran (1972) reported IAA production by *V. dahliae* on a medium containing L-tryptophan. Pegg (1987) described enzymes in the shikimate pathway, chorismate mutase and anthranilate synthetase, in the synthesis of IAA by *V. albo-atrum*. Ethylene was synthesized by *V. albo-atrum* via methionine added to the medium (Pegg and Cronshaw, 1976). Tzeng and DeVay (1984) found that ethylene production by *V. dahliae* grown on aqueous Czapek's medium was produced after addition of L-, D- or DL-methionine, DL-ethionine, or α-keto-γ-methyl butyric acid (KMBA). Exposure to light and brief heating or the addition of riboflavin stimulated ethylene release. Riboflavin in the light, but not in the dark, was fungitoxic. The P_2 (SS4) non-defoliating cotton strain produced substantially more ethylene than the P_1 defoliating strain – a converse reaction to *in vivo* ethylene production in cotton infected by the two strains. Ethylene at concentrations up to 35 v.p.m. had no apparent effect on growth sporulation or the production of microsclerotia *in vitro* (Ioannu *et al.*, 1977a).

Tomato plants infected by some isolates of *V. albo-atrum* may show a small but significant increase in stem height compared with healthy controls. Gibberellin-like activity has been shown by cultures of many isolates of *V. albo-atrum*, *V. dahliae* and *V. tricorpus*, which could be a basis for this effect (Aubé and Sackston, 1965).

Heteropolymers (see also Chapters 3 and 8)

Most studies have concerned extracellular substances related to cell wall synthesis, as either precursor molecules, analogues or degradation products. Cultures of *V. albo-atrum* and *V. dahliae* secrete large quantities of complex polymeric substances containing lipids, proteins and polysaccharides (Zel'tser and Malysheva, 1966; Malysheva and Zel'tser, 1968; Keen and Long, 1972; Keen *et al.*, 1972; Asamov *et al.*, 1975; Nachmias *et al.*, 1982b). The complex molecule (PLP) had a large molecular weight of approximately 3×10^6 consisting of 70% polysaccharide, 15% protein and 15% lipid (Keen and Long, 1972) and was unstable. Hydrolysis of the polysaccharide yielded galactose, mannose and galacturonic acid (Keen and Long 1972), similar in proportion to the alkali-soluble fraction of *Verticillium* cell walls (Wang and Bartnicki-Garcia, 1970). The *precise* role of the PLP in pathogenesis or its structure in relation to different culture substrates has yet to be determined. As with other molecules produced by *Verticillium*, definitive differences (or common identities) between different substances described in the literature have still to be established. Moreover, it is not clear whether the PLP is functional as an entire (stable) complex, or whether individual components are the active principals, or whether independent similar molecules are secreted as constitutive or media-dependent compounds. Thus Mussell (1972) described toxic proteins secreted by cotton isolates of *V. albo-atrum* [*sic*] (*V. dahliae*). Asamov *et al.* (1975) found as apparently separate molecules, a polypeptide and an oligosaccharide from *V. dahliae* which affected membrane function. Buchner *et al.* (1982) attributed the activity of 'the' PLP from *V. dahliae* from potato to a glycopeptide. Subsequent work by the Israel group (Nachmias *et al.*, 1985, 1987) considered the active toxic principle from culture fluids to be peptides which conferred specificity on races 1 and 2 of *V. dahliae*. To date, no-one has attributed a specific role to lipids other than as constituents of the PLP. Many reports of uncharacterized heteropolymers produced in *Verticillium* cultures have appeared in the literature (Porter and Green, 1952; Green, 1954; Caroselli, 1955; Kiessig and Haller-Kiessig, 1957; Talboys, 1957; Krasil'nikov *et al.*, 1965, 1969; Pegg, 1965). When *Verticillium* is grown on sucrose, oligosaccharides composed of fructose units are formed (Le Tourneau, 1961; Choy and Unrau, 1971). These compounds form from transglycosylation from sucrose but are not produced from glucose or fructose substrates. Polymeric fructosans have been reported from *V. albo-atrum* by Stoddart and Carr (1966) and Cronshaw and Pegg (1976), but would not survive the severe acid hydrolysis used in many carbohydrate studies. Le Tourneau (1958) isolated a glycoprotein from old *V. dahliae* cultures which on hydrolysis yielded glucose, galactose, mannose and arabinose in association with a protein content of 10–16%. This molecule, probably a cell wall precursor or degradation product, may well have been described by workers 15–20 years later and ascribed various properties.

Elicitors

Very few reports have occurred of characterized elicitors from *Verticillium*. Zaki *et al.* (1972b) showed that the toxin PLP from *V. dahliae* cultures elicited phloroglucinol-reactive compounds in cotton stem segments in amounts proportional to the concentration used. Crude culture filtrates of *V. albo-atrum* and extracts of cell walls and cytoplasm were used by Hutson (1978) to elicit the sesquiterpenoid rishitin in tomato. In a more comprehensive study, Woodward and Pegg (1986) found that different cultural conditions affected the type of elicitor produced. Large quantities of rishitin were induced in tomato stem segments by a high molecular weight fraction of *V. albo-atrum* culture filtrate from a 33-day shake culture. A different elicitor found in 14-day stirred cultures was shown to consist of approximately nine molecules of glucose, mostly 1,3-linked but with some 1,4-linkages. The low molecular weight elicitor induced six antifungal compounds in tomato, the principal one being rishitin. Homogenized mycelium of *V. dahliae* induced 5- to 500-fold increases in the benzophenanthridine alkaloid sanguinarine (23 mg l^{-1}) in cell suspension cultures of *Papaver bracteatum* (Cline and Coscia, 1988). A protein elicitor from *V. dahliae* at 10 μg per 15 ml of incubation cultures induced phytoalexin production in cotton cell suspension cultures. The addition of oxalate at 0.2–2.0 mM led to a tenfold stimulation of phytoalexin (Davis *et al.*, 1990). Koike *et al.* (1992) found that an elicitor from *V. albo-atrum* cultures with a molecular weight >12 kDa induced higher levels of medicarpin in resistant than in susceptible tissue-cultured calli. Low molecular weight fractions (<3.5 kDa) induced only low levels with no genotype specificity. A heat-stable glycoprotein from *V. dahliae* (*ex* sugar beet) was isolated from culture filtrates by Davis *et al.* (1998) which induced hemigossypol in cotton cell cultures. A 65-kDa glycoprotein purified by lectin (canavalin A) affinity chromatography and sodium dodecylsulphate–polyacrylamide gel electrophoresis (SDS–PAGE) gave a higher specific induction. Protease (pronase and papain) hydrolysis of crude filtrate and purified glycoprotein eliminated elicitor acitivity. Deglycosylation of the 65-kDa molecule by peptide *N*-glycosidase F resulted in no loss of activity and established the protein as the elicitor moiety. Some evidence was presented to suggest that a high molecular weight elicitor of >100 kDa was an aggregate of the 65-kDa molecule. It was claimed that the 100-kDa elicitor was unrelated to the 197-kDa lipoprotein polysaccharide of Meyer *et al.* (1994), one of five proteins which had a mass of 62 kDa. Reference was made to a further low molecular weight approximately 14-kDa elicitor protein. The authors, while emphasizing that the 65-kDa protein (53 kDa after glycosylation) unlike the crude filtrate did not cause an H_2O_2 oxidative burst in cotton cells, made little attempt to identify or contrast their results with other studies on *Verticillium*.

It is clear from the foregoing that several molecules of different size and composition with *inter alia* phytoalexin-inducing properties exist in *Verticillium* culture filtrates. The time is long overdue for a consolidation of elicitor studies to establish:

1. How substrate dependent are the molecules described?
2. How far do molecular sizes and composition reflect purification techniques or partial hydrolysis products? In this context, there is a need to account for molecules described by other work using the same species (but see 1, above).
3. Do differences exist between *V. albo-atrum* and *V. dahliae* or between strains of these species?
4. What are the multiple roles of culture filtrate metabolites – toxin versus elicitor, or both, depending on the assays used?

The question most in need of addressing is how far cultural metabolites reflect the pathogen–host reaction *in planta*?

Detoxification of plant secondary metabolites

Virulent cotton isolates of *V. dahliae* detoxify sanguinarine to dihydrosanguinarine (Howell *et al.*, 1973). A tomato isolate of *V. albo-atrum* was shown to possess an inducible β-glucosidase which hydrolysed a single glucose unit from the alkaloid α-tomatine to the less inhibitory β2-tomatine (Pegg and Woodward, 1986). This was confirmed for *V. dahliae* by Sandrock and Van Etten (1998) which was claimed to account for the high tolerance of this species to α-tomatine (ED_{50}300 μm). Ravisé and Chopin (1981) reported that one O-glucoside and five flavone C-glycosides were less toxic to *V. albo-atrum* producing a β-glucosidase than to other fungi lacking this enzyme.

Pathogenesis 8

Introduction

The unique lifestyle of the vascular pathogen, confined to the fluid environment of the xylem vessel and exerting its effect on the host's physiology indirectly, has been the stimulus for much research and scientific controversy over many years. Why the pathogen is confined to the xylem vessels and only rarely found in xylem parenchyma until after the death of the host is still not known, and opinions on this are largely speculative. Prokaryote vascular pathogens with no infective energy behave quite differently by breaking out of the vessels and spreading indiscriminately in root and shoot parenchyma. *Verticillium* spp. rarely leave the vessel until the death of the surrounding tissue or host. The classic debate on wilt pathogenesis has been concerned with the identification of the cause of wilting. In the extensive literature which has appeared over the last 50 years, two notable presumptions have persisted. It has been assumed by many writers that the whole spectrum of symptoms seen in the wilt syndrome are due to a single causal factor and that all wilt diseases, involving different pathogens and widely differing hosts, involve a common mechanism.

An addditional fundamental problem in a chapter on pathogenesis is whether some of the responses to infection by a virulent pathogen constitute essential features of pathogenicity or represent ineffectual non-specific resistance responses. Tyloses, for example, found commonly in the heartwood of healthy trees, might be seen as obstacles to systemic colonization by hyphae or spores, or alternatively to water conduction, or both! A similar ambiguity exists over the origin and function of vascular browning. The term *function* used in many publications implies a *teleological relationship* and would be better substituted by *effect*. The answer to these problems begs the whole question of the nature of wilt resistance.

Vascular Colonization

Colonization of the xylem is effected by mycelial growth but more rapidly by blastoconidia or phialoconidia from single-celled phialides (Talboys, 1962; Buckley *et al.*, 1969; Tolmsoff, 1973). In some instances, entry into the xylem is effected directly from the soil through broken major or minor roots, of which up to 20% may be found in a healthy plant. In this way, hyphae or conidia, together with a variety of prokaryotes, may enter the xylem passively (Pegg, 1985). Garber (1973) described in cotton the direct entry of *V. dahliae* hyphae through the bordered pit field from an adjacent pericyclic or cortical parenchyma cell. Penetration is associated with a callosity, and hyphae may pass from vessel to vessel via shared pit fields. Blackhurst and Wood (1963b) found similarly in tomato, the passage of *V. albo-atrum* hyphae through pits from vessel to vessel. As with cotton (Garber, 1973), hyphae occasionally entered paravascular parenchyma often stained with a phenolic deposit, but did not progress further. Rudolph (1931) reported that discoloration, 'plugging', preceded mycelial growth and could be found in potato from the root tip to the stem apex. The quantity of hyphae in the xylem system *in toto* or the number of infested bundles is not a good measure of the severity of disease. The numbers of invaded bundles is a measure, but probably an underestimate, of the number of root infections and penetrations from the root pericycle (Garber, 1973). Reinke and Berthold (1879) recorded two sizes of mycelium of *V. albo-atrum* in potato. Garber (1973) claimed that, dark, 4.0 μm hyphae (dauermycelien) of *V. dahliae* in cotton were derived from a population of xylem conidia measuring 5 × 10 μm, while hyaline, 1.5 μm hyphae arose from a population of 6 × 2 μm spores. Selman and Pegg (1957) reported the loss of all melanoid pigment in *V. albo-atrum in vivo* in isolations from plants inoculated with hyaline, dark pathotypes and found hyphae of different sizes (see Chapter 3). There was no loss of virulence.

Reinke and Berthold (1879) in the original description of *V. albo-atrum* on potato recorded the presence of spores in the vessels, but it was not until about 80 years later that their significance in rapid colonization was appreciated (Christie, 1956; Talboys, 1962). Schnathorst (1963) found higher conidial production of *V. dahliae* in cotton by the most virulent pathovars and the converse with pathovars of lower virulence. Conidia are free-floating in the tracheal fluid and can colonize a 115 cm tall plant by 24 h (Presley *et al.*, 1966). Dixon and Pegg (1969) also found conidia of *V. albo-atrum* in the extremities of tomato stems and petioles 24 h after inoculation of susceptible cultivars (see Chapter 9). Wright (1969) reported conidia in tobacco. In hop, *V. albo-atrum* conidial numbers reached 6870 ml^{-1}, an enormous inoculum for the host plant. The rate of spread was an order of magnitude greater than the hyphal growth rate *in planta*; this was also the same in cotton (Schnathorst *et al.*, 1968). Spore numbers *in planta* will vary considerably depending on the pathovar, host cultivar (and most importantly its plane of nutrition) and the general environmental conditions of temperature and water availability. Leydendecker (1950) failed to find *V. dahliae* conidia in cotton, while Garber observed 50–100 in the field of view of a single vessel. Brandt *et al.* (1984) found 100–15,000 conidia mm^{-1} of *V. dahliae* in the stems of resistant *Mentha crispa*, while only 10–57 spores mm^{-1} were found in the susceptible *M. peperita*. This reduced sporulation rate is not necessarily an explanation for resistance but may merely reflect an initial lower hyphal biomass for sporulation in *M. crispa*. The rate of spread in cotton root and stem appeared to be less than in leaves (Savov, 1978a). Pennypacker and Leath (1983) concluded that lucerne is colonized rapidly by internal conidiation of *V. albo-atrum* for which the age of the host is critical (Knoll, 1972). Heinz *et al.* (1998) reported on cyclical colonization of *V. albo-atrum* in R and S tomato Craigella cultivars following conidial inoculation at the four-leaf stage. Colonization was based on a PCR assay of the pathogen and *PAL*–t *PAL* 5 gene activity. The pathogen was systemic, reaching a peak at 2–4 days followed by a period of fungal elimination (see Dixon and Pegg, 1969) and a second peak 12–15 days post-inoculation.

The phenomenon of hyphal lysis is now well established; most authors, however, have not considered spatial growth (by conidiation) and partial elimination followed by growth, as a continuous process during the growth of the host. Such a process could be interpreted as the suppression of a natural (but inadequate) defence reaction in susceptible plants. In potato and presumably in most other hosts, conidia are swept up to perforated xylem end walls where, if they do not pass through, they germinate and grow through to sporulate again on the other side (Perry and Evert, 1983; Pegg, 1985). Mycelium encountered in distal parts of the xylem is more likely to be derived from spores than from a hyphal continuum from root infection. While the extent of mycelial invasion or conidial production in general is an unreliable estimate of pathogenicity (cf. the symptomless colonized host (Dixon and Pegg, 1969)), several workers have shown close correlation with the location of mycelium and/or vascular browning and the timing and appearance of symptoms. Hall and Busch (1971) claimed that chrysanthemum leaf petiole colonization with *V. dahliae* is essential before wilt symptoms becomes evident. Robb *et al.* (1975b) and Robb and Busch (1983) provided confirmatory fine structural evidence (see also Robb *et al.*, 1982). Leaf colonization was also studied in tomato (Dixon and Pegg, 1969) and hop (Talboys, 1958c, 1978). Tattar (1976) and Newbanks and Tattar (1982) measured electrical resistance in the xylem of *Acer* spp. to detect colonization of the stem before wilting symptoms had developed. Resistance to a pulsed electric current was significantly reduced in *V. albo-atrum*-infested xylem of *A. saccharinum*, *A. platanoides* and *A. rubrum* than in the xylem from corresponding healthy trees. Talboys (1958c) found that a highly virulent isolate of *V. albo-atrum* invaded roots, stems, petioles and leaves of hop. No penetration occurred in the terminal vascular elements of the interveinal areas of leaf. Only a small proportion of stem and petiolar vessels were occluded sufficiently to prevent water transport. Talboys concluded that foliar necrosis could not be attributed to vascular blockage and that normal post-infection (supernumary) secondary xylem was substantially in excess of the plant's water transport requirement. In potato with *V. dahliae*, Rudolph (1931) found petiolar and laminar xylem colonization but not in relation to symptom severity. Talboys (1958b), in a study of wilt-susceptible and wilt-tolerant hop cultivars, attempted to correlate low stem colonization with mild foliar symptoms and severe symptoms in susceptible cultivars with intensive stem xylem invasion. Harris (1953), however, had found mycelium in 'high resistance' hops which were symptomless. No immunity *per se* exists in any genetic lines of host species normally susceptible to *Verticillium*. Resistant infected hosts all become colonized in the root and hypocotyl region to a limited extent (see Chapter 9). Thus, while a good correlation may exist between the location and intensity of xylem colonization with symptom severity in susceptible hosts, the same does not apply to tolerant hosts – which may exhibit a polygenic, non-specific resistance to symptoms but not to hyphal colonization (see Non-hosts). Talboys, Beckman and others have considered the physical responses in the xylem and adjacent tissue to invasion, i.e. tylosis, hyphal and xylem wall coating, gummosis, gel formation, supernumerary cambial activity (xylem hyperplasia) and callose deposition, to be resistance functions and thus these are considered in Chapter 9. In an interesting experiment Robb and Busch (1982) detail ultrastructural changes found in *V. dahliae*-infected and drought-wilted leaves.

Studies on the nutritional conditions for the pathogen in the xylem are limited (see Pegg, 1985). The oxygen content of tomato bleeding sap was 0.6–0.9 p.p.m. at 24°C compared with the 8.5 p.p.m. saturation value (Dimond, 1962). Ioannou *et al.* (1977a), however, found that *V. dahliae* could grow and sporulate at 0.5% O_2 saturation. Tomato xylem fluid contains low concentrations of some sugars, and organic and amino acids (Wood, 1961). The actual volume of xylem fluid from a 30-cm stem segment was 0.1 ml, a considerable reduction on the volume of bleeding fluid released (Vessey and Pegg, 1973). Nearly all analytical studies and spore germination bioassays have been made on root xylem exudate from plants decapitated near the cotyledonary node. Amino acids and amides appear to be the major sources of nitrogen and carbon available to the pathogen (Wood, 1961; Dixon and Pegg, 1972). Kessler (1966) grew *V. dahliae* on sap from *Acer rubrum*, *Ulmus americana*, *Prunus serotina*, *Betula lenta* and *Diospyros virginiana* containing little or no sugar. Oxygen at 6% or less prevented microsclerotial production and reduced sporulation. CO_2 at 15% stopped microsclerotic formation but gave good sporulation. Infection with *V. albo-atrum* generally reduces xylem amino acid levels in tomato (Dixon and Pegg, 1972), and strawberry with *V. dahliae* (Springer, 1967). The claim by Springer that leaf stunting results from an amino acid deficit is dubious. Specific amino acids have been implicated in growth and resistance, e.g. proline (Dixon and Pegg, 1972) and alanine (Booth, 1969; Singh *et al.*, 1971). There is no clear-cut pattern of association between xylem amino acid or amides and the development of shoot symptoms. *V. dahliae*-discoloured xylem from sugar maple (*Acer saccharum*) had higher moisture content, total ash and concentrations of Ca and K (see Pegg, 1984) than adjacent uninfested wood and showed a lower electrical resistance (Malia and Tattar, 1978). The results are somewhat equivocal, since bacteria and other fungi were also isolated from discoloured wood.

A point of major concern regarding all cultural studies to date with *Verticillium* spp. involving artificial media is that the carbon source is frequently many orders of magnitude higher and/or is qualitatively different from that constituting the natural aqueous substrate.

Symptoms

The general pattern of symptoms found in most herbaceous and some tree hosts is a total or partial loss of turgor, originating as flaccidity of the lowest leaf, or a terminal leaflet in a compound leaf developing towards the stem and spreading acropetally until the whole plant is affected, in severe cases resulting in death. Tap roots and feeder roots of many infected plants show small brown lesions or larger brown patches representing major sites of infection (Rudolph, 1931). The initial flaccidity is followed in rapid sequence by chlorosis and necrosis. Associated with these symptoms, but depending on the type and age of host, epinasty, adventitious root production, petiolar abscission and stunting may occur. There is a general reduction in leaf area, and in stem, root, petiolar and laminar dry weights. Epinasty and stunting, the latter associated with reduced photosynthetic rates, were described in *V. albo-atrum*-infected tomato (Selman and Pegg, 1957). Stunting was found as a chronic symptom, in the absence of general wilting, followed seedling infection. Several authors have described foliar local lesions: potato and *V. albo-atrum* (Reinke and Berthold, 1879); beet and *V. albo-atrum* (Westerdijk and van Luijk, 1924); and tomato and *V. albo-atrum* (Bewley, 1921). The effect of a defoliating T_1 and T_9 (P_1) and non-defoliating SS4 (P_2) strains of *V. dahliae* on cotton have been described by Wiese and DeVay (1970), Garber (1973), Schnathorst and Mathré (1966b), Schnathorst (1981) and Friebertshauser and DeVay (1982). With the mild (SS4) strain, symptoms appeared 35–40 days after plant emergence. Cotyledons became chlorotic and quickly died, followed by interveinal and marginal chlorosis in the first leaf. Chlorotic tissue became necrotic and the condition spread up the plant. Moderate epinasty occurred and, depending on the cultivar and environmental conditions, especially temperature, the plant could die. Normally SS4 infection resulted in a non-lethal chronic infection. T_1-infection led to a rapid downcurling of the terminal leaf, followed in quick succession by severe epinasty, necrosis and an almost total abscission of fruit and bolls (Schnathorst, 1981). Symptoms of K deficiency as a foliar marginal chlorosis developed with some strains and cultivars (DeVay, 1990). Similar symptoms were described in hop cultivars infected by fluctuating (mild, V_1) and progressive (virulent, V_2 and V_3) pathotypes of *V. albo-atrum* on hop (Isaac and Keyworth, 1948). A distinct foliar symptom association with V_2 strain infection, tiger-striping, was described by Talboys (1978). Marginal and interveinal necrosis and chlorosis spread inwards to the laminar base leaving the main vein green for a long period; the leaf did not become flaccid but remained still and uncurled while progressively desiccating. Initially, in a heavily infected plant, stomata in the initial chlorotic zone remained closed but as the tissue became necrotic stomata opened and remained so. Stomata in the residual green lamina remained tightly closed. This condition – classic 'drought hardening' – was unaffected by water availability. Defoliation has also been described in blackcurrant with *V. dahliae* (Sherengovyi, 1968) and in apricot with *V. albo-atrum* (Vigouroux and Castelani, 1969). Foliar symptoms in raspberry infected by *V. dahliae* resemble tiger-striping in hop; stems often show a blue/purple longitudinal stripe (blue stem) on the side of an invaded bundle (Lawrence, 1912). Severe *V. dahliae* infection of apricot results in fruit with a black core (Phillips, 1916; Czarnecki, 1923). Potato tubers infested with *V. dahliae* or *V. albo-atrum* may show all shades of discoloration from yellow brown to black (Pethybridge, 1916; Rudolph, 1931). The localization of foliar symptoms and wilting or chlorosis corresponding to the juxtaposition of an isolated infested vein or veinlet in a half leaf has been described in tomato and other hosts by many authors. Examination by electron microscopy of *V. dahliae*-infected chrysanthemum leaves showing slight flaccidity of the terminal lobe revealed vessel lumina with a fibrillar material or occluded with solid plugs. Walls had a smooth or blistered coating. These conditions were unrelated to colonization but became more severe towards the flaccid region (Douglas and MacHardy, 1981) (see Chapter 9). A detailed chronological development of foliar symptoms in *V. dahliae*-infected sunflower, as seen in fine structure by T.E.M. was presented by Robb *et al.* (1977).

Kudela (1973) working with lucerne cv. Moravska Krajova reported an apparent correlation between *V. albo-atrum* infection and seed colour (brown). Stems which were dry before harvest also produced 3.5-fold more brown seeds than plants with green harvested stems. Seed colour ranged from yellow to dark brown, with intermediate shades. Eight lucerne cultivars infected with *V. albo-atrum* all showed reductions in height, dry weight and flowering in symptomatic and *symptomless* plants (Pennypacker *et al.*, 1988). Distinct yellow–brown foliar lesions described by Efimova (1978) in seven cotton cultivars jointly inoculated with *V. dahliae* and *F. oxysporum* f.sp. *vasinfectum* are almost certainly due to the latter pathogen.

The Involvement of Hydrolytic Enzymes

Within the non-living xylem vessel, the vascular pathogen lives effectively as a saprophyte and exerts its effect on the host plant passively through xylem occlusion and/or by the release of toxic metabolites into the transpiration stream (Pegg, 1985). How far the pathogen's sole carbon requirements are provided by solutes in the xylem fluid has not been established convincingly. The xylem vessel may be regarded as a membrane thin tube of pecto-cellulose – the primary cell wall – lined peripherally with a discontinuous covering of lignified cellulose in the form of annular, helical, reticulate thickened bands or perforate sheets. The middle lamella is exposed in simple or bordered pits. Pit fields shared by adjacent vessels (pit pairs) constitute the pathway for lateral water movement and nutrient enrichment or the efflux of toxic metabolites. Pits can provide nearly 50% of the vessel lumen surface area to which the pathogen has access (Van der Molen *et al.*, 1987). The composition of the membrane is complex (see Pegg, 1989) and would require an array of enzymes in sequence for complete dissolution. It is against this background that pectolytic enzymes must be considered.

The ability of extracellular, constitutive and inducible pectinolytic and cellulolytic enzymes to induce symptoms of wilt when introduced into plants has been known for 80 years. The pioneering work of Bewley (1920) on *V. albo-atrum* from tomato showed that culture filtrates contained a substance (cytase) causing young tomato shoots to wilt, which was thermolabile and could be precipitated in 95% ethanol. Kamal and Wood (1956) demonstrated the same effect in cotton shoots with culture filtrate from *V. dahliae* containing what they called protopectinase. It is now clear that Bewley's cytase and Kamal and Wood's protopectinase were most probably polygalacturonase (PG). Kamal and Wood attributed browning to PG but wilting to an unidentified thermostable substance. Scheffer *et al.* (1956) working on *V. dahliae* from tomato found a heat-labile vascular-browning factor. High PG activity was found, greater on Na polypectate than pectin. Low pectin esterase (PE) and β-glucosidase were also recorded. Blackhurst and Wood (1963a) showed that *V. albo-atrum* produced PG on pectin and was enhanced by an amino acid source. In a first crude attempt to detect pectolytic enzymes *in planta*, Wood (1961) found that *V. albo-atrum*-infected tomato plants contained 50% more PE than healthy plants. Blackhurst and Wood (1963a) found a small increase in tomato shoot PG following conidial inoculation – higher than in root-inoculated plants.

Deese and Stahmann (1962) described the production of PE and PG from *V. albo-atrum* on tomato stem and wheat bran and claimed that there was a 3–10 times higher enzyme activity on a susceptible tomato cultivar (Bonny Best) than on resistant cultivars (Loran Blood and Moscow). In contrast, Wiese *et al.* (1970) found only trace amounts of PG in *V. dahliae*-infected cotton plants. Deese and Stahmann (1962) interpreted their results in terms of enzyme inhibition in resistant plants by quinone-type substances, or the binding and inactivation of the enzyme on the resistant cell wall. No conclusive evidence was published for this, although resistant stem extracts had higher phenol-oxidizing properties than the susceptible extracts. Enzyme studies are subject to the same constraints as other metabolic studies, namely the differential growth rates of the pathogen on the two hosts. A more plausible explanation in view of Cooper *et al.*'s (1978) results is that the reduced enzyme activity in Loran Blood and Moscow reflected a smaller enzyme-producing biomass. Talboys (1958a) showed that the V_1 and V_2 (weak and strongly virulent) hop strains of *V. albo-atrum* produced C_1 and C_x cellulases *in vitro* as inducible enzymes on cellobiose and cellulose-containing media and were also capable of penetrating and degrading cellulose film. Neither the cellulases nor polymethylgalacturonase (PMG) activity detected on the same media could be correlated with strain virulence. Russel (1975a,b) confirmed cellulase production by *V. albo-atrum* from tomato. C_1 and three C_x enzymes, all pH stable and thermostable, were found in culture and in infected plants, more than in healthy plants. IAA and gibberellic acid (GA_3) were claimed to stimulate host cellulase production, but, whereas GA_3 decreased the resistance of resistant plants, IAA increased the resistance of suceptible plants. The possible mechanism for this is the reduced binding of PE to cell walls in the presence of high IAA concentrations (see Pegg, 1981a, for other references), thereby producing more free carboxyl radicals which bind Ca^{2+} to form rigid Ca pectates, which are more resistant to depolymerizing enzymes. Conversely, under appropriate conditions, IAA can be shown to stimulate an ATPase proton pump causing an efflux of hydrogen ions through the plasmalemma and a removal of Ca^{2+} from the wall. *V. albo-atrum* in tomato is known to induce a powerful redistribution of Ca^{2+} from the stem to the leaves (Pegg, 1985). Heale and Gupta (1972) using *V. albo-atrum* from lucerne found the constitutive production of endo-PG exo pectin lyase (pectin transeliminase) (exo-PL) and small amounts of PE. Cellulase was only induced by cellulose. PL, PG and C_x were all isolated and partially purified from infected lucerne. These authors implicated PL as a major enzyme in pathogenesis since its production in roots and stems coincided with symptoms and preceded PG and C_x, both of which were post-symptom in occurrence. Blackhurst and Wood (1963a) described the production by *V. albo-atrum* from tomatoes of PG, PME and cellulase (C). Whitney *et al.* (1969) showed carboxymethylcellulase activity by *V. albo-atrum* from lucerne. The repeated passage of a tomato isolate of *V. dahliae* into a resistant lucerne cultivar led to a progressive decrease in fungal protein synthesis and the ability to produce C_x. The loss of C_x activity was compensated by a gradual gain in PG activity (El Aissami *et al.*, 1998).

Early work on the constitutive or inducible nature of cellulolytic and pectolytic enzymes and the implications for their roles *in planta* failed to appreciate the subtlety of product, or catabolite concentration and stimulation, or repression of enzyme activity. This was demonstrated elegantly in *V. albo-atrum* and tomato by Cooper and Wood (1973, 1975, 1980) and Cooper *et al.*

(1978). Patil and Dimond (1967) found that *V. albo-atrum* PG from tomato on 1% Na polypectate increased two- to fourfold in the presence of 0.1 M glucose, sucrose or ribitol. On a 0.1 M glucose nitrate medium supplemented with 0.1 M α-amino-D-glucose, PG increased 20-fold. The specific activity of PG was reduced by 0.001 M galacturonic acid. Mussell and Green (1968, 1970) confirmed with *V. dahliae* from cotton that glucose and galacturonic acid acted as inducer and repressor of PG, unlike *F. oxysporum* f.sp. *lycopersici*, where glucose and galacturonic acid in unspecified concentrations acted oppositely. Cooper and Wood (1973, 1975) in a detailed re-examination of the problem with *V. albo-atrum* from tomato elucidated the importance of concentration, in which monomeric substrates from a complex polymer could act variously as specific enzyme inducers or repressors depending on the level used. Specific monomers of the various components found in the pectocellulosic membrane were slowly released into the culture medium through membrane-covered capsules. The sequence of enzymes produced, with their specific inducers (in parentheses) was: endo-PG (D-galacturonic acid); exo-arabinase (L-arabinose); endo-PL, (D-galacturonic acid); endo-xylanase (D-xylose); C_x (cellobiose); D-galactosidase and galactanase (D-galactose and L-arabinose); and D-glucosidase (D-glucose). Synthesis of endo-PG and endo-PL increased almost linearly when supplied with galacturonic acid from 0 to 10 mg 100 ml^{-1} h^{-1}, but showed a 90% decrease at 20 mg 100 ml^{-1} h^{-1}. Cooper and Wood (1975) found that the same range of enzymes capable of acting on cellulose, hemicellulose and pectin moieties in the primary wall and middle lamella was produced in *V. albo-atrum* cultures using tomato cell walls as the sole carbon source. Unlike other authors (Deese and Stahmann, 1962; Russel, 1975a,b), no differences in enzyme quantity or type were found using cell walls from resistant and susceptible lines. The results of Cooper and Wood clearly illustrate why earlier work showed conflicting results and why cultures in a simple sugar salts medium failed to produce the range of potential enzymes. It is now clear that cellulose and pectinaceous material in the pit membrane can be attacked by a sequence of enzymes where exo-enzymes release monomeric inducers of specific chain splitting enzymes, followed by succeeding enzymes as the next component of the matrix is exposed. Other examples of enzymes regulation are by Booth (1966, 1974), who found that L-alanine stimulated growth and PG production of *V. dahliae*, both of which were reversed by choline. Root exudates from a tolerant cotton cultivar contained more choline that those from a susceptible cultivar.

The establishment of a case for the involvement of pecto and cellulolytic enzymes in vascular wilt pathogenesis, as distinct from cataloguing inducible enzymes produced in culture, requires the application of modified Koch's postulates. These are: (i) *in vivo* detection in diseased plants; (ii) physical evidence of a degraded pit membrane; (iii) the presence of cleavage products in diseased plants; (iv) reduced disease severity following enzyme inhibitor treatment; and (v) greater production (or increased specific activity) of enzymes in susceptible compared with resistant plants (Pegg, 1981a). Rarely in any one investigation have more than two of these criteria been satisfied. Moreover, many studies have erected a case for or against enzyme involvement based on PG while excluding PL. Mussell and Strouse (1972) purified endo-PG from wild-type and hyaline variants of *V. dahliae* from cotton, using gel permeation chromatography and immunoelectrophoresis. Pure endo-PG from the T_1 strain in aqueous solution into cotton shoots (*G. hirsutum*) developed interveinal bronzing, necrosis and desiccation, but not chlorosis or wilt. PG was only effective in the presence of divalent cations Ca^{2+}, Mg^{2+} and Mn^{2+} (Mussell and Strand, 1974). Pegg (1985), however, has described similar symptoms in cotton and tomato shoots treated with 5 mM Ca^{2+} following an ethylene gassing pre-treatment (see Ethylene). Kamal and Wood (1956) lost protopectinase (PG) activity following dialysis of the culture medium, with the corresponding loss of ions. In contrast to the work of Mussell (1972, 1973) and Mussell and Strouse (1972), Keen and Erwin (1971) using the same host but different isolates of *V. dahliae* failed to produce symptoms using aqueous solutions of endo-PG. These studies are of particular importance, since PG was used in a purified form (Wang and Keen, 1970).

In all the foregoing work, however, no mention was made of vascular occlusion following enzyme treatment. Accounts in the literature from Bewley (1921) onwards, citing wilting following the bioassay of culture filtrates, while possibly implicating pectolytic enzymes, do not exclude many other probable wilt-inducing components of the culture. The presumed action of pectolytic enzymes is that carbon-containing oligosaccharide by-products of pit membrane degradation, further modified by other hydrolases, form gums occluding the vessel lumen or pit fields. Mussell (1972, 1973) presented evidence for a series of small (<10 kDa) molecular weight non-enzymic proteins capable of inducing wilting independently of enzymes produced in the same culture. Cooper (1975) found endo-PG and endo-PL in quantity in *V. albo-atrum*-infected tomato cuttings early after inoculation. PG but not PL was considered to be of host origin. Levels of PG and PL were comparable in resistant and susceptible tomato lines. PL failed to increase following root inoculation, while PG levels rose twofold. Cooper *et al.* (1978) resolved the PG into three isozymes with pIs of 5.0, 5.9 and 6.2, and PL into two isozymes with pIs of 5.0 and 5.9. PL had a partial requirement for Ca^{2+}. Plants treated with partially purified PL and PG at levels comparable with those in infected plants induced gels, vascular browning, chlorosis, wilting, necrosis and desiccation (Cooper, 1975).

The relationship between the enzyme-producing ability and virulence of isolates has been controversial. Leal and Villaneuva (1962) claimed that non-pathogenic isolates of *V. dahliae* failed to produce pectic enzymes in culture. Talboys and Busch (1970) found that 23 isolates of *V. albo-atrum*, 16 of *V. dahliae*, two of *V. tricorpus*, four of *V. nigrescens* and one of *V. nubilum* all produced PG, endo-PL and PME in amounts that varied as much within species as between them. Enzyme production and virulence were not closely correlated.

Wiese *et al.* (1970) and Wiese and DeVay (1970) were unable to correlate virulence of *V. dahliae* to cotton with PG production. Keen and Erwin (1971) similarly found no relationship between rates of production of extracellular endo-PG in culture and virulence of 13 cotton strains of *V. dahliae*, although Mussell (1973), working with the same fungus, observed a positive correlation. Puhalla and Howell (1975) used a technique for the isolation of UV mutants developed by Puhalla (1973a). Stable mutants of US cotton-defoliating strains were obtained deficient in endo-PG production and detected on a polygalacturonic acid medium. Virulence of the mutants which retained PG deficiency on reisolation was unimpaired. Symptoms identical to those produced by parent strains were obtained in *G. hirsutum*. Howell (1976) in an extension of this study identified two isozymes of PG with optimal activities at pH 4 and pH 6 (endo-PG_4, endo-PG_6). Using repeated induction on mutant strains, stable strains were isolated individually deficient for PG_4, PG_6, PL (exo-pectin transeliminase and exo-pectic acid transeliminase) and PME. Some strains were collectively deficient for PG_4, PG_6 and PL but not PME. Inoculation of cotton with mutant strains in all cases resulted in typical but slightly delayed symptoms – epinasty, chlorosis, necrosis, defoliation, as induced by wild-type strains. In contrast to the earlier work of Talboys and Busch (1970), Carder *et al.* (1987) found overall good correlation between the virulence of a large population V_1, V_2 and V_3 hop *V. albo-atrum* strains and PG and PL production. However, a large variation between PG and PL secretion and virulence in each population was not examined. Durrands and co-workers (Durrands and Cooper, 1988a,b,c; Durrands *et al.*, 1988; Cooper and Durrands, 1989), working on mutant strains of tomato *V. albo-atrum* produced strains that were variously and relatively (but not wholly) deficient in PL, PG, cellulase, β-galactosidase, β-glucosidase, L-leucine arylamidase, valine arylamidase, acid phosphatase, phosphoamidase, esterase (C4), esterase lipase (C8) and lipase (C14). The authors concluded that PL is an important determinant of virulence but not *pathogenicity* [*sic*] (infectivity) since symptomless plants showed colonization comparable with wild-type infection. These results, though of special interest, are nevertheless equivocal. Unlike most other studies, Durrands and Cooper identified a range of isozymes, six of PL (pIs 4.5–10.5) and 25 of PG (pIs 3–9.5). A secretory mutant C23 producing only 3, 9 and 3% PL, PG and cellulose, respectively, of the wild-type failed to induce more than mild basal chlorosis after 15 days. This was reflected by a greatly reduced ability to colonize the host plant. Thus C23 infection was comparable with that found in Ve-resistant, infected plants. In contrast, mutant 24d, which produced 43 and 95% PL and PG of the wild-type, similarly failed to induce symptoms while host colonization was equivalent to wild-type infection in susceptible plants. These results cannot be interpreted as *prime facie* evidence (as claimed by the authors) for the role of either PL or PG in virulence. The curious feature of this work is the absence of symptoms following inoculation with strains having only partial loss of PL or PG, unlike that of Howell (1976) who induced symptoms with *V. dahliae* strains wholly lacking pectolytic activity. As Durrands and Cooper (1988c) comment, the possibility of simultaneously neutralizing all genes for pectolytic isozymes is remote. There is thus a real possibility in Howell's (1976) work of isozymes of PL and PG functioning at pHs which were not examined. Conversely, the findings of Durrands and Cooper (1988c) and Cooper and Durrands (1989) with mutant 24d could represent the loss during mutagenesis of some virulence factor or recognition factor, other than enzymic, leading to virulence. Pegg (1981a, 1985) suggested that pectolytic enzymes could provide an essential source of carbon from middle lamella galacturonides. Mutant 34i, however, was unable to use saturated or unsaturated galacturonides *in vitro* yet succeeded in colonising the xylem system extensively *in vivo* (Durrands and Cooper, 1988; Cooper and Durrands, 1989).

Relatively little work has been done on enzyme inhibition other than catabolite repression. Gossypol, a substituted dinaphthalene carboxaldehyde, widely distributed in cotton as a normal metabolite and implicated in resistance, was shown by Sagdieva *et al.* (1974) to inbibit *V. dahliae* PL activity at 2×10^{-5} M. It was claimed that similar concentrations were found in cotton xylem fluid. Extracts from the xylem of resistant plants inhibited PL threefold compared with those from susceptible plants. Deese and Stahmann (1962) claimed a similar inhibition of PG from *V. dahliae* by quinone substances. While the results of both studies may represent the well-known phenolic-type inhibition of enzymes, the reduced PL and PG activities could also reflect reduced fungal growth. Ksaymova *et al.* (1990) purified an 18–20 kDa protein from cotton seed capable of inhibiting three native proteases from cotton and suppressing proteolytic activity in *V. dahliae*. The authors speculate that challenge infection of resistant plants increases the content of protease inhibitor suppressing pathogen growth. Acid phosphatase and DNase activity were reported to be greater in virulent cotton pathovars of *V. dahliae* than avirulent ones (Rubin and Tukeeva, 1978) (for reports on other hydrolases in *Verticillium* wilt, see Chapter 9). In summary, the evidence for the involvement of pectolytic and cellulolytic enzymes in *Verticillium* wilt pathogenesis is strong. The inconclusive results obtained from several reputable studies, however, beg the questions regarding the nature of pathogen nutrition and the precise determinant(s) of shoot symptoms. In particular, the carbon source requirements of the pathogen *in planta* remain unresolved for even one *Verticillium* strain and host. The use of enzyme-deficient mutants, one of the best techniques available prior to the advent of gene manipulation techniques, has given conflicting and equivocal results, emphasizing the great importance of substrate in determining isozyme patterns and the possible substitution (including *de novo* host induction) of one enzyme for another. Since several different agencies may give rise to the same symptoms, experiments are required in which the deletion of genes coding for specific enzyme proteins results in the loss of pathogenicity. The converse situation, where single or multiple enzyme-deficient strains still manifest pathogenicity, does not advance the role of enzymes more than circumstantially.

Toxins

The concept of a toxin basis of symptom induction in *Verticillium*-infected plants, although widely reported over 40 years as an alternative to vascular occlusion, was largely speculative and hypothetical until 1966 (Stoddart and Carr, 1966; Zel'tser and Malysheva, 1966). For convenience, toxins may be classified arbitrarily into small molecular weight (<1000 Da) and macromolecular weight compounds, the latter usually large polymers of polysaccharides and/or proteins. In this category, the distinction between toxin *sensu stricto* and some enzymes is mute; the former, however, would always act directly causing intrinsic damage to cell membranes, organelles and metabolic function. Toxic molecules as extracellular metabolites of the pathogen, are envisaged passing through xylem pit membranes into leaf (and other) parenchyma and there directly affecting permeability, causing cellular dysfunction and death. Few attempts have been made to identify *Verticillium* toxins *in planta*; the literature, therefore, is largely concerned with cultural metabolites and their use to simulate symptoms in bioassays.

Low molecular weight toxins

Reports of the involvement of low molecular weight substances in disease are limited. Talboys (1957) implied the existence of a low molecular weight toxin responsible for wilt in *V. albo-atrum*-infected hop. Subsequently, Talboys (1975) strikingly simulated 'tiger stripe' foliar symptoms in hop using a model toxin (polyethylene glycol) and ethylene. Pegg (1965), studying growth inhibition in tomato wilt, isolated growth inhibitors from ether-soluble neutral, basic and acidic fractions of *V. albo-atrum* culture filtrates. The chromatographic extracts were bioassayed *in vitro* on a tomato hypocotyl assay. Krasil'nikov *et al.* (1965, 1969) found similar low molecular weight substances in *V. dahliae* cultures, which were stable between pH 2 and 10 and were detoxicated by tetraene and tetracycline antibiotics. These toxins were active on cotton and legumes (see also Askarova *et al.*, 1971). Chepenko *et al.* (1969) also described five amino acid-containing toxic pigments, partially purified by chromatography and electrophoresis. These substances had chelating properties and were strong oxidizing agents in light. Ten *et al.* (1977a,b, 1981) described a range of small molecular weight metabolites from cotton strains of *V. dahliae* including pigments, pentaketides, neutral lipids and a peptide. The roles of these substances, if any, in pathogenicity is not clear. In general, no convincing evidence has been produced to support a small molecular toxin basis for *Verticillium* symptoms or specificity.

Macromolecular toxins

Early studies on wilt-inducing *Verticillium* culture filtrate components, which by deduction were high molecular weight compounds, were carried out by

Bewley (1922), Picado (1923), Porter and Green (1952), Green (1954), Caroselli (1955) and Threlfall (1959). In most of these studies, only a simple separation of the toxic moiety was achieved. Green (1954) precipitated a thermolabile protein–polysaccharide mixture from *V. dahliae* in 90% ethanol, which induced wilting in tomato shoots, unlike the crude filtrate which also induced chlorosis and necrosis. Green found that wilting shoots recovered after a section of the stem base was removed and concluded that wilting was due to vascular occlusion. Le Tourneau (1958) found 10–16% protein in ageing cultures of *V. albo-atrum* associated with a polymer containing glucose, galactose, mannose and arabinose. A two-component toxin was found in *V. albo-atrum* cultures from maple. The polysaccharide fraction caused wilting and slight leaf injury, which was largely reversible. A fraction identified as thiurea caused permanent wilting, chlorosis and necrosis (Caroselli, 1955). Whole filtrates were thermolabile but gave negative protein reactions. A thermostable (presumed to be non-enzymic) wilt-inducing factor was isolated from 11 to 15-day tomato *V. albo-atrum* cultures. Wilting and reduced transpirational activity, claimed to result from xylem occlusion, were reduced following dialysis (Threlfall, 1959). No differences were found between virulent and weakly virulent strains. In contrast, McLeod and Smith (1960) found that culture filtrates of *V. dahliae* caused wilting in susceptible tobacco cultivars but not in those possessing field resistance. A similar host-specific effect was described by Michail and Carr (1966), who proposed that cultures of *V. albo-atrum* could be used to screen for resistance in lucerne cultivars. In lucerne, Kiessig and Haller-Kiessig (1958) proposed a toxin basis for pathogenicity, which would be consistent with the bleached foliar desiccation found in the field. Reversible wilting was due to a thermolabile component, while necrosis was attributed to a thermostable fraction. Stoddart and Carr (1966), in a more critical study, purified two components from 21-day freeze-dried cultures of *V. albo-atrum*. Using gel permeation chromatography with water as a solvent, a protein with a molecular weight of about 1000 kDa and slight cellulase activity was isolated together with a fructosan of 5–10 kDa (4:1 fructose:glucose), both of which induced wilting in lucerne. These fractions, however, were not dialysed, and contamination with culture medium components could not be excluded.

Work on cotton *V. dahliae* toxins by Zel'tser and Malysheva (1966) has pioneered much of the later and current thinking about *Verticillium* toxins. Symptoms in cotton, especially resulting from T-strain infection, like hop and lucerne, present good circumstantial evidence for a foliar poison. The Russian groups Zel'tser and Malysheva (1966), Malysheva and Zel'tser (1968) and Krasil'nikov *et al.* (1969) isolated a protein–lipopolysaccharide (PLP) which reproduced many of the symptoms in cotton shoots and young plants. Keen and Long (1972) and Keen *et al.* (1972) found that the PLP isolated from log-phase 3-day cultures of *V. dahliae* had a molecular weight of 3000 kDa and consisted of 75% polysaccharide (yielding glucose, galactose, galacturonic acid and mannose on hydrolysis) and 15% each protein and lipid. Glucose, galactose and mannose were in the ratio 11:15:10 and the complex reached 1 mg ml^{-1} of the log-phase culture. This toxin was ionically unstable, precipitating at 1 mM KH_2PO_4 and degrading at 4 M Na acetate. Keen *et al.* (1972) claimed that cotton leaf symptoms could be induced in a solution of 5 μg PLP ml^{-1} from agarose gel, equivalent to 20 μg PLP per leaf. ^{14}C-Labelled PLP was found widely distributed in minor leaf veins and interveinal parenchyma, discounting a major xylem occlusion mechanism. Asamov *et al.*'s (1975) claims that a polypeptide and an oligosaccharide from *V. dahliae* altered the transmembrane flow of Na^+ and K^+ in cotton are consistent with repeated claims of increased permeability following PLP treatment. The interaction of ethylene (Cronshaw and Pegg, 1976) was not considered in any of these studies. A similar PLP complex from still cultures of *V. dahliae* from potato was isolated by Nachmias *et al.* (1982b). These authors obtained an acetone precipitate from 21-day cultures which was resolved into two fractions, a 2000 kDa PLP consisting of 64% polysaccharide, 7% lipid and 27% protein, and an 80,000 Da fraction containing 44% polysaccharide, 43% protein and 9% lipid. The general composition of the PLP suggested that the cotton and potato isolates were producing a similar molecule. Differences in composition can be attributed to medium, age of culture and pathotype. The lower lipid content would be due to acetone treatment (Buchner *et al.*, 1982; Nachmias *et al.*, 1982a, 1985). Nachmias and co-workers claim that toxin production is related to virulence as an innate character of a pathotype, not requiring the intervention of, or triggering by, host recognition. Similar 'toxin' patterns were described for cultures of *V. dahliae* isolates from aubergine, avocado, watermelon, olive, cotton and tomato. Both PLPs from potato *V. dahliae* were toxic to leaves of most host plants, but not those of non-hosts. Borodin and Shtok (1982) described the effect of cultural toxins inhibiting cotton seed germination. Similarly, compounds from *V. dahliae* were toxic on stone fruit trees (Kostroma and Kropis, 1972).

Buchner *et al.* (1982) and Nachmias *et al.* (1985) presented evidence to show that the active moiety of the PLP, based largely on gel permeation, amino acid analyses, high-performance liquid chromatography (HPLC) and pronase studies, was of the order of 2200 ± 500 Da. Earlier, Mussell (1972, 1973) described proteins <10 kDa from cotton *V. dahliae* capable of inducing wilt quite independently of pectolytic enzymes also present in the medium. Subsequently, Buchner *et al.* (1989), using a several-stage purification, claimed the toxin to be a peptide of 1000 Da molecular mass with an Asp-Thr-Ser-Glu-Gly-Ala-Val-Iso-Leu-Leu-Tyr-Phe-Lys-His composition. Arg-Cys and Pro could not be determined. The larger molecule, described earlier, itself part of a larger PLP complex, was attributed to aggregation due to the extraction and purification techniques. Strong evidence in support of a *V. dahliae* toxin in potato wilt was provided by immunofluorescent toxin antibody detection on xylem vessel walls of tubers and stems (Nachmias *et al.*, 1985) and the extraction and purification of a molecule from infected stem xylem fluid with similar but not identical amino acid composition, molecular mass, antigenic and biological properties. The *V. dahliae* toxins were claimed to induce symptoms (including half-leaf symptoms) at concentrations of 20 ng (2×10^{-6} M) injected into detached potato leaves. Unlike the PLP, the activity of the peptide was lost following pronase treatment (Nachmias *et al.*, 1985). The results of Mussell (1972, 1973) and Nachmias *et al.* (1985) are in contrast to those of Ten *et al.* (1977) who found that low molecular weight peptides were inactive following isolation from a high molecular weight toxin showing activity on cotton. The idea of differential host tolerance to a toxin was supported by the claim of Gour and Dube (1985) and Gour (1987) that culture fluids of the cotton strain of *V. dahliae* affected K^+ and Na^+ ion transport. A crude acetone precipitate (200 μg ml^{-1}) and 0.2 mM oubain (G-strophanthin), an ATPase ion pump inhibitor, both induced symptoms and K^+ and Na^+ leakage similar to those found in *V. dahliae*-infected cotton. The authors claim that oubain and the toxin were inactive on resistant cotton. Asamov *et al.* (1975) previously claimed that a polypeptide and oligosaccharide had increased permeability of an artificial phospholipid membrane to Na^+ and K^+. These results are different from those described by Cronshaw and Pegg (1976) and Pegg (1985), who found increased mobility and leakage of divalent but not monovalent cations following *V. albo-atrum* infection of tomato. Nachmias *et al.* (1987) reported that the *Ve* gene in tomato conferred tolerance on the peptide from race 1 cultures of *V. dahliae*. Harling *et al.* (1986), however, described PLP-complex toxins from virulent and avirulent strains of *V. albo-atrum* and *V. dahliae*, both of which were non-specific, causing symptoms on tomato and potato. Orenstein *et al.* (1994) using the PLP toxin of Buchner *et al.* (1982) and Nachmias *et al.* (1985) found that pollen germination and tube growth were inhibited only in the susceptible potato cv. Promesse and *Solanum phureja* line Iv P35 and not in the resistant-tolerant cvs Alpha and WA-85-4-5. However, the toxin concentration used, 10 mg ml^{-1} or greater, seems to be at an unreasonably high physiological concentration (see Meyer *et al.*, 1994). Nevertheless, the authors claim that pollen germination was only reduced at concentrations 100-fold higher than that required to cause damage to vegetative potato tissue. In a further reinvestigation of a PLP toxin, Meyer *et al.* (1994) isolated from aqueous 7 h culture filtrates of a virulent cotton isolate of *V. dahliae*, a complex phytotoxin which closely resembled that described by Keen and Long (1972), or more especially by Zel'tser and Malysheva (1966) and Malysheva and Zel'tser (1968), in chemical proportions; protein (15.7%), lipid (13%), carbohydrate (70%), phosphate (0.4%), if not in molecular mass (197 kDa). Gel permeation chromatography and preparative agarose electrophoresis of an acetone precipitate yielded 4.5 mg toxin l^{-1} of culture filtrate. The protein moiety could be dissociated into five protein-containing components, i.e. 78, 62, 48, 32 and 28 kDa. The 28-kDa fraction had PG activity, the 48-kDa fraction had cellulose activity and the 28- and 32-kDa fractions were associated with 1,3-β-glucanase activity. It seems most probable from this description that the 'toxin', or the specific binding fraction, had not been purified completely. Whether the active component(s) would be functional in a dissociated state is a moot point. The toxin complex induced wilt and necrosis in cotton seedlings at 2.5 μg ml^{-1}. An interesting feature of this work is that the PLP *toxin* also functioned as an *elicitor*, inducing 'pathogenesis related proteins' in the intercellular spaces of toxin-treated mature cotton leaf discs. In a subsequent paper (Dubery and Meyer, 1996), binding of ^{125}I-radiolabelled PLP to cotyledon protoplasts was saturable and reversible, K_d = 17.3 nM (NB, in Meyer and Dubery, 1993, K_d = 14.2 nM) with a single class of binding site estimated at 2.41×10^{-16} moles per protoplast with a binding affinity nearly identical to that of plasma membranes from roots (5.4×10^{-15} M μg^{-1} of protein; Meyer and Dubery, 1993). The authors claimed that no difference in binding affinity existed between resistant (cv. OR19) and susceptible (cv. Acala 1517-70) cotton, but there were five times the number of binding sites per protoplast and 16 times the number of sites per mg of membrane protein in cv. OR19.

Contemporary toxin work in China is still at a rudimentary level; Gu *et al.* (1995) reported studies on a crude *V. dahliae* toxin on aubergine with thermostability based on 60°C for 42 h. Guo *et al.* (1995) claim that the virulent cotton pathotype P_1 strain (T9 and GY) produces a glycoprotein with higher glycosylation that the P_2 non-defoliating strain (HT). Chen *et al.* (1998) working on the VD8 strain from cotton identified two proteins using fast liquid chromatography. One protein induced chlorosis on cotton seedlings after 2–10 h and the second induced wilt after 24–64 h. Both proteins were thermostable after 100°C, thus eliminating normal enzyme activity (cf. Mussell, 1972, 1973). Xu and Chen (2000) described the production of a polyclonal antiserum derived from cultural toxins (unidentified) from a cotton isolate of *V. dahliae* purified by ion chromatography and PAGE. No cross-reactivity was reported from other fungal species. The authors claim to detect the pathogen in cotton stems and seed in quantities down to 1.56 ng.

In addition to enzymes and toxins, culture filtrates of *V. albo-atrum* were reported to produce volatile compounds, such as methyl paraben, propyl paraben, tethane and xylene, all identified by HPLC and gas chromotography–mass spectroscopy (GC–MS) (El Aissami *et al.*, 1999). No role for these compounds was suggested and no indications were given as to whether they occurred *in planta* or were substrate specific.

The collected research on *Verticillium* spp. (largely *V. dahliae*) particularly over the last 35 years shows a repetition in support of a family of proteins and or glycoprotein toxins sometimes in association with lipid or other components. Notwithstanding the special claims for these toxins, many questions remain for which there has been a curious reluctance to provide answers. The key question presented is whether the molecule(s) described from culture filtrates is/are produced in the same or different form(s) in the infected plant? From our knowledge of other aspects of cultural biochemistry, many extracellular metabolites are substrate dependent. How far the range of substances and the reported variation of near similar molecules represent differences in culture medium composition is a question worthy of serious consideration. As mentioned elsewhere, the nutrition base of most cultures greatly exceeds in richness that of the vascular environment. Such media rich in carbohydrates, while promoting vegetative growth, may inhibit or transform those substances induced *in vivo* near the xylem wall from a relatively sparse mycelium which is bathed in a weak, low carbohydrate solution. Other variations between studies include fungal pathotype, culture age, temperature and oxygen content, all of which could produce quantitative and qualitative differences in cultural metabolites. There has been a signal lack of response to resolve these questions, each study purporting to describe the 'definitive' toxin. A further common presumption is that 'toxins' and 'elicitors' represent unique and very different molecules, the former responsible for pathogenesis and the latter uniquely involved in resistance. That the same molecule might function differently or at a different rate in different genotypes of the same host species is scarcely envisaged. In addition, the effects of predisposing molecules such as ethylene 'elicited' more in susceptible than resistant hosts (Cronshaw and Pegg, 1976; Pegg, 1985) or the nutritional status of the host (Englehard, 1989) have not been considered. Methods and conditions of toxin isolation and chemical purification vary enormously from one experimenter to another. How far reported differences in chemical proportion and molecular size (usually cited to justify the existence of a 'new' toxin) represent molecular aggregation, analytical shortcomings or a failure to separate independent but chemically bound molecules is an open question. Apart from limited immunological studies in potato, nothing has been achieved with *Verticillium* toxins *in planta* comparable with the work on ceratoulmin in Dutch elm disease. Whether a common toxin/elicitor exists for *V. dahliae* and *V. albo-atrum*, with variations appropriate to pathotype and host species, is a question deserving of priority for the future. This work involving an attempt to reproduce the specific host substrate as a nutrient source together with multiple bioassays for 'toxins' and 'elicitors' may resolve the question as to whether some or all of the toxins described to date represent cultural artefacts.

To date, the work on proteins and glycoproteins in cotton and potato provides the best evidence for a toxin basis for *Verticillium* pathogenesis. The evidence, however, is often contradictory and with many inconsistencies. The pioneering work of Zel'tser and Malysheva (1966) and Malysheva and Zel'tser (1968), though providing the basis for most of the modern work on wilt toxins, has received scant recognition from subsequent workers.

Physiological Aspects of Pathogenicity

Most of the work reviewed in this section, while involving *Verticillium* diseases, represents a selective and frequently discontinuous account of the developments in a particular field. Where a multiple role has been claimed or inferred for a particular topic, e.g. ethylene, the evidence is reviewed elsewhere in the appropriate section. In no instance has *Verticillium* wilt been the subject of a pioneering investigation. Much relevant work forming a continuum of discovery relates to studies on *Fusarium* and prokaryote pathogens and is not covered in this book.

Growth Substances

Talboys (1958b) and Selman and Pegg (1957) described symptoms in hop and tomato attributable to an imbalance in levels of host growth substances (see also reviews by Sequeira, 1963, 1973; Pegg, 1976a, 1981a). Pegg and Selman (1959) identified indol-3-yl-acetic acid (IAA) at 0.036 $\mu g\ ml^{-1}$ in cultures of *V. albo-atrum* grown on a tryptophan-free Czapek-Dox medium. Young leaves and stem apices of infected tomato plants showed a sevenfold increase over healthy plants. Young cut shoots treated with 1–5 p.p.m. of IAA in aqueous solution developed acute epinasty and hypertrophy. Older shoots became chlorotic and wilted. Dixon and Pegg (1972) failed to find the IAA precursor tryptophan in tomato bleeding sap, but this could have been degraded during the extraction and subsequent acidic analysis. Wiese and DeVay (1970) reported hyperauxiny in *V. dahliae*-infected cotton. No differences in auxin levels were found in plants infected with non-defoliating (S) and defoliating (T) isolates. Pegg (1959) induced tyloses in tomato shoots with exogenous IAA, and proposed that a similar mechanism existed in the *Verticillium*-infected host. Since all classes of plant growth substances, IAA, ethylene, gibberellins, cytokinins and abscisic acid (ABA), are known to interact and in some cases act as substitutes in a particular effect, it is difficult to attribute an exclusive role to an individual compound.

Ethylene has been shown to play a multiple and major role in wilt diseases (Pegg 1976a, 1981a); since it is IAA mediated, it is not easy to separate the effects of the two components. Wiese and DeVay (1969, 1970) showed a strong correlation between *V. dahliae*-evolved ethylene, host ethylene and defoliation of cotton. Ethylene from T9 (defoliating strain)-infected cotton increased fivefold, 13 days after inoculation, whereas only a twofold increase occurred with SS4 (non-defoliating) infection. Pegg and Cronshaw (1976) reported ethylene occurring as a pulse 9–12 days after inoculation of tomato with *V. albo-atrum* and coinciding with the onset of symptoms. No ethylene was detected from healthy or *Ve* gene-resistant-infected plants. Talboys (1978) reproduced in fine detail the 'tiger stripe' foliar symptoms of hop wilt, following uptake of 1500 p.p.m. polyethylene glycol (PEG) molecular weight 1500, and 10 p.p.m. of chloroethylphosphonic acid, an ethylene-generating compound. The essential direct and indirect involvement of ethylene in other aspects of *Verticillium* pathogenesis has been described by DeVay (1989) and Pegg (1989). The link between pectolytic enzymes and ethylene is complex (Pegg, 1981a), involving *de novo* synthesis of enzymic mRNA (Bennet and

Della Penna, 1987) and increased tissue sensitivity to the enzyme (Cronshaw and Pegg, 1976). The interaction with divalent ions is similarly complex; where these are necessary for PG activity (Mussell, 1973) or may simulate symptoms in ethylene-pre-treated tissue in the absence of enzymes (Pegg, 1985).

Although ethylene production has been seen as a function of loss of leaf turgor in *Fusarium* wilt, Pegg and Cronshaw (1976) found no increase in ethylene in water-deficient tomato leaves until irreversible cell death occurred. Although epinasty, adventitious root production, chlorosis, necrosis, tylosis, supernumerary xylem formation and patterns of stomatal behaviour can all be attributed to ethylene; IAA, giberellic acid (GA) and ABA can all accelerate abscission in cotton, and kinetin and benzyl adenine (BA) can increase symptom severity (Misaghi *et al.*, 1969). Mussell and Strand (1977) and Strand and Mussell (1975) have described the release of proteins containing peroxidase and IAA oxidase from cotton and tomato parenchyma cell walls by purified PG from *V. dahliae*. IAA oxidase was solubilized from cell wall preparations from susceptible but not from resistant plants. These authors claimed that the release of IAA oxidase was an early critical event in symptom formation and that application of IAA to susceptible infected plants delayed symptoms (see Talboys, 1972). This suggestion of intrinsic differences in cell wall composition in single-gene resistant plants has not been confirmed or supported by other studies. These results, while illustrating the association of pectin-depolymerizing enzymes and growth hormones, do not account for the hyperauxiny associated with susceptibility. Mussell *et al.* (1982) further describe an enzyme capable of generating ethylene from 1-amino cyclopropane carboxylic acid (ACC), liberated from tomato cell walls by a purified PG from *V. dahliae*. The ethylene-producing reaction mixture was stimulated by IAA, Mn^{2+} and *p*-coumaric acid. Ethylene was inhibited in the presence of a competitive inhibitor of IAA oxidase. Critical concentrations or the explanation for this apparent IAA paradox were not given. *V. dahliae*-infection of cocoa (*Theobroma cacao*) bears similarity to cotton wilt in that a highly virulent, defoliating pathotype has been shown by Resende *et al.* (1996a,b) to induce large quantities of ethylene in newly developed leaves as a prelude to defoliation. The case for ethylene involvement is strong since the process can be reversed by the application of the ethylene inhibitor, silver thiosulphate. These symptoms are not apparent in infection caused by a non-defoliating isolate where a rapid decrease in transpiration, stomatal conductance and midday leaf water potential is the prelude to wilt in the absence of defoliation. The results of Van der Molen *et al.* (1983) suggest that vascular blockage occurs as a result of melting of the primary cell wall by depolymerase enzymes induced by host ethylene. The available evidence suggests that ethylene may be both a cause and a result of pectolytic activity in a complex synergism (Abeles, 1973).

Stunting reported by many authors may result from a toxin, or a water or growth substance deficit. Misaghi *et al.* (1969) suggested that accelerated senescence in *V. dahliae*-infected cotton might be due to a reduction in endogenous cytokinins. Krikun *et al.* (1971) using a soybean callus assay reported reduced cytokinin and auxin, but increased ABA activities in *V. dahliae*-infected tomato. Reduced cytokinin activity in *V. albo-atrum*-infected tomato was confirmed in a more critical study by Patrick *et al.* (1977). Since cytokinin synthesis is in roots, a reduction in shoot supply would be expected from impeded vascular flow. Misaghi *et al.* (1972) found a reduction in one of three endogenous cytokinins in cotton infected with *V. dahliae*. The compound closely resembling zeatin was reduced by 28 and 39% in tracheal fluid and shoots, respectively. This reduction occurred after the appearance of symptoms and was considered to be a result and not a cause of disease. The role of growth inhibitors in *Verticllium* wilt is conjectural, with limited evidence. Pegg and Selman (1959) first reported increased growth inhibitor (inhibitor-β) levels in *V. albo-atrum*-infected tomato. This subsequently was confirmed as ABA (Pegg, 1981a). Although young leaves and shoot apices from diseased plants had a higher ABA content than those of healthy plants, the endogenous level could not account for the reduced leaf area. Since ABA is markedly affected by leaf turgor, the results were interpreted as a consequence of wilting rather than a cause of stunting. The role of ABA in foliar abscission is also uncertain. Wiese and DeVay (1970) found a good correlation between ABA content of cotton and foliar abscission caused by the T9 strain of *V. dahliae*. The evidence from ethylene studies suggests that this is a more likely cause of abscission, as well as stunting and other symptoms, while the major role for ABA may be stomatal regulation. No information is available on gibberellins and *Verticillium* disease. The findings of Aube and Sackston (1965) that gibberellin-like substances were produced by *V. dahliae* cultures may explain the stimulated growth of plants inoculated with avirulent strains of *Verticillium* spp.

Photosynthesis and Photoperiodicity

Selman and Pegg (1957) described dry weight losses of 72, 70 and 65% in tomato leaf stem and root, respectively, 8 weeks after inoculation with *V. albo-atrum*. Leaf photosynthesis was greatly impaired, with net assimilation rates, calculated on non-necrotic tissue, of 0.47 and 0.39 g DM^{-2} of leaf area $week^{-1}$ for healthy and infected plants. Mathre (1968) measured a twofold reduction in photosynthesis in cv. Acala 4-42 cotton by O_2 evolution and $^{14}CO_2$ fixation. Infected plants showed decreased O_2 evolution, lower starch content and reduced efficiency of the Hill reaction. Dark CO_2 fixation was reduced by between 10 and 30%. Kr'stev and Savov (1977) reported reduced photosynthesis and transpiration in cotton accompanied by an oxidative burst of increased peroxidase but decreased catalase activity. Leyendecker (1950) had earlier shown cotton leaf chloroplast degradation soon after infection with *V. dahliae*. Similar but somewhat equivocal work on cotton was carried out in the former USSR: Rubin *et al.* (1977) could find no change in the fatty acids of galactosyldiglycerides, sulpholipids or phospholipids during the first 3 weeks after infection. These results are at variance with those of Rakhmankulova *et al.* (1973) when increased photochemical activity of the chloroplast and increased binding strength between chlorophyll and the protein–lipid complex was recorded only 3 days after seedling inoculation. Fifteen days after inoculation, considerable destruction of chloroplasts occurred. Safiyazov (1978), also working on Uzbek cotton cultivars Tashkent-2 and 108-F, found a substantial reduction in photosynthesis leading to loss of assinilates in 2–3 leaf stage plants. The first pathological symptom of *V. dahliae* infection was an ultrastructural breakdown in the chloroplast which led progressively to the death of the plant. In plum, naturally infected with *V. dahliae*, chlorophyll-b decreased in green leaves in which oxidized carotenoids increased. As the disease progressed, chlorophyll and some yellow pigments were completely destroyed. In partially and wholly chlorotic leaves, carotin increased and violaxanthin decreased this effect (Shishkanu and Istrati, 1970). How far early changes to chloroplast structure and activity reflect the involvement of a toxin or incipient water loss is a question worthy of investigation. An acute water deficit, ABA closure of stomata, as well as toxaemia, would all lead to reduced photosynthesis and increased chlorophyllase activity. Salikhova *et al.* (1970) described the effect of *V. dahliae*-cultured 'toxins' on gas exchange in cotton leaves. A resistant clone of lucerne inoculated with *V. albo-atrum* showed reduced plant height, stem dry weight and aerial biomass at 40% photosynthetic photon flux density (PPFD) comparable with susceptible clones at 100% PPFD (Pennypacker *et al.*, 1994). Dark respiration in lucerne was suppressed under 40% PPFD, increased under 70% PPFD and unaffected (as was resistance) under 100% ambient PPFD. Total protein and *in vivo* ribulose-1,5-bisphosphate carboxylase/oxygenase (RUBISCO) activity were reduced in 40% illuminated plants only. The photosynthetic properties of inoculated resistant plants at 40% PPFD mimicked those for susceptible plants, but only dark respiration and the amount of RUBISCO were reduced by *V. albo-atrum*; other effects were reduced by low PPFD. The authors suggest that increased *in vitro* total RUBISCO and a decreased ribulose-1,5-bisphosphate (RuBP) regeneration rate in inoculated resistant plants were due to photosynthetic habituation, optimizing carbon assimilation (Pennypacker *et al.*, 1995a,b).

A study of the interaction of *V. dahliae* and *P. penetrans* using non-destructive, infrared gas analysis (IRGA) on potato plants was conducted by Saeed *et al.* (1997a,b), up to 45 days after planting and inoculating cv. Russett Burbank. IRGA analysis of a population on 1- to 25-day-old leaves recorded no effect of infection of either pathogen individually. Joint inoculation significantly reduced net photosynthesis, stomatal conductance and transpiration. Intercellular CO_2 greatly increased with both pathogens especially in older leaves. Assimilation rate and stomatal conductance were in a linear relationship. The reduction in gas exchange with conjoint infection without a concomitant fall in intercellular CO_2 suggested non-stomatal involvement. Assimilation rates were reduced by up to 40% in newly produced leaves before the appearance of visible symptoms (Guidi *et al.*, 1997). In a later study, Saeed *et al.* (1999) were able to show that concomitant infection of potato by *V. dahliae* and the nematode *P. penetrans* at ambient CO_2 concentration reduced carbon assimilation and light use efficiency (μmol CO_2 fixed per μmol of light used) and increased intercellular CO_2 (Ci) of young leaves at 42 days after infection, but not earlier. Significant reductions in both the initial slope of C-assimilation rate versus Ci curves and C-assimilation, at Ci = 500 μmol mol^{-1}, in jointly infected plants were apparent at 42 days. Infection by either individual pathogen did not change the initial slope at 16 days after infection. It was concluded that C-assimilation in late stage infection could be non-stomatal but leaf patchiness could be a contributing factor. Photosynthesis in *V. albo-atrum*-infected tomato was reduced in leaves above the colonization point, except the terminal youngest leaves, and was attributed largely to altered mesophyll activity (see also Tomescu *et al.*, 1997, on pepper). An IRGA study on field-grown whole plant aubergines in soil naturally infested with *V. dahliae* and fumigated control plots by Gent *et al.* (1995), using clear-walled gas-tight chambers, found that gas exchange differences and symptoms were only apparent after fruit set. Leaf area and photosynthesis were greatly reduced; when IRGA readings were expressed per unit leaf area, the reduced carbon assimilation was directly related to reduced plant leaf area rather than reduced photosynthetic efficiency.

Photoperiod has a profound effect on disease development in *Verticillium* wilts. McLeod and Thomson (1959) observed that late maturing tobacco cultivars were slower in developing symptoms from *V. dahliae* infection than early maturing cultivars. Tolmsoff (1960) and Harrison and Isaac (1968) found that *V. dahliae*-induced symptoms in potato appeared after tuber initiation during the late stage of maturity. Tuberization and *V. dahliae* symptoms are both controlled by day length (Busch and Edgington, 1967).

Plants grown in short photoperiods (SD) of 14 h dark and 10 h light developed tubers and moderate to severe symptoms, while plants in long photoperiods (LD) remained tuberless and symptomless. A similar relationship between flowering, symptoms and *V. dahliae* infection was found in *Chrysanthemum* (Busch and Schooley, 1970; Hall and Busch, 1971). Under SD (14 h dark, 10 h light), plants developed severe stunting at flower initiation which progressed to wilting necrosis and death at full bloom. Plants maintained in LD (16 h light, 8 h dark) remained vegetative and showed only mild symptoms in the lower leaves. Pegg and Jonglaekha (1981) failed to find marked differences in symptoms in LD- and SD-grown *Chrysanthemum* infected with *V. albo-atrum*. Stem colonization, however, measured by colony counts and chitosan analysis, was significantly greater in SD plants. Lucerne (*Medicago sativa*), a quantitative LD plant at high temperatures and SD plant at low temperature, showed 24% 'resistance' under LD condition (16 h light) but with SD of 12 or 8 h light, this was reduced to 12% (Kudela, 1975). Tomato, like tobacco (Wright, 1969), a day-neutral plant, developed acute symptoms following *V. dahliae* inoculation under a 4 h photoperiod (Jones *et al.*, 1975) with 98% mortality. Only a mean 5% of the plants died at photoperiods of 8, 12 and 16 h. *Ve* resistant cv. Tropic remained healthy under the 4 h SD photoperiod. (See Jones *et al.* (1978) on the interactions of temperature and short days on symptom expressions in R and S tomato cultivars.)

Sunflower (*Helianthus annuus*) had a different photoperiodic response to *V. dahliae* infection from other plants studied. Plants grown under LD (16 h light) developed more severe symptoms than under SD. No flowers were produced in inoculated SD or light-interrupted night plants (Sackston and Sheppard, 1973). This work would merit a careful re-examination since LD plants grew taller and more vigorously than SD plants. In all the reports of the response of infected plants to different photoperiods, no explanation for the mechanism has been forthcoming (Pegg, 1981a).

Water Relations

Many accounts of water relations in *Verticillium*-infected plants have represented only a minor part of the main study. Work discussed here is that in which water relations represents the main theme of the paper; see also Talboys (1968), MacHardy *et al.* (1972), Hall and MacHardy (1981), Hall (1986), and Jones (1986).

In young *V. albo-atrum*-infected tomato plants in sealed containers with a controlled water supply, transpiration was reduced by 10% with the appearance of symptoms; subsequently, transpiration fell by 20% of that of the controls. Dye studies showed petiolar and some stem blocking (Scheffer *et al.*, 1956). Selman and Pegg (1957) failed to find differences in water content between healthy and stunted but not wilting leaves. Threlfall (1959) confirmed Scheffer *et al.*'s (1956) findings. A rise in transpiration in infected plants 3–4 days after inoculation was followed by an 80% decline by day 13–15, when epinasty and chlorosis developed, immediately followed by wilt. Using aqueous 0.01% basic fuchsin under a pressure of 30.4 bars, it was concluded that stem resistance to water flow was increased 3–500 times. Threlfall (1959) demonstrated that 90% of the stem internode vessels were blocked, with hyphae found in 50% – much higher values than those recorded by others. Recovery of excised leaves in water was inversely related to the number of occluded vessels. The relative water content (RWC) of wilted leaves was 59–70% compared with 72–75% for healthy turgid leaves. Using simple techniques, no difference in osmotic values between healthy and infected wilting leaves could be found. Tomescu *et al.* (1997) also recorded a 75.4% increase in transpiration in pepper, 40 days after inoculation with *V. dahliae*; at 95 days, however, at fruiting, a major water sink, transpiration fell by 71% compared with controls. The relationship of leaf water potential, phenology and lint yield in cotton was described by Tzeng *et al.* (1985).

Other Effects

The content of cotton seed oil was markedly reduced in four cultivars infected with *V. dahliae*, in proportion to the degree of infection (Alimukhamedov, 1972). *V. dahliae* similarly depressed the yield of pelargonin oil by 64% from Egyptian-grown *Pelargonium zonale*, a disease largely transmitted in cuttings from infected stock plants (Mohamed *et al.*, 1987). Increased permeability, a common feature of infected plant tissue, was reported in cotton by Dube (1971).

Resistance 9

Resistance to *Verticillium* spp. is recognized by the total or partial absence of symptoms in comparison with other host species or cultivars similarly exposed to virulent pathotypes and showing severe damage or death. Throughout the voluminous literature on vascular wilts, there is a major underlying assumption that resistance is based on the exclusion or expulsion of the pathogen from the host or its restriction within the vascular system. Unlike most biotrophic and some necrotrophic reactions, however, there is no immunity to infection or anything strictly comparable with hypersensitivity. Thus monogenic and polygenic resistant hosts alike all show limited root or hypocotyl colonization. This presents a philosophical and a physiological problem.

The bases of resistance are seen as reduced inoculum potential in the rhizosphere, or rhizoplane (see Chapter 6); host physical restriction and limitation of the pathogen in the host, and host chemical inhibition of the pathogen. Since theories on pathogenesis have been dominated by vascular occlusion with its concomitant shoot water deficit, the alternative or complementary mechanism of host tolerance to pathogen metabolites has received scant attention. Most of the mechanisms referred to may contribute to a general expression of resistance rather than function alone. Above all, they represent *non-specific* responses, some of which can be found in susceptible as well as single R gene lines. (See reviews by Beckman and Talboys, 1981; Bell and Mace, 1981; Beckman, 1984, 1987.) The major aspect of resistance, the identification and selection of genetic resistance, will be considered under plant breeding and control.

Physical Aspects of Resistance

There are few reports of host morphology acting as a determinant of resistance, although a general unsubstantiated view exists that the absence of vascular pathogens in gymnosperms is related to the short length of tracheid elements. Rudolph (1968) claimed that the resistance of hop cultivars to *V. albo-atrum* was a function of the point of emergence of the first lateral root from the tip. Cultivars in which this interval was short were more tolerant to the disease. Phillip and Wilhelm (1971) showed a similar correlation in cotton, probably related to disease escape. *Gossypium barbadense* cv. Waukena White produced more lateral roots deep in the soil and was less susceptible to *V. dahliae* than *G. hirsutum* cv. SJ-1 with fewer deep laterals. While the obvious explanation that deep-originating lateral roots may escape the bulk of soil inoculum may be true, such a comparison in two highly polygenic *species* is not strictly valid when other resistance features in these plants are taken into account.

Cell wall thickening and lignitubers

Van der Meer (1925) first described conic protuberances (lignitubers) in potato and cucumber root cortical and pericyclic parenchyma walls in response to advancing hyphae of *V. albo-atrum*. The lignitubers were pierced by a fine (0.5–1.0 μm) tube through which the invading hyphae grew to swell when passing into the cell lumen. Dufrenoy (1927c) later reported the same phenomenon in pea roots infected with *V. dahliae*, but not as a resistance phenomenon. Isaac (1946) failed to find such structures in *V. dahliae*-infected sainfoin and described instead appressorial-like swellings on the tips of the penetrating hyphae. Much uncertainty has surrounded the role and chemical composition of lignitubers first observed by De Bary (1863) and named by Fellows (1928) for thickenings in wheat root in response to *Gaeumannomyces graminis* infection. Alternative names have been: infection pegs, papillae, or callosities. It is now thought that callose 1,3-β-glucan is the principal constituent, but Talboys (1958a) and Griffiths and Lim (1964, 1966) all found evidence of lignin deposition probably occurring as a secondary modification. Talboys (Beckman and Talboys, 1981) observed lignitubers in roots of 42 species of 13 dicotyledon families and in *Equisetum arvense* rhizoids infected with *V. albo-atrum*.

The role of lignitubers in resistance is equivocal. Callose deposition is wholly non-specific and may result from mechanical damage or a fungal glycoprotein induction (Beckman and Talboys, 1981). Talboys (1972) described lignitubers and other physical structures as *secondary determinants* of resistance. In a pioneering study, Selman and Buckley (1959) found lignitubers on all walls of cortical and stelar parenchyma cells in young roots, but not in the piliferous layer. In sterile solution at 200 or 2000 p.p.m. sucrose concentration, numerous lignitubers formed and invading mycelium was restricted to the hypodermis. At

20,000 p.p.m. sucrose, all root tissues were invaded and lignitubers were small, or absent in the dark-grown cultures. This result does not establish a *prima facie* case for lignitubers and resistance, since no account was taken of parallel changes in fungal enzyme induction or repression at the different sucrose levels. Griffiths and Isaac (1966b) obtained complete colonization of tomato seedling roots and no lignitubers by placing agar blocks with *V. albo-atrum* in contact with the roots in moist dishes. *V. dahliae*, *V. nigrescens*, *V. nubilum* and *V. tricorpus*, however, all induced cortical lignification and lignitubers. Root systems subsequently remained uncolonized or were invaded only slowly. High nitrogen stimulated root invasion by some species and suppressed the host response. Whether these isolates were non-, or only weakly virulent was not confirmed. In general, there has been no correlation between genetic resistance and lignituber production. Talboys (1958c), however, recorded penetration of lignitubers by *V. albo-atrum* mostly in low resistance hop cultivars. In pea roots invaded by *V. dahliae*, callosities formed from a coalescence of paramural vesicles on the plasmalemma after rupture and depletion of contents (Griffiths, 1971b). While most formed from apparent contact with invading hyphae on the opposite cell surface, lignituber formation started before penetration and contact with the plasmalemma. Griffiths (1971b) found no hyphal penetration of the lignituber tip and described the degeneration and death of the enveloped hypha. The frequency of papilla (lignituber) function was the same in Ve^{+} and Ve^{-} tomato cultivars infected with *V. dahliae* (Bishop and Cooper, 1983a,b). While individual structures were effective in restricting hyphae, they did not constitute a resistance barrier to colonization. The suberized endodermis similarly created a partial barrier, but penetration into stelar parenchyma occurred through endodermal cells. Hutson and Smith (1983) found a similar result with near-isogenic lines of tomato to *V. albo-atrum* and claimed that the only entry into the root was at the point of emergence of lateral roots. This was refuted by Gerik and Huisman (1985). In general, intact roots of tomato seem more resistant to penetration by *Verticillium* than by *Fusarium*. Transplanting injury and cutting tomato roots increased infection to *V. albo-atrum* by ten- and fourfold, respectively (Selman and Buckley, 1959). Any lignitubers or endodermal resistance mechanism functioning in intact seedling roots *in vitro* would be purely academic in relation to the naturally growing abraded root systems found in soil.

Tursunov and Dariev (1979) described the initial response of high resistant cotton, *G. australe* ('immune' [*sic*]) as browning of the paravascular parenchyma; partial destruction of starch grains in the lesser resistant cultivars C4727 and Uzbekistan 3 (an F_3 of C4727 × *G. mexicanum* [*sic*] = *G. mexicanum*). The authors discounted any resistance effect of wall thickening since vessel lumen penetration was exclusively via pit pores. A similar comparison of *G. hirsutum* ISA205 (susceptible) and the resistant *G. barbadense* Ashmounii line challenged by a P_1 strain isolate by Daayf *et al.* (1997) found that after 1–4 days, walls were strengthened with callose and cellulose accompanied by a strong production of terpenoids and phenolics. The interesting non-specific aspect of this study is that precisely the same reactions occurred in the susceptible cultivar, but later. An ultrastructural study of cotton (presumably a susceptible cultivar) showed that contact cells adjacent to initially infested vessels had the central vacuole broken up into smaller ones and the development of an apposition layer. Osmiophilic deposits were secreted through the plasmolemma and deposited on the apposition layer and pit fields. Other contact cells had a totally degenerate cytoplasm but no fungal presence. Contact vessels adjacent to secondarily infested vessels secreted the osmiophilic substance into the vessel lumina causing occlusion and embedding hyphae (see Peresse *et al.*, 1971; Mueller and Morghan, 1993). Benhamou (1995) identified the paravascular parenchyma cells, in this case of aubergine, as the site signalling the mobilization of a number of defence strategies, the earliest of which were the accumulation of electron-opaque, globular structures in paramural spaces and the intracytoplasmic vesicles filled with osmiophilic deposits. The globular structures were rich in callose, identified by gold labelling with a purified 1,3-β-glucanase, and contained glucosides and galactose residues. Lipids and phospholipids were detected in the vesicles, while chitinase and 1,3-β-glucanase were found in the vacuoles.

Vascular occlusion and restriction of fungal spread

The historic and traditional view of vascular occlusion, a prelude to a shoot water deficit, and essentially a feature of pathogenesis in susceptible plants, is interpreted by a number of authors as a resistance mechanism (see Beckman and Talboys, 1981), isolating the pathogen and restricting its spread. Vascular occlusion is attributed variously to blocking by gums, gels, tyloses, xylem vessel wall coatings and fungal mycelium.

Wall coating

The deposition of an amorphous layer on vessel walls and fungal hyphae should be distinguished from the grosser plugging of vessels and the infusion of phenolic compounds into vessels and adjacent xylem parenchyma. Cooper and Wood (1974) using scanning electron microscopy (SEM) described the 'blistered' appearance of *V. albo-atrum*-infected, tomato xylem walls, showing interconnecting outgrowths between hyphae and from hyphae to vessel walls. Pegg *et al.* (1976) using *V. albo-atrum*-infected tomato and the near-isolines Craigella Ve^+ and Ve^-, described under transmission electron microscopy (TEM) a deposit of electron-dense material, frequently layered around the invading hyphae and also on the vessel walls occluding paired pit fields. The deposit occurred in both resistant and susceptible cultivars and was considered to be identical to the 'outgrowths' described by Cooper and Wood. A similar, if not identical, deposit was described later for chrysanthemum and sunflower infected with *V. dahliae* (Robb *et al.*, 1978, 1979a,b). Street *et al.* (1986) found that a massive infusion of conidia of *V. albo-atrum* into the tomato xylem induced cell wall coating in both resistant and susceptible tomato plants. The coating was of two kinds, one, histochemically resembling lignin, associated with xylem parenchyma cells that had not yet formed secondary walls, and the other, histochemically resembling suberin, associated with the secondary walls of xylem parenchyma. The authors infer that the coating response is due to hypersecretion and/or modification of normal cell wall components induced by the presence of the pathogen.

Peresse *et al.* (1971), Catesson *et al.* (1976), Moreau *et al.* (1978) and Moreau and Catesson (1982, 1984, 1985) compared the responses of carnation xylem parenchyma to *V. dahliae* (a non-pathogen of carnation) and *Phialophora cinerescens*, a specific pathogen of this host. The secretory activity of xylem parenchyma mitochondria coincided with the appearance of *V. dahliae* introduced into the plant by wounding, but not following the introduction of *P. cinerescens*. The authors concluded that an efficient defence system existed in carnation against *V. dahliae* but the response to *P. cinerescens* was disorganized.

Newcombe and Robb (1988), working on the reaction to *V. albo-atrum* of three phenotypic classes of lucerne, highly resistant, moderately resistant and susceptible, induced host responses by infusing conidia into cut stems of healthy plants. Twenty-four hours after infusion, vascular coating at spore trapping sites varied inversely with the penetration of pit membranes; thus, highly resistant plants restricted pit penetration and lateral growth; moderately resistant plants failed initially but were resistant to secondary colonization; while the susceptible plants failed to restrict secondary colonization. The same authors (Newcombe and Robb, 1989a) traced the host cytological response to the initial accumulation of lipid bodies in vascular parenchyma cells close to *Verticillium* trapping sites. This was followed by the formation of intercellular coating material between the trapping sites and vessels, and finally by a suberin-like coating material in the lumina of the vessels providing a barrier to fungal penetration. The colonization ratio, the measure of pathogen invasiveness and host resistance in lucerne was defined by Newcombe *et al.* (1989).

Robb *et al.* (1989), working on *V. albo-atrum* colonization of tomato, claimed that vascular coating formed earlier in resistant than in susceptible plants, and treatment of inoculated resistant plants with a PAL inhibitor depressed the secretion of coating material, resulting in the plant behaving as a susceptible phenotype. These authors considered the timing of the coating response in vessels as an expression of the *Ve* gene for resistance. In a subsequent study (Gold and Robb, 1995a,b) on tomato cultivars Craigella (Ve^+), Craigella (Ve^-) with *V. dahliae* races 1 and 2, it was claimed that an aggressive pathogen was able to suppress coating, resulting in a susceptible reaction which was also associated with a high capacity for hoizontal spread. Robb *et al.* (1991) claimed that the wall and hyphal coating was suberin and could be induced by precursors of the phenylpropane pathway or ABA. Suberin was not identified unequivocally and the coating response, although appearing first in Ve^{+} cultivars, subsequently was far more intense in the susceptible cultivars. Essentially the same reaction was reported in two resistant lucerne clones, 1079 and WL-5, by Pennypacker and Leath (1993). In WL-5, *V. albo-atrum* was confined to the crown for 14 weeks after inoculation, by vascular occlusion before passing to stems; (16 weeks in the case of 1079). The authors interpret the development of narrow metaxylem vessels in clone 1079 as a cause of resistance rather than an associated response, and also hypertrophied xylem parenchyma in clone WL-5 which tested positively for suberin. *V. albo-atrum* frequently was found encased in a suberin- and lignin-positive substance in the vessels. While it is attractive to interpret hyphal coating as a defence response, it is not a ubiquitous feature of infection in resistant plants and is commonly seen in pathogeneic susceptible reactions. The total sealing of pit field and vessel end plates, however, would constitute one of the most effective blocks to the shoot water supply and could hardly be considered a primary product of the *Ve* gene.

Vascular gelation and gummosis

Beckman, in an extensive literature review (Beckman and Talboys, 1981; Beckman, 1987), has considered the role of gels and gums principally as a spore-trapping mechanism delaying or preventing fungal spread and development. The literature is largely confined to *Fusarium*, and the sequence of work on *Verticillium* is incomplete. No clear distinction is made in most publications between gels described by Beckman (1987) as weak, ephemeral spore-trapping entities which can shear under the transpiration pull, and the dense permanent phenol-infused plugs described by Moreau *et al.* (1978). Both occur in susceptible plants and cause severe disruption to the leaf water supply (Bewley, 1922; Threlfall, 1959; Talboys, 1968; Robb *et al.*, 1975a,b). Whether one is derived from the other or both arise by the same basic process in every host is not clear. Localization of infection is seen as a basic mechanism of resistance however, in cotton (Bugbee, 1970; Mace, 1978; Harrison and Beckman, 1982), hop (Talboys, 1958b) and tomato (Threlfall, 1959). Van der Molen *et al.* (1977) showed that pectinaceous gels arose from perforation plates, end walls and pit membranes by the distension of the primary and middle lamella following inoculation of nine species with a non-host-specific strain of *V. dahliae*. Beckman (1987) in a speculative scheme considered that elevated respiration and glycolysis normally associated with infection would provide enhanced levels of Kreb's cycle acids which had been shown, with alternating pH from <3 to <4.5, to cause middle-lamella membranes to swell. It was speculated that during the night, di- and tricarboxylic acids would accumulate to sufficiently high levels which, with reduced pH, would induce gelation. During the day, the transpiration flow would dilute Kreb's cycle acids, and previously formed gels would dissipate. The carboxyl groups on the uronide polymer of the middle lamella are bridged with Ca^{2+} forming a stable structure further protected by hemicellulose components which are hydrogen-bonded to the molecule. A possible predisposing environment to membrane attack was described in tomato affected by *V. albo-atrum* (Pegg, 1985), where calcium and other divalent ions moved from the stem and accumulated in the leaf lamina.

Moreau *et al.* (1978) studying the effect of the non-pathogen *V. dahliae* on carnation found evidence for *de novo* synthesized substances secreted through the plasmalemma of contact parenchyma cells, which moved through the pit membranes and accumulated as gums and plugs in the vessel lumina. This effect was seen as a feature of pathogenicity rather than resistance. Van der Molen *et al.* (1982, 1983) found that gels in banana contained predominantly arabinose, xylose, glucose and galacturonic acid. With the notable absence of cellulose, the gel constituents closely resembled the composition of the primary cell wall. Contrary to the conclusion of Beckman and Talboys (1981), many workers have associated vessel plugging and vascular browning with fungal pectolytic enzymes. McIntyre (1965), working with tomato cv. Bonny Best, infiltrated shoots with PG preparations from *V. albo-atrum*. Leaves showed some marginal necrosis but no wilt. No vessel plugs were observed in stem section. Van der Molen *et al.* (1983) induced gel plugs in castor bean (*Ricinus communis*) leaves by treating with fungal PG, PL and 1,4-β-xylanase. Gels stained with ruthenium red were observed in fresh sections, since fixing and dehydration caused shrinkage and loss. Mussell *et al.* (1982) and Pegg (1981a) have described the production of ethylene by pectolytic enzymes, and Abeles (1973) the role of ethylene in enzyme synthesis. Van der Molen *et al.* (1983) reported ethylene production and also gel formation following enzyme treatment when 3.0 p.p.m. of ethylene was used alone. Gel production, however, was blocked when leaves were treated with enzymes in the presence of an ACC synthetase inhibitor, preventing ethylene synthesis. The obvious conclusion from this work, that gels are induced by ethylene alone, is difficult to reconcile. A possible explanation is that *host enzymes* such as PG and/or hemicellulases may be specifically required for gel formation and are only synthesized following ethylene action. This is well documented for fruit ripening. Treatment of plants with fungal enzymes (Van der Molen *et al.*, 1983) would not preclude the involvement of host enzymes, in place of, or supplementing, fungal enzymes in the formation of gels.

The apparent paradox of gels functioning as defence agents and as a primary cause of wilting may depend on the precise timing of the event, as described in detail by Beckman (1987). Barchilon *et al.* (1987) found a positive correlation between the resistance of avocado cultivars inoculated with *V. dahliae* and the rapid formation and quantity of gel observed microscopically. Susceptible lucerne clones infected with *V. albo-atrum* showed intense vascular plugging and coating with gums and other materials which stained positively for lignin and suberin. These deposits and the crushing of metaxylem elements were seen as the principal cause of wilt (Bishop and Cooper, 1983b; Pennypacker and Leath, 1986). It is difficult to separate the physical effect of a gel in limiting a pathogen from its possible chemical antifungal role. Vascular browning is one of the most constant and stable features of *Verticillium* infection. It is seen as a phenolic infusion of primary and secondary walls and also of vascular gels or gum plugs. The literature and experimental work on phenolic staining is confined almost exclusively to *Fusarium* (see Wood, 1967; Beckman and Talboys, 1981; Pegg, 1981a). The dark brown colour of vascular plugs staining positively with nitroso reagents indicates the presence of orthodihydric phenols. Tannins and gallocatechins may also be present, depending on the host genus and species. The presence of additional molecules such as chemically bound oxidized phenols may explain why no evidence of hydrolysis or degradation of gums has been reported (Pegg, 1981a).

Tylosis

The extrusion of paravascular cell walls into the xylem lumen, often with nuclei and cell contents, is commonplace in higher plants (Reichenbach, 1845; Chattaway, 1949). Tyloses and gels frequently occur together in healthy and infected plants and can be induced in susceptible plants by mechanical damage. Since gummosis and tylosis involve removal of Ca^{2+} from the primary wall followed by 'acid growth' and the intrusion of protons (Rayle, 1973), both may arise from a common mechanism. Prior to evagination the parenchyma cell shows intense metabolic and particle activity. Numerous mitochondria and Golgi bodies appear and vesicles extend from the Golgi to and through the plasmalemma. The tylose wall is at least two-layered and the outer surface has a fibrillar appearance (Beckman and Talboys, 1981). The same process which affects the parenchyma primary wall must also affect the adpressed middle lamella which in some hosts is carried into the xylem. Bishop and Cooper (1984), using TEM, observed in tomato that the pit membrane was ruptured. Much controversy has centred on the role, if any, tyloses play in *Verticillium* resistance. Talboys, in a series of papers (1957, 1958a,b) based on hop cultivars differing in resistance to *V. albo-atrum*, found a correlation between high tylosis, sparse colonization and mild symptoms. Conversely, in susceptible cultivars, intense colonization was accompanied by few tyloses but severe symptoms. Talboys (1958b), on the exclusive basis of the hop results, proposed a hypothesis in which infection and disease could be considered in terms of a *determinative phase* during which resistance factors such as gels and tyloses determine the growth and spread of the pathogen. This is succeeded by an *expressive phase* in which symptoms develop in proportion to the restriction of the pathogen in the *determinative phase*. This concept was developed by Beckman (1987) and extended as a general model for all vascular pathogens. Suchorukov (1957) considered that tyloses sealed off areas of cotton xylem containing *V. dahliae*. Talboys (1964) extended the concept to involve the release during infection of a critical concentration of fungal metabolites which directly or indirectly induce tylosis or gums sufficient to occlude the pathogen. Such a mechanism could not operate in single-gene resistant host plants, unless the host determined the level of pathogen metabolite in the first instance.

Pegg and Dixon (1969) and Dixon and Pegg (1969) studied colonization, tylosis and hyphal lysis in four tomato cultivars: Bonny Best and Potentate susceptible to tomato isolates of *V. albo-atrum*, and Loran Blood and Moscow, both containing the *Ve* gene for resistance. In addition to the tomato isolate, two hop isolates causing fluctuating (HF) and progressive (HP) wilt in hop caused moderate to severe symptoms in all cultivars. Colonization was not always proportional to symptom severity as measured by the reduction in leaf area. Few tyloses were induced by the T isolate, and tylosis in Loran Blood and Moscow infected with HF and HP was proportional to symptom severity rather than resistance. Dimond (1966) considered tylosis a greater contributor to symptoms than to resistance. Sinha and Wood (1967a) found the correlation between symptoms in tomato and tylosis predicted by Talboys (1958b). Tylosis in susceptible plants was delayed and symptoms developed beforehand. Conversely, Dixon and Pegg (1969) showed that colonization of plants was complete before tyloses appeared. There was a progressive loss of mycelium, however, in all host–pathogen combinations, but this was greater in the less susceptible plants. Tjamos and Smith (1975), and Bishop and Cooper (1983b, 1984) all found evidence for tylosis in resistant tomato cultivars as a basis for the supression of *V. albo-atrum*. These authors conclude that the endodermis rather than cortical lignitubers provides a first barrier, and tylosis and gelation a second isolating barrier permitting chemical defences to operate.

Beckman (1987) claimed that tyloses only occur when pits in reticulated walls are 10 μM or greater. Pegg (1959) induced tylosis in tomato with exogenous IAA. Although tylosis is recognized as non-specific, treatment of Ve^+ and Ve^- Craigella isolines with 5 v.p.m. ethylene gas following inoculation with *V. albo-atrum* induced significant increases in tyloses in the hypocotyl and in the first and second internode region of the resistant plants only (Pegg, 1976a). Tylosis, unlike gels, gums and vascular coating, appears to be greater in the resistant host. However, intense tylosis, if effective in arresting the pathogen, would be expected to cause a concomitant limitation to the shoot water supply (Dixon and Pegg, 1969). Tylosis, therefore, under some circumstances may comprise one of a number of non-specific effects contributing to resistance. The apparent role of tylosis in single-gene resistant hosts may merely represent a stress reaction correlating or linked with a dominant gene for some other resistance factor. The growth response of the stele by the isodiametric expansion of xylem parenchyma resulting from enhanced growth substance activity is often sufficient to crush annular and spirally thickened elements. This has been interpreted by Beckman and Talboys (1981) as a pathogen-limiting defence mechanism.

Supernumerary tissue

Talboys (1958b) described the differentiation of new xylem vessels from interfascicular cambium following infection of wilt-tolerant hop cultivars by *V. albo-atrum*. The net increase in fungus-free xylem was seen as a compensatory resistance mechanism. Pegg and Selman (1959) similarly reported hyperplasia in tomato xylem parenchyma adjacent to *V. albo-atrum*-infected vessels. Treatment of Craigella Ve^- tomato plants with 5.09 v.p.m. ethylene before inoculation induced resistance comparable with that in Ve^+ cultivars (Pegg, 1976a). While the mycelial content was reduced throughout the plant's vascular system, ethylene also stimulated new xylem vessels in the secondary thickening zone at the crown of the plant. Adventitious roots, also initiated by ethylene and supplying newly differentiated, fungus-free xylem could also be envisaged as a resistance response. Pennypacker and Leath (1986, 1993) found reduced vessel diameters in *V. albo-atrum*-infected lucerne, which were associated with increased IAA oxidase and possible hyperauxiny reported for the same disease by Krakta and Kudela (1981), but the evidence for a direct role in resistance is equivocal.

The control of *Verticillium* diseases by grafting susceptible scions onto resistant rootstocks (often of a different species) has been analysed for the most part in terms of relative colonization in reciprocal root–shoot grafts. Zhou *et al.* (1998a) found no differences in either free proline or electrolytic activity between healthy, self-rooted and grafted aubergine plants. Following *V. dahliae* infection, however, free proline and electrolytic levels were found as disease indicators. PAL activity was higher in roots and leaves of grafted plants, an indication in part of root–shoot transmission.

Biochemistry and Physiology of Resistance

Phenol metabolism

Most phenolic compounds in healthy and *Verticillium*-infected plants originate from the shikimic acid pathway. Some are formed from the head-to-tail condensation of acetate and malonate units. Compounds such as gossypol and related forms from cotton, although phenol-like, are strictly terpenoids in view of their isoprenoid origin (Pegg, 1981a) but are considered in this section.

There are many examples of increases in polyphenols in cotton following *V. dahliae* infection. The major phenolics are thought to be tannins (Rubin and Perevyazkina, 1951; Avetisyan, 1960; Gubanov, 1962; Krasnoshchekova and Runov, 1962; Babaev, 1964; Bell, 1973; Mace *et al.*, 1978). Gubanov (1962) found an increase in lignins and quinones; condensation of the latter is considered to be the basis of classical vascular browning. Rubin and Perevyazkina (1951) showed that the leaf tannin content of resistant cotton fell relative to that in susceptible plants, while the root content of resistant plants only, increased. Bell (1973) claimed that only a few specific tannin-producing xylem parenchyma cells exist in healthy cotton, whereas under infection stress most cells have this propensity. This was later confirmed by Howell *et al.* (1976), Mace *et al.* (1978) and Bell and Stipanovic (1978). The origin of vascular browning is thought to be condensation of oxidation products of dopamine in banana, catechin in cotton, and dihydroxy phenols in tomato (Bell and Mace, 1981; see also Bell, 1991b). Phenol concentration has been correlated with resistance of cotton species. *V. dahliae*-resistant *G. barbadense* cotton contains higher levels of total and dihydroxy phenols than the susceptible *G. hirsutum* (Bhaskaran and Muthusamy, 1974). Similarly, *V. dahliae*-resistant tobacco (Sheppard and Peterson, 1976) and potato (McLean *et al.*, 1961) have greater amounts of chlorogenic acid than corresponding susceptible plants. The decline in resistance of potato with age also correlates with falling chlorogenic acid content (Patil *et al.*, 1964). In all these examples, infection leads to increased levels of phenols in both resistant and susceptible hosts. In strawberry, the phenol content of healthy susceptible and resistant plants, however, is identical, but increases more in the resistant plant following infection (Okasha *et al.*, 1968a). Pollock and Drysdale (1976) measured the total soluble phenolic content of tomato, resistant and susceptible to *V. albo-atrum* and found little difference initially. Subsequently, there was a greater build up in the S cultivar. In a similar exercise on tomato cultivars, Khatib *et al.* (1974) demonstrated an inverse relationship between susceptibility and phenolic content. Of particular interest was the correlated decline in phenolics and increased susceptibility after flowering. This post-flowering increase in susceptibility to *Verticillium* has also been recorded in potato, cotton and chrysanthemum plants (Robinson *et al.*, 1957). The relationship between peroxidase (PO) activity and resistance of tomato to *V. dahliae* was described by Reuveni and Ferreira (1985). One of the few contributions on the possible role of wood tannins was made by Somers and Harrison (1967) studying *V. dahliae* infection of apricot. Pegg (1981a) listed the roles of phenolic compounds in host physiology as: uncouplers of respiration from oxidative phosphorylation, enzyme inhibitors, host growth inhibitors, antifungal compounds and enzyme stimulants.

Phenol-oxidizing enzymes, oxidases and peroxidases are widespread in healthy plants (Waggoner and Dimond, 1956) and rise substantially in infected plants. Phenol oxidases oxidize monophenols which, by dehydrogenation, form flavonoids, tannins, lignins and quinones. The source of enzymes in diseased plants is most probably of host origin. Neither *V. dahliae* nor *V. albo-atrum* cultures contained constitutive peroxidase, but culture filtrates strongly stimulated horseradish peroxidase (Suchorukov, 1957). The interaction of POs with the growth substances IAA and ethylene was described by Pegg (1981a). Hydrogen peroxide derived from oxidative reactions is a potent toxin in its free state. Since peroxidases utilize 1 mol and catalases utilize 2 mol of H_2O_2 in an oxidative reaction, their activity in infected plants has been considered part of a resistance response. Ethylene, which has a role in resistance, may be self-generating through the action of PO utilizing H_2O_2 produced by wall-bound phenol oxidases. These with peroxidases are released by pectolytic enzyme activity. Gordon and Paleg (1961) showed a link between IAA production and polyphenol oxidase (PPO) activity. The action of PPO on tryptophan produced an indole quinonimine which was converted to indolylpyruric acid and IAA. Pegg (1976a) found that PPO, PO and PAL all increased in tomato following *V. albo-atrum* inoculation, more so in susceptible than resistant plants; only PPO, however, increased following ethylene treatment. Ortho-dihydroxy phenols: chlorogenic acid, caffeic acid, DOPA and catechol, and their derivatives, scopoletin and vanillyl alcohol, all act as inhibitors of IAA oxidase, which would lead to hyperauxiny. Monohydric phenols, such as tyrosine, *p*-hydroxybenzoic acid and *p*-, *m*- and *o*-coumaric acid all act as stimulants of IAA oxidase depending on concentration (Pegg, 1976a). While phenolic acids and their oxidation products may function as antifungal compounds, quantities required are relatively high in cellular physiological terms. Lee and Le Tourneaux (1958) and McLean *et al.* (1961) showed an excellent correlation between the *o*-dihydric phenol (measured as chlorogenic acid) content of six potato cultivar tissues and their field resistance to *V. dahliae*. The LD_{100} for spore germination, however, was between 1000 and 2500 p.p.m.

As plants age, the phenol content and resistance fall; levels of phenols fall more rapidly and to a lower level in susceptible cultivars. This effect coincided with shortening photoperiod and lower light intensity. A fundamental relationship exists between phenol content (lignification), auxin content, light and disease resistance (Pegg, 1981a). Thus PAL is light induced, which leads to an increase in chlorogenic acid, itself inhibiting IAA oxidase. Lee *et al.* (1994) characterized the upstream regulatory region of the PAL gene in tomato (*PAL5*) by DNA sequencing and S_1 nuclease protection studies. The multiple regulatory sites were suggested to provide for constitutive PAL production and a differential expression in response to different stresses, in this case light, wounding and *V. albo-atrum* infection. Resistance in aubergine by grafting was associated with greater PAL activity in leaves and roots and lower proline and electrolytic leakage than in self-rooted plants (Zhou *et al.*, 1998a). See Powell *et al.* (1987) on the induction of PAL in *Verticillium*-infected tomato plants. Poletaeva and Bekker (1986) examined the balance of dehydrogenases and oxidases from two cotton species challenged by different wilt pathogen species.

Recent interest in different oxidases has come from cotton studies in China and the CIS. Vlasova (1993) studying the early phase of infection (4 and 7 days) of a P_1 strain isolate on a highly susceptible cv. S-4727 and the relatively tolerant cv. Tashkent-1, found a common increase in PPO and PO in detatched roots (root culture?). Only in R plants did total phenols increase. Leaves from susceptible cotton plants infected with *V. dahliae* showed greatly reduced activities of superoxide dimutase (SOD) and PO; 89 and 82%, respectively of healthy leaves, with a concomitant increase in malondialdehyde (MDA) (Guo *et al.*,

1995). It is now fashionable to refer to the long-established increase in oxidative reactions as the 'oxidative burst'. Part of the product of this is thought to consist of free oxygen radicals which are considered to be as damaging, if not more so, than H_2O_2. SOD utilizes free oxygen and is therefore assumed to be part of a defensive response. Zhu *et al.* (1995) claimed that infected susceptible *G. hirsutum* cotton plants in the pre-symptom period displayed higher SOD, PO, membrane lipid peroxidation and lower soluble proteins than tolerant cultivars. Symptom appearance was accompanied by an oxidative burst with greatly increased SOD and PO activities. Soluble protein decreased in R cultivars with increased dielectric leakage and MDA levels in S cultivars. Conflicting results were presented by Li *et al.* (1995) who claimed that while PO and SOD increased in both S and R cotton cultivars following inoculation, SOD activities were lower in R (tolerant) cultivars and higher in S cultivars. Xia *et al.* (1994) found essentially the same results in cotton. Peroxidase activity was positively correlated with symptom intensity, but where esterase isoenzyme levels were higher in resistant plants, SOD bands were 1–2 fewer in R and S cultivars. These results do not accord with general findings where SOD is regarded as an attribution of resistance. The PO activity of resistant *G. hirsutum* cotton cv. Zhong 3474 increased rapidly after infection, while the susceptible cv. Litai 8 increased more slowly. Large differences were observed at 5 days (Ji *et al.*, 1995). Wu *et al.* (1997) rightly conclude that the role of active oxygen species in plant pathogen resistance is not yet understood. Transgenic potato cv. Russet Burbank normally susceptible to *V. dahliae*, encoding glucose oxidase, produced several-fold increases in total salicylic acid in leaves. No change was found in free salicylic acid. RNAs of anionic peroxidase and acidic chitinase were also introduced. Several isoforms of extracellular PO including a new induced one were detected. The authors suggest that elevated sublethal levels of H_2O_2 may activate an array of defence mechanisms. The results, however, are too incomplete to make fair speculations.

Shadmanov (1995) presented an interesting but as yet unsubstantiated hypothesis on cell surface resistance. In this, primary defence from fungal infection is conditioned by two chains of respiration, which are non-mitochondrial, do not form ATP and in which the terminal oxidases are PPO or ascorbatoxidase. Penetration of surface cell layers by a vascular pathogen are thought to invoke intense oxidative processes which result in localized fungitoxic activity. In tolerant reactions, Shadmanov envisages the pathogen compelled to enter the more neutral environment of the xylem vessel but in a very weakened state and to have no great effect on the host. This hypothesis is suggested to explain why there is often mass clustering of *Verticillium* on the root surface, but little penetration. The general effect of light and day length on lignification is well known. The involvement of phenols in this process has a more probable role in resistance than phenols acting as fungistatic or fungicidal substances. A further major important role for free or bound phenols and the associated lignification is as blocks or inhibitors of pectolytic enzyme activity.

Antifungal Chemical Resistance

This section is concerned with naturally-occurring chemicals which act directly in a fungistatic or fungicidal manner.

Phenolics have been considered previously from a physiological rather than a fungicidal role. Catechin in cotton is a potent antisporulent at 10^{-5} M but $>10^{-3}$ M is required to inhibit mycelial growth. Catechin and its proanthocyanidin derivatives are confined mostly to the endodermis (Mace and Howell, 1974). Extracts and tracheal fluids from resistant cotton cultivars frequently inhibit *Verticillium* spore germination (Bell, 1969) but may have little effect on mycelial growth *in planta*. Excluding phenols, there are few identified pre-infectional antifungal compounds involved in vascular wilts. Pegg and Woodward (1986) using a [^{14}C]cholesterol assay found lower levels of the steroidal glycoalkaloid tomatine in roots, stems and leaves of *V. albo-atrum*-resistant tomato compared with susceptible plants. Wounding constituted a greater enhancer of tomatine levels than infection. ED_{50}s for germination and germ tube growth were 132 and 66 μg ml^{-1}, respectively. *V. albo-atrum* was shown to produce an inducible, extracellular 1,2-β-glucosidase which hydrolysed tomatine to the weakly active β-2-tomatine with an ED_{50} for germ tube growth of 390 μg ml^{-1}. Pegg and Woodward (1986) concluded that α-tomatine played no role in *Ve* gene resistance. Sandrock and Van Etten (1998) showed that *V. dahliae* too was able to degrade α-tomatine to the less toxic β-2-tomatine and tomatidine. *V. dahliae* was tolerant of α-tomatine, with an LD_{50} of >300 μM.

Phytoalexins *sensu stricto* represent stress metabolic compounds which are either absent in healthy plants or are present only at very low and biologically inactive concentrations. The term, however, has been used incorrectly by some authors to include compounds which are active at pre-infection levels. Phytoalexins in *Verticillium* wilt have been studied mostly in cotton, tomato, aubergine and lucerne. Sesquiterpenoid naphthaldehydes are found in cotton, sesquiterpenoids and polyacetylenes in tomato and aubergine, and isoflavanoid pterocarpans in lucerne.

In cotton, the principal terpenoid is gossypol found in plants and cotton seed oil as a natural constituent. 6-Methylgossypol ether (MG) and the 6,6′-dimethyl ether (DMG) are major constituents of the root epidermis and the lysigenous pigment gland on the phloem (Mace *et al.*, 1974; Bell *et al.*, 1978). The major derivatives of gossypol, acting as phytoalexins, are hemigossypol (HG) and hemigossypol-6-methyl ester (MHG), especially in resistant cultivars (Bell *et al.*, 1975). Desoxyhemigossypol (DHG) and desoxyhemigossypol-6-methyl ether (DMHG) are precursors found 12–48 h after infection (Stipanovic *et al.*, 1975). Vergosin (Zaki *et al.*, 1972a) and isohemigossypol (Sadykov *et al.*, 1974) were later identified correctly as DMHG (Stipanovic *et al.*, 1975) and HG (Veech *et al.*, 1976). Two other derivatives, isohemigossypol (Sadykov *et al.*, 1974) and gossyvertin, were isolated from stem xylem extracts of *V. dahliae*-infected cotton. Little gossypol was detected in these extracts (Coxon, 1982).

Bell (1969) and Bell and Presley (1969b) first implicated phytoalexins in resistance of cotton to vascular pathogens. Another early contribution to cotton phytoalexins was by Metlitskii *et al.* (1971). Using live and heat-treated conidia of a defoliating strain of *V. dahliae*, gossypol and related compounds increased more in resistant *Gossypium barbadense* cultivars during the first 48 h after inoculation, but subsequently increased amounts were found in five *G. hirsutum* cultivars. Resistance was believed to function by an initial containment, similar to the spore-trapping gel concept. In general, resistant cultivars contain phytoalexins missing or present in only low concentration in susceptible cultivars (Bell *et al.*, 1975; Mace *et al.*, 1976).

Resistant *G. barbadense* synthesize large amounts of gossypol-type phytoalexins in response to heat-killed cells, whereas susceptible *G. hirsutum* do not (Bell and Presley, 1969b; Bell and Stipanovic, 1978). Similarly, moderately resistant [*sic*] (tolerant) cultivars of *G. hirsutum* produce phytoalexins more rapidly than susceptible cultivars in response to avirulent or weakly virulent, but not highly virulent isolates of *V. dahliae* (Bell, 1969; Zaki *et al.*, 1972b; Mukhamedova *et al.*, 1975). Cotton mutants selected for increased resistance to *V. dahliae* showed a more rapid and intense phytoalexin synthesis than their progenitors (Gaibullaev *et al.*, 1975; Mukhamedova *et al.*, 1975; Avazkhodzhaev *et al.*, 1976). Young tissue in susceptible *G. hirsutum* cultivars synthesizes HG more slowly than older and more resistant tissue (Bell, 1969). Ageing cotton leaves, however, showed increased susceptibility and had reduced phytoalexin activity (Howell *et al.*, 1976). Bell and Presley (1969a) in a complex experiment studied the effect of temperature and a defoliating strain of *V. dahliae* on *G. barbadense* cv. Seabrook Sea Island (resistant), *G. hirsutum* cv. Acala 4-42-77 (tolerant) and cv. Stardel (susceptible). At 22°C, all cultivars were susceptible; at 25–29°C, resistance and susceptibility were expressed. Cv. Stardel was susceptible at 25°C, but tolerant at 27°C. At 32°C, all cultivars were resistant, correlating with near maximum phytoalexin synthesis. However, at this temperature, conidial production declined almost to zero (Bell and Presley, 1969a). Mace *et al.* (1976) and Mace (1983) found that the gossypol family of phytoalexins are formed mainly by paratracheal parenchyma and exude into the xylem lumen to be absorbed by cell walls, tyloses, hyphae and conidia. A localized green staining product of DHG was found in the vessel lumen and in specialized, often solitary paravascular cells (Mace *et al.*, 1989). Mace (1978) considered that physical occlusion of the vessels permitted the accumulation of phytoalexins which could thus act on the isolated pathogen. Bell (1969) claimed that 50% of the induced terpenoid aldehydes could be recovered in ethanolic eluates of vessels compared with <5% of induced flavanols (Bell and Stipanovic, 1978).

HG, gossypol and DMHG had ED_{50} values of 10–60, 100 and 6 μg ml^{-1} for inhibition of *V. dahliae* mycelial growth (Zaki *et al.*, 1972b; Sadykov *et al.*, 1974). The ED_{50} for gossypol against conidial germination was given by Bell (1967) as 50 μg ml^{-1}. Raj (1974), however, claimed that only a 25% inhibition of *V.*

dahliae was obtained at 2740 μg ml^{-1}. Bell and Mace (1981) emphasized the need for adequate solubilization and stabilization in the test solution. In later studies (Mace, 1978; Mace *et al.*, 1985), it was found that concentrations of HG, MHG, DHG and DMHG between 15 and 45 μg ml^{-1} killed *V. dahliae* conidia over 18–40 h and mycelium at 48 h. Curiously, only HG had the requisite water solubility to attain fungicidal concentration.

Garas *et al.* (1986) and Garas and Waiss (1986), studying the effect of strongly and weakly virulent isolates of *V. dahliae* on resistant and susceptible cultivars, concluded that resistance was directly correlated with levels of DMHG and MHG induced by infection. Portenko *et al.* (1985) found similar supporting evidence in cross-protection experiments. Cultivars were claimed to be resistant and susceptible to races 1 and 0 of *V. dahliae*. When a cultivar susceptible to race 1 was inoculated with both races, accelerated levels of gossypol-type phytoalexins were associated with less severe symptoms than when inoculated with race 1 alone. The race concept in cotton isolates of *V. dahliae* has not been resolved, especially in the CIS, and the 'races' 1 and 0 of Portenko *et al.* (1985) most probably compare with strains such as the P_1 defoliating and P_2 non-defoliating strains as described in the USA.

Mace *et al.* (1990a,b) presented revised data for phytoalexin induction in cotton. Ten days after inoculation of resistant *G. barbadense*, concentrations (μg ml^{-1}) of stelar extracts were DHG 25, HG 26, DMHG 57 and MHG 79. The LD_{100} for mycelium was DHG 15 μg ml^{-1}, DMHG 25 μg ml^{-1}, HG 35 μg ml^{-1} and MHG 45 μg ml^{-1}. The authors concluded that only DHG, which solubilized in water at 50 μg ml^{-1} at pH 6.3, had a role *in planta*. Mace *et al.* (1990b) speculated that by a process of proton exclusion from DHG, a series of fungitoxic free radicals could be produced. DHG was found in *V. dahliae*-infected cotton roots and stems at 6.1 μg ml^{-1} and reduced *V. dahliae* growth by 75%. DHG appeared to be decomposed in an H_2O_2-mediated process. Its activity was reduced by catalase at 500 U ml^{-1}, glutathione at 0.1 mM and SOD at 1700 U ml^{-1}, but the latter gave the smallest effect (Mace and Stipanovic, 1995). The first stage in the gossypol and lacinilene phytoalexin synthetic pathway was elucidated by Chen *et al.* (1995). Using a *G. arboreum* cell-suspension culture initiated by a *V. dahliae* elicitor preparation, a 1.9 kb mRNA increased tenfold. This was used to isolate two, cDNA clones coding for a protein which was characterized by GS–MS, chiral phase gas chromatography (GC) and nuclear magnetic resonance (NMR) as (+)δ-cadinene synthase (CAD). A second CAD synthase, similarly elicited in cell culture by a *V. dahliae* elicitor was described by Chen *et al.* (1996). Heinstein *et al.* (1996) found by differential induction of the two CAD mRNAs after eliciting with *V. dahliae* and using reverse transcription PCR, that CAD A and CAD C were differentially expressed in cotton roots and leaves leading to phytoalexin induction. Transcription of CAD A is induced preferentially by *V. dahliae* in roots of *G. arboreum* seedlings, whereas *Xanthomonas campestris* pv. *malvacearum* induces CAD A and CAD B in leaf tissue. The enzyme had a K_m of 7 mM faresyl diphosphate and a k_{cat} of 0.039 s^{-1} at 30°C. Benedict *et al.* (1995) confirmed that cotton stele tissue extracts incubated after inoculation with *V. dahliae* synthesized (+)δ-cadinene from farnesyl diphosphate. A detailed metabolism of farnesyl diphosphate was proposed by Alchanati *et al.* (1998). In soluble preparations of *V. dahliae*-infected stele tissue, (1-RS)-[1-^{2}H]E,E-farnesyl diphosphate was converted to [5-^{2}H] and [11-^{2}H]δ-cadinene, and [4,4,13,13,13-^{2}H5]nerolidyl diphosphate was converted to [8,8,15,15,15-^{2}H5]δ-cadinene. It was suggested that nerolidyl diphosphate is an intermediate in the enzymic cyclization of the natural substrate E,E-farnesyl diphosphate to nerolidyl diphosphate followed by cyclization to *cis*-germacradienyl cation, a 1,3 hydride shift, a second cyclization to cadinanyl cation and deprotonation to δ-cadinene. Twelve hours after inoculation, δ-cadinene synthase mRNA was maximal. Water controls showed only trace activity. After 12 h, δ-cadinene synthase was 54% but no phytoalexins were detected. A methyl transferase (*S*-adenosyl-L-methionine:desoxyhemigossypol-6-*O*-methyltransferase), which specifically methylates the 6-position of DHG to form DMHG was purified and characterized from *V. dahliae*-infected cotton steel tissue. The enzyme had a K_m of 4.5 μM for DHG and a k_{cat}/K_m of 5.08×10^4 s^{-1} thin (mol l^{-1}). The methyl ether leads to methylated HG, gossypol, hemigossypolone and heliosides with lower effectivness. The proposal for cotton phytoalexins synthesis at about 24 h is a signal from *V. dahliae* inducing a steady-state level of δ-cadinene synthase, converting E,E farnesyl diphosphate→nerolidyl diphosphate→δ-cadinene→DHG, DMHG, HG and MHG.

As a prelude to resistance genetic engineering in cotton, C.J. Liu *et al.* (1999) isolated from *G. arboreum* a full-length cDNA encoding farnesyl diphosphate synthase (FPS). This was identified by *in vitro* assay of the enzyme heterologously expressed in *Escherichia coli*. Treatment of a suspension of *G. arboreum*-cultured cells with a *V. dahliae* elicitor induced transcription of FPS and CAD. A wild Australian cotton *G. australe* cell culture was also induced to produce FPS, but only a low rate of CAD was found due to a constitutive expression of the sesquiterpene cyclase gene in the cell-suspension culture. Two transcripts and proteins of FPS were recovered from elicited *G. australe* cells. One of the difficulties in the interpretation of the role of phytoalexins in cotton wilt is that they accumulate in both susceptible and resistant cultivars and species. Joost *et al.* (1995) examined conserved regions in the cotton genome for 3-hydroxy-3-methylglutaryl coenzyme A reductase (HMGR), the first enzyme in the terpene biosynthctic pathway. These were used to design a PCR primer for cloning a fragment of a cotton HMGR gene. The clone was used as a probe on Northern blots to show that following inoculation, the induction of HMGR mRNA and the specific activity of HMGR was more rapid in the resistant *G. barbadense* cv. Seabrook Sea Island than in the susceptible *G. hirsutum* cv. Rowden. The level of mRNA returned to that of healthy controls after 4 days in the resistant cultivar but continued to increase in the susceptible cultivar. The inference drawn from this and the succeeding work by different authors is partly compromised by the use of different polygenic host species. The work of Bianchini *et al.* (1999) nevertheless adds support to the phytoalexin argument in cotton.

The resistant (tolerant) cv. Seabrook Sea Island 12 B2 (SBSI) from *G. barbadense* showed earlier HG and DHG synthesis (12 h after inoculation) than in the susceptible *G. hirsutm* cv. Rowden. By 48 h, levels of HG and DHG in cv. SBSI were 23.9 and 10.5 μg^{-1} of fresh tissue, respectively, sufficient to suppress conidial germination completely. This level in cv. Rowden was not attained until 96 h. As might be expected, these changes were reflected in the relative activities of CAD in the two cultivars. Using the same cultivars, Cui *et al.* (2000) developed cloned probes to code for enzymes of the phenylpropanoid pathway. PAL mRNA was constitutively expressed. Levels of mRNA transcripts coding chalcone synthase for caffeic-*O*-methyltransferase showed a greater increase or initiation in cv. SBSI than in cv. Rowden. Equal levels of 1,3-β-glucanase mRNA were present in *V. dahliae*-inoculated and water controls. Chitinase nRNA sythysis was strongly induced by inoculation. Curiously, mRNA levels in the susceptible cv. Rowden continued to increase but diminished rapidly in cv. SBSI after 48–60 h. This was interpreted by the authors as a reflection of the host response to pathogen spread – greater in Rowden than SBSI. A question of equal validity is why cv. Rowden does not become more resistant as hostile levels of glucanohydrolases increase. The results also beg the question of interpretations based on experiments on non-isogenic host lines. A study by Xia and Achar (1999) on eight cotton cultivars representing three resistance types found no correlation with reducing sugar or tannin concentration of leaves and roots but a significant linear regression between gossypol concentration and disease indicies. Similarly, PO activity was positively related to disease severity. It is not clear whether this work is supportive of phytoalexins in resistance or merely illustrates markers of disease severity.

Whereas phytoalexins have been considered essentially as fungistatic or fungicidal compounds, Sagdieva *et al.* (1974) demonstrated an enzyme inhibitor role for gossypol comparable with phenolic inhibitors. Gossypol at 2×10^{-5} M inhibited PL from *V. dahliae* cultures. A similar role for DHG or other derivatives at sub-fungitoxic levels is not precluded. Avazkhodzhayev and Zeltser (1990) claim that the suppression of resistance in *G. hirsutum* cultivars is partly by a block to phytoalexin synthesis but also by phytoalexin metabolism by virulent strains of *V. dahliae*. DHG, DMHG and hemigossypolone induced in *G. hirsutum* and *G. barbadense* stems were shown to be highly toxic to zoopathogenic fungi such as *Candida albicans* and *Cryptococcus neoformans* (Mace *et al.*, 1993). With the prospect of genetically engineered cotton by inserting genes for novel phytoalexins (Bell *et al.*, 1994), Bell *et al.* (1998) isolated from *V. dahliae*-infected *Hibiscus cannabinus* (Kenaf or Deccan Hemp) two trinorcadelene phytoalexins, hibiscanal (2,8-dihydroxy-4,7-dimethoxy-6-methyl-1-naphthaldehyde) and *O*-hibiscanone (3,8-dimethyl-1,2-naphthoquinone). The BD_{50} values of hibiscanal and *O*-hibiscanone against *V. dahliae* were 25.83 and 1.18 $\mu g\ ml^{-1}$, respectively. All propagules were killed with *O*-hibiscanone at 8 $\mu g\ ml^{-1}$ which was more toxic than DHG or mansonone-C (3,8-dimethyl-5-isopropyl-1,2-naphthoquinone) found in other malvaceous spp.; see also Stipanovic *et al.*, 1997.

Puckhaber *et al.* (1998) demonstrated by means of ^{13}C-NMR experiments that O-hibiscanone was reduced to is hydroquinone as an apparent detoxification mechanism by conidia of *V. dahliae*. The hydroquinone was substantially less toxic than the quinone.

Smith *et al.* (1971) first identified a phytoalexin, the isoflavanoid pterocarpan, medicarpin, in *Medicago sativa* (lucerne). Khan and Milton (1975) claimed that an avirulent strain of *V. albo-atrum* induced more phytoalexin in leaf drop diffusates than a virulent strain. When conidia from both strains were used, phytoalexin induction by the non-virulent pathovar was reduced. This was interpreted by Bell and Mace (1981) to be due to a lower sensitivity of virulent than non-virulent pathovars to medicarpin and sativan. Khan and Milton (1978) claimed a direct relationship between the rates and amounts of phytoalexin synthesis by lucerne stem segments and the degree of resistance of the plant. No such relationship was observed between cultivar resistance and leaf phytoalexin production. Flood *et al.* (1978b) using TLC, found five phytoalexins of which medicarpin and sativan were present in greatest quantity. The three others were not identified. Flood and Milton (1982), studying the effect of different *V. albo-atrum* pathovars on excised lucerne leaves, found that attempted penetration by an avirulent *V. albo-atrum* isolate led to rapid death of the host cells and accumulation of medicarpin and sativan, both of which restricted hyphal growth. In a compatible interaction, only low levels of phytoalexins formed and unrestricted hyphal growth occurred, which was not halted subsequently when high levels of medicarpin and sativan accumulated.

Latunde-Dada and Lucas (1985) investigated the *in vitro* response of callus lines showing varying degrees of wilt resistance to a lucerne strain of *V. albo-atrum*. Medicarpin accumulated in both resistant and susceptible cell lines challenged with conidia or agar blocks with mycelium, but higher levels formed when the callus was derived from a resistant than a susceptible cultivar. The lowered sensitivity of callus lines to wilt toxins might serve as a useful pointer when selecting for field resistance. Latunde-Dada and Lucas (1988) subsequently concluded that since seed progeny of plants derived from second-generation tissue cultures reverted to susceptible reaction types, somaclonal selection offered no advantage over conventional methods of selective breeding. Latunde-Dada and Lucas (1986), in *in vitro* studies of colonization of callus lines by *V. albo-atrum*, showed that medicarpin accumulated more rapidly at 28°C than at 20°C and with an increased sensitivity of the fungus to the phytoalexin at the higher temperature. Lignin deposition was also shown to be initiated earlier and occur more rapidly in resistant than susceptible lines. Enzymes associated with the phenylpropanoid and isoflavanoid pathways, PAL, chalcone synthetase and chalcone isomerase, responded variably. In general, activity of all three enzymes was higher in resistant than susceptible callus following inoculation. At 28°C, however, activity was reduced except for chalcone synthetase and isomerase which had higher activity than at 20°C, but only in the susceptible callus (Latunde-Dada *et al.*, 1986).

Several, chemically different phytoalexins associated with resistance to *Verticillium* have been identified in tomato, the *nor*-sesquiterpenoid rishitin (Tjamos and Smith, 1974; Grzelinska and Sierakowska, 1978), the polyacetylenes, falcarindiol, falcarinol and *cis*-tetradeca-6-ene-1,3-dyne-5,8 diol (Elgersma and Overeem, 1981; Elgersma *et al.*, 1984) and the triterpene lupeol (Grzelinska, 1980). Tjamos and Smith (1974) found only low levels of rishitin in cultivars Gem resistant (3.5 $\mu g\ g^{-1}$) and Gem susceptible (1.0 $\mu g\ g^{-1}$) following spore infiltration of stem segments. These results contrasted with those from *V. albo-atrum*-injected fruit, where the rishitin content of resistant fruit was 14-fold higher (20 $\mu g\ g^{-1}$) than that of susceptible fruit. Using a stem incision, inoculum droplet uptake technique, Elgersma (1980) found no difference in colonization between cv. Multicross (resistant) and Maascross (susceptible) 4 days after inoculation, but subsequently, colonization increased in susceptible plants. Similarly, at 11 days, the rishitin content which had risen to 40 $\mu g\ g^{-1}$ was the same in resistant and susceptible plants alike. Hutson and Smith (1980), comparing cultivars of Craigella showing reciprocal resistance and susceptibility to *Verticillium* and *Fusarium*, found six antifungal compounds in roots and stems resulting from *V. albo-atrum* infection, regardless of whether the plant reaction was resistant or susceptible. Rishitin was a major constituent of these. Using a terpenoid-specific stain, $SbCl_5$, terpenoids were found restricted to the vessel walls and a few layers of xylem parenchyma in resistant plants, with tyloses giving intense reactions. In susceptible plants, staining was diffuse over the xylem parenchyma. The ED_{50} values for *V. albo-atrum* conidial germination (30–100 $\mu g\ ml^{-1}$) and germ tube growth (3.0 $\mu g\ ml^{-1}$) (Tjamos and Smith, 1974) suggest that unless localized concentrations occurred, the levels detected *in planta* by Tjamos and Smith (1974) and Hutson and Smith (1980) would have been too low to be effective. Elgersma and Overeem (1981) and Elgersma *et al.* (1984) confirmed the existence of polyacetylinic phytoalexins in tomato and showed that resistance could not be interpreted solely in terms of rishitin.

In a definitive paper, Elgersma and Liem (1989) described rishitin and *cis*-tetradeca-6-ene-1,3-dyne-5,8-diol as the two main phytoalexins induced by *V. albo-atrum* infection. Using two different tomato isolines, resistant and susceptible to *V. albo-atrum*, both phytoalexins accumulated at the same rate in susceptible and resistant plants alike for the first 7 days. At 11 days, 100% more rishitin (110 $\mu g\ g^{-1}$ of fresh weight) and polyacetylene (18$\mu g\ g^{-1}$ of fresh weight) was present in susceptible compared with resistant plants. Plants were inoculated by inoculum droplet uptake through a stem incision just above the first leaf. Elgersma and Liem (1989b) demonstrated that conidial numbers in 4-cm stem segments cut above the inoculation point were the same in susceptible and resistant plants. Moreover, the accumulation of each phytoalexin was proportional to the quantity of fungus present. The concentration of phytoalexin extracted was clearly related to the inoculation method and was higher than would be expected from an undamaged root-dip method. However, the concentration in susceptible diseased plants, if available to the pathogen, was well in excess of a lethal dose. This experiment, while resulting in higher levels of phytoalexins than would occur in natural infection, discredits any role for rishitin and polyacetylenic phytoalexins in *Verticillium* wilt resistance in tomato, or an explanation for *Ve* gene resistance. Little work has been done on other post-infection compounds in tomato (Grzelinska, 1980), but an explanation of resistance based on phytoalexins exclusively would seem improbable.

Emmanouil and Wood (1983) found several antifungal compounds in *Solanum melongena* (aubergine) roots following inoculation with *V. dahliae* or treatment with benzimidazole fungicides. One sesquiterpenoid, lubimin, was identified. Yoshihara *et al.* (1988) identified by IR, NMR and GC–MS, three sesquiterpenes: solavetivone, lubimin and epilubimin, and five phenolic compounds: vanilin, ethyl ferulate, ethyl *p*-coumarate, ethyl caffeate, *p*-hydroxybenzoic acid and 3 hydroxy-1-(4-hydroxy-3-methoxy phenyl)-1-propanone from the roots of *S. melongena*. A healthy field-grown hybrid of *S. melongena* and *S. integrifolium* resistant to *V. dahliae* had 5.6- and 1.6-fold increases in solavetivone and lubimin, respectively, over the wild-type. In glasshouse-grown plants, however, absolute amounts of these compounds were 20- to 80-fold lower and only small increases could be attributed to the *R* gene. No unequivocal or convincing case for a phytoalexin-mediated resistance mechanism to *V. dahliae* was presented. Tyuterev *et al.* (1980) reported that the use of *V. dahliae* as a challenge organism on wheat resulted in a 40% decrease in infection when subsequently inoculated with *Puccinia triticina*.

An interesting study on the resistance of cacoa (*Theobroma cacoa*) to *V. dahliae* by Cooper and co-workers in the UK (Resende *et al.*, 1994, 1995a,b, 1996; Cooper *et al.*, 1995, 1996) identified four phytoalexins, two of which were directly associated with resistance. Two compounds, 4-hydroxyacetophenone and 3,4-dihydroxyacetophenone, both reported as pre-infectional antifungal substances in a range of hosts, were present in healthy plants in trace amounts. The ED_{50}s for germ tube length for 4-hydroxyacetophenone and 3,4-dihydroxyacetophenone were 4.4 and 95.5 $\mu g\ ml^{-1}$, respectively, but the accumulated stem levels 15 days post-inoculation in the resistant cv. Pound-7 were only 2.0 and 4.9 $\mu g\ g^{-1}$ fresh weight, respectively. A novel finding was the post-infectional accumulation of elemental cyclo-octasulphur (S_8). In this case, the ED_{50} for germ tube length was 1.7 $\mu g\ ml^{-1}$, whereas the stem level in cv. Pound-7 was 51–116 $\mu g\ g^{-1}$ fresh weight. Sulphur identified by GC–MS and X-ray crystallography was shown by SEM and X-ray microanalysis to be present in vessel walls, scattered parenchyma cells and vascular gels. A fourth phytoalexin, the triterpenoid arjunolic acid, was also induced in stems in the resistant cultivar at 28 times the required concentration for the ED_{50} for germ tube growth (6.0 $\mu g\ ml^{-1}$). The authors could find no positive association with levels of condensed tanins or tylosis with resistance in cacoa.

With the exception of cotton, lucerne and cacoa there is no convincing evidence for a role for phytoalexins in *Verticillium* wilt resistance. In these hosts, phytoalexins are not seen functioning in isolation. Most authors envisage their antifungal role in conjunction with some localization and restriction of the pathogen and only then at localized relatively high concentrations.

Lectins

A preliminary account by Ten *et al.* (1995) poorly translated from Russian to English, described the isolation, from resistant cotton cv. Mexicanum and susceptible cv. C-4727, of lectin proteins with molecular weights of 5–45 kDa able to agglutinate erythrocytes and differentially agglutinate conidia of races 1 and 2 of *V. dahliae*. The authors also speculate that lectins may play a role in pathogen recognition and phytoalexin synthesis. This little known area of resistance in *Verticillium* studies merits further detailed examination. Zhang *et al.* (1996) published a preliminary account of breeding cotton for *Verticillium* resistance using biochemical parameters.

Sources of novel proteins and miscellaneous antifungal substances

The advent of genetic engineering combined with limitations in the availability of natural resistance genes in some crop genera has stimulated the search for sources of anti-*Verticillium* substances, usually from genetically incompatible genera. The wheat or barley thionin LD_{50} concentration for *inter alia V. dahliae* mycelial growth was lowered 2- to 73-fold when combined with 2S-albumins from radish or rape (Terras *et al.*, 1993a). The thionin fungal toxicity was enhanced 2- to 33-fold by the small or large subunit from 2S radish albumins. Two other 2S-albumin-like proteins, the barley trypsin inhibitor and two barley Bowman–Birk-type trypsin inhibitor isoforms also acted synergistically with the thionins (2- to 55-fold). K^+ leakage and SEM confirmed that thionins and 2S-albumins each alone and in synergistic combination increased hyphal plasmalemma permeability. Terras *et al.* (1993b) identified a family of basic, cysteine-rich antifungal proteins, comprising seven substances isolated from the seed of four brassicae species. The proteins, multimers of a 5-kDa polypeptide, inhibited *inter alia V. dahliae* (see also de Samblanx *et al.*, 1996). An experimental 'membrane-interactive' peptide D5-C (source undisclosed) described by Qui *et al.* (1995) was toxic to pollen and protoplasts of rape (*Brassica napus*) at 1–5 μM. At 5.0 μM, germination of *V. albo-atrum* conidia was completely inhibited. Two isoforms I and II of a thaumatin-like protein NP24 isolated from tomato fruit cv. Better Boy and differing only in two amino acids of the N-terminal sequence, were toxic *inter alia* to *V. dahliae* at IC_{50} 0.8–6 μg ml^{-1}. NPI, which was crystallized, increased approximately eightfold during ripening (Pressey, 1997). Gupta (1975) described electrophoretic differences in proteins from R and S cultivars of lucerne.

Ajoene (4,5,9-trithiadodeca-1,6,11-trien-9-oxide) synthesized, but also naturally occurring in garlic (*Allium sativum*) was active *inter alia* to *V. dahliae* within the range 2–200 μg ml^{-1} (Reimers *et al.*, 1993). Okamoto *et al.* (1993) found two novel and a known chlorinated archinol derivative from the orchid *Hericium etinaceum* (*H. erinaceus*). One of the new structures 4-chloro-3,5-dimethoxybenzaldehyde was active against *V. dahliae* as well as other fungi. Ouf *et al.* (1994) identified six saponins from roots and aerial shoots of *Zygophyllum coccineum*, *Z. album* and *Z. dumosum*. Quinovic acid was the dominant aglycone. Kaempferol was also isolated from the flavonoid fraction of *Z. dumosum*. All compounds were fungistatic to *V. albo-atrum*. Some saponins and kaempferol gave control against *V. albo-atrum* when used as a seed protectant in glasshouse tests. Since *V. albo-atrum* infects mostly by hyphal invasion, a mycelial inoculum would have provided a more realistic challenge to the compounds. Decomposing rapeseed meal (RSM) in soil releases volatile anti-*Verticillium* compounds from glucosinolates present at 36 μM g^{-1}. RSM at 4 g kg^{-1} of soil gave 65% inhibition of mycelial growth, and at 60 g kg^{-1} soil growth was completely inhibited. Melanin synthesis was reduced by 99% in *V. dahliae* cultures exposed to the lowest rate of RSM (20 g kg^{-1} of soil; Melouk *et al.*, 1995).

The antifungal terpenoids cupressotropolone A (6-isopropyltropolone-β-glucoside) and cupressotropolone B ((5-l3-hydroxy-3-methyl-trans-1-butenyl)-6-isopropyl-tropolone-β-glucoside) isolated from the bark of *Cupressus sempervirens* were shown by Madar *et al.* (1995) to be active against conidial germination in *V. dahliae*. The search for novel anti-*Verticillium* compounds had extended to lichens. (–)Usnic acid isolated from *Alectoria ochroleuca* by Proska *et al.* (1996) inhibited *V. albo-atrum* growth at 100 μg per disc. Two synthesized derivatives of (–)usnic acid showed potent antifungal properties. Additionally, two hydrazones from (–)usnic acid reduced fungal elastase and trypsin activities more than (–)usnic acid. Four weed species which had not been recorded as symptomless hosts of *V. dahliae*, *Euphorbia helioscopia*, *Galium tricornutum*, *Sisymbrium irio* and *Ranunculus asiaticus*, were toxic to *V. dahliae*, significantly reducing colony growth after 4 days when tested as aqueous extracts (Qasem and Abu-Blan, 1995). A merediterpnoid, methoxybifurcarenone, isolated from the brown alga *Cystoseira tamariscifolia* from Rabat, Morocco by Bennamara *et al.* (1999) was antifungal to tomato pathogens, *inter alia V. albo-atrum* [*sic*]. *Artemesia annua* from China contains a cadinane derivative, arteannuin B, with a minimum inhibitory concentration of 100 μm l^{-1} against *V. dahliae* (Tang *et al.*, 2000).

The increasing culture of oilseed rape (*Canola*) and a threat from stable diploid isolates of the pathogen *V. longisporum* Karapapa Stark (*V. dahliae* var. *longisporum*) has stimulated a search of the genus *Brassica* for genotypes producing antifungal chemicals for use in breeding programmes or for disease-suppressing green manure cultivars. To this end, Heale *et al.* (1991) found that gluconasturtin, a product resulting from myrosinase hydrolysis of oilseed rape glucosinolates, was highly inhibitory to stable diploid isolates of the pathogen *V. dahliae longisporum*. No difference in resistance, however, was found between high and low glucosinolate cultivars. A screen of 28 genotypes of *B. nigra* (black mustard) and 35 of *B. juncea* (Indian mustard) for the release of allylisothiocyanate (AI) following the enzymic hydrolysis of the allylglucosinolate sinigrin by Olivier *et al.* (1999) found compounds in all genotypes which suppressed *V. dahliae*. Antifungal activity and a.i. release varied. Plants releasing >1.2 mg a.i. g^{-1} of tissue (a high level in physiological terms) were fungicidal. Sinigrin was the major glucosinolate in all genotypes of *B. nigra*, together for the most part with 2-phenylethylisothiocyanate (PI). *B. juncea* genotypes produce variable quanities of a.i. together with, in most types, PI, benzyl-I and 3-butenyl-I. The significance of this work was relevant possibly only for green manure cultivars for the control of potato wilt since *V. dahliae* from potato was used exclusively in the assay. A screen of 25 flavanoid compounds for antagonism to *V. albo-atrum* carried out by Picman *et al.* (1995) found that unsubstituted flavanoids were the strongest growth inhibitors in the range 1–5 p.p.m. Hydroxylation, methoxylation and glycosylation resulted in loss of activity. Karapapa *et al.* (1997b) examined glucosinolate metabolism in winter oilseed rape (*Brassica napus* ssp. *oleifera*) cv. Cobra in compatible, (cv. Cobra × *V. longisporum*) and incompatible (cv. Cobra × non-pathogenic *V. dahliae*) reactions. While glucosinolate breakdown products following enzymic hydrolysis with myrosinase are fungicidal to *Vericillium* spp., their role in resistance is debatable. Total glucosinolates (aromatic, indolic and aliphatic) increased in all plant parts in the incompatible reaction compared with controls and the compatible reaction. The aromatic glucosinolate, gluconasturtin (phenylethylglucosinolate, GLNS) was greater in roots and hypocotyls in an incompatible reaction (cv. Cobra inoculated with an avirulent *V. dahliae*). In cotyledons and leaves, however, GLNS levels were greater in the pathogenic reaction cv. Cobra inoculated with *V. longisporum*. Indole and aliphatic indoles varied in levels between different plant parts in the two host–fungus interactions. Of five different glucosinolates, progoitrin, gluconapin, gluconasturtin, glucobrassicin and neoglucobrassicin, in the presence of myrosinase, only progoitrin (PI), at pH 7 was more toxic than propenyl nitrate (fungicidal concentration >20 μg ml^{-1}). The non-pathogenic *V. dahliae* was more sensitive to PI than *V. longisporum*. The evidence to date for a role for glucosinolates in resistance is mostly circumstantial; though largely in favour, it is not overwhelming.

Elicitors

There is limited information available on elicitors of antifungal compounds in *Verticillium* diseases. Solutions of cupric and mercuric ions induced gossypol and gossypol-type compounds in cotton stem segments (Bell, 1967) and medicarpin and sativan in lucerne leaflets (Khan and Milton, 1979). A PLP complex which induced disease symptoms in detached cotton leaves (Keen and Long, 1972) was shown by Zaki *et al.* (1972a,b) to elicit phloroglucinol-reactive compounds in cotton stem segments in amounts proportional to the PLP concentration. Greater quantities of phloroglucinol-reacting compounds, mostly HG and DMHG, were formed in tolerant cultivars inoculated with a mild strain of *V. dahliae* rather than in those inoculated with a virulent defoliating strain. Smit and Dubery (1997) also found increased lignification in cotton hypocotyl tissue following treatment with a PLP from *V. dahliae*. This was accompanied by transitory increases in PAL, cinnamyl alcohol dehydrogenase and cell wall peroxidase activities. Resistant cotton cv. Acala 1517-70 exhibited earlier and higher levels of lignification, wall thickening and enzyme activities than the susceptible cv. OR19. Stationary and staling cultures contained larger quantities of related low molecular weight compounds believed to be breakdown products of PLP (Keen and Long, 1972). The composition of the PLP is similar to that of the outer glycan heteropolymer of *V. dahliae* (Wang and Bartnicki-Garcia, 1970). The actual heteropolymer may be more active and specific than the PLP since washed, heat-killed conidia elicited stronger synthesis of phytoalexins in resistant *G. barbadense* than in susceptible or tolerant *G. hirsutum* (Bell and Presley, 1969a). Avazkhodzhaev *et al.* (1984) described the characteristics of a *V. dahliae* metabolite inducing phytoalexins in cotton.

The concept of fungal heteropolysaccharides as elicitors is supported in part by common antigen studies. Both resistant and susceptible cotton cultivars have at least one antigen common to pathogenic *V. dahliae* and *Fusarium*, but not to a non-pathogen. Davis *et al.* (1998) describe the purification by SDS–PAGE of a glycoprotein elicitor of cotton phytoalexins from *V. dahliae*. There is little attempt in this and other recent studies to compare purification protocols with earlier studies and to eliminate the possibility of 'new' higher molecular weight moieties being aggregations of earlier described smaller molecules, or peculiarities of cultural and/or extraction methods. Altman *et al.* (1985) showed that culture filtrates of *V. dahliae* inhibited growth and embryo development of *G. hirsutum* callus cultures in direct proportion to the content of phytoalexin elicitors. Apostol *et al.* (1987) found that citrate, inhibited elicitor-induced phytoalexin production in cotton and soybean. Wall extracts from both virulent and avirulent strains of *V. dahliae* induced gossypol-type phytoalexins in stem segments of resistant and susceptible cotton cultivars. Techniques used for the induction of sanguinarine in *Papaver bracteatum* were applied by Cline *et al.* (1993) to *Sanguinaria canadensis* (bloodroot), a host of *V. dahliae*. When 2,4-D was withheld from suspension cultures and a *V. dahliae* elicitor added, dopamine, an alkaloid precursor, increased to more than 20% of the culture's dry weight. Chelirubine increased 0.1–1.3% dry weight; chelerythrine 0.01–0.10% dry weight and sanguinarine 0–0.02% of the dry weight. Treatment of aubergine, pepper and tomato with non-phytotoxic levels of benomyl, carbendazim or thiobendazole induced the same seven phytoalexins as *V. dahliae* infection (Emmanouil and Wood, 1983). Three only were identified on the basis of co-chromatography and staining on TLC. These were the sesquiterpenoids lubimin from aubergine, capsidiol from pepper and rishitin from tomato. The unsaturated fatty acids oleic C18:1, linoleic

C18:2 and linolenic C18:3 were found in three isolates of *V. dahliae* in the polar lipids and were double in content of free fatty acids. Moderate levels of lubimin were elicited by linoleic acid (the predominant fatty acid). A higher induction occurred when tissue browning was inhibited by phenylthiourea (Castoria *et al.*, 1995). Low and Heinstein (1986) described the use of fluorescent dyes to monitor elicitor-stimulated defence responses in cultured plant cells.

Culture filtrates of *V. albo-atrum* yielded two elicitors of rishitin (Woodward and Pegg, 1986). In staling, 33-day-old shake cultures, a high molecular weight fraction induced higher levels in susceptible than resistant tomato isolines. Stirred, 14-day-old 5 l cultures yielded a 3-β-glucan of six glucose units which elicited similar quantities of rishitin (12–19 $\mu g\ g^{-1}$ of fresh weight) in both susceptible and resistant plants. The work illustrates the important influence of the culture conditions on quantitative and probably qualitative aspects of elicitor biosynthesis. The authors concluded, however, that this elicitor activity was unrelated to single-gene resistance in tomato. Onuorah (1987) found that culture filtrates of pathogenic and non-pathogenic isolates of *V. albo-atrum* elicited phytoalexins in lucerne. Smith *et al.* (1971) isolated a glycopeptide from cultures of a virulent strain of *V. albo-atrum* which elicited medicarpin and sativan in lucerne. Only the carbohydrate moiety acted as the elicitor. Koike *et al.* (1995a) used heat-released cell wall elicitors from a *V. albo-atrum* isolate from lucerne to select all lines from cv. *Vertus* with high PAL activity, lignin deposition and phytoalexin accumulation. A glycoprotein elicitor from *V. albo-atrum* was shown by Cooke *et al.* (1994) to induce synthesis of medicarpin in lucerne cv. Kabul. Treatment of seedlings of the same cultivar with the cell-permeating cyclic AMP analogue dibutryl cyclic AMP also induced medicarpin and stimulated PAL activity. The interesting feature of this work was that treatment of cell suspension cultures with the *Verticillium* elicitor led in a few minutes to a transitory increase in intracellular cyclic AMP together with a pulse of adenylyl cyclase and a subsequent increase in cyclic AMP phosphodiesterase. Incubation of a membrane fraction from *M. sativa* cells with the glycoprotein elicitor resulted in a dose-dependent stimulation of adenylyl cyclase activity. These results strongly suggest that cyclic AMP may be a second messenger in the lucerne defence response. Robinson and Smith (1995) showed that fluxes of Ca^{2+} into lucerne protoplasts were implicated in the intracellular signal transduction mechanism that mediates between interaction of elicitor and receptor and the activation of gene transcription involved in phytoalexin production. The rate of efflux of $^{45}Ca^{2+}$ from elicitor-challenged protoplasts also occurred. A plasma membrane Ca^{2+}-ATPase was involved in the efflux.

Elicitors play an important role in the stimulation of oxidative enzymes, a topic which has received new interest in recent years. Davis *et al.* (1993) maintained that only the protein moiety in a *V. dahliae* glycoprotein was the specific extracellular signal for phytoalexin induction in cotton and soybean cell suspension cultures. The carbohydrate fraction was responsible for the H_2O_2 burst. Similarly, polygalacturonic acid (a 20-mer) elicited oxidative activity but not antifungal substances. The role of elicitors and *inter alia* a *V. dahliae* glycoprotein in the oxidative burst has been reviewed by Low and Merida (1996). Tomato anionic peroxidase (PO) genes *tap*1 and *tap*2 are induced by wounding and pathogen attack. The 5′-flanking region of *tap*1 controls wound- and pathogen-inducible β-glucuronidase (GUS) in tobacco plants transformed with a *tap*1–*GUS* chimaeric fusion gene construct. When tobacco protoplasts with −358 bp and larger *tap*1 constructs were incubated with a *V. albo-atrum* [*sic*]? elicitor, GUS activity was increased twofold over a high constitutive level (Mohan *et al.*, 1993). Low *et al.* (1993), in a comparison of biotin and a *V. dahliae* glycoprotein, found that uptake rates based on molecular size, the degree of saturation of ligand binding and kinetics were similar in endocytic pathways in cultured soybean cv. Kent cells. However, the elicitor was transported mainly to the vacuole, while biotin moved to the cytoplasm. The involvement of phospholipase A (PLase A) in plant signal transduction for auxin-stimulated growth is well documented but not associated with a plant defence response, while oligogalacturonic acid, a potent elicitor of the soybean H_2O_2 burst, did not stimulate PLase A; harpin and a *V. dahliae* glycoprotein elicitor stimulated both (Sreeganga *et al.*, 1996). A further indication of the unique channel of signalling by *V. dahliae* is that the pathogen-stimulated oxidative burst was blocked by pre-treatment of cells with chlorpromazine-HCl (a PLase A inhibitor) at the same concentration that inhibited PLase A, but chlorpromazine-treatment of harpin- or oligogalacturonic acid-elicited cells had no effect on the H_2O_2 burst.

A comparison of the *V. albo-atrum*-stimulated oxidative burst in Ve^+ and Ve^- tomato cell lines by Liu and Kolattukudy (1996) showed that the time course in each cultivar was identical, peaking at 20 min, but was elicitor concentration dependent. At 1 $\mu g\ ml^{-1}$, the oxidative burst was low but was maximum at 25–30 $\mu g\ ml^{-1}$ of cell suspension. In the Ve^+ line, the luminol chemiluminescence was 100–1000 greater than in the susceptible line. Trevisan *et al.* (1997a) described the PO activity in cell cultures of five hop cultivars elicited with *V. albo-atrum* culture filtrate and a crude hyphal homogenate. Trevisan *et al.* (1997b) also demonstrated a two- to threefold initial increase in PAL activity in resistant (cv. Wye Target) cell cultures compared with susceptible cultures (cv. Hallertaner Mittelfruh) after the addition of culture filtrate. Flocco *et al.* (1998) detected maximum (100% increase) PO activity in cell cultures of horseradish (*Armoracia lapathifolia*) transformed by *Agrobacterium rhizogenes* LBA 9402 at the end of the exponential phase. Elicitors included a *Verticillium* preparation (unspecified) and the abiotic elicitors $AgNO_3$ and $CuSO_4$. Park and Lee (1996) demonstrated that *V. albo-atrum* elicitors induced the cross-hybridization of 3-hydroxy-3-methylglutaryl (HMG)coenzyme A DNA to 2.7-kb mRNA in tomato cells. Wounding stems and leaves or roots also induced synthesis of HMG reductase (a regulatory enzyme of isoprenoid compound biosynthesis including phytoalexins) in a biphasic mode (Heesung and Yongse, 1996). The effect was transient, with maximum levels from 9 to 12 h after the addition of the elicitor.

Fungal culture filtrates and *inter alia V. dahliae* were used to elicit taxol (the anticancer drug) and other taxanes in *Taxus baccata* cell suspension cultures. Arachidonic acid at 1 mg l^{-1} raised taxol production by 150%, but oxidative stress induction and recognized abiotic elicitors had no effect.

Tissue Culture and Fungal Extract Experiments

Protoplasts of six lucerne genotypes were incubated with *V. albo-atrum* culture filtrates and cell wall components (Koike *et al.*, 1993). Those from susceptible lines reacted more to a low molecular weight fraction of the filtrate while protoplasts from resistant lines were affected more by high molecular weight fractions and cell wall components. A subsequent study by Katsumata *et al.* (1993) attempted to screen 65 cell lines for high PO activity (and presumably resistance) using a heat-released cell wall elicitor. From the susceptible cv. Vertus treated with *N*-methyl-*N*-nitro-*N*-nitrosoguanidine, four stable high PO cell lines were obtained; these, however, failed to show higher PO activities than the parental lines following normal inoculation with *V. albo-atrum*. Murakami *et al.* (1993) failed to identify *V. dahliae* resistance in hypocotyl fragments and calli derived from Ve^+ and Ve^- genotypes using culture filtrates in *in vitro* bioassays. Subsequently, Murakami *et al.* (1996) refuted earlier findings in favour of *in vitro* screening of calli and hypocotyl explants with ethanol and acetone extracts of *V. dahliae* culture filtrates. Tissue from a Ve^+ genotype was more resistant to ethanol extracts and ouabain (G-strophanthin, an ion transport inhibitor). Acetone extracts however are notoriously toxic from fusil oil contamination unless subjected to multiple fractional distillation immediately prior to use in extraction. The authors conclude that *in vitro* bioassay could be used to identify resistant genotypes. This approach, however, would only be convincing if extended to derived whole plants which were shown to survive normal infection (see also Ireland and Leath, 1987; Ji *et al.*, 1995).

Single cells derived from resistant, susceptible, aubergine callus lines were identified successfully by de Melo and Brar (1995a,b) using culture filtrates from *V. dahliae* isolate VD-02. Rapid cell death greater in R than S lines was identified by fluorescein diacetate. Culture filtrate selections closely paralleled glasshouse infectivity screenings. A similar scheme to select *Verticillium*-resistant potato was used by Koike *et al.* (1996b) in which tuber and callus were inoculated with *V. albo-atrum* or grown on a medium containing culture filtrate of the pathogen. Tolerant, cv. Triumph suppressed fungal growth in culture, and callus tolerated and grew in the culture medium. Nachmias *et al.* (1985a) were able to identify tolerance to *V. dahliae* in pollen germination and tube growth. Cultivars Alpha and WA-85-4-5 successfully tolerated a PLP toxin (Buchner *et al.*, 1982) while the susceptible cv. Promesse and *Solanum phureja* lv P35 did not.

Induced Resistance

Investigations into 'acquired immunity' or cross-protection against *Verticillium* wilt are few in number. More comprehensive work has been carried out on *Fusarium* (see Matta, 1989) but the principles involved seem common to both. Pathogens, non-pathogens (cross-protection) and abiotic factors (induced resistance) have all conferred protection when applied as a pre-inoculation treatment.

Roots of cotton cv. Acala 4-42 sprayed with conidia of a mild (non-defoliating) strain of *V. dahliae* (SS4) protected 66% of the population from a defoliating strain (T1) inoculated 7 days later. When field soil containing an estimated 5×10^3 microsclerotia of T1 g^{-1} of soil was supplemented with 1×10^5 SS4 microsclerotia g^{-1}, 94% of the plants were protected compared with 100% in non-supplemented soil (Schnathorst and Mathré, 1966b; Barrow, 1969). Equal numbers of mild and severe strain propagules only resulted in defoliating wilt. Soil inoculum in the ratio 1×10^5 microsclerotia SS4:1×10^4 T1 protected 50% of plants from severe wilt. Bell and Presley (1969a) achieved small increases in resistance in cotton by cross-protecting roots with heat-killed conidia. Live cells of avirulent or weakly virulent *V. dahliae* were more effective. Muslimov (1974) reduced cotton wilt by 80% after treating plants with culture filtrates of *V. dahliae* (see also Fedotova and Guseva, 1970; Sidorova, 1981). The role of silica in induced resistance is usually associated with foliar pathogens and the penetrative resistance of epidermal cells. Colson (1998) found that two foliar sprays of silicic acid and benzothiadiazole (BTH) applied to cotton cv. Siokra 1-4 in the pre-wilt period of the Australian spring (January) significantly decreased *V. dahliae* wilt severity in March. It is not clear from this study whether BTH and silicic acid were applied separately or in combination, and what role, if any, silicic acid played in symptom alleviation.

A measure of crop-protection control in cotton was achieved by Paplomatas and DeVay (1990) in plot experiments, treating soil infested with *V. dahliae* with the non-pathogen *V. tricorpus*. A 14% reduction of *V. dahliae* foliar symptoms resulted from root colonization with *V. tricorpus*. *Verticillium* wilt in tomato was reduced by simultaneous inoculation with *F. oxysporum* or *V. albo-atrum*. The degree of acquired resistance was proportional to the inoculum concentration of the non-pathogen challenge fungus (Tigchelaar and Dick, 1975). A curious feature of *Verticillium–Fusarium* mixed inoculum reported by Pineau and Besri (1991) was that *V. dahliae* resistance in *Fusarium*-resistant tomato increased while resistance to *V. dahliae* in *Verticillium*-resistant tomato decreased. Amemiya *et al.* (1989) used three non-pathogenic isolates appplied to roots at 1×10^7 conidia ml^{-1} or to soil pre-inoculated with *V. dahliae*. Protection was achieved when the challenge fungus numerically exceeded the pathogen. Plants inoculated with both organisms produced more tyloses (also reported for cotton by Suchorukov, 1957), and the antifungal activity of root exudates was comparable with that of single-gene resistant plants. Significantly, inoculation by the challenge organism alone did not induce these changes. Earlier, Amemiya *et al.* (1985, 1986) had shown that while pre-inoculation of tomato with *F. oxysporum* f.sp. *cucumerinum* and a non-pathogenic isolate of *V. dahliae* protected against *F. oxysporum* f.sp. *lycopersici*, so did a cell-free *F. oxysporum* f.sp. *lycopersici* conidial germination fluid, but only transitorily. Somewhat similarly, Fernandes *et al.* (1975) showed that *V. albo-atrum* cross-protected tomato against *Pseudomonas solanacearum* when inoculated more than a day in advance of the bacterium. The protection was proportional to the inoculum concentration of the protector. No protection was conferred in simultaneous inoculation and no antagonism was observed in mixed culture. A further example of multiple pathogen interaction was provided by Page and Gray (1988) experimenting with mixed inoculation of lucerne with *V. albo-atrum* and *Phytophthora megasperma* f.sp. *medicaginis* (PMM) in cultivars Ladak (susceptible to both) and WL316 (resistant (tolerant) to both). In cv. WL316, mortality to a single pathogen equalled that of combined inoculation, whereas in cv. Ladak *Verticillium* wilt was suppressed by the presence of PMM. In third harvest of cv. WL316, *V. albo-atrum* alone and the combined pathogens caused a significant reduction in yield compared with PMM alone and the control. Matta (1989) discounted the role of tyloses or vascular occlusion in cross-protection.

Treatment of *Cucumis sativa* (cucumber) roots with filtrates of *V. albo-atrum* 3–5 days before inoculation with the same isolate significantly reduced vascular colonization and wilt symptoms compared with non-treated, inoculated controls (Tjamos, 1979). Similar results were obtained by spraying the two basal leaves with a *V. albo-atrum* conidial suspension. A similar, acquired, 'auto immunity' in cucumber, with translocation from the challenge leaf, has been variously achieved with *Sphaerotheca*, *Collototrichum* and virus infection, which may represent a phenomenon unique to this host. The minimal response time between challenge and pathogen inoculations was 72 h (Zaki *et al.*, 1972b; Melouk and Horner, 1975). Skotland (1971) protected *Mentha piperita* (peppermint) and *M. cardiaca* (spearmint) against a virulent isolate of *V. dahliae* by pre-inoculating with an isolate of *V. nigrescens*, weakly pathogenic to these species and to tomato, pepper and aubergine. Melouk and Horner (1975), using the same combination, observed a maximum response with pre-inoculation 7–9 days before the pathogen. Four weeks after the pathogen inoculation, wilt severity was greatly reduced, with many plants remaining symptomless. Microsclerotial formation in cross-protected plants was only 5–7% of that in *V. dahliae* controls. The same response time was reported for cotton (Bell and Presley, 1969a).

Both natural and synthetic growth-regulating substances act as inducers of resistance and vascular wilts (Bell and Mace, 1981). Many studies have been conducted on *Fusarium*, and the picture with *Verticillium* spp. is incomplete. Ethylene plays a multiple role in wilt disease (see Chapter 8). Whole young tomato plants exposed to a 5 p.p.m. ethylene:air mixture for 24 h remained symptomless after inoculating with a virulent strain of *V. albo-atrum* (Pegg,

1976a). The numbers of colonized vessels were greatly reduced in gassed plants compared with controls. Conversely, the numbers of vessels differentiated in the secondary thickened zone of the lower stem increased significantly, resulting in a substantial net increase in fungus-free xylem. Unlike reports on *Fusarium*-infected plants, in which excised parts were gassed, ethylene treatment of whole plants had no effect on PO activity, and PAL was reduced. PPO activity, however, was greatly stimulated by ethylene. Ethylene acts as a powerful stimulant of mRNA synthesis, affecting enzyme induction and antifungal metabolites, both of which could be involved in the resistance response (Abeles, 1973). Pegg (1976a), however, found no increase in antifungal activity in ethylene-treated tomato.

The effect of the non-fungicidal dinitroaniline herbicides in conferring resistance to *V. dahliae*-inoculated aubergine and tomato may be interpreted similarly. 4-(Methyl sulphonyl)-2,6 dinitro-*N,N*-dipropylaniline (nitralin), and α,α,α-trifluro-2,6-dinitro-*N,N*-dipropyl-*p*-toluidine (trifluraline), both non-fungitoxic to *V. dahliae*, induce endogenous ethylene in seedlings grown in soil containing 1 μg a.i. g^{-1} of soil. Seedlings transplanted to *V. dahliae*-infected soil showed 95% (tomato) and 75% (aubergine) survival compared with controls (Grinstein *et al.*, 1976).

Growth retardants blocking mevalonate synthesis of gibberellins or blocking gibbberellin-induced enzyme activity (see Chapter 10, Chemical methods) affect *V. dahliae* colonization and microsclerotial production (Buchenauer, 1971; Buchenauer and Erwin, 1972, 1973a; Erwin *et al.*, 1974, 1976). Two growth retardants, *N,N*-dimethylpyrrolidinium iodide (DP11) and *N,N*-dimethylpiperidinium iodide (DPY1), with no antifungal activity induced resistance to *V. dahliae* in treated cotton plants which showed increased hemigossypol levels compared with controls (Buchenauer and Erwin, 1973b). In contrast, 4-hydroxy-3,6-dioxo-hexa-hydropyridazinyl-4-acetic acid (Pynadon), a growth retardant inducing resistance to growth and reproduction of *V. dahliae*, had no effect on host antifungal activity.

Hyphal Lysis and Host Hydrolytic Enzymes

A reduction in the number of infested vessels or reduced mycelial density following infection has been reported by various workers (Blackhurst and Wood, 1963b; Wilhelm and Taylor, 1965; Sinha and Wood, 1967a; Taylor and Flentje, 1968; Dixon and Pegg, 1969; Pegg, 1978). In olive trees (Wilhelm and Taylor, 1965), apricot (Taylor and Flentje, 1968; Harrison and Clare, 1970) and peach (Ciccarese *et al.*, 1990) infected with *V. dahliae*, a seasonal decline in the level of hyphae is reached to a point where no pathogen could be isolated. These authors considered that perennial infection was maintained by annual re-infection. Photoperiod has a profound effect on resistance, reflected in levels of mycelial colonization. Field resistance to *V. dahliae* and *V. albo-atrum* in potato

(*Solanum tuberosum*) declines with shortening days. Plants grown in a controlled environment under a long (15 h) photoperiod showed a large reduction in mycelial colonization after 2 weeks, declining to 3% of that of short-day plants by 5 weeks (Pegg, 1977). Chrysanthemum (*Chrysanthemum morifolium*) only develops severe infection under short (flowering) photoperiods. Pegg and Jonglaekha (1981) similarly found a rapid decline in mycelial colonization in long-day plants. The same phenomenon occurs in tomato, a day-neutral plant, regardless of photoperiod (Blackhurst and Wood, 1963b; Sinha and Wood, 1968). Dixon and Pegg (1969) found a 70% reduction in *V. albo-atrum*-colonized vessels in compatible and incompatible interactions alike. In a near-isogenic resistant cultivar, the pattern of hyphal lysis paralleled the Ve^- cultivar, commencing 8 days after inoculation and reaching 90% of the susceptible level by 14 days.

Pegg (1977) compared the loss of mycelium in *Verticillium* infection with mycophagy found as a constitutive activity of orchid mycorrhizal infection. An enzymic basis for heterolysis was proposed by Pegg and Vessey (1973). Chitinase (endo-1,4-β-*N*-acetylglucosamine hydrolase) activity was found as a constitutive activity of healthy plants, which increased following infection by hop and tomato isolates of *V. albo-atrum*. Pegg and Young (1981, 1982) subsequently characterized different constitutive chitinases, an *N*-acetylglucosaminidase from the pathogen and tomato. Since chitin is not found in tomato, in common with other higher plants, the enzymes were considered to be non-specific, possibly acting on bound *N*-acetylglucosamine (NAG) as glycoproteins in cell membranes. The tomato chitinase had a molecular weight of 27–31 kDa, a K_m of 10.46 mg chitin ml^{-1} and a V_{max} of 55.6 μg NAG h^{-1} ml^{-1}. The kinetics, pH and temperature responses of the tomato enzyme distinguished it from a *V. albo-atrum* chitinase of 63 kDa. It was concluded that increased chitinase associated with hyphal lysis was attributable largely to host activity. Young and Pegg (1981) characterized three isozymes of a constitutive endo-1,3-β-glucanase from tomato. Exo-, but no endo-glucanase was found in *V. albo-atrum*. Large increases in glucanase activity were found in infected plants coincident with increased hyphal lysis (Pegg and Young, 1981). Activity was higher in susceptible than resistant lines, reflecting the greater *in vivo* mycelial biomass in the former. In an attempt to simulate hyphal lysis *in vitro*, Young and Pegg (1982) found that hydrolysis of purified *V. albo-atrum* hyphal wall required a combination of host 1,3-β-endo-glucanase and endo-chitinase together with either fungal exo-glucanase or 1,4-β-glucosidase. Ethylene-treated whole tomato plants led to increased synthesis of 1,3-β-glucanase in leaves and decreased synthesis in stems (Pegg, 1976a). Chitinase was inhibited in stems and leaves.

The results of Pegg and Young (1981) suggested that the Ve^- isoline (Craigella S) in *in vivo* experiments, recognized *V. albo-atrum* cell wall components, resulting in a linear increase in chitinase and glucanase activity over 21 days. While the results are consistent with a non-specific defence reaction in tomato, no role could be found for glucan or chitin hydrolases in single-gene resistance mechanisms.

Strong evidence was found in potato (Pegg, 1977) and chrysanthemum (Pegg and Jonglaekha, 1981) for a correlation between chitinase and glucanase activities and hyphal lysis in relation to photoperiod. Enzyme levels were low in long- and short-day control plants, but increased in long-day infected plants in parallel with the loss of invading mycelium. Cotton seeds inoculated with VAM mycorrhizal fungi *Glomus mosseae*, *G. versiforme* or *Sclerocystis sinuosa* with or without *V. dahliae* revealed that VAM reduced the incidence and severity of *Verticillium* (Liu *et al.*, 1993, 1995). Both VAM and *V. dahliae* infection induced ten new proteins, one of which had chitinase activity. The authors claim that the proteins (one, some or all?) retard hyphal growth and kill *V. dahliae* conidia, but no indication is given as to whether the chitinase protein is of constitutive host origin. El-Quakfaoui *et al.* (1995) reported for the first time growth inhibition of *V. albo-atrum* by a purified chitosan gene from *Streptomyces* sp. strain N174 which was used to transform tobacco cv. Xanthi. A plasmid containing the chitosan gene with its own signal peptide under the control of the cauliflower mosaic virus (CaMV) 35S promoter was used to transform cv. Xanthi via *Agrobacterium tumefaciens* mediation. Three chitosanases were recovered from leaves and intercellular fluids by SDS–PAGE, one of which resembled the *Streptomyces* enzyme but with a different signal peptide cleavage site. Cotton leaf tissue treated with heat-released soluble *V. dahliae* cell wall fragments induced a biphasic synthesis, or increase of three chitinases with pIs of 3.7–4.2, and a 1,3-β-glucanase with a pI of 4.5, in addition to synthesis of the antifungal metabolites lasinilene and cadalene. The hydrolases reached a peak at 12 h followed by a gradual increase up to 120 h. Differences were found between constitutive and inducible enzymes (Dubery and Slater, 1997). Kawaradani *et al.* (1998) described a new, simpler and quicker method for determining 1,3-β-glucanase using *p*-nitrophenyl-β-D-laminaritetraoside as a substrate. Using this method, the authors found 1.5-fold increased enzyme activity in aubergine 4 days prior to and 1 day after symptom appearance.

Protein protease inhibitors have been isolated from dormant cotton seed by Mezhlum'yan *et al.* (1994) and Kasymova and Yuldashev (1995), who discuss a possible protective role against *V. dahliae*. A mycelial proteolytic extract of 10 mg protein ml^{-1} was completely suppressed by a 1 mg of protein ml^{-1} inhibitor preparation. Mirzaazimova *et al.* (1994a,b, 1995) reported differences in free lipids and lipids weakly and strongly bound to proteins in seed of resistant cv. Tashkent-1 and susceptible cv. 175-F cotton; lipids and fatty acids varied according to the degree of resistance. Xing and Chin (2000) suggested increasing production of 16:1 fatty acids in aubergine as an approach to resistance against *V. dahliae*. Infection increased the lipids strongly bound to proteins in resistant cotton cultivars with a concommitant reduction of weakly bound lipids. The significance of these lipid changes in resistance is not clear and may merely represent a consequence of infection with no particular defence role (Gusakova *et al.*, 1995).

An interesting experiment breaking single-gene resistance in tomato was described in an incomplete experiment by Jones and Overman (1976b). Two-week-old seedlings of cv. Tropic (with Ve resistance) were root-dip inoculated before incubation under a 4-h photoperiod and at temperatures of 20, 24, 28 and 32°C. The respective wilt percentages were 10, 32, 61 and 80%, but surprisingly the surviving percentages were 40, 68, 79 and 92%. While the authors conclude that the *Ve* gene is broken by high temperatures and low photoperiod, this must represent metabolic stress, since 32°C is not normally limiting for the *Ve* gene and the low survival rate correlating with a low wilt incidence is not consistent with a temperature lethality.

Control 10

1 Physical Methods

Heat

Various temperatures have been cited for the thermal death point of *Verticillium* spp. which naturally depend upon duration and other factors in the particular experiment. Solovyeva and Polyarkova's (1940) claim, quoted by Domsch *et al.* (1980a,b), that *V. dahliae* survives up to 80°C in soil demands verification and explanation. The control of *V. albo-atrum* in tomato achieved by Bewley (1922) by raising the glasshouse temperature and ridging soil around the lower stems of plants resulted in symptom-free growth rather than elimination of the pathogen.

Nelson and Wilhelm (1958) claimed that the thermal death point (TDP) for hyphae and conidia of *V. dahliae* was a minimum of 5 min at 47°C in water, and 10 min in soil at 50°C for microsclerotia. A slight variation of 30 min in soil at 57.5°C was given by Bollen (1969) and 30 min in soil at 52.5°C for *V. albo-atrum*. In a subsequent study, Bollen (1985) found different values for different strains of *V. dahliae*: 45–47.5°C for potato and tomato strains and 40–45°C for a sugarbeet strain, values in agreement with Pullman *et al.* (1981a,b) – LD_{90} exposure values for two strains of *V. dahliae* were 23 and 27 min at 50°C (Katan, 1981). Attempts by Gilad and Borochov (1993) to eliminate *V. dahliae* 'spores' [*sic*] (microsclerotia) from tubers of *Liatris spicata* (gayflower or blazing star) by postharvest hot water treatment was successful at 49°C for 40 min. At 53°C, however, tuber membrane integrity was impaired as evidenced by enhanced electrolyte leakage and a decline in membrane protein thiols. Takano *et al.* (1985) advocated the use of dry heat to disinfest seed of *Verticillium*. Staffeldt (1959) and Sterne *et al.* (1979) eliminated *V. dahliae* from cotton waste by composting, during which the temperature reached 50°C in 4 days and 68°C in 12 days, thereafter declining. Yuen and Raabe (1984) similarly eliminated *V. dahliae* in turned plant residue compost. Effective temperatures of >45°C lasted for 10–14 days with peaks at 70°C. Corners of the compost heap failed to reach lethal temperatures. Van Wambeke *et al.* (1987) described the novel use of microwaves for sterilizing crop growing substrates which could also apply on a small scale to highly infective diseased crop residues. Castejon-Munoz and Bollen (1993) found that exposure of *V. dahliae* microsclerotial suspensions to 45°C for 30 min gave a survival of <0.01% of the population; when suspensions were subjected to a non-lethal pre-conditioning exposure of 40°C, survival at 45°C was 0.4%. The duration of thermal pre-conditioning depended on the degree of melanization of the microsclerotia. Exposure of PDA cultures of *V. dahliae* from olive at 55, 50, 47, 45, 42 and 40°C for 15 min, 45 min, 2, 4, 24 and 24 h, respectively, all proved lethal. Similarly *V. dahliae* could not be cultured from diseased olive branches subjected to 55, 50, 47, 45 and 40°C for 15 min, 60 min, 2, 6 and 30 h, respectively. The work on which this study was based (Al Ahmad *et al.*, 1997a) led to a whole-tree, clear plastic sheet-covered chamber, placed over an orchard tree, previously irrigated and with soil covered by a black tarp. Trees were covered for 10, 15 and 20 days when chamber temperatures reached 55°C. Soil temperatures at 5 and 15 cm were 55 and 45°C. After 15–20 days, the fungus could not be isolated from the tree. This temperature exposure had no deleterious effect on the tree (Al Ahmad, 1993).

The burning of plant residues in the field to eliminate pathogenic inoculum has been used for several crops. Powelson and Gross (1962) reported a successful reduction of *V. dahliae* in surface potato haulm debris following harvest. Using temperature, melting-indicator tubes inserted into the pith of diseased stems and a tractor-drawn propane-flamer boom, speeds of 1.5 m.p.h. eliminated *V. dahliae* in 95% of the stems but left unburnt root stubs. At this speed the percentage of 125°F, 150°F and 200°F tubes melted was 100, 90 and 53%, respectively, while at 3 m.p.h. only 60% of the 125°F and none of the 200°F tubes melted. The same technique was used by Horner and Dooley (1965) to control *V. dahliae* in peppermint stubble. Easton *et al.* (1972a) in a 3-year study on potato found that stubble burning following frosting of the haulm only successfully reduced soil inoculum after 3 years. The authors concluded that inoculum production from buried haulm compensated for that killed by flaming. In the absence of fumigation, burning in the third year reduced inoculum and plant infection, and led to a 16% increase in yield. Largest yields, however, resulted from a combination of flaming and fumigation, which reduced soil inoculum in some samples to zero. In a subsequent 2-year study (Easton *et al.*, 1975), haulm burning in the absence of fumigation led to a substantial reduction in soil propagules (65–82% reduction in year 1 and 15–94% reduction in year 2), with 25 and 20% increases in tuber yield in successive years. Curiously, these results were not mirrored in wilt incidence in the field or in the number of microsclerotia found in stem tissue. Ausher *et al.* (1973) showed that flaming of cotton waste in the field effectively reduced the number of propagules per g of soil from 2×10^6 to 5×10^3. Heating stem tissue to 70°C for 30 s or to 100°C for 25 s eliminates *V. dahliae* even in wet tissue. The need for such treatment was emphasized by Krikun and Susnoski (1971) who foresaw the spread of the pathogen into newly cultivated areas by the use of cotton waste in cattle bedding. A novel application of heat sterilization was described by Runia *et al.* (1988) for the control of *V. dahliae* infection in recirculating soil-less culture nutrient solutions. Raising the temperature of the recycled nutrient briefly to 90°C before cooling via heat exchangers and passing to root systems effectively controlled *V. dahliae.*

Solarization

The use of polyethylene or PVC sheeting as a ground cover to trap solar radiation and retard the loss of convective heating is termed solarization. This technique pioneered by Katan *et al.* (1975) raises soil temperature in surface layers to the TDP of *Verticillium* spp. thereby eliminating inoculum not only of *Verticillium* but also of other pathogens, nematodes and weed seeds. Where soil tarping is carried out around a standing crop, e.g. plantation crops such as pistachio or olive, the technique, often referred to as mulching, may affect crop growth in ways other than reduction of the pathogen, e.g. by water retention. The term solarization should therefore be restricted to the treatment of fallow ground or if in the presence of crop plants when the result is assessed in terms of inoculum reduction. In other circumstances, the term plastic mulching is preferred.

Katan and co-workers (Katan *et al.*, 1975, 1976a,b) in the Jordan Valley, Israel, covered irrigated land with a 0.03 mm polyethylene sheet during July and August. *V. dahliae* inoculum was buried at various depths in sheeted and unsheeted soil. Temperatures of 45–48°C were reached at 5 cm and of 40–48°C at 15 cm below the surface. After 2 weeks, the pathogen was eliminated to a depth of 25 cm. The authors found that weed seeds were killed also and potato yield was enhanced by 300%. Abu-Blan and Abu-Gharbieh (1994) tested 12 weeks of solarization in the Jordan Valley with clear and black plastic followed by winter crops of potato, cauliflower and cucumber. Soil pathogens and *inter alia V. dahliae* populations were greatly reduced. Both tarps were retained as a mulch, with the transparent sheet covered with a perforated black one. Yield increases, transparent versus black, were: potato 31 and 29%; cauliflower, 48 and 10%; and cucumber 10 and 0%, respectively. The cucumber results most probably reflect the resistance of *Sclerotinia sclerotiorum*, the major pathogen of cucumber, to solarization temperatures. Wilt in aubergine was reduced by between 25 to 95%. Subsequently, Grinstein *et al.* (1979a,b) described the successful application of the technique to tomato (see Bourbos and Skoudridakis, 1996), potato, cotton and groundnut. Yield increased on average by 100% and the effect persisted for 3 years. Aloj and Noviello (1982) in Italy, achieved temperatures of 44–57°C at 5 cm and 36–45°C at 20 cm 4 weeks after sheeting wet soil with polyethylene; higher temperatures were produced under PVC film. Wilt of potato, tomato and aubergine were effectively controlled.

Partial control of *V. dahliae* in potato was reported by Davis and Sorensen (1986) over 2 years. While soil temperatures in surface layers rose to 41°C from 26°C, roots growing below 15 cm still became infected. The use of black plastic sheeting had no effect on infection rate or yield of aubergine (Moorman, 1982). This was illustrated in comparative studies by Stapleton *et al.* (1989) using black or transparent sheeting on apple, pecan and grapevine plantations in the San Joaquin Valley, California. While soil temperatures at 15–23 cm under clear plastic were raised by 10–18°C, under black plastic the increase was only 8–12°C. In these experiments, *V. dahliae* was reduced by 55–97% in depths from 0 to 23cm under clear plastic during winter or summer. Weeds were reduced by 82% under either clear or black sheeting. Neither trees nor vines sustained injury, but all showed increased negative growth and fruiting. Pre-irrigated pecan trees were grown for an entire season without further irrigation. The optimal thickness for maximal solar transmission is 25–30 μm (Pullman *et al.*, 1979; Katan, 1981). Since hydrated microsclerotia and resting mycelium are thermally more sensitive than when dry, solarization in irrigated soil is more effective (Macfayden, 1967; Katan *et al.*, 1976a,b; Katan, 1981). This has been achieved by pre-irrigation or trickle irrigation under the sheeting (Katan, 1981) or by furrow irrigation under the sheeting (Pullman *et al.*, 1979). Good control of *V. dahliae* inoculum was achieved in Italy (Cartia, 1989) with benefits at soil depths beyond the enhanced temperature zone of solarization. Increased levels of nitrates, Ca and Mg were found. In plastic tunnels, temperatures of 46.9°C at 5 cm and of 42.3°C at 15 cm were recorded (Cartia *et al.*, 1991). Tjamos and Faridis (1981) and Tjamos *et al.* (1989) similarly recorded excellent control of *V. dahliae* by ground tarping in plastic tunnels. Botseas and Rowe (1994) described the effectiveness of Cretan summer sun for solarizing plastic tunnel soil where *V. dahliae* of early potatoes is a major problem. After 10 weeks' cover with transparent polyethylene during June, July and August, inoculum was reduced from 1378–1806 microsclerotia g^{-1} of soil in controls to undetectable levels. This was reflected in a tuber increase of 112% in plants with only 0.3–0.4% root infections compared with 67% in non-solarized soils. Solarization of Cretan tomato field soil by Ioannou (1999) gave effective control of *V. dahliae* and other pathogens, including a variety of weeds: *Malva, Aramanthus, Chrysanthemum, Chenopodium, Urtica urens, Lolium rigidum* and *Calendula arvensis*, a number of which are recognized as non-hosts of *Verticillium*. Soil, in 0.8-m strips, was solarized under transparent polyethylene sheets for 7–8 weeks in July–August, raising the soil temperature by 10–11°C. Tomato cv. Marmande planted in mid-September and harvested in March showed increased fruit yield by 60–135%. Glasshouse solarization in Murcia, Spain reduced the incidence of wilt in aubergine from 23 to 1.8% (Cenis *et al.*, 1988). The use of plastic tunnels or glasshouses offers potential for solarization in northern latitudes where solar radiation is lower and of shorter duration. In Australia, Hardy and Sivasithamparam (1985) obtained satisfactory control of aubergine wilt, but Porter and Merriman (1985) reported inconclusive results for *V. dahliae* in tomato.

The interaction of solarization and mineral availability (see later) described by Cartia (1989) was investigated by Stapleton *et al.* (1991). Soil heated in an incubator for temperatures and periods achieved by solarization showed increased concentrations of NH_4-N, NO_3-N, P, K, Ca and Mg. Field application of 305 kg of NH_4-N ha^{-1} sometimes reduced soil populations of *V. dahliae*, but growth and yield were always stimulated more by solarization than NH_4 fertilizer. A similar threefold increase in soil N was detected in the top 10 cm of Ontario potato field soil after 8 weeks solarization. *V. dahliae* propagules were reduced by 73 and 54% in the upper 15 and 30 cm soil horizons, respectively. The nematode *P. penetrans* was also reduced by 50% (Lazarovits *et al.*, 1991). Yields of potato, soybean, maize and aubergine planted the following season all showed reduced wilt and enhanced growth.

Solarization has been effective in erradicating *V. dahliae* from irrigated cotton fields (Pullman *et al.*, 1979, 1980, 1981b; Conway and Martin, 1983; Stapleton and DeVay, 1986). Agar disc experiments showed that the exposure time at 37°C for LD_{90} was 29 min, and 23 min at 50°C. Most propagules were killed after 14–66 days under clear polyethylene and Californian conditions (Pullman *et al.*, 1979, 1981b). The mycorrhizal fungus *Glomus fasciculatus* with a high thermal tolerance survived these conditions. Weir *et al.* (1987) reported that solarization ameliorated the potassium–*Verticillium* disease complex, increasing cotton yields by >60%. Melero-Vara *et al.* (1990) claimed cotton yield increases in Spain of 230% (see also Jimenez-Diaz *et al.*, 1991, who claimed beneficial effects (possibly nutritional) in the absence of major pathogens). Basallote *et al.* (1994) and earlier workers on cotton, while claiming a substantial reduction in inoculum density following treatment were equivocal about solarization in cotton (a relatively low-value crop) as a cost-effective control measure, since with the replenishment of soil inoculum it would need to be repeated on a regular or fairly frequent intermittent basis. While solarization originated as a fallow ground (pre-plant) treatment, its successful use on growing crops, especially plantation trees, has been widespread. Ashworth and Gaona (1982) using clear polyethylene sheets, controlled *V. dahliae* wilt in established pistachio nut groves. Ashworth *et al.* (1982) claimed the eradication of *V. dahliae* to a depth of 4 feet in pre-irrigated soil, which was reflected in a 75% reduction of wilt the following year; it is difficult, however, to see how convected heat could transfer so deeply and be effective. Similar control programmes have been carried out on established avocado, almond and olive. Trees sustained no damage except for transient Fe deficiency in avocado. In the first year, the disease percentages in control and solarized avocado and almond plantations were 30 and 18%, and 22 and 4%, respectively. Second year data for avocado were 13 and 17% (Shabi *et al.*, 1986, 1987). In a subsequent study (Shabi *et al.*, 1990), control of the pathogen to a depth of 70 cm was described. A severely diseased olive showed greatly improved growth after 2 years. Tjamos *et al.* (1986) also reported the recovery after solarization of 10- to 15-year-old infected olive trees (see also Tjamos, 1991). Wilt in newly-planted olive orchards from carry-over inoculum from other crops is a major Andalucian problem (Lopez Escudero and Blanco Lopez, 1997). Single or double season solarization under clear 50 μm polyethylene sheets, placed either side of the tree line, reduced inoculum to 0–0.2 propagules g^{-1} of soil. Disease severity was greater and control marginally less effective in irrigated trials. Natural recovery after solarization was claimed to be remarkable, supporting the hypothesis that disease in olive is the result of continuous re-infection. Tjamos *et al.* (1991) report similar recovery in olive. Despite obvious problems with cultural and harvest practices, Morgan *et al.* (1991) found that polyethylene sheeting of growing field cherry, tomato and aubergine was as effective at controlling *V. dahliae* as on fallow ground. Both crops showed yield increases the following year.

One of the recurrent findings in solarization studies is that control is not always directly attributable to the thermal killing of the pathogen, particularly in deep soil levels or when the temperature only marginally approaches the pathogen's TDP (Pullman *et al.*, 1981b; Pinkas *et al.*, 1984; Katan *et al.*, 1988). Solarization temperatures and conditions seem to favour microbial antagonists such as *Trichoderma*, *Talaromyces flavus* and lytic bacteria (Elad *et al.*, 1980; Tjamos and Paplomatas, 1986, 1987a,b, 1988; Greenberger *et al.*, 1987), and *T. flavus* and *Aspergillus terreus* (Tjamos and Skretis, 1990). Examination of pre- and post-solarization soil by Stapleton and DeVay (1982) showed that thermophilic and thermotolerant fungi and bacteria species populations increased contrary to other microorganisms. Stapleton and DeVay (1986) considered that these saprophytes benefited by comparison with soil pathogens by virtue of their less specific and limiting requirements. Soils near Gaza City, Egypt heavily infested with *Verticillium* and other root-invading pathogens had greatly reduced microbial populations including *V. dahliae*, following solarization for 75 days. Actinomycetes and fluorescent *Pseudomonas* spp. quickly recolonized the treated ground. Fruit yields increased by 70% (Wadi, 1999).

A number of workers have combined the effect of solarization with soil fumigation with mixed success. Grinstein *et al.* (1978, 1979a,b) claimed that solarization for 31 days reduced *Pratylenchus thornei* by 80–100% and the complete control of *V. dahliae* microsclerotia, reducing disease in potato by 66%. Ethylene dibromide and chloropicrin fumigation by comparison scarcely affected *V. dahliae*, and the 55% disease reduction was entirely due to a 90% reduction in the eelworm. The yield increases were 35% after solarization and 30% after fumigation. Tjamos and Faridis (1981) found that solarization for tomato was comparable with methyl bromide fumigation. Glasshouse experiments comparing solarization with metham sodium fumigation found that solarization in each experiment gave superior control over chemical fumigation. The percentage infections in aubergine planted after treatment were: control, metham Na and solarization expt I, 23, 2.4 and 1.8%, respectively; and expt II, 9, 23.8 and 6.5%. Ben-Yephet *et al.* (1986) found a synergistic effect of solarization and metham sodium fumigation. Overman and Jones (1986), in a detailed study of solarization and chemical fumigation in Florida, prepared field tomato beds by either cover cropping with *Sesbania macrocarpa* (a forage legume) or solarizing under polyethylene from July to September. Beds were then injected with a mixture of isothiocyanate (17%) + 1,3-dichloropropene (34%) + chloropicrin (15%), or methyl bromide (98%) + chloropicrin (2%). Beds were mulched and planted with tomato cv. Sunny in November and replanted in mid-February. Solarization reduced the incidence of *V. albo-atrum* race 2 [*sic*] (*V. dahliae*) during both seasons. Yield increased 26% in the autumn crop but not in the spring. *Meloidogyne incognita* was controlled only during the autumn. Both fumigants only improved yield in non-solarized plots. The fumigation of strawberry in California pioneered by Wilhelm and Ferguson (1953) and Wilhelm and Koch (1956) required injected field soil to be covered with a continuous sheet of polyethylene to prevent escape of the volatile chloropicrin–methyl bromide fumigant mixture. Fumigation has been used on all strawberry crops since 1965 with exponential increases in yield. While the sterilizing effect on weed seeds, *V. dahliae*, nematodes and other pathogens has been attributed to the gas fumigant, no account has been taken of the solarization effect which must have occurred under the polyethylene tarp. It is known that the killing of the soil microflora releases large quantities of N locked up in protein which is bound electrostatically to clay constituents. There is also a concomitant reduction in biological oxidation of NH_3 and NO_3 which in field practice has led to a 50% reduction in applied N. These effects were attributed wholly to chloropicrin–methyl bromide, but were also described by Cartia (1989), Stapleton *et al.* (1991) and Lazarovits *et al.* (1991) in the absence of chemical fumigants. It is likely, therefore, that at least some of the benefit of soil fumigation as practised in California and elsewhere can be attributed to solarization, depending on the length of time the soil was covered. With the growing concern regarding bromine residues in plants, especially in rotation leaf crops such as lettuce, and the likelihood of more restrictive legislation on the use of methyl bromide fumigation, solarization alone will assume a greater importance (see Imaizumi and Tairako, 1996).

Evidence for the effect of black plastic sheet mulching on *Verticillium* infection and on crop growth is limited and equivocal (Moorman, 1981). Basoccu and Garibaldi (1970) reported that mulching with black polyethylene increased *Capsicum* yield and fruit number, but only in comparison with shaded and irrigated plants. No comment was made specifically on the effect of mulching on disease incidence or yield. Glasshouse tomatoes grown under black PVC developed *V. dahliae* symptoms earlier than non-mulched plants (Basoccu and

Garibaldi, 1974). This effect was confirmed by Elmer and Ferrandino (1991) on *V. dahliae* wilt of aubergine, where plants under black plastic mulch developed symptoms 13 days earlier than non-mulched plants. The effect of black mulching on disease incidence and severity is not clear from the description given. A critical study of the effect of clear, black, or no plastic mulching, on a newly established plantation of apricot (*Prunus armeniaca*) and almond (*P. dulcis*) showed a remarkable penetration of temperature into surface layers. Under clear, black or no plastic, the respective temperatures after 19 weeks at 18 cm were 46, 41 and 33°C . At 30 cm, temperatures were still 41, 37 and 32°C. *V. dahliae* was reduced by 86–100% in both trees, but trees mulched in clear plastic from planting time either failed to survive or did not grow as well as black mulched trees. Continuous mulching in black polyethylene gave better control of *V. dahliae* than conventional solarization and was effective in water conservation. For a chronology of solarization studies to 1986, see Katan *et al.* (1987). The introduction of new types of ground surface film based on low-cost sprayed covers of bitumen, bitumen latex or latex offer new possibilities for disease control. Bochow (1989) found that sprays of these emulsions were as effective as plastic sheeting in sealing the ground and preventing water loss or the escape of metham sodium after injection. Good control of *V. dahliae* was obtained on potato haulm buried in soil which was injected with metham sodium and covered for 7 days with a sprayed latex emulsion sheet. The development of a transparent sprayed sheet of sufficient durability would be invaluable for soil solarization.

2 Chemical Methods

Chemical control methods against *Verticillium* species may be directed against inoculum levels in the soil or growing substrate (soil fumigants), or chemotherapy, based on the containment or elimination of the pathogen *in planta* by chemicals sprayed on aerial parts or applied as drenches or granular preparations. A third category defined by Langcake (1981) as *alternative chemical agents* includes compounds which are not directly fungicidal or fungistatic but which alleviate symptoms. The subject has been reviewed by Erwin (1981) and Talboys (1984).

Soil (Substrate) Fumigants

Fungicidal fumigants act as general biocides, the most common of which are described below.

1. Methyl bromide (CH_3Br) is a broad-spectrum biocide injected 15–20 cm deep, permeating soil pore space as a gas. It is highly water soluble and works most effectively in moist soil, but water balance is critical to prevent pore blockage. As with other fumigants, soil cover with an impermeable polyethylene tarp for 48 h minimum is necessary to maintain effective concentrations. Using a range of polyethylene sheet thicknesses from 0.0254 to 0.1524 mm, Voth *et al.* (1973) found that CH_3Br retention was proportional to film thickness. A major concern with the use of CH_3Br is the danger of toxic bromine residues accumulating in plant tissues, especially leaf crops grown in rotation with strawberry. Voth *et al.* (1973) claimed that strawberry fumigated for 1, 6 or 7 successive years with CH_3Br did not accumulate bromine residues. Hutchinson *et al.* (2000) substituted methyl iodide for CH_3Br and found a synergistic 2.7-fold increase in control of all microorganisms with chloropicrin compared with a conventional CH_3Br–chloropicrin mixture.

2. Chloropicrin (trichloronitromethane) was developed as tear gas based on its powerful lachrymatory and toxic action. Large stocks remaining after World War 1 led to its experimental use as a soil fumigant by Russell (1920). Although non-soluble in water, the same principles apply to its use as for CH_3Br. The pioneering use of chloropicrin as a *Verticillium* fumigant was by Wilhelm and Koch (1956). Later, Wilhelm *et al.* (1961) reported on the superior effect of chloropicrin–CH_3Br mixtures (192 kg ha^{-1} of chloropicrin:106 kg ha^{-1} of CH_3Br). Wilhelm *et al.* (1974c) referred to this as a 'synergistic' mixture.

3. Methylisothiocyanate (MIT) compounds. These general (mostly water-soluble) biocides include Vapam or methamsodium (sodium *N*-methyldithiocarbamate). This breaks down to MIT (methyl isothiocyanate) (Vorlex) which is most probably the active principle of Vapam. MIT (Vorlex) may be used in its own right. Dazomet (Mylone, Basamid) (3,5-dimethyltetrahydro-1,3,5-2H-thiadiazine-2-thione) also releases MIT. Vapam has been applied effectively against *Verticillium* in sprinkler irrigation (Gersti *et al.*, 1977; Erwin, 1981; Krikun and Frank, 1982) but seems only effective in arid regions (Talboys, 1984).

4. Nematicides. The intimate association of nematodes and *Verticillium* spp. in a disease complex is illustrated by the use of nematicides. The most commonly used have been: Telone (1,3-dichloropropene); DD or Trappex (1,3-dichloropropene + 1,2-dichloropropane), often combined with MIT or chloropicrin; Nemagon (1,2-dibromo-3-chloropropane); and EDB (ethylenedibromide).

Although soil fumigation may be effective on a wide range of arable and plantation crops, in practice it is confined to intensively grown, high-value crops, for the most part with shallow root systems. Soil fumigants are limited by cost since high concentrations are required to achieve, by diffusion, concentrations lethal to microsclerotia and melanized mycelium (Wilhelm and Ferguson, 1953; Henis and Bar, 1973; Ben-Yephet and Frank, 1984).

Strawberry

Since the early experiments with chloropicrin–methyl bromide mixtures as soil fumigants by Wilhelm and Koch (1956), the control of *V. dahliae* by regular fumigation has transformed the Californian strawberry industry. The history of the disease and the fumigation technology was described by Wilhelm (1961), Wilhelm *et al.* (1974c), Wilhelm and Paulus (1980) and Pegg (1984). Approximately one-third of the US strawberry crop is grown in California, mostly on muck (black, highly organic) soils. The yield from cultivars derived from cultivars Shasta and Lassen increased from 6000–10,000 kg ha^{-1} to 40,000–60,000 kg ha^{-1} largely due to the control of *V. dahliae*. Soils are fumigated prior to planting with mixtures of 57 parts CH_3Br and 43 parts chloropicrin, and applied at rates of 364–504 kg ha^{-1} through soil chizels to a depth of 15–20 cm. Immediately after injection, soil is covered with a tarp of clear polythene, 0.025 mm thick, glued at the margins to give a complete gas-tight cover to the whole plantation for a minimum of 48 h. In addition to *V. dahliae*, weed seeds are eliminated; and it provides a valuable control of alternate (non-) hosts of *V. dahliae*. Nematodes and other root rot organisms are also killed (Wilhelm, 1961, 1962, 1966; Wilhelm *et al.*, 1974c; Pegg, 1984). Talboys (1984) discussed some of the limitations of the method including the unsuitability of many soils, including soils of deep-rooted plants, for gas diffusion, the extreme toxicity of the chemicals and the extremely high cost. A further disadvantage is the loss of the endophyte *Endogone* (*Glomus fasciculatus* Thaxter). Additional benefits to pathogen control include nitrogen release from killed bacteria, a reduction in NH_3 and NO_3 oxidation and enhanced uptake of K^+, Cl^- and PO_4^{3-} by strawberry roots (Pegg, 1984).

Cassini *et al.* (1971) using methyl bromide reported excellent control of *V. dahliae*, nematodes and weeds, but 25% of plants subsequently were killed by *Phytophthora cactorum*. Wilhelm (1971) found, contrary to Cassini *et al.* (1971a), that CH_3Br alone at doses up to 400 lb $acre^{-1}$ was ineffective, but combined with chloropicrin at 1:1 was excellent. Similar results were described by Zinkernagel (1971) in Germany using CH_3Br at 50 ml m^{-2}, CH_3Br–chloropicrin mixtures and Di-Trappex. The latter was most effective at 74 ml m^{-2} with two- to threefold yield increases. Treatment with methyl bromide alone very occasionally gave yield increases, but chloropicrin was better (Gilles, 1971). In New Zealand, Tate and Cheah (1983) found dazomet more effective than chloropicrin–methyl bromide mixtures at comparable rates applied as a ridge-only fumigant for a crop lasting one or two seasons. Spring application of CH_3Br (Terabol) was ineffective against *V. albo-atrum* [*sic*] (*V. dahliae*) microsclerotia with a soil temperature of 6–8°C at 20 cm. In mid-May with a soil temperature of 16°C, the treatment was effective and extended to the second year (Hoffman and Zinkernagel, 1972). Di-Trappex (DD + MIT) at 300 or 600 l ha^{-1} gave better control of *V. dahliae* in Poland than 0.2% benomyl at 250–600 ml per plant, with the benefit of weed control for 3 years (Szczygiel and Rebandel, 1981, 1982); yields of three strawberry cultivars; Senga Sengana, Midway and Surprise des Halles, increased from 15.6, 5.2 and 2.3 t ha^{-1} in the controls to 54.6, 31.8 and 36.6 t ha^{-1}, respectively. Wilt incidence was closely associated with *Pratylenchus penetrans* populations which were effectively reduced by the DD–MIT mixture. Benomyl had no effect on the nematode. Harris *et al.* (1986) described details of pre-planting fumigation with Basami (dazomet), chloropicrin and formalin.

Harris (1989) retained good control of strawberry wilt in cv. Hapil over three seasons using chloropicrin at 210 or 420 l ha^{-1}, or dazomet at 450 kg ha^{-1} and sheeting over with a polyethylene–nylon sheet, resulting in a two- to threefold yield increase. In a subsequent more detailed study, Harris (1990a,b) compared the pre-planting effects of dazomet (380 and 570 kg ha^{-1}) and chloropicrin (75, 150 and 300 l ha^{-1}) in a planting of cv. Elsanta grown on raised beds and mulched with polyethylene. Vapam (260 kg ha^{-1}) and Telone (210 l ha^{-1}) were also included for observation. In the first season, wilt in control beds was 4.7 and <1% in all main treatments. Wilt incidence following Vapam and Telone was 1.6 and 0%. All treatments reduced nematode numbers, but only dazonet under UK conditions and Kent brick-earth soils killed weed seeds. This did not occur in the following year (Harris, 1991). In spite of the control of pathogens, significant yield increases were not obtained in this low wilt trial. Yields following Telone were reduced. All treatments increased soil mineral nitrogen. In the following season, Harris (1991) claimed that dazomet and CH_3Br treatments increased yield by 40–60%, while chloropicrin at 150 l ha^{-1}, which reduced nematode numbers and *V. dahliae* to indetectable levels, had no effect on yield. This result raises the question of possible deleterious residual effects of fumigants on plant growth which is scarcely considered by the various workers.

Tomato

Soil fumigation against root-knot nematodes and *Verticillium* wilt of tomato has been used mainly in North America. McKeen and Sayre (1964) controlled *V. dahliae* in glasshouse tomato soil using Vorlex (MIT) and EP-201. Methyl bromide–chloropicrin mixtures and Vorlex were effective against root-knot nematode and *Verticillium* (Overman and Jones, 1977). An evaluation of broad-spectrum fumigants versus nematicides was carried out by Jones and Overman (1978). Carbofuram, EDB and 1,3-D, and NaN_3 alone or in combination with carbofuram all decreased the incidence of *V. dahliae* in tomato and increased yields. Broad-spectrum fumigants were most effective, a result confirmed in field trials (Overman and Jones, 1980). Overman (1982) used off-centre, 5 cm deep buried, irrigation tubes to deliver pre- and post-plant tomatoes cv. Flora-Dade with Vapam and EDB to control *Meloidogyne incognita* and *V. albo-atrum* [*sic*] (*V. dahliae*). While the report claimed that yield was improved by a single pre-plant injection of Vapam into the drip system and best root-knot nematode control from a single pre-plant injection compound, neither treatment affected the incidence of *V. dahliae* in the crop. Noling (1987) recommended methyl bromide–chloropicrin for *Verticillium*, weed seed and nematode control. Methyl bromide was also used effectively by Matta (1976) at 70 g m^{-2} and by McSorley *et al.* (1985). An early report by Gindrat *et al.* (1973) showed that while CH_3Br fumigation gave good control of *V. dahliae* and *Meloidogyne* spp. in Switzerland, the accumulation of bromine residues in tomato (up to 1130 p.p.m.) was unacceptable; the cost was also excessive. DD and dazomet controlled *Meloidogyne* but not *Verticillium*. In Greece, Thanassoulopoulos and Kitsos (1973) reported that dowfume MC_2 and Di-Trappex were most effective against *V. albo-atrum* [*sic*] (*V. dahliae*). Treatments were not recommended in the first year of cultivation (see also Jones *et al.*, 1973). Bourbos (1986) found that while *V. albo-atrum* [*sic*] (*V. dahliae*) was sensitive to CH_3Br; live pathogen below the depth of effective fumigation re-infested the upper 30 cm soil layer. Trappex 40 (SN556) gave good control of *V. dahliae*, and Terr-o-gas 67 best control of *M. incognita* (Shoemaker, 1985). Van Wambeke *et al.* (1984) used the oxidase inhibitor sodium azide (NaN_3) as a successful soil fumigant against *V. albo-atrum* in tomato. In Tunisia, triple-yield increases in tomato in *V. dahliae*-infested soil followed 1,3-D or MIT fumigation (Moens and Ben Aicha, 1986). McSorley *et al.* (1985) claimed that under Florida conditions and soils, MIT, or chlorinated C_3 compounds containing fumigants, performed as well as CH_3Br–chloropicrin combinations. Methamsodium and dazomet had intermediate effects.

Potato

Soil fumigation against *V. dahliae* and nematodes in a *Verticillium*–nematode early dying complex (Guthrie, 1960) has been used with some success in the USA and Israel, but the treatment is restricted by cost. One of the pioneering studies on potato soil fumigation using Vapam was by Young and Tolmsoff (1958). Easton *et al.* (1972a, 1975) in a critical study found that annual springtime field fumigation for 5 successive years with Telone and chloropicrin, and DD and chloropicrin, reduced propagule numbers between March and May which later built up to a half or one-third of numbers in non-fumigated soil. Fumigation in the absence of burning increased tuber yields by 40% (Telone + chloropicrin) and 38% (DD + chloropicrin). Following late autumn haulm burning (see Physical methods) and spring fumigation, yield increases were 59 and 71%, respectively. The propagules as determined by Easton *et al.* (1969) included both conidia and microsclerotia. Since fields were irrigated with water containing 15,410 propagules l^{-1}, annual fumigation was necessary to control seasonal re-inoculation. This finding also explained the late season build up of inoculum following fumigation. Notwithstanding the late availability of inoculum, plants in fumigated land were symptomless compared with 47% disease incidence in control plots. Easton (1976) fumigated Washington potato soil with Telone C (1,3-dichloropropane, MIT and trichloronitromethane) from August to March with a corresponding range of soil temperatures of 20, 16.1, 7.2, 5.6 and 3.9°C. The yield in quintals (100 kg ha^{-1}) with the percentage wilted plants were 586 (57%), 618 (57%), 659 (35%), 683 (22%) and 773 (13%), respectively. The non-fumigated control plot yielded 488 quintals and 70% wilted plants. Subsequent laboratory tests found no difference in propagule numbers at different soil temperature fumigations. In Oregon, Powelson and

Carter (1973) found fumigation in autumn with MIT and Telone or chloropicrin alone or in combination to be as effective as in spring. Potato early death syndrome (PED) was delayed and yields improved. Trichloronitromethane (chloropicrin) was more effective as a spring application. Workman *et al.* (1977) confirmed the Telone C results. O'Sullivan and Reyes (1980) found Telone C fumigation or rotation of potatoes with maize to have comparable effects on wilt decrease and yield increases. In general, 1,3-D was adequate for control based on the high cost of MIT and chloropicrin (Erwin, 1981). Kotcon *et al.* (1984) found a correlation between PED and reduced root growth and surface area. Methyl bromide was only partially effective in reducing the effect of *V. dahliae* on root extension and root surface area. Davis (1985) listed chloropicrin, Vapam, Vorlex, Telone C and Terr-o-cide (mixtures of EDB and chloropicrin) as the most effective fumigants available to control *Verticillium* and nematodes. Easton *et al.* (1972b) mixed potential tuber disinfectants with soil containing *V. dahliae*. The organomercurial Semasan Bel (1.3 g l^{-1}) killed all propagules. Captan, polyram and $NaOCl_3$ solutions were all effective, but not dusts of captan and polyram. The partial sterilization of soil by the use of Vapam, Vorlex, Telone and chloropicrin gave sigificantly increased potato yields for 2 years following treatment (Kunkel and Weller, 1965). This appeared to be due to the control of PED, but the Telone results were inexplicable since Telone is largely nematocidal and the nematode levels were considered to be too low to cause symptoms.

The penetration and effectiveness of Vapam (metham sodium) has been studied by Ben-Yephet *et al.* (1981, 1983, 1984) and Ben-Yephet and Frank (1984, 1985, 1989). The technique of applying Vapam as a dilute solution by irrigation sprinkler against *V. dahliae* wilt of groundnut (Krikun and Frank, 1982) was modified by Ben-Yephet *et al.* (1984, 1989). Application of Vapam (32% a.i.) at 600, 800, 1000 and 1200 l ha^{-1} by combining half the dose in the first 12.5% of the irrigation water followed by the remainder in 87.5% of the water gave a far deeper penetration of the toxicant. Seventeen days after treatment, potatoes were planted and microsclerotia sampled. Dilute application killed all microsclerotia to 10–20 cm, but the split application killed all microsclerotia to 30–40 cm with a concomitant reduction in plant infection. Vapam at 225 kg ha^{-1} applied before planting in Oregon controlled PED and doubled the yield. The increased yield, however, did not compensate for the cost of treatment (Young, 1956).

Ben-Yephet *et al.* (1986, 1989) described the dramatic synergistic effect of soil solarization of Vapam applied as a split dose in irrigation water. In a July experiment, solarization alone killed 61% of microsclerotia in 4 weeks; Vapam at 25 ml m^{-2} killed 70% of microsclerotia. A combination of solarization and Vapam resulted in 100% death of microsclerotia in 1 week. Experiments started later in the year were less effective. In laboratory studies designed to optimize Vapam effectiveness in the field, Ben-Yephet and Frank (1978) found that microsclerotia stored prior to fumigation at field capacity were less affected than those stored in air-dry soil.

Notwithstanding the dangers of bromine residues, fumigation against PED in Israel using methyl bromide is limited largely by cost. In the Negev where *V. dahliae* is now endemic, CH_3Br kills fungi, bacteria, nematodes and weed seeds. The loss of phosphate from the killing of *Endogone* spp. is compensated for by supplying phosphoric acid through an irrigation line (Maharshak *et al.*, 1995b). Fumigation experiments with methyl bromide on potato soils in the Negev desert between 1983 and 1992 by Nachmias *et al.* (1995b) led to an increased yield in the spring season in a loess-type soil by 23%. In a sandy soil, the beneficial effect was only 13%, presumably due to greater volatility of the gas. Notwithstanding the high cost of treatment, the authors claimed that the treatment was cost-effective in a 3-year crop rotation. Gamliel *et al.* (1995) demonstrated that the effective levels of CH_3Br used in soil fumigation are necessary to compensate for high losses from low density polyethylene sheeting. Using co-extruded multilayer films with polyamide or ethylene vinyl alcohol as the barrier layer for post-fumigation soil sheeting, *V. dahliae* was controlled to a depth of 40 cm at 25 g m^{-2} compared with 50 g m^{-2} using standard polyethylene. Control plants collapsed 4 weeks before harvest. The percentage of large tubers (745 g) under impermeable film was 50–60% compared with 30% in non-fumigated controls. The use of such films offers great saving on fumigation costs in general, with considerable environmental advantages, especially in bromine residue-sensitive crops such as carnations and antirrhinum (Baker, 1970). Gamliel *et al.* (1997) again reported the effects of reduced CH_3Br fumigation in both experimental and commercial plots or fields. Formaldehyde at application rates of 2500 or 5000 l ha^{-1} has been used to control *V. dahliae* and *Pratylenchus mediterraneus*, the two major pathogens in PED, and has been effective subsequently in wilt control in succeeding crops of potato or groundnut (Maharshak *et al.*, 1995a).

Cotton

Chemical control of cotton wilt was reviewed by Minton (1973) and Bell (1992a). Fumigation with 55% chloropicrin–45% methyl bromide delivered at 280–400 kg ha^{-1} (Wilhelm *et al.*, 1966, 1972b) or a 1:1 mixture by weight at 225–375 lb $acre^{-1}$ (Wilhelm *et al.*, 1967) eradicated *V. dahliae* but stunted cotton (from bromine residues) and severely reduced yield. In California, McClellan *et al.* (1955) fumigated cotton land with the nematicide EDB at 40.7 l ha^{-1}. Although nematodes were controlled and yield increased, no effect was apparent on the fungus, which gave a wilt incidence of 77–81%. Wiles *et al.* (1968) found that fumigation-induced symptoms of stunting and reduced leaf area with mottled chlorosis could be reversed by foliar sprays of 0.5% $ZnSO_4$ but not with other macro or trace elements. Minton (1973) induced similar abnormal symptoms after soil fumigation with high rates of MIT–chlorinated 3C hydrocarbons. Such high rates controlled wilt at the expense of yield. In one of the early studies, Waddle and Fulton (1956) found that CH_3Br, chloropicrin and EDB all reduced cotton wilt, but of these CH_3Br was most effective; conversely, DD fumigation led to increased wilt. Chilton (1956) recognized the effectiveness of chloropicrin, whereas the use of EDB, Vapam and/or thironitrochlorobenzene led to only a slight (non-cost-effective) reduction in wilt. Blank (1971) cited in Minton (1973) failed properly to control cotton wilt with 1,2-dibromo-3-chloropropane, chloropicrin, 1,3-D, used alone and/or in, combination with chloropicrin or propargyl alcohol. Dose rates and condition were not given.

Where cotton is grown in sequence to potato as in Israel, Vapam fumigation reduces the disease in both cases (Ben-Yephet *et al.*, 1989). The effectiveness of Vapam increases with increasing temperature from 15 to 35°C and with slower rather than faster water application (Ben-Yephet *et al.*, 1981). The overall picture with cotton is that while fumigation may be cost-effective where a high-value crop preceding cotton is first protected, in many cases the cost of field treatment is not justified in relation to yield returns. This is particularly true considering the scale of field plantations and the more obvious advantage of using resistant high-yielding cultivars.

Other crops

1. Artichoke.
Ciccarese *et al.* (1985) and Cirulli *et al.* (1987) obtained good control of *V. dahliae* at 80 g m^{-2} fumigation with dazomet.
2. Aubergine.
Skotland (1964) in the USA and McKeen and Mountain (1967) in Canada reported effective control of wilt and *P. penetrans* using Vorlex, while Vigouroux (1972), in France, found CH_3Br slightly more effective than chloropicrin.
3. Chinese cabbage.
Fujinaga *et al.* (1999) controlled *V. dahliae* yellows by Vapam injection into a mulched ridge and spraying prior to cultivation.
4. Groundnut.
Krikun and Frank (1982) used Vapam applied by sprinkler line as a pre-plant sterilant against microsclerotia. Thanassoulopoulos and Kitsos (1973) found that EDB and Telone C both increased crop yields.
5. Watermelon.
Yield was increased by 280% following Edigan (metham sodium) injection into the irrigation line (Krikun *et al.*, 1976). Gersti *et al.* (1977) found additionally that a 32% aqueous solution at pH 9.1 in irrigation water reduced wilt by 68%.
6. Peppermint.
Faulkener and Skotland (1963) and Green (1964) were early advocates of soil fumigation against *V. dahliae* in mint using chloropicrin, trizone (chloropicrin, methyl bromide and propargyl bromide) and DD, all of which controlled *Verticillium* and nematodes.

7. Avocado.
Munnecke *et al.* (1978) used CH_3Br to sterilize roots of avocado infected with *V. albo-atrum* [*sic*] (*V. dahliae*). *V. dahliae* microsclerotia were more resistant than mycelium, but the pathogen was the least sensitive of ten soil-borne pathogens to CH_3Br.
8. Pistachio nut.
Where pistachio followed cotton, Ashworth and Zimerman (1976) recommended pre-plant fumigation with a methyl bromide–chloropicrin mixture.
9. Tobacco.
Taylor and Canter-Visscher (1970) obtained significant control of wilt of tobacco in New Zealand in the subsequent 3 years with a single fumigation of chloropicrin or chloropicrin–CH_3Br. Telone PBC was ineffective against wilt but controlled *Pratylenchus*, as did the other fumigants.

Plant Growth Regulators

The role of growth-regulating compounds in the alleviation of plant disease symptoms has long been recognized (Wain and Carter, 1977). Multiple applications of 2-chloroethyl-trimethyl ammonium chloride, 2(2 methyl-4-chlorophenoxy) propionic acid and the K salt of 2-benzothiozyl-carboxymethyl sulphide as soil drenches, retarded *V. dahliae* in cotton following root ball inoculation (Erwin *et al.*, 1966). Ranney (1959) reported that plant growth regulators reduced wilt incidence with concomitant changes in soluble N, reducing sugars, sucrose, starch, leaf, root and fruit weights. Blank, however (reported in Minton, 1973), found no effects of 2,4-D, methoxane and K-gibberellate on wilt intensity or cotton yield. Working on lucerne, Kratka and Kudela (1982) found that 11-fold pre- and post-inoculation, or fivefold post-inoculation sprays of 10^{-5} M IAA, markedly reduced *V. albo-atrum* infection in R and S cultivars. Pre- and post-sprays of 10^{-2} M IAA induced only a slight wilt reduction, but was effective as a post-inoculation treatment. Similarly, kinetin sprays at 10^{-4} and 10^{-8} M significantly reduced infection. Khasanov *et al.* (1986) achieved a 6.4% reduction in cotton wilt accompanied by increased leaf weight and boll number following application of mepiquant chloride ('Pix') at 1 l a.i. ha^{-1}; yield was increased by 0.25 t ha^{-1}.

Chernyaeva *et al.* (1989) claimed that application of calcium carbide to cotton soils inhibited nitrification and increased the coefficient of N utilization by increasing soil ethylene production. Disease incidence was also reduced. A comprehensive study of the effects of growth regulators on potato yield and *V. dahliae* incidence was conducted by Corsini *et al.* (1989). IAA was only partially effective due to its lability. The synthetic auxin NAA consistently reduced wilt and stem colonization. This was recorded on cultivars Norgold Russet (highly susceptible), Russet Burbank (moderately susceptible) and the highly resistant selection A66107-51. Although NAA effected a 50% reduction in wilt severity, the reduction in stem colonization was not as great as between cv. A66107-51 and cv. Norgold Russet. Wilt supression was not accompanied by an increase in total yield, but the treatment led to some tuber malformation. Other growth regulators, benzyladenine and gibberellic acid (GA_3), gave little or no consistent effect or even *increased* wilt and stem colonization.

Buchenauer and Grossman (1969, 1970) found that several morphactins applied to tomato roots and leaves reduced both *Verticillium* symptoms and the level of colonizing mycelium. Subsequently, the dwarfing compounds chlorocholine chloride (CCC, cyclocel, chlormequat) and 2 (4-dichlorobenyl-tributyl phosphonium) (Phosphon-D) and *N*-dimethyl succinamic acid (B995) were shown to reduce symptom severity (Buchenauer, 1971). Other gibberellin-inhibiting growth retardants, *N-N*-dimethyl piperidinium iodide (DPII) and *N-N*-dimethyl pyrrolidinium iodide (DPYI) delayed symptoms of *V. dahliae* in cotton by 2–3 weeks (Buchenauer and Erwin 1973a,b). Since all of these synthetic compounds and the morphactins showed no fungitoxicity up to 10^{-3} M and the reduction in symptoms appeared to be directly related to a supression of the pathogen *in planta*, the mode of action must lie in the stimulation of host defence mechanisms. Similar results were obtained by Buchenauer and Erwin (1976) and Erwin (1977a) in field experiments on cotton using CCC and the chloride salt of DPYI (DPIC) as foliar sprays at 10–25 g a.i. ha^{-1}. A marked reduction of petiolar vascular fungus was accompanied by only a slight diminution of foliar symptoms. Erwin (1977a) again confirmed that CCC and DPIC were non-fungitoxic. Both substances were shown to increase cotton seed and lint in uninfected plants; their role in symptom alleviation must therefore be regarded as conjectural. The plant growth regulator mepiquat (Pix, in China) used at various application rates was claimed to control *V. dahliae* in cotton by Jian and Ma (1999). Best results (0.9–9.6% increase) were obtained by foliar sprays of 40–60 mg kg^{-1} (60 mg kg^{-1} = 45 g ha^{-1}) once or twice during blooming. In another attempt to control cotton wilt by chemicals, Yuan *et al.* (1998) listed in descending order of effectiveness, Kangkuwi (40% KKK, Chinese); Veratrine, 0.5%; Lufeng, 95%; BFA; Mepiquat and Carbendizim, 50%. Kangkuwi was claimed to effect 71.3% control, but essential details including the identity of Kangkuwi were not identified. Pacbutrazol, the gibberellin inhibitor, used in trees and shrubs to restrict shoot growth and reduce pruning, was shown to inhibit by 25–100% a range of soil-borne tree and shrub pathogens including *V. dahliae*. The compound was used at ×200 dilution of the shoot retardant dose (Jacobs and Berg, 2000). See also Dutta (1980, 1981) and Dutta and Isaac (1981) on the use of growth substances and antibiotics for control of tomato wilt.

Miscellaneous Compounds with Fungicidal Activity

A range of diverse compounds has been investigated *in vitro* and *in vivo* with variable and occasionally converse results. A 100% control of *V. dahliae* from ornamental species *in vitro* using the quaternary ammonium compounds Dimani A and Menno Terforte at dilutions up to 1:100 was claimed by Brielmaier (1985). Of a range of compounds tested *in vitro* including ethers, thio-ethers and amides of N-phosphorylated carbamic acid, quaternary ammonium salts containing polyvalent iodine as anions were highly fungicidal (Andreeva *et al.*, 1976). Pulido (1969) described the effective elimination of *V. dahliae* from cotton seed using 2-(thiocyanomethyl-thio) benzothiazol at 225 ml 100 kg^{-1} of machine-linted and 200 ml 100 kg^{-1} of acid-linted seed. No phytotoxicity was observed. Triforine was partially effective against *V. dahliae* and *V. albo-atrum in vitro* but not *in vivo* (Fuchs and Drandarevski, 1973). Similar success was reported by Buchenauer (1979) when the triazoles, triadimefon, triadimenol, fenarimol, nuarimol, imazalil and fluotrimazole were all shown to be active against *V. dahliae* and *V. albo-atrum in vitro* only. Abdugafurov *et al.* (1990) described a novel triazol 1-phenyl-4 (5)-(x-aryloxymethylene) 1,2,3 triazol active on *V. dahliae*. The 'x' refers to substituents such as: 2-F; 4-F; 4-C1; 2, 4, 5-C1; 3-Br; 4-I; 2-NO_2; 2-CH_3; and 4-C_3H_7. Gangopadhyay and Kapoor (1975) claimed *inter alia* that *V. albo-atrum* [*sic*] (*V. dahliae*) infection of tomato, sweet pepper and aubergine seed could be controlled by soaking seed in 0.2% a.i. solutions of captofol and aureofungin in succession. In Tashkent (CIS), application of 2 t ha^{-1} [*sic*] paraformaldehyde on soil 20 days prior to sowing decreased cotton infection from 90 to 30% and increased seed yields by 1 t ha^{-1} (Yunusov *et al.*, 1975). Popov *et al.* (1976) in an unexplained result claimed that higher doses (unstated) of urea partially or completely disinfested cotton soil of microsclerotia. A similar claim was made by Maksudov *et al.* (1975) using 0.005–0.1% isatin, 5-methylisatin or isonitrosoaceto-*p*-chloramin. Hops were successfully protected against *V. albo-atrum* using two drenches each of 1 l per hill of fentin acetate at early and mid-growth stages (Leibelt and Senser, 1972). Surprisingly, no subsequent reports on the merits or disadvantages of this treatment were forthcoming. The partial success of unspecified semicarbazides ohlosom, semicar, trilan and BMK, against *V. dahliae* in cotton following autumn and spring soil application at 50 and 100 kg ha^{-1} was described by Andreeva (1977). Gafurov *et al.* (1990) reported the alleviation of cotton wilt symptoms and increased yield following sprays of c-fluron (unspecified) (see also Buïmistru, 1977a,b; Khalitov *et al.*, 1977).

Several authors have reported antifungal responses of nematicides, a result not always easy to evaluate in the field. Direct nematode control was practised effectively against *Pratylenchus* and *Helicotylenchus* in potato using aldicarb (Davis *et al.*, 1981) and the elimination of *Hopolaimus* sp. in Indian cotton infected with *V. dahliae* using Nemagon and Dasanit (Shanmugam *et al.*, 1976). Sakhacheva *et al.* (1988), however, described a direct antifungal and nematicidal role for a series of 2,2′ and 4,4′ dipyridyls with metal salts, and 2,6 dichloro-4, 4′; 2,6,2′,6′ tetrachloro 4,4; octochloro-4,4, dipyridyls; 2, chloro-4-trichloromethyl pyridine and 2,3,5-trichloro-4-trichloromethyl pyridine. These compounds were all highly effective against *Aphelenchoides besseyi* and some were 100% effective as fungicides against *V. dahliae*. Some foliar scorching occurred following sprays at 0.1% concentration. An unusual effect of the aphicide, acephate (*O,S*-dimethyl acetylphosphoramidothioate) applied to cut potato seed tubers at 1–2 oz a.i. cwt^{-1} of tubers was noted by Krause-Huether *et al.* (1975) in the USA. While protecting against the green, peach aphid, the compound also delayed symptoms of *V. albo-atrum* and *V. dahliae*, but did not affect the rate of disease development. It was suggested that acephate may protect against initial root infection possibly indirectly, perhaps acting as a nematicide.

A novel control of *V. dahliae* of tomato *in vitro* and partial control *in vivo* by metabolic blocking was achieved by Mussell *et al.* (1987). Difluoromethylornithine (DFMO), a specific inhibitor of fungal polyamine synthesis via ornithine decarboxylase, inhibited mycelial growth at concentrations as low as 5 fM. The inhibition could be reversed by putrescine – co-incidentally a product associated with foliar necrosis in acute K deficiency in mineral nutrition and some *Verticillium* symptoms. Dogo *et al.* (1999) claimed similar success for a fluorocompound, 5-fluoroindole, as a broad-spectrum *in vitro* antifungal chemical, affecting among other pathogens *V. dahliae*. The concentration of 100 $\mu g\ ml^{-1}$ is relatively high, and while the authors suggest the possible use of 5-fluoroindole, no indication was given as to how such a poisonous substance might be deployed. Hall (1975) recorded small but statistically significant differences in the *in vitro* sensitivity of a collection of *Verticillium* isolates to benomyl. *V. dahliae* (21 isolates), *V. albo-atrum* (nine isolates) and *V. nigrescens* (eight isolates) gave mean effective dose responses (ED_{50s}) of 0.36, 0.26 and 0.19 $mg\ l^{-1}$, respectively. The mean specific values for the slope of probit inhibition–10 g dose curve were: 10.55, 12.55, not significantly different, and 6.87, significantly different. Identification based on this response, however, was not possible because of overlap of values within a single species (Hall, 1975).

Naturally occurring saponins of lucerne incorporated into culture media were shown to be fungistatic (Manninger *et al.*, 1978). No correlation was found between a high plant saponin content and resistance to *V. albo-atrum*. Tannic acid was strongly inhibitory *in vitro* to *V. dahliae* (Cheo, 1982). Picman *et al.* (1990) working on sunflower resistance in Canada found that five terpenoids, two sesquiterpene lactones and three diterpene acids were potent inhibitors of *V. dahliae* in the range 10–100 p.p.m. It was proposed that selection of high terpenoid sunflower lines could improve natural resistance to the pathogen. In a subsequent study, Picman and Schneider (1993) found that pathogens *inter alia V. albo-atrum* [*sic*] were most affected by the sesquiterpene lactones, alantolactone and isoalantolactone, causing significant inhibition at the minimum inhibitory concentration (MIC) of 1–5 p.p.m. Leptocarpin at MIC 1–5 p.p.m. was also effective but the related hydroxyleptocarpin had an MIC 10- to 50-fold higher. A third pair, parthenin and coronophilin, were partially active on a weakly virulent isolate but less so on a highly virulent one.

A number of conventional fungicides have been shown to be ineffective against *Verticillium* or in some cases to enhance the disease. Blank (see Minton,

1973) found zineb, lithium carbonate, Bordeaux mixture, cycloheximide and pentachlorophenol to be inactive against *V. dahliae* on cotton. The herbicides 3-(β-chlorophenyl)-1,1-dimethyl urea (monuron) and 3 (3,4-dichlorophenyl)-1,1-dimethyl urea (diuron) actually increased disease severity. A similar result was obtained by Nilsson (1976). Trifluralin increased the susceptibility of oilseed rape and eliminated resistance in resistant cultivars. Sprays of chlorothalonil and fentin hydroxide applied to potato for the control of *Alternaria solani* and *Phytophthora infestans* led to a substantial increase in microsclerotial inoculum – 7–14% increase in 1 year and 20–26% in another (Easton, 1990). The efficacy of three ligands, thiolactic acid (TLA), thiosalicylic acid (TSA) and bipyridyl (BY), and their metal sulphates (Mn, Fe, Co, Ni and Zn) was tested *inter alia* on *V. albo-atrum* [*sic*] (*V. dahliae*) by Garg *et al.* (1995). All complexes affected hyphal growth, but the TSA ligand complexes were more toxic than TLA complexes. Ternary complexes were more toxic than binary complexes.

The advent of genetic engineering and the possibility of inserting into plant genomes, resistance genes from unusual sources has stimulated a search for anti-fungal activity in a wide range of plant and animal genera and species. One such candidate is the peptide magainin-2 from the African clawed frog. Concentrations in aqueous culture of 0.0005 p.p.m. completely inhibited a number of pathogens, *inter alia V. dahliae*. Leading edges of colonies on PDA treated with 0.01 p.p.m. magainin-2 showed degradation of mitochondria and cytoplasmic matrix and a reduction in ribosomes, all after 12 h of treatment (Kristyanne *et al.*, 1997). Extracts of rue (*Ruta graveolans*) containing the psoralens (furanocoumarins) 8-methoxypsoralen and 5-methoxypsoralen caused significant inhibition of various pathogens, *inter alia V. dahliae*. A mixture of these chemicals at 64.8 p.p.m. inhibited mycelial growth (Oliva *et al.*, 1999). Four, thio-functionalized glucosinolate enzyme-derived products, glucoiberin, glucocheirolin, glucoerucin and glucoraphenin, showed high toxicity to *V. dahliae* and other fungi as potential alternatives to commercial soil fumigants (Manici *et al.*, 1999) (see also usinic acid, chlorinated orcinol, ajuene and cypress terpenoids in Chapter 9).

Systemic Fungicides

Since 1968 with the introduction by Du Pont of benomyl, a family of systemic compounds has been available for *Verticillium* control, all of which generate in the plant, benzimidazol-2-yl carbamate, with the trivial name carbendazim (MBC).

The systemic fungicides are: benomyl (methyl-1-[butylcarbamoyl] benzimidazol-2-yl-carbamate), (BEN); thiabendazole (2-[thiazol-4-yl] benzimidazole), (TBA); and thiophanate-methyl (dimethyl 4,4′-[*o*-phenylene] bis [3-thioallophanate]), (TPM). These fungicides are sold under various trade names, as is MBC which is used in its own right. They may be applied as root drenches, as stem injections, or as foliar sprays. As a group, they exhibit low water solubility which usually prevents soil penetration deeper than 100 mm (Talboys, 1984).

Formulation is all important for the effectiveness of a fungicide for a particular application. Benzimidazoles are not foliar absorbed unless in an acidified solution (Buchenauer and Erwin, 1971) or by adding paraffin oils to spray formulations (Erwin *et al.*, 1974). Relatively high dose rates are required to control herbaceous plants, requiring large volumes of irrigation water. MBC and TBA applied in drip irrigation were not satisfactory (Solel *et al.*, 1979).

Within the plant, the chemical is transported apoplastically in the transpiration stream. It suppresses the pathogen in part in shallow roots and stem vessels but accumulates rapidly in leaves far from the required site of action and not at all in deep roots harbouring the fungus (Talboys, 1984). MBC is nonfungicidal and is only fungistatic while the vascular concentration remains sufficiently high. BEN, TAB and TPM generally have failed to control tree diseases when applied as soil drenches. Even at high concentrations, the xylem fluid content is low, possibly due to binding (particularly TBA; Wang *et al.*, 1971) or inactivation (Talboys, 1984) in the plant. The acropetal movement of systemic fungicides represents a fundamental problem in chemotherapy. To date, no basipetally, phloem-translocated fungicides have been found (Edgington, 1981). A principal concern with the benzimidazoles and especially BEN has been mutagenesis. This problem is exacerbated in fungistasis, where the pathogen survives and tolerant strains can be selected. BEN and MBC inhibit mitosis (Davidse, 1973), and DNA production is suppressed (Hammerslag and Sisler, 1973). In MBC-sensitive *Verticillium*, the molecule binds to the protein tubulin, a precursor of microtubules essential for the spindle fibres of mitosis. MBC acts in a similar way to colchicine. In BEN-resistant strains, a tubulin was formed which did not bind to MBC, allowing mitosis to proceed normally (Davidse, 1975; Davidse and Flach, 1977). Talboys and Frick (1974) and Talboys and Davies (1976a) described the *in vitro* production of BEN-tolerant *V. dahliae* strains by subculturing on to progressively higher BEN–agar concentrations. Strains tolerant to 0.5 p.p.m. yielded strains tolerant to 3, 5 and 10 p.p.m. Some of these were highly virulent on strawberry and were insensitive to BEN drenches. Such variants, however, have not been found in the field as a result of BEN treatment of crops. Karoleva and Kas'yanenko (1978) described a BEN-resistant *V. dahliae* mutant which retained its resistance after passage through cotton. Emmanouil and Wood (1982) obtained a BEN-tolerant strain of *V. dahliae* from a wild-type BEN-sensitive isolate. The mutant retained its resistance after continuous growth on BEN-free agar for 15 months. It was also resistant to MBC and TPM, but not to TBA. Locke and Thorpe (1976) reported a *V. dahliae* isolate from a commercial tomato crop which tolerated 500 p.p.m. BEN; this was later (Locke and Thorpe, 1977) shown to be a strain of *V. tricorpus*. Sensitive strains of *Verticillium* respond to as little as 0.3–0.4 mg BEN causing inhibited growth, but not inhibited conidial germination (Erwin, 1969). Neither BEN nor MBC are mobile in soil, but BEN may persist in soil for as long as 1 year (Erwin, 1973). *In vitro* studies of the effect of BEN on growth and development of conidia of *V. dahliae* strains of differing virulence were reported by Atakuzieva and Safiyazov (1980). The tolerance of several strains was listed by McHugh and Schreiber (1984) and the effect of systemic fungicides on microsclerotia in columns of different soil types was described by Ben-Yephet (1979).

Strawberry

This crop has perhaps profited most from the application of systemic fungicides. Lockhart *et al.* (1969) in Canada found that BEN controlled strawberry wilt when applied as a root-dip but was ineffective as a foliar spray. Talboys and Frick (1972, 1973) showed that three BEN or TPM soil drenches gave 97–100% control of *V. dahliae* in cv. Cambridge Vigour. A low wilt incidence persisted as a residual effect into the second year. Effective control was also achieved by a late autumn and spring drench, and BEN was marginally better than TPM. Jordan (1972, 1973, 1974, 1975) obtained similar results using BEN, TBA and TPM. Fourteen isolates were tested on ten standard commercial cultivars. A single field application of 3000 l ha^{-1} of 1000 p.p.m. protected only the first two runners. A subsequent treatment protected all runners. Since MBC is transported apoplastically, treatment of mother plants failed to protect roots of runner plants (Jordan, 1975); Nicholson *et al.* (1972) concluded that the dose rate affected the duration but not the degree of protection. Van der Sheer *et al.* (1975) controlled wilt for up to 5 months with 0.1 g of BEN but viable fungus could still be isolated from petioles. Plants treated with BEN or TPM regardless of the degree of infection were all larger than non-treated plants, reflecting the cytokinin-like nature of the benzimidazole molecule. Talboys *et al.* (1975, 1976) showed that drenching runners in a high (0.075% a.i.) BEN concentration and a large (600 ml per plant) volume and planting out into chloropicrin-fumigated soil previously infested with *V. dahliae* controlled the disease for the following two seasons. Rebandel (1981) reported similar results. Field increases of 25% in cv. Surprise des Halles and 299% in cv. Senga Precosa may in part reflect the control of other diseases as well as *V. dahliae*.

Van der Sheer *et al.* (1975) in The Netherlands and Profic-Alwasiak *et al.* (1977) in Poland found that both BEN and TPM gave good control of strawberry wilt when applied as drenches during or after planting. An important and salutary finding by Van der Sheer *et al.* (1975) was that while *V. dahliae* was controlled by TPM, there was a concomitant increase in the incidence of *Phytophthora cactorum*.

Cotton

Considerable work has been carried out in the laboratory and glasshouse (Erwin and DeWolfe, 1968; Erwin *et al.*, 1968a,b; Buchenauer and Erwin, 1971; Knoll, 1978) to show that benzimidazole fungicides can control *V. dahliae* in cotton, but little of this knowledge has been applied commercially because of the cost-effectiveness of treatment. From work initiated in 1964, Erwin and DeWolfe (1968) and Erwin *et al.* (1968a,b) conducted the earliest studies on the uptake, movement and effectiveness of benomyl (Du Pont 991) and TBA in cotton seedlings. Leach *et al.* (1969) could find no fungicide in root tips or apical meristem from benomyl soil drenches. Of the systemics tested, BEN was the most effective on cotton (Erwin *et al.*, 1969; Ranney, 1971) and was readily taken up as an active fungicide from the soil (Erwin *et al.*, 1968a; Locke and Green, 1971). The capacity for acropetal movement of BEN is limited; Hine *et al.* (1969) found downward movement to root tissues in 8-week-old 'Deltapine Smooth Leaf' cotton seedlings, but not in 4- or 12-week-old plants foliar sprayed with 1000 and 2000 p.p.m. BEN in polyhydric alcohol esters, surfactant wetters. Sims *et al.* (1969) established in cotton that the active breakdown principle *in planta* was MBC. Erwin (1970) and Erwin *et al.* (1971) using a ^{14}C-label showed that TBA applied to cotton roots decreased in concentration with increasing plant height. TBA was shown to be absorbed and transclocated as such, mostly in the xylem but with some lateral movement into the phloem (Erwin *et al.*, 1971). The effectiveness of BEN and TBA is greatly enhanced by the type of adjuvant used, presumably affecting uptake and, possibly translocation (Rawlins and Booth, 1968; Booth, 1970; Booth and Rawlins, 1970; Booth *et al.*, 1971; Buchenauer and Erwin, 1971). No foliar absorption of benzimidazole occurs unless the chemicals are acidified (Buchenauer and Erwin, 1971); thus BEN and TBA need to be adjusted to pH 1.7 and 2.7, respectively, with HCl, but even when applied twice as a foliar spray at 1–3 day intervals, wilt was not competely controlled. Erwin *et al.* (1974) subsequently found that foliar uptake was enhanced by the addition of paraffin oils to the formulations. Minton in 1970 (1973) showed that BEN applied with Esso oils Orchex N795, 696 and 3408 with emulsifiers and Tween-20 reduced early wilt more effectively than BEN or BEN + Tween-20. No preparation, however, reduced late wilt or increased yield. Buchenauer and Erwin (1971) reported that a 10–12 day post-inoculation spray of 5000 p.p.m. acidified BEN had a curative effect on early disease symptoms. BEN and TBA have been shown to have long persistence in soil (Erwin *et al.*, 1968a,b, 1969, 1971; Hine *et al.*, 1969; Leach *et al.*, 1969). Erwin *et al.* (1968, 1969) using a microbial bioassay technique found evidence of high concentrations of BEN in six replicate cotton plants 6 months after soil treatment. Only two out of six plants in TBA-treated soil were inhibitory in bioassays. Erwin *et al.* (1978) found that BEN application to plants led to reduced microsclerotial production in tissues.

BEN and TPM break down very slowly in soil to MBC (Erwin *et al.*, 1969; Hine *et al.*, 1969) which is taken up by roots and translocated in the xylem to upper leaves and stems (Erwin *et al.*, 1968a,b; Hine *et al.*, 1969; Booth and Rawlins, 1970; Booth *et al.*, 1971) but not to cotton seed embryos (Erwin and De Wolfe, 1968; Erwin *et al.*, 1968a,b; Ashworth and Hine, 1971); pot experiments employed 20 mg of BEN per pot containing 800 g of soil equivalent to 2.4

kg ha^{-1} in the field. Leach *et al.* (1969) explained the difference in response between plants in field- or glasshouse-treated soils. In the field, only a proportion of feeder roots were absorbing in BEN-treated soil, whereas in the confined space of a plant pot, all roots would be in the BEN zone.

A similar experiment in the field incorporated BEN at 24.3 kg ha^{-1} into a 464.5 cm^2 or 232.2 cm^2 profile. The greatest accumulation of BEN was in plants grown in the larger profile. Using ^{32}P, Bassett *et al.* (1970) showed that the soil profile explored by cotton roots increased from 161.3 cm^2 on 1 May, to 14,194 cm^2 on 18 July. The 232.2 cm^2 experimental profile used by Leach *et al.* (1969) only represented 1.5% of the cotton roots' exploratory zone in mid-July. Most experiments on wilt control in the glasshouse using BEN and TBA were highly successful (Erwin *et al.*, 1968a,b,c; Erwin *et al.*, 1969; Leach *et al.*, 1969), whereas attempts at field control achieved only fair or poor results (Erwin *et al.*, 1966, 1968a,b, 1969; Hefner, 1969; Hine and Simbwa-Bunnya, 1969; Leach *et al.*, 1969; Booth, 1970; Booth and Rawlins, 1970; Ranney, 1971). Multiple point (widespread) distribution of the fungicide in field soils was more effective than single point placement (Erwin *et al.*, 1969; Booth, 1970). Variable results have been described; Ranney (1971) found that moderate rates of BEN and TBA applied to cotton seed, soil and seedlings reduced loss and increased yield. BEN was found to be more effective on a w/w a.i. basis than TBA. Erwin (1972, 1977a) reported experiments where BEN and TPM were sprayed behind soil cultivation implements. Rates of 25 kg ha^{-1} applied before and after planting led to increased seed and lint yields. Control depended on the extent of the cotton root system exposed to the fungicide since increasing the dose to 112 kg ha^{-1} (an economically unrealistic level) did not consistently increase control or yield. Hefner (1969) found that soil or foliar application made no difference in wilt control in cv. Hopicala grown under natural field infestation of *V. dahliae*. Booth and Rawlins (1970) found that row dressing with BEN and inter-row dressing with BEN and Tween-20 (polyoxyethylene sorbitan monolaurate) more effectively reduced wilt than row dressing alone; subsequent experiments by these authors using wilt-susceptible and tolerant cotton cultivars showed no effect on wilt incidence. Erwin *et al.* (1968a) similarly recorded negative results. Yunusov *et al.* (1973) in Uzbekistan, however, obtained almost complete control of field wilt and restored yield with 50–100 kg ha^{-1} BEN or uzgen.

Reports from the CIS are generally more enthusiastic about prospects for field control than the more pragmatic findings from the USA, regardless of cost-effectiveness. The incorporation of 50 kg of BEN ha^{-1} at autumn ploughing decreased cotton wilt by 3.1% at flowering and 10.8% at maturation, and increased seed yields by 390 kg ha^{-1}. When applied in spring ploughing, no effect on wilt was found, but curiously yield increases of 310 kg ha^{-1} were claimed (Khasanov, 1975a). Similar results were claimed for a compound uzgen (unspecified) (Khasanov, 1975b). In Azerbaijan field trials, best results were obtained by seed treatment with 1:1000 antifungin solution (unspecified) followed by two sprays at the cotyledon and 4–5 leaf stage (Panteleev and Bagdasaryan, 1972).

Application of 100–250 kg ha^{-1} of BEN, uzgen (possibly a Russian BEN preparation) or olgin (unspecified) increased seed cotton yields from 3.7–3.8 t ha^{-1} to 4.0–4.4, 4.2–4.7 and 4.5–5.3 t ha^{-1}, respectively (see also Yuzbash'yan, 1976; Kurbanov and Nasridinov, 1980). Yunusov *et al.* (1979) claimed improved yield from *V. dahliae* infection and higher fibre quality from BEN and uzgen (100–200 kg ha^{-1}), Topsin NF44 (thiophanate at 200 kg ha^{-1}) and ammonia at 1500 kg ha^{-1}. Shibkova and Tkachenko (1971) reported the transfer of successful glasshouse experiments to the field. Best results were obtained when BEN was introduced into the soil with superphosphate after autumn ploughing. Yunusov and Sultanov (1978) recommended spraying with BEN, uzgen or algin at the 2–3 leaf stage, at budding and fruiting, when the fungicides prevented microsclerotia on fallen leaves from germinating. Sboeva *et al.* (1977) claimed to have reduced *Verticillium* wilt from 82% to 6–7% by one soil application at 50–100 kg a.i. ha^{-1} with protection lasting for two or three seasons. Similar results were also claimed by Khamidov and Umarov (1980) and by Yunusov *et al.* (1980). Success of this magnitude has not been reported from the USA.

Koroleva and Kasyanenko (1977, 1978) and Koroleva *et al.* (1978) found BEN-resistant mutant clones in natural populations of *V. dahliae*. BEN resistance persisted through passage on cotton and artificial culture and was highly aggressive to cultivars 108-F and Tashkent 1. In strongly infested fields with benomyl in the soil, it was postulated that >1000 mutant clones kg^{-1} of dry soil could develop. Pulido (1969) described the antifungal properties of Busan (2-thiocyanomethylthio) benzothiazole used as a seed treatment against *V. dahliae* and other cotton pathogens. Atakuzieva and Safiyazov (1980) found that the leaves of two *V. dahliae*-infected cotton cultivars had lower oxidoreductase activity than healthy leaves. Application of 150 mg BEN kg^{-1} of soil prior to sowing 'infected' seed prevented symptom appearance and restored enzyme levels to those in healthy plants. BEN at 50 mg kg^{-1} delayed symptoms and had only a partial effect on enzyme activity (see also Popov, 1979).

Other crops

1. Artichoke.
BEN and chinosol (unspecified) used as a root-dip reduced *V. dahliae* infection (Cirulli *et al.*, 1987).
2. Chrysanthemum.
Busch and Hall (1971) sprayed individual leaves infected with *V. dahliae* weekly with 20 p.p.m. BEN in 0.25% aqueous Tween-20. Local inhibition of colonization and symptom development was achieved but not in adjacent unsprayed leaves. The authors concluded from this that wilt symptoms result from the activity of the pathogen in the leaves and not in roots and stems. Comparative glasshouse tests on chrysanthemum cuttings with 70% TPM at 2.4–14.4 g l^{-1}, 50% BEN at 4.8 g l^{-1} and MBC at 2.4 g l^{-1} and inoculated with *V. albo-atrum* [*sic*] (*V. dahliae*) showed

that MBC caused stunting but gave complete control. TPM at 12 or 14.4 g l^{-1} and BEN were both satisfactory. In field tests, TPM as a root-dip at 7.2 or 9.6 g l^{-1} or sprayed at 7.2 g l^{-1} a week after planting gave good control. BEN was less satisfactory than in the glasshouse test (McCain and Farnham, 1974).

3. Clover.

Pot experiments with *Trifolium pratense* showed that BEN reduced *V. dahliae* infection but increased seed mortality (Skipp, 1986).

4. Apricot.

Severe losses due to *V. dahliae* occur in Hungary. Best control was found in soil treatment with 50% Ferbam or Heteron granules (75% dazomet). An injected BEN suspension gave 95% control but was ineffective as a soil treatment; presumably at the concentrations used (Kovacs, 1976).

5. Cucumber.

Costache and Tomescu (1980) described the effect of BEN, TPM and MBC on glasshouse cucumber (see also Costache *et al.*, 1978a).

6. Cacao.

From a screen of 15 fungicides to control *V. dahliae* in *Theobroma cacao* L., MBC induced complete inhibition at 1 p.p.m. followed by BEN, TBA and TPN each at 10 p.p.m. (de Oliveira, 1983).

7. Aubergine.

BEN as a soil drench to control *V. dahliae* has been used with various success by Sivaprakasam and Rajagoopalan (1974b) and Thorat *et al.* (1976), both in India, by Costache and Raicu (1976) in Romania, and by McKeen (1972) in Canada also on tomato, together with TPN. Svampa (1974), in Italy, achieved *moderate* control from large doses sprayed at 5-day intervals [*sic*]. The author claimed some phloem transport presumably from an unnatural saturation of the plant tissues. Tomatoes too were included in the experiment (see Marchoux *et al.*, 1973, for the result of early studies of benzimidazole fungicides on a range of solanaceous crops against *Verticillium* diseases).

8. Flax and *Impatiens sultani.*

Coosemans (1979) obtained only partial control of *V. albo-atrum* and *V. dahliae* and of *Pratylenchus penetrans* using a combination of BEN and aldi carb.

9. Hop.

The incidence of *V. albo-atrum* wilt was reduced in pot experiments using TBA as a soil drench (Chambers *et al.*, 1983). It is difficult to imagine the successful use of systemic fungicides in the field in relation to the massive perenial root stock and multiple stems of *H. humulus*, and certainly not as a permanent measure.

10. Lucerne.

Dixon (1972) reported favourable results with TBA *in vitro* and in glasshouse experiments as a soil drench of 50 p.p.m. a.i. A field experiment showed that successive treatments resulted in a cumulative wilt reduction. Higher doses led to phytotoxicity. Veverka and Kudela (1980, 1983) evaluated BEN and trimorphamid. Tu (1982) and Antoun *et al.* (1984) tested several fungicides including BEN against *V. albo-atrum* in seedlings. Knoll (1978) reported the presence in all parts of tomato and lucerne of 'phytogenic inhibitors' – non-specified thermolabile compounds – which increased following infection but not after application of carboxin, TBA or BEN.

11. Pepper (Capsicum).
D'Armini *et al.* (1976) examined the relationship between *V. dahliae* wilt control and the effect of the soil microflora by *inter alia*, benzimidazole compounds. Aleksic *et al.* (1976) claimed that a single BEN spray suppressed disease development for 60 days. Best results came from a root dip in a 2.5% suspension followed by 0.1 g per plant drench once or twice at a 30-days interval. Costache (1974, 1977) and Costache and Stoenescu (1978b) reported successful control using BEN, TPM and MBC. Mancini *et al.* (1973) described the *in vitro* effect of BEN and other fungicides on a *Capsicum* isolate of *V. albo-atrum*, probably *V. dahliae* Kleb.
12. Potato.
No successful application of systemic fungicides has been described for this crop. Celetti and Platt (1987a) recommended TPM as the most acceptable of the fungicides tested in glasshouse and field experiments. Partial success with BEN to control *V. dahliae* in the presence of *H. rostochiensis* was not reflected in reduced soil inoculum for a succeeding potato crop (Hide and Corbett, 1973). Frank *et al.* (1973) established an interactive relationship between *V. dahliae* and pink-eye disease caused by *Pseudomonas fluorescens*. When BEN was applied 2 or 4 weeks after planting *Verticillium*-inoculated seed pieces, pink-eye developed to varying degrees depending on the cultivar or line. No pink-eye developed in seedlings treated with BEN before *V. dahliae* innoculation.
13. Sesame (*Sesamum indicum*).
El-Deeb *et al.* (1985) successfully protected seed against *V. albo-atrum* [*sic*] (*V. dahliae*) with BEN and chlorothalonil.
14. Soybean (*Glycine max*).
V. nigescens causing pod infection was successfully controlled by BEN (Vesper *et al.*, 1983).
15. Tobacco.
BEN applied as a soil drench at 2.5 kg a.i. ha^{-1} greatly reduced leaf loss from *V. dahliae* (Hartill, 1971).
16. Tomato.
Bochow (in Fuchs, 1977) reported the effect of BEN residues in roots of treated tomatoes persisting for two seasons and protecting succeeding crops against *V. albo-atrum*. Growth regulators and BEN interacted in *Antirrhinum* infected with *V. albo-atrum*. Matta and Garibaldi (1970) and Garibaldi and Lamonarca (1974) treated tomato with BEN thiophanate (TP) and TPM at 4 g m^{-2} 20 days after transplanting, with a marked reduction in *V. dahliae* wilt. Plants immersed in a 0.1% suspension followed 20 days later with a 2 g m^{-2} drench gave good results but with some phytotoxicity. Pepper was treated with 2, 4 and 6 g m^{-2} BEN and TP. The incidence of *V. dahliae* initially was reduced at the highest concentration, but this effect was later reduced and then cancelled. No difference in effectiveness was found with the different compounds. Garibaldi (1976) subsequently showed that foliar sprays of BEN and TPM applied to tomatoes led to substantially reduced *V. dahliae* wilt symptoms. Formigoni *et al.* (1973) achieved similar results treating glasshouse tomatoes in Sicily with granular TPM.
17. Woody plants.
Benzimidazoles have been found to be of little use in controlling *Verticillium* wilt

in trees. Barak *et al.* (1981) suggested that this was due to binding onto xylem lignin. Petsikos-Panayotarov (1981) found that MBC hydrochloride injected into olive tree trunks could be detected by *Penicillium italicum* bioassay 1–2 weeks later in branches 4.25 m distance from the injection point. The distribution was variable and toxicity declined to zero at 24 weeks. In contrast to the negative results of Petsikos-Panayotarov (1981) in Greece, more recent experiments by Fodale and Mule (1999) in Italy claimed an 83% cure of olive cv. Nocellara del Belice following trunk injection of benzimidazoles, benomyl, thiophanate methyl and carbendazim. The authors also claimed control with fosetyl-Al, copper sulphate and dodine. More information on the persistence and cost-effectivness of the treatment will be required before the authors' proposal of trunk injection as the only efficient way to control olive wilt could be adopted. Soil drenches of BEN and TBA delayed and reduced *V. dahliae* wilt symptoms in sugar maple (*Acer saccharum*) and redbud (*Cercis canadensis*) following the infection of young plants (Born, 1974). Davis and Peterson (1973) described an asphalt tree-wound dressing containing *inter alia*, 50 p.p.m. BEN which controlled *V. dahliae* and other tree pathogens.

Notwithstanding some of the enthusiastic claims for the successful control of *Verticillium* spp. by fungicides, the reality is somewhat different. While a number of compounds have worked *in vitro*, few, if any, have shown any promise for control *in planta*. Difficulties of application, and/or cost-effectiveness (a problem with systemic fungicides) have prevented any practical alternatives to soil fumigation (with selected crops), solarization or varietal resistance.

3 Biological Control

The term *biological control* in this chapter is used specifically in the sense of the antagonism of an organism against a wilt pathogen as distinct from the addition of organic amendments to soil. Early studies reviewed by Baker (1981) and Marois *et al.* (1982) attempted field control through the incorporation of antagonistic microorganisms into the soil using relatively crude methodology and often with variable, inconsequential results. The subject did not become established until more precise results from laboratory experiments were forthcoming. During the late 1980s and progressively throughout the 1990s, public hostility to fungicides and especially chemical soil sterilization led to increased activity to find antagonistic microorganisms and develop preparations for field use, and to seek novel genes for antifungal molecules from animals and plants with the potential for introduction into crop plants.

Soils showing a natural suppressiveness to the pathogen or decreased infection by comparison with a control were described by Nelson (1950), while Born (1971) showed that this 'resistance' could be destroyed by steaming. Similarly the results of incorporating soil amendments (see Chapter 6) leading to a subsequent disease reduction have been interpreted in terms of an increase in the population of bacterial (McKeen, 1943; Marupov, 1990; Strunnikova *et al.*, 1990) or actinomycete (Dutta and Isaac, 1979b) antagonists. Little experimental evidence was presented in early studies to confirm these claims which disregarded other possibilities such as direct inhibiting or stimulating effects of decomposing residues. Lucerne distillates applied to soil containing *V. dahliae* microsclerotia eliminated the pathogen (Gilbert and Griebel, 1969). Berestetskii *et al.* (1982) found that volatiles from decomposing residues (presumably in excess), methanol from pea and maize, butanol from pea, and acetaldehyde from lupin completely suppressed spore germination in *V. dahliae*. Melouk *et al.* (1995) produced very similar results with volatiles from decomposing rape seed meal in soil. These results and those of Grishechkina (1988) showed that ground residues of barley, wheat, maize, lucerne, clover, tomato, cucumber, mustard, pepper and aubergine actually stimulated growth of *V. dahliae* microsclerotia, which as a fallow treatment could reduce the inoculum potential for succeeding crops. Conversely, Melouk *et al.* (1995) showed that volatile compounds from rape seed meal in soil (60 g kg^{-1} of soil) completely inhibited growth of *V. dahliae*. These foregoing results illustrate the necessity for a careful experimental interpretation of soil amendments and disease alleviation even when there is an apparent *prima facie* case for antagonism by microorganisms.

The Role of Bacteria

Klingner *et al.* (1971) observed that *Erwinia carotovora* was present (presumably in enhanced numbers) in the rhizosphere of cotton which escaped *Verticillium* wilt, and showed that isolated *E. carotovora* antagonized *V. dahliae in vitro*. Seed coating with a bacterial suspension or soil amended with a pre-plant bacterium–carrot mixture reduced wilt incidence. Other reports of *in vitro* inhibition include *Rhizobium* and 15 strains of *Agrobacterium tumefaciens* on *V. dahliae* (Kerr *et al.*, 1978), metabolites of actinomycetes from a cotton rhizosophere (Ezrukh, 1978) and fungistatic and fungicidal effects of *Streptomyces rimosus* on *V. albo-atrum* and *V. dahliae* (Chi, 1963) and of *Bacillus* spp. (more prevalent in wilt-resistant than wilt-susceptible rhizospheres), *Pseudomonas*, *Gluconobacter*, *Flavobacterium* and *Streptomyces* on *V. dahliae* (Azad *et al.*, 1987). *Bacillus subtilis* isolated from healthy maple stems reduced wilt symptoms when introduced into stem wounds in Norway and in silver maple seedlings prior to inoculation. Leben *et al.* (1987) successfully increased potato plant height and shoot weight and reduced *V. dahliae* propagules by treating seed tubers with strain M-4 of *Pseudomonas fluorescens* before planting in infested soil. No increase in yield was achieved, however, in comparable field experiments. Success similar to the laboratory experiments of Leben *et al.* (1987) was achieved by Wadi (1985) using a streptomycin sulphate mutant (No. 45) of a bacterial isolate from potato rhizospheres used to coat potato roots. Fu *et al.* (1999a) claimed that unnamed endophytic bacteria isolated from cotton tissues [*sic*] suppressed toxin production in culture and in *in vivo* wilt tests on cotton. This report in the absence of any identification, or definition, or measurement of a toxin does not inspire confidence. Similarly, in Fu *et al.* (1999b), the same strains are reported to promote cotton shoot growth and produce extracellular proteins in culture, but the significance of these and any convincing details are lacking.

Berg (1997), Berg and Ballin (1994) and Berg *et al.* (1996) reported a screen of more than 5000 rhizosphere bacteria for antifungal activity against *V. longisporum* (*sensu* Karapapa *et al.*, 1997c) on oilseed rape. Seventeen species shared antifungal activity *in vitro*; all the Gram-positive bacteria were *Bacillus cereus* or *B. subtilis* (see also Li and Zhang, 1995) but the majority of antagonists were pseudomonads, *P. aureofaciens*, *P. chlororaphis*, *P. fluorescens* and *P. putida* (fluorescent types) (see also Solarska, 1995). Other pseudomonads were *Chrysomonas luteola*, *Sphingomonas* (*Pseudomonas*) *paucimobilis*, *Stenotrophomonas maltophilia* and *Burkholderia cepacia*. Other types included (enterics) *Pantoaea agglomerans*, *Serratia liquefasciens*, *S. plymuthica* and *S. rubideae*, and the remainder were *Agrobacterium radiobacter*, *Chromobacterium violaceum* and *Flavobacterium indologenes*.

Strains of *Stenotrophomonas maltophilia* (*Xanthomonas maltophilia*, Berg and Ballin, 1994) were shown to produce the antifungal macrocyclic lactam antibiotics, maltophilin and alteramid-A, probably supported by chitinases and 1,3-β-glucanases. In field trials, one isolate (R. 3089) reduced wilt by 23% leading to an increased yield of 9% (see also Berg *et al.*, 1998). Berg and Ballin (1994) and Berg *et al.* (1994) reported that *B. subtilis*, *P. fluorescens* and *S. maltophilia* were the strongest antagonists against *V. dahliae*, inducing ultrastructural and morphological changes in the pathogen. *S. maltophilia* was isolated by Berg *et al.* (1996) from the rhizosphere of oilseed rape using a selective medium. The average number of bacteria of root fresh weight was 10^7, representing 3.7% of the total bacterial population. Strains differed in their ability to cause *in vitro* inhibition. Selected strains significantly reduced disease incidence in glasshouse tests. In a subsequent study (Berg *et al.*, 1998; Lottmann and Berg, 1998), the population of antifungal bacteria, fluorescent pseudomonads, *Bacillus* spp. and *S. maltophilia* all increased during the plant's life cycle. Population densities based on c.f.u.s were higher in spring and summer than in autumn and winter.

A stable average of 7% of all bacterial isolates were antifungal to *V. longisporum*. Using transposon T*n*5 mutants of *S. maltophilia*, a maltophiin-negative R3089:T*n*5-1 derivative was less antagonistic to *V. longisporum* than the wild-type maltophiin-producing R3089. In other isolates, antibiotics, lytic enzymes and siderophores appeared to be involved (Lottmann *et al.*, 1997). The diazotroph *Alcaligenes paradoxus* strain 1AG4 isolated from the rice rhizoplane by Sun *et al.* (1999) showed *in vitro* suppression of *V. albo-atrum* [*sic*]. Another species, *A. fecalis* No. 4, inhibited 13 pathogens among which was *V. dahliae* (Honda *et al.*, 1999). This isolate produced hydroxylamine which inhibited growth when added to pathogen cultures. A transposon-non-hydroxylamine-producing mutant was non-suppressive, suggesting a dominant role for it in biological control.

In other studies, Berg *et al.* (1997) and Berg and Bahl (1997) isolated 16 active strains of *Serratia plymuthica*, all of which were antagonistic to *V. longisporum* the pathogen of oilseed rape, and some of which produced the antifungals prodigiosin and pyrrolnitrin, siderophores, and chitin and glucan hydrolases.

Chitinolytic activity alone appeared to be the basis of antifungal control *in vitro*, in strain C48. Frankowski *et al.* (1998) found *N*-acetyl-β-D-glucosaminidase, chitobiosidase and endochitinase as constituents of the chitinolytic activity. Mutants with reduced chitin lyase activity gave reduced protection to oilseed rape cv. Lambada. Other bacterial antagonists of *V. dahliae* have been described by Tjamos *et al.* (1997a) (*Bacillus* spp. and *Paenibacillus alvei*) with an attempt to find specific PCR primers for their soil and plant detection; Tjamos *et al.* (1997b) with unidentified isolates of *Bacillus* and *Pseudomonas*, with antifungal and chitinolytic activity, and Wang *et al.* (1997) (*P. fluorescens* and *B. subtilis*) with claims of reduced wilt incidence on cotton in pot experiments. Three *Streptomyces* (*S. pulcher*, *S. canescens* and *S. citreofluorescens*) were active *inter alia* against *V. albo-atrum in vitro* and *in vivo* after coating tomato seed with spores of the antagonists (E1-Abyad *et al.*, 1993, 1996). A novel extension of the *in vitro* antagonism of *Streptomyces* sp. strain N174 against *inter alia V. albo-atrum* was demonstrated by El-Quakfaoui *et al.* (1995) by transferring a plasmid containing the chitosanase gene with its own signal peptide under the control of the CaMV 35S promoter into tobacco via an *A. tumefaciens* transformation. Three chitosanases were expressed *in planta*, one of which resembled the *Streptomyces* enzyme. *In vivo* inhibition of *V. albo-atrum* was observed. Eshita *et al.* (1995) reported a new antibiotic (bacillomycin Lc) of the inturin class from *B. subtilis*, active *inter alia* against *V. dahliae*. Bacillomycin Lc was isolated as a set of five congeners and differed from bacillomycin L by the change from Asp1 to Asn1 and from Gln5 to Glu5. The congeners differed only in the structure of the aliphatic side chain of the constituent β-amino acid. Increased hydrophobicity of the β-amino acid was associated with increased antifungal activity. Safiyazov *et al.* (1995) found that isolates of *B. subtilis* (23) and *P. fluorescens* (41) were active *inter alia* against *V. dahliae* on cotton. The authors claimed that diluted culture filtrate applied in field tests inhibited disease development and stimulated seed germination.

Xia *et al.* (1996) claimed that *V. dahliae*-antagonistic bacteria (unnamed) were more prevalent *in* cotton plant tissues (6.8%) than in rhizophere soils (1.4%). When introduced into cotton plants, induced resistance to *V. dahliae* occurred which was proportional to bacterial mobility within the tissues. A most interesting corollary to this work (Xia *et al.*, 1997) showed that bacterial inoculation of cotton stems followed by *V. dahliae* inoculation resulted in higher peroxidase, superoxide dismutase (SOD) and esterase isoenzymes compared with the non-bacterial controls. Further new isozymes developed 4 days after inoculation. Li and Zhang (1995) claimed to control *inter alia V. albo-atrum* [*sic*] (*V. dahliae*) of cotton using *Bacillus* spp. and *B. cereus*, but no details were presented.

Solarska (1994a, 1995) and Solarska *et al.* (1996) have reported the beneficial effects of intercropping rye with hops on disease improvement and the increase in fluorescent pseudomonads antagonistic to *V. albo-atrum* in *in vitro* tests. It has been known since the 1960s that rhizosphere bacteria and in particular IAA- and gibberellin-producing *Pseudomonas* spp. ameliorate crop growth in the absence of any demonstrable role in pathogen control (Leben *et al.*, 1987). In a study on potato, Lottmann *et al.* (1997) found that of a population of 384 isolates from root and tuber, 27% produced IAA compared with 8% which were antagonists to *V. dahliae*. Of a total of 18 different antifungal species, 14 were found on roots and eight on tubers. Some antifungal species also produced IAA. New antifungal bacteria were: *Actinobacillus ligniersii*, *Comammonas acidovorans*, *Enterobacter intermedius*, *Paenibacillus macerans*, *Serratia grimesii*, *Sphingobacterium heperinum*, *S. maltophilia* and *Yersinia frederiksenii*. Sharma and Nowak (1998) studied the effect of a selected strain (PsJN) of a plant growth-promoting *Pseudomonas* sp. on the wilt-susceptible tomato cv. Bonny Best. Two methods of bacterial inoculation were achieved: *in vitro*, in which tissue-cultured plantlets were inoculated (systematically colonized) by co-culturing; and *in vivo*, 3 weeks after bacterial inoculation of seedlings in a greenhouse. *In vitro* inoculated transplants showed significant supression of *V. dahliae* infection and stimulated growth, compared with non-inoculated controls. The most effective protection was achieved at low (1×10^3 conidia ml^{-1}) inoculum and least at the highest inoculum level (1×10^6 conidia ml^{-1}). *In vivo* bacterially colonized seedlings, while not exhibiting stimulated growth, showed significantly less infection than controls after pathogen inoculation. Unlike the *in vitro* transplants, *in vivo*-treated seedlings did not show delayed symptom appearance, only reduced intensity. The authors suggested that endophytic bacterial colonization of tomato is a prerequisite for effective *Verticillium* protection and that growth promotion precedes the disease resistance response.

Fungal Antagonists

Askarova and Mamadaliev (1974) claimed that greater numbers of fungi than bacteria and actinomycetes were antagonistic to *V. dahliae*. In *in vitro* tests, Sezgin *et al.* (1982) found that *Aspergillus ochraceus*, *A. sulphureus*, *A. terreus*, *Gliocladium roseum*, *G. virens*, *Myrothecium roridum*, *M. verrucaria*, *Penicillium patulum*, *Trichoderma harzianum* and *T. viride*, were all active against *V. dahliae*. Similarly, Babushkina (1982) reported that *Aspergillus flavus*, *A. fumigatus*, *A. lutescens*, *A. nidulans*, *A. terreus*, *Fusarium equiseti*, *Penicillium cyclopium*, *P. claviforme*, *P. notatum*, *P. roquefortii*, *P. rubrum* and *T. viride* were highly active against one cotton strain of *V. dahliae* and moderately so to another. *V. dahliae* was more highly resistant than 14 other fungi in antagonism tests against *Neocosmospora vasinfecta* var. *africana* (Turhan and Grossmann, 1988). Marois *et al.* (1982) in a glasshouse experiment tested 34 soil isolates for their activity against *V. dahliae* from aubergine. The most active were: *Aspergillus alutaceus*, *Gliocladium virens*, *Paecilomyces lilacinus*, *Talaromyces flavus* and *Trichoderma viride*. When introduced into *V. dahliae*-infested soil, these were able to reduce wilt to 0–20% compared with 90% in the controls. Other minor fungi antifungal to *Verticillium* spp. are *Tuber melanosporum* on *V. albo-atrum in vitro* (Bonfante *et al.*, 1972), *Zygorhynchus moelleri* lysing hyphal walls of

V. albo-atrum (Brown, 1986), *Hypochytrium catenoides* parasitizing microsclerotia of *V. dahliae* (Tolmsoff and Wheeler, 1977), *Phaeotheca dimorphospora* isolated from elm wood and fungicidal *in vitro* to *V. albo-atrum* causing bursting and destruction of mycelia (Yang *et al.*, 1993), and *Chaetomium* sp. Most *Chaetomium* spp. isolated from soil were active on *V. dahliae* but one of two antifungal substances from *C. globosum*, chaetoglobosin-A, completely inhibited spore germination of *V. dahliae* at 32 μg ml^{-1} – a high concentration in biological terms (Amemiya *et al.*, 1994). Tian *et al.* (1998c) in Nanjing, China found that *Stemphylium* sp. inhibited *V. dahliae in vitro* and *in vivo*. Mycelial growth and microsclerotial formation were inhibited by a mycelial culture or culture filtrates. The active component precipitated from culture filtrates by ethanol or $(NH_4)_2SO_4$ was a heat-labile (100°C for 10 min), non-protein secondary metabolite. The authors claimed that inoculation of cotton with *Stemphylium* pre- or post-*V. dahliae* inoculation reduced the disease index. No details were provided of the site of *Stemphylium* growth or of its effects on cotton. *Pythium oligandrum* is parasitic on *V. dahliae* when grown in dual culture (Al-Rawahi and Hancock, 1998). Isolates of *P. oligandrum* differed in their ability to reduce growth and microsclerotial numbers, as did the sensitivity of isolates of *V. dahliae* to the parasite. Temperature and matrix potential both affected the response. Populations of *V. dahliae* were lower in the rhizosphere of sweet pepper in the presence of *P. oligandrum* than in *Pythium*-free soil. In greenhouse studies, growth of pepper and fruit yield were higher in *V. dahliae*-infested soil in the presence of *P. oligandrum* compared with *Pythium*-free controls. However, in the absence of *Verticillium*, plants grown in the presence of *P. oligandrum* were 40–50% taller, suggesting a growth-promoting role similar to the fluorescent pseudomonads. Ultra structural and cytochemical details of *P. oligandrum* mycoparasitism was described by Benhamou *et al.* (1999).

Thielaviopsis basicola and *V. dahliae* frequently occur together in US cotton-growing soils. Schnathorst (1964) considered that the sudden wilt syndrome of cotton in California could involve both *T. basicola* and *V. dahliae*. This was investigated by Mathre *et al.* (1967) using cv. Acala 4-42. Results were variable depending on the population of *T. basicola*. Low levels of *T. basicola* had no effect on either the P_1 (T1) or P_2 (SS4) strains of *V. dahliae*. With an increasing level of *T. basicola*, infection by the P_1 strain increased compared with P_1 only. However, with high levels of *T. basicola*, wilt symptoms from P_1 and P_2 infection were greatly delayed and associated with reduced disease incidence 4 months later. Schnathorst (1981) suggested that *T. basicola* which is infective in cotton (black root rot) in its own right may destroy infective sites favourable to *V. dahliae* since cotton with a high proportion of dead roots was difficult to inoculate in glasshouse experiments.

Of various fungi examined as potential biocontrol agents, three genera, *Talaromyces*, *Trichoderma* and *Gliocladium*, have dominated the more recent literature. Henni (1987) found *T. harzianum* and *Penicillium griseo-fulvum* to inhibit microsclerotia of *V. dahliae* from aubergine.

Gliocladium

Muromtsev (1980) showed in both laboratory and field experiments that *Giocladium* spp. were able to suppress cotton wilt. Isolates of *G. roseum* reduced the viability of *V. dahliae* microsclerotia on nylon mesh by 70% after burying in soil for 2 weeks. Of seven isolates, two were ineffective (Keinath *et al.*, 1990). Globus and Muromtsev (1990) used a powder containing 2–3 × 10^8 propagules of *G. roseum* g^{-1} at a rate of 5–10 kg ha^{-1} to alleviate *V. dahliae* wilt of cotton by 12–34%.

Trichoderma

Trichoderma commonly occurs in soil and the plant rhizosphere (Domsch *et al.*, 1980a,b) where it fulfils an antagonistic role. *T. viride* was reported to be antagonistic to *V. albo-atrum* (Aubé, 1967) and *V. dahliae* (Catani and Peterson, 1966). Of 12 fungal antagonists tested *in vitro* against *V. dahliae* from Mexican cotton, *Acremonium* sp., *Trichoderma* sp. and *Chaetomium globosum* reduced radial colony growth on PDA by 65–75%. *Alvephoma* sp. and *Trichoderma* sp. colonized 100 and 20% of the mycelium, respectively, while *Aspergillus* sp. and *C. globosum* produced inhibitory haloes (Castrejon Sanguino, 1994). Under laboratory conditions, the activity of *T. viride* against *V. dahliae* increased with temperature from 24 to 27°C.

Wilt suppression by *Trichoderma* spp. has been largely confined to the former USSR. *T. viride* (syn *T. lignorum*) grown on peat or lignin, supplemented with nutrients and called trichodermin, has been used as a cotton seed coating, as a pre-planting amendment and as a side dressing with fertilizer (Fedorinchik, 1964). The antagonist is sometimes grown on barley and broadcast in the soil, or on oats (Marupov, 1976). *T. viride* has also been applied in COMF, an organic mixture containing *Chlorella vulgaris*, hydrolytic lignin and inorganic fertilizer and used extensively in Uzbekistan. Wilt was reduced in grey, semi-desert soils by 22–28%, increasing yield by 0.3–0.5 t ha^{-1} (Azimkhodzbayeva and Ramasanova, 1990). The use of ploughing-in green manure such as mustard acts as a sustrate for *T. viride* endemic in soil (Marupov, 1976; see also Wilson and Porter, 1958). Mirzabaev (1977) and Mirzabaev *et al.* (1979) introduced *Trichoderma* sp. in combination with glauconite sand into soil, reducing wilt incidence. Shadmanova *et al.* (1981) examined the effectiveness of *T. viride* against *V. dahliae*, while Khakimov and Ynusov (1980) found that a *T. viride*–lignin mixture applied at 120 kg ha^{-1} reduced cotton wilt by half; at 2000 kg ha^{-1}, wilt was reduced by two-thirds. Khakimov *et al.* (1981) claimed that the antibiotic sesquiterpenes, trichodermin-2 and trichodermin-5 from *T. viride* (Tillaev, 1980), are used in many parts of the former USSR to reduce cotton wilt and increase yield. Notwithstanding the glowing claims for the aleviation of cotton wilt in the former USSR by *T. viride* in various formulations, it is by no means clear whether it is always a reliably successful control measure and, more especially, whether it is practical in commercial culture and cost-effective viz à viz the use of partially-resistant cultivars.

Accounts of the use of *Trichoderma* to control wilts of other hosts include the small-scale experiments of Czaplinska (1973) finding inhibition of *V. albo-atrum* on lucerne roots; reduced wilt of aubergine by the incorporation of mycelium to soil (Buimistru, 1976; D'Ercole *et al.*, 1997); and a similar study on tomato using *T. viride* by Sportelli *et al.* (1983). Rhizosphere soils from *Verticillium*-susceptible tomato cultivars were found to be dominated by *Fusarium* (unidentified), whereas *T. viride* was dominant in resistant cultivars. When roots of susceptible cultivars were injected with *V. dahliae* and *T. viride*, disease intensity was reduced compared with inoculation by the pathogen alone (Subbarao and Bailey, 1961). Such experiments however, would bear only a slight relationship to the balance of pathogen–antagonist populations present in a natural soil environment. Moreover, single-gene resistance is not decisively operative at the root–soil boundary. D'Ercole *et al.* (1984) isolated from soil *T. viride* (29 strains), *T. harzianum* (14), *T. koningii* (six), *T. hamatum* (two), *T. polysporum* (two) and *T. pseudokoningii* (one), all active against *V. dahliae*. D'Ercole and Nipoti (1986) claimed successful control of tomato wilt using the first three species. Jordan and Tarr (1978) dipped rooted strawberry runners in suspensions of *Coniothyrium* sp., *Penicillium* sp. and *Trichoderma* sp. before planting in *V. dahliae*-infested soil, obtaining a significant increase in plant size the following spring compared with untreated controls. The literature to 1985 on *Trichoderma* and *Gliocladium* as biological control agents *sensu lato* is reviewed by Papavizas (1985). Kowalik (1994) claimed to control *inter alia*, *V. albo-atrum* in tomato hydroponic culture in peat and mineral wool by adding *T. aureoviride*, *T. harizianum*, *T. piluliferum*, *T. pseudokoningii* and *T. viride* to the circulating solution. To be effective, the antagonists would need to adapt to the aquatic environment and colonize the support substrates for the extended life of the crop, a minimum of 8 months. In Brazil, Martins-Corder and de Melo (1998) screened 47 strains of four species of *Trichoderma* for antagonistic activity against *V. dahliae*. Seven isolates were selected for volatile and non-volatile anti-*V. dahliae* metabolites and for induced hyphal malformation of the pathogen, i.e. *T. viride* (T15P, Tal-1), *T. koningii* (TW6, CNP311A), *T. harzianum* (CNP17, TC11) and *T. aureoviride* (Tal-10). Hyphae responded by coiling and hook-like formations.

Talaromyces flavus

T. flavus is a teliomorph from the Eurotiales with anamorph *Penicillium dangeardii* (Pitt, 1980; Fravel and Adams, 1986) widely distributed throughout temperate regions. Using isolate Tf1 ascospore pre-plant drenches in soil, Marois *et al.* (1982) in a pioneer study with this antagonist, suppressed *V. dahliae* wilt of aubergine, with a concomitant yield increase comparable with chemical control. This was confirmed by Henis and Fahima (1990). Similarly, Wilderspin and Heale (1984) in laboratory experiments obtained a 50% reduction in *V. albo-atrum* infection in *Antirrhinum*. One reason for the success of *T. flavus* is its ability to survive in different soils for long periods, e.g. 16 months in Idaho field soil (Fravel *et al.*, 1986), and in 25 different soils (Fravel and Marois, 1986). Some *T. flavus* isolates suppress *Verticillium* by hyperparasitism (see Fravel, 1989), but not isolate Tf1 which kills microsclerotia in soil (Marois *et al.*, 1982).

Isolates of *T. flavus* with the *Penicillium vermiculatum* anamorph produce a range of antibiotics, vermistatin, vermicillin and vermiculine with cytotoxic effects (see Fravel, 1989), but isolate Tf1 was shown by Fravel *et al.* (1987) to produce a compound with a low molecular weight which was able to kill microsclerotia. This was later shown by Kim *et al.* (1988) to be glucose oxidase (GO), and the killing factor was the oxidative product H_2O_2. Gluconic acid, the second product, may also be used as a carbon source by *V. dahliae*. Kim *et al.* (1988) established that H_2O_2 is deactivated rapidly in soil and, while glucose and H_2O_2 applied singly were ineffective, glucose and GO resulted in 100% killing of microsclerotia, presumably by H_2O_2 acting locally as synthesized. Culture filtrates of *T. flavus* grown on glucose produced high levels of GO and are able to suppress *in vitro* germination of *V. dahliae* microsclerotia, whereas xylan-based cultures produced only negligible amounts of GO and were non-inhibitory. Both cultures revealed complex protein profiles, but only one band (71 kDa) from the glucose medium culture, detected in Western blots, responded to a polyclonal antiserum PABGO-1 raised against and specific for GO. Cultures from which GO was removed by immunoprecipitation were no longer inhibitory. Conversely, xylan culture filtrates which were amended solely with GO from *T. flavus* were able to kill microsclerotia in culture. The authors, Stosz *et al.* (1996), concluded that GO was the sole protein responsible for the antagonism of *T. flavus* to *V. dahliae*. Using a GO polyclonal antibody and immunostaining, Stosz *et al.* (1998) were able to reveal that GO was both intra- and extracellular. Old and young cells contained GO but labelling of the cell wall enzyme decreased with age. GO release was by exocytosis rather than cell lysis. The enzyme was stable for 2 weeks at 25°C and remained active for several days at up to 50°C.

For inoculum production, ascospore numbers were highest when grown on an oligosaccharide carbon source and a C:N ratio of 30:1, but greatest control (as a drench) of aubergine wilt was achieved using ascospores from PDA rather than the more productive medium with hypoxanthine plus lactose or maltose (Engelkes *et al.*, 1997). Using complex organic 'alginate' prill mixtures, Fravel *et al.* (1995) found that different formulations responded differently in different soil types. Colonization of aubergine was greater on *V. dahliae*-infected roots than healthy controls and was proportional to the severity of infection (Fahima and Henis, 1995). While ascospore concentrations of 10^5 and 10^6 reduced *V. dahliae* by 51 and 69%, respectively, there was no direct correlation with reduced disease severity. On aubergine, tomato and potato, *T. flavus* inoculum applied as an ascospore drench, or as a mycelial alginate prill, localized more on roots than on root hairs. Populations were lower in the rhizosphere than on roots and were greater from drenches than from the solid substrate (Tjamos and Fravel, 1995a, 1997). Colonization of potato and aubergine roots proceeded passively from ascospore-inoculated tubers or seed (aubergine) and decreased, log-linearly with distance from the inoculum source (Nagtzaam and Bollen, 1997; Nagtzaam *et al.*, 1997), though results for potato and aubergine described by Nagtzaam (1998) were often inconsistent. Both Zeise (1997) and Solarska (1997) working with tomato, oilseed rape and hops found greater disease control was obtained by inoculating soil before sowing or planting. Two benomyl-resistant mutants, BenR and TF1-R6, showed enhanced production of chitinase, 1,2-β-glucanase and cellulase compared with ten wild-type isolates. Madi *et al.* (1997) postulated that these enzymes acted in conjunction with antifungal substances and GO in controlling *Verticillium*. Using IG and T4 mutant strains of *T. flavus*, Saito *et al.* (1995) inhibited *V. dahliae* by 73% *in vitro* and 44% in soil; an ethylacetate-soluble metabolite extracted from *T. flavus* culture filtrates was a potent inhibitor of the pathogen.

Some of the subtlety of the pathogen–antagonist relationship was illustrated by Nagtzaam (1995) in a comparison of the effect of *T. flavus* on old and newly established inoculum. Experimental field potato plots were inoculated with microsclerotia (50 microsclerotia g^{-1} of soil) in autumn 1992 and spring 1993 and seed tubers coated with ascospores of three *T. flavus* strains in a talc–wheat bran–alginate powder, were planted, while simultaneously amending soil with 43 kg ha^{-1} of the same material. Control, 40 and 136 days after planting, measured by decreased stem colonization and increased yield, was only achieved with the autumn-applied pathogen inoculum. *T. flavus* did not reduce stem colonization or microsclerotial numbers in plants which had received haulm-killing herbicides. Spink and Rowe (1989), however, reported no effect of the antagonist on the *V. dahliae* population in potato plant rhizospheres or of disease levels in plants. In a further study on *T. flavus* in greenhouse and small plot experiments, Nagtzaam *et al.* (1998) confirmed that *T. flavus* applied as ascospores in carboxymethylcellulose or talc reduced the viability of *V. dahliae* on senescent potato haulm collected from the field. When a preparation of *T. flavus* in an alginate–wheat bran substrate was applied at 0.5% (w/w) to soil, the population of *V. dahliae* at either 15 or 25°C was reduced by 90%. Population densities of *V. dahliae* were negatively correlated with those of *T. flavus* ($r = -0.50$, $P = 0.001$). However, alginate and wheat bran alone led to some reduction of *V. dahliae* and also reduced infection. The authors claimed that combinations of *T. flavus* with other antagonists (*B. subtilis*, *F. oxysporum* or *G. roseum*) each at half strength gave control results equivalent to a single antagonist. With our existing knowledge of the subtlety of biological antagonism, it is difficult to envisage a simple additive role for each of these very different antagonists with no problems of compatibility arising. Biological control of cotton pathogens and *inter alia V. dahliae* was reviewed by Larkin and Fravel (1998). Soesanto (2000) developed a model system with the short life cycle *Arabidopsis thaliana* using *T. flavus* and *Pseudomonas fluorescens* strain P60 to study the dynamics of biological control in *V. dahliae*. Applications of *T. flavus* to fresh organic debris containing microsclerotia at 15 or 25°C was followed after 3 weeks by an increase in the population density. The density of soil microsclerotia was greatly reduced, especially at 25°C, and symptoms on *A. thaliana* were delayed. *P. fluorescens* P60 induced similar effects; inoculated, treated, *A. thaliana* plants appeared the same as uninoculated controls and with reduced microsclerotial formation.

In the light of the foregoing, it seems most unlikely that a commercial preparation based on a *single* strain and *one* formulation for different crops, grown on widely differing soil types, would offer any possibility of success (see also Bilai, 1963).

Mycorrhiza

Preliminary attempts to utilize vesicular–arbuscular mycorrhizal fungi (VAMF) as biocontrol agents have met with limited success. Simultaneous inoculation of cotton plants (*G. hirsutum* and *G. barbadense*) with the VAMF *Glomus mosseae*, *G. versiforme*, or *Sclerocystis sinuosa* and *V. dahliae* led to mutually reduced percentage infection. VAMF colonized root cap cell, meristem and differentiating and elongating zones of the root tip, while *V. dahliae* infections occurred from the root tip up to 2 cm. A reduction in disease development was correlated with increased arbuscule production in cortical cells. *G. versiforme* was the most effective antagonist (Liu, 1995). Similar results were achieved on aubergine by Matsubara *et al.* (1995). Plants inoculated with *Glomus tunicatum* and *Gigaspora margarita* and transplanted in *V. dahliae*-contaminated field soil exhibited delayed and reduced wilt symptoms. *G. tunicatum* was most successful and led to increased and less deformed fruit. The authors reported increased lignification in secondary cortical cell walls. A degree of control of *V. dahliae* wilt of olive was achieved by Karajeh and Al-Raddad (1999) inoculating seedlings with *G. mosseae*. VAMF-inoculated plants were significantly bigger and showed reduced disease severity. Mycorrhizal colonization of roots was not retarded by *Verticillium* infection. However successful this greenhouse experiment may have been, the problems of stabilizing a *G. mosseae* population on mature tree roots under olive grove conditions are immense and such that the control may only be of academic interest.

While competition for infection sites may play a role in disease alleviation, Liu *et al.* (1993, 1995) showed that ten novel proteins, one exhibiting chitinase activity, were induced in cotton roots and leaves following seed inoculation with *Glomus mosseae*, *G. versiforme*, *G. hoi* and *Sclerocystis sinuosa*, or *V. dahliae* infection. These so-called 'pathogenesis-related proteins' (although chitinase is recognized as a constitutive enzyme in healthy cotton) were shown *in vitro* to retard hyphal growth and kill conidia. Joint infection with VAMF and pathogen led to increased protein levels. Other organisms attacking *Verticillium* are: the amoeba *Thecamoeba granifera* subsp. *minor*, penetrating hyphal and conidial walls (Alabouvette *et al.*, 1979); the nematode *Aphelenchus avenae* which feeds on *V. albo-atrum* (Mankau and Mankau, 1962); and soil insects *Collembola*, *Proisotoma minuta* and *Onychiurus encarpatus* which reduced germination of microsclerotia by consuming germ tubes (Curl and Gudauskas, 1985). A previously undescribed fungal endophyte – the hyphomycete *Heteroconium chaetospira* – was inoculated into Chinese cabbage to control yellows (*V. dahliae*) and club root. *H. chaetospira* colonizes cortical cells in the root tip region. After 3 months, *V. dahliae* yellows was reduced by 49–67% compared with non-mycorrhizal controls.

With growing pressure for non-chemical, environmentally acceptable control measures, there is an increasing search for novel sources of naturally-occurring anti-*Verticillium* molecules, the genes for which could be transferred to a susceptible host or to a vector organism. A number of these have unusual sources. Examples active on *Verticillium* spp are as follows: 3-formyl-7,11 dimethyl-docecatrien-1-al in the mandibula gland of *Lasius fuliginosus* (Akino *et al.*, 1995); *Streptomyces* chitinase introduced as a plasmid with its own signal peptide under the control of the CaMV 35S promoter into tobacco cv. Xanthi via *A. tumefaciens* transformation (El-Quakfaoui *et al.*, 1995); chlorinated orchinol from the fungus *Hericium erinaceum* (Okamoto *et al.*, 1993); terpenoids, 6 isopropyltropolone-β-glucoside and 5-(3-hydroxy-3-methyl-trans-1-butenyl-6-isopropyl-tropolone-β-glucoside) (cupressotropolones A and B) from *Cupressus sempervirens* bark (Madar *et al.*, 1995); usnic acid from the lichen *Alectoria ochroleuca* (Proska *et al.*, 1996); and 4,5,9-trithiadodeca-1,6,11-trien-9-oxide (ajoene) (Reimers *et al.*, 1993). The widespread release of transgenic potato plants with the *Bacillus thuringiensis* var. *tenebrionis* insecticidal endotoxin led *inter alia* to concern about microbial ecological changes. Donegan *et al.* (1996), however, could find no effect of such plants on plant and rhizosphere populations of *V. dahliae*.

Verticillium spp. as Antagonists

In a reversal of the foregoing, species of *Verticillium* may themselves assume the role of antifungal antagonist. Paplomatas *et al.* (1997) found that *V. tricorpus* effected biological control over *Rhizoctonia solani* attacking cotton in field and greenhouse experiments. The authors invoke catalase as a *Rhizoctonia* virulence factor, since the addition of catalase at 1.1 U ml^{-1} converted an avirulent isolate to weakly virulent and increased the activity of a virulent isolate. Ascorbate, a catalase inhibitor at 100 mg l^{-1}, reduced virulence, leading to the suggestion that *V. tricorpus* produced a catalase inhibitor. *V. tricorpus* encapsidated on cotton seed protected seedlings against *R. solani* damping-off. *V. biguttatum* Gams (a soil saprophyte) was observed by Nicoletti and Lahoz (1997) to parasitize cultures of *R. solani*. Root-dips of tobacco seedlings in 2×10^6 ml^{-1} *Verticillium* conidial suspensions of two strains gave partial protection. The most effective strain had higher endochitinase and glucosaminidase than the other. Another soil *Verticillium*, *V. chlamydosporium*, was shown by Bharadwaj and Trivedi (1997) to reduce population levels of the cyst nematodes *Heterodera avenae* and *H. cajani* on wheat and cowpea, respectively. It was claimed that wheat bran inoculum applied pre-sowing increased wheat yield by 76% and reduced reproduction in *H. cajani* by 90–92%; no control mechanism was suggested.

The dubious practice of disseminating wilt pathogens to control other plant diseases has been proposed by some workers. An early example of cross-protection involving two different pathogens but with a similar host–pathogen environment was provided by Grasso and Tirro (1982). Symptoms of mal secco disease on sour orange due to *Phoma (Deuterophoma) tracheiphila* were less severe and delayed if plants were pre-inoculated with *V. dahliae* below the point of *P. tracheiphila* introduction. Pre-inoculation above the *P. tracheiphila* inoculum site merely delayed mal secco symptoms. Injection of *V. dahliae* into field elm (*Ulmus procera*) and *U. hollandica* cv. Commelin suppressed *Ophiostoma ulmi* symptoms almost totally (Scheffer and Elgersma, 1990). In a re-examination of this work, Sutherland *et al.* (1995) found that *V. dahliae* had no effect on Dutch elm disease in the highly susceptible *U. procera*. Symptom expression was reduced progressively with increased resistance of *Ulmus* clones. This reaction suggests a 'cross-protection' role for *V. dahliae*, enhancing native defence reactions rather than involving antibiosis or hyperparasitism. In partial contrast to the result on *U. procera*, Gazquez *et al.* (1998) reported significant but variable control on *Ulmus minor* var. *minor* in Granada, Jaen, Salamanca and Alicante in Spain. Using *V. dahliae* strain WC5850 from Arcadis Hidemij Realisatie, The Netherlands, the authors found that elm species, cultivar, age, size, soil type and weather conditions all affected the results. At the present state of our knowledge, the future for *Verticillium* bio-control of Dutch elm disease appears problematic. *V. dahliae* capable of infecting lupin (*Lupinus polyphyllus*) was proposed as a 'bioherbicide' for this plant colonizing braided (multichannel) riverbeds in South Island, New Zealand (Harvey *et al.*, 1996). The prospect of microsclerotial formation on a large scale in a wild habitat with an aquatic dispersal medium however, is not to be envisaged. Similarly the use of *V. dahliae* to control the weed, velvet leaf (*Abutilon theophrasti*) could similarly be fraught with problems (Green and Wiley, 1987). The control of *V. albo-atrum* of hop by composting infected bine (haulm) (Sewell *et al.*, 1962), may be interpreted as biological control in its widest sense.

Control by Grafting on Resistant Stocks

Where wilt resistance is lacking or only weakly expressed in a relatively high-value crop, the alternative to soil fumigation, of grafting on to resistant rootstock, may be a cost-effective alternative. Ginoux (1974) described the results of a 4-year grafting programme on solanaceous plants and *inter alia*, aubergine against *V. dahliae*. The author found the cost then, (1974) comparable with soil fumigation (see also Beyres, 1974). Canevascini and Caccia (1976) claimed that aubergine cv. Mariner produced 3128 kg of fruit per plant [*sic*] (= 31.28?) when grafted on rootstock hybrid KNVF compared with 1592 kg (= 15.92?) on its own roots. Ginoux and Dauple (1982) described the costs and technique of grafting and the benefits of grafting aubergine cv. Giniac on tomato rootstock KNVF (see also T.-Q. Li *et al.*, 1999). Zhou *et al.* (1998) also comparing the effectiveness of grafting aubergine, examined consequent changes in physiological and biochemical characteristics. Cirulli (1975) advocated olive grafting on to resistant rootstock as the best control against *V. dahliae* (see also Hartmann and Whisler, 1970). Vine (*Vitis vinifera*) as a plantation crop which is normally grafted benefits from the use of *V. dahliae*-resistant stock. Nipoti and D'Ercole (1982) reported the results of a 3-year trial of 11 *V. dahliae*-resistant rootstock clones, the most resistant of which was 57R.

Ushiki *et al.* (1996) screened root extracts of 53 species of medicinal plants in Japan against six soil-borne pathogens including *V. dahliae*. *Geranium pratense*, *Eupatorium fortunei* and *Sanguisorba officinalis* were particularly active (inhibitory).

4 Integrated Control

The principles of integrated control were outlined by Talboys (1984), who diagramatically presented the potential interactions of chemical agents, microorganisms, the physical environment and wilt resistance, indicating that two or more of these factors selected to operate in close sequence should result in lasting control. The larger the number of control methods integrated, the greater the possibility of synergism. Most strategies for integrated control (IC) presented here on a crop basis reveal little subtlety of choice and fall far short of Talboy's idealized concept.

Cotton

The concept of IC in the cotton crop was reviewed by El-Zik (1985). Most work was carried out in the USSR (now CIS). Yarovenko *et al.* (1976) combined the application of fertilizers and pesticide in conjunction with benomyl to reduce infection by 30–50%. No comment, however, was made on yield or cost–benefit analysis – a common omission in most studies. Rotation of cotton with lucerne and maize with a *Trichoderma*–mineral soil treatment combined with a carbamide foliar spray at the 2–3 leaf stage resulted in 15–20% reduced disease expression (Alimukhamedov *et al.*, 1977). Yuldashev (1978) and Ubaĭdullaev (1980) introduced *T. viride* in rotation with resistant cultivars with the use of quintozene as a soil fumigant, and benomyl or 'uzgen' used as foliar sprays. The use of fungicides and defoliators to reduce microsclerotial formation was advocated by Sboeva *et al.* (1980) and Tellyaev and Ligai (1981). The value of integrating row spacing with irrigation scheduling and nitrogen applications was discussed by El-Zik and Yamada (1981). Mineral fertilizer requirements essential for cotton yield often have deleterious effects on wilt severity. Mamadaliyev and Boirov (1990) in Andizhan (CIS) reversed this effect by applying lignine [*sic*] in conjunction with phosphate–potassium fertilizer in triple doses and reduced wilt from 95.8 to 33.3%. Lignin applied alone reduced yield compared with the appropriate control. Azimkhodzbayeva and Ramasanova (1990) in Tashkent achieved similar results incorporating into soil a complex organic–mineral fertilizer with *Trichoderma lignorum* and a unicellular alga (*Chlorella vulgaris*), which increased yield under wilt conditions by 0.3–0.5 t ha^{-1}, an effect which persisted for 2 years. No information is forthcoming on the role of the alga in this method. Field solarization in Andalucian clay soils infested with the P_1 strain of *V. dahliae* combined with a resistant (tolerant) Acala GC 510 cotton reduced wilt incidence to less than 13% compared with controls. Bearing in mind that inoculum reduction is a transitory effect and considering the field scale of cotton production and crop value, it seems unlikely that large-scale solarization would be economically feasible (Melero *et al.*, 1988; Melero-Vara *et al.*, 1995a,b). Neither carbendazim nor *Pseudomonas fluorescens* strain P32-treatments of *V. dahliae*-inoculated cotton were effective when applied singly (Niu *et al.*, 1999); however, when carbendazim was applied to soil at 10 μg^{-1}, the *P. fluorescens* population of soil and rhizosphere rose by 11.6- and 12.8-fold, respectively. When P32 and carbendazim were applied simultaneously, cotton wilt incidence was reduced to 82.2% compared with 40.4% in the absence of carbendazim. Seed dressed with fungicide and bacterium showed a 72.7% reduced disease rate. Since no synergistic effect of the dual treatment was found in sterilized soils, the authors interpreted the results on the basis of the elimination of P32 competitors. Proposals for the integrated management of *Verticillium* and *Fusarium* wilts in China were reported by Shen (1985).

Potato

Problems associated with early attempts at IC of potato were described by Cole *et al.* (1972). Grinstein *et al.* (1979a,b) successfully combined polyethylene mulching with chemical fumigation, incorporating a nematicide. Elad *et al.* (1980) used the same control but incorporated *Trichoderma harzianum*. Yield increases in the presence of *V. dahliae* have been achieved by rotating potatoes with maize and fumigating with 1,3-dichloropropene and chloropicrin (O'Sullivan and Reyes, 1980). A variety of soil fumigant, systemic fungicide and resistant cultivar combinations were described by Hide *et al.* (1984) to combat potato early dying syndrome (PED). However, the shortage of effective resistance in potato (Rowe *et al.*, 1987) and the longevity of microsclerotial inoculum in soil (5–10 years) (Davis and McDole, 1979) make control of PED difficult. Krikun *et al.* (1984) outlined the criterion employed in integrated control in Israel where *Pratylenchus thornei* predisposed potato to early infection. The main features included metham sodium soil sterilization delivered through the irrigation system and the use of partially tolerant cultivars.

Many accounts of 'integrated control' merely report the effect of two (rarely three) control methods, with no attempt at integration either between treatments or with environmental or other factors. A factorial field experiment comparing continuous cropping with alternating crops of lucerne, yellow sweet clover, maize and sudax (*Sorghum halepense* × *S. sudanense*) and potato in the presence of *V. dahliae* and *Pratylenchus penetrans* was conducted by J. Chen *et al.* (1995). One year of any of the break crops followed by 2 of potato, showed no difference in yield compared with 3 of potato. Two years of sweet clover or lucerne followed by potato gave significantly higher yields and lowest *P. penetrans* population densities (J. Chen *et al.*, 1995). Nachmias *et al.* (1997b) described a four-component integrated system to control PED in Israel involving soil fumigation (metham sodium, methyl bromide and formaldehyde); *Verticillium*-free seed tuber stock, tolerant cultivars (Shepody, Lady Rosetta and Nicola); and *Verticillium*-free compost – tested prior to use. Simple greenhouse experiments on pre-harvest treatment of haulm to control microsclerotial development, but with important field implications, were conducted by Mol and Scholte (1995b). Early and late pre-harvest haulm, killed by chemical treatment, burning, cutting at different lengths and buried, or left on the soil surface, was air-dried after 4 weeks and microsclerotia determined. The early chemical killing yielded more microsclerotia than cutting. Covering haulm with non-sterile soil was partially inhibitory to microsclerotial growth. At the late treatment, short-chopped haulm kept on the soil surface was most inhibitory to microsclerotial (and soil inoculum) density. A large variation was recorded in all treatments.

The above experiment with slight variation, was continued by Lamers and Termorshuizen (1998) in The Netherlands between 1993 and 1997 to investigate microsclerotial establishment in *V. dahliae*, *V. tricorpus* and *V. albo-atrum* [*sic*]. Mechanical chopping of leaves followed by 'green' lifting (presumably green haulm) compared with chemical killing and/or normal lifting gave an 'apparent' reduction in microsclerotia, but results were non-significant. It is surprising at this late stage in our knowledge of *Verticillium* control that more precise data are not available to establish the most effective protocol for inoculum limitation.

Strawberry

Talboys and Frick (1975) advocated the integration of disease avoidance, soil fumigation, wilt-resistant cultivars and systemic fungicides with adjustment relative to prevailing conditions. Although methyl bromide–chloropicrin soil fumigation has been the mainstay against *V. dahliae* and other soil-borne problems in everbearing strawberries, especially in the USA, the growing awareness of bromine residue problems (especially in rotation with leaf crops) and increasing political pressure for non-chemical control methods has led to the use of solarization in conjunction with fumigation with the object of replacing or reducing chemical levels. A comparison of solarization alone or combined with metam sodium with methyl bromide–chloropicrin in California (Hartz *et al.*, 1993) showed that solarization and chloropicrin reduced inoculum in 1989/90 but chloropicrin was superior in 1990/91. Solarization was less effective than chloropicrin in controlling annual weeds. Solarization alone increased strawberry yields by 12%, but combined with metam sodium gave a 29% increase equal to that achieved by chloropicrin. Similar results were obtained in Mexico by Cebolla *et al.* (1995), no significant difference in control was found between solarization combined with low doses of methyl bromide and methyl bromide at standard doses alone. In Chile, control of soil inoculum at depths of 10, 20 and 30 cm by solarization was 98, 85 and 70%, respectively, whereas methyl bromide at 68 g m^{-2} gave 100% at all levels. Nematode results were inconsistent (Aballay *et al.*, 1996). Gordon *et al.* (1997) similarly found that chloropicrin:Telone (1,3-dichloropropene) 70:30 fumigation gave absolute control of inoculum survival and less than 1% plants showed wilt symptoms (presumably from infected runners or from soil microsclerotial levels too low to detect statistically). Plots cropped to rye and tarped for 2 months during the summer had an average 32 microsclerotia g^{-1} of soil surviving in the soil, compared with 45 in control plots.

Cover crops were not seen by Gordon *et al.* (1997) as a credible alternative to fumigation. In contrast, in a different soil with different ambient temperatures, Termorshuizen *et al.* (1997) obtained effective control of soil microsclerotia using a combination of solarization and organic amendments. The soil tarp, Hytilene, was a three-layered 0.135 mm polyethylene sheet. Organic soil amendments were grass or broccoli (40 t ha^{-1}) rotavated to 25 cm and irrigated overnight before covering. Organic amendment in combination with solarization led to a decline of *V. dahliae* of >90%. Broccoli was more suppressive than grass, but organic matter or solarization alone had little effect. There was no effect on microsclerotia below the level of broccoli residue incorporation. Under the Hytilene cover, O_2 and the redox potential declined and persisted several weeks under broccoli where the maximum soil temperature remained <37°C. The authors hypothesized that anaerobic fermentation of the organic matter produced compounds toxic to microsclerotia. Blok *et al.* (2000) found a redox potential reduced to –200 mV and since *V. dahliae* and other pathogens were not, or hardly, inactivated by the broccoli treatment in the absence of tarping, or under tarping in the absence of the amendments, the control success of the combined treatment could not be attributed to thermal inactivation. Background for the Danish integrated fruit production programme for strawberry was described by Daugaard *et al.* (1997).

Other Crops

Hops

As early as 1964 Sewell and Wilson (1964b) recommended a combination of field sanitation, use of tolerant cultivars and weed control to limit non-host inoculum build up, and grassing over between gardens to restrict movement and carry-over. Subsequently, nitrogen limitation was added to the programme. These methods combined with strict legislation have greatly reduced the problem which threatened the survival of the industry in the 1950s and 1960s.

Melon

Pardossi and Tognoni (1985) proposed an integrated scheme using solarization, resistant cultivars, grafting on to resistant stocks and growing in nutrient film culture.

Tomato

In Florida, soils of low pH, Jones and Overman (1986) used fumigation with a polyethylene tarp. Raising the pH to 7.5 with hydrated lime discouraged *Fusarium* crown rot, but enhanced *V. albo-atrum* [*sic*] (*V. dahliae*) wilt which was controlled by fumigation. Accati *et al.* (1974) described the effect of irrigation, mulching and plastic shading on tomato and bell pepper production and *V. dahliae* severity.

Aubergine

The combination of soil fumigation with grafting onto resistant tomato rootstock, while improving yield, was limited by cost. An interesting development on the integration of solarization and the bio-control agent *Talaromyces flavus* by Fravel and Tjamos (1995) and Tjamos and Fravel (1995a), simulated solarization experiments on microsclerotia by heating them in water at temperatures sublethal to *T. flavus*. Microsclerotia thus treated had a lower germination rate and slower melanization, an effect which interacted synergistically with *T. flavus* to supress wilt of aubergine. Microsclerotia in non-sterile soil were affected by metam sodium at 93.5 l ha^{-1} but, in sterile soil, the growth rate was unaffected at rates of 1871 l ha^{-1}. *Gliocladium roseum* but not *T. flavus* (both in an alginate prill) had a reduced growth rate at 187 and 935 l metam sodium ha^{-1}. In a greenhouse test in field soil, wilt of aubergine was unaffected by 187 l ha^{-1}, but was reduced when either fungus was combined at this level (Fravel, 1996).

Earlier Welvaert and Poppe (1986) had shown that *Verticillium* was almost eliminated by methyl bromide (100 mg m^{-2}) or dazomet (80g m^{-2}) with formaldehyde, all under a transparent tarp. Soil fungi generally were reduced by 80–100% and 75% by the treatments, respectively. However, there was a clear shift to *Trichoderma* in the survivors. Ben-Yephet *et al.* (1987) found the lethal effect of solarization and metam sodium on microsclerotia to be 81–100% compared with 61% for solarization and 70% for metam sodium (see also Cartia *et al.*, 1996).

The control option of soil fumigation alone for *V. dahliae* wilt of apricot or in combination with solarization is discussed by Stapleton and Duncan (1999).

An analysis of the effects of combining soil amended with cabbage residues, or chicken manure and solarization, by Gamliel and Stapleton (1995) showed that heated soil liberated fungitoxic volatile compounds such as aldehydes, and sulphur compounds including isothiocyanates. After 1 week, microbiological activity of the soil was reduced rapidly. Solarization of soil amended with ammonium phosphate also reduced *Verticillium*. Colonization of roots by *Bacillus* spp. and fluorescent pseudomonads was unaffected by the combined treatment. Wilt of cauliflower (*Brassica oleracea* var. *botrytis*) is a recent and serious disease in California, especially in the Salinas Valley and other coastal regions. In the absence of resistant cultivars and the uneconomic methyl bromide–chloropicrin fumigation, control measures are required to reduce the 93 microsclerotia g^{-1} of soil, which leads to a wilt incidence of 90%. When broccoli, *B. oleracea* var. *italica* (a non-host of *V. dahliae*), residues were incorporated in soil and covered with a plastic tarp for 2 weeks, control of soil microsclerotia was equal to either methyl bromide–chloropicrin or Telone fumigation. Fresh broccoli residues at 20–30°C were more effective than dried (Koike *et al.*, 1997). In a subsequent study (Subbarao *et al.*, 1999), no differences were obtained between tarped or non-tarped plots with either broccoli incorporation into soil or controls. In fumigated plots, propagules increased to pre-treatment levels at the end of the cropping season. The best control was obtained by metham sodium fumigation followed by broccoli treatment and chloropicrin over 2 years. The authors claim that the control achieved by broccoli residues in reducing inoculum and cauliflower wilt, equalled or exceeded soil fumigation, with concomitant economic and environmental advantages.

A comprehensive study on *V. dahliae* inoculum density control in peppermint plots by Crowe *et al.* (1997) established that cumulative seasonal wilt incidence and poor winter survival (winter kill), were directly proportional to soil microsclerotial density and presumably the continuous infection of new rhizomes. Treatments were: postharvest flaming without tillage, or tillage and no flaming, followed by killing the mint crop and planting with irrigated wheat and peas, before replanting with nematode-free peppermint. Flaming without tillage was shown to be highly advantageous for sustained cropping. It was suggested that tillage may lead to winter kill and a high inoculum density. An IC system for Mediterranean olive culture was described by Tjamos (1993) involving controlled irrigation (preventing microsclerotial dissemination), pruning prior to defoliation, solarization for individual trees and the testing of two Californian potentially wilt-resistant rootstocks. Herbicide control of non-host, inoculum-supporting weeds was also advocated. A laboratory study by Saeed *et al.* (1997c) investigated the influence of soil water content and temperature on metam sodium (metham sodium) fumigation and its degradation product methyl isothiocyanate. Temperatures were 2 or 22°C and the soil water contents expressed as soil matrix potential were −23 (wet), −113 (moist) and −2845 J kg^{-1} (dry). Fumigation had greatest efficacy against *V. dahliae* in wet cold soils, which led to the recommendation of metam sodium fumigation in late autumn, or early spring. This result would clearly need confirming in soils of different composition and pH (see also Reed and Verhalen, 1970). Since *T. flavus* damages and kills microsclerotia partly through the indirect production of H_2O_2, Fravel (1997) included H_2O_2 in an interactive study of sublethal stress with heat and metam sodium on microsclerotial viability. When microsclerotia were treated with heat then metam sodium, or metam sodium then heat, the two stresses were additive for microsclerotial mortality. However, when microsclerotia were treated with H_2O_2→metam sodium, or metam sodium→H_2O_2, there was interaction, such that at moderate levels of metam or H_2O_2 there was significantly less mortality than with low or high rates of metam sodium and H_2O_2 on successive generations of microsclerotia, showing that third-generation progeny were less susceptible to stress than their parents.

5 Legislation and Quarantine

The history of *Verticillium* disease has been one of spread within and from crop to crop and the infestation of land previously free from the pathogen. It is surprising, therefore, especially with high-value plantation crops where seed-borne transmission may be unimportant, that quarantine measures or legislative attempts to contain wilt disease have received so little interest.

An exception to this has been the involvement of government legislation, combined with plant breeding, integrated with nutrition and sanitation to control *V. albo-atrum* wilt of hops in the UK (Ebbels and King, 1979; Pegg, 1984). The success of this approach has provided a model system for crops in other countries but has also illustrated the inertia associated with the introduction of what was later, clearly seen as an 'obvious' solution. The initial discovery of the disease (Harris, 1925a) was in Penshurst, Kent in 1924; 6 years later, a second outbreak was found in Hertfordshire (Harris, 1936); both isolates were of the M or mild type. In 1933, a more aggressive 'progressive' strain V_1 was isolated at Paddock Wood in Kent (Harris and Furneaux, 1938; Keyworth, 1945). This was the forerunner of the 'V' strains each overcoming the resistance of a succession of erstwhile tolerant cultivars. The rapid spread of V_1 over large distances (see Pegg, 1984) led Keyworth (1939) to identify propagation of rooted strap cuts from the bases of diseased stock plants as one of the prime causes and to inaugurate a certified scheme, following voluntary Ministry of Agriculture inspection of hop propagation gardens. Only gardens found to be disease-free could sell certified propagation material; layered bines and softwood cuttings (Keyworth, 1945). The purchase of such material became almost mandatory in 1947, with the government 'Progressive *Verticillium* Wilt of Hops Order 1947' making *Verticillium* wilt of hops a notifiable disease and prohibiting the sale of hop plants from land known to be affected by the disease. In 1950, the Hop Research Department of Wye College (University of London) in a then disease-free area in Kent, started a breeding programme to select tolerant cultivars (in the absence of true resistance), possessing good brewing qualities, which were screened by East Malling Research Station and such material propagated by carefully selected growers. Simultaneously, a growers association, the Hops Marketing Board, organized the distribution of healthy cultivars, the first being OR.55 (Keyworth's Midseason). From 1955 to 1975, a series of certified schemes (A, B, AA) was established and introduced to provide a gradual improvement in plant health, pedigree and the discontinuation of earlier grades (Keyworth, 1951; Ebbels, 1979). During this time, new 'progressive' races of *V. albo-atrum* were identified on their reaction with tolerant cultivars in current use (Sewell and Wilson, 1976), with moderate symptoms occurring on cv. Wye Target (the most tolerant cultivar available) infected with race V_3. To limit the possible spread of susceptible cultivars and to contain the movement of new pathogenic races (see Neve, 1979), the 'Progressive *Verticillium* Wilt of Hops Order 1978' was imposed by the Government Ministry. This scheme required Ministry of Agriculture licensed breeding material produced at Wye College, tested at East Malling Research Station, to be propagated in East Anglia and the West Midlands by licensed propagators; such material was then to be bulk produced by selected growers in different parts of the UK, followed by inspection and licensing for commercial growing in strictly specified areas of Kent and Sussex, the principal hop growing areas affected by *Verticillium* wilt (Ebbels, 1979). The important principle established here was that only Ministry-approved cultivars were available to growers, and the only source of these was from outside the disease area. These somewhat draconian regulations were relatively easy to enforce since the planting area quota and the marketing of hops were both under the control of the Hops Marketing Board. *V. albo-atrum* is still a major problem in hop cultivation, but much less so than heretofore. This is due to a control policy in which breeding, sanitation and husbandry practices have all been integrated and made effective through legislation. A Ministry of Agriculture advisory leaflet 413 (Ebbels and Dickens, 1981) summarized the position for growers – see also the European Plant Protection Organisation guidelines for good plant protection practice for the hop (Anon, 1996).

The problems facing an isolated cotton-producing country such as Australia in the absence of strict legislative measures are outlined by Allen (1995b). An Italian Ministry of Agriculture and Forestry decree of 1993 formally established a voluntary certification scheme for the propagation of olive to include *inter alia*, *V. dahliae*. Until legislation is introduced to restrict *new* olive grove plantings to certificated pedigree plants with associated control recommendations, there is little likelihood of success (Martelli *et al.*, 1995; Martelli and Prota, 1997). The limited administrative control of *Verticillium* wilt of lucerne in the UK is outlined by Heale *et al.* (1979). Regular summaries on the quarantine requirements of lucerne wilt in Canada are produced by the Plant Health Division of the Canadian Department of Agriculture (Anon, 1986).

From the many examples in this monograph, there is clearly enormous scope in many countries for the introduction, for certain crop plants, of quarantine measures combined with a rigorously-applied plant health scheme to exclude all but certified pathogen-free seed or plant stock. While there is a limited amount of previously cropped land which is free from *Verticillium*, there is still a pressing need to exclude, by regional or national legislation if necessary, the introduction of specific virulent pathotypes. With hindsight, in areas of earstwhile pathogen-free soil such as the Negev in Israel, the imposition of simple quarantine rules to exclude or restrict the planting of potato, a known major carrier of *Verticillium* spp., would have prevented or greatly lessened the perpetual and costly problem of wilt disease in other crops.

6 Breeding For Resistance

For the most part, diseases of crops of economic importance are caused by *V. dahliae* and *V. albo-atrum*. Both species are genetically plastic, and strains of a given species assume a greater significance in a particular host than does infection by one or other species (Pegg, 1984). International trade and crop movements, pioneering colonization, explorers and invading armies have all contrived to spread host plants and their pathogens worldwide, but not always simultaneously. The introduction of crop monocultures often of non-diverse genotype has provided the selection pressure for the evolution of new strains and pathotypes of increased virulence, with a concomitant breakdown of existing resistance and the need to find or develop new resistant lines.

Many of the early (and some contemporary) reports failed to distinguish between *V. dahliae* and *V. albo-atrum*. Thus unless there is reference to microsclerotia or a geographical distribution, or ambient temperature at which *V. albo-atrum* would not survive, the true identity of the pathogen in some accounts is conjectural. Detailed reports have been published by Wilhelm (1975, 1981) for field and fruit crops; Goth and Webb (1981) for vegetable crops; Schreiber and Townsend (1981) for shade trees; and Bell (1992a) for cotton.

The literature on wilt resistance breeding and the cataloguing of natural resistance is extensive; for simplification, research is described on a crop basis under: field and plantation crops, both major and miscellaneous; bush and tree fruit crops; and ornamental trees.

Major Field and Plantation Crops

Cotton (*Gossypium* spp.)

Cultivated cotton is derived from four species of *Gossypium*; two diploids ($2n = 26$), *G. arboreum* L. and *G. herbaceum* L., and two allotetraploids with chromosome complements $2n$–$4\times = 52$, *G. hirsutum* L. and *G. barbadense* L. The upland cottons derived from cultivars of *G. hirsutum* grown in central Asia, the CIS (formerly USSR), China, India, USA, Mexico, Australia and also Africa and South America comprise 95% of the world production. *G. barbadense* is native to South America and the West Indies, cultivars of which are grown in South America, western USA, CIS, India, Sudan and Egypt. Comprising approximately less than 5% of the world production, the drought-resistant *G. arboreum* is grown locally in India and South East Asia. *G. herbaceum* is grown in Africa and parts of Asia; it hybridizes with *G. arboreum*, and together these make up less than 1% of world cotton production. All species are attacked by strains and pathotypes of *V. dahliae*.

History of the disease

The first account and accurate identification of the pathogen was by Carpenter (1914) in *G. hirsutum* in Arlington, Virginia. Later, Carpenter (1918) found that *V. dahliae* from okra (*Hibiscus esculentus*) caused wilt of cotton, and Bewley (1922) repeated this with *V. dahliae* from tomato on Asiatic cotton (*G. herbaceum*). The disease subsequently spread to Tennessee (Sherbakoff, 1928, 1929) and extensively in Mississippi (Miles and Persons, 1932), Arizona (Taubenhaus, 1936) and California (Shapovalov and Rudolph, 1930). Thereafter, *V. dahliae* was found in all cotton-growing areas of the USA. Simultaneously with the early reports in the USA, Butler (1933) and Miles (1934) described the disease in central Asia, Bulgaria and Greece (Pandian and Isaac, 1971). Subsequently, it was confirmed in Brazil, China, Peru, Uganda and the central Asian republics of the former USSR (Barducci and Rada, 1942; Mukhamezhanov, 1966). By 1940, cotton wilt was a limiting factor in all the above regions.

Losses

Verticillium wilt is the major source of yield loss in the three major cotton-producing countries, China (including *Fusarium*), the CIS and the USA, and to a lesser extent in Australia, Greece, Syria, Zimbabwe, Peru, South Africa and Spain (Bell, 1992a). Peak losses in the USA in 1961 accounted for the loss of 580×10^3 bales, and 4.4% in 1967. During the periods 1971, 1973 and 1975, the average losses in 14 cotton-producing states were 1.95, 2.87 and 2.9%, respectively (Anon, 1974, 1976a,b). The subsequent decline has been due largely to improved management (reduced irrigation and nitrogen, and new resistant lines). The average annual yield loss over the last 40 years was 1.7%, with New Mexico and California losing 5.73 and 3.87%, respectively (Bell, 1992a). Ranney (1995) estimated national yield loss between 1961 and 1965 to have reduced by 44%. On some plantations, however, losses of 80%, and in some cases of the total crop, have been recorded. Losses of 500×10^3 t ($\sim 76 \times 10^4$ bales) in the CIS in 1966 were estimated (Mukhamedzhanov, 1966), 80% of which were in Uzbekistan. From its origins in China in 1935 from cotton cultivars introduced from the USA, the disease has spread dramatically. In the 1950s, wilt was reported in only a few counties in four provinces; by 1964, the disease was present in 18 provinces and 372 counties. In 1982, 12% of the country's cotton land surveyed (82.7%) had the disease (Liu, 1985; Shen, 1985). In 1993, 270×10^6 ha, 50% of the cotton-growing area of China, was affected by both P_1 and P_2 strains, with annual losses of lint cotton estimated at 16×10^4 t (Ma and Shezeng, 1997). Wei *et al.* (1999) reported losses of 1500 t year^{-1} for the combined effects of *Fusarium* and *Verticillium* in Zhenping, Henan province. The effect of *Verticillium* wilt on cotton yield, fibre properties and seed quality was described by Bugbee and Sappenfield (1970). The importance of time of wilt appearance on subsequent cotton yield was stressed by Lu *et al.* (1985).

Sources of resistance

Although *V. dahliae* is the major pathogen of cotton, as with tomato, *Fusarium* wilt is also important (*Fusarium oxysporum* f.sp. *vasinfectum*). Thus work on resistant lines frequently has involved the search for linked resistance to both pathogens. No immunity to *V. dahliae* exists in *Gossypium* germplasm, and progress generally has been made by introducing resistant or tolerant cultivars while eliminating older highly susceptible ones. Bell (1992a) reported that nine out of ten planted cultivars in 1954 were highly susceptible to the P_1 strain, whereas in 1990 the ten leading cultivars all had greater resistance than seven of the 1954 cultivars. The main problem is on *G. hirsutum* but, while *G. barbadense*, *G. arboreum* and *G. herbaceum* show some resistance, all are susceptible to some pathotypes of the P_1 strain.

Various early breeding programmes to incorporate *inter alia*, resistance to *Verticillium* in cotton have been undertaken (Harrison and Brinkerhoff, 1947; Leyendecker, 1949; Sherbakoff, 1949; Harland, 1957; Voitenok, 1958, 1960; Solovyeva, 1959; Wiles, 1960, 1963, 1971; Nazirov, 1962; Cooper, 1963; Lehman *et al.*, 1963; Pudovkina and Alikhodzhaeva, 1963; Malinin, 1964, 1968; Mirakhmedov and Adylkhodzhaev, 1964; Polyanichko and Runov, 1964; Straumel, 1964; Ter-Avanesyan, 1964; Zunnunov, 1964; Cotton, 1965; Kratirov, 1965; Adylkhodzhaev, 1966; Barducci, 1966; Bird, 1966; Presley, 1966; Mathre and Otta, 1967; Fisher, 1968; Wiles *et al.*, 1968; Bell, 1973).

MEXICAN RESISTANCE A selection from *G. hirsutum* cultivated near the town of Acala, Mexico, was introduced into the USA in 1906 and became the parent stock of many 'Acala' cultivars. Ten years after planting 'Acala' cottons in California, any existing resistance broke down (Wilhelm, 1981). 'Acala 442' originated as a resistant single-plant selection from 'Acala 5-12' made by Harrison in 1939. 'Acala 5-12' (later known as Acala 1517) was wilt susceptible, but developed resistance after six inbred generations. This reached its acme in the 'Acala 4-42-77' family in 1966 (Cooper *et al.*, 1967). Resistance in 'Acala' cultivars has been progressively broken and improved over many years, with new pathotypes of *V. dahliae* arising on new plant selections.

PERUVIAN RESISTANCE This is based on the selection of a single healthy plant of *G. barbadense* in a crop of severely diseased cv. Suave plants, between 1907 and 1917 in Peru, by a grower called Tanguis. Many 'Tanguis' resistant plants were derived from the original selection, replacing the upland susceptible cultivars in Peru and subsequently transferring to the USA (Barducci and Rada, 1942; Baker, 1966).

SOVIET SOURCES OF RESISTANCE Shadmanov and Saranskaya (1990) claimed that all cotton cultivars in the CIS after planting sustain 10–15% wilt, which may rise to 95% after 2–3 years of production, leading to their being discarded. Some 200 wilt-resistant cultivars grown in the cotton republics of the CIS were derived from a single wilt tolerant type of *G. hirsutum* subsp. *mexicanum* var. *nervosum* collected in Mexico. From this have developed the Tashkent series of cultivars and, more recently, Andizhan 9 and 60, Kirgizsky 3, etc. (Wilhelm 1974a,b; Kalandarov, 1990; Kravtsova, 1990). These cultivars have shown high resistance to race 1 (P_2 strain) of *V. dahliae* but only slight resistance to race 2 (P_1 strain). Resistance from *G. barbadense* has been released in cv. Termez 7 (see review by Maksumov, 1978).

HYBRID RESISTANCE – MOSTLY IN THE USA *G. hirsutum* and *G. barbadense* cross readily, yielding fertile F_1 offspring. Wilhelm *et al.* (1972a) transferred resistance from *G. barbadense* to upland cotton. From 1967, complex triple hybrids (Coker 100W *G. hirsutum* × *G. arboreum* × *G. thurburi*) from crosses and backcrosses with the Georgia cv. Early Fluff have produced the San Joaquin (SJ) family of resistant plants grown principally in California. Harrison (1955) introduced a novel source of resistance in *G. hopi* (now probably *G. hirsutum* var. *punctatum*). Crossing Arizona Queen Creek Acala, *G. hopi*-76 and Acala 1517, the AHA and C6 line families were developed with high resistance and have been planted throughout the US cotton belt. Resistance to *Fusarium* and *Verticillium* in *G. hirsutum* cultivars was derived by introgression from the Sea Island, and Mit-Afifi cultivars of *G. barbadense* (Sherbakoff, 1949; Wiles, 1963) which yielded cv. Seabrook Sea Island. Wiles (1963), also identified *Verticillium* resistance in *G. arboreum* and *G. herbaceum* in the upland cv. MVW 26, a derivative of cultivars Auburn 56 and Rex and other lines. Shen (1985) described the selection of the first wilt-resistant cultivars in China from the adopted cultivars Delfos 531 and Deltapine 15.

Hybrid F_1 cotton has been produced by hand emasculation and hand pollination since 1971 (Wilhelm, 1981). The discovery of cytoplasmic male sterility and fertility-resisting genes leads to the prospect of large-scale F_1 seed production and the synthesis of F_1 hybrids with high wilt resistance.

Inheritance of resistance

From crosses between wilt-susceptible and wilt-tolerant Acala cottons, Barrow (1970a) suggested that tolerance to the P_2 (mild SS4, non-defoliating) strain of *V. dahliae* was controlled by a dominant V^t gene. Bell (1973) considered that a second gene for tolerance and a number of modifier genes were present, suggesting polygenic inheritance. Al-Rawi and Ahmed (1984) claim that dominance is of greater importance than additive effects. A detailed analysis of wilt tolerance using pedigree data from three crosses is given by Devey and Roose (1987). The inheritance of the wilt-resistant Acala 4-42 as a true breeding line from the susceptible Acala 5-12 was indicative of a recessive polygenic inheritance (Harrison, 1955); this was supported by the stability of Acala 4-42 for 15 years.

Bell and Presley (1969b) and Barrow (1970b) have shown that the expressions of resistance to a given strain can vary with the ambient temperature. The temperature range for host reaction to the P_1 strain from susceptible to immune is 4°C. The critical temperature ranges are: genetically susceptible, 27–31°C; moderately resistant, 25–29°C; and highly resistant cultivars 23–27°C. While a mean temperature of 27°C is the best temperature for screening resistance levels to P_1, 22.5°C is best for the P_2 strain (Barrow, 1970b). Barrow (1970a) found that temperature was critical for selecting true breeding lines in cultivars heterozygous for resistance. Other environmental factors complicate the expression of, and selection for, resistance. Thus, in the field, a moderately resistant cultivar can appear resistant or susceptible with low or high inoculum or low or very high nitrogen (Bell, 1992b). Another problem identified by Khaidarov (1990) is the reaction of genotypes to inoculum of mixed (especially *Fusarium* and *Verticillium*) pathogens. Plant susceptibility increases with the onset of flowering. Rysbayeva and Urunov (1990) illustrated this by inducing resistance in *G. hirsutum* cultivars by removal of flower buds.

In general, resistance in *G. hirsutum* is thought to be multigenic, with pooled additive and dominant effects (Verhalen *et al.*, 1971; Devey and Roose, 1987). The former authors concluded that resistance was quantitatively inherited but with occasional dominance towards susceptibility. Barnes and Staten (1961) had earlier found transgressive segregation towards either resistance or susceptibility but that most resistance was quantitative.

The resistance of *G. barbadense* from cv. Seabrook Sea Island 12-B-2 to mixed strains of *V. dahliae* was inherited as a single dominant gene in progeny from crosses with *G. hirsutum* cv. Hartsville or the Alabama line H257, both already containing resistance to *Verticillium* and *Fusarium* by the previous introgression of one or more genes from *G. barbadense* (Smith and Dick, 1960; Wiles, 1963).

Resistance to the P_2 strain by various Acala lines was dominant at least in the F_1 progeny (Barrow, 1973). The incomplete dominance of resistance in F_1 hybrids from *G. hirsutum* (S) × *G. barbadense* (R) and indistinct segregation in the F_2 and F_3 progeny indicate multigenic quantitative inheritance. Incomplete dominance of resistance in F_1 progeny derived from *G. hirsutum* ssp. *mexicanum* var. *nervosum*, originally thought to be dominantly monogenic (Wilhelm *et al.*, 1974b), is now known to be polygenic (Makhubov, 1990). The genetics of *Verticillium* resistance [*sic*] (tolerance) were reviewed by Barrow (1973).

USA In field and glasshouse experiments, Wilhelm *et al.* (1970, 1971) detected high levels of resistance in 16 accessions of *G. barbadense*, *G. armourianum*, *G. thurberi*, *G. arboreum* and *G. hirsutum* races *yucatense* and *mexicanum* (Leningrad strain). The partial resistance exhibited by strains of *G. hirsutum* was described by Wilhelm *et al.* (1970) as wilt tolerance.

Over the last 35 years, a number of successful *Gossypium* cultivars have been released into commercial production, many with resistance not only to *V. dahliae* but also to *F. oxysporum* f.sp. *vasinfectum*, *Xanomonas malvacearum*, *Meloidogyne incognita* and other pathogens, e.g. cv. Hancock, free from wilt for 12 years (Hoskinson and Hancock, 1972); Tamcot SP21, Tamco SP23 and Tamcot SP37 (Bird, 1976); Tamcot SP21S derived from Tamcot SP21X Tamcot SP37 (Bird 1979a); Tamcot SP37H (Bird, 1979b), and Tamcot HQ95 (El-Zik and Thaxton, 1990). These hybrids developed by Texas A & M University represent a continuing programme of improved and linked resistance, also incorporating many other desirable features including lint quality and yield. Many lines of Acala cotton have been developed and released, based on Acala 1517 and also show moderate to good resistance to *V. dahliae* (Davis *et al.*, 1978, 1980; Malm *et al.*, 1978a,b; Barnes *et al.*, 1980). Other lines of *V. dahliae*-resistant *G. hirsutum* incorporating other resistance to other pathogens have been described by Chambers and McCutchen (1978), Sappenfield (1980, 1981, 1985), Jenkins *et al.* (1979) and Stokes and Sappenfield (1981). The problem of pathotypes was described by Batson and Blasingame (1988) when wilt symptoms developed 70 days after planting ten mid-south adapted cotton cultivars in *V. dahliae*-infested Mississippi field soils. Wilhelm *et al.* (1976) reported a short-branching upland cotton which was wilt tolerant by virtue of withstanding dense low planting. Temperature within specific limits is critical for pathogen survival and disease expression. Foliar characters known as Okra leaf and Super Okra leaf, raise stem and canopy temperatures by approximately 2–3 and 3–4°C, respectively, and (presumably depending on the ambient temperature) reduce wilt in the susceptible cv. Stoneville 7A by 26 and 33%. Schnathorst (1975) envisaged the incorporation of these characters into commercial cultivars to enhance resistance. Wilhelm *et al.* (1978) advocated resistance breeding combined with dense planting to maximize yields. Phenotypic modification was developed for dense plant populations (Wilhelm *et al.*, 1985).

Until plant resistance genes or gene clusters from cotton species can be identified reliably and transgenic plants become a working reality, conventional breeding for multiline crosses, of which *V. dahliae* tolerance is but one part of multi-adversity resistance (MAR), will continue to be the main approach. More recent breeding work and the release of new lines, albeit incorporating *Verticillium* resistance from old established sources, include: Davis (1993), cultivars Laguna 89 and CIAN 95, Palomo Gil and Godoy Avila (1994), Palomo Gil *et al.* (1993a,b, 1996) five upland-type A and B hybrid cotton cultivars, Cantrell *et al.* (1995); a new line 86-6 Cun and Cuikang (1995); a new Acala release, El-Zik and Thaxton (1996), a *G. hirsutum* MAR cv. Tamcot sphinx, Chambers and Bush (1998); evaluation of 16 cultivars for wilt resistance, Wheeler and Gannaway (1998); cultivars showing good resistance in the Texas High Plains, Paymaster 330, H526, Tamcot sphinx, All-Tex Atlas and Stoneville 887, Adair *et al.* (1995); review of cotton cultivar tolerance, Thaxton and El-Zik (1998); breeding for resistance in Tamcot cultivars; and wilt resistance breeding in Arkansas, Bourland and Oosterhuis, (1999).

CIS (COMMONWEALTH OF INDEPENDENT STATES, FORMERLY USSR) Early work by Adylkhodzhaev (1966) studied the partial resistance of subspecies of *G. hirsutum* to *V. dahliae*. Straumal (1964) working in Tashkent outlined the principles for the selection of wilt-'resistant' cotton. Ter-Avanesian (1976) outlined the difficulty of accurate resistance selections at temperatures above 27°C. This author listed *G. hirsutum* ssp. *punctatum*, ssp. *polycarpum* and ssp. *rupestre* as potential sources of wilt resistance. The cv. Tashkent I, while resistant to the P_2 strain, was completely susceptible to P_1 (race 2 in the CIS). Kasyanenko *et al.* (1976) working in Tajikistan found resistance to the P_1 strain in three accessions of *G. arboreum* from India and Korea, and *G. arboreum* ssp. *neglectum* from India. Various accessions of *G. herbaceum* from Iraq, Afghanistan and the USSR were also resistant to P_1. Teshabaeva *et al.* (1973) and Teshabaev and Teshabaeva (1974) described white [*sic*] and black forms of *V. dahliae* both apparently of the P_2 strain. Several accounts indicate that cv. Tashkent I was only moderately resistant to pathotypes of the P_2 strain (Usmanov, 1974; Straumal and Tishin, 1976). Tolerance (partial resistance) to P_1 and immunity to P_2 were obtained in line 777, an interspecific hybrid of *G. hirsutum* cv. Tashkent × *G. tricuspidatum* ssp. *purpurascens* [*sic*] from 02800 (Makhbubov *et al.*, 1985). The problems of pathogen nomenclature in the CIS have been confused by a failure to compare isolates on a worldwide basis. Some authors refer to five races 0, 1, 2, 3 and 4 (Kasyanenko, 1990), while not distinguishing between pathovars = pathotypes of defoliating or non-defoliating strains. The picture is confused further by reference to A and B races (Sattarov, 1981a,b).

One indication that there is comparability of the major strains P_1 and P_2 in the CIS and USA is in the response of US resistant cultivars in the CIS. In general, cultivars such as Acala, Coker, Deltapine, Locket, Stoneville, Kemp and Doubled Hope were resistant under Azerbaijan and Uzbekistan conditions. In many instances, US resistant cultivars outyielded those from the CIS under wilt conditions (Kandov and Bakharamov, 1976; Rakhimzhanov, 1977, 1987; Ter-Avanesyan *et al.*, 1977; Sherzhanov, 1987, 1989). Indian and Mexican lines were also resistant under CIS conditions (Abdullaev and Kiktev, 1979; Guseinov *et al.*, 1979; see also Mirakhmedov *et al.*, 1969). From a screen of the world *G. hirsutum* germplasm collection in the Vavilov Institute, CIS, Atlantov and Rakhimzhanov (1990) claim that no absolute resistance to *V. dahliae* exists in this genus. These authors maintain that the narrow resistance gene base of CIS cultivars of Tashkent and most other former Soviet cultivars based on a single selection of *G. hirsutum* ssp. *mexicanum* var. *nervosum* has led to an overall breakdown to a range of P_1 pathovars. The claim that selection pressure from Tashkent monocultures led to the evolution of race 2 (USA P_1 strain) is unlikely and, as with other countries, was most probably introduced from the USA; see also Makhubov (1990). Attempts to improve resistance to race 2 have involved *G. hirsutum* cultivar crosses with *G. hirsutum* ssp. *yucatense*, ssp. *punctatum* and ssp. *purpurescens* (Bolbekov, 1987). Kalandarov (1990) recognized four races in Tajikistan and Uzbekistan, of which races 2 (P_1) and 3 are most serious. Ruderal forms of *G. hirsutum* from Mexico were crossed with the best local cultivars and selected, resistant F_3, F_4 and F_5 plants were crossed with Acala and Tamscot cultivars to give polygenic resistance (see Dotsenko, 1971). The use of colchicine treatment on triploid *Gossypium* hybrids ($2n = 39$) has led to nine new amphidiploids ($2n = 78$) combining cultivated and wilt-tolerant characters from the diploid species, *G. raimondii*, *G. trilobum*, *G. sturtii*, *G. aridum*, *G. harknessii*, *G. lobatum*, *G. armourianum*, *G. triphyllum* and *G. thurberi*. High wilt resistance was obtained from amphidiploids and introgressive lines obtained from *G. trilobum*, *G. thurberi*, *G. aridum*, *G. lobatum* and *G. sturtii* (Egamberdiyev, 1990).

The problems of finding wilt resistance sources was outlined by Lemeshev *et al.* (1987). See Makhmadaliev (1984) on the inheritance of wilt resistance in interspecific hybrids. Lemeshev and Shakhmedova (1990) in a complex breeding programme, converted sterile triploid F_1 hybrids to hexaploids and backcrossed these to produce highly resistant tetraploids derived from *G. thurberi* and *G. aridum*. Negmatov *et al.* (1990) found resistance to three 'physiological races' of *V. dahliae*; 1, 2 and 3 phenotypically with black-, brown- or cherry coloured microsclerotia, in a dwarf line of *G. barbadense*; unfortunately, much resistance was lost in backcrosses with the F_1 hybrid of this species. Similar results were reported by Akhmedov *et al.* (1991). The importance of linking drought-stress tolerance with disease resistance particularly for Uzbekistan conditions was demonstrated by Akkuzhin *et al.* (1990) with the introduction of a selection of N510 which had good (but not absolute) resistance to P_1. Gorkovtseva and Azizova (1997) claimed to be able to differentiate resistant and susceptible reactions of breeding stock using fungal infection of stem callus cultures *in vitro*. Rakhimzhanov (1980) made five resistant selections from Indian forms of *G. hirsutum* (see also Mannanov *et al.*, 1971).

OTHER COUNTRIES The pattern of cultivar susceptibility and pathogen race distribution described for the USA and CIS appears similar in all cotton-growing regions of the world, probably reflecting the widespread distribution of contaminated seed over the last 70 years. Screening programmes for resistance have been carried out in: Peru (Baker, 1966); Syria (Al-Hamidi, 1985); Iraq, with resistance from Acala F1517 and Giza 45 (*G. barbadense*) (Khalida *et al.*, 1983); India (Siddiqui and Deshprabhu, 1970); Iran (Hamdollah-Zadeh, 1993); Israel (Tsror and Nachmias, 1990); Bulgaria (Savov, 1986b); Turkey, where the Uzbekistan Tashkent cultivars are the main sources of resistance (Karcilioglu *et al.*, 1982; Dolar, 1984); South Africa (Swanepoel and deKock, 1996); Spain (Gutierrez *et al.*, 1994; Bejarano-Alcazar *et al.*, 1996); Mexico (Godoy Avila *et al.*, 1994, 1995); China (Li and Yang, 1989); and Brazil (Cia, 1977).

China, particularly during the last decade, has emerged as one of the major centres of *Verticillium* research in the world. This reflects the rapid development of Chinese science; it also reflects the widespread culture and importance of cotton for the national economy and the equally widespread distribution of *V. dahliae* pathotypes. In the Henan province of China alone, 4×10^6 ha of cotton were severely damaged from the P_1 strain (Wang *et al.*, 1997). Attempts at fungicidal control using Zhikuling, Duojunlins and Huangfusuanyan were only moderately successful (Wei *et al.*, 1999). In the Yellow River Valley, *Fusarium* and *Verticillium* predominate; cv. Jinmian, while highly resistant to *Fusarium*, was only moderately tolerant to *V. albo-atrum* [*sic*] (*V. dahliae*). Yingfan *et al.* (1994) reviewed the characteristics of cotton germplasm showing *Verticillium* resistance. The results of 16 years of wilt-resistance breeding in this area are reviewed by Li and Yang (1989). Gao and Gu (2000) recorded lowest yields and highest wilt indices in cv. Sumian-8 derived from a susceptible *G. hirsutum* and the highest yield and *Verticillium* resistance in cv. Changkangmian, 53.3 kg mu^{-1} and 56.1 kg mu^{-1}, respectively (mu = 0.067 ha). Wu (1990) described a *G. hirsutum* cv. – Xinluzao (Xin Upland Early) with multiple pathogen resistance but only partial resistance to *V. dahliae*. Over the period 1974–1994, *Verticillium* tolerance genes were introduced into Chinese upland cotton cultivars from *G. thurberi*, *G. bickii*, *G. sturtianum* and *G. raimondii*. From these, 15 lines were selected, the most resistant of which was Shiyuan 185 incorporating a gene from *G. sturtianum* (Sui, 1995). Other studies on resistance included Cai *et al.* (1994) describing progeny from a diallel cross between resistant cultivars Chuan 739 and Chuan 2808 and wilt-susceptible cultivars Yu 86-1 and TX-5 (American). Cao *et al.* (1988), Pan *et al.* (1994), Zhao *et al.* (1994), Shuxiang (1995), Ma and Jian (1995), Shijie *et al.* (1996), and Han *et al.* (1996) were all involved in the derivation and screening of new lines. Ouyang and Huang

(1996) developed cv. Xinluzao 5, an early variety incorporating cold hardiness and wilt tolerance. Similarly a very early tolerant cv. Xinluzao 5 was introduced by Benlian and Dingyuan (1996). These Chinese cultivars like all tolerant selections may anticipate a limited life as more virulent pathotypes emerge and/or new cultural characters supersede old. Li *et al.* (1996) described lines Chuan 737 and Chuan 2802 both with resistance to three races of 11 strains and Cheng and Zhang (1985) cv. Jinmian 7. A new wilt-resistant cv. Shan 2234 has been developed by the Shaanxi Institute of Cotton as a multihybrid, incorporating American multi-resistance genes (SP21) combined with Chinese agronomic characters from Liao 756-57 and Laio 756 (Jing *et al.*, 1999). Wang *et al.* (1999) described the crosses 93K12, Pengze 70 and JS2028 with resistance to two strains. From an examination of F_1, F_2 and the BC_1 and BC_2 segregation ratios, the authors concluded that resistance (tolerance) was determined by two dominant complementary genes. The genetic male-sterile line Dong A and ten other tolerant cultivars were used to construct a recurrent selection base population from which two selection pools were derived (Min *et al.*, 1999). A high yielding (1560 kg ha^{-1}) cv. Su Mian9, derived from Su Mian2 × Zhong Mian12 with combined *Fusarium–Verticillium* tolerance, was described by Gu *et al.* (1998) (see also Qian *et al.*, 1987; Qiongfang *et al.*, 1996; Xiaoxuan *et al.*, 1997) (see also reviews by Shi *et al.*, 1998; Tan, 1998). Three types of pathogen were described by Song *et al.* (1997) – a filamentous, defoliating, highly pathogenic type corresponding to the P_1 strain, and nucleate and intermediate types more closely resembling P_2 strain pathotypes.

With the exception to date of South Africa, all upland cotton-growing regions of the world have reported the existence of P_1 and P_2 strains. There is general agreement that apart from pathogenicity (and unlike hop strains), P_1 and P_2 can be distinguished readily; on microsclerotial size, $P_1 = 22 \times 95$ μm and $P_2 = 32 \times 55$ μm; on temperature, 27 and 23°C, respectively; while *Verticillium*-resistant tomato cv. Pearson with P_1 is symptomless, it wilts in the presence of P_2 (Hamdollah-Zadeh, 1993). P_1 isolates produced round and elongate microsclerotia on water agar while P_2 produce only round. On sanguinarine agar, P_1 was able to grow but P_2 was greatly inhibited; similarly, P_1 unlike P_2 fluoresces under UV (Bejarano-Alcazar *et al.*, 1996). The distribution of *V. dahliae* pathotypes in Andalucia southern Spain was described by Blanco-Lopez *et al.* (1987).

The development of multiple disease resistance (including *V. dahliae*) in advanced cotton lines in Brazil was described by Cia *et al.* (1999). Onan and Karcilioglu (1998) reviewed Turkish and Aegean *V. dahliae* pathotypes and wilt tolerance in Nazilli 84 cotton stocks. Australian cultivars Sicala V-2RR and Sicot 189RR were released by the CSIRO Cotton Research Unit (Anon 2000a,b). Both were glyphosate tolerant and combined multiple fungal and bacterial resistance together with high-yielding, quick-maturing and other desirable agronomic characters.

The future for *Verticillium* resistance plant breeding in cotton (and other crops) must lie in the development of genetically engineered plants. Since pathogen virulence and host resistance are both so multifaceted, the problem will be which genes for which particular proteins should be incorporated. In a pioneering approach in Australia by McFadden *et al.* (1997), genes for chitinase, glucanase and glucose oxidase were used to transform independent cotton lines. If successful, this approach would be expected to enhance existing polygenic resistance of cotton; immunity however, will be more elusive. Hill *et al.* (1999) constructed a cDNA library with mRNA from root tissue of a superior *G. hirsutum* wilt-tolerant cv. Sicala V-1. Using a differential screening technique and differences in gene expression between healthy and *V. dahliae*-infected root, the temporal expression of nine, putative defence-response genes was examined by Northern blot at intervals after infection. The rapid increase in mRNA transcripts corresponding to each clone consequent to infection, strengthened their possible role in resistance. Genes for a 14-3-3 protein and pathogenesis-related proteins were identified, together with presumed novel genes for which, at present, no role can be asscribed. Upland cotton was screened against eight wild-type strains of *V. dahliae* and two diploid strains by Guozhong *et al.* (1994). Differences in *V. dahliae* wilt incubation time and severity in *Bt* transgenic cotton and CRI controls were recorded by Sun *et al.* (1998b). The peak occurrence in *Bt* cotton in a wheat–cotton intercropping system was delayed by at least 10 days. A significant correlation coefficient was obtained for *Bt* cotton wilt index and hours of sunshine.

Solanaceous plants: potato (*Solanum* spp.)

Wilt is caused both by *V. albo-atrum* and *V. dahliae*, depending on geographical location. It is present in all potato-growing areas; *V. albo-atrum* is widespread in the cooler, mild-temperate regions and may occur there with *V. dahliae*, the latter, however, extends into countries with warm temperate and subtropical summer temperatures.

Although resistance in *Solanum* spp. and hybrids has been examined by Webb and Hougas (1959), the genetic basis of resistance is still not well known. Resistance breeding in *S. tuberosum* is complicated by tetrasomic inheritance and the difficulty of complete recovery of resistant progeny in backcrosses. Hougas *et al.* (1958) described the production of polyhaploids; pollinations of *S. tuberosum* $2n = 48$ by the diploids *S. phureja*, *S. simplicifolium* and *S. verrucosum* ($2n = 24$) resulted in some crosses inducing the parthenocarpic development of the *S. tuberosum* egg cell to yield diploid plants. Such plants might be used in crosses with wild-type resistant plants, followed by conventional backcrosses to desirable cultivars with subsequent restoration of tetraploidy with colchicine. Much of the breeding and selection has been done on an empirical basis. In North America, wilting is followed by chlorosis and premature senescence, hence the name potato early dying syndrome (PED). In the cooler eastern and north-western regions, the pathogen is *V. albo-atrum*; elsewhere, for the most part, *V. dahliae* prevails (Beckman, 1973). In Idaho, in one of the earliest references to *Verticillium*, Stevenson and Jones (1953) reported a number of cultivars with partial resistance including cvs Populair Menominee and Sequoia. McClean (1955) found two types of resistance in Idaho breeding stock, one (the most prevalent) associated with late maturity of the cultivar. This author found that selfed crosses of susceptible cultivars often yielded progeny with substantial wilt resistance. S × R and R × R crosses gave increasing numbers of resistant progeny but, as Goth and Webb (1981) point out, in the absence of genetic analysis, the 'resistance' may reflect late maturity, i.e. partial disease escape. Akeley *et al.* (1956) working on *V. albo-atrum* resistance in Maine found no immunity in several crosses with 33 cultivars including the common cvs Houma and Russet Burbank. Some lines were high yielding and wilt tolerant; several, however, were high yielding despite high levels of resistance. In a major *V. dahliae* wilt breeding project in Peru (International Potato Centre) involving 236 accessions in 55 species and 376 plants from five crosses of *S. phureja* × *S. tuberosum*, both diploid and tetraploid, the pathogen was recovered from resistant, tolerant and susceptible plants. Diploids showed higher resistance than tetraploids and hexaploids. Resistance was detected in *S. ambinosum*, *S. berthaultii*, *S. bulbocastanum*, *S. chacoense*, *S. gandarilasii*, *S. microdontum*, *S. mochicense*, *S. oplocense*, *S. pamplasense*, *S. polyadenium*, *S. spegazzinii*, *S. tarijense* and *S. vernei*. Resistance was found in 32 and tolerance in 41 accessions. Segregation of resistance in more than 60% of the tetraploids was in a 1:2:1 ratio. Resistance was inherited as a single incompletely dominant gene segregating in a disomic or tetrasomic ratio (Malamud and O'Keefe, 1976). Concibido *et al.* (1994) maintain that the difficulty in *Verticillium* resistance breeding is due to a dearth of genetic and breeding information, particularly the obscure parentage of standard cultivars and complex segregation at the tetraploid level. Six 2*n* wild interspecific *Solanum* hybrids were derived from R × R and S × R crosses. Results of *S. gourlayi* × *S. chacoense* and the reciprocal cross indicated that resistance was dominant and polygenic. This cross showed the best potential as a source for wilt resistance.

Other reports of resistance in *Solanum* species were in the diploid *S. chacoense* and *S. chacoense* × *S. connersonii* hybrids (Webb and Buck, 1955), and in *S. chacoense*, *S. tarijense*, *S. acaule*, *S. microdontum* ssp. *gigantophyllum*, *S. kurzianum*, *S. maglia* and *S. vernei* (Rosse and Rowe, 1965). Bazan de Segura and Ochoa (1959) found 11 resistant and five moderately resistant varieties of *S. tuberosum* ssp. *andigena* in 100 varieties tested in Peru. High resistance to *V. dahliae* was also found in *S. berthaultii*, *S. chacoense*, *S. sparsipilum*, *S. sucrense* and *S. tarijense* (Corsini *et al.*, 1987) and in *S. aculeatissimum*, *S. scabrum*, *S. sisymbriifolium* and *S. toruum* (Alconero *et al.*, 1988). Highest resistance was found in the poor tuberizing series, Commersonia, Tarijensa and Transequatorialia originating in Bolivia and north-western Argentina, all of which are diploid (except *S. sucrense* which is tetraploid) and hybridize readily with *S. tuberosum* (Corsini *et al.*, 1986). Nieto (1989) working with the Columbian Central Potato Collection in Bogota screened 21 commercial cultivars and 326 clones for resistance to *V. dahliae* and *V. albo-atrum* at 2600 m above sea level. Of these cultivars, ICA Tequendama, Parda Pastusa and 15 clones were 'resistant' to *V. dahliae*. No true resistance, but some slight tolerance, to *V. albo-atrum* was found. Screenings of 398 genotypes from accessions of *S. berthanetii*, *S, chacoense* and *S. tarijense* found tolerance more commonly than resistance to *V. albo-atrum* in all but two (Lynch *et al.*, 1997). The inheritance of resistance in two genotypes identified in *S. chacoense* P_1 472189 was studied in crosses with a susceptible *S. chacoense* P_1 472810. Segregation ratio in F_1 and F_2 and backcross populations in an R genotype (18-21 R) showed that resistance was controlled by *Vc*, a single dominant gene, which, if transferred to tetraploid germplasm, could provide effective and stable control.

As a result of reciprocal grafting experiments between resistant and susceptible potato rootstocks and scions, Tsror and Nachmias (1995) concluded that resistance resided mainly in the rootstock, a susceptible rootstock (cv. Nicola) permitting enhanced colonization in a resistant (cv. Desiree) scion. The potato genotype did not lead to a differential expression of races 1 and 2 as seen in tomato. The authors conclude that general defence mechanisms in addition to the root barrier exist in potato. Nachmias *et al.* (1997a) discuss the Israel strategy to minimize potato wilt; the indiscriminate references to 'tolerance' and 'resistance' by this group, often relating to the same cultivar, however, leads to much confusion. Nachmias and Orenstein (1986, 1987) described a peptidal toxin produced by *V. dahliae* which induced symptoms similar to those produced in natural infection, in detached leaflets of susceptible but not tolerant cultivars. This was proposed as a basis for selecting tolerant germplasm. Similar differential results were obtained from the toxin, derived *ex planta* on susceptible but not tolerant protoplasts. The osmoticum, pH and hormonal status of the assay were critical.

Most disease resistance breeding and the release of new cultivars has been in the USA and here largely confined to the northern and north-western (especially Idaho) states. *V. dahliae* and more rarely *V. albo-atrum* resistance, although an important feature, is only one of many desirable attributes incorporated into new genotypes (Pavek *et al.*, 1973; Twomey *et al.*, 1977; Johansen *et al.*, 1981; Yarris, 1981; Davis, 1984; Reeves *et al.*, 1985; Azad *et al.*, 1987; Mohan *et al.*, 1987; Rowe *et al.*, 1987; Plaisted *et al.*, 1990). Canadian breeding work has been directed similarly to multiple disease resistance, including *V. albo-atrum* as well as other desirable cultural characters (Anon, 1972, 1973, 1975; Davies *et al.*, 1977; Johnston and Rowberry, 1980; Young *et al.*, 1988). More recent North American releases with *inter alia* moderate to high resistance to *V. albo-atrum* and/or *V. dahliae* include: Hoyos *et al.* (1993), Johansen *et al.* (1993), Reeves *et al.* (1994, 1996, 1997), Davis *et al.* (1994a), Goth *et al.* (1994b), Lynch *et al.* (1995) and Tarn *et al.* (1995). The practical importance of gross yield in the field versus *Verticillium* resistance was emphasized by Corsini and Pavek (1996) in summarizing a 3-year study comparing yield of highly-resistant clones and their progeny against standard cultivars. Crosses of high-yielding highly resistant cultivars A66107-51 × A68113-4 yielded 29 resistant from 125 clonal progeny. Of these, 11 had lower yields than cv. Russett Burbank and none outyielded the parents. Similarly, of 15 interspecific hybrids – crosses between *Solanum* spp. and $2n$ and $2n = 4x$ cultivars – only four resistant progeny had sufficient yield to produce seed for yield trials. The authors conclude that a primary selection for resistance is inefficient and yield and other agronomic criteria should be the primary objective for planting in *Verticillium*-infested fields. This argument, while perhaps succeeding in the early years, ignores the inevitable progressive increase in soil inoculum from infected susceptible haulm. The old cultivar Russet Burbank still provides a basic genotype for modern hybrids. Such cultivars, although moderately or highly resistant to *Verticillium*, also carry genes conferring variable resistance to *Phytophthora infestans*, *Fusarium* spp., *Streptomyces scabies*, potato leaf-roll virus, PVX, PVY and net necrosis viruses, golden cyst nematode (*Globodera rostochiensis*) and physiological disorders such as hollow heart, brown centre and growth cracks. Mosley *et al.* (2000a,b,c) described three such recent releases, from joint collaboration of Agricultural Experimental Stations in Oregon, Idaho, Washington, Colorado, California, Texas and the USDA: cv. Russet Legend (= Century Russet × WN672-2); cv. Umatilla Russet (= Butte × A77268-4) and Century Russet. Above all, these cultivars, not surprisingly, greatly out-yield Russet Burbank, e.g. Russet Legend (47.3 t ha^{-1} = 23.65 t ha^{-1}), Russet Burbank (32.4–36.5 t ha^{-1}). A multiline germplasm release by Corsini *et al.* (1999) in Washington State USA was AWN 86514-2 with high *Verticillium* resistance ♂ cv. Ranger Russet × ♀ KSA195-96 from Idaho with Polish germplasm which also incorportated genes from *S. acaule*, *S. demissum*, *S. phureja*, *S. simplicifolium*, *S. stolonifera* and *S. tuberosum* sp. *andigena*. Wheeler *et al.* (1994b) demonstrated in microplots what had been known from glasshouse studies, that a *V. dahliae* × *Pratylenchus penetrans* interaction was associated with earlier and a higher incidence of PED in cultivars Kennebee (S) and Russette (R). In 2 of 3 years, *V. dahliae* alone affected yield in cvs Kennebee and Superior (also S) but not the resistant cultivars Reddale and Russette. Regression equations showed that the product of V_d and P_p density terms can describe yield losses in R cultivars. With S cultivars, a quadratic function with V_d and either $V_d \times P_p$ or P_p alone was the best descriptor of yield loss. In a resistance trial in northern Maine, where *V. dahliae* and *V. albo-atrum* are found in conjunction, Goth and Haynes (2000) confirmed the resistance of cultivars Reddale and Russette and also cvs Abnaki and B0169-56. Infection in susceptible cultivars significantly reduced the number and weight of US No1 and B-sized tubers. The study was used to enhance the *Verticillium* resistance breeding programme. A study of 1330 geneotypes from 30 potato families found that, in general, there was no consistent relationship between the severity of *V. albo-atrum* or *V. dahliae* and incidence of pink-eye in segregating families (Goth *et al.*, 1994a). Resistance to *V. albo-atrum* was claimed in the New Zealand cv. Iwa (Butcher, 1975) and in cv. Waitangy, a cultivar grown from seed (Anon, 1981). Resistance claimed for *V. albo-atrum* in

Uttar Pradesh (Achal-Shah *et al.*, 1985) must refer to *V. dahliae*. A list of *V. dahliae*-tolerant cultivars currently in use in Israel was given by Nachmias *et al.* (1997b). The role of *V. dahliae* in the UK potato crop is unknown (Monnington, 1996), but UK isolates of *V. dahliae* from potato and linseed all induced wilt symptoms in 2-week-old potato plantlets under controlled conditions.

Sebastiani *et al.* (1994) found resistance to *V. dahliae* in somaclonal variation from stem explant callus cultures from cv. Desiree. Regenerated plants from one clone showed resistance comparable with resistant cv. Kondor. Chromosome manipulation techniques inducing structural or ploidy changes are described by Frusciante *et al.* (1996) for the transfer of *V. dahliae* resistance from 2*n* wild *Solanum* into 4*n* cultivated potatoes via haploids and meiotic mutants.

Tomato

Brief surveys of *Verticillium* wilt in tomato have been presented by Goth and Webb (1981) and by Beckman (1987). Massee (1896) gave the first description of *V. albo-atrum* in tomato, but erroneously called it *Fusarium oxysporum* f.sp. *lycopersici*. Jagger and Stewart (1918) may be credited with the first accurate account of a *Verticillium* sp. (*V. dahliae*) as a pathogen of tomato; Bewley (1922) subsequently showed that Massee's pathogen was *V. albo-atrum sensu stricto* but added confusion by describing 'microsclerotia'. Bryan (1925) confirmed *V. dahliae* in field tomatoes. The tomato was reviewed by Rick (1978).

Both species are responsible for wilt in both glasshouse and field crops, and their relative importance depends entirely on ambient summer temperature and hence on geographic location. Isolates of both species cross-infect and induce disease to varying extents in tomato, potato, pepper, okra and aubergine. Dixon and Pegg (1969) further showed that hop strains of *V. albo-atrum*, albeit under experimental conditions, induce symptoms in both conventionally resistant and susceptible tomato cultivars depending on the particular combination and inoculum level. To date, *V. albo-atrum* is not separated into races; the pathogen exists as a population of isolates of varying pathogenicity, but not segregated by host cultivars. However, *Verticillium* taxonomy is often still confused in plant breeding literature, for e.g. where *V. dahliae* used in screening trials is referred to as '*V. albo-atrum* race 1'.

Bryan (1925) reported resistance to *V. dahliae* in Ohio, USA. The first resistant cultivars were delveoped by H.L. Blood in 1932 using a small fruited wilt-resistant wild Peruvian species called 'Peru Wild' (accession 665, Californian Experimental Station) crossed with the wilt-susceptible cv. Century. Shapovalov and Lesley (1940) subsequently showed that a large fruited selection W6 had *V. dahliae* resistance controlled by a single dominant gene. Schaible *et al.* (1951) used Peru Wild as a source of resistance to race 1 of *V. dahliae*. Cannon and Waddoups (1952) produced Loran Blood and V. R.

Moscow, two very long surviving cultivars which have been used as a source of the *Ve* gene for other resistant hybrids (Denby and Wooliams, 1962). The single-gene (*Ve*) dominance resistance has been confirmed by Kosuge *et al.* (1977) in Japan, Matta *et al.* (1980) in Italy and Petrovskaya (1985) in Russia.

Sidhu and Webster (1979) in an early attempt to elucidate the genetics of resistance to *Fusarium* (I genes) and *Verticillium* (*Ve*) considered them to be located respectively on chromosomes 11 (I_1 and I_2) and 12 (Rick, 1978). Later studies by Diwan *et al.* (1999), with the advantage of molecular techniques, used three different mapping populations: a recombinant inbred line population; an F_2 population segregating for *V. dahliae* resistance; and a population of 50 introgression lines to give an unambiguous map location of *Ve*. In all three populations, the *Ve* gene was located on the short arm of chromosome 9, tightly linked to the RFLP marker GP39. The linkage was confirmed in screens for GP39 in different breeding lines with known resistance or susceptibility to race 1. A perfect match was found in each case, supporting the GP39 marker as a starting point for map-based cloning of the *Ve* gene. Kawchuk *et al.* (1994) produced a population segregating for the Ve allele from cvs Craigella (R) × Ailsa Craig (S). Analysis of parental DNA using PCR and 400 (10-nucleotide) random primers produced 1880 amplified DNA fragments. Of the four polymorphisms obtained between R and S genotypes, only one was linked to the *Ve* gene. The linkage between the marker and the *Ve* locus was not particularly close at <3.5 ± 2.7 centiMorgans (cM). The marker detected *Ve* and *ve* alleles producing amplified DNA fragments of 1350 and 1300 bp, respectively. The primer sequence was 5′CTCACATGCA3′. Another analysis of the sequence linked to *Ve* and *ve* for codominant markers by Kawchuk *et al.* (1998) revealed a homologous sequence with a central polymorphic region comprising 79 nucleotide substitutions, insertions and deletions. High-resolution linkage analysis using F_2-progeny segregation for resistance and marker-assisted selection showed that the linkage between the genetic marker and the *Ve* locus was <0.67 ± 0.49 cM.

In Wisconsin 1957 (see Nachmias *et al.*, 1987), a new pathovar (race 2) of *V. dahliae* was reported which was virulent against cultivars carrying the *Ve* gene and became a serious problem in California and North Carolina. Race 2 occurred in the Mediterranean region in 1969 (Cirulli, 1969), Greece in 1980 (Tjamos, 1980) and in Queensland, Australia, in 1977, when the *Ve*-bearing cv. Tropic became infected (O'Brien and Hutton, 1981). Paternotte and Van Kesteren (1993) described a new aggressive strain of *V. albo-atrum* in *Verticillium*-resistant cultivars of tomato in The Netherlands.

Some confusion has surrounded the nature of Ve resistance which has not been helped particularly by plant breeders using loose terminology and giving poor descriptions of the pathogens and the precise sources of resistance. Ve resistance is always associated by some slight colonization. This varies in non-isogenic cultivars and particularly with the pathogenic isolate (Dixon and Pegg, 1969), such that Blackhurst and Wood (1963) suggested that the 'resistance' of cultivars Loran Blood and Gem which originated in 'Peru Wild' should be termed tolerance. Okie and Gardner (1982a,b) screening seedlings for R_1 and R_2 inferred that the *Ve* gene was incompletely dominant, and Matta and Garibaldi (1984) described an outbreak of wilt on Ve-carrying cultivars. Resistance screening using *V. dahliae* isolates of races 1 and 2 from the USA, Europe and Australia, O'Garro and Clarkson (1988b) found that race 2 isolates from the USA were more virulent than those from Australia. This latter problem compounds the difficulties encountered with tolerance-like responses of a 'resistant' geneotype.

In addition to major gene resistance, polygenic resistance exists particularly in cultivars described as resistant prior to 1940. Bewley (1921) described the cv. Manx Marvel which Walter (1967a,b) showed to possess polygenic resistance. Major screenings for *Verticillium* resistance have been conducted by Govorova (1979, 1989) at the Vavilov Institute Leningrad using 2000 sources from 35 countries. Pugacheva (1981, 1982), Vlasova *et al.* (1979) and Vlaslova and Pugacheva (1984) found resistance in *L. peruvianum*, *L. peruvianum* var. *dentatum*, *L. chilensei* and *L. pinpinellifolium*; de Badilla (1974) in Chile, including polygenic resistance; and Tjamos (1981) in Greece. Govorova *et al.* (1987) derived mutant lines of *L. pimpinellifolium* by γ-irradiation which were resistant to *V. albo-atrum* [*sic*] *V. dahliae*? Beye and Lafay (1988b) discussed selection pressure on the *V. dahliae* population leading to pathovars of increased virulence. *L. peruvianum* var. *humifusum* is a rich source of resistance genes, albeit from a reproductively isolated species. Vulkova and Sotirova (1993) overcame this incompatibility by using *L. chilense* as an intermediary before deriving a triple hybrid with *L. esculentum*. After a series of backcrosses of the three-genome hybrid, several lines were obtained with resistance *inter alia* to *V. dahliae* and *Meloidogyne incognita*.

Tomatoes grow as intermediate types (non-determinative), usually (but not exclusively) under protective structures, while determinate types (bush) are grown as field crops. The fruit are produced variously for the fresh market or for processing as canned fruit, paste, ketchup, concentrates and juice. With these uses in mind and the necessity for mechanical harvesting for field crops, plant breeders look at many plant attributes in the development of new cultivars. *Verticillium* resistance is usually included with that for *Fusarium* and also other diseases and nematodes. Most wilt resistance breeding has been carried out in the USA and Canada as part of a continuous programme of new cultivars, with resistance *inter alia* to *V. dahliae* race 1.

Dobinson *et al.* (1996) in Essex County, Ontario, found that *V. dahliae* race 2 was establishing as the dominant race. An average of 31–41% of field tomatoes were infected with *V. dahliae*, and of 126 isolates tested 122 were race 2. Gold and Robb (1995a) described partial resistance to race 2 in cv. IRAT. While sporulation in this host is greatly reduced, the authors suggest an alternative resistant mechanism, operative to some degree against R_1 and R_2. A comparison of R_2 isolates from the USA (North Carolina), Brazil and Spain showed no difference in virulence. Cultivars IRAT L3, Morden Lac, Okitsu Sozai and UC82 showed a tolerance to R_2 quite different from the single *Ve* gene resistance to R_1 isolates. At present, what appears to be polygenic partial resistance (tolerance) is the best defence against R_2. One of the few published accounts of attempts to introduce transgenic resistance to *Verticillium* is by Tabaeizadeh *et al.* (1999). An acidic endochitinase gene (*pcht 28*) isolated from *L. chilense* was introduced into *L. esculentum* through *Agrobacterium*-mediated transformation using the CaMV 35S promoter, inducing a high level of constitutive chitinase enzyme activity. Kanamycin-resistant R_1 and R_2 plants resulting from self-pollinated transformants were tested against races 1 and 2 of *V. dahliae*. In each case, the transformed plants exhibited a higher level of tolerance than the appropriate untransformed controls. The mechanism by which endochitinase affects resistance is not known, unless evidence for enhanced hyphal lysis is forthcoming. The enzyme in question must be non-specific since there is no record of a chitin substrate in higher plants.

Some of the important or more recent breeding work against *V. dahliae* R_1 or *V. albo-atrum* includes: in the USA, Crill *et al.* (1971, 1977), Jones and Crill (1973), Tigchelaar *et al.* (1974), Stoner (1977), Volin *et al.* (1977), Berry and Gould (1979, 1981, 1986), Jones *et al.* (1980), Augustine *et al.* (1981a,b,c), Gardner (1982, 1993a,b), Lambeth (1982), McFerran *et al.* (1982), Jones and Millet (1984), Baggett and Kean (1986), Scott *et al.* (1989), Ashcroft *et al.* (1993), Baergen *et al.* (1993), Berry *et al.* (1993, 1995); Prashar and Enevoldsen (1993), Nguyen (1994), Baggett *et al.* (1995), Scott *et al.* (1995) and Summers (1996a,b,c); in Canada, Kerr and Cook (1979, 1982, 1983a,b) and Metcalf *et al.* (1984, 1986); in Italy, Casarini, (1970), Casarini and Antonellini (1971), Matta and Garibaldi (1974), Cirulli *et al.* (1985) and Ferrari *et al.* (1990); in Bulgaria, Iordanov and Yordanov (1978, 1979); in Australia, Sumeghy (1979) and Herrington and Saranah (1985); in Morocco, Besri *et al.* (1976); and in Switzerland, Granges (1981).

A review of tomato breeding for low temperature fruit setting and disease resistance is presented by Cuartero (1994).

Aubergine (*Solanum melongena*)

V. dahliae causes a serious disease of aubergine (*Solanum melongena* L.) resulting in worldwide dramatic crop losses. The pathogen has been referred to erroneously as *V. albo-atrum* by a number of authorities (e.g. Goth and Webb, 1981) long after the individual identity of each species was universally recognized. The virtual absence of resistance in aubergine has led to it being used as the universal suscept in isolate and pathovar studies. Perhaps because of this, relatively little work has been carried out on resistance breeding (Goth and Webb, 1981).

Macit and Saydam (1970) in Turkey, found that two of 30 cultivars tested had partial resistance. A screen of cultivars from different parts of the world by

Khar'kova (1975) using a Russian isolate found none to be resistant. A different result was reported by Dikii and Neklyodova (1975, 1977, 1979) in Krasnodar (CIS) who found in a screen of 615 cultivars from 64 countries, 73 cultivars with moderate resistance and two with high resistance. All Russian cultivars were susceptible. Similar results were published by Nothman and Ben-Yephet (1979) and Bletsos *et al.* (1998), where no resistance but different degrees of tolerance were found. Nipoti (1982a,b) found good tolerance in one line. O'Brien (1983) in the USA evaluated 473 plant introductions and ten cultivars by grading resistance on a ten-point scale. Thirty-six introductions graded 5+ were regarded as tolerant. Likewise, Sukhanberinda *et al.* (1986) in Russia found three of 164 cultivars with partial tolerance. Since host cultivars rather than pathogen isolates were the subjects of tests, it is not possible to make valid comparisons or consider results in absolute terms. Sivaprakasam and Rajagopalan (1974a,b, 1975), using a Madras isolate of *V. dahliae*, found all cultivars tested to be susceptible. In China, Xiao and Lin (1995) and Lin and Xiao (1995) report a range of reactions from moderately resistant to tolerant in wild *Solanum* spp. and *S. melongena* cultivars. A total of 1031 accessions of *Solanum* including 983 landraces and local cultivars, 26 foreign accessions and 15 wild or semi-wild species were root-dipped in a 1×10^7 ml^{-1} suspension of propagules of a single (Beijing) isolate of *V. dahliae*. A Beijing landrace Jiuyeqie (S) and an improved commercial cv. Qiqie (R) were used as comparative controls. No accession was completely resistant; on a disease scale of 0 = symptomless, 5 = death, the most resistant accessions and disease scores were: *S. aethiopicum* II6B0301, 2.1; *S. melongena* II6B0506, 2.4; *S. sisymbrifolium* II6B0980, 2.7; and *S. coaguulans* IIB0345, 3.0. Thirty-three accessions were tolerant, most of which were long-fruited (see Jarl *et al.*, 1999). Bletsos *et al.* (1998) found *S. sisymbrifolium* to be more susceptible than the two commercial cultivars Tsakoniki and Langada.

From pioneering attempts at resistance (tolerance) breeding in Puerto Rico in 1944 (Goth and Webb, 1981), most progress has been made in Italy by Cirulli, Ciccarese and co-workers. Restaino (1986) working in Sicilly derived three new cultivars with field tolerance by treating seed with the mutagen ethyl methanesulphonate at 0.8%; cultivars Picentia and Floralba were derived from the susceptible cultivars Lunga Violetta and Florida Market, respectively. Cultivar Macla was a mutant selection from Florida Market crossed with an Asian cv. Nagaoka. In 1981, Cirulli and Ciccarese (1984) and Cirulli *et al.* (1990) screened 116 accessions under field conditions and found no major gene resistance. Some accessions showed a reduced rate of disease development by comparison with a standard cultivar used in commercial culture – Florida Market – which were used in subsequent breeding experiments. Four lines were selected designated R_1, R_2, R_3 and R_4, all characterized by a low wilt rate. Over 2 years, losses in two susceptible cultivars were 54–57% in cv. BA-12 and 62–85% in cv. Florida-Market; of the R lines and their self-pollinated progeny, losses of 12–42% occurred in R_3 and R_4, with no losses in R_1 and R_2. The results of diallele crosses with the susceptible Florida Market with F_1 selfed hybrids backcrossed on to R and S parents, showed that partially-dominant additive genes were involved as well as non-allelic epistatic effects which limited the diallele analysis (Ciccarese *et al.*, 1994). Cirulli *et al.* (1997) concluded that the predominance of additive gene effects offered progress in further aubergine selections. de Melo and da Costa (1985) also found promising resistance from backcrossed hybrids (see Yi and Chen (1998) for a review of *Verticillium*-resistance breeding in China).

The reaction of mixoploid plants regenerated from leaf tissue cultures and screened against *V. dahliae* culture filtrates was described by Alicchio *et al.* (1984). Rotino *et al.* (1987) similarly tested 23 resistant cell lines by exposing leaf-disc callus embryoids to *V. dahliae* culture filtrates. No information is forthcoming from either of these studies on the fate of mature plants derived from filtrate screening to conventional infection. Little information is available on pathogen isolates. Porta-Puglia and Montorsi (1982) reported a range of virulence in isolates obtained from aubergine seed. Cristinzio and Scala (1994) proposed electrolyte leakage as a measure of aubergine resistance to *V. dahliae*. *Solanum torvum* was shown by Jarl *et al.* (1999) to exhibit tolerance to *V. dahliae*. This was transferred to *S. melongena* by protoplast fusion. Putative hybrids were regenerated and screened against *V. dahliae* and *V. albo-atrum*. Twelve plants with a high degree of field *Verticillium* tolerance and aubergine characteristics and normal seed set were selected for a breeding programme. One plant showed changed genomic DNA, with some changes corresponding to specific characteristics of *S. torvum*.

Pepper (*Capsicum annuum*)

The central organisms responsible for sweet pepper (*C. annuum* var. *grossum*) and chilli pepper (*C. annuum* var. *longum*) wilt are listed as *V. albo-atrum* and *V. dahliae*, although the limited literature on disease resistance breeding rarely includes a description of the pathogen. *C. pendulum*, *C. frutescens* and *C. chinense* have been reported as sources of horizontal resistance when hybridized with *C. annuum* (Saccardo and Sree Ramulu, 1977).

No vertical resistance has been reported. Several, mostly Eastern European, accounts of resistant hybrids refer to tolerance rather than resistance *per se* (Aleksic *et al.*, 1975; Milkova *et al.*, 1988, 1989 for *V. albo-atrum*, and Suta and Stoenescu, 1986; Gil-Ortega *et al.*, 1986 for *V. dahliae*). Aleksic *et al.* (1976), Sutic *et al.* (1981) and Marinkovic *et al.* (1984) in the former Yugoslavia found a stronger resistance in hybrids derived from Colombian lines. The authors reported virulence in terms of host stunting, with seven of 10 pathogenic isolates being extremely virulent. More recent *V. dahliae*-resistant sweet pepper releases showing various degrees of tolerance are described from Bulgaria by Todorov *et al.* (1995) and from Romania by Suta *et al.* (1996) and Tanasescu *et al.* (1996). Similar resistance was reported in chilli pepper (*C. annuum* var. *longum*) accession PBC408 by Riley and Bosland (1995). No pepper cultivars with *Verticillium* resistance were found in China (Lu *et al.*, 1984).

Large-scale screenings for resistance were carried out by Iglesias-Olivas *et al.* (1987) of 325 accessions of *C. annuum*, *C. baccatum*, *C. chaconense*, *C. chinense*, *C. frutescens* and *C. pebescens*; but only partial resistance was found in one New Mexico line. Porcelli *et al.* (1987) similarly tested 480 lines at seedling and adult stages, concluding that seedling resistance was controlled by one or a few genes while adult resistance was polygenic. Kharkova (1990) described a successful *V. dahliae* resistance breeding programme in Moldavia (CIS) since 1970, using a standard hybrid backcrossing technique. The polygenic resistance was only preserved in F_3 and F_4 crosses when both parents were resistant. Germplasm from 24 of 264 local populations of *C. annuum* (sweet pepper) collected throughout Bulgaria was identified by Todorova *et al.* (1997) as 'resistant'. Gonzalez-Salan and Bosland (1993) raised a 68% *Verticillium*-resistant population of *C. annuum* by four cycles of selfing of PI215699; segregation analysis suggested that resistance was controlled by additive and epistatic genes. Grafting susceptible cultivars on to resistant root stocks provides an alternative control method, albeit more costly. Nervo *et al.* (1999) successfully protected four commercial F_1 hybrids from *V. dahliae* by grafting onto rootstock line P_5, but one cv. P5358 graft combination showed reduced yield irrespective of the presence of the pathogen.

Tobacco (*Nicotiana tabaccum* L.)

Wilt caused exclusively by *V. dahliae* is a major problem in New Zealand, now limiting tobacco production following its first observation in 1944 (Thompson and McLeod, 1959). The only other country where tobacco wilt constitutes a serious threat is in the province of Quebec, Canada. Wright and Sackston (1973) showed that while mixed pathogen isolates from tomato only induced 0.83% infection in tobacco, reisolates from infected plants gave 100% infection in field tests the following year, highlighting the possible danger in crop rotation of residual *V. dahliae* inoculum from former potato crops. Occasional reports have come from Chile (Latorre *et al.*, 1989), the USA and Africa (Wilhelm, 1981). Wright (1968) found no true immunity *per se*, but described host reactions in New Zealand cultivars from extreme susceptibility to high polygenic resistance. During rapid growth, cultivars such as McNair 121 (Wright, 1968) resisted symptom expression even with extensive vascular colonization (tolerance); others such as T.I. 448 A (a Columbian cultivar) and Florida 301 showed high resistance associated with an exclusion of the pathogen from the leaves (Wright, 1969). Hartill and Campbell (1976) also found different levels of tolerance to the disease, with cv. Waimea showing marked resistance (Wright, 1973). Beatson (1973), however, considered cv. Kuaka 427 a more highly resistant plant. Beatson *et al.* (1990) using standard pedigree breeding techniques released cv. Moskuaka 85 as an F_{13} line with resistance to *V. dahliae*, tobacco mosaic virus and *Thielaviopsis basicola*. Latorre *et al.* (1987) confirmed outbreaks of tobacco wilt in Chile in 1985 and 1986, but like Wright and Sackston (1973) predicted an increased incidence from a build up of microsclerotial inoculum in other crop residues.

The work of Murray *et al.* (1999) on genetic constructs is a salutory lesson against the concept of genetic engineering as a universal and instant solution to the problem of disease resistance. The glucose oxidase (GO) gene from *T. flavus* was modified with a carrot extensin signal peptide and fused to either a constitutive or root-specific plant promotor. T_1 tobacco plants expressing GO constitutively protected against *Rhizoctonia solani*. Selected T_3 homozygous lines showed some protection against *V. dahliae*, but not against *F. oxysporum*. However, high levels of GO expressed in cotton roots were associated with reduced plant height, seed set, germination rate and the formation of lateral roots. Clearly, even when a gene for, or associated with, resistance can be identified, much work is necessary to control its expression, in order to eliminate deleterious phenotypes.

Lucerne (*Medicago sativa*)

Lucerne is an important deep-rooted forage legume in largely temperate regions of the world, Europe, North America and Asia, but it is also grown in South America, South Africa and Oceania – mostly Australasia. Wild types of *M. sativa* and *M. falcata* Lam. are indigenous to Iran, Anatolia and Trans Caucasia, (a Vavilov centre of plant diversity). From these native areas, the crop was distributed around the ancient world associated with horse breeding – Greece (500 BC), China (200 BC), Spain (AD 1500) – and subsequently as cattle food to the New World, Peru, Mexico, the USA and Canada. The crop preceded by many centuries the development of *Verticillium* wilt (Pegg, 1984). The principal pathogen is *V. albo-atrum*, but *V. dahliae* has also been reported as a minor pathogen and remains a potential threat where lucerne is grown in hotter climates; although the lucerne strain of *V. albo-atrum* has a higher optimal temperature (25°C) than other *V. albo-atrum* strains. The disease has been reviewed by Kreitlow (1962), Wilhelm (1981) and Pegg (1984).

V. albo-atrum was first recorded in Sweden in 1918 (Hedlund, 1923) and subsequently in Germany (Richter and Klinkowski, 1938), the UK (Noble *et al.*, 1953; Isaac, 1957c), Europe and the former USSR (Kreitlow, 1962), Canada (Aubé and Sackston, 1964), New Zealand by 1965 (Crooks, 1975; Hawthorne, 1987) Washington State, USA (Graham *et al.*, 1977) and South California (Erwin *et al.*, 1988; Howell and Erwin, 1990). In the Pacific North West USA, the disease is well established on high-rainfall coastal land and in California on irrigated desert land; in 1981, involving a total of 202,000 ha (Christen and

Peaden, 1981). Although a drought-resistant crop, lucerne does not occur on non-irrigated dryland (Pegg, 1984). Aubé and Sackston (1964) described a *Verticillium* wilt in *M. sativa* and other forage legumes in Canada including *Trifolium repens* L., *T. pratense* L., *Lotus corniculatus* L., *Onobrychis viciifolia* Scop., *Lupinus alba* L. and *Melilotus alba* Desr. In British Columbia (Sedun, 1980) the cost–benefit of growing high, wilt-resistant cultivars over moderate, wilt-resistant cultivars in western Canada was calculated at \$21 ha^{-1}, and over low resistant cultivars at \$44 ha^{-1} (Smith *et al.*, 1995). At 1995 values, this equated to a regional western Canadian benefit of \$2.2 million. The comparable benefit from growing regionally adapted cultivars was \$26.6 million per annum (see also Arny and Grau, 1985; Viands *et al.*, 1988).

Although Aubé and Sackston (1964) regarded lucerne pathovars as broadly host specific, Müller (1969) showed that seven of 26 weed species growing in association with lucerne were susceptible to the lucerne strain. Busch and Smith (1982) also claim to have infected a number of leguminous and non-leguminous plants with an Ontario lucerne strain. While most species exhibited mild or no symptoms, the pathogen could be re-isolated from 18 of 21 species tested. Christen and French (1982) obtained symptoms on aubergine, cantaloupe and watermelon, but not potato or tomato, using eight lucerne isolates. Notwithstanding this result, the authors maintain that isolates from lucerne are limited to this host. Isaac (1949) and Heale and Isaac (1963) claimed that the UK lucerne strain was mildly pathogenic to potato, tomato, other crops and weeds. Lucerne is a polyploid – an autotetraploid $2n = 4x = 32$. Zaleski (1957), Isaac and Lloyd (1959) and Roberts and Large (1963) found that all the cultivars grown in Britain and France in the late 1950s including cultivars Dupuits and Grimm were susceptible. Panton (1965, 1966, 1967a,b,c) in Sweden found highly resistant offspring in Dupuits × Grimm hybrids indicative of transgressive segregation associated with polypoids. Lucerne hybridizes with other species of *Medicago*, a feature examined by Carr (1960) who found a great variation in species resistance, and in a *M. sativa* × *M. falcata* hybrid. The resistant cv. Maris Kabul was derived originally from a *M. sativa* × *M. hemicycla* cross (Head and Finch, 1975; Rogers, 1976). Other resistant cultivars which served subsequently as sources of resistant germplasm were Vertus (Lundin and Jonsson, 1975; Viands, 1985) and Sabilt (Carr *et al.*, 1972; Davies and Young, 1975).

The outstanding contribution to *Verticillium* resistance breeding was carried out by Panton (1964, 1965, 1966, 1967a,b,c) from first studies in 1962 in Sweden. Diallel crosses of resistant plants and susceptible clones showed that the basis of resistance was polygenic, with additive and possibly multiplicative gene effects. The highest resistance was from R × R hybrids. Steuckardt and Hempel (1976), Hawthorne and Triggs (1986) and Dixon *et al.* (1989) all confirmed the polygenic nature of *Verticillium* resistance. Latunde-Dada and Lucas (1983) found that protoclones from mesophyll proplasts of cv. Europe were highly tolerant to *V. albo-atrum*, which was attributed to their higher ploidy.

Various screenings of lucerne have highlighted the most resistant cultivars, some of which have remained durable for many years, e.g. Germany, cv. Luna (Scheller, 1971); Czechoslovakia, cv. Hybride de Crecy (Kudela and Rezac, 1972), cv. Palava (Gubis, 1979) and cv. Vertibenda (Gubis *et al.*, 1990); New Zealand, cultivars Saranac and Iroquois (Dunbier *et al.*, 1976); Hungary, cultivars Furez and Vertibenda (Malatin and Manninger, 1980); and Switzerland, cultivars Europe, Euver, Everest, Franken Ness, Resis, Vela and Vertus (Joggi *et al.*, 1982a,b) and cv. Vetus (Charles *et al.*, 1989). Schiefer *et al.* (1987) described glasshouse trials in Germany, as did Tomaszewski *et al.* (1983) with 49 Polish and other cultivars. Kudela (1974) shows the comparison between disease ratings for natural and artificial infection. Five test cultivars and one locally adapted cultivar were grown on each of 12 sites in eight European countries as part of a cooperative study on lucerne wilt. The prominent resistant cultivar was Vertibenda (Bocsa and Raynal, 1984). The results of a long-term Slovakian study of the resistance of 21 progenies of selected species of lucerne to various root pathogens, *inter alia*, *V. albo-atrum* were presented by Mistinova (1999).

While breeding programmes in Europe concentrated co-equally on *Verticillium* resistance and agronomic qualities, North American lucerne breeding, following the introduction of wilt in 1976, involved multiple disease and pest resistance, including *Clavibacter michiganense* sp. *insidiosus*, *Phytophthora megasperma* f.sp. *medicaginis*, *Colletotrichum trifolii*, *Fusarium oxysporum* f.sp. *medicaginis*, *Ditylenchus dipsaci*, *Therioaphis maculata* and *Acyrthosiphon pisum*, in addition to agronomic characters. Such cultivars may represent a synthesis of 400 (Stratton, 1994) or more clones with varying degrees of germplasm from *M. falcata*, *M. varia* and Flemish, Ladak, Peruvian, Chilean, Turkistan and Persian sources (Fox and Knipe, 1987; Hanna and Huang, 1987; Woodward *et al.*, 1987, 1990; Viands *et al.*, 1990). Typical sources of multiple resistance in a synthetic cultivar would be: *M. falcata* (8%), Kadak (12%), *M. varia* (22%), Turkestan (5%), Flemish (47%) and Chilean (6%) (Kugler *et al.*, 1995). As mentioned above, *V. albo-atrum* resistance is but one part of a multiple defence strategy; see, for example, Peaden *et al.* (1985, 1995) where *Verticillium* resistance (tolerance) is a mere 16% of the total. Moreover, new lucerne releases in the USA and Canada where the main breeding programmes are in progress are designed specifically for regions of these countries, e.g. cv. Victory in northern USA (Viands *et al.*, 1995a); cv. Multistar, a 146 clone synthetic in upper midwest to mid-south production areas (Stratton, 1997); and cv. '5715' a 102 plant synthesis designed for Argentina, Chile, Spain, Australia and south-western USA (Woodward *et al.*, 1993).

Much of the resistance (tolerance) bred into contemporary north Amerian cultivars, e.g. AC Blue Jay (Acharya *et al.*, 1995), is derived from the old-established, European resistant cultivars Vertus, Sverre, Maris Kabul and Lutece, with yield increases based on other genetic characters. Losses from *V. albo-atrum* and *Phytophthora megasperma* in the USA were estimated at 3.3 t ha^{-1} (Gray and Page, 1988), while the use of a *Verticillium*-resistant cultivar e.g. cv. Oneida, yielded 3.3 t ha^{-1} more than susceptible cultivars over three harvests. In comparison with the susceptible cv. Saranac-AR, cv. Oneida produced 4.5 t ha^{-1} more in the third year (Murphy and Lowe, 1989). New resistant cultivars in the USA with combined resistance characters were reported by Beard *et al.* (1982), Hanson *et al.* (1987), Murphy and Lowe (1989), Murphy *et al.* (1989) and Sorensen *et al.* (1989). Julen (1980) developed an intensive selection method within several cultivars through to F_3, combining lines and reselecting; the highly resistant cv. Sverre was developed by this method. Chloupek (1996) and Chloupek *et al.* (1996) combined *Verticillium* resistance with selections for higher nitrogenase activity. The progeny of a diallele cross of four germplasms selected for *V. albo-atrum* resistance and producing symptomless inoculated plants, were analysed together with the parents for *Verticillium* resistance. Cultivar symptom expression interacted with inoculation age but not assessment age. Resistance was conferred by additive gene action. No heterosis was detected and the general combining ability mean square was ×3.75 larger than the specific combining ability mean square (Martin *et al.*, 1993).

In British Columbia, yield is reduced by poor disease resistance, winter hardiness or both, leading to the strategy of increasing persistence by sowing mixtures of both types. In the absence of a severe winter, wilt-resistant cv. Beaver yielded 2 t ha^{-1} more than the winter hardy cv. Beaver (Stout *et al.*, 1998). Erwin and Khan (1993), Viands *et al.* (1995b), Bouton *et al.* (1997) and Yamaguchi *et al.* (1995) described new lucerne releases with *Verticillium* resistance in the USA and Japan, respectively. Typical of modern cultivars registered, Cash *et al.* (1998a,b) developed cv. Parade derived from 634 parental plants selected for vigour and resistance to pests and pathogens; similarly, cv. Tahoe, derived from 782 parents. These references represent a small part of the literature relating to new cultivars mostly from the USA.

Flood *et al.* (1978a) considered that US isolates were more virulent than those from Europe. This was refuted by Busch and Smith (1981) who found in reciprocal experiments with US and European isolates and four resistant European cultivars in America and Britain, that a wide range of reaction from very susceptible to highly resistant, was unrelated to the geographic area of the pathogen isolate. Christen *et al.* (1983) in fact concluded that the North American epidemic resulted from the import of a European pathovar. Joint experiments between the UK, France, Canada and America, testing North American and European isolates on European cultivars in Europe and North America have shown, with minor exceptions, that the pattern of resistance and susceptibility remains the same regardless of the origin of the *Verticillium* isolate (Pegg, 1984). Hawthorne (1983) found much variation in pathogenicity from isolates from different parts of New Zealand. The proposal that both resistance and pathogenicity are under polygenic control and are part of a gene for gene relationship (Hawthorne and Triggs, 1986) is not supported by hard evidence. Gagne and Richard (1982) in a list of principal cultivated and weed hosts of *V. albo-atrum* concurred with Wilhelm (1981) that the lucerne strain is highly specific, this with its high temperature optimum make it a probable candidate for a new specific ranking. Christie *et al.* (1985) reported on studies on both European and North American cultivars. Pennypacker *et al.* (1985) highlighted a little recognized flaw in the argument for breeding tolerance, as distinct from pathogen exclusion, by identifying tolerant, infected hosts as *symptomless carriers* with the potential risk of a build up of inoculum in the soil. A glasshouse assessment method for resistance was described by Hawthorne (1987). Pi-Azek (1996a) described a comparison of laboratory and greenhouse methods for the evaluation of *Verticillium* resistance.

An American *V. albo-atrum* resistance trial by Huang *et al.* (1999) on 25 Chinese lucerne cultivars and one *M. falcata* cultivar in comparison with North American and European cultivars showed that only one, cv. Qi, exhibited any resistance which was comparable with that of the Swedish cv. Vetus but much lower than that of the Canadian cv. Barrier. Since there is no recorded *Verticillium* wilt in lucerne in China, this study highlights the necessity for rigid quarantine restriction in China and Taiwan for an absolute exclusion of infested seed, or hay, or potentially infected breeding plants. Elite seed stocks should be grown under strict observation in isolated quarantine areas.

Strawberry (*Fragaria ananassa* Duch.)

The origins and cultivated history of the strawberry are complex. Varieties in cultivation today are derived from the octoploid 'Pinapple' or Pine ($2n = 8x = 56$) originating in 1750 (hence the specific epithet *ananassa*). 'Pine' was derived from the North American *F. virginiana* and the Californian beach strawberry *F. chiloensis* which were introduced into Britain by Elizabethan explorers and subsequently by botanists, where hybrids were made. North American cultivars were all derived from hybrids including 'Pine' germplasm introduced from Europe. The history of the strawberry is reviewed superbly by Wilhelm and Sagen (1974); strawberries may be short-term fruiting (syn June-bearing) cultivars found in Europe and the UK and everbearing cultivars with genes for extended fruiting from *F. chiloensis*, the mountain strawberry *Fragaria vesca* and other species, e.g. *F. ovalis* (Draper *et al.*, 1981). Everbearing types dominate the North American industry, especially California where the climate is congenial to extended fruiting.

Strawberry wilt is caused by both *V. albo-atrum* and *V. dahliae* (Talboys *et al.*, 1961, 1977), but the latter represents the major pathogen even in temperate regions. Little information exists on strains of *V. dahliae*, but at least two are thought to exist. The strawberry strains also infect potato, cotton, tomato and various weeds.

Hitherto, the success of soil fumigation in the USA as a control measure has given little impetus to disease resistance breeding. Most resistance breeding has been done in Europe and Canada where soils are heavier and less conducive to the diffusion of fumigants. Public resistance to the use of bromine compounds may enhance work on plant breeding, but cultivars in the USA would need to reach the present very high yields achieved by fumigation (Pegg, 1984). Resistance and tolerance to wilt are both expressed, with the latter in most of the UK cultivars. Newton and Van Adrichem (1958) derived wilt-resistant and/or tolerant seedlings from susceptible plants of *F. yukonensis*, *F. ovalis*, *F. ovientalis*, and Verney *et al.* (1959) from a susceptible clone of *F. chiloensis*. Wilhelm (1955b) found evidence for resistance to be dominant and quantitatively inherited, although dominance was not always complete. This has been confirmed by later studies. Wilhelm (1955b) also found transgressive segregation – a feature common to octoploids. This was confirmed by Shaw *et al.* (1997a) who found some susceptible progeny from resistant parents while five of 21 resistant selections showed greater resistance than parental types. Shaw *et al.* (1996) advocated a continuous selection of transgressive segregants from superior parents. D'Ercole *et al.* (1980) in Italy reported 94 of 154 *Fragaria* lines inoculated with *V. dahliae* to be resistant (used in a generic sense to include tolerance), of which 38 plants remained symptomless. Faedi (1983) conducted a similar screen for resistant cultivars suitable for the Trentino and Piedmont mountainous regions. Popova *et al.* (1975), Andreeva (1975) and Govorova (1983, 1985) identified sources of resistant germplasm suitable for Russian isolates and conditions.

The following *Verticillium*-resistant cultivars have been bred and released in the countries indicated:

1. Canada.
Everbearing, cv. Centennial, third-generation inbred of Century × Gem seedling (Evans, 1975); June-bearing, cv. Micmac (Craig *et al.*, 1978); cv. Vantage with high resistance (Ricketson *et al.*, 1986) and cv. Secord (Dale *et al.*, 1986).
2. USA.
Everbearing, cultivars Tribute, *V. dahliae* tolerant and cv. Tristar resistant. The everbearing trait was derived from a wild clone of *F. ovalis* (Draper *et al.*, 1981). June-bearing, cv. Allstar with field resistance (Galletta *et al.*, 1981), cv. Scott, tolerant (Galletta *et al.*, 1980) and cv. Scarlet, wilt resistant (Moulton and Anderson, 1978). Work in the University of California strawberry-breeding programme by Shaw *et al.* (1996, 1997b) described geneotypes as relatively susceptible, intermediate and wilt resistant. The relative resistance categories were distinguished at inoculum levels of 10^6 and 10^5 conidia ml^{-1}.
3. Europe.
European cultivars, all short season fruiting, include: UK – cv. Pandora = Humbolt × Redstar × Merton, tolerant (Simpson and Blanke, 1989) and Pegasus = Redgauntlet × Gorella – moderate resistance (Simpson and Blanke, 1991), 'high level of field resistance' (Simpson *et al.*, 1994). Cultivars Calypso and Tango = (Rapella × Selva) × Rapella with agronomic advantages but only moderate resistance to *V. dahliae* (Simpson and Blanke, 1994); Italy – for Po valley conditions, cv. Dana (Rosati *et al.*, 1982) and cv. Etna (Faedi *et al.*, 1986). Some stunting was encountered in all levels of resistance; France – three new cultivars, Cirafine, Cijosee and Cirano combining various improvement in yield,

quality and disease resistance including *Verticillium* (presumably *V. dahliae*) were described by Markocic *et al.* (2000).

4. Other countries.

Russia – cv. Purpurovaya (Shakova, 1974); Brazil – cv. Sierra (Bastos-Cruz *et al.*, 1972). Govorova and Govorov (1997a,b) described the development of 30 hybrid cultivars including, for example, the super early, day-neutral cv. Vechnaya Vesna combining *high* resistance to *V. dahliae* and 3–6 other diseases together with drought resistance and other desirable agronomic characters.

Possible limitations in the use of tissue-cultivated stock were highlighted by Shoemaker *et al.* (1985) who found that such plants exhibited greater susceptibility to *V. dahliae* than runner-propagated plants (see also Basu, 1961).

Hop (*Humulus lupulus* L.)

The hop plant is grown as a plantation (hop garden) perennial above 32° latitude in the northern hemisphere, especially in the UK, Northern Europe and North West America, but also in Australasia, South Africa and South America.

The principal pathogen is *V. albo-atrum*, with *V. dahliae* as an occasional and minor problem. The first account of a mild wilt in cv. Fuggle, in Kent, UK, was by Harris (1925b); thereafter, with the occurrence of new pathovars, the disease spread through Kent and East Sussex, reaching epidemic proportions and peaking in the 1960s. In subsequent years, infection was contained, largely by: (i) the introduction of resistant cultivars; (ii) a national certification scheme for disease-free clonally propagated hop plants; (iii) legislative quarantine requirements and restricted movement of propagation material; (iv) fallowing of infested gardens; (v) greatly improved field hygiene; and (vi) controlled nutrition. The programme adopted for the successful containment of hop wilt in the UK could serve as a model for other wilt disease problems in the world (see Pegg, 1984).

Four strains of hop *V. albo-atrum* have been recognized and designated M = mild, causing fluctuating wilt, and V_1, V_2 and V_3 = virulent or progressive strains. The original outbreak (Harris, 1925a) caused a mild disease with fluctuating wilt and recovery on cultivars Fuggle and Whitbread Golding. The development of the 1930's epidemic was due to a new strain which induced rapid and severe terminal symptoms on all cultivars then in existence. This was designated the progressive strain (V_1) (Keyworth, 1942; Isaac and Keyworth, 1948); subsequently (Sewell and Wilson, 1975, 1976, 1978, 1979) two further progressive (super-progressive) strains, V_2 and V_3 were identified.

Disease resistance breeding in the UK was carrried out at Wye College (University of London) [now Imperial College at Wye] and the former East Malling Research Station (now Horticulture Research International). Early breeding studies by E.S. Salmon were based on visual assessment and clonal selection with no knowledge of the genetics of *Humulus*. Even today, relatively little is known about the inheritance of resistance, and breeding in the UK at least is made difficult by hybridizing pollen from the wild hop.

Keyworth (1946) selected two clones (OR55 and OJ47) from a New Mexico hop, which were tolerant (root and stem colonization with mild symptoms) to the V_1 strain which caused severe symptoms on cv. Fuggle. In the 1950s, cultivars Whitbread Golding, Progress and Alliance were released, all with tolerance to the V_1 strain (Keyworth *et al.*, 1953; Pegg, 1984). Coincident with the emergence of strain V_2 causing severe wilt on all previously tolerant cultivars, Neve (1972, 1976, 1979) and Thompson and Neve (1972) introduced cv. Wye Target which was tolerant to V_2. This strain, however, was severe on cv. Whitbread Golding and on the general suscept cv. Fuggle. With the emergence of strain V_3, cv. Wye Target succumed to moderate, but debilitating wilt symptoms. The position in the UK is fluid, with no immunity to *V. albo-atrum* and with great selection pressure on a narrow range of tolerance in the cultivars under production.

German hop breeders using mainly British germplasm and with the pathogenic strains extant in German hop gardens (=M and V1) found cultivars Northern Brewer, Record and Star to be tolerant, and cultivars Hersbrucker spät, Spalter, Deutscher Fruhhopfen and Hull to be moderately susceptible (Rintelen, 1974; Scott, 1977). Subsequently, with the emergence of strain V_2 and V_3, these cultivars became vulnerable. The modern approach to hop breeding has been towards seedless hops with a high α-acid content. In the context of English hop culture, with strict confinement of infection by quarantine of land and plants and with a rigorously imposed plant health certification scheme, it has been possible to grow erstwhile hop cultivars susceptible to V_2 and V_3 in parts of the country by 'disease escape', while taking advantage of new high-yielding and high α-acid content cultivars, but with no tolerance to the virulent pathogenic strains (Pegg, 1984). Chambers and Darby (1995) report on the selection of several lines at Wye College, UK, which are more *Verticillium* resistant than the current leader cv. Wye Target. Line TC105 was undergoing farm trials.

The *Verticillium* strains in Oregon and Idaho USA are quite different from those in the UK, and the potato strain of *V. dahliae* has been reported as a hop problem (Romanko *et al.*, 1979). Curiously, English germplasms for *Verticillium* tolerance and α-acid content have been used in local breeding programmes resulting in *V. dahliae*-tolerant cultivars such as cv. Cascade (Brooks *et al.*, 1972), cv. Galena (Romanko *et al.*, 1979) and cv. Eroica (Romanko *et al.*, 1982). Cv. USDA 21055 was described by Haunold *et al.* (1978) as having good 'resistance' to *V. albo-atrum* in Oregon. Derived from seed of open-pollinated cv. Brewers Gold, cv. Banner was released in 1994 (P1558759) with tolerance to *V. dahliae* combined with a high α-acid content and multiple disease resistance for cultivation in Washington and Idaho (Romanko *et al.*, 1996). Cv. Crystal, a femal triploid ($2n = 3x = 30$) bred for the Pacific Northwest, remains seedless even in the presence of fertile ($2n = 2x$) males. Its composition is 2/3 Hallertauer mittelfruh, 1/6 Cascade, 1/12 German aroma hop, 1/24 Brewers Gold and 1/24 Early Green. During the 4 years of pre-release trials, *V. albo-atrum* symptoms were not observed. The authors, Haunold *et al.* (1995), failed to comment on whether the trial was on infested soil or whether other cultivars succumbed. Piotrowski and Milczek (1982) published biochemical indices of resistance to *V. albo-atrum*.

Chambers *et al.* (1986) described various techniques for assessing the virulence of hop pathovars but without success for the separation of isolates of progressive and fluctuating strains. Connel and Heale (1985) and Heale and Connel (1987) subjected hop mesophyll protoplasts and suspensions of callus cells with field resistance to culture filtrates. The method was considered too unreliable to replace the conventional root challenge from soil inoculum on hybrid plants. A modified technique might have merit in future, in a primary screen of cell cultures using genetically engineered cells.

Mint – peppermint (*Mentha piperita* L.) and spearmint (*M. spicata*)

Wilt in both species is caused by a fairly host-selective pathovar of *V. dahliae*, but spearmint is slightly less susceptible than peppermint. Originally Old World garden plants, they are now grown commercially, in the US mid-west and north-west states, Ukraine, Moldavia and the Balkans under cool–warm temperate conditions and long days, as valuable essential oil crops (Nelson, 1926, 1950). Spearmint is a diploid ($2n = 48$). Nelson (1950) considered peppermint an allo-hexaploid ($2n = 6x = 72$) to be a natural hybrid of *M. aquatica* ($2n = 96$) × *M. spicata* probably of English origin, where both occur as garden plants, or escapes, and where *M. peperita* was grown as an oil crop for the northern chocolate industry. Prior to the introduction of resistant cultivars, *Verticillium* wilt destroyed a large part of the US mint industry and, in Michigan from 1938 to 1970, mint oil was reduced from 50 to 1% of US production (Nelson, 1950).

No immunity to *V. dahliae* exists in the genus *Mentha*, but Nelson (1950) found that *M. arvensis* var. *canadensis*, *M. arvensis* var. *piperascens*, *M. gentilis*, *M. pulegium*, *M. rotundifolia*, *M. rubra* and *M. sylvestris* showed varying degrees of resistance. *M. crispa* had the highest resistance. A screen of 2000 hybrid seedlings from *M. piperita* cv. Mitcham (S) × *M. crispa* against *V. dahliae* ranged from extreme susceptibility to resistance higher than that of *M. crispa*. Genetic analysis revealed two dominant R genes. Murray (1969, 1971) in Michigan produced 100,000 X-ray and neutron mutants of cv. Mitcham which over 6 years of rigorous wilt screening and quality selection resulted in the release of the resistant cv. Todd's Mitcham. A subsequent clone cv. Murray Mitcham produced by the same technique was also released (Todd *et al.*, 1977). Horner and Melouk (1976) and Yarris *et al.* (1977) had similar success radiating rhizome pieces of *M. cardiaca* (Scotch spearmint) with γ-rays from a ^{60}Co source, with 15 of 15,000 resistant plants being selected for field testing and release. Also working with ^{60}Co mutants, Johnson and Cummings (2000) derived by irradiation, *M. piperita* and *Mentha* clones, which were evaluated with *M. piperita* seed-fertile clones and hybrids for *V. dahliae* resistance. One *M. piperita* mutant, three *M. piperita* conventional hybrids and four *M. piperita* fertile clones exhibited low and variable resistance, based on a wilt index correlated with c.f.u. isolations from inoculated stems. Native *Mentha* and native *Mentha* mutants, however, showed a high level of *Verticillium* resistance. The pathogenicity of *V. dahliae* isolates from 11 hosts to peppermint was described by Horner (1954). The selection pressure on *V. dahliae* from the near monoculture of mint crops resulted in increased pathogenicity of the isolates (Green, 1977). From a comparison of disease assessment using symptoms and fungal invasion of the stem (Melouk and Horner, 1976) and Wilhelm (1981) concluded that two resistance mechanisms were involved.

An attempt by Krasnyanski *et al.* (1998) to produce somatic hybrids with improved oil yields and *V. dahliae* resistance met with only partial success. Protoplast fusion was achieved between the high quality oil *M. peperita* cv. Black Mitchum and spearmint, *S. spicata* cv. Native Spearmint, also of high oil quality. Cell division was inhibited by iodoacetic acid and fusion aided with polyethylene glycol (PEG) and dimethylsulphoxide (DMSO). Iodoacetic acid-treated peppermint protoplasts were unable to divide, and cell division in the hybrid, resulting in callus and shoots, occurred as a consequence of the untreated spearmint protoplast component. Hybrids were identified by RAPD profiles, chromosome counts and Southern hybridization patterns. Unfortunately, 18 somatic hybrids were more susceptible than native hybrid spearmints. GC oil analysis of the hybrids was typical of spearmint.

Sunflower (*Helianthus annuus*)

This is a most important, worldwide grown, oil seed crop originally native to North America. It is also grown as a green fodder and silage crop. In addition to the original open-pollinated varieties, since 1969, F_1 hybrid cultivars are available through the use of cytoplasmic male-sterile, seed-producing lines discovered in progeny of the cross *H. petiolaris* × *H. annuus* and pollinator lines that restore fertility to the offspring (Wilhelm, 1981). Wilt or leaf mottle caused by *V. dahliae* was discovered by Sackston (1949) in southern Manitoba, Canada (Sackston *et al.*, 1957). These authors also found that the pathogen was seedborne (Sackston and Martens, 1959). Hoes (1972) described losses due to *V. dahliae*. It is now a serious problem in the USA, Canada and Europe where disease-resistant cultivars offer the only feasible control measure. Putt (see Wilhelm, 1981), in a study of inbred lines, suggested that resistance and susceptibility were based on stable heterozygosity, with resistance inherited dominantly. Crosses of resistant lines by Putt and by Rava (1974) showed that resistance was governed by a single dominant gene V_1. Resistance modifier genes were also postulated since the resistance of some *progeny* excluded that of either parent. Little is known about strains of *V. dahliae* pathogenic to sunflower, however, Klisiewicz (1981) found that a highly resistant line was susceptible to an isolate (T_1) from safflower. Sedun and Sackston (1982) described two other virulent isolates in Canada. A pathovar appeared in Argentina in 1983 which overcame the V_1 resistance lines in the breeding programme (Bruniard *et al.*, 1983, published 1984).

Verticillium-resistant hybrid F_1 and inbred lines have been registered for use in the USA by Fick *et al.* (1979a,b), Roath *et al.* (1981), Miller and Gulya (1985) and Miller (1986). Miller (1993) and Miller and Gulya (1995) described F_6-derived F_7 maintainer lines converted to male sterility by backcrossing, which were homozygous for resistance to North American races of *V. dahliae*. In Canada, lines have been registered by Dedio *et al.* (1983), Dedio and Hoes (1983, 1988) and Hoes and Dedio (1988). A wide range of sunflower cultivars and hybrids are grown in Russia, but yield losses of 60% or more are attributed to a range of fungal pathogens of which *V. dahliae* is not regarded as a major problem (Yakutkin and Tavolzhanskri, 1999). The authors discuss various control options. The distribution of *V. dahliae* and other diseases on sunflower in Yugoslavia was described by Acimovic (1977), in Romania by Hulea *et al.* (1973) and in Bulgaria by Khristov (1977).

Cruciferae

*Oilseed rape (*Brassica napus *s.sp.* oleifera*)*

Until the widespread cultivation of oilseed rape as a source of vegetable oil, *Verticillium* wilt was a minor disease. The crop has been grown intensively in Sweden since the 1960s and, not surprisingly, the first accounts of wilt as a problem in 1965 originated there (Kroeker, 1976). Yield losses have been estimated generally at 5–10%, but exceptionally as high as 50% (Kroeker, 1976). In northern Germany, where the crop over-winters following a September sowing, infection increased from 2 to 56% in land continuously cropped for 5 years (Krüger, 1987, 1989; Seidel and Zeise, 1990). The disease has now spread to Poland, southern Russia and the Ukraine (Wilhelm, 1981). Nilsson (1985) used remote sensing to study the scale of the problem.

Little work has been done on plant resistance breeding, and the disease has not been reported in the USA or the UK. Nilsson (1977) noted that cv. Norde appeared more resistant or tolerant than cv. Panter. The application of the herbicide trifluralin as a routine practice severely enhanced wilt and eliminated any resistance in cv. Norde (Nilsson, 1977). Krüger (1991) and Busch (1991) reported preliminary experiments in Germany on resistant cultivar breeding. In Poland, of 55 breeding lines of winter rape selected in 1989 for resistance to *Leptosphaeria maculans*, all but one were highly susceptible to *V. dahliae* (Grzybowska, 1993). Disease resistance screening in the short term is difficult since inoculated winter rape seedlings do not develop symptoms until shortly before flowering (Zeise *et al.*, 1990). Zielinski and Sadowski (1994) observed that no complete resistance existed in 1993; natural infection in 24 cultivars and lines was 1.5–5% – the most resistant were BOH1582 and cv. Liradette. Grzybowska (1996) and Grzybowska and Olechnowicz (1996) crossed five maternal and two paternal 'resistant' lines, selecting 26 F_2 hybrids for further breeding. Of 3861 seedlings tested in the glasshouse, 48% were healthy in the first year and 40% healthy and 22.8% tolerant in the second. Under field conditions, infection occurred late and infected plants were described as 'very tolerant with no premature ripeness'.

The pathogen previously described as a diploid, long-spored variety of *V. dahliae* (*V. dahliae* var. *longisporum* Stark 1961) was considered to be fairly host specific. Karapapa *et al.* (1995, 1997a) and Heale (1997), however, have now presented convincing evidence based on *inter alia*, morphology, DNA microdensitometry, DNA 'fingerprinting' using RAPD and molecular karyotyping using pulsed-field gel electrophoresis (PFGE), for the erection of new species, *V. longisporum* comb nov. Heale (1997) has speculated that the sole recording of *Verticillium* infection of brussels sprout by Isaac (1957a) was most probably caused by *V. longisporum*. Evidence from DNA 'fingerprinting' suggested a possible hybrid origin of *V. longisporum* from a *V. albo-atrum* × *V. dahliae* cross. In Russia, rape seed is grown as a winter crop from *B. napus* ssp. *oleifera* and as a spring crop from turnip (*B. rapa*). Portenko (1997), at the Institute of Rapeseed Research Lipetsk, described four strains of *V. longisporum* on winter rape seed and one haploid strain of *V. dahliae* on spring rape seed in the Krasnodar region, the latter corresponding to VCG-2B of the US OARDC system. All strains caused wilt on aubergine and no resistance was recorded. Tabrett *et al.* (1995) showed that the cruciferous weed *Arabidopsis thaliana* was not host specific for *V. longisporum* and could be infected by haploid *V. dahliae* and *V. albo-atrum* isolates.

*Cabbage (*Brassica oleracea*)*

Yui (1986) in Taiwan screened 249 genotypes of *Brassica campestris*, *B. chinensis*, *B. pekinensis* sspp. *narinosa*, *japonica* and *rapifera*, and five cultivars of *B. napus* for resistance to *V. dahliae* (presumed to be haploid). The latter and *B. pekinensis* var. *rapifera* were least susceptible. The most tolerant were the Kanamachi cultivar group (small white turnips) (see also Williams, 1979 and Yui *et al.*, 1985).

*Cauliflower (*B. oleracea *var.* botrytis*).*

The first outbreak of wilt in cauliflower (also referred to as *vein disease*) was recorded in the Salinas valley, California, in 1990 (Koike *et al.*, 1994; Koike and Subbarao, 1995). The latter authors also found cabbage and Chinese cabbage susceptible to the same isolate. The description of the pathogen (Subbarao *et al.*, 1995) confirmed it to be *V. longisporum* which also infected Japanese radish, Chinese cabbage, turnip, broccoli (presumably cauliflower) and cabbage. The inability of *V. longisporum* to attack broccoli was exploited by Koike *et al.* (1997)

by incorporating non-host broccoli residues in cauliflower field soils to reduce microsclerotial inoculum levels. Since infection can reach >90% in a valuable and expanding crop, the need for resistant hybrid cultivars is imperative. Studies in Belgium in 1999 of 20 cauliflower cultivars (Callens *et al.*, 2000) showed that *V. longisporum*-vein disease was the major limiting problem. The preferred cultivars Fremont and Somerset, superior in all other qualities, were suceptible to the pathogen. The disease resistance breeding in swede rape and turnip rape in Sweden to 1995 is reviewed by Happstadius and Melander (1995).

*False flax (*Camelina sativa *L.)*

Camelina sativa was grown under trial as a new oil seed crop in Germany in 1997. As a member of the same family as oilseed rape (*Brassicae*), Henneken *et al.* (1998) and Foller *et al.* (1998) investigated its susceptibility to *V. longisporum*. Ten cultivars or breeding lines were inoculated under laboratory and field conditions with *V. longisporum* isolated from winter oilseed rape, all of which were susceptible to varying degrees.

Melon – muskmelon (*Cucumis melo*)

This is a vine of central Asian origin, cultivated worldwide, with many cultivars. Little has been written about the pathogen (*V. dahliae*) which is capable of infecting watermelon (*Citrullus lanatus*), preserving melon-citron (*C. lanatus* var. *citroides*), cucumber, pumpkin and squash (Middleton and Bohn, 1953), or the disease. Wilt was recorded in the UK in 1926 (Moore, 1959) and in the USA in 1953 when no resistant cultivars were available (Middleton and Bohn, 1953). An extensive breeding programme has been in existence for melon cultivars resistant to the major pathogen *F. oxysporum* f.sp. *melonis* and for watermelon resistant to *F. oxysporum* f.sp. *niveum*. The limited work on *Verticillium* resistance has had for the most part to consider linked resistance to *Fusarium* in addition to other diseases.

In the CIS, Mukhamedalieva (1972) screened five early and nine mid-season local cultivars against different races of *V. dahliae*. Only one cultivar, Taslaki 862, showed some tolerance. This cultivar was found to be tolerant in a subsequent examination of 17 cultivars inoculated with a cotton strain of *V. dahliae* and showed 62.5% resistance under mixed infection with *F. oxysporum* f.sp. *melonis* (Mukhamedalieva, 1973, 1977). In California, lines developed from several commercial types and cv. Big River Bush produced U.C. SR-91 Bush, U.C. Top Mark Bush and U.C. Perlita Bush all with resistance to *Verticillium* and *Fusarium* (Zink, 1978). An eighth generation cultivar from cv. Honey Ball × (SR91 × Honey Dew), cv. Honeyloupe was resistant to *V. dahliae* (Zink, 1979). Cv. Crenshaw Bush, a SR91 × Big River Bush cross, which was backcrossed three times to SR91, the progeny of which were crossed three times to cv. Crenshaw, incorporated resistance to *Verticillium* (Zink, 1980). Many *Fusarium*-

resistant cultivars are wholly susceptible to *V. dahliae*. Zink and Gubler (1986, 1987) developed cv. U.C. PMR-45, a three-parent compound cultivar with *Verticillium* resistance combined with single-gene resistance to race 1 of *Fusarium* and race 1 of *Sphaerotheca fuliginea*. The genetics of *Verticillium* resistance, unlike those of *Fusarium*, have not been described. Isolates and strains of *V. dahliae* from many vegetables, cotton, soft fruit and ornamentals are capable of causing wilt in melon.

Guayule (*Parthenium argentatum*)

Guayule is a shrub native to the arid plains of Mexico and south-western USA producing a high-quality hypoallergenic latex used in the medical industry. Campbell *et al.* (1943) first described wilt in California caused by *V. dahliae*. The disease reduced rubber yield by up to 25%, limiting crop growth in California (Schneider, 1945; Campbell and Presley, 1946). Of four morphological types of guayule, Schneider (1948) found that type 4 including strains 405, 407 and 416 showed most resistance. Gerstel (1950) found that in a polyploid series of strains, resistance increased with the number of chromosome sets: thus diploid clones ($2n = 2x = 36$) were extremely susceptible whereas the multiplication of a resistance factor in triploid, tetraploid and hexaploid plants gave increasing resistance. Tysdal *et al.* (1983) described four resistant hybrids: Cal-1 derived from open-pollinated seed of F_2 and BC1 plants from a *P. argentatum* × *P. tomentosum* cross; Cal-2, which was similar but from a *P. argentatum* × *P. fruticosum* hybrid; Cal-3 which was developed from 12 diploid *P. argentatum* crosses; and Cal-4 which was a composite of open-pollinated, resistant, diploid *P. argentatum*.

No details have been published of the genetic nature of resistance or of pathogen strains and alternate hosts. In response to an anticipated increased demand for guayule latex, Ray *et al.* (1995) established a new breeding programme. A diploid ($2n$) population was subject to three cycles of selection screening for increased tolerance to *V. dahliae* and the highest latex yield. Among 18 improved lines (13 of which had not been evaluated previously), a screened $2n$ line (418-6) had the highest tolerance but was not significantly better than four unscreened polyploids. Cuttings were more tolerant than seedlings, which increased with the age of plant at the time of inoculation. Guayule resistance breeding may be expected to assume greater importance when planting pressure will require the crop to follow cotton and other *V. dahliae*-susceptible crops.

Chrysanthemum (*Chrysanthemum morifolium*)

In addition to its effects on the cultivated decorative species (*C. morifolium*), *V. dahliae* causes wilt in a range of species including ox-eye daisy (*C. leucanthemum*), marguerite (*C. frutescens*) and Italian chrysanthemum (*C. indicum*) (Engelhard, 1957).

In the cultivated species, wilt may be severe, especially under the short days required for flowering. In the absence of specific breeding programmes, the finding of resistance (nature unknown) appears to be largely fortuitous. McCain and Welch (1976) and McCain (1975) reported on a resistance screen of 55 cultivars, 73% of which were resistant to wilt and 39% to both wilt and *Puccinia chrysanthemi* (see also Welch and McCain, 1977). Fifty-three cultivars were recorded as resistant to *V. dahliae* including cultivars Artemis, Dixie and Gay Ann (Byrne *et al.*, 1980).

Olive (*Olea europaea*)

The olive (*O. europaea*) of Mediterranean origin has been cultivated since ancient times for oil and edible fruit. Wilt caused by *V. dahliae* was unknown as a serious problem until its first reported occurrence in Sicily in 1946 (Ruggieri, 1946, 1948) in cultivars Bella di Spagna, Nocellara and Ogliarola. Shortly afterwards, it was found in California, in young groves planted after tomato (Snyder *et al.*, 1950). Since then, the disease has been a serious problem, destroying established groves in Greece, Spain, Italy, the Middle East and western USA.

Isolates of the pathogen from cotton, tomato, peppers, aubergine and the weeds cocklebur (*Xanthium canadense*) and tomatillo (*Physalis ixocarpa*) are all capable of infecting olive. The association of olive wilt following planting after pepper, aubergine and tomato was noticed by Ruggieri (1948), and in California after planting in old cotton soils by Snyder *et al.* (1950). Greek olive isolates were confined to a specific RAPD (random amplified polymorphic DNA) group together with potato, melon, pistachio and Dimorphotheca isolates, tomato and cotton – non-specific members were also in the same group (Paplomatas and Lampropoulos, 1997).

In the largest screening experiment to date, Wilhelm and Taylor (1965) screened large numbers of Mediterranean and Californian seedlings in old cotton-field soil as a source of resistant rootstock. Since olive is an open wind-pollinated crop, much variability was found in the seedlings. Much resistance was on the basis of tolerance but some showed true resistance in the exclusion of pathogen from the roots, even from erstwhile susceptible cultivars. The most susceptible cvs were Manzanillo, Chemlalie and Mission and the most resistant Frantojo and Arbequina. A clone selection of Arbequina was registered as cv. Allegra; Hartman *et al.* (1971) also described cv. Oblonga as a recommended resistant rootstock. The most susceptible of Wilhelm and Taylor's (1965) cultivars, cv. Manzanillo, was found to be the most resistant of five cultivars studied in Crete (Linardakis *et al.*, 1980), emphasizing the importance of pathovars in resistance screening. No information is available from either study of the precise source or virulence of the pathogen(s). From an investigation of resistance in Syrian olive cultivars since 1985, seven lines have emerged designated 'yarmouk lines', individual cultivars which may be used as resistant rootstocks (Al-Ahmad *et al.*, 1997b). The 1993 status of olive wilt in newly established plantations in Andalucia, Spain was reviewed by Rodriguez Jurado *et al.* (1993). A resistant (tolerant) cv. Oblonga inoculated with a P_1 cotton strain of *V. dahliae* remained symptomless but showed evidence of extensive colonization from cultured stem. Cirulli and Montemurro (1976) compared pathogenic isolates of *V. dahliae* and sources of resistance in Puglia, Italy.

There appears to be sufficient resistance in *O. europea* to justify an extensive breeding programme using a wide range of field isolates, particularly for Mediterranean countries where the problem is serious, and this should be combined with good hygiene and quarantine practice and the introduction of certified, elite, pathogen-free propagation stock.

Pistachio (*Pistacia vera* L.)

This is a native plant of western Asia and is cultivated there and in the Mediterranean countries. Following its introduction into the USA at the turn of the century, it has been grown in California since 1950. Infection by *V. dahliae* has been described in Greece by Papaioannou (1956).

An extended glasshouse and field screening programme conducted by Raabe and Wilhelm (1978) over 14 years, included 108 trees of *Pistacia* spp. Young trees showed severe symptoms. Two species, *P. terebinthus* and *P. integessima*, were more resistant than other species (*P. atlantica*, *P. mutica* and *P. chinensis*). *P. integessima* is now used as a rootstock in areas of severe wilt infestation. The trial ground heavily infested with *V. dahliae* was cropped previously with tomato. An evaluation of rootstocks of *Pistacia* spp. budded with *P. vera* cv. Kerman and tested in field soil with an inoculum concentration of 13–25 microsclerotia g^{-1} of dry soil showed that during the first two growing seasons, two trees on *P. atlantica* and one each on *P. integerrima* and *P. integerrima* × *P. atlantica* died. Subsequently, *Verticillium* wilt was present in 25% of *P. integerrima* × *P. atlantica*, 3% on *P. atlantica* and 2% on *P. integerrima* × *P. atlantica* and *P. atlantica* × *P. integerrima*, the latter hybrid showing that tolerance was inherited differently in the reciprocal cross (Teviotdale *et al.*, 1995).

Apricot (*Prunus armeniaca* L.)

No sources of genetic resistance to *V. dahliae* in apricot have been reported and contemporary control measures are limited to hygiene and husbandry practices, and solarization.

Blackberries (*Rubus* spp.) and raspberries (*Rubus idaeus*)

Cultivars of the genus *Rubus* widely dispersed in the northern hemisphere, especially North America, represent complex hybrids of cultivated and wild species which range in ploidy from diploid ($2n = 14$) to dodecaploid. Many are apomictic. The disease on raspberries, known as 'blue stripe', was first reported in the USA by Lawrence (1912) and in the UK in 1923 (Harris, 1925b). In the USA, the disease, especially on blackberry species and cultivars, may be devastating in the absence of resistant lines, causing a total failure of the crop. *V. dahliae* has not been reported as a serious problem on black-fruited trailing species (brambles) outside the USA (see Rudolph, 1931). Wilhelm *et al.* (1965), in a signal contribution, reviewed the resistance of *Rubus* spp. based on a single isolate of *V. dahalie* originally isolated from infected tomato plants growing between rows of boysenberries, which was also pathogenic to strawberries. Resistant species included *R. ursinus*, from which cultivars Logan Mammoth, Marion and Olallie were derived in part; *R. procerus* in the Himalaya yielding in part, cultivars Chehalem and Marion; *R. lacinatus* – evergreen blackberry; *R. ulmifolius* var. *inermis* – Burbank thornless; *R. ulmifolius* var. *inermis* (tetraploid) – Merton thornless. Seedlings from selfed cv. Olallie segregated 60.6% resistant to 39.4% susceptible; in a subsequent test on 209 seedlings, R:S segregation was approximately 1:1 and clearly defined, with no tolerance. This study still represents the most comprehensive examination of wilt resistance, but is weakened by the use of only a single pathogenic isolate. Little or no information is available on the genetic basis of resistance or its inheritance. Similarly, the spectrum of pathovars is unknown.

Cultivars of *R. idaeus* (raspberry) are mostly susceptible to *V. dahliae*. Boysenberry (susceptible) of unknown parentage probably had *R. idaeus* as one progenitor species. Ourecky (1977) achieved tolerance in cv. Brandywine, with purple fruit, derived from cv. New York 631 (purple) × cv. Hilton (red). Screening for resistance in UK cultivars was carried out by Montgomerie and Kennedy (1982) and Montgomerie *et al.* (1983). A high level of *V. dahliae* tolerance was found by Kennedy (1985) in a red × black-fruited raspberry cross. The inheritance of *V. albo-atrum* tolerance in raspberry *Rubus* subgenus *Idaeobatus* was studied by Fiola and Swartz (1994) in progeny of an incomplete partial diallel of two black, purple and red cultivars. Fourteen weeks after root inoculation, symptom severity and black raspberry parentage showed a significant correlation ($r^2 = 0.90$) as did fungus re-isolation ($r^2 = 0.66$). Tolerance in some genotypes was indicated by colonization of symptomless plants. While the results indicated an additive inheritance of resistance, the authors postulate a gene for gene system for symptom expression, with partial dominance of resistance alleles.

Chickpea (*Cicer arietinum*)

A native of western Asia, chickpea or garbanzo is grown in Asia, Mediterranean regions and the Americas. Wilt caused by *V. dahliae* is severe in Spain and North America. Little work has been done specifically in wilt resistance breeding or selections. Phillips (1979) evaluated 279 accessions of *C. arietinum*, 34 of which were reported free from *Verticillium*, presumably after screening in infested field soil.

Groundnut (*Arachis hypogaea*)

The genus with 40–70 species is a native of South America, with Brazil as the centre of diversity. Wilt caused by *V. dahliae* was reported in New Mexico by Smith (1960), who noted an earlier outbreak in Tashkent (CIS). Purss (1961) also described the disease in Queensland, Australia. Relatively little is known about the provenance of pathogenic isolates, but it is significant that isolates from cotton, chilli pepper and *Solanum elaeagnifolium* all caused wilt to varying degrees in groundnut. Curiously, cultivars became more severely infected in the field compared with pot experiments in a glasshouse (Smith, 1961), presumably due to the presence of pathovars more virulent than those used in controlled inoculation. The same author found high wilt resistance in one of several cultivars tested, cv. Georgia Bunch 182-28, and what appeared to be tolerance in others. The highly resistant cv. Georgia Bunch matured too late and lacked other commercially desirable attributes and thus was not suitable for development. A resistant line P1295233 was compared with the susceptible commercial cv. Tamnut 74 by Erikson *et al.* (1986). The effect of wilt on cv. Tamnut 74 was also described by Melouk *et al.* (1983).

Frank and Krikun (1968, 1969), in Israel, screened for resistance to *V. dahliae*, bacterial wilt and *Pythium* root rot. The selection 65–121 from cv. Schwarz-21 was found to combine resistance to all three.

Safflower (*Carthamus tinctorius*)

Safflower is an ancient crop grown for food and textile dyes extracted from yellow–red flower pigments and also for edible seed oil. Wild species of *Carthamus* are found in Afghanistan, Pakistan, Ethiopia and the Middle East, and, in the latter, *C. palaestinus* and *C. flavescens*, both morphologically similar to the cultivated diploid species ($2n = 2x = 24$). *C. tinctorius* is grown in the USA, Australia, Mediterranean countries and Mexico.

Wilt caused by *V. dahliae* was first reported on safflower in Nebraska, USA by Schuster and Nuland (1960). Almost simultaneously, severe losses were reported in California on a crop succeeding cotton (Urie and Knowles, 1972). The implications for crop rotation are not clear since isolates from safflower are pathogenic to a wide range of crops and vice versa (Wilhelm, 1981).

Disease resistance breeding has been conducted almost exclusively in the USA. Early experiments by Urie and Knowles (1972) screened 1300 plant introductions (P_1) and 1100 commercial breeding lines for *Verticillium* resistance. P_1

222-240 and P_1 250-922 from Iran, and P_1 253-385 from Israel were highly resistant. In 1971, cultivars AC1 and AC1W bred in Texas were released as wilt-resistant cultivars (Prato, 1971) as was cv. VFR-1 with triple resistance to *Verticillium*, *Fusarium* and *Phytophthora drechsleri* (Wilhelm, 1981). Thomas (1976) showed that this unique triple resistance was controlled by a single dominant gene. Other resistant releases were made by Abel and Lorance (1975) with cv. Dart, of moderate resistance and by Abel (1977) with cv. B68195-25, an F_5 selection from a composite of an F_4 derived from a double cross. This cultivar is resistant to *P. drechsleri*, *Puccinia carthami* and *V. albo-atrum* [*sic*] (*V. dahliae*). *Verticillium* resistance was incorporated into safflower releases with resistance to other pathogens by Urie *et al.* (1976, 1980) and Thomas *et al.* (1978). A 10-year evaluation of safflower in Texas, Arizona and California in heavily infested *V. dahliae* soils found that no cultivars were resistant at all sites; cultivars VFR1 and LMVFP1 were resistant to the cotton P_1 strain, and P1249081 to the P_2 strain.

Artichoke (*Cynara scolymus*)

Severe outbreaks of *V. dahliae* wilt were reported in artichoke in southern Italy in the early 1980s, and the disease is now established (Ciccarese *et al.*, 1995). Although methyl bromide or dazomet fumigation provides almost complete control, cost and environmental constraints are providing the impetus for a resistance breeding programme. From a preliminary screen of 130 clones from different countries, 11 were selected for a 3-year perennial screening trial with three susceptible clones. Three categories emerged: 'resistant', <30% of diseased plants and severity index <20%; moderately 'resistant' 30–80 and 20–60%, respectively; and susceptible >80 and >60%, respectively. The most resistant (tolerant) clone (76) had only 19.8 and 19.2% plants showing external symptoms and vascular discoloration, respectively, with severity indices for these of 9.2 and 10.3%. By comparison, a susceptible clone (125) had 100% plants showing symptoms of 79% severity and 97.5% showing vascular discoloration of 78.7% severity. Some clones showed a higher percentage discoloration over wilt symptoms suggesting a different form of tolerance. The third year of cropping was essential to confirm resistant lines (Cirulli *et al.*, 1994).

Horseradish (*Armoracia rusticana*)

Little is known of the nature or inheritance of resistance to *V. dahliae* in horseradish. A collection of 113 cultivars in a University of Illinois germplasm collection held at Urbana could provide a source for material for disease resistance breeding. Root discoloration is a major disadvantage to this crop. Six cultivars, 635A, 1236A, 769A, 125A, 761A and 28A, were regarded as resistant, but this was not complete. On a root cross-section discoloration index of 0 = symptomless, 1 = trace to <10%, 2 = 10–50% and 3 = >50% resistant (tolerant cultivars), these were in the range of 0.2 (Atibalentja and Eastburn, 1998).

Cacao (*Theobroma cacao*)

T. cacao clones from the Parinari (PA) group originating in Peru provide an important source of resistance *inter alia* to *V. dahliae* (Yamada *et al.*, 1996). All clones studied were self-incompatible and shared interclonal relationships with 11 clones from the cacao germplasm collection at CEPEC Bahia, Brazil. Four compatibility groups were identified.

Sea buckthorn (*Hippophae rhamnoides*)

These willow-like trees which are grown in the Baltic states and central Asia as a plantation bush crop and are a rich source of vitamin C (ascorbic acid, 368 mg 100 g^{-1}) and oil from the fruits. Siberian cultivars grown in Estonia produced high yields of good quality fruit in the first cropping years, but all subsequently died from *V. dahliae* infection (Siimisker, 1996). Piir (1996) reported on a long-term evaluation of 33 sea buckthorn genotypes from the Polli Experimental Station Estonia, 16 cultivars from the Siberian Research Institute of Horticulture, Barnaul in the Altai district and 12 cultivars from Moscow State University. All cultivars of Altai origin were very susceptible to *V. dahliae*. Baltic × Siberian ecotypes have been produced in Moscow, but no details are available.

Stone fruit

Economic losses in two Californian almond orchards were evaluated over 6 years. Trees of cultivars Carmel and Nonpareil showed considerable mortality in the first growing season. The disease was active throughout the summer months, but could be re-isolated all year. Losses over 5 years were estimated at US$9000 ha^{-1} (Asai and Stapleton, 1994). In Italy, isolates of *V. dahliae* from almond, tomato, peach, apricot, cherry and olive were inoculated on all these hosts, including plum and aubergine. Almond, cv. Don Carlo and the rootstock hybrid almond × peach 677 were highly susceptible to all isolates. Almond cv. Filippo Ceo showed some tolerance. The cherry isolate was weakly aggressive on all species, but the olive isolate, while virulent on olive, was moderately aggressive on other species and avirulent on tomato. The remaining isolates were aggressive on all the tested species (Luisi *et al.*, 1994). Of six *Prunus* rootstocks tested by Corazza and Chilosi (1986), only Pixi was immune to *V. dahliae* (*ex* peach). Zamani-Zadeh and Zakii (1995) reported a serious decline of apricot in Shahrood, Iran due to *V. dahliae*. *V. dahliae* wilt of peach is a serious problem in southern Italy. Ciccarese *et al.* (1990) described natural recovery from infection over 3 years in cultivars Andros, Spring Gold, Coronet, June Gold and May Flower. The percentage plants showing external symptoms over years 1, 2 and 3 were 41, 14 and 0.8%, respectively. This was not reflected in discoloured wood which was, correspondingly, 65, 62 and 70%. Fungal biomass calculated from cultured wood chips correlated with symptoms but not vascular discoloration, and was 72, 18 and 1.7% for years 1–3, respectively. This result poses a serious problem for evaluating breeding lines for resistance. Gavrilenko *et al.* (1977) described the situation on plum in Moldavia.

Ornamental trees

Wilt of ornamental trees and shrubs is caused largely by *V. dahliae*, especially in countries with continental summer temperatures, but *V. albo-atrum* is also important where temperature is not limiting. For the most part, the literature pertaining to shade trees and shrubs consists of host lists of susceptible genera and species, with very few studies on breeding for resistance. The host range is extensive (see Rudolph, 1931; Engelhard, 1957; Himelick, 1969; Piearce and Gibbs, 1981; Phillips and Burdekin, 1983), with some 70 tree species and many shrubs. The most commonly infected tree genera include *Acer, Ulmus, Fraximus, Catalpa, Liriodendron, Magnolia, Juglans, Tilia, Cercis, Rhus, Cotinus, Robinia, Ailanthus* and *Koelreuteria*. In Poland, Lukomski (1962) reported a heavy loss of oak seedlings due to *Verticillium*, but the disease has not been reported in mature trees. Piearce and Gibbs (1981) claim that conifers and monocotyledonous plants are unaffected and that species of *Alnus, Betula, Carpinus, Fagus, Populus* and *Platanus* are very resistant.

Maple (*Acer* spp.) is severely attacked by *Verticillium* especially in nursery beds (Caroselli, 1957; Pirone, 1959). In New Jersey, Dochinger (1956) found during 1954/55 *A. saccharinium* (silver maple) to be most affected (8%), followed by *A. platanoides* (Norway maple) (2.8%), *A. rubrum* (red maple) (1.4%) and *A. saccharum* (sugar maple) (1%). Between 1969 and 1972, 20% of all maple trees were lost to *V. dahliae* in Indianapolis (Schrieber and Townsend, 1981). Powell (1977) considered *A. platanoides, A. saccharum* and *A. palmatum* (Japanese maple) to be more susceptible than *A. campestre* (field maple), *A. tartaricum* (Tartarian maple), *A. rubrum* and *A. saccharinum*. In one of the few experimental studies, Townsend and Hock (1973) stem-inoculated *A. rubrum* seedlings grown from seed from six provenances, Arkansas, Illinois, Minnesota, New Jersey, Ohio and Pennsylvania, with a homogenized miscrosclerotial, conidial inoculum from five different sources. Seedlings of Arkansas and Illinois provenance were consistently most resistant, even after re-inoculation, whereas those from Minnesota and Pennsylvania were most susceptible, with symptoms increasing following re-inoculation. Since *V. dahliae* could be re-isolated from

Arkansas and Illinois seedlings, the resistance exhibited was referred to as 'tolerance'. A preliminary report on the resistance of maple cultivars was published by Hoitink *et al.* (1979). Townsend *et al.* (1990) screened 13 cultivars of grafted *A. platanoides* (Norway maple) for *V. dahliae* tolerance by stem puncture inoculation. Two cultivars 'Jade Glen' and Parkway showed greatest tolerance, and two, Crimson King and Greenlace, the least. Six of the 13 cultivars were then grown on their own roots and root-inoculated. The results confirmed that the same cultivar tolerance was exhibited regardless of a grafted rootstock or inoculation method. Cultivars Crimson King and Greenlace showed up to an 80% mortality after root inoculation. Howard (1993) found that *V. dahliae*-infected rootstock depressed chip-bud take of the red-leaved scion, Crimson King but did not explain large differences in budding success from field to field. In Holland, Hiemstra (1997) made a preliminary screen of 20,000 Norway maple seedlings from different provenances for *Verticillium* resistance by root-dipping in a *V. albo-atrum* suspension and assessing after 4–6 weeks. On average, approximately 1% of plants remained symptomless, with no clear pattern of parental resistance inheritance. Selected plants were propagated for further screening. All commercial cultivars tested were susceptible.

While the major wilt disease of *Ulmus* is caused by *Ophiostoma ulmi*, species of elm may be seriously infected with *Verticillium* (Van der Meer, 1926). Holmes (1967) compared the resistance of five Dutch elm clones to *O. ulmi* and *V. albo-atrum* [*sic*] (*V. dahliae*). Symptoms of *Verticillium* wilt were later and milder than those from *O. ulmi* but, in general, resistance to *Ophiostoma* seen in *U. hollandica* 'Groenveld' and *U. hollandica* N148 also applied to *V. dahliae*. *U. carpinifolia* 'Schuurhoek' and *U. hollandica* 'Belgica' were susceptible. Using growth reduction as a measure of disease severity, Rauscher *et al.* (1974) inoculated variously, three elm species and two elm clones from each of five elm species or hybrid families. *U. americana* was most affected and *U. pumila* least, with *U. japonica* intermediate. When clones of species or families were examined, those of *U. pumila* × *U. japonica* showed the most resistance (or least growth reduction). *U. americana* clones were most susceptible. Symptoms were related to inoculum concentration. Each species had an inoculum threshold, above which symptoms were severe. In a subsequent study, Lester (1975) inoculated 594 seedlings of *U. americana*, resulting in a series of symptoms. Clones selected from a range of seedlings showing different symptom severities, also showed significant variation, but not proportional to the original seedling symptoms. There is here a clear indication of clonal variation in tolerance which could be exploited in a vegetative reproduction programme irrespective of an apparent lack of genetic (chromosomal) tolerance. The correlation of *O. ulmi* resistance and *V. dahliae* tolerance reported by Holmes (1967) was demonstrated by Smalley and Lester (1973a) in their elm 'Sapporo Autumn Gold' grown from open pollinated seed from *U. pumila* in Hokkaido, Japan. They considered it to be an *U. pumila* × *U. japonica* hybrid. Another released elm cv. 'Regal', also showed high *O. ulmi* resistance as well as high tolerance to *V. dahliae* (Smalley and Lester, 1983b).

Verticillium wilt of ornamental plants is more widespread than is generally recognized by foresters and horticulturalists. Before a meaningful plant breeding programme could be undertaken, much preliminary work would be required on genetic typing of pathogenic strains and to determine the range of pathotypes capable of infecting a particular species. Above all, in view of the time scale of potential benefits, a comprehensive disease resistance/tolerance breeding programme would need to be justified in terms of the economic value of the planting and the perceived seriousncss of the disease in the treescape or in nursery production. Species of *Acer* would make a suitable subject based on European and North American requirements alone (see Chapter 11).

Hosts 11

Verticillium wilts are overwhelmingly diseases of dicotyledenous plants in temperate regions. The extent to which disease is expressed is a complex function of the species and strain of *Verticillium* responsible, the degree of resistance of the host and the environmental conditions.

Various authors have attempted to record all instances of wilting attributable to *Verticillium* at that time, and an amalgamation of these is given in Table 11.1.

Woolliams (1966) recorded *V. dahliae* from *Equisetum arvense* L. in Canada, the sole cryptogam host. In The Netherlands, Van der Lek (1918) quoted Shoevers as isolating *Verticillium* spp. from *Thuja* sp. while, in France, Végh (1987) reported *V. dahliae* from *Chamaecyparis* sp., the only gymnosperm hosts recorded.

Monocotyledonous plants very occasionally have been cited as true hosts for *Verticillium*. Sherrod and Elliott (1967) inoculated '*Sorghum vulgare*' by root-dipping in a culture isolated from cotton and obtained similar symptoms to those in cotton, but *Verticillium* cultures isolated from the *Sorghum* never formed microsclerotia. Mathré (1986, 1989) observed symptoms of '*Cephalosporium* stripe' in *Hordeum vulgare* L. growing in a field previously cropped with potatoes. Isolations consistently yielded *V. dahliae* which caused comparable symptoms when inoculated into barley, wheat (*Triticum aestivum* L.) and oats (*Avena sativa* L.). Some monocotyledonous plants including wheat and other cereals (Martinson and Horner, 1962), and cereals, onions and tulips (Malik and Milton, 1980) have been shown to harbour *V. albo-atrum* or *V. dahliae* in the superficial tissues of their roots. Where microsclerotia are formed, they pose a source of infection for susceptible dicotyledonous crops grown in the same soil.

Host lists

Early host lists (Van der Meer, 1925; Wollenweber, 1929; Rudolph, 1931) predated the recognition of *V. albo-atrum* and *V. dahliae* as individual species, with the result that many references to *V. dahliae* are listed as *V. albo-atrum*. Some confusion exists in later accounts where compilers have not checked source material and where the original author used the earlier convention of combining micro- and non-microsclerotial pathogens under *V. albo-atrum*. In the present work, in some instances, the author's name for the causal organism has been corrected while still giving the original attribution in parentheses.

Van der Meer (1925) listed hosts of *V. albo-atrum* and *V. dahliae* in The Netherlands under group, order, family, genus and species, with reference to the published source. Rudolph (1931) tabulated by family, genus and species, those plants recorded in world literature as susceptible to attack by *V. albo-atrum* 'and related forms', which included *V. dahliae*, with 134 references dating from 1879 to early 1929; i.e. for the first 50 years of recognition of *Verticillium* as a cause of wilting. He included descriptions of symptoms, where known, for those wilts of 'truck crops', fruit trees, nuts, bush fruits and ornamentals, but did not distinguish host plants from symptomless 'non-hosts', which yielded the fungus on sampling. The list prepared by Wollenweber (1929) of *Verticillium* wilts of woody plants and trees under the country of each report appeared after Rudolph concluded his search of world literature, and thus includes some records not included in his account.

Engelhard (1957) tabulated records in alphabetical order of host, genus and species under general headings of trees, shrubs, field crops, vegetables and small fruits, ornamentals, flowers and weeds. He gave the reference (frequently without citing the author), country, if it was the first record for the species on that host, and whether it involved *V. albo-atrum* or *V. dahliae*. Parker (1959) reviewed *Verticillium* infection of deciduous fruit trees, taking his selections from world literature. The list of species and varieties of tree and shrub hosts compiled by Himelick (1969) included both species of *Verticillium* under *V. albo-atrum*, while Stark (1961) gave records of both species on garden plants in the Hamburg region of Germany. Smith (1979) reviewed the literature of *V. albo-atrum* and *V. dahliae* wilts of landscape trees in the USA, while Carter (1975) listed *Verticillium* wilts of trees in the mid-west USA, all as *V. albo-atrum*. Phillips and Burdekin (1983) gave a list of tree hosts of *V. albo-atrum* and *V. dahliae*, while Slawson (1987) summarized reports of isolation of *V. dahliae* from various woody ornamental plants. Végh (1987) listed French ornamental tree and shrub hosts to *V. dahliae*. Devaux and Sackston (1966) gave those species of horticultural crops from which *V. albo-atrum*, *V. dahliae* or *V. nigrescens* were isolated in Quebec, Canada.

Weed Non-hosts

Many plants growing as weeds in an apparently disease-free environment harbour *Verticillium* in their vascular system without showing detectable parasitic symptoms, disease only manifesting itself with the introduction of a susceptible crop. Such symptomless carriers are referred to as 'non-hosts' and act as reservoirs of infection: Sewell and Wilson (1958) in the UK, Oshima *et al.* (1963) in the USA, Skadow (1969a) in Germany, Evans (1971a) in Australia, Thanassoulopoulos *et al.* (1981) in Greece and Vargas-Machuca *et al.* (1987) in Peru. The importance of microsclerotial build up in weeds of potato was highlighted by Busch *et al.* (1978) and in cotton field weeds by Johnson *et al.* (1980). Graminaceous plants, non-hosts of *Verticillium*, have also served as sources of microscleratia on roots (Kirikun and Bernier, 1990). The weeds of south-western Ontario serving as non-hosts to *V. dahliae* and *V. albo-atrum* in potato growing areas were listed by McKeen and Thorpe (1973) (see also Pegg, 1974).

While no comprehensive list of host plants has been published since that of Engelhard (1957), individual records abstracted from the world literature appeared in the *Review of Applied Mycology* from Vol. 27 (1948) to Vol. 48 (1969), and thereafter in the *Review of Plant Pathology* from Vol. 49 onwards, recently both as new records for the host and for the relevant country.

In Table 11.1, records are listed in alphabetical order of host family, updating the host name where known, but quoting the name given in the original reference as well. The relevant literature is by now so large that no claim is made as to the comprehensiveness of the list. Some mention of hosts not included in the table occurs in other chapters.

Table 11.1. Abbreviations used are as follows: auctt. = of authors; *V. a-a* = *Verticillium albo-atrum; V. d.* = *Verticillium dahliae; V.* sp. = *Verticillium* sp.; isol. ex = isolated from a plant; isol. from = isolated from the environment, i.e. the soil, etc; (h.) = host (only used where there is ambiguity); (n.h.) = non-host or carrier; re-inoc. on to = re-inoculated on to the same host or, if followed by a different host name, on to the latter.

Host	Verticillium	Reference
Aceraceae		
Acer campestre L.	*V. a-a* isol. ex, in the USA & re-inoc. on to	Himelick (1969)
A. ginnala Maxim.	*V. a-a* isol. ex, in the USA & re-inoc. on to	Himelick (1969)
A. macrophyllum Pursh.	*V. a-a* on, in the USA	Mielke (1935)
A. mono Maxim.	*V. a-a* on, in the USA	Felix (1955)
A. negundo L.	*V.* sp. on, in Canada	Drayton (1924)
	V. a-a. on, in the USA (1950)	Felix (1955)
A. nigrum Michx. f.	*V.* sp on, in the USA	Carter (1938)
A. palmatum Thunb. var. *rubrum* Schwerin.	*V.* sp. isol. ex, in the USA & re-inoc. on to *Acer* spp.	Zimm (1918)
	V. a-a (? *V. d.*) isol. ex, in the USA	Wilhelm *et al.* (1955)
	V. d. isol. ex & re-inoc. on to, in Japan	Harada *et al.* (1997)
A. pennsylvanicum L.	*V. a-a* on, in the USA	Hibben (1959)
A. platanoides L.	*V.* sp. isol. ex, in the USA & re-inoc. on to *Acer* spp.	Zimm (1918)
	V. d. on, in Italy	Goidanich (1934a,b)
	V. a-a on street trees in The Netherlands	Alkemade (1979)
A. pseudoplatanus L.	*V.* sp. isol. ex, in the UK & re-inoc. on to	Bewley (1921)
	V. d. on, in Italy	Goidanich (1934a,b)
	V. sp. on, in the USA (1940)	Engelhard (1957)
	V. a-a on, in Canada	Anon (1961a)
A. rubrum L.	*V.* sp. isol. ex, in the USA & re-inoc. on to	Zimm (1918)
	V. a-a isol. ex, in the USA	Carter (1938)
	V. a-a. on, in Canada	Anon (1961a)
A. saccharinum L.	*V. a-a* on, in the USA (1950)	Engelhard (1957)
	V. a-a on, in the USA & re-inoc. on to	Himelick (1969)

A. saccharum Marshall	*V.* sp. isol. ex, in the USA & re-inoc. on to	Zimm (1918)
	V. sp. (as *Acrostalagmus* sp.) on, in the USA	Rankin (1914)
	V. a-a on, in the USA	Martin (1926)
	V. d. colonization of leader shoots correlated with N concentration	Smith (1983)
Acer sp.	*V.* sp. (as *Acrostalagmus* sp.) on, in the USA	Rankin (1914)
	V. sp. on, in The Netherlands	Anon (1920)
	V. a-a isol. ex, in the USA	Zeller (1936)
	V. spp. on, in the USA	Gravatt (1926)
Acer spp.	*V. d.,* symptoms of, in nurseries, UK	Piearce & Gibbs (1981)
	V. d. on, general review	Caroselli (1957)
	V. d. inoc. on to, in the USA, movement in stems	Sinclair *et al.* (1981)
	V. d. growth on sap	Regulski & Peterson (1983)
Aizoaceae		
Cryptophytum crystallinum N.E. Br.	*V. d.* on, in California, USA	MacDonald *et al.* (1984)
Tetragonia expansa Thunb.	*V. a-a* (? *V. d.*) isol. ex, in the USA	Wilhelm *et al.* (1955)
Amaranthaceae		
Amaranthus retroflexus L.	*V. a-a* isol. ex, in the USA	Rudolph (1931)
A. viridis L.	*V. d.* on, in the Aegean region	Kocaturk & Karcilioglu (1976)
Celosia cristata L. (as *C. argentea* var. *cristata*)	*V. a-a* (? *V. d.*) isol. ex, in the USA	Baker & Locke (1946)
Gomphrena celosoides Mart.	*V. d.* isol. ex, in Australia (n.h.)	Evans (1971a)
Anacardiaceae		
Cotinus coggygria Scop.	*V. a-a* on, in The Netherlands	Westerdijk & van Luijk (1924)
	V. a-a isol. ex, in the USA	Fowler (1937)
	V. a-a & *V. d.* isol. ex, in the Ukraine	Trotzenko (1958)
Mangifera indica L.	*V. d.* on, in Israel	Pinkas & Chorin (1970)
Pistacia vera L.	*V. a-a* (? *V. d.*) isol. ex, in the USA	Snyder *et al.* (1950)
	V. d. on, control of, in the USA	Ashworth & Zimmerman (1976)
	V. d. affecting commercial pistacio, in Greece	Chitzanidis (1995)
	V. tricorpus isol. ex trees affected by *Xanthomonas* sp. in Australia	Edwards & Taylor (1998)

Continued

Table 11.1. *Continued*

Host	Verticillium	Reference
Anacardiaceae *Continued*		
Rhus aromatica Ait.	*V. a-a* isol. ex, in the USA & re-inoc. on to (1943)	Himelick (1969)
(as *R. canadensis* Marsh.)	*V. a-a* on, in the USA	Dufrenoy (1927a), citing Barrus, unpublished
R. glabra L.	*V. a-a* on, in The Netherlands	Westerdijk & van Luijk (1924)
	V. a-a isol. ex, in the USA	Carter (1938)
R. trilobata Nutt.	*V. a-a* on, in the USA (1950)	Engelhard (1957)
R. typhina L.	*V. a-a* on, in The Netherlands	Westerdijk & van Luijk (1924)
	V. d. on, in The Netherlands	Van der Meer (1925)
	V. d. on, in the USA	Ludbrook (1933)
	V. a-a isol. ex, in the USA (1960)	Himelick (1969)
Schinus terebinthifolius Raddi	*V. sp.* on, in the USA (1950)	Engelhard (1957)
Apocynaceae		
Vinca major L.	*V. a-a* isol. ex & re-inoc. on to, in the USA	Koike (1988)
Araliaceae		
Aralia cordata Thunb.	*V. a-a* on, in the USA	Vaughan (1924)
	V. d. isol. ex & re-inoc. on to, in Japan	Sawamura & Soma (1976)
A. racemosa L.	*V. d.* isol. ex, in the USA	Dufrenoy (1927a)
Panax quinquefolius L.	*V. a-a* on, in the USA (1950)	Engelhard (1957)
Balsaminaceae		
Impatiens balsamina L.	*V.* sp. on, in Italy	Curzi (1930)
	V. a-a isol. ex, in The Netherlands	Van den Ende (1958)
	V. a-a on, in the USA (1950)	Engelhard (1957)
I. walleriana Hook.	*V. d.* isol. ex & re-inoc. on to, in the USA	Taylor (1993)
Basellaceae		
Basella rubra L.	*V. d.* isol. ex & re-inoc. on to, in Japan	Kanno & Horiuchi (1999)
Ullucus tuberosus Caldas	*V. d.* (*V. nubilum*?) isol. ex & re-inoc. on to & on to potato, in Peru	Otazu *et al.* (1998)

Begoniaceae		
Begonia spp.	*V. a-a* on, in the USA (1950)	Engelhard (1957)
Berberidaceae		
Berberis thunbergii DC.	*V.* sp. isol. ex & re-inoc. on to, in the USA	Jagger & Stewart (1918)
	V. a-a isol. ex & re-inoc. on to, in the USA	Boyd (1930)
B. vulgaris L.	*V. a-a* on, in the USA (1950)	Engelhard (1957)
Mahonia spp.	*V. d.* on, in the UK	Slawson (1987)
Bignoniaceae		
Campsis radicans (L.) Bureau (as *C. radicans* (L.) Seem)	*V. a-a* on, in the USA	Carter (1945)
Catalpa bignonioides Walter	*V.* sp. on, in the USA	Engelhard (1957)
C. speciosa (Warder ex Barry) Engelm. (as *C. speciosa* Warder)	*V. d.* on, in Italy	Bisiach *et al.* (1982)
	V. a-a isol. ex & re-inoc. on to, in the USA	Carter (1938)
	General review of, in Europe	Smith *et al.* (1988)
Boraginaceae		
Heliotropum arborescens L.	*V. a-a* on, in the USA (1950)	Engelhard (1957)
Heliotropum sp.	*V. a-a* isol. ex & re-inoc. on to, in the USA	Bryan (1928)
Lithospermum arvense L.	*V. a-a* & *V. d.* on, in Germany	Skadow (1969a)
Silene noctiflora L. (as *Melandrium noctiflora*)	*V. d.* on, in Germany	Skadow (1969a)
Buxaceae		
Buxus microphylla koreana Nakai	*V. a-a* isol. ex, in the USA & re-inoc. on to	Himelick (1969)
B. sempervirens L.	*V.* sp. on, in the USA	Gravatt (1926)
	V. sp. isol. ex & re-inoc on to, in the USA	Harrar (1937)
Campanulaceae		
Campanula isophylla Moretti	*V. a-a* on, in the UK	Beaumont (1951)
C. macrantha Fisch.	*V. d.* on, in The Netherlands	Van der Meer (1925)
Platycodon grandiflorus (Jacq.) A.DC.	*V. d.* isol. ex, in Japan	Takeuchi & Horie (1996)

Continued

Table 11.1. *Continued*

Host	Verticillium	Reference
Cannabidaceae		
Cannabis sativa L.	*V. d. isol.* ex & re-inoc. on to, in the USSR	Vasilieff (1933)
Humulus lupulus L.	*V. a-a* is the main cause of wilt in hops, *V. d.* and *V. tricorpus* minor causes	
	V. a-a isol. ex, in the UK in 1924	Harris (1925a)
	'progressive' str. of *V. a-a* described in the UK	Keyworth (1942)
	V. a-a from 'fluctuating' & 'progressive' outbreaks compared	Isaac & Keyworth (1948)
	Soil conditions in Bavaria	Wittman (1971)
	V_1, V_2 & V_3 races differentiated	Sewell & Wilson (1978)
	Control	Sewell & Wilson (1974)
	Review	Rintelen (1974)
	Impact on agriculture	Pegg (1984)
	V. a-a on, in the USA (1952)	Engelhard (1957)
	V. a-a on, in Oregon, USA	Horner (1965)
	Distribution in Germany	Schmidt & Skadow (1967)
	V. a-a and *V. d.*, control of, in Germany	Kohlmann *et al.* (1974)
	V. a-a & *V. d.* on, in New Zealand	Christie (1956)
	V. a-a on, in Bulgaria	Ĭotov (1962)
	V. a-a & *V. d.* isol. ex in Slovenia (1974)	Cerenak *et al.* (1999)
	V. tricorpus isol. ex, in New Zealand	Taylor (1968)
Caprifoliaceae		
Sambucus racemosa L.	*V. d.* isol. ex, in The Netherlands	Van der Meer (1925)
Viburnum × *burkwoodii* Burkwood	*V. a-a* isol. ex, in the USA & re-inoc. on to	Himelick (1969)
V. lantana L.	*V. a-a* isol ex, in the USA	Carter (1938)
V. lentago L.	*V. a-a* on, in the USA (1950)	Engelhard (1957)
V. plicatus Thunb. (as *V. tomentosum* (Thunb.) Rehder)	*V. a-a* on, in the USA	Baines (1945)
	V. a-a isol. ex, & re-inoc. on to, in the USA	Himelick (1969)

V. tinus L.	*V. a-a* on, in the USA (1950)	Engelhard (1957)
	V. d. isol. ex, in the UK	Slawson (1987)
	V. d. on, in Italy	Garibaldi *et al.* (1990)
V. tomentosum (Thunb.) Rehder	*V. a-a* on, in the USA	Baines (1945)
	V. a-a isol. ex & re-inoc. on to, in the USA	Himelick (1969)
Viburnum spp.	*V. a-a* isol. ex, in the USA	Himelick (1969)
Caricaceae		
Carica papaya L.	*V. d.* isol. ex, in Australia	Burden (1968)
Caryophyllaceae		
Dianthus caryophyllus L.	used as 'non-host'	Catesson *et al.* (1976)
Dianthus sp.	*V.* sp. isol. ex, in France	Dufrenoy (1928)
Lychnis chalcedonica L.	*V. a-a* on, in The Netherlands	Van der Lek (1918)
Chenopodiaceae		
Beta vulgaris L. ssp. *vulgaris* (as *B. vulgaris* L.)	*V. a-a* isol. ex & re-inoc. on to, in The Netherlands	Westerdijk (1918)
	V. d. on, in The Netherlands	Van der Meer (1925)
	V. a-a on, causing root rot (?) in the USA	Anon (1943)
	V. d. on, in Greece	Karadimos *et al.* (2000)
Chenopodium album L.	*V. d.* on, in The Netherlands	Van der Meer (1925)
	V. a-a on, in Germany	Goetz (1935)
	V. d. isol. ex, in New Zealand	Cunningham (1949)
	V. a-a isol. ex, in the UK (n.h.)	Sewell & Wilson (1958)
C. berlandieri	*V. a-a* isol. ex, in the USA (n.h.)	Oshima *et al.* (1963)
Rhagodia nutans R.Br.	*V. d.* isol. ex, in Australia, (n.h.)	Evans (1971a)
Cistaceae		
Cistus purpureus Lam.	*V. a-a* (? *V. d.*) isol. ex, in the USA	Wilhelm *et al.* (1955)
Cistus sp.	*V. d.* isol. ex, in the UK	Slawson (1987)
Compositae		
Anaphalis margaritacea (L.) Benth. (as *Gnaphalium margaritaceum* L.)	*V. d.* on, in The Netherlands	Van der Meer (1925)

Continued

Table 11.1. *Continued*

Host	Verticillium	Reference
Compositae *Continued*		
Argyranthemum frutescens (L.) Schultz-Bip	*V. d.* on, in Italy	Garibaldi *et al.* (1990)
Artemisia absinthium L.	*V. d.* isol. ex & re-inoc. on to, in Italy	Garofalo (1970)
Aster amellus L.	*V.* sp. on, in Denmark	Dowson (1922, 1923)
A. cordifolius L.	*V.* sp. on, in Denmark	Dowson (1922, 1923)
A. diffusus Ait.	*V.* sp. on, in Denmark	Dowson (1922, 1923)
A. ericoides L.	*V.* sp. on, in Denmark	Dowson (1922, 1923)
A. novae-angliae L.	*V.* sp. on, in Denmark	Dowson (1922, 1923)
A. novi-belgii L.	*V.* sp. on, in Denmark	Dowson (1922, 1923)
A. paniculatus Lam.	*V.* sp. on, in Denmark	Dowson (1922, 1923)
A. tradescantii L.	*V.* sp. on, in Denmark,	Dowson (1922, 1923)
	'*V. vilmorinii*' on, in Denmark	Gram & Rostrup (1925)
A. vimineus Lam.	*V.* sp. on, in Denmark	Dowson (1922, 1923)
Aster sp.	Wilt of Michaelmas daisy in the UK	Wiltshire (1920)
	V. sp. isol. ex Michaelmas daisy & re-inoc. on to, in the UK	Dowson (1922)
	'*Cephalosporium asteris*' described from	Dowson (1923)
Bidens subalternans DC.	*V. d.* isol ex, in Australia (n.h.)	Evans (1971a)
Callistephus chinensis Nees	*V. a-a* isol. ex, in France and described as *Acrostalagmus vilmorinii* n.sp.	Guéguen (1906)
	V. a-a isol. ex, in the USA	Baker *et al.* (1940)
Carthamus lanatus L.	*V. d.* isol ex, in Australia (n.h.)	Evans (1971a)
C. tinctorius L.	*V. d.* on, in Australia	Oxenham (1963)
	V. d. on, in Australia	Anon (1966)
	V. d. isol. ex, in Morocco: seed transmission	Goethal (1971)
	V. d. isol ex & inoc. on to, in the USA: seed transmission	Klisiewicz (1974a,b, 1975)
	V. d. isol. ex & re-inoc. on to, in Italy	Fiori & Carta (1982)
	V. d. on, in Spain	Jimenez-Diaz *et al.* (1985)

	V. d. isol. ex, in Japan	Takeuchi & Horie (1996)
Centaurea cyanus L.	*V. a-a* (? *V. d.)* isol. ex, in the USA	Baker & Locke (1946)
	V. d. on, in Australia	Anon (1966)
C. imperialis Hausskn. *ex* Bomm.	*V. a-a* (?*V. d.)* isol. ex, in the USA	Baker & Locke (1946)
C. solstitialis L.	*V. d.* isol ex, in Australia (n.h.)	Evans (1971a)
	Susceptibility to *V. d.*	Klisievicz (1986)
Chrysanthemum frutescens	*V. d.* on, in Italy	Gullino & Garibaldi (1998)
C. × *grandiflora* auct.	*V. a-a* on, in Bulgaria	Khristova (1958)
C. indicum Cass.	*V. d.* on, in The Netherlands	Van der Meer (1925)
C. leucanthemum L.	*V. d.* on, in The Netherlands	Van der Meer (1925)
C. maximum auct. var. Shastra daisy	*V. a-a* (? *V. d.)* isol. ex, in USA	Baker & Locke (1946)
C. morifolium (Ramat.) Hemsl.	*V. a-a* isol. ex, in Germany	Donant (1932)
	V. sp. on, in the USA	Huber and Jones (1934)
Chrysanthemum sp.	*V.* sp. on, in the USA	Haensler (1924)
	V. a-a inoc. on to, in the USA	Haensler (1925)
Chrysanthemum spp.	*V. d.* on, in Italy	D'Aulerio & De Polzer (1982)
	V. a-a isol. ex, in the USA (1950)	Engelhard (1957)
	V. d. isol. ex, in Argentina	Alippi & Ronco (1980)
Coreopsis lanceolata L.	*V. a-a* on, in the USA, 1950	Engelhard (1957)
Cosmos spp.	*V. d.* on, in Turkey	Sezgin *et al.* (1985)
Cynara scolymus L.	*V. d.* on, in France	Dufrenoy (1927a,b)
	V. d. isol ex & re-inoc. on to, in Chile	Fernandez & Tobar (1989)
	V. d. isol. ex & re-inoc. on to, in the USA	Bhat *et al.* (1999)
Dahlia pinnata Cav. (as *D. variabilis* Desf.)	*V. d.* isol. ex & re-inoc. on to, in Germany & *V. d.* described	Klebahn (1913)
	V. d. on, in The Netherlands	Van der Meer (1925)
Dahlia sp. (as *D. cultorum*)	*V. d.* on, in Moldavia (CIS)	Bukar and Salinskya (1986)
	V. d. isol. ex & re-inoc. on to, in Japan	Takeuchi & Hori (1998)
Dimorphotheca sinuata DC. (as *D. aurantiaca* DC.)	*V. a-a* on, in the USA (1950)	Engelhard (1957)

Continued

Table 11.1. *Continued*

Host	Verticillium	Reference
Compositae *Continued*		
Echinacea purpurea (L.) Moench.	*V. d.* isol. ex & re-inoc. on to, in the USA	Putnam & Crowe (1999)
Echinops banaticus Rochel *ex* Schrader (as *E. ritro*)	*V. d.* isol. ex & re-inoc. on to, in Japan	Takeuchi & Hori (1998)
Erigeron canadensis L.	*V. d.* on, in The Netherlands	Van der Meer (1925)
Erigeron spp.	*V. a-a* on, in the USA (1950)	Engelhard (1957)
Gerbera jamesonii Bolus *ex* Hook. (as *G. jamesonii* Bolus)	*V. a-a* (? *V. d.)* isol. ex, in the USA	Snyder *et al.* (1950)
	V. sp. on, in Italy	Ragoni Galluci (1959)
	V. a. & *V. d.* on, in the USSR	Kulibaba (1972)
Gnaphalium margaritaceum L.	*V. d.* on, in The Netherlands	Van der Meer (1925)
Helianthus annuus L.	*V. d.* on, in the USSR	Zelle (1932)
	V. d. on, in Canada	Sackston *et al.* (1957)
	V. d. on, in Uruguay	Sackston (1957)
	V. d. on, in France	Alabouvette & Bremeersch (1975)
	V. d. on, in Spain	Jimenez Diaz *et al.* (1980)
	V. d. on, in China	Yang & Wei (1988)
	V. d. on, in Japan	Sumino & Abe (1999)
	V. d. isol. ex & re-inoc. on to, in the UK	Church & McCartney (1995)
	V. d. isol. ex seed, in Nigeria	Ataga & Akueshi (1996)
Helichrysum bracteatum (Vent.) Andrews	*V. a-a* (? *V. d.*) isol. ex, in the USA	Baker & Locke (1946)
Ixodia achillaeoides R.Br.	*V. d.* on, in Australia	Hall *et al.* (1996)
Lactuca sativa L. var. *longifolia* Lam.	*V. d.* on, in Greece (Crete)	Ligoxigakis & Vakalounakis (1997)
Liatris spicata (L.) Willd.	*V. d.* isol. ex & re-inoc. on to, in Japan	Takeuchi & Hori (1998)
Matricaria chamomilla sensu auct.	*V. a-a* on, in Germany	Skadow (1969a)
Parthenium argentium A. Gray	*V. a-a* isol. ex, in the USA,	Campbell *et al.* (1943)
Pericallis hybrida R. Nordenstam	*V. a-a* on, in Bulgaria	Khristova (1958)
Rudbeckia hirta L.	*V. a-a* on, in the USA (1950)	Engelhard (1957)

Senecio cineraria DC. (as *S. bicolor*)	*V. d.* isol. ex & re-inoc. on to, in the USA	Koike (1991)
S. cruentus DC.	*V.* on, in the USA	White (1928) reported by Rudolph (1931)
	V. a-a on, in the USA (1950)	Engelhard (1957)
	V. d. on, in Italy	Aloj & Garibaldi (1984)
S. vulgaris L.	*V. d.* on, in The Netherlands	Van der Meer (1925)
	V. a-a on, in the USA	Zeller (1936)
	V. a-a & *V. d.* isol. ex, in the UK (n.h.)	Harrison & Isaac (1969)
Stevia rebaudiana	*V. d.* on, in the USA (2000)	Farrar *et al.* (2000)
Tagetes spp.	*V. d.* on, in the USA	Hine and McCain (1984)
Tanacetum cinerariifolium (Trev.) Schultz-Bip. (as *Chrysanthemum cinerariifolium*)	*V. a-a* on, in India	Shivanandappa & Govindu (1965)
Taraxacum officinale Wigg	*V.* sp. on, in the USA	Rudolph (1926b)
(as *T. officinale* Weber)	*V. d.* inoculated on to, from raspberry, in the USA	Rudolph (1931)
Tragopogon porrifolius L.	*V.* sp. on, in the USA	Jagger & Stewart (1918)
	V. a-a on, in the USA (1950)	Engelhard (1957)
Verbesina encelioides Benth. & Hooker f. ex A. Gray (as *V. encelioides* (Cav.) A. Gray)	*V. d.* isol ex, in Australia (n.h.)	Evans (1971a)
Xanthium strumarium L. (as *X. pungens* Wallr.)	*V. d.* isol. ex, in Australia (n.h.)	Evans (1971a)
X. strumarium L. (as *X. spinosum* L.)	*V. d.* isol. ex, in Australia (n.h.)	Evans (1971a)
	V. d. on, in Turkey	Kocaturk & Karcilioglu (1976, 1980)
Xanthium sp.	*V.* wilt reported, USA	Carpenter (1914)
Cornaceae		
Aucuba japonica Thunb.	*V. a-a* on, in the USA (1950)	Engelhard (1957)
	V. d. on, in Italy	Lorenzini *et al.* (1983)
Cornus florida L.	*V. a-a* on, in the USA	Holmes (1957)
Convolvulaceae		
Ipomoea batatas (L.) Lam.	*V. a-a* on, in the USA (1950)	Engelhard (1957)
I. lonchophylla J.M. Black	*V. d.* isol. ex, in Australia (n.h.)	Evans (1971a)

Continued

Table 11.1. *Continued*

Host	Verticillium	Reference
Crucifereae		
Armoracia rusticana P. Gaertner, Meyer & Scherb.	*V. a-a* isol. ex & re-inoc. on to, in Germany	Pötschke (1923)
(as *Radicula armoracea* (L.) Robbins	*V. d.* isol. ex & re-inoc. on to, in Germany	Böning (1936)
	V. a-a on, in the USA (1950)	Engelhard (1957)
	V. d. isol. ex & re-inoc. on to, in Italy	D'Ercole *et al.* (1989)
	V. d. from; host range	Chang & Eastburn (1994)
	V. d. on, in Illinois, USA	Eastburn & Chang (1994)
Brassica napus L., Napobrassica group		
(as *B. napobrassica* Mill.)	*V. a-a* on, in the USA (1950)	Engelhard (1957)
Brassica napus L.	*V. d.* on, in Germany	Daebeler *et al.* (1985)
		Daebeler *et al.* (1988)
	Spread of *V. d.* to other *Brassica*	Ohlsson (1988)
	Influence of herbicides on *V. d.*	Nilsson (1977)
	Effects of *V. d.* on	Svensson and Lerenius (1987)
	Remote sensing of infection	Nilsson (1985)
	Survey in western Germany	Krüger (1989)
	V. longisporum described from, in the UK	Karapapa *et al.* (1997a,b)
B. oleracea Botrytis group	*V. d.* isol. ex & re-inoc. on to, in the USA	Koike *et al.* (1994)
Capitata group	*V. a-a* isol. ex, in the USA	Snyder *et al.* (1950)
Gemmifera group	*V. a-a* isol. ex, in the USA	Snyder *et al.* (1950)
	V. d. isol. ex, in the UK	Isaac (1957a)
	History & description	Isaac (1957a)
B. pekinensis Skeels	*V. a-a* isol. ex & re-inoc. on to, in Japan	Watanabe *et al.* (1973)
B. rapa L.	*V. d.* on, in Germany	Daebeler *et al.* (1985)
	V. d. on, in Italy	Ciccarese *et al* (1987)
	V. d. on, in Japan, control	Hirota & Miyagawa (1988)
Brassica spp.	250 vars. infected with *V. d.* disease assessment	Yui *et al.* (1985)

Camelina sativa (L.) Crantz	*V. longisporum* ex oilseed rape inoc. on, in Germany	Henneken *et al.* (1998)
		Foller *et al.* (1998)
Capsella bursa-pastoris Medik.	*V. a-a* isol. ex, in The Netherlands. Re-inoc. on to lucerne	Kort & van Rheenen (1959)
	V. a-a & *V. d.* isol. ex, in the UK (n.h.)	Harrison & Isaac (1969)
	V. d. isol. ex, in Australia (n.h.)	Evans (1971a)
Cichorium intybus L.	*V. d.* isol. ex, in Belgium	Vanderwalle & Parmentier (1951)
	V. d. on, in Italy	Ciccarese *et al.* (1987)
Hesperis matronalis L.	*V. d.* isol. ex & re-inoc. in Italy	Frisullo *et al.* (1987)
Matthiola sp.	*V. d.* isol ex, in New Zealand	Cunningham (1949)
M. incana (L.) R.Br. (as *M. incana* R.Br.)	*V. a-a* (?*V. d.)* isol. ex, in the USA	Snyder *et al.* (1950)
Raphanus sativus L.	*V. a-a* isol. ex, in the USA	Snyder *et al.* (1950)
	V. d. isol. ex, in Japan. Path. to turnip, aubergine, strawberry, tomato & cucumber	Kitazawa & Suzui (1980)
Rapistrum rugosum (L.) J.P. Bergeret (as *R. rugosum* (L.) All.)	*V. d.* isol. ex, in Australia (n.h.)	Evans (1971a)
Sisymbrium irio L.	*V. a-a* on, in the USA	Leyendecker (1951)
Cucurbitaceae		
Citrullus lanatus (Thunb.) Matsum. & Nakai (as *C. vulgaris* Schrad.)	*V. a-a* on, in the USA	Wollenweber (1922)
	V. a-a isol. ex, in the USA	Rudolph & Snyder (1936)
	V. d. on, in Australia	Anon (1964)
	V. d. on, in Japan,	Kondo *et al.* (1987)
Cucumis melo L.	*V. a-a* on, in The Netherlands	Van der Leck (1918)
	V. d. on, in The Netherlands	Van der Meer (1925)
	in the UK	Bewley (1928)
	in Denmark	Gram & Rostrup (1925)
	V. a-a isol. ex, in France	Dufrenoy (1927b)
	V. sp. on, in the USA	Kendrick & Schroeder (1934)
	V. a-a isol. ex, in the USA	Rudolph & Snyder (1938)
	V. d. on, in Turkey	Saydam & Copcu (1973)

Continued

Table 11.1. *Continued*

Host	Verticillium	Reference
Cucurbitaceae *Continued*		
	V. d. on, in Turkey	Ecvil and Yalcin (1977)
	V. d. on, in Romania	Costache & Tomescu (1987)
C. melo, inodorus group	*V. a-a* isol. ex, in the USA	Rudolph & Snyder (1938)
Cucumis sativus L.	*V. a-a* isol. ex & re-inoc. on to, in Sweden	Lindfors (1917)
	V. a-a isol. ex, in The Netherlands	Van der Meer (1925)
	V. d. isol. ex & re-inoc. on to, in Italy	D'Ercole (1976)
	V. a-a on, in Republic of Ireland	Clancy (1986)
	V. a-a on, in South Africa	Visser (1980)
	V. a-a more severe than *V. d.* in the USA	Gubler *et al.* (1978)
	V. d. (race 2) on, in Greece (Crete)	Ligoxigakis & Vakalounakis (1997)
Cucurbita pepo L.	*V. d.* (race 2) on, in Greece (Crete)	Ligoxigakis & Vakalounakis (1997)
Cucurbita sp.	*V.* sp. on, in The Netherlands	Gram & Rostrup (1925)
Cupressaceae		
Thuja sp.	*V.* sp. isol. ex, in The Netherlands	Shoevers in Van der Lek (1918)
Ebenaceae		
Diospyros kaki L.	*V. a-a* on, in the USA (1942)	Engelhard (1957)
D. texana Scheele	*V. a-a* on, in the USA (1950)	Engelhard (1957)
D. virginiana L.	*V. a-a* isol. ex, in the USA	Carter (1945)
Elaeagnaceae		
Elaeagnus angustifolia L.	*V. a-a* isol. ex & re-inoc. on to, in the USA	Rogerson (1957)
Hippophaë rhamnoides L.	*V. d.* isol. ex, in the USSR	Petrova (1982)
	V. d. isol. ex, in the UK	Kennedy (1987)
	V. d. on, in the USSR	Chikrizova (1987)
	V. d. on, in Estonia, on plants of Siberian origin	Piir (1996)
Ericaceae		
Erica australis Hort.	*V. a-a* (? *V. d.*) isol. ex, in the USA	Snyder *et al.* (1950)

E. milii Des Moul.	*V. a-a* on, in the USA (1950)	Engelhard (1957)
E. persoluta L.	*V. a-a* (*V. d.*) isol. ex, in the USA	Snyder *et al.* (1950)
Rhododendron molle G. Don	*V. a-a* isol. ex & re-inoc. on to, in the USA	Himelick (1969)
Rhododendron spp.	*V. a-a* on, in the USA	Edson & Wood (1936)
Vaccinium angustifolium Aiton	*V. d.* on, in the USA	Brisson *et al.* (1974)
Euphorbiaceae		
Codiaeum variegatum (L.) Blume	*V. d.* on, in Italy	Garibaldi & Gullino (1976)
Euphorbia drummondii Boiss.	*V. d.* isol. ex, in Australia (n.h.)	Evans (1971a)
E. pulcherrima Willd. *ex* Klotzsh	*V. d.* on, in Italy	Grasso & Pacetto (1969)
Manihot esculenta Crantz		
(as *M. utilissima* Pohl)	*V. a-a* on, in Brazil	Averna-Saccá (1922)
	V. d. on, in Rwanda	Maraite & Meyer (1976, 1977)
Fagaceae		
Castanea sativa L.	*V. a-a* on, in the USA	McCain (1963)
	V. d. isol. ex, in the UK	Slawson (1987)
Quercus agrifolia Née	*V. a-a* artificially inoc. on to	Thomas & Boza (1984)
Q. cerris L.	*V. a-a* on, in Hungary	Georgescu *et al.* (1959)
Q. lobata Née	*V. a-a* inoc. on to, in the USA	Thomas & Boza (1984)
Q. palustris Muench.	*V. a-a* isol. ex & re-inoc. on to, in the USA	Himelick (1969)
Q. petraea (Mattuschka) Liebl.	*V. d.* isol. ex, in Hungary	Vajna (1999)
Q. robur L.	*V. a-a* on acorns, in Czechoslovakia	Urosevic (1987)
Q. rubra L.	*V. a-a* isol. ex & re-inoc. on to, in the USA	Himelick (1969)
Quercus sp.	*V. d.* on, in Moldavia, USSR	Krangauz (1958)
Geraniaceae		
Pelargonium domesticum L. Bailey	*V. a-a* isol. ex, in the USA	Baker *et al.* (1940)
P. grandiflorum (Andrews) Willd.	*V. d.* on, in Italy	Garibaldi & Gullino (1973)
P. graveolens l'Hérit.	*V. a-a* on, in India	Ranganathiah & Swarmy (1969)
P. hortorum L. Bailey	*V. a-a* isol. ex, in the USA	Baker *et al.* (1940)
	V. a-a isol. ex & re-inoc. on to, in the USA	Torgeson (1951)

Continued

Table 11.1. *Continued*

Host	Verticillium	Reference
Geraniaceae *Continued*		
Pelargonium sp.	*V. a-a* & *V. d.* isol. ex & re-inoc. on to, in the UK	Fletcher & Griffin (1972)
	V. a-a on, in Denmark	Hellmers (1975)
	V. a-a on, in Poland	Orijkowski (1995)
Grosulariaceae		
Ribes nigrum L.	*V. d.* on, in The Netherlands	Van der Meer (1925)
R. rubrum L.	*V. d.* on, in The Netherlands	Van der Meer (1925)
R. sanguineum Pursh var. *lombartii*	*V. d.* isol. ex, in The Netherlands,	Van der Meer (1925)
R. uva-crispa L. & *Ribes* sp.	*V.* sp. on & inoc. on to currants in Germany	Aderhold (1907)
R. uva-crispa (as *R. grossularia* L.)	*V. d.* on, in The Netherlands	Van der Meer (1925)
R. uva-crispa	*V. d.* isol. ex, in France	Dufrenoy & Dufrenoy (1927)
Ribes spp.	*V. d.* on, in eastern Germany	Muller (1990)
Hippocastanaceae		
Aesculus hippocastanum L.	*V. a-a* isol. ex, in Germany	Wollenweber (1929)
	V. a-a on, in the USA	Armstrong (1941)
	V. a-a on, in the UK	Greig *et al.* (1982)
Juglandaceae		
Juglans regia L.	*V. d.* on, in France (1930)	Engelhard (1957)
Juglans sp.	*V.* sp. isol. ex, in France	Dufrenoy (1927c)
Labiatae		
Agastache rugosa Kuntze	*V. d.* isol. ex, in the USA	Block *et al.* (1989)
	Possible specificity of *V. d.* strain	Fuentes-Granados & Widrlechner (1995)
Lamium amplexicaule L.	*V. d.* isol. ex, in Australia (n.h.)	Evans (1971a)
Mentha × *piperita* L.	*V.* sp. on, in the USA	Nelson (1926)
	V. d. isol. ex, in the USA	Post (1935)
	General	Nelson (1950)

Mentha × *spicata* L.	*V. a-a* on, in the USA (1952)	Engelhard (1957)
Mentha sp.	*V. d.* on, in Bulgaria	Tanev *et al.* (1979)
Monarda didyma L.	*V. d.* isol. ex & re-inoc. on to, in The Netherlands,	Van der Meer (1925)
M. fistulosa L.	*V. a-a* var. *menthae* on, in the USA (? experimental)	Engelhard (1957)
Perilla ocymoides L.	*V. d.* isol. ex & re-inoc. on to in Japan	Kanno & Horiuchi (1999)
Salvia farinacea Benth.	*V. a-a* isol. ex, in the USA	Baker & Locke (1946)
S. reflexa Hornem.	*V. d.* isol. ex, in Australia (n.h.)	Evans (1971a)
Satureja hortensis L.	*V. d.* isol. ex, in Canada	Gourley (1979)
Lauraceae		
Cinnamomum camphora (L.) J. Presl	*V. a-a* (? *V. d.*) isol. ex, in the USA	Wilhelm *et al.* (1955)
Laurus nobilis L.	*V. d.* on, in Italy	Gullino (1983)
Persea americana Miller	*V. a-a* on, in the USA (1952)	Engelhard (1957)
	V. a-a on, in Chile & Equador	Zentmeyer (1949)
	V. a-a on, in Argentina	de Ramallo (1972)
	V. d. on, in Mexico	García *et al.* (1984)
	V. a-a on, in Mexico	Morales-Garcia (1989)
	V. d. on, in Morocco	Vanderweyen (1978)
	V. d. on, in Chile	Latorre & Allende (1983)
Sassafras albidum (Nutt.) Nees	*V. a-a* isol. ex, in the USA	Carter (1945)
Leguminosae		
Arachis hypogaea L.	*V. d.* isol. ex, in Israel	Krikun & Chorin (1966)
	V. d. on, in Turkey	Kocaturk & Karcilioglu (1976a,b)
Ceratonia siliqua L.	*V.* sp. on, in the USA (1952)	Engelhard (1957)
Cercis canadensis L.	*V. a-a* isol. ex & re-inoc. on to, in the USA	Carter (1945)
	V. d. isol. ex & re-inoc. on to, in Iran	Ershad (1972)
	V. d. on, in Canada	Thorpe & Jarvis (1978)
C. chinensis Bunge	*V. d.* (isol. ex *Acer* sp. etc.) inoc. on to	Jacobs *et al.* (1994)
C. siliquastrum L.	*V. a-a* on, in Italy	Engelhard (1957)
C. yunnanensis Hu & Cheng	*V. d.* (isol. ex. *Acer* sp. etc.) inoc. on to	Jacobs *et al.* (1994)

Continued

Table 11.1. *Continued*

Host	Verticillium	Reference
Leguminosae *Continued*		
Cicer arietinum L.	*V. a-a* [*sic*] (*v. d.*) on, in Pakistan	Bhatti *et al.* (1985, 1987)
	V. a-a [*sic*] (*v. d.*) on, in Tunisia	Halila & Harrabi (1987)
	V. d. on, in Turkey:seed-borne	Maden (1987)
	V. d. on, in Italy	Infantino & Porta-Puglia (1990)
	V. d. on, in Greece (Crete)	Ligoxigakis & Vakalounakis (1997)
	V. d. on, in the USA	Erwin (1958)
Cladrastis lutea (Michx.) K. Koch	*V. a-a* isol ex, in the USA	Carter (1940)
Crotalaria juncea L.	*V. d.* on, in South Africa	Wickens (1949)
Glycine max (L.) Merr.	*V. d.* isol. ex & re-inoc. on to, in the USSR	Vasilieff (1933)
	V. d. isol. ex & re-inoc. on to, in Japan	Ohto *et al.* (1993)
Gymnocladus dioica (L.) Koch	*V. a-a* isol. ex & re-inoc. on to, in the USA	Himelick (1969)
Hedysarum coronarium L.	*V. d.* on, in USA, (h.) to pea, sweet pea & *Onobrychis* sp. (n.h.) to lucerne, potato, *Antirrhinum*, chrysanthemum & tomato	Milton & Isaac (1976)
Lathyrus odoratus L.	*V. a-a* isol ex & re-inoc. on to, in the UK	Bewley (1922)
	V. a-a on, in the USA (1952)	Engelhard (1957)
	V. d. on, in Australia	Anon (1963)
Lotus corniculatus L.	*V. a-a* isol. ex (n.h.) in Canada	Aubé & Sackston (1964)
Lupinus albus L.	*V. a-a* on, in the UK	Griffiths & Isaac (1963)
	V. a-a & *V. d.* on, in Canada	Aubé & Sackston (1964)
L. polyphyllus Lindley	*V. d.* on, in The Netherlands	Van der Meer (1925)
	V. a-a more virulent than *V. d.*	Griffiths & Isaac (1963)
	V. a-a on, in NZ	Harvey *et al.* (1996)
Medicago hispida Gaertn.	Soil infection of, with *V.* strains *ex* raspberry, tomato & apricot	Rudolph (1931)
M. polymorpha L.	*V. d.* isol. ex tap-root in Australia (n.h.)	Evans (1971a)

M. sativa L.	*Verticillium* wilt of lucerne (alfalfa) caused by *V. a-a*, though mild attacks by *V. d.* also	Isaac (1957c, 1959)
	From the first report of *V. a-a* in Sweden	Hedlund (1923)
	Germany	Richter & Klinkowski (1938)
	UK	Noble *et al.* (1953)
	Poland	Truszkowska & Legiec (1973)
	Verticillium wilt was considered as a European disease	Kreitlow (1962)
		Holliday (1980)
	V. a-a & *V. d.* had been reported in Canada	Aubé & Sackston (1964)
	but only as a dangerous wilt by and its spread as	Sheppard (1979)
	lucerne acreage increased noted by	Atkinson (1981)
	V. a-a on, in the USA	Graham *et al.* (1977)
	V. a-a on, in Wisconsin, USA	Grau *et al.* (1981)
	V. a-a on, in Wyoming, USA	Gray & Roth (1982)
	Spread in Pacific NW of USA	Christen & Peaden (1981, 1982)
	V. a-a on, in Kansas, USA	Stuteville *et al.* (1986)
	V. a-a on, in Maryland, USA	Grybauskas & Dutky (1987)
	Spread in South California, USA	Erwin *et al.* (1989)
	Spread in California, USA	Gordon *et al.* (1989)
	Spread in USA	Leath (1984)
	V. a-a outbreak in Canada	Richard (1987)
	V. a-a on, in Quebec, Canada	Nicholls *et al.* (1987)
	New Zealand	Sanderson (1976)
	Japan	Kitazawa & Sato (1981)
	Effect of *V.* wilt on world agriculture	Pegg (1984)
	Outline of current (1985) research	Heale (1985)
Neptunia gracilis Benth.	*V. d.* isol. ex, in Australia (n.h.)	Evans (1971a)
Onobrychis viciifolia Scop.	*V. a-a* isol. ex & re-inoc. on to, in Germany	Richter & Klinkowski (1938)
	V. d. isol. ex, in the UK	Isaac (1946)
	V. a-a & *V. d.* isol. ex, in Canada	Aubé & Sackston (1964)

Continued

Table 11.1. *Continued*

Host	Verticillium	Reference
Leguminosae *Continued*		
Phaseolus vulgaris L.	*V. a-a* on, in Brazil	Averna-Sacca (1922)
	V. a-a on, in Chile	Mujica (1957)
Pisum sativum L.	*V. d.* isol. ex, in the UK	Isaac & Rogers (1974)
Robinia pseudoacacia L.	*V. a-a* on, in The Netherlands	Buisman (1932)
	V. a-a isol. ex, in the USA	Carter (1938)
	V. a-a on, in Italy	Goidanich (1935)
Sophora japonica L.	*V. a-a* on, in Italy	Goidanich (1935)
	V. a-a isol. ex & re-inoc. on to, in the USA	Himelick (1969)
Tephrosia sp.	*V. d.* isol. ex & re-inoc. on to, in Uganda	Hansford (1939)
Trifolium hybridum L.	*V. a-a* & *V. d.* isol. ex, in Canada	Aubé & Sackston (1964)
T. pratense L.	*V. a-a* on, in Canada	Sackston (1959)
	V. a-a & *V. d.* isol. ex, in Canada	Aubé & Sackston (1964)
T. repens L.	*V. a-a* & *V. d.* isol. ex, in Canada (n.h.)	Aubé & Sackston (1964)
Trigonella foenum-graecum L.	*V. a-a* on, seed-borne	Hashmi (1988)
Vicia faba L.	*V. d.* isol. ex, in Greece (Crete)	Ligoxigakis & Vakalounakis (1994)
Vigna radiata (L.) R. Wilczek	*V. a-a* on, in Pakistan	Ehteshamul-Haque & Ghaffar (1994)
V. unguiculata (L.) Walp. (as *V. sinensis* Endl.)	*V. a-a* isol. ex, in the USA	Rudolph & Snyder (1936)
	V. d. on, in Turkey	Kocaturk & Karcilioglu (1976a)
V. unguiculata sp. *sesquipedalis* (L.) Verdc. (as *V. sesquipedalis* W.F. Wright)	*V. a-a* (? *V. d.*) isol. ex, in the USA	Wilhelm *et al.* (1955)
Linaceae		
Linum usitatissimum L.	*V. d.* on flax, in Belgium	Marchal (1939)
	V. d. on linseed in the UK and Germany	Fitt *et al.* (1992)
	effect of *V. d.* on linseed crops in the UK	Fitt *et al.* (1998)
Magnoliaceae		
Liriodendron tulipifera L.	*V. d.* on, in France (1931)	Engelhard (1957)

	V. a-a isol. ex & re-inoc. on to, USA	Waterman (1956)
Magnolia grandiflora L.	*V. a-a* isol. ex, in the USA	McCain (1963)
M. soulangeana Soul.	*V. a-a* isol. ex & re-inoc. on to, in the USA	Schreiber *et al.* (1961)
M. tomentosa Thunb. (as *M. stellata* (Sieb. & Zucc.) Maxim.)	*V. a-a* isol. ex, in the USA	Himelick (1969)
Magnolia sp.	*V. a-a* isol. ex, in the USA	Engelhard & Carter (1956)
Malvaceae		
Abelmoschus esculentus (L.) Moench	*V. d.* isol. ex & re-inoc. on to, in the USSR	Vasilieff (1933)
	V. d. on, in Turkey	Esentepe *et al.* (1972)
(as *Hibiscus esculentus* L.)	*V.* sp. isol. ex plants from the USA & re-inoc. on to, in Germany	Wollenweber (1913)
(as *Hibiscus esculentus* L.)	*V. a-a* on, in the USA	Carpenter (1918)
Abutilon theophrasti Medik.	*V.* wilt of in the USA	Carpenter (1914)
	V. d. on, in the USA	Kirkpatrick & Harrison (1979)
	V. d. on, in the USA	Sickinger *et al.* (1987)
	V. d. Use of as control of velvetleaf	Wiley *et al.* (1985)
		Hartzler (1996)
(as *A. avicennae* Gaertn.)	*V. d.* on, in USSR	Vasilieff (1933)
Alcea rosea L. (as *Malva alcea* L.)	*V. d.* isol. ex & re-inoc. on to, in the Netherlands	Van der Meer (1925)
Callirhoë papaver (Cav.) A. Gray	*V. a-a* on, in the USA (1952)	Engelhart (1957)
Gossypium L. (Cotton)		
G. arborium L.	India, West & South-East Asia	
G. barbadense L.	South America, California, CIS, India, Sudan & Egypt	
G. herbaceum L.	Africa, West & Central Asia	
G. hirsutum L. (Upland Cotton)	Central Asia, China, the USA, India, Mexico, Central & South America, Africa, Australia	
	V. d. from, defoliating isolates = P_1 (CIS 2 & 3; formerly T1 & T9)	
	Less virulent, non-defoliating isolates = P_2 (CIS 1 & 4; formerly SS4)	
	V. a-a on, in the USA (probably *Fusarium* sp.)	Carpenter (1914)

Continued

Table 11.1. *Continued*

Host	Verticillium	Reference
Malvaceae *Continued*		
G. hirsutum L. (upland Cotton) *Continued*	Inoc. with *V. d.* (as *V. a-a*) from okra	Carpenter (1918)
	Inoc. with *V.* sp. from tomato	Bewley (1921)
	V. a-a (? *V. d.*) on cotton, in the USA	Sherbakoff (1928)
	V. a-a (? *V. d.*) isol. ex & re-inoc. on to cotton, in the USA	Miles & Persons (1932)
	V. d. isol. ex & re-inoc. on to cotton, in Uganda	Hansford (1939)
	V. d. on cotton, in Mozambique	Teixeira & Teixeira (1953a,b)
	V. d. on cotton, in Australia	Anon (1959)
	on, in Turkey	Kamal and Saydam (1970)
	V. d. isol. ex & re-inoc. on to, in India	Isaac *et al.* (1972)
	V. d. on, in India	Kannan and Srinivasan (1984)
	V. d. isol. ex, in Madagascar	Cauquil (1973)
	V. d. on, in Iraq	Mamluk (1974)
	strains of	Mamluk (1975)
	V. d. on, in Iran	Moshirabadi (1978)
	V. d. on, in Bulgaria	Savov (1979)
	Virulence of Greek isolates	Tjamos and Kornaros (1978)
	V. d. on, in Israel	Tsror *et al.* (1990)
	V. d. on, in China	Zhang *et al.* (1981)
	V. d. & *V. a-a* on, in Egypt	Michail (1989)
	V. d. on, in Turkey	Esentepe *et al.* (1986)
	V. d. on, in South Africa	Swanepoel & de Kock (1996)
	V. nigrescens isol. ex, in the USA	Bugbee (1967)
Hibiscus cannabinus L.	*V. d.* on, in Italy	Frisullo *et al.* (1995)
	V. d. on, in Italy	Di Corato *et al.* (1996)
H. trionum L.	*V. d.* isol. ex, in Australia, (n.h.)	Evans (1971a)
Lagunaria patersonii (Andrews) Don (as *L. patersonii* L.)	*V. d.* isol. ex and re-inoc. on to, in Italy	Polizzi (1996)

Malva neglecta Wallr.	*V. d.* on, in Germany	Skadow (1969a)
M. rotundifolia L.	*V.* sp. on, in the USA	Rudolph (1926b)
Sida spinosa L.	*V. d.* on, in South Africa	Wickens (1948)
Moraceae		
Maclura aurantiaca L.	*V. a-a* on, in Italy	Goidánich (1934b)
	V. a-a on, in the USA	Waterman (1941)
Nyssaceae		
Nyssa sylvatica Marsh.	*V. a-a* isol. ex, in the USA	Engelhard & Carter (1956)
Oleaceae		
Fraxinus americana L.	*V. a-a* isol. ex & re-inoc. on to, in the USA	Himelick (1969)
	V. d. isol. ex & re-inoc. in the USA	Heffer & Regan (1996)
	V. d. affecting nursery trees in the USA	Worf *et al.* (1994)
F. angustifolia	*V. d.* causing leaf scorch on, in the USA	Heffer & Regan (1996)
F. excelsior L.	*V. a-a* isol. ex & re-inoc. on to, in the USA	Himelick (1969)
	V. d. affecting young plantations in The Netherlands	Schuring & van der Schaaf (1999)
F. nigra Marshall	*V. a-a* isol. ex, in the USA	Himelick (1969)
F. pensylvanica Marshall var. *subintegerrima* (Vahl) Frenald	*V. a-a* isol. ex & re-inoc. on to, in the USA	Thomas (1950)
	V. d. isol. ex, in the USA	Heffer & Regan (1996)
F. quadrangulata Michx.	*V. a-a* isol. ex & re-inoc. on to, in the USA	Himelick (1969)
Ligustrum amurense Carr.	*V. a-a* isol. ex, in the USA	Carter (1945)
Olea europaea L.	*V.* sp. on, in Italy	Ruggieri (1946)
	V. d. on, review of, in Italy	Cirulli (1981)
	V. d. on, in Greece	Demetriades *et al.* (1958)
		Zachos (1963)
	review of, in Greece	Thanassoulopoulos *et al.* (1979)
	spread in nurseries, in Greece	Thanassoulopoulos (1993)
	V. a-a [*sic*] (*V. d.*) isol. ex and re-inoc. on to, in the USA	Boyle (1963)

Continued

Table 11.1. *Continued*

Host	Verticillium	Reference
Oleaceae *Continued*		
Olea europaea L. *Continued*	*V. d.* (race 2) on, in Greece (Crete)	Ligoxigakis & Vakalounakis (1997)
	V. sp. on, in Turkey	Saydam & Copcu (1972)
	V. d. on, in France (1969)	Vigouroux (1975)
	V. a-a (? *V. d.*) isol. ex, in the USA	Snyder *et al.* (1950)
	Review of *V.* wilt in California, USA	Wilhelm et al. (1962)
	V. d. on, in Australia	Anon (1961a)
	V. d. on, in Spain	Caballero *et al.* (1981)
	update in Spain	Rodriguez Jurado *et al.* (1993)
	V. d. on, survey of, in Mediterranean	Ciccarone (1974)
	V. d. on, survey of, in Syria	Al-Ahmad (1988)
Osmanthus ilicifolius Standish	*V. a-a* on, in the USA	Gruenhagen & Fordyce (1963)
Syringa vulgaris L.	*V. d.* isol. ex & re-inoc. on to, in The Netherlands	Van der Meer (1925)
Onagraceae		
Clarkia elegans Dougl.	*V. a-a* on, in the USA (1952)	Engelhart (1957)
Fuchsia hybrida Hort. *ex* Vilm. (as *F. hybrida* Voss)	*V. a-a* isol. ex, in the USA	Baker *et al.* (1940)
Paperveraceae		
Eschscholzia californica Cham.	*V. d.* isol. ex & re-inoc. on to, in The Netherlands	Van der Meer (1925)
Papaver bracteatum Lindl.	*V. a-a* on, in Germany	Donant (1932)
P. nudicaule L.	*V. d.* on, in Italy	Garibaldi & Gullino (1976)
P. orientale L.	*V. d.* isol. ex & re-inoc. on to, in The Netherlands	Van der Meer (1925)
	V. d. on, in the UK	Beaumont (1953)
	V. a-a on, in the USA (1952)	Engelhard (1957)
P. rhoeas L.	*V. d.* isol. ex & re-inoc. on to, in The Netherlands	Van der Meer (1925)
Papaver sp.	*V. a-a* on, in USA (1952)	Engelhard (1957)

Pedaliaceae		
Proboscidea louisianica (Mill.) Thell.	*V. d.* isol. ex, in Australia (n.h.)	Evans (1971a)
Sesamum indicum L.	*V. d.* isol. ex & re-inoc. on to, in the USSR	Vasilieff (1933)
	V. d. on, in Turkey	Esentepe *et al.* (1972)
	V. a-a (?*V. d.*) isol ex & re-inoc. on to, in USA	Chilton (1957)
(as *S. orientale*)	*V. d.* isol. ex & re-inoc. on to, in Uganda	Hansford (1939)
	V. d. seed-borne in Uganda	Mathur & Kabeere (1975)
Piperaceae		
Peperomia obtusifolia var. *variagata*	*V. d.* on, in Italy	Grasso & La Rosa (1982)
Pittosporaceae		
Pittosporum tobira (Thunb.) Aiton	*V. a-a* isol. ex, in the USA	McMurphy via Rudolph (1931)
Plantaginaceae		
Plantago lanceolata L.	*V. a-a* & *V. d.* isol. ex, in the UK (n.h.)	Harrison & Isaac (1969)
P. major L.	*V. a-a* isol. ex, in The Netherlands & re-inoc. on to *Medicago sativa*	Kort & van Rheenen (1959)
Polemoniaceae		
Phlox carolina L.	*V. a-a* on, in the USA (1953)	Engelhard (1957)
P. decussata Lyon ex Pursh	*V.* sp. isol. ex & re-inoc. on to, in The Netherlands	Van der Meer (1925)
	V. a-a (? *V. d*) isol. ex, in Hungary	Husz (1936)
P. drummondii Hook.	*V.* sp. isol. ex, in The Netherlands	Van der Meer (1925)
	V. a-a on, in the USA (1953)	Engelhard (1957)
P. paniculata L.	*V. a-a* on, in the USA (1953)	Engelhard (1957)
Polemonium spp.	*V. a-a* on, in the USA (1953)	Engelhard (1957)
Polygonaceae		
Fallopia convolvulus (L.) A. Löve (as *Polygonum convolvulus* L.)	*V. d.* isol. ex, in Australia (n.h.)	Evans (1971a)
Polygonum persicaria L.	*V. a-a* & *V. d.* isol. ex, in the UK	Harrison & Isaac (1969)
Rheum rhaponticum L.	*V. a-a* (? *V. d.*) isol. ex, in the USA	Wilhelm *et al.* (1955)

Continued

Table 11.1. *Continued*

Host	Verticillium	Reference
Portulacaceae		
Portulaca oleracea L.	*V. a-a* isol. ex, in the USA (n.h.)	Oshima *et al.* (1963)
	V. d. isol. ex, in Australia (n.h.)	Evans (1971a)
Proteaceae		
Banksia victoria	infected with *V. d.* isol. ex *Leucospermum cordifolium*	Koike *et al.* (1991)
Leucospermum cordifolium	*V. d.* isol. ex & re-inoc. on to, in the USA	Koike *et al.* (1991)
L. salignum	not infected when inoc. with *V. d.* ex *L. cordifolium*	Koike *et al.* (1991)
Protea spp.	not infected when inoc with *V. d.* ex *L. cordifolium*	Koike *et al.* (1991)
Pyrolaceae		
Pyrola elliptica Nutt.	*V. a-a* on, in the USA (1953)	Engelhard (1957)
P. rotundifolia L. (as *P. asarifolia* Michx.)	*V. a-a* on, in the USA	Engelhard (1957)
Ranunculaceae		
Aconitum napellus L.	*V. d.* isol. ex, in The Netherlands	Van der Meer (1925)
	V. a-a on, in France	Desjobert & Herisset (1972)
Clematis sp.	*V. a-a* on, in Belgium	Marchal (1939)
Consolida ajacis (L.) Schur (as *Delphinium ajacis* L.)	*V. a-a* on, in the USA (1953)	Engelhard (1957)
Paeonia lactiflora Pallas	*V. d.* on, in the USA (1953)	Engelhard (1957)
P. officinalis L.	*V. d.* on, in the USA (1953)	Engelhard (1957)
Paeonia sp.	*V.* sp. isol. ex, in the USA	Dufrenoy (1929)
	V. a-a on, in USA	Martin (1926)
Resedaceae		
Reseda odorata L.	*V. a-a* on, in the USA (1953)	Engelhard (1957)
Rhamnaceae		
Ceanothus sp.	*V. a-a* on, in the USA	Harrington & Cobb (1984)
Rosaceae		
Cydonia oblonga Miller	*V. d.* on, in UK (1936)	Engelhard (1957)

	V. d. on, in France (1941)	Engelhard (1957)
Fragaria × *ananassa* Duch. (commercial strawberry)		
	V. d. on, in the USA	Thomas (1931)
	distribution of, in the USA	Strobel (1963)
	V. d. on, in Florida, USA	Howard and Albregts (1982)
	V. d. (most) & *V. a-a* isol. ex & re-inoc. on to, in Poland	Leski (1974)
	V. d. on, in South Africa	Visser & Kotze (1974)
	V. d. & *V. a-a* on, in the UK	Anon (1979)
	V. d. on, in Italy	D'Ercole (1975)
	Agricultural impact	Pegg (1984)
Malus pumila Miller	*V. a-a* on, in Bulgaria	Atanazoff *et al.* (1932)
	V. a-a isol. ex, in the UK	Sewell & Glasscock (1958)
	V. tadzhikistanicum (? *V. d.*) isol. ex, in Tadzhik, SSR	Osmolovskaya & Zlotina (1969)
	V. nigrescens on, in Tadzhik, SSR	Osmolovskaya & Zlotina (1969)
Prunus armeniaca L.	*V.* sp. on, in the USA	Czarnecki (1923)
	V. d. on, in France	Dufrenoy (1927c)
	V. sp. isol. ex & re-inoc. on to, in Canada	Berkeley (1927)
	V. d. on, in France	Vigouroux (1984)
	V. a-a on, in Hungary	Husz (1947)
	V. a-a isol. ex, in the USA	Goheen (1949)
	V. d. on, in NZ	June (1950)
	V. d. on, in Turkey	Saribay *et al.* (1973)
	Decline disease and control, Greece	Rumbos and Karayiannis (1999)
	V. d. on, in Moldavia, USSR	Popushoi & Kulik (1976)
		Popushoi (1977)
	V. d. isol. ex & re-inoc on to, in Iran	Zamani-Zadeh & Zakii (1995)
P. avium (L.) L.	*V.* sp. on, in The Netherlands	Van der Meer (1925)
	V. sp. on, in Czechoslovakia	Anon (1926)
	V. a-a (? *V. d.*) isol. ex, in the USA	Wilhelm *et al.* (1954)

Continued

Table 11.1. *Continued*

Host	Verticillium	Reference
Rosaceae *Continued*		
P. cerasus L. (as *P. cerasus acida* Ehrh.)	*V. a-a* isol. ex, in The Netherlands	Van der Lek (1918)
P. cerasus L. (as *P. ceracus austera*)	*V. a-a* isol. ex, in The Netherlands	Van der Meer (1925)
	V. a-a on, in the USA	Parker (1959)
P. ceracifera Ehrh.	*V.* sp. on, in France	Dufrenoy & Dufrenoy (1927)
	V. a-a on, in Hungary	Husz (1947)
P. domestica L.	*V.* sp. on, in France	Dufrenoy & Dufrenoy (1927)
	V. a-a on, in the UK	Keyworth (1944b)
	V. d. on, in New Zealand	June (1950)
	V. a-a on, in the USA (1953)	Engelhard (1957)
	V. a-a on, in Canada	Anon (1961a)
	V. d. & '*V. tadzhikistanicum*' on, in the CIS	Dermovskaya (1972)
	V. d. on, in Moldavia (CIS)	Gavrilenko *et al.* (1977)
	V. d. on, in Yugoslavia	Arsenijevic & Sevar (1981)
	V. d. on, in Turkey	Saribay & Demir (1984)
	V. d. on, in Iran	Kamran (1985)
P. dulcis (Miller) D. Webb (as *P. communis* Fritsch)	*V.* sp. isol ex, in the USA	Czarnecki (1923)
(as *P. amygdalus* Batsch.)	*V. a-a* isol. ex, in the USA	Carter (1938)
	V. d. on, in France	Anon (1932)
	V. d. on, in Bulgaria	Vitanov *et al.* (1972)
	V. d. isol. ex & re-inoc. on to, in Italy (1991)	Luisi & Sícoli (1993)
P. laurocerasus L.	*V. a-a* on, in the USA (1953)	Engelhard (1957)
P. lusitanica L.	*V. a-a* on, in the USA (1953)	Engelhard (1957)
P. mahaleb L.	*V. d.* isol. ex, in The Netherlands	Van der Meer (1925)
	V. a-a (? *V. d.*) isol. ex, in the USA	Wilhelm *et al.* (1954)
P. mume Siebold & Zucc.	*V. a-a* on, in the USA	Engelhard (1957)
P. persica (L.) Batsch. (as *P. persica* Sieb. & Zucc.)	*V. a-a* isol. ex, in the USA	Haensler (1922)

	V. d. on, in New Zealand	June (1950)
	V. a-a on, in Chile	Cancino *et al.* (1971)
	V. d. isol. ex, in Turkey	Saydam *et al.* (1971)
	V. d. isol. ex, in the CIS	Azimdzhanov (1971)
	V. d. isol. ex, in Yugoslavia & re-inoc on to pepper & cotton (symptoms) aubergine & tomato (symptomless n-h)	Arsenijevic (1977)
	V. d. on, in Spain	Jimenez-Diaz & Montes-Agusti (1974)
P. persica var. *nectarina* (Aiton) Maxim.	*V. d.* on, in Australia	Anon (1962)
P. nigra Aiton	*V. a-a* on, in Canada	Anon (1961a)
Pyrus communis L.	*V. d.* on, in The Netherlands	Anon (1951)
Rhaphiolepis indica Lindley	*V. a-a* isol. ex, in the USA	Wilhelm *et al.* (1955)
R. umbellata var. *integerrima* Rehd.	*V. a-a* isol. ex, in the USA	Wilhelm *et al.* (1955)
Rosa multiflora Thunb. *ex* Murray (as *R. multiflora* Thunb.)	*V. a-a* isol. ex, in USA	Engelhard & Carter (1956)
R. rugosa Thunb.	*V. d.* on, in Lithuania	Shal'tyanis & Saltenis (1974)
Rosa spp.	*V. a-a* on, in the USA (1953)	Engelhard (1957)
	V. d. on, in Jersey (Channel Islands)	Collingwood (1965)
	V. d. on, in Italy	Garibaldi & Gullino (1973)
	V. d. on, in the CIS	Khripunova (1976)
	Review of *V.* on, in the UK	Thomas, B.J. (1981)
	V. d. on: General review of	Gullino & Garibaldi (1996)
Rubus allegheniesis Porter	*V.* sp. isol. ex, in the USA	Lawrence (1912)
	V. a-a (? *V. d.*) isol. ex, in the USA	Snyder *et al.* (1950)
R. idaeus L.	*V.* sp. on, in the UK	Harris (1925b)
	'*V. ovatum*' isol. ex, in the USA	Berkeley & Jackson (1926)
	V. d. isol. in The Netherlands	Karthaus (1927)
	V. a-a on, in Germany	Goetz (1935)
	V. a-a & *V. d.* on, in Germany	Kuhne (1980)
R. occidentalis L.	*V.* sp. isol. ex (as '*Acrostalagmus caulophagus*') in the USA	Lawrence (1912)

Continued

Table 11.1. *Continued*

Host	Verticillium	Reference
Rosaceae *Continued*		
R. occidentalis L. *Continued*	*V. a-a* isol. ex, in the USA	Zeller (1936)
	V. a-a isol. ex, in the USA & Canada	Hockey (1923)
R. ursinus Cham. & Schldl.	*V. a-a* on, in the USA	Wilhelm & Thomas (1950)
Rubus sp. (as *R. procumbens* Muhl. var. *roribaccus* Bailey)	*V.* sp. isol. ex, in the USA	Rudolph (1931)
Rubus spp. (dewberry)	*V. a-a* on, in the USA (1953)	Engelhard (1957)
Rutaceae		
Citrus limon (L.) Burm.	*V. d.* on, in Tadzhik, SSR	Zlotina & Urakov (1977)
Phellodendron amurense Rupr.	*V. a-a* isol. ex & re-inoc. on to, in the USA	Himelick (1969)
Salicaceae		
Populus tremula L.	*V. a-a* on, in Germany	Liese (1934)
Sapindaceae		
Koelreuteria paniculata Laxm.	*V. a-a* isol. ex & re-inoc. on to, in the USA	Engelhard & Carter (1956)
Scrophulariaceae		
Antirrhinum majus L.	*V.* sp. on, in The Netherlands	Van der Meer (1925)
	V. a-a (? *V. d.*) isol. ex, in the USA	Baker *et al.* (1940)
	V. d. isol. ex & re-inoc. on to, in Argentina	Merlo *et al.* (1985)
Antirrhinum sp.	*V. a-a* isol. ex & re-inoc. on to, in the USA	Brown (1914)
Calceolaria spp.	*V. a-a* on, in the USA (1953)	Engelhard (1957)
Digitalis purpurea L.	*V. a-a* on, in the USA (1953)	Engelhard (1957)
Hebe elliptica (Forster) Pennell (as *Veronica elliptica* Forster)	*V.* sp. isol. ex, in the USA	McMurphy via Rudolph (1931)
Simaroubaceae		
Ailanthus altissima (Miller) Swingle	*V. a-a* on, in the USA (1953)	Engelhard (1957)
A. glandulosa Desf.	*V. a-a* isol. ex, in the USA	Martin (1929)
	V. a-a on, in Italy	Goidanich (1935)

	V. d. on, in France (1933)	Engelhard (1957)
	V. d. isol. ex, in Greece	Skarmoutsos & Skarmoutsou (1998)
	V. sp. on, in Austria	Cech (1998)
Simmondsiaceae		
Simmondsia chinensis (Link) C. Schneider	*V. d.* on, in the USA	Orum *et al.* (1981)
	V. d. on, in Israel	Tsror & Erlich (1996)
Solanaceae		
Atropa belladonna L.	*V. a-a* (? *V. d.*) on, in the USA	Tervet (1944)
Capsicum annuum L.	*V. tracheiphilum* described & re-inoc. on to, in Italy	Curzi (1925)
		Snyder and Rudolph (1939)
	V. a-a isol. ex, in the USA	Rudolph & Snyder (1937)
	Capsicum strain of *V. d.* isol. in Canada	Evans and McKeen (1975a,b)
	V. d. on, in Turkey	Saydam & Kamel (1970)
	V. d. on, in Portugal	Lucas & Borges (1972)
	V. d. isol. ex, in Israel; highly virulent strain, Vdp.	Tsror *et al.* (1998)
Capsicum sp.	*C.* inoc. with *V. a-a* ex tomatoes	Bewley (1921)
	V. a-a on, in Germany	Wollenweber (1922)
	V. d. on, in Uganda	Hansford (1939)
Datura stramonium L.	*V. d.* isol. ex, in Australia (n.h.)	Evans (1971a)
Lycopersicon esculentum Miller	*V. d.* (as *Fusarium lycopersici* Sacc.) on, in the UK	Massee (1896)
	V. d. (as *V. a-a*) on, in the UK	Bewley (1922)
(as *Lycopersicum esculentum* Miller)	*V.* sp. on, in the USA	Jagger & Stewart (1918)
(as *Solanum esculentum* Trn.)	*V. a-a* isol. ex & re-inoc. on to, in The Netherlands	Van der Lek (1918)
	V. a-a & *V. d.* on, in The Netherlands	Van der Meer (1925)
	Aggressive str. on, in The Netherlands	Paternotte & Van Kesteren (1993)
	V. a-a on, in the USA	Bryan (1925)
	V. a-a (? *V. d.*) isol. ex, in the USA	Snyder *et al.* (1950)
	V. sp. on, in the USA	Rudolph (1926a)
	V. d. on, in Turkey	Saydam & Kamel (1970)

Continued

Table 11.1. *Continued*

Host	Verticillium	Reference
Solanaceae *Continued*		
L. esculentum Miller *Continued*	*V. a-a* on, in Moldavia	Khar'kova & Okhova (1972)
	V. a-a & *V. d.* on, in Poland	Glaser (1977)
	V. d. on, in Portugal	Lucas & Borges (1972)
	V. a-a & *V. d.* on, in Canada	McKeen (1972b)
	V. d. on, in Jordan	Mamluk & Skaria (1979)
	V. d. on, in Morroco	Pineau (1977)
	V. d. on, in South Africa (race 2)	Ferreira *et al.* (1990)
	V. d. on, in the Korean Republic	Park *et al.* (1995)
	V. d. (race 2) on, in Greece (Crete)	Ligoxigakis & Vakalounakis (1997)
	V. nigrescens on, in Greece	Thanassoupoulos & Kitsos (1972)
	V. d. on, in New Zealand	Hill (1982)
L. esculentum var. *cerasiforme* (Dun.) A. Gray	*V. a-a* on, in the USA (1953)	Engelhard (1957)
L. esculentum var. *pyriforme* Alef. (as f. *pyriforme* (Dun.) C.H. Mull.)	*V. a-a* on, in the USA (1953)	Engelhard (1957)
Nicotiana suaveolens Lehm.	*V. d.* isol. ex, in Australia (n.h.)	Evans (1971a)
N. tabacum L.	*V. a-a* (? *V. d.*) isol. ex, in the USA	Sherbakoff (1934)
	V. d. on, in NZ (1944)	Thomson & McLeod (1959)
	V. d. on, in Chile	Latorre *et al.* (1987)
	V. tricorpus on, in NZ	Taylor (1968)
	V. d. & *V. nigrescens* on, in Canada	Sheppard & Viswanathan (1974)
Petunia hybrida Hort. Vilm.	*V. a-a* (? *V. d.*) isol. ex, in the USA	Baker & Locke (1946)
Physalis alkekengi L. (as *P. francheti*)	*V. a-a* on, in the USA	Nance (1939)
P. angulata L.	*V. d.* isol. ex, in Australia (n.h.)	Evans (1971a)
Salpiglossis sinuata Ruiz & Pavón	V. *a-a* (? *V. d.*) isol. ex, in the USA	Baker & Locke (1946)
Schizanthus pinnatus Ruiz & Pinón	*V. a-a* (? *V. d.*) isol. ex, in the USA	Baker & Locke (1946)
Solanum carolinense L.	*V.* sp. isol. fr. tomato-planted field, USA (n.h.)	Jagger & Stewart (1918)

	V. a-a on, in the USA (1953)	Engelhard (1957)
S. eleagnifolium Cav.	*V.* sp. isol. fr. tomato-planted field, USA (n.h.)	Jagger & Stewart (1918)
	V. a-a on, in USA (1953)	Engelhard (1957)
S. integrifolium Poir.	*V.* sp. isol. fr. tomato-planted field, USA (n.h.)	Jagger & Stewart (1918)
S. marginatum L.	*V.* sp. isol. fr. tomato-planted field, USA (n.h.)	Jagger & Stewart (1918)
S. melongena L.	*V.* wilt of, in Japan, reported in 1931	via Hashimoto (1989)
	V. tricorpus isol. ex, in NZ	Taylor (1968)
	V. d. on, in India	Sivaprakasam & Rajagopalam (1971)
	V. a-a & *V.* d. on, in Canada	McKeen (1972b)
	V. d. on, in Jordan	Mamluk & Skaria (1979)
S. nigrum L. *sensu lato*	*V. d.* isol. ex, in The Netherlands	Van der Meer (1925)
	V. a-a on, in Germany	Goetz (1935)
	V. d. isol ex, in New Zealand	Cunningham (1949)
	V. a-a (? *V. d.*) isol. ex, in the USA	Snyder *et al.* (1950)
	V. a-a isol. ex, in the UK (n.h.)	Sewell & Wilson (1958)
	V. d. isol. ex, in Australia (n.h.)	Evans (1971a)
	V. d. (race 2) on, in Greece (Crete)	Ligoxigakis & Vakalounakis (1997)
S. rostratum Dunal	*V.* sp. isol. from tomato-planted field, USA (n.h.)	Jagger & Stewart (1918)
	V. a-a on, in the USA (1953)	Engelhard (1957)
S. torvum Sw.	*V.* sp. isol. from tomato-planted field, USA, (n.h.)	Jagger & Stewart (1918)
	General in the USA	Engelhard (1957)
S. tuberosum L.	*V. a-a* isol. ex & described, Germany	Reinke & Berthold (1879)
	V. a-a isol. & re-inoc. on to, in the UK	Pethybridge (1916)
	V. a-a (as *V. duboys* Foëx) in France	Foëx (1923)
	V. a-a & *V. d.* on, in The Netherlands	Van der Meer (1925)
	V. a-a on, in the USA (1953)	Engelhard (1957)
	V. spp. on seed stocks, in the UK	MacGarvie & Hide (1966)
	V. d. on, in Turkey	Saydam & Kamel (1970)
	V. d. on, in Canada	McKeen (1972b)

Continued

Table 11.1. *Continued*

Host	Verticillium	Reference
Solanaceae *Continued*		
S. tuberosum L. *Continued*	*V. d.* on, in Colombia	Turkensteen and Eduardo Nieto (1987)
	V. d. on, in Jordan	Mamluk & Skaria (1979)
	V. d. isol. ex, in Peru	Torres & Gutierrez (1981)
	V. d. on, in Peru	Martin (1985)
	V. d. on, in Peru	Pinillos *et al.* (1987)
	Epidemic in Piedmonte, Italy	Gullino *et al.* (1987)
	V. a-a & *V. d.* isol. ex, in Pakistan	Turkensteen (1988)
	V. a-a on, in Poland, in 1989	Malek *et al.* (1991)
	V. d. on, in South Africa	Denner & Millard (1997)
	V. d. review of in Israel	Nachmias and Krikun (1985)
	V. nigrescens isol. ex, in South Africa	Denner & Millard (1997)
	V. a-a on, in Australia	Sampson (1980)
Sterculiaceae		
Fremontodendron californicum (Torrey) Cov. (as *Fremontia californica* Torr.)	*V. a-a* isol. ex, in USA	Baker *et al.* (1940)
Theobroma cacao L.	First report on	Turconi (1920)
	V. d. on, in Uganda	Leakey (1965)
	V. a-a & *V. d.* isol. ex & re-inoc. on to, in Uganda	Trocmé (1972)
	V. d. isol. ex and re-inoc. on to, in Uganda	Emechebe *et al.* (1971)
	Factors affecting inoculation	Emechebe *et al.* (1975)
	Aetiology	Resende *et al.* (1995c)
Thymelaeaceae		
Daphne cneorum L.	*V. a-a* isol. ex & re-inoc. on to, in the USA	Himelick (1969)
D. mezereum L.	*V. d.* on, in The Netherlands	Van der Meer (1925)
Tiliaceae		
Corchorus capsularis L.	*V. d.* isol. ex & re-inoc. on to, in the USSR	Vasilieff (1933)

C. oritorius L.	*V. d.* isol. ex & re-inoc. on to, in Japan	Kanno & Horiuchi (1999)
Tilia americana L. (as *T. glabra* Vent.)	*V. a-a* on, in the USA	Carter (1940)
T. cordata Mill.	*V. a-a* isol. ex, in Germany	Wollenweber (1929)
	V. a-a isol. & re-inoc. on to, in USA	Himelick (1969)
T. × *euchlora* K. Koch	*V. a-a* on, in Germany	Wollenweber (1929)
T. japonica (Miq.) Simonkai	*V. d.* isol. ex & re-inoc. on to, in Japan	Harada *et al.* (1997)
Umbelliferae		
Apium graveolens L. var. *dulce* DC.	*V. a-a* on, in the USA (1953)	Engelhard (1957)
Ulmaceae		
Ulmus americana L.	*V. a-a* on, in the USA	Carter (1938)
	V. a-a isol. ex & re-inoc. on to, in the USA	Himelick (1969)
	V. a-a on, in Canada	Anon (1961a)
U. campestris L.	*V. a-a* on, in The Netherlands	Van der Meer (1926)
	V. a-a on, in Italy	Goidanich (1935)
U. glabra Hudson (as *U. montana* With.)	*V.* sp. isol. ex, in Germany	Stapp (1928)
U. glabra Hudson (as *U. montana*)	*V. a-a* isol. ex, in Germany	Donant (1932)
U. parvifolia Jacq.	*V. a-a* isol. ex, in the USA	Engelhard & Carter (1956)
U. procera Salsb. (as *U. campestris* L. var. *wheatlei* Simon-Louis)	*V.* sp. on, in The Netherlands	Anon (1920)
U. procera Salsb. (as *U. campestris* L.)	*V.* sp. on, in The Netherlands	Anon (1920)
	V. a-a isol. ex, in USA	Kelsheimer & May (1940)
U. rubra Muhl.	*V. a-a* on, in the USA	Kelsheimer & May (1940)
Ulmus spp.	Varied response of hybrid clones to inoculation with *V. a-a*	Rauscher *et al.* (1974)
Urticaceae		
Urtica dioica L.	*V. a-a* on, in Germany	Goetz (1935)
U. urens L.	*V. d.* isol. ex, in The Netherlands	Van der Meer (1925)
	V. a-a isol. ex, in the UK (n.h.)	Sewell & Wilson (1958)
Valerianaceae		
Valeriana officinalis L.	*V. d.* on, in Germany	Gerlach & Franz (1973)

Continued

Table 11.1. *Continued*

Host	Verticillium	Reference
Verbenaceae		
Verbena bonariensis L.	*V. d.* isol. ex, in Australia (n.h.)	Evans (1971a)
V. officinalis L.	*V. d.* isol. ex, in Australia (n.h.)	Evans (1971a)
Vitaceae		
Parthenocissus tricuspidata (Siebold & Zucc.) Planchon (as *Ampelopsis veitchii* Hort.)	*V.* sp. isol. ex, in The Netherlands	Van der Lek (1918)
Vitis vinifera L.	*V.* sp. on vines with American rootstocks in Germany (1953)	Böning *et al.* (1960)
	V. d. on, in Germany	Thate (1961)
	V. d. ? (as *V. a-a*) on, in Greece	Zachos & Panagopoulos (1963)
	V. d. isol. ex & re-inoc. on to, in NZ	Canter-Visscher (1970)
	V. d. on, in Chile	Alvarez & Sepulveda (1977)
	V. d. on, in the USA	Schnathorst & Goheen (1977)
	V. d. on, in Italy	Egger (1973)
	V. d. on, in Portugal	Kuhn (1981)
	V. a-a on, in Austria	Nieder (1980)
	V. d. on, in Turkey	Kapkin & Ari (1982)
	General	Pearson and Goheen (1988)

Table 11.2. Common names of hosts.

Host	Common name
Aceraceae	
Acer campestre L.	Field or hedge maple
A. ginnala Maxim.	Amur
A. negundo L.	Box elder
A. nigrum Michx.	Black maple
A. palmatum Thunb.	Japanese maple
A. pennsylvanicum L.	Moosewood
A. platanoides L.	Norway maple
A. pseudoplatanus L.	Sycamore, great or Scottish maple
A. rubrum L.	Red maple
A. saccharinum L.	Silver or white maple
A. saccharum Marshall	Sugar or striped maple
Aizoaceae	
Tetragonia expansa Thunb.	New Zealand spinach
T. tetragonoides (Pallas) Kunze	New Zealand spinach
Cryptophytum crystallinum N.E.Br.	Ice plant
Amaranthaceae	
Amaranthus retroflexus L.	Pigweed
A. viridis L.	—
Gomphrena celosoides Mort.	—
Anacardiaceae	
Cotinus coggygria Scop.	Smoke tree or bush, wigtree
Mangifera indica L.	Mango
Pistacia vera L.	Pistachio
Rhus aromatica Ait. (prob. syn. of *R. canadensis*)	Fragrant sumac
R. canadensis Marsh.	Fragrant sumac
R. glabra L.	Smooth sumac
R. trilobata Nutt.	Skunkbush sumac
R. typhina L.	Staghorn or velvet sumac
Schinus terebinthifolia Raddi	Brazil pepper tree
Apocyanaceae	
Vinca major L.	Greater periwinkle
Araliaceae	
Aralia cordata Thunb.	Udo
A. racemosa L.	American spikenard
Panax quinquefolius L.	American ginseng
Balsaminaceae	
Impatiens balsamina L.	Balsam
I. walleriana Hook.	Busy Lizzie
Basselaceae	
Ullucus tuberosus Caldas	Ulluco, oca quina
Berberidaceae	
Berberis thunbergii DC.	Japanese barberry
B. vulgaris L.	European barberry

Continued

Table 11.2. *Continued*

Host	Common name
Bignoniaceae	
Campsis radicans (L.) Bureau	Trumpet vine
Catalpa speciosa (Warder ex Barry) Engelm.	Cataba, cigar tree
Boraginaceae	
Heliotropium arborescens L.	Heliotrope, cherry pie
Lithospermum arvense L.	Bastard alkanet
Silene noctiflora L.	Night-flowering catchfly
Buxaceae	
Buxus microphylla koreana Nakai	Korean boxwood
B. microphylla Siebold & Zucc.	Japanese box
Campanulaceae	
Campanula isophylla Moretti	Bellflower
C. macrantha Fisch.	Bellflower
Platycodon grandiflora (Jacq.) A. DC.	Balloon flower, Chinese bellflower
Cannabidaceae	
Cannabis sativa L.	Hemp
Humulus lupulus L.	Hop
Caprifoliaceae	
Sambucus racemosa L.	Red-berried elder
Viburnum lantana L.	Wayfaring tree, twistwood
V. lentago L.	Nannyberry
V. tinus L.	Laurestinus
V. tomentosum (Thunb.) Rehder	Japanese snowball, doublefile
Caricaceae	
Carica papaya L.	Papaya, pawpaw, melon tree
Caryophyllaceae	
Dianthus caryophyllus L.	Carnation, clove pink
Lychnis chalcedonica L.	Maltese cross, cross of Jerusalem
Chenopodiaceae	
Beta vulgaris L. ssp. *vulgaris*	Beetroot, sugarbeet, mangel-wurzel, mangold
Chenopodium album L.	Goosefoot, fat-hen
C. berlandieri Moq.	Pitseed goosefoot, lambsquarter
Rhagodia nutans R. Br.	—
Cistaceae	
Cistus purpureus Lam.	Rockrose
Compositae	
Anaphalis margaritaceum (L.) Benth. (as *Gnaphalium margaritaceum* L.)	Pearly everlasting
Argyranthemum frutescens (L.) Schultz-Bip.	—
Artemisia absinthium L.	Absinth, wormwood
Aster amellus L.	Italian aster, Michaelmas daisy
A. cordyfolius L.	Blue wood aster
A. diffusus Ait.	—
A. ericoides L.	Heath aster
A. novae-angliae L.	New England aster

Table 11.2. *Continued*

Host	Common name
Compositae *Continued*	
A. novae-belgii L.	New York aster
A. paniculatus Lam.	—
A. tradescantii L.	—
A. vimineus Lam.	—
Bidens subalternans DC.	Bur-marigold
Callistephus chinensis (L.) Nees	The aster or China aster of gardens
Carthamus lanatus L.	—
C. tinctorus L.	Safflower, kurdee, false saffron
Centaurea cyanus L.	Cornflower, bachelor's buttons
C. solsitialis L.	—
Chrysanthmum × *grandiflora* auct.	—
C. indicum Cass.	—
C. leucanthemum L.	Oxeye daisy, whiteseed
C. maximum auct. var. Shastra daisy	Max daisy
C. morifolium (Ramat.) Hemsl.	Florists' chrysanthemum
Cosmos spp.	Cosmos
Cynara scolymus L.	Globe or French artichoke
Dahlia pinnata Cav. (as *D. variabilis* Desf.)	Gave rise to garden dahlias
Dimorphotheca sinuata DC.	Cape marigold
Echinops banaticus Rochel *ex* Schrader	Small globe thistle
Erigeron canadensis L.	Fleabane
Gerbera jamesonii Bolus *ex* Hook. (as *G. jamesonii* Hook.)	Barbeton or Transvaal daisy
Helianthus annuus L.	Sunflower
Helichrysum bracteatum (Vent.) Andrews	Strawflower
Ixodia acchillaeoides L.	—
Lactuca sativa L. var. *longifolia* Lam.	Lettuce, Romaine lettuce
Liatris spicata (L.) Willd.	Gayfeather
Matricaria chamomilla sensu auctt.	Sweet, false chamomile
Parthenium argentium A. Gray	Guayule
Pericallis hybrida R. Nordestam	Florists' cineraria
Rudbeckia hirta L.	Black-eyed Susan
Senecio cineraria DC.	Dusty-miller
S. cruentus DC.	Parent of florists' cineraria
S. vulgaris L.	Groundsel
Stevia rebaudiana	Stevia
Tagetes spp.	Marigold
Tanacetum cinerariifolium (Trev.) Schultz-Bip (as *Chrysanthemum cinerariifolium*)	Pyrethrum
Taraxacum officinale Wigg	Dandelion
Tragopogon porrifolius L.	Salsify, vegetable oyster
Verbesina encelioides (Cav.) Gray	—
Xanthium strumarium L. (as *X. pungens* Wallr.)	Cocklebur
Xanthium strumarium L. (as *X. spinosum* L.)	Clotbur

Continued

Table 11.2. *Continued*

Host	Common name
Convolvulaceae	
Ipomoea batatas (L.) Lam.	Sweet potato
I. lonchophylla J.M. Block	—
Cornaceae	
Aucuba japonica Thunb.	Japanese laurel
Cruciferae	
Armoracia rusticana P. Gaertner, Meyer & Scherb. (as *Radicicola armoracia* (L.) Robbins)	Horseradish
Brassica napus L.	Rape
B. napus L. Napobrassica group	Rutabaga, swede
B. oleracea L.	
Botrytis group	Broccoli, cauliflower
Capitata group	Cabbage
Gemmifera group	Brussels sprout
B. pekinensis Skeels	Pe-tsi, wong-bok, chili, Chinese cabbage
B. rapa L.	Sarson, turnip
Camelina sativa L.	False flax
Capsella bursa-pastoris Medik.	Shepherd's purse
Chichorium intybus L.	Chicory, succory, witloof
Hesperis matronalis L.	Dame's violet, sweet rocket, damask
Raphanus sativus L.	Radish
Rapistrum rugosum (L.) All.	Turnip weed
Sisymbrium irio L.	London rocket, blue-eyed grass, satin flower
Cucurbitaceae	
Citrullus lanatus (Thunb.) Matsum. & Nakai (as *C. vulgaris* Schrad.)	Watermelon
Cucumis melo L.	Cantaloupe, honeydew & muskmelon, etc.
C. sativus L.	Cucumber
Curcurbita pepo L.	Summer & autumn squash, pumpkin, vegetable marrow, summer squash
Cupressaceae	
Thuja sp.	Cedar, arbor-vitae
Ebenaceae	
Diospyros kaki L.	Japanese persimmon
D. texana Scheele	Texas persimmon
D. virginiana L.	Persimmon
Elaeaginaceae	
Elaeagnus angustifolia L.	Oleaster, Russian olive, Trebizond date
Hippophaë rhamnoides L.	Sea buckthorn
Ericaceae	
Erica spp.	Heaths (heathers)
Erica australis Hort.	—
E. persoluta L.	—
Rhododendron molle G. Don	Azalea
Vaccinium angustifolium Aiton	Blueberry

Table 11.2. *Continued*

Host	Common name
Euphorbiaceae	
Codiaeum varigatum (L.) Blume	Croton
Euphorbia drummondii Boiss.	—
E. milii Des Moul.	Crown of thorns
E. pulcherrima Willd. *ex* Klotsch	Poinsettia
Manihot esculenta Crantz	Cassava, manioc, mandioca, tapioca, gari, yuca
Fagaceae	
Castanea sativa L.	Sweet chestnut
Quercus agrifolia Née	—
Q. lobata Née	California oak
Q. palustris Moench	Pin oak
Q. petraea (Mattuschka) Liebl.	English, French, Polish, Slavonian, sessile oak
Q. robur L.	Pedunculate oak, English oak, etc.
Q. rubra L.	Red oak
Geraniaceae	
Pelargonium domesticum L. Bailey	Show geranium, pelargonium
P. grandiflorum (Andrews) Willd.	—
P. graveolens l'Herit.	Rose geranium
P. hortorum L. Bailey	Fish geranium
Grossulariaceae	
Ribes sanguineum Pursh. var. *lombartii*	Lombart's flowering currant
R. uva-crispa L.	Gooseberry
Hippocastanaceae	
Aesculus hippocastanum L.	Horse chestnut
Juglandaceae	
Juglans sp.	Walnut
J. regia L.	English walnut
Labiatae	
Agastache rugosa Kuntze	Giant hyssop
Lamium amplexicaule L.	Henbit
Mentha piperita L.	Black peppermint
M. spicata L. ×	Spearmint
Monarda didyma L.	Oswego tea, bee balm, fragrant balm, bergamot
Salvia reflexa Hornem.	—
Satureja hortensis L.	Summer savory
Lauraceae	
Cinnamomum camphora (L.) J. Presl.	Camphor
Laurus nobilis L.	Bay, laurel
Persea americana Miller	Avocado
Sassafras albidum (Nutt,) Nees	Sassafras

Continued

Table 11.2. *Continued*

Host	Common name
Leguminosae	
Arachis hypogaea L.	Peanut, groundnut, monkeynut, earthnut
Ceratonia siliqua L.	Carob, St John's bread
Cercis canadensis L.	Redbud
C. siliquastrum L.	Judas tree
Cicer arietinum L.	Chickpea, garbanzo bean
Cladrastris lutea (Michx) K. Koch	Yellow-wood
Glycine max (L.) Merr.	Soy or soybean
Gymnnocladus dioica (L.) Koch	Kentucky coffee tree
Hedysarum coronarium L.	French honeysuckle
Lathyrus odoratus L.	Sweet pea
Lotus corniculatus L.	Bird's-foot trefoil, eggs and bacon
Lupinus albus L.	Egyptian or white lupin
L. polyphyllus Lindley	Cultivated, Russell lupin
Medicago hispida Gaertn.	Bur clover
M. polymorpha L.	—
M. sativa L.	Alfalfa, lucerne
Neptuna gracilis Benth.	—
Onobrychis viciifolia Scop.	Sanfoin, holy clover
Phaseolus vulgaris L.	Kidney, French or dwarf flageolet, string, haricot bean, navy bean
Pisum sativum L.	Pea
Robinia pseudoacacia L.	False acacia, black locust
Sophora japonica L.	Pagoda tree, Chinese Scholar tree
Trifolium hybridum L.	Alsike
T. pratense L.	Red or purple clover
T. repens L.	White or Dutch clover
Trigonella foenum-graecum L.	Fenugreek
Vicia faba L.	Broad, field or horse bean
Vigna radiata (L.) Wilczek	Mung bean, green or golden grass
V. unguiculata (L.) Walp.	Cowpea, yaya, chowlea, gubgub
V. unguiculata ssp. *sesquipedalis* (L.) Verdc.	Chinese yard-long bean
Linaceae	
Linum usitatissimum L.	Flax, linseed
Magnoliaceae	
Callirhoe papaver (Cav.) A. Gray	Poppy mallow
Liriodendron tulipifera L.	Tulip tree
Magnolia grandiflora L.	Bull bay, loblolly magnolia
M. soulangeana Soul.	Saucer magnolia
M. tomentosum Thunb.	—
Malvaceae	
Abelmoschus esculentus (L.) Moench.	Okra, lady's fingers
Abutilon theophrasti Medik.	Chingma, Chinese jute or hemp, Indian mallow, flowering maple, Manchurian jute, velvetleaf

Table 11.2. *Continued*

Host	Common name
Malvaceae *Continued*	
Alcea rosea L.	Hollyhock
Gossypium arborum L.	Tree cotton
G. barbadense L.	Sea Island cotton
G. herbaceum L.	Commercial cotton
G. hirsutum L.	Upland cotton
Hibiscus cannabinus L.	Kenaf, Ambari or Deccan hemp, Bimlipatum jute
H. trionum L.	Flower-of-an-hour
Lagunaria patersonii (Andrews) G. Don	Pyramid tree
Lavateria olbia L.	Tree mallow
Malva spp.	Mallow
M. neglecta Wallr.	—
M. rotundifolia L.	—
Malvaviscus arboreus Cav.	Wax mallow
Moraceae	
Maclura aurantiaca L.	Osage-orange
Nyssaceae	
Nyssa sylvatica Marsh.	Black gum, cotton gum, sour gum, pepperidge
Oleaceae	
Fraxinus americana L.	American, Canadian or white ash
F. angustifolia	—
F. excelsior L.	Common or European ash
F. nigra Marshall	American, black, brown, Canadian or white ash
F. pennsylvanica Marshall	American, Canadian green, red or white ash
F. quadrangulata Michx.	Blue mountain ash
Ligustrum amurense Carr.	Privet, amur
Olea europaea L.	Olive
Osmanthus ilicifolius	—
Syringa vulgaris L.	Lilac
Onagraceae	
Fuchsia hybrida Hort. *ex* Vilm.	Garden fuchsia
Papaveraceae	
Eschscholzia californica Cham.	California poppy
Papaver nudicaule L.	Iceland poppy
P. orietale L.	Oriental poppy
P. rhoes L.	Common corn or field poppy
Pedaliaceae	
Proboscidia louisianica (Mill.) Thell.	Unicorn plant
Sesamum indicum L.	Sesame, simsim, gingelly
Piperaceae	
Peperomia obtusifolia var. *varigata*	—

Continued

Table 11.2. *Continued*

Host	Common name
Pitosporaceae	
Pitosporum tobira (Thunb.) Aiton	Tobira
Plantaginaceae	
Plantago lanceolata L.	Ribwort
P. major L.	Common plantain
Polemoniaceae	
Phlox decussata Hort.	—
P. drummondii Hook.	Annual or Drummond phlox
P. paniculata L.	Perennial garden phlox
Polygonaceae	
Polygonum angulata L.	—
P. persicaria L.	Redleg
Rheum rhaponticum L.	Rhubarb
Portulacaceae	
Portulaca oleracea L.	Purslane
Proteaceae	
Banksia victoria	—
Leucospermum cordifolium	—
L. salignum	—
Protea spp.	Sugar bushes
Pyrolaceae	
Pyrola asarifolia Michx.	Wintergreen
P. elliptica Nutt.	Shinleaf
Ranunculaceae	
Aconitum napellus L.	Monk's hood, wolfsbane
Consolida ajacis (L.) Schur	Rocket larkspur
Paeonia spp.	Peony
Resedaceae	
Reseda odorata L.	Mignonette
Rhamnaceae	
Ceanothus sp.	Californian lilac
Rosaceae	
Cydonia oblonga Miller	Quince
Fragaria × *ananosa* Duchesne	Commercial strawberry
Malus pumila Miller	Apple
Prunus armeniaca L.	Apricot
P. avium (L.) L.	Gean, mazzard, wild cherry, hagberry
P. cerasifera Ehrh.	Myrobalan, cherry plum
P. cerasus L.	Morello or amorella cherry
P. domestica L.	Plum
P. dulcis (Miller) D. Webb	Almond
P. laurocerasus L.	Cherry laurel
P. lusitanica L.	Portugal laurel
P. mahaleb L.	Mahaleb cherry
P. mume Siebold & Zucc.	Japanese apricot

Table 11.2. *Continued*

Host	Common name
Rosaceae *Continued*	
P. persica (L.) Batsch.	Peach
P. persica var. *nectarina* (Aiton) Maxim.	Nectarine
Rhaphiolepis indica Lindley	India hawthorn
R. umbellata var. *integerrima* Rehd.	Yeddo-hawthorn
Rosa multiflora Thunb. *ex* Murray	Rose
R. rugosa Thunb.	Japanese rose
Rubus allegheniesis Porter	Alleghany or Crandall blackberry
R. ideus L.	Raspberry
R. occidentalis L.	Blackberry
R. ursinus Cham. & Schldl.	Dewberry, trailing blackberry (USA)
Rhaphiolepsis sp.	—
Rutaceae	
Citrus limon (L.) Burm.	Lemon
Phellodendron amurense Rupr.	Cork tree
Salicaceae	
Populus tremula L.	Aspen
Sapindaceae	
Koelreuteria paniculata Laxm.	Golden rain
Scophulariaceae	
Antirrhinum majus L.	Snapdragon
Calceolaria spp.	Slipperwort
Digitalis purpurea L.	Foxglove
Hebe elliptica (Forster) Pennell	Veronica
Simaroubaceae	
Ailanthus altissima (Miller) Swingle	Tree of heaven
A. glandulosa Desf.	Tree of heaven
Simmondsiaceae	
Simmondsia chinensis (Link) C. Schneider	Jojoba, goat nut, pig nut
Solanaceae	
Atropa belladonna L.	Belladonna
Capsicum annuum L.	Paprika, cultivated pepper
Datura stramonium L.	Thorn apple, Jimson or Jamestown weed
Lycopersicon esculentum Miller	Tomato
L. esculentum Miller var. *cerasiforme* A. Gray	Cherry tomato
f. *pyriforme* (Dun.) C.H. Mull.	Pear tomato
Nicotiana suaveolens Comes.	—
N. tabacum L.	Tobacco
Petunia hybrida Hort. Vilm.	Garden petunia
Physalis alkekengi L.	Chinese lantern
P. angulata L.	—
Salpipiglossis sinuata Ruiz & Pavón	Painted tongue
Schizanthus pinnatus Ruiz & Pinnóe	Poor man's orchid, butterfly flower
Solanum carolinense L.	Carolina horse-nettle
S. eleagnifolium Cav.	Silverleaf nightshade, white horse-nettle

Continued

Table 11.2. *Continued*

Host	Common name
Solanaceae *Continued*	
S. integrifolium Poir.	Scarlet, or tomato eggplant
S. marginatum L.	—
S. melongena L.	Aubergine, eggplant, Jew's apple, brinjal, badinjam
S. nigrum L.	Black nightshade
S. rostratum Dunal	Kansas thistle, bufalo bur
S. torvum Sw.	Berenjena, cimarrona
S. tuberosum L.	Potato
Sterculiaceae	
Fremontodendron californicum (Torrey) Cov.	Flannel bush
Theobroma cacao L.	Cacao, cocoa
Thymelaeaceae	
Daphne cnorum L.	Rose daphne
D. mezereum L.	Mezereon
Tiliaceae	
Corchorus capsularis L.	Jute
Tilia americana L.	American linden, basswood, whitewood
T. cordata Mill.	Small leaved lime
T. japonica (Miq.) Simonkai	Japanese lime
Ulmaceae	
Ulmus americana L.	American, Littleford, or Moline elm
U. glabra Hudson	Dutch elm, wych elm
U. parvifolia Jacq.	Chinese elm
U. procera Salisb.	English elm
U. rubra Muhl.	Red elm, slippery elm
Umbelliferae	
Apium graveolens L. var. *dulce* DC.	Celery
Urticaceae	
Urtica urens L.	Small or dog nettle
Valerianaceae	
Valeriana officinalis L.	Valerian
Verbenaceae	
Verbena bonariensis L.	—
V. officinalis L.	Vervain, Juno's tears
Vitaceae	
Parthenocissus tricuspidata (Siebold & Zucca) Planchen	Boston ivy
Vitis vinifera L.	Grape vine

Techniques and Methodology 12

The references included in this chapter comprise a non-exhaustive compendium of methods and techniques used in wilt pathology. Some early ones have now been superceded but are included for general information. Other references to techniques are to be found in the text under different chapter headings.

Assessment of the Pathogen in the Soil

Direct observation

Sewell (1959) used glass-walled seedling germination boxes and a wide-field dissecting microscope to observe early stages of tomato root infection with *V. albo-atrum*. Mol and Van Riessen (1995) and Mol (1995a) used a similar chamber with microsclerotia coated on the glass walls in a water–agar film. A similar system was employed by Tjamos and Vellios (1997) to observe biological control directly on roots (see Harley and Waid, 1955). A variation of this method used two membrane filters covered with porous glass discs, kept vertically in soil for up to 5 months after which germinating microsclerotia were viewed on the filters after removal (Muromtsev *et al.*, 1977; Vishnevskaya *et al.*, 1990). Transparent polyester growth pouches were used by Elango *et al.* (1986) for infection studies in lucerne roots, a technique that could be adapted for scanning or transmission EM. Using a variation of the Rossi–Cholodny technique, Jordan *et al.* (1972) and Galanopoulos and Tribe (1974) studied the development of conidia smeared on glass slides and coverslips, respectively, after burial in soil for different times and with different treatments. The simple technique of allowing infected, surface-sterilized root and stem fragments to germinate on and penetrate into sterile filter paper (Whatman 1–4) permitted inversion of the Petri dish and direct observation of *V. albo-atrum, V. dahliae, V. nigrescens, V. nubilum* and *V. tricorpus* (McKeen and Thorpe, 1971). Condensation on Petri lids was prevented by spraying 'Duron', antistatic textile spray (Huang & Kokko, 1983), a valuable aid to observation and photography in all chamber experiments.

Retrieval of the pathogen

Filter paper

Isaac *et al.* (1971) successfully separated conidia and chlamydospores of *V. dahliae* and *V. nigrescens* from microsclerotia by filtering a ×10 dilution through Whatman No. 5 paper and plating the filtrate on soil-extract agar.

Nylon filtration

Lumsden (1981) described the use of 30-μm pore and 1-μm pore, nylon monofilament meshes for separating *V. dahliae* microsclerotia and conidia, respectively.

Wet-sieving soil

Several variations on the standard wet-sieving technique for microsclerotial recovery have been employed: the use of 37-, 53- and 74-μm sieves followed by plating sievings on low-sugar Czapeks agar with added streptomycin sulphate and a cellophane cover as additional C (Ashworth *et al.*, 1972b); the replacement of cellophane by polygalacturonate (pectate) as a principal C source (Huisman and Ashworth, 1974a,b); and the method of Smith and Rowe (1984) in potato field soils. Harris (1990b) concluded that the critical factors in producing accurate reliable results were removal of soil particles <20 μm in size, applying an optimal quantity to each plate and drying the soil suspension on plates before incubation. Using these methods and 20 replicate plates, infestation levels of 1 c.f.u. in 10 g of soil could be detected. The pathogen level for 5% disease in susceptible strawberry was 1 c.f.u. in 2 g of soil. The need for standardization of methods was illustrated by Termorshuizen (1997a) following the distribution of 15 identical soil samples for a sixfold replicated analysis to 13 different workers and the extraordinary variation in the results produced.

Andersen sampler

Using a modified method, Butterfield and DeVay (1977) increased the accuracy and sensitivity for *V. dahliae* microsclerotial detection in soil and retained 2.8 times more propagules per gram of soil than using the wet-sieving technique of Huisman and Ashworth (1974a,b).

Water column

Alternate washing and decanting soil samples in a water column separated microsclerotia in the denser samples (Evans *et al.*, 1967).

Density flotation

A flotation technique developed by Ben-Yephet and Pinkas (1976) involved shaking 5 g of soil in a separating funnel with 20 ml 1:1 (w/v) caesium chloride, utilizing its high specific gravity and low viscosity. Microsclerotia appeared in the upper layer, but recovery experiments retrieved only 55% of added microsclerotia. Harris (1990b) achieved only half the microsclerotial numbers isolated by wet-sieving.

Nicot and Rouse (1983, 1987a) in comparative studies of microsclerotial assessment methods concluded that wet-sieving was the most precise, but time consuming and biased; Andersen sampling was unbiased but least precise, and dilution plating was the least time consuming and least biased, giving 100% recovery. A statistical evaluation of sampling variation for field-soil sampling was presented by Evans and Gleeson (1980). In an important study, the efficiency of c.f.u. recovery from 21 soil samples from commercial potato fields in relation to soil characteristics was tested by Wheeler and Rowe (1995). Samples were infested with two levels of *V. dahliae* or left with natural levels of *V. dahliae* and dried for 0, 2, 4 or 6 weeks before analysing for c.f.u. cm^{-3} of soil. Recovery from naturally-infested fine-textured soils was similar for all drying periods. Two to four weeks drying yielded best recovery in peaty or coarse-textured soils. Recovery following artificial infestation was variable at all times. Recovery was highest at 2 and 4 weeks of drying for peat and least for all times with fine-textured soils. Organic matter was not normally associated with low recovery, while gravel content and pH were inversely correlated with recovery. Curiously, and importantly, glass Petri plates yielded greater numbers of colonies in two of three soils than plastic plates. Soil dilution did not influence recovery.

For many years the problems associated with accurate determinations of inoculum levels in soil and differences in results recorded by different workers in various parts of the world have dominated soil ecological discussions at International *Verticillium* Symposia (IVS). Termorshuizen (1995) at the sixth symposium in Israel initiated collective experiments involving IVS members to compare findings from a number of worldwide laboratories using identical blind samples. He postulated that previously published differences using the same methods (Powelson and Rowe, 1993) could be due to: drying period (Wheeler and Rowe, 1995); medium composition (Harris *et al.*, 1993); air humidity during incubation; or water quality. Termorshuizen (1995) used a microsclerotial recovery method similar to that of Pegg (1978) who used conidia to measure the recovery of *V. albo-atrum* and *V. dahliae* from host tissue. Recoveries on sandy soils were higher than clayey. In some experiments, recovery of added microsclerotia was <1% while, in the absence of the soil suspension, germination was 80%. The possibility that microsclerotia could be damaged by soil particles, or affected by the residence time in the soil suspension, was considered. Harris *et al.* (1993) proposed that the behaviour of plate-grown microsclerotia (i.e. those used in recovery experiments) might be very different from those constituting part of the soil biosphere.

The collective experiment was conducted on 14 'blind' samples in which 13 research groups in seven countries analysed the microsclerotial content each using their preferred method. Twelve samples were naturally-infested and two had added microsclerotia. An additional control was included to eliminate transport effects. The results (Termorshuizen *et al.*, 1998) did not lead to a general confidence in published studies. There was a 118-fold difference between groups with the lowest and highest mean estimates. Recovery of *V. dahliae* added to samples varied from 0 to 59%, with most variability occurring at the Petri dish level. It was concluded that wet plating (soil suspension dilutions) led to underestimates of a population and with greater variability than dry plating assays. This important experiment emphasizes the need still for a critical examination of the parameters of different assays to achieve a more acceptable conformity of soil population numbers by different workers. Until this has been achieved, attempts to quantitate disease incidence with soil inoculum levels will be questionable or futile. The authors advocate the need for a standardization of results from different laboratories by comparing recoveries with a common set of soil samples.

Plant indicators

Soil samples in the laboratory were seeded with tomato which subsequently were sectioned and plated (Schreiber and Green, 1962). In Israel, aubergine seedlings were regularly used as soil test 'bait' plants and also used in the field to survey for pathogen-free land. Symptomless rotation crops were also sampled (30 ha^{-1}) for plating. This method identifies soil suitable for susceptible crops or requiring fumigation (Tsror and Nachmias, 1990). Harris and Yang (1996) recorded the incidence of wilt in runners and fruiting plants in 13 strawberry cultivars at 72 locations over 2 years with accompanying *V. dahliae* analyses of the soil at each site. A successful linear regression of wilt incidence was obtained for cv. Elsanta runners but not cv. Elsanta fruiting plants. Data were used to estimate an inoculum concentration which corresponded to a 5% wilt incidence for cv. Elsanta (IC_5). For the 44 sites for each of which a 15-year cropping history was known, there was no clear correlation between any crop and soil ICs at or above IC_5. Perhaps not surprisingly, *V. dahliae* was more common where soil had supported vegetatively propagated crops rather than crops from seed (see also Harris *et al.*, 1993). A rapid (10 min) tomato seedling wilt assay for *V. dahliae* cell-free culture filtrates was described by Madhosingh (1996) by growing partially etiolated seedlings (under 2.2×10^3 lux) and a twofold concentration of the filtrate. Rapid wilt was enhanced by conducting the assay in light at 23.8×10^2 lux in a wind stream of 125–150 m min^{-1} and relative humidity of 28% at 30°C. The author claimed that the assay distinguished between the relative wilt capacities of races 1 and 2 and the comparative resistance of four cultivars. Details of the medium were not provided and, it should be noted that constituents of some *uninoculated* media are notorious as 'wilt agents'; moreover, unless host cells were used as carbon source, there is no reason why cultural metabolites of races 1 and 2 should differ. Using potato cv. Element, aubergine cv. Black Beauty and thorn apple (*Datura stramonium*), Nagtzaam *et al.* (1997) obtained linear correlations between soil inoculum density, root population density and, for aubergine, between soil inoculum density and sap population density (presumably conidia): R^2 0.45–0.99 for root soil inocula and R^2 of only 0.04–0.26 for stem densities based on plating sap on a solidified ethanol medium. The latter result is not surprising, since, as has been discussed elsewhere, sporulation in xylem vessels may be unrelated to the density of overall colonization. There is, additionally, the real possibility that conidiation would be localized in blocked vessels, reflected in much lower sap yields usually found in infected plants. The pathozone (= infection court) was calculated at <300 µm, indicating the close proximity of microsclerotia to roots.

Soil DNA assessment

Using *V. dahliae* and DNA as a target sequence, soil suspensions are ground in liquid N_2 with added skim milk powder and the DNA from disrupted cells extracted in sodium dodecyl sulphate–phenol. Following dilution, the DNA extract was assayed by PCR amplification. With internal controls, the method was claimed to be rapid and cost-effective (Volossiouk *et al.*, 1995). A modified system using a two-step PCR system, samples first amplified with general primers followed by specific primers (nested-PCR), readily detected *V. dahliae*, *V. albo-atrum* and *V. tricorpus* (Platt, *et al.*, 1997b). The use of internal controls was essential. Problems identified included: inhibition by soil compounds, variable DNA extraction rates and low fungal populations (but see Heinz and Platt, 2000a,b).

Selective Media

The alcohol–agar medium of Nadakavukaren and Horner (1959) as modified by Easton *et al.* (1969), involved a 1:50 soil dilution, plated on 15 ml of ethanol–streptomycin agar with an additional 50 p.p.m. penicillin G. Taylor (1969) obtained a high recovery of *V. tricorpus* utilizing cellulose and biotin as sole carbon sources. A variation (Camporota and Rouxel, 1977) of the medium described by Isaac *et al.* (1971) increased quintozene to 100 p.p.m. and included 0.1% streptomycin, chlortetracycline and chloramphenicol. This medium was similar to one developed and used successfully by Evans *et al.* (1972, 1974) and was slightly modified for use on Chinese soils by Lu *et al.* (1990). Ethanol agar and a pectate-based agar were used by Goud and Termorshuizen (1997) for easy identification of *V. dahliae* and *V. tricorpus* based on coloration of the medium, colony shape, and shape and size of the microsclerotia. The selective medium of Komada for *F. oxysporum* combining L-sorbose and L-asparagine, with salts and streptomycin sulphate gave good recovery of *V. albo-atrum* from hop and lucerne soils (Christen, 1982a).

Microsclerotial Inoculum Production and Storage Viability

Henni (1987b) treated a suspension of the dark surface of a *V. dahliae* culture with ultrasound for 30s followed by filtration on a 37-μm mesh. A variation on this technique provided viable microsclerotia stored for a long period (Hawke and Lazarovits, 1994, 1995b). Semi-solid cultures were blended and wet-sieved to isolate microsclerotia from hyphae and conidia which were mixed with acid-washed quartz sand to prevent clumping and subsequently separated from the sand by dry-sieving. The authors found that 76–106 μm microsclerotia thus produced gave 94% germination after 59 weeks at 24°C. Little difference in viability was found between microsclerotia stored between 24°C and −70°C, but intermittent cooling and partial heating, conditions found in frost-free refrigerators, was damaging, as was small size (mesh size <75 μm).

In early studies on the viability of preserved inoculum, Wilhelm (1955a) was able to resuscitate 17 of 51 13-year-old agar slants of *V. dahliae* by covering with cooled molten agar. Pathogenicity to tomato was retained. Microsclerotia added to a mineral soil, sand, peatmoss (4:1:1 (by vol.)) mixture and kept at 50% water-holding capacity, retained 100% viability to infect tomato for 82 weeks, whereas conidium-infested soil declined to zero infectivity after 21 weeks (Schreiber and Green, 1962). Microsclerotial storage in distilled water has been reported by many workers to be superior to periodic transfer and with extended longevity, i.e. 7 years, with pathogenicity to aubergine after 5 years (Pimentel *et al.*, 1980); and 15–17 years, with pathogenicity retained to a range of crop plants (Russomanno *et al.*, 1995).

Greenhouse-grown lucerne stems infected with *V. albo-atrum* buried in sterile and non-sterile soils were examined monthly for 3 years. Storage at −5°C to 5°C led to the longest viability; survival decreased with increased temperature (Basu, 1987). Lucerne stubble in the laboratory sporulated after 1 year's storage but not after 2 years (Jiminez-Diaz and Millar, 1987). The genus *Verticillium* was reported by Smith and Onions (1983) as surviving for >10 years on silica gel and 13–15 years when freeze-dried. The genus is not generally recommended for mineral oil or liquid N_2 storage, although Talboys and Wilson (1954) maintained *V. albo-atrum* under paraffin oil on prune–lactose–yeast for a short time. In a comprehensive study of 3000 'strains' [*sic*] (= isolates) of *V. dahliae*, 2000 of *V. tricorpus*,

400 of *V. nigrescens* and 50 isolates of other species, Molchanova and Portenko (1990) reported that cultures maintained on a dry nutrient medium under a coating of vaseline oil at room temperature with periodic transfer to new medium survived for 3–5 years. Those kept on agar nutrient medium under oil at 4–6°C retained viability for 8–12 years. Auxotrophic mutants and those not forming resting bodies were sensitive to storage. Frequent transfers onto fresh medium led to loss of virulence. A procedure for the electrophoretic analysis of proteins from *V. dahliae* microsclerotia treated with metham sodium and with altered viability was described by Engelkes and Fravel (1998).

Inoculation Methods

Stem injection of lower cotton stems by 5-ml syringe using a 2.3×10^6 ml^{-1} conidial suspension was effective in evaluating resistance (Bugbee and Presley, 1967). The number of cells retained by the host, however, is notoriously variable. Panella *et al.* (1969) found root-dipping in a conidial suspension best for young plants, and spraying more suitable for older ones (see also Saydam *et al.*, 1973). A glasshouse experimental comparison of different methods on lucerne found that immersing cut roots in a conidial suspension led to 98–99% infection in S cultivars and 78–79% in R cultivars (Moller and Andreasen, 1971). The importance of the correct method to evaluate effective resistance in the field was highlighted by these experiments. Stem injection or surface application of inoculum to seeds or soil proved quite ineffective. The most severe (and earliest) symptoms on strawberry were produced after root-dipping in comminuted mycelium (Maas *et al.*, 1985).

The danger of loss of viability through enhanced temperature during ultrasonic treatment of inoculum was highlighted by van Wambeke *et al.* (1985). On maple (*Acer platanoides*) seedlings, Yang and Harris (1994) found that roots soaked in a conidial suspension of *V. dahliae* accurately reflected natural field disease development. In a subsequent experiment (Chambers and Harris, 1997) using a standard 10^6 conidia ml^{-1} inoculum concentration, found that dilution of inoculum only became limiting at 4×10^4 conidia ml^{-1}. A 20% reduction of the standard inoculum had no effect on disease. The efficacy of root-dipping was confirmed by Atibalentja and Eastburn (1997). Four inoculation methods: colonized oatseed, root-dip, infested soil and set-piece dip, on susceptible and resistant horseradish cultivars were all effective, although with much variation in differentiating between S and R cultivars. Root-dipping gave the best statistical results, the highest incidence of foliar symptoms and the shortest incubation time. For controlled inoculation with field incubation, Mol and Scholte (1996) used the simple expedient of sinking in the field 75-l containers filled with steam-sterilized soil and measured microsclerotial inoculum. A system of assessing hop *V. albo-atrum* strain pathogenicity under controlled soil temperature in concrete tanks, based on 'Wisconsin tanks' was described by Talboys and Wilson (1954). This method was used to identify M and V strains of the pathogen.

Fungal Identification

References to fungal identification may be found in other chapters, especially Chapters 3 and 4.

A simple culture method for distinguishing *V. albo-atrum* from *V. dahliae* was the use of prune–lactose–yeast agar (Talboys, 1960). Cultures incubated at 22°C produced conidiophores in 3–4 days while microsclerotia formed in *V. dahliae* at 5–6 days. The latter had a 'gritty' texture and *V. albo-atrum* a 'fibrose' appearance. Durrands and Cooper (1988b) described two pectate agar media to detect PG and PL enzymes used in detecting pectinase-deficient mutants of *V. albo-atrum*. Platt and Bollen (1995) screened potato stem segments for *inter alia V. albo-atrum, V. dahliae* and *V. tricorpus* on different media. Pectate–NPX medium (Huisman and Ashworth, 1974a, used to isolate *V. dahliae* from soil; subsequently modified by Huisman, 1988b), successfully selected *V. dahliae* but suppressed recovery of *V. albo-atrum* and *V. tricorpus*. Detection of these was greater on PDA. However, while *V. albo-atrum* and *V. tricorpus* sporulated profusely on pectate-PPX agar, *V. dahliae*, while growing vigorously, did not.

Using RAPD on rape seed fungi, polymorphic fragments were selected in Southern hybridization experiments. Amplified taxon-specific DNA fragments were obtained for *V. dahliae* (Schleier *et al.*, 1997; see also Paplomatas and Elena, 1995; Karapapa *et al.*, 1997b,c; Koike *et al.*,1997; Paplomatas and Lampropoulos, 1997; Pramatefraki and Typas, 1997; Typas, 1997). Konnova *et al.* (1995) used hyphal wall polysaccharides to identify *inter alia V. dahliae* on cotton. Nitrate-non-utilizing (*nit*) mutants commonly used to determine VCG compatibility appeared as chlorate-resistant sectors growing out of partly restricted wild-type colonies on chlorate-amended media. An improved water–agar–chlorate (WAC) medium comprising 2% agar, 0.02% glucose and 2–5% potassium chlorate strongly inhibited growth of most *V. dahliae* isolates. Of the chlorate-resistant sectors, 66–100% were *nit* mutants, most of which were *nit* 1 and approximately 6% as *nit* M (Korolev and Katan, 1997). Nazar *et al.* (1991) and Robb *et al.* (1994) described a PCR-based assay using primers derived from internal transcribed spacer regions of ribosomal RNA genes which was capable of distinguishing *V. albo-atrum* and *V. dahliae*. A sequel to this work (Moukhamedov *et al.*, 1994) was a PCR assay to identify *V. tricorpus* from potato. The 18–28S rDNA intragenic region was obtained by PCR amplification of the pathogens genomic DNA (see Robb and Nazar, 1996). The internal transcribed spacer regions of *V. tricorpus* were clearly divergent from those of *V. dahliae* and *V. albo-atrum*. Carder *et al.* (1994) described in detail RFLP and PCR techniques to identify subspecific groups (host- and non-host-based specific isolates) of *V. dahliae* and *V. albo-atrum*. Primer pairs were identified which were used to distinguish all isolates of *V. dahliae*, the non-lucerne isolates of *V. albo-atrum* and the A group of *V. dahliae*. Other primers amplified DNA from a wide range of isolates of both species. Amplified taxon-specific fragments (RAPD) were obtained as hybridization probes to identify *V. longisporum* and *V. lateriticum* and distinguish them from nine other pathogens of oilseed rape (*Brassica napus*) (Schleier *et al.*, 1997). Dobinson (1995) successfully transformed DNA from *V. dahliae*, conferring resistance to hygromycin B, which was integrated into the *V. dahliae* genome. Sphaeroplasts were transformed with the cosmid vector DNA and single and multiple copies, in a tandem assay were incorporated into the genome. Transformation efficiencies ranged from three to five tranformants μg^{-1} of vector DNA. Direct evidence of chromosome rearrangements was seen in two tranformants using electrophoretic karyotype analysis. The technique was seen as a prelude for the facilitation of cloning and characterizing genes for pathogenicity. Using PCR fragments of known sequences from *V. dahliae* and *V. albo-atrum* and simple submarine electrophoresis in agarose gels containing a bisbenzimide–PEG conjugate, Muller *et al.* (1997) found that DNA fragments 567 bp long differing by as little as a single base change could be distinguished. Only changes affecting bisbenzimide-binding sites (consisting of at least four consecutive A/T bases) altered mobility.

Assessment of Colonization

Staining methods

Chlorazol black-E was used to stain mycelium and conidia in sections without staining host tissues (Newcombe and Robb, 1989b). Fluorescent staining of *V. dahliae* in cotton using fluorescein isothiocyanate (FITC) was used by Khalkhodzhaev and Akhmedzhanov (1971) and Akhmedzhanova and Makhmudov (1978). While it is possible to quantify these methods using microdensitometry, the distribution in the plant is so irregular that multiple replication of many areas in the plant would be necessary, perhaps giving no better estimate of colonization than an arbitrary scoring system as used by Talboys (1958b) and Pegg and Dixon (1969). An indirect enzyme-linked immunosorbent assay (ELISA) system was used on *V. dahliae* in cotton plants (Gerik *et al.*, 1987). Rabbit antiserum derived from *V. dahliae* was coupled with naphthyl phosphate which, reacting with a diazonium salt, appeared as a coloured hyphal precipitate.

Plating method

Davis *et al.* (1983a) used a novel method whereby potato stems containing *V. dahliae* were air-dried, powdered, passed through an Andersen sampler and plated on a nutrient pectate medium. The authors found a close correlation between the recovery of fungus from susceptible and resistant cultivars and foliar symptoms.

Measurement of vascular flow

The indirect method of measuring a consequence of infection, i.e. restricted vascular flow, rather than fungal invasion *per se* was used by Street and Cooper (1984). Flow in excised tomato petioles infected with *V. albo-atrum* as measured in a modified Scholander pressure bomb was affected before the appearance of foliar symptoms.

All the above methods fail to distinguish between mycelial colonization and sporulation, the latter reflecting colonizing potential or metabolically active cells contributing to pathogenicity.

Isolation of the pathogen *in planta*

Direct isolation of mostly conidia from *V. dahliae*-infected grape vine was achieved by pumping 2–3 ml of sterile water under low pressure through a 30–40 cm root segment and plating on agar (von Tiedemann *et al.*, 1983). Depending on vessel size, the degree of occlusion from mycelium, gums and tyloses and the delicacy of the tissue, it is unlikely that this technique would be effective on many plants. See also Sewell and Wilson (1964a) who also used a woody plant (hop).

Tissue comminution and plating

Tissue comminution and plating as used by Busch and Schooley (1970), Busch and Hall (1971), Pegg and Jonglaekha (1981) (chrysanthemum infected with *V. dahliae*); Pegg (1981a) (tomato and non-hosts infected with *V. albo-atrum*); Pegg and Street (1984), hop infected with *V. albo-atrum*, is the only method capable of measuring the total (mycelium and conidia) *viable* fungal biomass. Pegg (1981a) demonstrated that there is not a simple relationship between mycelial and tissue fragments and colonies on a spread-plate. Using recovery experiments with different host tissues and conidia or mycelium, he found that some tissues, e.g. chrysanthemum, were stimulatory, enhancing colony numbers, while others, e.g. tomato root and leaf, were wholly inhibitory. Pegg and Street (1984) showed that NaClO used for surface sterilizing was absorbed by open xylem vessels and severely inhibited colonies. These authors advocated stripping extrastelar pericycle and homogenizing the infected vascular system in sterile, distilled water.

Image analysis

A method for the quantitation of microsclerotia colonizing potato stem using an image analyser was described by Mol and Meijer (1995). The system (an alternative to counting by eye) overestimated infection on most outdoor crops due to plant and soil particles that did not discolour in boiling NaOH. The method was claimed to be reliable only for potato haulm obtained from glasshouse-grown plants inoculated with *V. dahliae* under controlled conditions.

DNA technology

Robb *et al.* (1990) pioneered DNA hybridization probes for the identification of *V. dahliae* and *V. albo-atrum* which, following amplification, could be used to measure fungal DNA *in planta*. Using 5.8S rRNA probes of genomic libraries, two bands, a 1.9- and a 2.6-kb fragment, were identified for each species. Differential oligonucleotide probes were synthesized based on two intervening transcribed sequences (ITS1 and ITS2). By varying the hybridization conditions, it was possible to distinguish between the two species and also between fungal genomic and lucerne DNA. The divergent region was amplified by PCR to detect small quantities of pathogen *in planta* (see also Robb *et al.*, 1994 for a detailed protocol). Using a similar technique with PCR amplification, *V. albo-atrum* biomass and colonization patterns in stem and roots of susceptible and resistant lucerne cultivars were determined, which showed close correlation with aerial symptoms (Hu *et al.*, 1993, 1995). The system was affected by host compounds requiring recovery experiments and careful interpretation. While it was capable of detecting very small quantities of fungal DNA, there was no indication that it superseded the comminution–plating method or involved any less preparation. Aspromougos and Schlösser (1997) designed primers based on differences in the 5.8S rRNA ITS1 and ITS2 sequences of *V. dahliae* (Robb *et al.*, 1990). An homologous DNA template was used as a competitor to the target pathogen DNA template. PCR amplification products were analysed by HPLC (see also Platt *et al.*, 1997b).

Carder *et al.* (1994) described in detail the protocol for PCR amplification of subspecific groups of *V. dahliae* and *V. albo-atrum* and DNA extraction from host tissue. The authors claim that isolates may be identified successfully in tomato, lucerne and *Arabidopsis* to <1 pg (<45 genome copies) of fungal DNA (200 pg of total DNA from infected plants) but not in hop or strawberry. No details of *in planta* quantitative studies are given, but the authors stress the importance of sterility in the PCR mixture and note artefacts encountered in the technique. Few primers were wholly specific for individual RFLP groups. While the technique is extremely sensitive and would be valuable for strain identification *in planta*, the host tissue requirement is only 2 mm^3 and it is difficult to imagine how valuable the technique would be for colonization studies – calculating the spatial fungal biomass of a replicated plant experiment – without enormous industry. Moukhamedov *et al.* (1994) claimed that the PCR assay based on sequence differences in rDNA used to identify *V. tricorpus* was also valuable for the quantitation of the pathogen in diseased field plants. *V. tricorpus* primers amplified a specific 337-bp DNA fragment in culture extracts or from infected potato stems. No amplification occurred with DNA from *V. dahliae* or *V. albo-atrum*. The addition of a *V. tricorpus* internal control DNA template allowed the quantitation of the pathogen in diseased field plants. The value of the technique in measuring relative colonization due to treatment or isolate beyond a precise identification and presence or absence of the pathogen remains to be demonstrated.

One of the limiting factors in DNA extraction from plants and especially soils is inhibition of the PCR technique. With this in mind, Heinz and Platt (2000a) developed an improved technique for large-scale studies using PCR-based methods. The extraction buffer contained proteinase K, and further DNA purification was achieved with added ammonium acetate. In soils, a nucleic acid carrier was utilized with the proteinase K–ammonium acetate method. Since organic solvents are not required, DNA can be extracted from small volumes; an advantage when large-scale samples are involved. Comparative tests with the SDS buffer–phenol protocol confirm the reduction in PCR inhibition. The authors (Heinz and Platt, 2000b) confirmed the effectiveness of the technique in recovery experiments involving competitive PCR on *V. tricorpus* in soils. Co-amplification of *V. tricorpus* with competitor DNA gave accurate quantitation in the range 102–106 conidia and 1–500 microsclerotia. The correlation of spores added to *V. tricorpus*-free soil and the spores estimated by PCR was $r = 0.99$. Propagules detected in Prince Edward Island potato field soils range from 0.16 to 19.2 microsclerotia g^{-1} of soil. Propagule numbers at harvest time were not correlated with P_i at planting time. This method deserves critical examination in quantative studies *in planta*.

Serology

Monoclonal antibodies (mAbs) were prepared against purified *V. dahliae* mycelial protein which was made unimmunogenic by conjugating it to keyhole limpet haemocyanin. Three hybridomas were cloned from mouse, all producing immunoglobulin G1-type mAbs that recognized a 60-kDa protein. Specificity of the mAbs was tested using indirect ELISA immunoblots and indirect competitive ELISA (IC-ELISA). A quantitative immunoassay based on the latter showed a linear response between fungal protein (4–500 ng) and absorbance at 405 nm. Accurate quantitative colonization was measured between R and S greenhouse-grown potatoes (Plasencia *et al.*, 1996). There was, however, some cross-reactivity with *V. albo-atrum* which would present problems in mixed infections. In a subsequent study involving 14 potato cultivars in Minnesota field trials, Plasencia and Banttari (1997) were able to rank cultivars as susceptible, intermediate and resistant to *V. dahliae* with good correlation coefficients. The authors claim that the IC-ELISA method is as good as the tissue comminution–plating technique, with the advantage that results are available in 3 days and at reduced cost (see also Nachmias and Krikun, 1984b).

A detailed protocol for a double monoclonal antibody sandwich (DAS)–ELISA detection of *V. dahliae* (Van de Koppel and Schots, 1994, 1995) was developed to provide a clean health scheme for symptomless glasshouse-grown roses. Symptoms in roses may be intermittent (Visser and Kotze, 1975; Horst, 1983) or may be confused with premature senescence. Using surface washings and culture supernatants as antigens on mice, nine hybridoma cell lines were produced which recognized *V. dahliae* in roses and reacted with washings and culture supernatant of *V. dahliae* but not with common antigens from other fungi or from roses. A DAS-ELISA system used one clone 7E12 (of the IgG2a subclass). Plates were coated with MA6-7E12, and alkaline phosphatase conjugates of MA6-7E12 used as the second antibody. It failed on some tests, however, to confirm positively the presence of the pathogen. This the authors attributed to using whole-plant extracts which diluted pathogen antigens at the site of location. Moreover, the MA6-7E12 was non-specific since *V. dahliae* and *V. albo-atrum* in chrysanthemum were both detected. ELISA results were always positive when *V. dahliae* was identified by plating and always negative when platings for *V. dahliae* were negative. In some instances, the mAbs reacted with common antigens from other fungi on platings which were negative for *V. dahliae*. Auger *et al.* (1995) used ELISA and TIBA (tissue immuno-binding assay) to determine the presence of *V. dahliae* in vine (*Vitis vinifera*) and apricot (*Prunus armeniaca*). Rabbit antibodies were obtained from a soluble protein, from *V. dahliae* from infected grape and apricot, respectively, separated by PAGE. Soluble proteins from mycelium were also separated by differential centrifugation. Immunoglobulin dilution end points of 1:10,000 were achieved with both antisera, but no details of the specificity of the assay were presented.

Assessment of Resistance

Glasshouse and field tests

Most of the methods of evaluating susceptibility, or its converse, resistance, have been used routinely, frequently with slight modifications, since the early studies on wilt diseases. For this reason, therefore, the originator of the technique is difficult to identify and only a brief review of the literature is given. Moreover, notwithstanding the title of this monograph, wilt, is not always the exclusive symptom of infection by the vascular verticillia.

The resistance of tomato seedlings to *V. dahliae* was assessed at 25 days by estimates of stunting, cotyledonary flaccidity and counts of chlorotic spots on lower leaves (Ben-Yephet and Pilowsky, 1979). Visser and Hattingh (1980) used vascular discoloration of tomato, but substituted the degree of wilting to avoid a destructive harvest. De Leeuw *et al.* (1986) inoculated tomato leaves and found in susceptible lines that hyphae entered xylem vessels and spread to adjacent mesophyll, while in resistant plants hyphae became blocked with gums and tyloses restricting the pathogen to the mesophyll of the inoculation area. Thanassoulopoulos and Kitsos (1974) regarded stem height in tomato as extremely variable and hence too unreliable to be used as a criterion of infection or, conversely, resistance.

Many wilt indices have been used with little appreciation of the form of the progress curve of the disease in either a single organ or the whole plant. Michail and Carr (1966) employed a wilt scale of 0–5 on 6-week-old lucerne plants with 0, in the absence of symptoms and 5, leaves wilting (beyond recovery). The authors used culture filtrates on seedlings as a good simulation of the effect of *V. albo-atrum* on mature plants. The use of unpurified culture filtrates, however, is fraught with danger since some uninoculated media are good 'wilt inducers' and many wilt-inducing substances produced *in vitro* and in particular inoculum itself are greatly in excess of those produced *in vivo* or may be cultural artefacts. Panella *et al.* (1969) found wilt scores unreliable, substituting a scale of vascular discoloration in the tap root from 1 = healthy to 6 = total discoloration, with the plant dying or dead. Dixon and Doodson (1971) described keys for seedlings and mature field plants. In the former based on 50 replicates, the scale ranged from 0 = symptomless to 7 = dead. In mature plants, an 11-point scale was used. Both scales were based on leaf chlorosis, leaf necrosis, leaf rolling and xylem staining. Kudela (1970) used a similar 6-point scale while Kratka and Kudela (1987) compared vascular discoloration with the histological estimation of hydroxyproline; this was shown by Dixon and Pegg (1972) to rise in susceptible but remain the same as uninoculated controls in resistant plants. Harvey (1982) used a 9-point key with illustrations on plants selected by quadrat, cut across the crown and scored on vascular discoloration. The colonization ratio (CR) of lucerne was determined by infusing disease-free cuttings with a mixture of conidia and red vinyl particles which accumulated at primary spore trapping sites. Plants sectioned 168 h after infusion were scored for the number of secondary colonization sites established per primary trapping site. The resulting CR values correlated with the level of resistance as measured by other methods. The method appeared equally valuable for determining pathovar virulence. Pepper (Johnson and Bosland, 1987) and potato (Gotz *et al.*, 1982) have also been assessed by disease rating. The relationship between field and glasshouse ratings of wilt tolerance in cotton were outlined by Devey and Ressielle (1986).

The assessment of resistance (tolerance) in clonally propagated hop against *V. albo-atrum*, using field scale pot or infection bed experiments, takes 2 years to complete; (see MacNeill (1972) for a description of the Wisconsin temperature-controlled soil tanks used originally for hop wilt testing at East Malling Research Station, UK.). Chambers *et al.* (1997) described preliminary experiments on a hop seedling screen which would facilitate an accelerated breeding programme. A seedling screen has also been developed to detect resistance to *V. dahliae* in *Acer platanoides* (Hiemstra, 1997). Seedling roots (4 to 6 weeks old) dipped in a conidial suspension proved most effective, if severe, compared with other methods, with account taken of over-wintering deaths. Curiously, roots damaged prior to inoculation led to disease escape. From a preliminary screen of 20,000 seedlings, only 200 remained symptomless. Piazek (1996b) claimed that inoculation of detached lucerne leaves (presumably through the petiole), differentiated between susceptible and resistant reactions as seen in whole-plant inoculation. Wilting, however, is notoriously unreliable in detached plant organs and it is essential to detect a tissue response (chlorosis, necrosis) before natural premature senescence occurs. Stem inoculation of sunflower hybrids with a conidial suspension was shown by Bertero de Romano *et al.* (1994) to be statistically effective in glasshouse and field tests to evaluate resistance to *V. dahliae*. Horseradish resistance to *V. dahliae* was assessed by four inoculation methods, i.e. colonized oatseed, root-dip, infested soil and set dip. Using R and S cultivars, root-dip inoculation was the most effective, with the lowest coefficient of variation, the shortest incubation time and giving the highest incidence of foliar symptoms. Inoculum density was not a good predictor of horseradish wilt susceptibility (Atibalentja and Eastburn, 1997). A comparison with resistance assessment, using root conidial-dip versus root injection on five mint cultivars showed comparable results (Sink and Grey, 1999). Plant nutrition from a soil–peat medium not surprisingly gave a higher growth rate than perlite–vermiculite. Cv. Black Mitchum was claimed to have the highest disease rating at 104 or 106 conidia ml^{-1}, a distinction which would be hard to achieve in practice. *Mentha spicata* (native spearmint) was resistant, *M. crispa* (*M. spicata*) was moderately resistant, cv. Murray Mitchum peppermint (*M.* × *piperita*) and Scotch spearmint [*sic*] (*M.* × *gracilis*) were moderately susceptible. Using the inoculation tests, seven somaclones derived from 743 Black Mitchum were found to be resistant (see also Chapter 10).

Physiological tests

Many of the methods appropriate to this section have been cited in Chapters 8 and 9. An assay for resistance of aubergine to *V. dahliae* based on electrolyte leakage from callus tissue was developed by Cristinzio *et al.* (1995). When callus was incubated with a tenfold dilution of culture filtrate, electrolyte loss from resistant tissue was significantly lower than from susceptible tissue. Similarly, the most resistant line had the lowest leakage. Electrolyte leakage on stem segments failed to distinguish between resistance and susceptibility. Tree breeding and resistance screening is a protracted exercise. To facilitate a rapid preliminary screen for symptom appearance, Malia and Tattar (1975) devised an electrical resistance assay on *Acer rubrum* (red maple) inoculated with *V. albo-atrum*. Four-year-old trees grown for 2 months in aerated Hoaglands solution were stab-inoculated in 1.0-mm diameter roots. Platinum electrodes 0.3 mm in diameter were inserted tangentially 4 mm deep either side of the incision, 3 cm apart. Plants were washed in distilled water, dried and electrical resistance measured on a Shigometer. In inoculated roots, a fall of 50 kOhms was detected after 2 days and 1 day before symptom appearance. In healthy control trees, wounded roots showed a resistance drop of 20 kOhms under identical conditions. Inoculated trees that died showed no fluctuation in electrical resistance after the initial drop. Trees that recovered from infection exhibited increased resistance to values higher than the initial readings. Madhosingh (1996) claimed to be able to distinguish between race 1 and race 2 resistance in a 10 min assay on tomato seedlings. To achieve this, the incubation conditions were contrived by producing etiolated seedlings under 2.2×10^3 lux, concentrating culture filtrates $\times 2$, at very low (28%) relative humidity, in a wind stream of 125–150 m min^{-1} and at a temperature of 30°C. It is claimed that under these stress conditions, the single-gene resistance to the appropriate race was expressed over susceptibility. Electrophoresis of polyphenol oxidase (catechol oxidase) on chalazal samples from cotton seeds showed that varieties with a high level of resistance had higher numbers of stained bands than susceptible seeds, which had few. Seeds with excised chalaza produced normal seedlings whose resistance could be verified in the field using conventional inoculation. This technique used in China by Zand *et al.* (1996) accelerated breeding selection and at lower cost. Platt and Bollen (1993) advocated the use of potato-tissue culture plantlets in tubes for the study of *V. dahliae* and *V. albo-atrum*.

Measurement of Wilt Intensity

Talboys (1955) was one of the earliest to use dyes to measure vascular leakage. Acidic dyes such as acid fuchsin, light green and eosin were unsuitable since, although showing vascular mobility, they diffused from vessels to surrounding stelar tissue. Basic dyes, e.g. Basic fuchsin and crystal violet, stained lignin well but showed poor mobility unless decolorized with sodium metabisulphate, after which they travelled rapidly through non-obstructed vessels thereby giving a measure of the extent of wilt progression. Robb *et al.* (1983) used the same technique.

Potential yield reduction in cotton (an inverse of wilt intensity) was estimated by a formula in which the sum of the number of affected plants multiplied by a coefficient in each of five symptom classes (0–4) was divided by the total number of plants in the crop sample (Shami, 1971). Lynov *et al.* (1985) found that leaves from 6-year-old *Hippophae rhamnoides* (sea buckthorn) infected with *V. dahliae* exhibited 'delayed' fluorescence. The authors were able to identify uninfected or resistant plants and distinguish these from those showing early changes associated with infection. Preliminary attempts to develop a field diagnosis computer system to predict the severity of *Verticillium* yellows in *Brassica pekinensis* (Chinese cabbage) by Itoh *et al.* (1989) achieved only partial success. Functions included the number of successful infections, the ratio of infected to healthy plants, disease intensity, microsclerotial colonies on roots, production of microsclerotia on plants, a time constant for the decrease in density of *V. dahliae*, temperature, soil organic matter, density of *Pratylenchus penetrans* and a constant for the decrease in the nematode. The model erected with these data attempted to evaluate the past management of the field site and predict future disease severity assuming the adoption of certain management techniques. Using data from four farms, however, the accuracy of prediction was not high. Future attempts will most certainly refine predictive models and provide more accurate input which would be especially valuable for diseases such as early dying in potato (Rouse, 1985).

Establishment of *Verticillium*-free Clones

Since *Verticillium* infection proceeds acropetally, stem cuttings of perennial plants had basal sections plated aseptically and incubated while the cuttings were kept in cold storage. After 20 days, only fungus-free cuttings were retained (Dimock, 1943). This method prevailed until Quak (1957) developed meristem culture which was combined with heat treatment.

Remote Sensing

Nilsson (1985) claimed that *V. dahliae*-infected rape plants exhibit leaf temperatures elevated 5–8°C above those of healthy controls, probably due to restricted transpiration and the absence of evaporative cooling. Although the experiments were glasshouse based, possibilities exist for field IR photographic inspection. Harris (1985) described such sensing for *Verticillium* wilt in potato, where aerial colour photography distinguished chlorosis in wilted crops from those with late blight. *V. albo-atrum* [*sic*] (*V. dahliae*). Infection of *Erythrina* spp. was detected in Uttar Pradesh, India, by Misra and Shedha (1990) using aerial multispectral analysis on a scale of 1:10,000 with blue, red and green filters in combination, or blue and red. These techniques have now been superseded in some countries by routine crop surveillance by satellite.

Bibliography

Aballay, E., Jardel, A. and Montealegre, J. (1996) Efectos de la solarización y bromuro de metilo sobre *Verticillium dahliae* Kleb. y nematodos fitoparasitos asociados a monocultivos de frutilla. *Fitopatologia* 31(3), 230–240.

Abdel-Raheem, A. and Bird, L.S. (1967) Effect of nutrition on resistance and susceptibility of cotton to *Verticillium albo-atrum* and *Fusarium oxysporum* f. *vasinfectum*. *Phytopathology* 57, 457.

Abdugafurov, I.A., Madikhanov, R.A., Mukhammedova, R.A. and Mahsumov, A.G. (1990) On fungicidal activity of 1-phenyl-4(5)-(x-Aryloxymethylene)-1,2,3 triazoles. *Proceedings of the Fifth International* Verticillium *Symposium*. Leningrad, p. 46.

Abdukarimov, A.A., Mukhamedov, R.S. and Safiyazov, Zh. (1990) On the probable role of mycovirus *Verticillium dahliae* pathogenic fungus–cotton relationships. *Proceedings of the Fifth International* Verticillium *Symposium*. Leningrad, p. 98.

Abdullaev, A.A. and Kiktev, M.M. (1979) [New wilt-resistant and leaf-shedding cotton lines]. *Khlopkovodstvo* 11, 28.

Abduraupov, K. and Kichanova, I.M. (1977) [Cotton wilt and glauconite fertilizers]. *Zashchita Rastenii̊* 1, 51.

Abduraupov, K. and Urunov, I. (1976) [Glauconite and cotton wilt]. *Khlopkovodstvo* 10, 45.

Abel, G.H. (1977) Registration of B68195-25 safflower germplasm (Reg.No. GP 13). *Crop Science* 17, 189–190.

Abel, G.H. and Lorance, D.G. (1975) Registration of Dart safflower (Reg.No. 7). *Crop Science* 15, 100.

Abeles, F.B. (1973) *Ethylene in Plant Biology*. Academic Press, New York.

Abrarov, A.A. and Gubanov, G.Ya. (1973) [Trials of stand density]. *Khlopkovodstvo* 9, 40–41.

Abrarov, A.A. and Ibragimov, Sh.I. (1976) [New sketches for spacing of cotton]. *Sbornik Nauchnykh Trudov, Sredneaziatskogo Otdela VASKhNIL* 1, 40–52.

Abu-Blan, H.A. and Abu-Gharbieh, W.I. (1994) Effect of soil solarization on winter planting of potato, cauliflower and cucumber in the central Jordan Valley. *Dirasat. Series B, Pure and Applied Sciences* 21(3), 203–213.

Accati, E.G., Basoccu, A. and Garibaldi, A. (1974) Effect of irrigation, mulching and plastic shading on tomato and bell pepper production and on *Verticillium* wilt disease severity. *Plastics in Agriculture. Proceedings of the 5th International Colloquium.* Agroinform, Budapest, Hungary, pp. 382–393.

Achal-Shah, Srivastava, K.K. and Bora, S.S. (1985) Reaction of some potato cultivars to *Verticillium* wilt *in vitro*. *Progressive Horticulture* 17, 155–156.

Acharya, S.N., Huang, H.C. and Hanna, M.R. (1995) AC Blue J alfalfa. *Canadian Journal of Plant Science* 75(2), 469–471.

Acimovic, M. (1977) Distribution of important sunflower diseases in Yugoslavia. *Report of the Institute of Field Vegetable Crops, Novi Sad, Yugoslavia.*

Adair, S., Wheeler, T.A. and Gannaway, J.R. (1995) Screening cotton for tolerance to *Verticillium dahliae*: subjective versus objective methodology. *1995 Proceedings Beltwide Cotton Conferences.* San Antonio, Texas, January 4–7, 1995, Vol. 1, pp. 217–218.

Aderhold, R. (1907) Über eine Thrombose der Johannis-und Stachelbeere. *Mitteilungen aus der K. Biologischen Anstalt für Land-und Forstwirtschaft* 4, 26–27.

Adylbekov, M.T., Sattarova, R.S., Mubarakova, F.Kh., Umarova, G.K. and Khalmuradov, A.G. (1991) β-1,3-Glucanase of *Verticillium dahliae*. *Proceedings of the Fifth International* Verticillium *Symposium, Leningrad*, p. 47.

Adylkhodzhaev, I.N. (1966) [Study of resistance in cotton of sspp. of *G. hirsutum* to *Verticillium* disease]. *Voprosy Genet. Embr. Khlopchatniki, Tashkent* 121–124.

Ageeva, I.V. (1999) Effect of photodynamic dyes as universal damaging agent on *Verticillium dahliae* lipid composition. In: Lye, H., Russell, P.E., Dehne, W. H. and Sisler, H.D. (eds) *Modern Fungicides and Antifungal Compounds II.* 12th International Reinhardsbrunn Symposium, Friedrichroda, Thuringia, Germany, 24–29 May 1998. Intercept Limited Ltd, Andover, UK, pp. 137–141.

Ahmad, I. and Malloch, D. (1995) Interaction of soil microflora with the bioherbicide phosphinothricin. *Agriculture, Ecosystems and Environment* 54(3), 165–174.

Akeley, R.V., Stevenson, F.J., Folsom, D. and Bonde, R. (1956) Breeding varieties of potato resistant to *Verticillium* wilt in Maine. *American Potato Journal* 33, 15–21.

Akhmedov, Kh., Zel'tser, S., Azimov, A. and Nuritdinova, Kh. (1991) [Wilt-resistant lines]. *Khlopkovodstvo* 4, 54–55.

Akhmedzhanov, I.G., Dzholdasova, K.B., Gussakovsky, E.E., Zeltzer, S.S. and Avazkhodzhayev, M.K. (1990) The influence of red light upon toxic properties of the pathogen of *Verticillium* wilt in cotton. *Proceedings of the Fifth International* Verticillium *Symposium.* Leningrad, p. 48.

Akhmedzhanova, A.N. and Makhmudov, K. (1978) [The use of fluorescence method for detecting latent infection in cotton seedlings]. *Byulleten' Vsesoyuznogo Nauchno-Issledovatel'skogo Instituta Zashchity Rasteniĭ* 42, 62–64.

Akimov, G.I. (1997) Determination of vegetative compatibility in strains of *Verticillium dahliae* from Tajikistan and Uzbekistan. *Proceedings of the Seventh International* Verticillium *Symposium.* Cape Sounion, Athens, Greece, p. 96.

Akimov, G.I. and Portenko, L.G. (1996) Studying vegetative compatibility in *Verticillium dahliae* Kleb. with non-nitrate-utilizing mutants. *Genetika Moskva* 32(2), 211–217.

Akino, T., Tsurushima, T. and Yamaoka, R. (1995) Antifungal and antibacterial activity of 3-formyl-7,11-dimethyl-(2E,6Z,10)-dodecatrien-1-al in the mandibular gland of *Lasius fuliginosus* Latreille. *Japanese Journal of Applied Entomology and Zoology* 39(4), 329–333.

Akkuzhin, D., Nurmatov, R., Kuznetsova, O., Usmanova, K. and Turdyaliev, R. (1990) [Way ahead for a new variety]. *Khlopkovodstvo* 6, 37.

Alabouvette, C. and Bremeersch, P. (1975) Deux maladies nouvelles en France dans les cultures de tournesol. *Compte Rendus des Séances de l'Academie d'Agriculture de France* 61, 626–636.

Alabouvette, C., Pussard, M. and Pons, R. (1979) Isolation and characterization of a mycophagous amoeba, *Thecamoeba granifera* subsp. *minor*. In: Schippers, B. and Gams, W. (eds) *Soil-borne Plant Pathogens. Proceedings of the 4th International Symposium on Factors determining the Behaviour of Plant Pathogens in the Soil.* Munich, Germany, August 1978. Academic Press, New York, pp. 629–633.

Al-Ahmad, M.A. (1988) [Quantitative survey of *Verticillium* wilt of olive in southern Syria]. *Arab Journal of Plant Protection* 6, 27–32.

Al-Ahmad, M.A. (1993) The solar chamber: an innovative technique for controlling *Verticillium* wilt of olive. *Bulletin OEPP* 23(3), 531–535.

Al-Ahmad, M.A. and Mosli, M.N. (1993) *Verticillium* wilt of olive in Syria. *Bulletin OEPP* 23(3), 521–529.

Al-Ahmad, M.A., Duski, A. and Nasan, H. (1997a) Effect of dry heating on growth of *Verticillium dahliae* on PDA media and on its survival in diseased olive branches. *Proceedings of the Seventh International* Verticillium *Symposium.* Cape Sounion, Athens, Greece, p. 85.

Al-Ahmad, M.A., Mosli, M.N. and Duksi, A. (1997b) The performance of local olive varieties against *Verticillium* wilt and selection of resistant Yarmouk lines. *Proceedings of the Seventh International* Verticillium *Symposium.* Cape Sounion, Athens, Greece, p. 43.

Alchanati, I., Patel, J.A.A., Liu Jinggao, Benedict, C.R., Stipanovic, R.D., Bell, A.A., Cui Yunxing and Magill, C.W. (1998) The enzymatic cyclization of nerolidyl diphosphate by δ-cadinene synthase from cotton stele tissue infected with *Verticillium dahliae*. *Phytochemistry* 47, 961–967.

Alconero, R., Robinson, R.W., Dicklow, B. and Shail, J. (1988) *Verticillium* wilt resistance in eggplant, related *Solanum* species, and interspecific hybrids. *HortScience* 23, 388–390.

Aleksic, Z. (1976) [Effect of some systemic fungicides on the control of *Verticillium* wilt of pepper]. *Poljoprivreda* 23(253), 41–49.

Aleksic, Z., Aleksic, D. and Miladinovic, Z. (1975) [Diseases of pepper and the possibilities of their control by cultivating resistant varieties]. *Agronomski Glasnik* 37, 73–82.

Aleksic, Z., Aleksic, D. and Sutic, D. (1976) Evaluation de la resistance du piment au *Verticillium albo-atrum* Reinke et Berth. et determination de la virulence des souches de parasite. *Poljoprivredna Znanstvena Smotra* 39, 63–70.

Alexander, L.J. (1962) Susceptibility of certain *Verticillium* resistant tomato varieties to an Ohio isolate of the pathogen. *Phytopathology* 52, 998–1000.

Alexander, S.J. and Hall, R. (1974) *Verticillium* wilt of chrysanthemum: anatomical observations on colonization of roots, stem, and leaves. *Canadian Journal of Botany* 52, 783–789.

Al-Hamidi, M.D. (1985) Untersuchungen zur Baumwollwelke in der Syrischen Arabischen Republik. Thesis, Justus Liebig Universität, Giessen, Germany.

Alicchio, R., Antonioli, C. and Palenzona, D. (1984) Karyotic variability in plants of *Solanum melongena* regenerated from callus grown in presence of culture filtrate of *Verticillium dahliae*. *Theoretical and Applied Genetics* 67, 267–271.

Alimov, A. (1980) [Response of cotton cultivars to herbicides and changes in soil nutrient regime]. *Trudy Sredneaziatskii Nauchno Issledovatel'skii Institut Zashchity Rastenii* 14, 23–25.

Alimukhamedov, Kh.A. (1972) [A study on seed-oil contents in cotton]. *Sbornik Rabot po Genetike, Selektsii i Semenovodstvy Khlopchatnika* 87–93.

Alimukhamedov, S.N., Mirpulatova, N.S., Kichanova, I.M., Tursunov, F., Akhunov, T. and Mirzaev, B. (1977) [The effectiveness of cotton–lucerne rotations]. *Zashchita Rasteniĭ* 2, 23.

Alippi, H.E. and Ronco, B.L. (1980) Marchitamiento del crisantemo (*Chrysanthemum morifolium* (Ramat.) Hemsl.) ocasionado por *Verticillium dahliae* Klebahn en la República Argentina. *Revista de la Facultad de Agronomía, Universidad Nacional de la Plata* 56, 5–9.

Alkemade, J.P.F. (1979) Oculeren van *Acer platanoides* cultivars op onderstammen, afkomstig van percelen met verschillende voorvrucht. *Groen* 1979(4), 150–151.

Allen, S.J. (1995a) The effects of various cultural practices on the epidemiology of *Verticillium* wilt of cotton in New South Wales. 6th International *Verticillium* Symposium, Dead Sea, Israel, 1994. *Phytoparasitica* 23, 51.

Allen, S.J. (1995b) Plant quarantine and diseases of cotton. *Australasian Plant Pathology* 24(1), 70–73.

Aloj, B. and Garibaldi, A. (1984) La cineraria (*Senecio cruentus* DC.), nuovo ospite di *Verticillium dahliae* Kleb. *Informatore Fitopatologico* 34, 55–56.

Aloj, B. and Noviello, C. (1982) Riscaldamento solare e controllo degli agenti fitopatogeni terricoli. *Annali della Facoltà di Scienze Agrarie della Università degli Studi di Napoli Portici* 16, 151–158.

Al-Rahawi, A.K. and Hancock, J.G. (1998) Parasitism and biological control of *Verticillium dahliae* by *Pythium oligandrum*. *Plant Disease* 82, 1100–1106.

Al-Rawi, K.M. and Ahmed Abdul-Jawad Ahmed (1984) A quantitative genetic analysis of the inheritance of resistance to *Verticillium* wilt of Upland cotton. *Iraqi Journal of Agricultural Science 'ZANCO'* 2, 13–23.

Altman, D.W., Stipanovic, R.D., Mitten, D.M. and Heinstein, P. (1985) Interaction of cotton tissue culture cells and *Verticillium dahliae*. *In Vitro* 21, 659–664.

Alvarez, A.M. and Sepulveda, R.G. (1977) *Verticillium dahliae* asociado con sintomas de 'amarillamiento' en vid. *Agricultura Técnica* 37, 138–140.

Amemiya, Y., Kirano, K. and Iida, W. (1985) [Induction of resistance to *Verticillium* wilt in tomato]. *Technical Bulletin of Faculty of Horticulture, Chiba University* 36, 135–139.

Amemiya, Y., Yamaguchi, K., Hirano, K. and Iida, W. (1986) [Suppression of *Fusarium* wilt of tomato by use of cross protection]. *Technical Bulletin of Faculty of Horticulture, Chiba University*, 79–83.

Amemiya, Y., Kondo, A., Hirano, K., Hirukawa, T. and Kato, T. (1994) Antifungal substances produced by *Chaetomium globosum*. *Technical Bulletin of Faculty of Horticulture, Chiba University* 13–18.

Andreev, L.N. and Serebryakova, V.N. (1974) [Electrophoretic analysis of proteins of some species of the genus *Verticillium* Nees ex Wallr.]. Микология и Фитопатология [*Mycology and Phytopathology*] 8, 3–8.

Andreeva, E.I. (1977) [Investigation of the activity of the preparations ohlosom, semicar, trilan and BMK against *Verticillium* wilt of cotton in Central Asia]. *Sbornik Nauchnykh Trudov VNIIKHSZR* 8, 93–98.

Andreeva, E.I., Dregval', G.F., Smirnova, K.F., Verbovskaya, T.M. and Pronchenko, T.S. (1976, published 1977) [Fungicidal activity of some derivatives of thiazole]. *Referativnyi Zhurnal* 2.55, 1177.

Andreeva, N.F. (1975) [Behaviour of varieties of strawberry towards *Verticillium* wilt]. *Byulletin' Vsesoyuznogo Nauchno Issledovatel'skogo Instituta Zashchity Rastenii* 34, 45–46.

Anon (1920) Verslag over de werkzaamheden van den phytopathologischen dienst in het jaar 1919. *Mededeelingen van den Phytopathologische Dienst Wageningen* 12, 18.

Anon (1926) [Verticilliosis of the cherry]. *Ochrana Rostlin* 6, 32.

Anon (1932) Rapports sommaires sur les travaux accomplis dans les laboratoires en 1931. *Annales des Épiphyties* 18, 54–96.

Anon (1943) Reports on diseases of sugarbeets. *Plant Disease Reporter* 27, 645.

Anon (1951) [Some diseases and pests of cultivted crops in 1950 and their control]. *Versl. PlZiekt. Dienst, Wageningen* 118, 1–120.

Anon (1959) Plant disease survey for the 12 months to June, 1959. *Twenty-ninth Annual Report of the New South Wales Department of Agriculture. Biological Branch Division of Science Services.*

Anon (1961a) Noms Francais des maladies des plantes au Canada. *Publication 263 of the Ministere de l'Agriculture de la Province de Québec.*

Anon (1961b) *Agricultural Gazette.* New South Wales 72, 155–156.

Anon (1962) *32nd Annual Report of the New South Wales Department of Agriculture, Biology Branch, Division of Science Services.*

Anon (1963) Plant disease survey for the twelve months ending June 1963. *Thirty-third Annual Report of the New South Wales Department of Agricultural Science Services, Biology Branch.*

Anon (1964) *Report on Primary Industry, Queensland, 1963–64. Division of Plant Industry, Science Branch. Plant Pathology Section.* pp. 24–25.

Anon (1966) New hosts for *Verticillium dahliae. Report of the Department of Agriculture for New South Wales, Division of Science Services, Biology Branch, Plant Pathology Section 1964–65.* pp. 75–80.

Anon (1972) Potato. *Canada Department of Agriculture: Research Report. 1971.*

Anon (1973) Potato. *Canada Department of Agriculture: Research Branch Report 1972. I Crop plants.*

Anon (1974) Reduction in yield of cotton caused by parasitic diseases in 1973, Cotton Disease Council. *Plant Disease Reporter* 58, 373.

Anon (1975) Potato. *Canada Department of Agriculture: Research Branch Report 1974. I Crop Plants.*

Anon (1976a) Report for 1975, Cotton Disease Loss Estimate Committee, Cotton Disease Council. *Plant Disease Reporter* 60, 862.

Anon (1976b) Report of the Cotton Disease Loss Estimate Committee, Cotton Disease Council. *Plant Disease Reporter* 60, 862.

Anon (1979) Strawberries. *Report. Scottish Horticultural Research Institute, 25th Annual, for the year 1978.*

Anon (1981) 'Waitangy' the world's first true potato seed variety. *New Zealand Potato Bulletin* 81, 12–13.

Anon (1986) *Summary of Plant Quarantine Pest and Disease Situation in Canada 1984–85.* Agriculture Canada, Plant Health Division, Ottawa, Canada, pp. 38–63.
Anon (1996) Guideline on good Plant Protection practice, No. 5. *Hop Bulletin OEPP* 26, 295–309.
Anon (2000a) Variety 'Sicala V-2RR'. Application No. 1999/036. *Plant Varieties Journal* 13(1) 38.
Anon (2000b) Variety 'Sicot 189-2RR'. Application No. 1999/037. *Plant Varieties Journal* 13(1), 38–39.
Antoun, H., Couture, L. and Guilmette, H. (1984) Effect of some systemic and protectant fungicides on growth of *Verticillium albo-atrum, Rhizobium meliloti* and lucerne seedlings in the laboratory. *Protection Ecology* 6, 279–286.
Apostol, I., Low, P.S., Heinstein, P., Stipanovic, R.D. and Altman, D.W. (1987) Inhibition of elicitor-induced phytoalexin formation in cotton and soybean cells by citrate. *Plant Physiology* 84, 1276–1280.
Arbogast, M. and Powelson, M.L. (1997) Response of potato cultivars to plant available water and *Verticillium dahliae. Proceedings of the Seventh International* Verticillium *Symposium.* Cape Sounion, Athens, Greece, p. 45.
Armstrong, N.G. (1941) *Verticillium* wilt of horsechestnut. *Arborist's News* 6, 82–83.
Arny, D.C. and Grau, C.R. (1985) Importance of *Verticillium* wilt of alfalfa in North America. *Canadian Journal of Plant Pathology* 7, 187–190.
Arsenijević, M. (1977) [*Verticillium dahliae* as a parasite of peach in Yugoslavia]. *Zaštita Bilja* 28(140), 131–142.
Arsenijević, M., and Ševar, M. (1981) [*Verticillium dahliae* Kleb. as a pathogen of plum]. *Zaštita Bilja* 32, 339–351.
Artykbaev, P.K. and Urunov, I.S. (1978) [Copper effect on nitrogen and nucleic acids metabolism in wilt-infected plants of cotton sown by different methods]. *Sbornik Nauchnykh Trudov, Sredneaziatskii Nauchno issledovatel'skii Institut Zashchity Rastenii* 12, 11–16.
Asai, W.K. and Stapleton, J.J. (1994) The economic impact of *Verticillium* wilt on the establishment of a young almond orchard. In: Barbera, G. (ed.) *Acta Horticulturae First International Congress on Almond.* Agrigento, Italy, 17–19 May 1993. pp. 333–337.
Asamov, D.K., Tursunkulova, R.K., Isaev, P.I., Ostroshchenko, O.S. and Stepanichenko, N.N. (1975) [Action of phytotoxic substances of *Verticillium dahliae* on the permeability of an artificial phospholipid membrane]. *Khimiya Prirodnykh Soedinenii* 5, 670.
Ashcroft, W.J., Gurban, S., Holland, R.J., Waters, C.T. and Nirk, H. (1993) 'Arcadia' and 'Goulburn': determinate fresh-market tomatoes for arid production areas. *HortScience* 28(8), 857–858.
Ashworth, L.J. (1983) Aggressiveness of random and selected isolates of *Verticillium dahliae* from cotton and the quantitative relationship of internal inoculum to defoliation. *Phytopathology* 73, 1292–1295.
Ashworth, L.J. and Gaona, S.A. (1982) Use of clear polythene mulch for control of *Verticillium* wilt in established pistachio nut groves. *Phytopathology* 72, 243–246.
Ashworth, L.J. and Hine, R.B. (1971) Structural integrity of the cotton fruit and infection by microorganisms. *Phytopathology* 61, 1245–1248.
Ashworth, L.J. and Huisman, O.C. (1980) New hope for *Verticillium* control in cotton. *California Agriculture* 34(10), 19–20.
Ashworth, L.J. and Zimmerman, G. (1976) *Verticillium* wilt of the pistachio nut tree:

occurrence in California and control by soil fumigation. *Phytopathology* 66, 1449–1451.

Ashworth, L.J., McCutcheon, O.D. and George, A.G. (1972a) *Verticillium albo-atrum*: the quantitative relationship between inoculum density and infection of cotton. *Phytopathology* 62, 901–903.

Ashworth, L.J., Waters, J.E., George, A.G. and McCutcheon, O.D. (1972b) Assessment of microsclerotia of *Verticillium albo-atrum* in field soils. *Phytopathology* 62, 715–719.

Ashworth, L.J., Harper, D.M. and Andris, H.L. (1974a) The influence of milling of air-dry soil upon apparent inoculum density and propagule size of *Verticillium albo-atrum*. *Phytopathology* 64, 629–632.

Ashworth, L.J., Huisman, O.C., Harper, D.M. and Stromberg, L.K. (1974b) Free and bound microsclerotia of *Verticillium albo-atrum* in soils. *Phytopathology* 64, 563–564.

Ashworth, L.J., Huisman, O.C., Grogan, R.G. and Harper, D.M. (1976) Copper-induced fungistasis of microsclerotia of *Verticillium albo-atrum* and its influence on infection of cotton in the field. *Phytopathology* 66, 970–977.

Ashworth, L.J., Huisman, O.C., Harper, D.M. and Stromberg, L.K. (1979a) *Verticillium* wilt disease of tomato: influence of inoculum density and root extension upon disease severity. *Phytopathology* 69, 490–492.

Ashworth, L.J., Huisman, O.C., Harper, D.M., Stromberg, L.K. and Basset, D. (1979b) *Verticillium* wilt disease of cotton: influence of inoculum density in the field. *Phytopathology* 69, 483–489.

Ashworth, L.J., Morgan, D.P., Gaona, S.A. and McCain, A.H. (1982) Polyethylene tarping controls *Verticillium* wilt in pistachios. *California Agriculture* 36(5/6), 17–18.

Ashworth, L.J., Galanopoulos, N. and Galanpoulov, S. (1984) Selection of pathogenic strains of *Verticillium dahliae* and their influence on the useful life of cotton cultivars in the field. *Phytopathology* 73, 1637–1639.

Ashworth, L.J., Gaona, S.A. and Surber, E. (1985) *Verticillium* wilt of pistachio: the influence of potassium nutrition on susceptibility to infection by *Verticillium dahliae*. *Phytopathology* 75, 1091–1093.

Ashworth, L.J., Morgan, D.P. and Surber, E. (1986) *Verticillium* wilt of pistachio. *California Agriculture* 21–24.

Askarova, S.A. and Mamadaliiev, A.K. (1974) [Dynamics of the content of microbial antagonists of *Verticillium dahliae* in the rhizosphere of cotton]. *Uzbekskii Biologicheskii Zhurnal* 18, 67–68.

Askarova, S.A., Ioffe, R.Ya. and Mamadaliev, A.Kh (1970) [Effect of trace elements on the resistance of cotton to wilt]. *Khimiya v Sel'skom Khozvaĭstve* 8, 27–30.

Askarova, S.A., Khodzhibaeva, S.M. and Azimkhodzhaeva, M.N. (1971) [The effect of polyene antibiotics on the formation of toxic substances by *Verticillium dahliae*]. *Uzbekskii Biologicheskii Zhurnal* 15, 60–61.

Askarova, S.A., Ioffe, R.Ya. and Mamadaliev, A.Kh. (1973) [*Microelements and the Resistance of Cotton to Wilt*]. FAN UzbekSSR, Tashkent, Uzbek SSR.

Aspromougos, I. and Schlösser, E. (1997) A competitive PCR assay to quantify *Verticillium dahliae* on infected resistant and susceptible tomato hybrids using HPLC. *Proceedings of the Seventh International* Verticillium *Symposium*. Cape Sounion, Athens, Greece. p. 15.

Ataga, A.E. and Akueshi, C.O. (1996) Fungi associated with sunflower seed in Nigeria. *Seed Research* 24(1), 64–65.

Atakuzieva, R.A. and Safiyazov, Zh.S. (1979) [Sensitivity of different strains of *Verticillium dahliae* to benzimidazole derivatives]. *Uzbekskii Biologicheskii Zhurnal* 3, 20–23.
Atakuzieva, R.A. and Safiyazov, Zh.S. (1980) [Oxidoreductase activity of cotton leaves]. *Khlopkovodstvo* 34–35.
Atanasoff, D., Dodoff, D.N., Kovacevski, I.C., Martinoff, S.I., Trifonova, V. and Christoff, A. (1932) [Parasitic fungi new to Bulgaria]. *Yearbook, University of Sofia Faculty of Agriculture* 10, 341–366.
Atibalentja, N. and Eastburn, D.M. (1997) Evaluation of inoculation methods for screening horseradish cultivars for resistance to *Verticillium dahliae*. *Plant Disease* 81(4), 356–362.
Atibalentja, N. and Eastburn, D.M. (1998) *Verticillium dahliae* resistance in horseradish germ plasm from the University of Illinois collection. *Plant Disease* 82(2), 176–180.
Atkinson, T.G. (1981) *Verticillium* wilt of alfalfa: challenge and opportunity. *Canadian Journal of Plant Pathology* 3, 266–272.
Atlantov, A.V. and Rakhimzhanov, A.G. (1990) [New sources and donors of wilt resistance genetically based in *Gossypum hirsutum* L. spp.]. *Proceedings of the Fifth International* Verticillium *Symposium*. Leningrad, p. 68.
Atmar, V.T., Throneberry, G.O. and Kuehn, G.D. (1976) Effects of adenine and cytokinins on growth and protein kinase activity of *Verticillium albo-atrum*. *Mycopathologia* 59, 171–174.
Aubé, C. (1967) Antagonisme des champignons du sol envers *Verticillium albo-atrum* R & B agent de la fletrissure verticillienne de la lucerne. *Canadian Journal of Microbiology* 13, 227–233.
Aubé, C. and Pelletier, G. (1968) Formation of endospores in *Verticillium albo-atrum* R. and B. *Canadian Journal of Microbiology* 14, 606–607.
Aubé, C. and Sackston, W.E. (1964) *Verticillium* wilt of forage legumes in Canada. *Canadian Journal of Plant Science* 44, 427–432.
Aubé, C. and Sackston, W.E. (1965) Distribution and prevalence of *Verticillium* species producing substances with gibberellin-like biological properties. *Canadian Journal of Botany* 43, 1335–1342.
Auger, S., Esterio, G.M. and Jurgensen, E.E. (1995) Immunodiagnosis de *Verticillium dahliae* Klebahn en vid (*Vitis vinifera* L.) y damasco (*Prunus armeniaca* L.). *Fitopatología* 30(3), 138–142.
Augustine, J.J., Volin, R.B., Bryan, H.H., Everett, P.H., Burgis, D.S. and Gull, D.D. (1981a) Hayslip, a jointless fresh-market tomato. *Circular, Agricultural Experiment Station, University of Florida* S-278, 1–7.
Augustine, J.J., Volin, R.B., Burgis, D.S., Hayslip, N.C., Bryan, H.H., Gull, D.D. and Crill, J.P. (1981b) Florida 1A, Florida 1B, Florida 1C. Three jointed fresh market tomato selections of Florida 1 resistant to race 1 and 2 of *Fusarium* wilt and *Verticillium* wilt. *Circular, Agricultural Experiment Station, University of Florida* S-282, 1–7.
Augustine, J.J., Volin, R.B., Everett, P.H., Bryan, H.H., Hayslip, N.C., Gull, D.D. and Crill, J.P. (1981c) Burgis, a jointless uniform green-fruited fresh-market tomato. *Circular, Agricultural Experiment Station, University of Florida* S-279, 1–7.
Ausher, R., Krikun, J., Frankel, H., Katan, J. and Rubin, D. (1973) Destroying *Verticillium* inoculum by flaming plant residues. *Phytoparasitica* 1, 131–132.
Ausher, R., Katan, J. and Ovadia, S. (1975) An improved selective medium for the isolation of *Verticillium dahliae*. *Phytoparasitica* 3, 133–137.
Avazkhodzhaev, M.K. and Zel'tser, S.S. (1990) Physiological principles of cotton varietal

immunity to *Verticillium* wilt and the ways of its inducing. *Proceedings of the Fifth International* Verticillium *Symposium*. Leningrad, p. 50.

Avazkhodzhaev, M.K., Egamberdiev, A.E., Gaibullaev, I.K. and Zhumaniyazov, I.Zh. (1976) [Nature of the wilt resistance of chemical mutants of cotton]. *Uzbekskii Biologicheskii Zhurnal* 3, 60–62.

Avazkhodzhaev, M.K., Zel'tser, S.S. and Adylova, A.N. (1984) [Characteristics of the metabolite of *Verticillium dahliae* inducing phytoalexins in cotton]. *Uzbekskii Biologicheskii Zhurnal* 1, 3–5.

Averna-Sacca, R. (1922) Algumas das molestias cryptogamicas do tobacco (*Nicotiana tabacum*). *Boletim de Agricultura São Paulo* 23a, 201–268.

Averre, C.W. (1973) Vegetable crop disease reports. In: Zehr, E.I. (ed.) *Fungicide and Nematicide Tests. Results of 1972*, Vol. 28. American Phytopathological Society, Winchester, Virginia, pp. 63–94.

Avetisyan, A.D. (1960) [On tannins, carbohydrates, and carbohydrates of cotton in relation to wilt resistance]. *Izvestiya Akademii Nauk Armyan SSR* Biological Science 13(7), 35–45.

Avrarov, A., Gubanov, G. and Faizillaev, M. (1976) [Wilt and disease sowing]. *Khlopkovodstvo* 10, 29–30.

Azad, H.R., Davis, J.R., Schnathorst, W.C. and Kado, C.I. (1987) Influence of *Verticillium* wilt resistant and susceptible potato genotypes on populations of antagonistic rhizosphere and rhizoplane bacteria and free nitrogen fixers. *Applied Microbiology and Biotechnology* 26, 99–104.

Azimdzhanov, I.M. (1971) [*Verticillium* wilt of peach in the Tashkent district]. *Uzbekskii Biologicheskii Zhurnal* 15(2), 13–14.

Azimkhodzbayeva, N.N. and Ramasanova, S.S. (1990) Application of organomineral fertilizers with lignin in *Verticillium* wilt control. *Proceedings of the Fifth International* Verticillium *Symposium*. Leningrad, p. 88.

Baard, S.W. and Pauer, G.D.C. (1981) Effect of alternate drying and wetting of the soil, fertilizer amendment, and pH on the survival of microsclerotia of *Verticillium dahliae*. *Phytophylactica* 13, 165–168.

Baard, S.W. and Pauer, G.D.C. (1982) Effect of acid soils on survival of *Verticillium dahliae*. *Transactions of the British Mycological Society* 79, 513–517.

Bååth, E. and Hayman, D.S. (1983) Plant growth responses to vesicular–arbuscular mycorrhizas XIV. Interactions with *Verticillium* wilt on tomato plants. *New Phytologist* 95, 419–426.

Babaev, F.A. (1964) [Role of tannic substances in the resistance of cotton varieties to *Verticillium* wilt]. *Transactions of the All-Union Symposium on Cotton Wilt Control*. Tashkent, Izdat. pp. 63–66.

Babaev, F.A. (1969) [Some control measures against *Verticillium* wilt]. *Khlopkovodstvo* 19, 31–33.

Babaev, F.A. and Bagirov, M. (1967) [On the control of *Verticillium* wilt of cotton]. *Zashchita Rasteniĭ Moscow* 12, 21–22.

Babaev, F.A., Mukhtarova, L.S. and Kembel', M.E. (1979) [Wilt resistance of cotton and metabolism characteristics under different irrigation conditions]. *Vestnik Sel'skokhozyaĭstvennoĭ Nauki Azerbaidzhana* 1, 72–79.

Babushkina, I.N. (1971) [The variability of *Verticillium* wilt causal agents under the influence of rhizosphere fungi]. Микология и Фитопатология [*Mycology and Phytopathology*] 5, 345–351.

Babushkina, I.N. (1974) [Soil microscopic antagonistic fungi of the pathogens of plant

Verticillium wilt]. Микология и Фитопатология [*Mycology and Phytopathology*] 8, 26–31.

Babushkina, I.N. (1979) [Some physiological features of *Verticillium dahliae* Kleb.; *V. tricorpus* Isaac and their strains induced by the action of antagonistic fungi]. Микология и Фитопатология [*Mycology and Phytopathology*] 13, 203–208.

Babushkina, I.N. (1982) [Some biological features of strains of *Verticillium dahliae* Kleb., *V. tricorpus* Isaac and their variants]. Микология и Фитопатология [*Mycology and Phytopathology*] 16, 95–99.

Baergen, K.D., Hewitt, J.D. and St Clair, D.A. (1993) Resistance of tomato genotypes to four isolates of *Verticillium dahliae* race 2. *HortScience* 28(8), 833–836.

Baggett, J.R. and Kean, D. (1986) 'Oregon Spring' and 'Santiam' parthenocarpic tomatoes. *HortScience* 21, 1245–1247.

Baggett, J.R., Mansour, N.S. and Kean, D. (1995) 'Oregon Star' and 'Oregon Pride' parthenocarpic tomatoes. *HortScience* 30(3), 649–650.

Bailey, J.A. (1986) *Biology and Molecular Biology of Plant–Pathogen Interactions. NATO ASI Series H: Cell Biology*, Vol. I, Springer Verlag, Berlin.

Bailey, R. (1971) Ecology of the cotton wilt fungus *Verticillium dahliae. Rhodesia Science News* 5, 11–14.

Baines, R.C. (1945) *Verticillium* wilt and die-back of *Viburnum. Phytopathology* 35, 145–147.

Baker, K.F. (1970) *Root Diseases and Soil-borne Pathogens. Second International Symposium on Factors Determining the Behaviour of Plant Pathogens in Soil.* Imperial College, London. University of California Press, Berkeley, California, pp. 234–239.

Baker, K.F. (1981) Biological control. In: Mace, M.E., Bell, A.A. and Beckman, C.H. (eds) *Fungal Wilt Diseases of Plants.* Academic Press, New York, pp. 523–561.

Baker, K.F. and Locke, W.F. (1946) An epiphytotic of *Verticillium* wilt in southern California. *Plant Disease Reporter* 30, 210.

Baker, K.F., Snyder, W.C. and Hansen, H.N. (1940) Some hosts of *Verticillium* in California. *Plant Disease Reporter* 24, 424–425.

Baker, R. (1966) New developments in the field of cotton breeding for disease resistance. Work done in Peru. *Proceedings, Seminar on Cotton Production Research.* Lima, Peru. pp. 7–16.

Bakhali, A.H. (1987) Degradation of date palm (*Phoenix dactylifera* L.) cell walls by extracellular pectolytic enzymes produced by *Verticillium albo-atrum* and *Verticillium dahliae. International Journal of Tropical Plant Diseases* 5, 165–172.

Bakhali, A.H. (1989a) Specificities of pectolytic enzymes produced by *Verticillium tricorpus. Transactions of the Mycological Society of Japan* 30, 161–167.

Bakhali, A.H. (1989b) Comparative studies on cellulolytic activities of species of *Verticillium. Phyton Buenos Aires* 50, 25–29.

Bakhali, A.H. (1991) Production of xylanase by *Verticillium dahliae. Microbiologica* 14, 363–366.

Balandina, I.D. and Shvetsova, L.P. (1977) [Features and strains of *Verticillium dahliae* Kleb. and *Fusarium oxysporum* f. *vasinfectum* differing in the degree of virulence]. *Trudy Sredneaz. Nii Zashchity Rast.* 11, 16–20.

Banfield, W.M. (1941) Distribution by the sap stream of spores of three fungi that induce vascular wilt diseases of elm. *Journal of Agricultural Research* 62, 637–681.

Bao, J.R., Katan, J., Shabi, E. and Katan, T. (1998) Vegetative-compatibility groups in *Verticillium dahliae* from Israel. *European Journal of Plant Pathology* 104, 263–269.

Barak, E., Dinoor, A. and Jacoby, B. (1981) Factors affecting the apoplastic translocation of systemic fungicides in plant stems. *Phytoparasitica* 9, 223.

Barasubiye, T., Parent, J.-G., Hamelin, R.C., Laberge, S., Richard, C. and Dostaler, D. (1995a) Discrimination between alfalfa and potato isolates of *Verticillium albo-atrum* using RAPD markers. *Mycological Research* 99(12), 1507–1512.

Barasubiye, T., Richard, C., Dorstaler, D., Parent, J.-G., Hamelin, R.C. and Laberge, S. (1995b) Physiological, pathological and molecular characterisation of *Verticillium albo-atrum* strains isolated from potato and alfalfa. 6th International *Verticillium* Symposium, Dead Sea, Israel. *Phytoparasitica* 23.

Barbara, D.J., Ashby, S.C. and Morton, A. (1987) Double-stranded RNA in *Verticillium albo-atrum*. *Plant Pathology* 36, 229–231.

Barbara, D.J., Carder, J.H., Morton, A. and Tarbrett, A.M. (1995) Molecular studies and relationships between isolates of *Verticillium dahliae* and *Verticillium albo-atrum*. 6th International *Verticillium* Symposium, Dead Sea, Israel, 1994. *Phytoparasitica* 23, 43.

Barchilon, V., Pinkas, Y. and Sztejnberg, A. (1987) Characteristics of an inoculated avocado rootstock resistant to *Verticillium* wilt. *Canadian Journal of Plant Pathology* 9, 78.

Barducci, T.B. (1966) New developments in the field of cotton breeding for disease resistance: work done in Peru. *Seminar on Cotton Production Research*. Lima, Peru, 1966, pp. 7–16.

Barducci, T.B. and Rada, G.G. (1942) *Verticillium* wilt del algodonero. *Bulletin of the Agricultural Experimental Station, Molina, Peru*, No. 23.

Barnes, C.E. and Staten, G. (1961) The combining of ability of some varieties and strains of *Gossypium hirsutum*. *Bulletin, New Mexico Agricultural Experimental Station*, p. 457.

Barnes, C.E., Davis, D.D., Malm, N.R., Roberts, C.L. and Wood, R.L. (1980) Registration of Acala 1517-77 Upland cotton (Reg.No.77). *Crop Science* 20, 113.

Barron, G.L. and Fletcher, J.T. (1970) *Verticillium albo-atrum* and *V. dahliae* as mycoparasites. *Canadian Journal of Botany* 48, 1137–1139.

Barrow, J.R. (1969) Cross-protection against field infestation of *Verticillium* wilt. *Proceedings of the Beltwide Cotton Production Research Conference*. New Orleans, Louisiana. pp. 32–33.

Barrow, J.R. (1970a) Heterozygosity in inheritance of *Verticillium* wilt tolerance in cotton. *Phytopathology* 60, 301–303.

Barrow, J.R. (1970b) Critical requirements for genetic expression of *Verticillium* wilt tolerance in Acala cotton. *Phytopathology* 60, 559–560.

Barrow, J.R. (1973) Genetics of *Verticillium* tolerance in cotton. Verticillium *Wilt of Cotton. Proceedings of a Work Conference of the National Cotton Pathology Research Laboratory*. College Station, Texas 1971. ARS-S19, pp. 42–62.

Baryshnikova, T.K. (1990) Complementation analysis of adenine dependent mutations of *Verticillium dahliae*. *Proceedings of the Fifth International* Verticillium *Symposium*. Leningrad, p. 100.

Basallote, M.J., Bejarano, J., Blanco, M.A., Jimenez-Diaz, R.M. and Melero, J.M. (1994) La solarización del suelo: una estrategia para controlar enfermedades causadas por organismos de suelo y reducir las alternativas de cultivo. *Investigación Agraria, Producción y Protección Vegetales* 207–220.

Basoccu, L. and Garibaldi, A. (1970) Influenza di alcune pratiche conturali sulla produzione del peperone in assenza o in presenza di attacchi di *Verticillium dahliae* Kleb. *Annali della Facoltà di Scienze Agraria della Università degli Studi di Torino* 6, 141–162.

Basoccu, L. and Garibaldi, A. (1974) Effetti della pacciamatura, del'irrigazione e della verticilliosi sulla coltura del pomodoro in serra. Proceedings of the First National Conference on Methods of Glasshouse Cultivation of Market Garden Plants. Ragusa, 22–24 November, 1971. *Tecnica Agricola, Italy* 26, 799–814.

Bassett, D.M., Stockton, J.R. and Dickens, W.L. (1970) Root growth of cotton and measurement by P32 uptake. *Agronomy Journal* 62, 200–203.

Bastos-Cruz, B.P., da Silveira, A.P. and Camargo, L. de S. (1972) Comportamento de algumas variedades e hibridos de morangueiro (*Fragaria* hibridos) com relacao a murcha verticilar (*Verticillium dahliae* Kleb.). *Archivos do Instituto Biológico* São Paulo 39, 221–229.

Basu, P.K. (1961) *Verticillium* disease of strawberries. *Canadian Journal of Botany* 39, 165–196.

Basu, P.K. (1987) Longevity of *Verticillium albo-atrum* within alfalfa stems buried in soil or maintained without soil at various temperatures. *Canadian Plant Disease Survey* 67, 15–16.

Basu, P.K. and Butler, G. (1991) Colony growth response to temperature and pathogenicity of isolates of *Verticillium albo-atrum* from alfalfa in Canada. *Canadian Journal of Plant Pathology* 13, 112–115.

Batirov, A.A. (1986) [Effect of lignin on microbiological processes in newly reclaimed soils, cotton productivity and the incidence of *Verticillium* wilt in cotton]. *Doklady Vsesoyuznoi Ordena Lenina i Ordena Trudovogo Krasnogo Znameni Akademii Sel'skokhozyaisvennykh Nauk Imeni V. I. Lenina* 5, 46–47.

Batson, W.E. and Blasingame, D.J. (1988) Field performance of selected mid-south cotton cultivars in *Verticillium dahliae* infested soil in Mississippi. *Proceedings of the Beltwide Cotton Production Research Conference, 1988*. pp. 30–31.

Baudin, M. and Lepoivre, P. (1984) Risques potentiels de transmission de champignons phytopathogenes par les boues d'epuration. *Annales de Gembloux* 90, 183–191.

Bazán de Segura, C. and Ochoa, C. (1959) *Informe Mensual Estación Agricola de la Molina* 33, 1–7.

Beard, D.F., Graham, J.H., Kugler, J.L. and Kawaguchi, I.I. (1982) Registration of WL 316 alfalfa (Reg.No. 101). *Crop Science* 22, 1257.

Beatson, R.A. (1973) Kuaka 427: a new *Verticillium* wilt resistant variety. *New Zealand Tobacco Growers' Journal* November, 9.

Beatson, R.A., Classon, A.P. and Jackson, D. (1990) Breeding and development of flue-cured tobacco cultivar 'Moskuaka 85'. *New Zealand Journal of Crop and Horticultural Science* 18, 11–15.

Beaumont, A. (1951) *Campanula* diseases. *Gardeners' Chronicle Series 3* 129(3358), 157.

Beaumont, A. (1953) Diseases of poppies. *Gardeners' Chronicle Series 3* 134(3476), 71.

Bechtol, M.K. and Thorneberry, G.O. (1966) Characteristics of respiration by conidia of *Verticillium albo-atrum*. *Phytopathology* 56, 963–966.

Beckman, C.H. (1973) The incidence of *Verticillium* species in soils, vines and tubers of Rhode Island-grown potatoes. *Plant Disease Reporter* 57, 928–932.

Beckman, C.H. (1984) Modelling the succession of host–parasite interactions in vascular diseases of plants. *Phytopathologia Mediterranea* 23, 109–129.

Beckman, C.H. (1987) *Nature of Wilt Diseases of Plants*. American Phytopathological Society, St Paul, Minnesota.

Beckman, C.H. and Talboys, P.W. (1981) Anatomy of resistance. In: Mace, M.E., Bell, A.A. and Beckman, C.H. (eds) *Fungal Wilt Diseases of Plants*. Academic Press, New York, pp. 487–521.

Beckman, C.H., Stessel, G.J. and Howard, F.L. (1969) *Verticillium* spp. and associated fungi from certified potato seed tubers. *Plant Disease Reporter* 53, 771–773.

Bedlan, G. (1987) Die Paprika-Welke. *Pflanzenschutz* 3(3–4), 8.

Bejarano-Alcazar, J. and Jimenez-Diaz, R.M. (1997) Analysis of the spatial pattern of *Verticillium dahliae* in soil of cotton fields. *Proceedings of the Seventh International* Verticillium *Symposium*. Cape Sounion, Athens, Greece. p. 87.

Bejarano-Alcazar, J., Blanco-Lopez, M.A., Melero-Vara, J.M. and Jimenez-Diaz, R.M. (1995a) Influence of crop rotation on population of defoliating and non-defoliating pathotypes of *Verticillium dahliae* in field soils. 6th International *Verticillium* Symposium, Dead Sea, Israel. *Phytoparasitica* 23, 49.

Bejarano-Alcazar, J., Melero-Vara, J.M., Blanco-Lopez, M.A. and Jimenez-Diaz, R.M. (1995b) Influence of inoculum density of defoliating and nondefoliating pathotypes of *Verticillium dahliae* on epidemics of *Verticillium* wilt of cotton in Southern Spain. *Phytopathology* 85(12), 1474–1481.

Bejarano-Alcazar, J., Blanco-Lopez, M.A., Melero-Vara, J.M. and Jimenez-Diaz, R.M. (1996) Etiology, importance, and distribution of *Verticillium* wilt of cotton in southern Spain. *Plant Disease* 80(11), 1233–1238.

Bejarano-Alcazar, J., Termorshuizen, A.J. and Jimenez-Diaz, R.M. (1999) Single-site root inoculations on eggplant with microsclerotia of *Verticillium dahliae*. *Phytoparasitica* 27(4), 279–289.

Bell, A.A. (1967) Formation of gossypol in infected or chemically irritated tissues of *Gossypium* species. *Phytopathology* 57, 759–764.

Bell, A.A. (1969) Phytoalexin production and *Verticillium* wilt resistance in cotton. *Phytopathology* 59, 1119–1127.

Bell, A.A. (1973) Nature of disease resistance. Verticillium *Wilt of Cotton. Proceedings of a Work Conference of the National Cotton Pathology Research Laboratory*. College Station, Texas 1971. ARS-S19. pp. 47–62.

Bell, A.A. (1989) Role of nutrition in diseases of cotton. In: Engelhard, A.W. (ed.) *Management of Diseases with Macro- and Microelements*. American Phytopathological Society, St Paul, Minnesota, pp. 167–204.

Bell, A.A. (1991a) Accumulation of ammonium ions in *Verticillium*-infected cotton and its relation to strain virulence, cultivar resistance and symptoms. In: Braun, J.M. and Richter, D.M. (eds) *Proceedings of the Beltwide Cotton Production Research Conference*. National Cotton Council, Memphis, Tennessee, p. 186.

Bell, A.A. (1991b) Tannin concentrations in *Verticillium*-infected cotton: relationships to strain virulence and cultivar resistence. In: Braun, J.M. and Richter, D.M. (eds) *Proceedings of the Beltwide Cotton Production Research Conference*. National Cotton Council, Memphis, Tennessee, p. 187.

Bell, A.A. (1992a) *Verticillium* wilt. In: Hillocks, R.J. (ed.) *Cotton Diseases*. CAB International, Wallingford, UK, pp. 87–122.

Bell, A.A. (1992b) Biology and ecology of *Verticillium dahliae*. In: Lyda, S.D. (ed.) *Comparative Pathology of Sclerotial-forming Plant Pathogens: A* Phymatotrichum omnivorum *Symposium*. Texas A & M University Press, College Station, Texas.

Bell, A.A. and Mace, M.E. (1981) Biochemistry and physiology of resistance. In: Mace, M.E., Bell, A.A. and Beckman, C.H. (eds) *Fungal Wilt Diseases of Plants*. Academic Press, New York, pp. 431–486.

Bell, A.A. and Presley, J.T. (1969a) Heat-inhibited or heat-killed conidia of *Verticillium*

albo-atrum induce disease resistance and phytoalexin synthesis in cotton. *Phytopathology* 59, 1147–1151.
Bell, A.A. and Presley, J.T. (1969b) Temperature effects upon resistance and phytoalexin synthesis in cotton inoculated with *Verticillium albo-atrum. Phytopathology* 59, 1141–1146.
Bell, A.A. and Puhalla, J.E. (1974) Cyclic DOPA a natural precursor in melanin of *Verticillium dahliae. Proceedings of the American Phytopathological Society* 1, 53–54.
Bell, A.A. and Stipanovic, R.D. (1978) Biochemistry of disease and pest resistance in cotton. *Mycopathologia* 65, 91–106.
Bell, A.A. and Wheeler, M.H. (1986) Biosynthesis and function of fungal melanins. *Annual Review of Phytopathology* 24, 411–451.
Bell, A.A., Simpson, M.E., Marsh, P.B. and Howell, C.R. (1971) *Proceedings of the Beltwide Cotton Production Research Conference.* National Cotton Council, Memphis, Tennessee, p. 84.
Bell, A.A., Stipanovic, R.D., Howell, C.R. and Fryxell, P.A. (1975) Antimicrobial terpenoids of *Gossypium. Phytochemistry* 14, 225–231.
Bell, A.A., Puhalla, J.E., Tolmsoff, W.J. and Stipanovic, R.D. (1976a) Use of mutants to establish (+)-scytalone as an intermediate in melanin biosynthesis by *Verticillium dahliae. Canadian Journal of Microbiology* 22, 787–799.
Bell, A.A., Stipanovic, R.D. and Puhalla, J.E. (1976b) Pentaketide metabolites of *Verticillium dahliae.* Identification of α-scytalone as a natural precursor to melanin. *Tetrahedron* 32, 1353–1356.
Bell, A.A., Stipanovic, R.D., O'Brien, D.H. and Fryxell, P.A. (1978) Sesquiterpenoid aldehyde quinones and derivatives in pigment glands of *Gossypium. Phytochemistry* 17, 1297–1305.
Bell, A.A., Stipanovic, R.D., Mace, M.E. and Kohel, R.J. (1994) Genetic manipulation of terpenoid phytoalexins in *Gossypium*: effects on disease resistance. In: Ellis, B.E., Kuroki, G.W. and Stafford, H.A. (eds) *Genetic Engineering of Plant Secondary Metabolism.* Plenum Publishing Corporation, New York, pp. 231–249.
Bell, A.A., Stipanovic, R.D., Zhang Jiuxu, Mace, M.E. and Reibenspies, J.H. (1998) Identification and synthesis of trinorcadalene phytoalexins formed by *Hibiscus cannabinus. Phytochemistry* 49, 431–440.
Bender, C.G. and Shoemaker, P.B. (1984) Prevalence of *Verticillium* wilt of tomato and virulence of *Verticillium dahliae* race 1 and race 2 isolates in western North Carolina. *Plant Disease* 68, 305–309.
Benedict, C.R., Alchanati, I., Harvey, P.J., Liu, J.G., Stipanovic, R.D. and Bell, A.A. (1995) The enzymatic formation of δ-cadinene from farnesyl diphosphate in extracts of cotton. *Phytochemistry* 39(2), 327–331.
Benhamou, N. (1988) Ultrastructural localization of carbohydrates in the cell walls of two pathogenic fungi: a comparative study. *Mycologia* 80(3), 324–337.
Benhamou, N. (1989) Ultrastructural study of galacturonic acid distribution in some pathogenic fungi using gold-complexed *Aplysia depilans* gonad lectin. *Canadian Journal of Microbiology* 35, 349–358.
Benhamou, N. (1995) Ultrastructural and cytochemical aspects of the response of eggplant parenchyma cells in direct contact with *Verticillium*-infected xylem vessels. *Physiological and Molecular Plant Pathology* 46(4), 321–338.
Benhamou, N., Chamberland, H., Noel, S. and Ouellette, G.B. (1990) Ultrastructural localization of β-1,4-glucan-containing molecules in the cell walls of some fungi: a

comparative study between spore and mycelium. *Canadian Journal of Microbiology* 36(3), 149–158.

Benhamou, N., Rey, P., Picard, K. and Tirilly, Y. (1999) Ultrastructural and cytochemical aspects of the interaction between the mycoparasite *Pythium oligandrum* and soilborne plant pathogens. *Phytopathology* 89(6), 506–517.

Benlian, O. and Dingyuan, H. (1996) A very early quality cotton variety – Xinluzao 5. *China Cottons* 23(12), 19.

Bennamara, A., Abourriche, A., Berrada, M., Charrouf, M., Chaib, N., Boudouma, M. and Garneau, F.X. (1999) Methoxybifurcarenone: an antifungal and antibacterial meroditerpenoid from the brown alga *Cystoseira tamariscifolia*. *Phytochemistry* 52(1), 37–40.

Bennet, A.B. and Della Penna, D. (1987) Polygalacturonase, its importance and regulation in fruit ripening. In: Thomson, W.W., Northnagel, E.A. and Huffaker R.C. (eds) *Plant Senescence: Its Biochemistry and Physiology*. American Society of Plant Physiology, Rockville, Maryland, p. 98.

Benson, D.M. and Ashworth, L.J. (1976) Survival of *Verticillium albo-atrum* on nonsuscept roots and residues in field soil. *Phytopathology* 66, 883–887.

Ben-Yephet, Y. (1979) Penetration of metham and Mylone into soil columns as measured by their effect on viability of microsclerotia of *Verticillium dahliae*. *Plant and Soil* 53, 341–349.

Ben-Yephet, Y. and Frank, Z.R. (1978) Toxicity of metham towards *Verticillium dahliae* microsclerotia. *Plant Disease Reporter* 62, 119–121.

Ben-Yephet, Y. and Frank, Z.R. (1984) Optimization of the metham-sodium dose in controlling *Verticillium dahliae* in potato. *Phytoparasitica* 12, 203–205.

Ben-Yephet, Y. and Frank, Z.R. (1985) Effect of soil structure on penetration by metham-sodium and of temperature on concentrations required to kill soil-borne pathogens. *Phytopathology* 75, 403–406.

Ben-Yephet, Y. and Frank, Z.R. (1989) Factors affecting the efficiency of metham-sodium in controlling *Verticillium dahliae*. *Acta Horticulturae* 255, 227–234.

Ben-Yephet, Y. and Pilowskey, M. (1979) A method for determining susceptibility of tomatoes to *Verticillium dahliae* in the greenhouse. *Plant Disease Reporter* 63, 66–69.

Ben-Yephet, Y. and Pinkas, Y. (1976) A cesium chloride flotation technique for the isolation of *Verticillium dahliae* microsclerotia. *Phytopathology* 66, 1252–1254.

Ben-Yephet, Y. and Szmulewich, Y. (1985) Inoculum levels of *Verticillium dahliae* in the soils of the hot semi-arid Negev region of Israel. *Phytoparasitica* 13, 193–200.

Ben-Yephet, Y., Letham, D. and Evans, G. (1981) Toxicity of 1,2-dibromoethane and 1,3-dichloropropene to microsclerotia of *Verticillium dahliae*. *Pesticide Science* 12, 170–174.

Ben-Yephet, Y., Siti, E. and Frank, Z. (1983) Control of *Verticillium dahliae* by metham-sodium in loessial soil and effect on potato tuber yields. *Plant Disease* 67, 1223–1225.

Ben-Yephet, Y., Siti, E. and Frank, Z. (1984) Penetration of metham sodium into loess soil as measured by its effect on the viability of *Verticillium dahliae* microsclerotia. In: Citrulli, M. (ed.) Pathobiology of *Verticillium* Species. Third International *Verticillium* Symposium, Bari, Italy. *Phytopathologia Mediterranea* 23, 99.

Ben-Yephet, Y., Melero, J.M. and DeVay, J.E. (1986) Synergistic interaction of solarization and metham sodium in controlling microsclerotia of *Verticillium dahliae*. *Proceedings of the Fourth International* Verticillium *Symposium*. Guelph, Canada.

Ben-Yephet, Y., Melero, J.M. and DeVay, J.E. (1987) Synergistic interaction of solariza-

tion and metham-sodium in controlling microsclerotia of *Verticillium dahliae*. *Canadian Journal of Plant Pathology* 9, 78.
Ben-Yephet, Y., Frank, Z.R., Melero-Vara, J.M. and DeVay, J.E. (1989) The effect of crop rotation and metham sodium on *Verticillium dahliae*. In: Tjamos, E.C. and Beckman, A.A. (eds) *NATO Symposium*. pp. 543–555.
Berestetskiĭ, O.A., Kravchenko, L.V. and Makarova, N.M. (1982) [The effect of volatile products of decomposed plant residues on the development of fungal spores]. Микология и Фитопатология [*Mycology and Phytopathology*] 16, 126–129.
Berg, G. (1997) Bacterial antagonists of *Verticillium dahliae* var. *longisporum* Stark in the rhizosphere of oil seed rape. *Proceedings of the Seventh International* Verticillium *Symposium*. Cape Sounion, Athens, Greece. p. 64.
Berg, G. and Bahl, H. (1997) Charakterisierung der nutzlichen Rhizobakterien von Winterraps fur einen Einsatz in der biologischen Kontrolle der *Verticillium* Welke. *Gesunde Pflanzen* 49(3), 76–82.
Berg, G. and Ballin, G. (1994) Bacterial antagonists to *Verticillium dahliae* Kleb. *Journal of Phytopathology* 141(1), 99–110.
Berg, G., Knappe, C., Ballin, G. and Seidel, D. (1994) Biological control of *Verticillium dahliae* by naturally occurring rhizosphere bacteria. *Archives of Phytopathology and Plant Protection* 29, 249–262.
Berg, G., Marten, P. and Ballin, G. (1996) *Stenotrophomonas maltophilia* in the rhizosphere of oilseed rape – occurrence, characterization and interaction with phytopathogenic fungi. *Microbiological Research* 151(1), 19–27.
Berg, G., Frankowski, J. and Hubert, B. (1997) Interactions between *Serratia plymuthica* and the soilborne pathogen *Verticillium dahliae*. *Proceedings of the Seventh International* Verticillium *Symposium*. Cape Sounion, Athens, Greece. p. 70.
Berg, G., Marten, P. and Bahl, H. (1998) Population dynamics of bacteria including antifungal species in the rhizosphere of oilseed rape during its life cycle. *Archives of Phytopathology and Plant Protection* 31, 215–224.
Bergeson, G.B. (1972) Concepts of nematode–fungus associations in plant disease complexes: a review. *Experimental Parasitology* 32, 301–314.
Berkeley, G.H. (1927) Report of the Dominion Field Laboratory of Plant Pathololology, St Catherines. *Canada Department of Agriculture: Dominion Experimental Farms, Report of the Dominion Botanist* 1926, pp. 59–102.
Berkeley, G.H. and Jackson, A.B. (1926) *Verticillium* wilt of the red raspberry. *Science and Agriculture* 6, 261–270.
Berkeley, G.H., Madden, G.O. and Willison, R.S. (1931) *Verticillium* wilts in Ontario. *Science and Agriculture* 11, 739–759.
Berry, S.Z. and Gould, W.A. (1979) 'Ohio 7663' tomato. *HortScience* 14, 550–551.
Berry, S.Z. and Gould, W.A. (1981) 'Ohio 7681' tomato. *HortScience* 16, 790–791.
Berry, S.Z. and Gould, W.A. (1986) 'Ohio 832' tomato. *HortScience* 21, 334.
Berry, S.Z. and Thomas, C.A. (1961) Influence of soil temperature and methods of inoculation on resistence of mint to *Verticillium* wilt. *Phytopathology* 51, 169–174.
Berry, S.Z., Wiese, K.L. and Aldrich, T.S. (1993) 'Ohio 8556' processing tomato. *HortScience* 28(7), 751.
Berry, S.Z., Aldrich, T.S., Wiese, K.L. and Bash, W.D. (1995) 'Ohio OX38' hybrid processing tomato. *HortScience* 30(1), 159.
Bertero de Romano, A., Vazquez, A., Piubello, S. and Sala, C. (1994) Quantifying the relationship between *Verticillium* wilt intensity and yield loss of sunflower

(*Helianthus annuus* L.), and feasibility of utilizing the hypodermic inoculation technique as a selection method. *Helia* 17(20), 49–54.

Besri, M. (1978) Phases de la transmission de *Fusarium oxysporum* f.sp. *lycopersici* et de *Verticillium dahliae* par les semences de quelques varietiés de tomate. *Phytopathologische Zeitschrift* 93, 148–163.

Besri, M. (1980a) Influence de la température du sol et de l'air sur la fusariose et la verticilliose de tomates d'automne de printemps au Maroc. *Phytopathologische Zeitschrift* 98, 97–107.

Besri, M. (1980b) Influence du potential osmotique de l'eau sur la croissance de *Fusarium oxysporum* f.sp. *lycopersici* et de *Verticillium dahliae*. *Phytopathologische Zeitschrift* 99, 1–8.

Besri, M. (1981) Qualité des sols et des eaux d'irrigation et manifestation des tracheomycoses de la tomate au Maroc. *Phytopathologia Mediterranea* 20(2–3), 107–111.

Besri, M. (1990) Effect of salinity on the development of tomato *Verticillium* wilt in Morocco. *Proceedings of the Fifth International* Verticillium *Symposium*. Leningrad. p. 61.

Besri, M. and Zrouri, M. (1983) Température et manifestation des tracheomycoses de tomate cultivée en plein air et sous tunnel plastique au Maroc. *EPPO Bulletin* 13, 127–131.

Besri, M., Ouddaf, M. and Khibabi, L. (1976) Comportement des principales varietés des tomate utilisées au Maroc vis à vis de *Fusarium oxysporum* f.sp. *lycopersici* et de *Verticillium dahliae*. *Poljoprivredna Znanstvena Smotra* 39, 89–99.

Besri, M., Zrouri, M. and Beye, I. (1984) Appartenance raciàle et pathogenie comparée de quelques isolats de *Verticillium dahliae* Kleb. obtenus a partir de tomates résistants au Maroc. *Phytopathologische Zeitschrift* 109, 289–294.

Bewley, W.F. (1920, publ. 1921) 'Sleepy disease' or wilt of the tomato. *Annual Report, Cheshunt Experimental and Research Station*, pp. 26–38.

Bewley, W.F. (1922) 'Sleepy disease' of the tomato. *Annals of Applied Biology* 9, 116–134.

Bewley, W.F. (1928) *Diseases of Glasshouse Plants*. Ernest Benn, London.

Beye, I. and Lafay, J.F. (1988a) Verticilliose de la tomate: analyse des interactions entre l'agressivité de l'agent pathogène et la résistance générale de l'hôte. *Agronomie* 8, 351–361.

Beye, I. and Lafay, J.F. (1988b) Verticilliose de la tomate: conséquences des interactions entre l'agressivité de l'agent pathogène et la résistance de l'hôte au niveau des populations. *Agronomie* 8, 435–439.

Beyres, A. (1974) Le greffage des solanacées maraichères. *Pepinieristes Horticulteurs Maraichers* 152, 27–32.

Bharadwaj, P. and Trivedi, P.C. (1997) *Verticillium chlamydosporium* a propitious biocontrol agent for the management of cyst nematode. *Proceedings of the Seventh International* Verticillium *Symposium*. Cape Sounion, Athens, Greece. p. 72.

Bhaskaran, R. (1972) *In vitro* production of indole-3-acetic acid by *Verticillium dahliae*. *Science and Culture* 38, 523.

Bhaskaran, R. and Muthusamy, M. (1974) Role of phenolics in *Verticillium* wilt of cotton. *Madras Agricultural Journal* 61(6), 160–164.

Bhat, R.G., Subbarao, K.V. and Bari, M.A. (1999) First report of *Verticillium dahliae* causing artichoke wilt in California. *Plant Disease* 83(8), 782.

Bhatti, M.A., Ali, S. and Khan, I.U. (1985) Comparative pathogenicity of fungi associated with wilt of chickpea, with special reference to *Verticillium albo-atrum* R. and B. *Tropical Pest Management* 31, 324–326.

Bhatti, M.A., Ali, S. and Khan, I.U. (1987) Pathogenicity of fungi associated with wilt of chickpea with special reference to *Verticillium albo-atrum*. *Pakistan Journal of Agricultural Research* 8, 39–42.

Bhutta, A.R., Bhatti, M.H.R., Nizamani, S.M. and Ahmad, I. (1997) Studies on effect of seed-borne fungi on germination of sunflower. *Helia* 20(26), 35–42.

Bianchini, G.M., Stipanovic, R.D. and Bell, A.A. (1999) Induction of delta-cadinene synthase and sesquiterpenoid phytoalexins in cotton by *Verticillium dahliae*. *Journal of Agricultural and Food Chemistry* 47(10), 4403–4406.

Bidochka, M.J., Leger, R.J.S., Stuart, A. and Gowanlock, K. (1999) Nuclear rDNA phylogeny in the fungal genus *Verticillium* and its relationship to insect and plant virulence, extracellular proteases and carbohydrases. *Microbiology Reading* 145, 955–963.

Bilai, V.I. (1963) *Antibiotic-producing Microscopic Fungi*. Elsevier Publishing Company, Amsterdam, The Netherlands.

Bird, L.S. (1966) Interrelation of resistance and escape of cotton from five major diseases. *Proceedings, 26th Cotton Disease Council*, pp. 92–107.

Bird, L.S. (1976) Registration of Tamcot SP21, Tamcot SP23 and Tamcot SP37 cottons (Reg.Nos. 61, 62 and 63). *Crop Science* 16, 884.

Bird, L.S. (1979a) Registration of Tamcot SP37H cotton. *Crop Science* 19, 412.

Bird, L.S. (1979b) Registration of Tamcot SP21S cotton. *Crop Science* 19, 410–411.

Bishop, C.D. and Cooper, R.M. (1983a) An ultrastructural study of root invasion in three vascular wilt diseases. *Physiological Plant Pathology* 22, 15–27.

Bishop, C.D. and Cooper, R.M. (1983b) An ultrastructural study of vascular colonization in three vascular wilt diseases I. Colonization of susceptible cultivars. *Physiological Plant Pathology* 23, 323–343.

Bishop, C.D. and Cooper, R.M. (1984) Ultrastructure of vascular colonization by fungal wilt pathogens II. Invasion of resistant cultivars. *Physiological Plant Pathology* 24, 277–289.

Bisiach, M., de Martino, A. and Vercesi, A. (1982) La verticilliosi della *Catalpa*. *Informatore Fitopatologico* 32(11), 59–62.

Blackhurst, F.M. and Wood, R.K.S. (1963a) *Verticillium* wilt of tomatoes – further experiments on the role of pectic and cellulolytic enzymes. *Annals of Applied Biology* 52, 89–96.

Blackhurst, F.M. and Wood, R.K.S. (1963b) Resistance of tomato plants to *Verticillium albo-atrum*. *Transactions of the British Mycological Society* 46, 385–392.

Blanco-López, M.A., Jiménez-DÍaz, R.M. and Caballero, J.M. (1984) Symptomatology, incidence and distribution of *Verticillium* wilt of olive trees in Andalucía. *Phytopathologia Mediterranea* 23, 1–8.

Blanco-López, M.A., Melero-Vara, J.M., Bejarano-Alcazar, J. and Jiménez-Díaz, J.M. (1987) Distribution and pathogenicity of *Verticillium dahliae* isolates infecting cotton in Andalucia, southern Spain. *Canadian Journal of Plant Pathology* 9, 78.

Blanco-López, M.A., Bejerano-Alcazar, J., Malero-Vera, J.M. and Jiménez-Díaz, R.M. (1989) Current status of *Verticillium* wilt of cotton in southern Spain: pathogen variation and population in soil. In: Tjamos, E.C. and Beckman, C.H. (eds) *NATO ASI Series H Cell Biology. Vascular Wilt Diseases of Plants*, Vol. 28, Springer-Verlag, Berlin, pp. 123–132.

Blanco-Lopez, M.A., Rodrigues-Jurando, D. and Jimenez-Dias, R.M. (1990) Incidence and seasonal variation of *Verticillium* wilt in olive orchards. *Proceedings of the Fifth International* Verticillium *Symposium*. Leningrad, p. 5.

Blank, L.M. and Leyendecker, P.J. (1951) *Verticillium* spread to disease-free soil by infected cotton stalks. *Plant Disease Reporter* 35, 10–11.

Blank, L.M., Leding, A.R. and Nakayama, R.M. (1953) *Bulletin of the New Mexico Agricultural Experimental Station*, p. 1081.

Bletsos, F.A., Thanassoulopoulos, C.C. and Roupakias, D.G. (1998) Sensitivity of two Greek egplant varieties and the wild species *Solanum sisymbriifolium* Lam. to *Verticillium* wilt. *Journal of Genetics and Breeding* 52, 99–102.

Bletsos, F.A., Thanassoulopoulos, C.C. and Roupakias, D.G. (1999) Water stress and *Verticillium* wilt severity on eggplant (*Solanum melongena* L.). *Journal of Phytopathology* 147(4), 243–248.

Block, C.C., Senechal, N.P. and Widrlechner, M.P. (1989) First report of *Verticillium* wilt of *Agastache rugosa* caused by *Verticillium dahliae*. *Plant Disease* 73, 1020.

Blok, W.J., Lamers, J.G., Termorshuizen, A.J. and Bollen, G.J. (2000) Control of soilborne plant pathogens by incorporating fresh organic amendments followed by tarping. *Phytopathology* 90(3), 253–259.

Bochow, H. (1977) After-effects of systemic fungicides in the control of soil-borne pathogens, and the interaction between systemic fungicides and growth regulators. In: Fuchs, A. (ed.) *Internal Therapy of Plants. Netherlands Journal of Plant Pathology* 83(Suppl. 1), 305–316.

Bochow, H. (1989) Enhanced disinfection effects of soil fumigation with sealing by film forming emulsions. *Acta Horticulturae* 255, 255–260.

Bocsa, I. and Raynal, G. (1984) Tentative d'appreciation de l'importance locale des principales maladies de la luzerne dans 8 pays d'Europe par l'emploi de varietes resistantes. *Proceedings of the* Medicago sativa *group of Eucarpia*. August 1984, Brno, Czechoslovakia. pp. 151–165.

Boesewinkel, H.J. (1976) Storage of fungal cultures in water. *Transactions of the British Mycological Society* 66, 183–185.

Boisson, C. and Lahlou, H. (1980) Étude du polymorphisme intraclonal chèz le *Verticillium albo-atrum*, forme à microsclerotes. I La morphogenèse des thalles à partir des microconidies et ses variations. *Canadian Journal of Botany* 58, 2407–2419.

Boisson, C. and Lahlou, H. (1983) Étude du polymorphisme intraclonal chez le *Verticillium albo-atrum*, forme à microsclérotes. III. Morphologie et morphogénèse comparées de deux isolats de phenotype sauvage et de leurs principaux variants. *Canadian Journal of Botany* 61, 188–196.

Bolbekov, M.E. (1987) [Analysis of wilt resistance in intraspecific distant hybrids of cotton in the F_3 involving *Gossypium hirsutum* subsp. *yucatenense* (Hutch.)]. *Nauchno-Tekhnicheskii Byulleten' Vsesoyuznogo Ordena Lenina i Ordena Drushby Narodov Nauchno-Issledovatel'skogo Instituta Rastenievodstva Imeni N.I. Vavilova* 175, 42–45.

Bollen, G.J. (1969) The selective effect of heat treatment on the microflora of a greenhouse soil. *Netherlands Journal of Plant Pathology* 75, 157–163.

Bollen, G.J. (1985) Lethal temperatures of soil fungi. In: Parker, C.A., Rovira, A.D., Moore, K.G., Wong, P.T. and Kollmorgen, J.F. (eds) *Ecology and Management of Soil-borne Plant Pathogens*. American Phytopathological Society, St Paul, Minnesota, pp. 191–193.

Bonfante, P.F., Ceruti Scurti, J. and Obert, F. (1972) Interazione di *Tuber melanosporum* Vitt. con miceli di altri funghi. *Allionia* 18, 53–59.

Bonifacio, A. (1974) Seccume delle branche di olivo. *Informatore di Ortoflorofrutticoltura* 15(4), 7–8.
Bonifacio, A. and Parrini, C. (1975) Presenza di *Verticillium dahliae* Kleb. in oliveti della Toscana. *Informatore Fitopatologico* 25(12), 21–25.
Böning, K. (1936) Untersuchungen über Meerrettichkrankheiten und deren Bekämpfung. *Angewandte Botanik* 18, 482–494.
Böning, K., Mallach, N., Spraw, F. and Wagner, F. (1960) [Contribution to 33rd German Plant Protection Conference, Freiburg]. *Pflanzenschutz* 12, 11–12.
Booth, J.A. (1966) Enhancement of *Verticillium albo-atrum* pectolytic enzyme production and activity by L-alanine. *Phytopathology* 56, 871.
Booth, J.A. (1969) *Gossypium hirsutum* tolerance to *Verticillium albo-atrum* infection. I Amino acid exudation from aseptic roots of tolerant and susceptible cotton. *Phytopathology* 59, 43–46.
Booth, J.A. (1970) A comparison of various surfactants on benlate fungicide activity. *Proceedings of the Beltwide Cotton Production Research Conference*. Houston, Texas. p. 25.
Booth, J.A. (1974) Effect of cotton root exudate constituents on growth and pectolytic enzyme production by *Verticillium albo-atrum*. *Canadian Journal of Botany* 52, 2219–2224.
Booth, J.A. and Rawlins, T.E. (1970) A comparison of various surfactants as adjuvants for the fungicidal action of benomyl on *Verticillium*. *Plant Disease Reporter* 54, 741–744.
Booth, J.A., Rawlins, T.E. and Chew, C.F. (1971) Field treatments for control of cotton *Verticillium* wilt with benomyl–surfactant combinations. *Plant Disease Reporter* 55, 569.
Born, G.L. (1971) Heat treatment of soil enhances *Verticillium* wilt infection of barberry and redbud. *Plant Disease Reporter* 55, 996–997.
Born, G.L. (1974) Root infection of woody hosts with *Verticillium albo-atrum*. *Bulletin, Illinois Natural History Survey* 31, 207–249.
Borodin, G.I. and Shtok, D.A. (1982) [Effects of toxic substances of fungi inhibiting cotton seeds]. Микология и Фитопатология [*Mycology and Phytopathology*] 16, 338–342.
Botseas, D.D. and Rowe, R.C. (1994) Development of potato early dying in response to infection by two pathotypes of *Verticillium dahliae* and co-infection by *Pratylenchus penetrans*. *Phytopathology* 84, 275–282.
Bourbos, E.A. (1986) Étude de l'evolution de la mycoflore du sol de tomate en serre non chaufée après désinfection. *EPPO Bulletin* 16, 321–325.
Bourbos, V.A. and Skoudridakis, M.T. (1996) Soil solarization for the control of *Verticillium* wilt of greenhouse tomato. *Phytoparasitica* 24(4), 277–280.
Bourland, F.M. and Oosterhuis, D.M. (1999) Breeding and evaluation of cotton genotypes. *Proceedings of the 1999 Cotton Research Meeting*. Special Report of the Arkansas Agriculture Experiment Station 193, 85–88.
Bouton, J.H., Gates, R.N., Wood, D.T. and Utley, P.R. (1997) Registration of 'ABT 805' alfalfa. *Crop Science* 37(1), 293.
Bowers, J.H., Nameth, S.T., Riedel, R.M. and Rowe, R.C. (1996) Infection and colonization of potato roots by *Verticillium dahliae* as affected by *Pratylenchus penetrans and P. crenatus*. *Phytopathology* 86(6), 614–621.
Boyd, O.C. (1930) Some diseases of shrubs and ornamentals in Massachusetts. *Plant Disease Reporter* 14, 201.

Boyle, A.M. (1963) Pathogenicity of an isolate of *Verticillium albo-atrum* from olive. *Phytopathology* 53, 242.

Brandt, W.H. (1962) Manganese stimulation of melanin synthesis and microsclerotia development in *Verticillium*. *Plant Physiology* 37(Suppl.), 30.

Brandt, W.H. (1964a) Morphogenesis in *Verticillium*: a self-induced, non-hereditary variation in colony form. *American Journal of Botany* 51, 820–824.

Brandt, W.H. (1964b) Morphogenesis in *Verticillium*: effects of light and ultraviolet radiation on microsclerotia and melanin. *Canadian Journal of Botany* 42, 1017–1023.

Brandt, W.H. (1967) Influence of near ultra-violet light on hyphal elongation in *Verticillium*. *Mycologia* 59, 736–739.

Brandt, W.H. and Reese, J.E. (1964) Morphogenesis in *Verticillium*: a self-produced diffusible morphogenetic factor. *American Journal of Botany* 51, 922–927.

Brandt, W.H. and Roth, J.N. (1965) Loss of melanin-containing structures by *Verticillium* cultures. *Phytopathology* 55, 1200–1202.

Brandt, W.H. and Wang, C.H. (1960) Catabolism of C^{14} labelled glucose, gluconate and acetate in *Verticillium albo-atrum*. *American Journal of Botany* 47, 50–53.

Brandt, W.H., Lacy, M.L. and Horner, C.E. (1984) Distribution of *Verticillium* in stems of resistant and susceptible species of mint. *Phytopathology* 74, 587–591.

Brielmaier, U. (1985) Wirkung von Desinfektionsmitteln auf pilzliche Krankheitserreger, die im Zierpflanzenbau von Bedeutung sind. *Mededelingen van de Faculteit Landbouwwetenschappen Rijksuniversiteit Gent* 50(3b), 1235–1242.

Brinkerhoff, L.A. (1969) The influence of temperature, aeration, and soil microflora on microsclerotial development of *Verticillium albo-atrum* in abscised cotton leaves. *Phytopathology* 59, 805–808.

Brinkerhoff, L.A. (1973) Effects of environment on the pathogen and the disease. Verticillium *Wilt of Cotton. Proceedings of a Work Conference of the National Cotton Pathology Research Laboratory*. College Station, Texas 1971, ARS-S19. pp. 98–104.

Brinkerhoff, L.A., Samayoa, E. and Murray, J.C. (1967) *Verticillium* wilt reactions in cotton as affected by post-inoculation temperatures. *Phytopathology* 57, 805.

Brisson, J.D., Pauze, J.F. and Lavoie, G. (1974) Mycological and histopathological study of the stem dieback in lowbush blueberry (*Vaccinium angustifolium* Ait.). *Proceedings of the Third North American Blueberry Research Workers' Conference*. East Lansing, Michigan, 6–7 November, 1974. pp. 63–80.

Brisson, J.D., Robb, J. and Busch, L. (1978) Ultrastructure of wilt syndrome caused by *Verticillium dahliae*. IV. Evidence for α- and β-conidial forms. *Canadian Journal of Botany* 56, 1458–1465.

Brody, A.K., Karban, R. and Schnathorst, W.C. (1990) Inverse relationship between cotton plant density and *Verticillium* wilt incidence and severity: evidence for an alternative hypothesis. *Crop Protection* 9, 174–176.

Brooks, S.N., Horner, C.E., Likens, S.T. and Zimmermann, C.E. (1972) Registration of Cascade hop. *Crop Science* 12, 394.

Brown, A.E. (1986) Ferrous complexes in fungal disease suppression. *Annual Report on Research and Technical Work of the Department of Agriculture for Northern Ireland 1985*, pp. 139–140.

Brown, F.H. and Wiles, A.B. (1970) Reaction of certain cultivars and weeds to a pathogenic isolate of *Verticillium albo-atrum* from cotton. *Plant Disease Reporter* 54, 508–512.

Brown, M.F. and Wyllie, T.D. (1970) Ultrastructure of microsclerotia of *Verticillium albo-atrum*. *Phytopathology* 60, 538–542.

Brown, N.A. (1914) A snapdragon wilt due to *Verticillium*. *Phytopathology* 4, 217.
Bruniard, J.M., Ludueña, P.M. and Ivancovich, A. (1983, published 1984) Selección de líneas de girasol con resistencia a *Verticillium dahliae*. *IDIA* 413/416, 96–98.
Bruschaber, J.A., Wilson, C.L. and Aist, J.R. (1967) Asexual nuclear behaviour of some plant pathogenic fungi. *Phytopathology* 57, 43–46.
Bryan, M.K. (1925) Phytopathological notes. *Verticillium* wilt of tomato. *Phytopathology* 15, 187–188.
Bryan, M.K. (1928) *Verticillium* wilt of *Heliotrope*. *Phytopathology* 18, 246.
Buchenauer, H. (1971) Einfluss einiger Wuchshemmstoffe (CCC, CMH, Amo-1618, Phosfon-D, B-995) auf die *Fusarium* und *Verticillium* Welke der Tomate. *Phytopathologische Zeitschrift* 72, 53–66.
Buchenauer, H. (1979) Comparative studies on the antifungal activity of triadimefon, triadimenol, fenarimol, nuarimol, imazalil and fluotrimazole *in vitro*. *Zeitschrift für Pflanzenkrankheiten und Pflanzenschutz* 86, 341–354.
Buchenauer, H. and Erwin, D.C. (1971) Control of *Verticillium* wilt of cotton by spraying foliage with benomyl and thiabendazole solubilized with hydrochloric acid. *Phytopathology* 61, 433–434.
Buchenauer, H. and Erwin, D.C. (1972) Control of *Verticillium* wilt of cotton by spraying with acidic solutions of benomyl methyl 2-benzimidazole carbamate and thiabendazole. *Phytopathologische Zeitschrift* 75, 124–139.
Buchenauer, H. and Erwin, D.C. (1973a) Reduction of the severity of *Verticillium* wilt of cotton by the growth retardants *N,N*-dimethyl-pyrrolidinium iodide and *N,N*-dimethylpiperidium iodide associated with induction of antifungal compounds. *Zeitschrift für Pflanzenkrankheiten und Pflanzenschutz* 80, 576–586.
Buchenauer, H. and Erwin, D.C. (1973b) *N,N*-dimethylpyrrolidinium iodide and *N,N*-dimethylpiperidinium iodide as new plant growth retardants. *Zeitschrift für Pflanzenkrankheiten und Pflanzenschutz* 80, 567–575.
Buchenauer, H. and Erwin, D.C. (1976) Effect of the plant growth retardant pydanon on *Verticillium* wilt of cotton and tomato. *Phytopathology* 66, 1140–1143.
Buchenauer, H. and Grossmann, F. (1969) Einfluss von Morphaktinen auf verschiedne Tomatenkrankheiten. *Deutsche Botanische Gesellschaft Neue Folge* 3, 149–159.
Buchenauer, H. and Grossmann, F. (1970) Therapeutic effects of fluorine, phenanthrene and xanthene derivatives on fungal diseases of tomato. *Nature* 227, 1267–1268.
Buchner, V., Nachmias, A. and Burstein, Y. (1982) Isolation and partial characterization of a phytotoxic glycopeptide from a protein–lipopolysaccharide complex produced by a potato isolate of *Verticillium dahliae*. *FEBS Letters* 138, 261–264.
Buchner, V., Burnstein, Y. and Nachmias, A. (1989) Comparison of *Verticillium dahliae*-produced phytotoxic peptides purified from culture fluids and infected potato stems. *Physiological and Molecular Plant Pathology* 35, 253–269.
Buckley, P.M., Wyllie, T.D. and DeVay, J.E. (1969) Fine structure of conidia and conidium formation in *Verticillium albo-atrum* and *V. nigrescens*. *Mycologia* 61, 240–250.
Bugbee, W.M. (1967) *Verticillium nigrescens* from cotton. *Plant Disease Reporter* 51, 760–761.
Bugbee, W.M. (1970) Vascular response of cotton to infection by *Fusarium vasinfectum* f.sp. *vasinfectum*. *Phytopathology* 60, 121–123.
Bugbee, W.M. and Presley, J.T. (1967) A rapid inoculation technique to evaluate the resistance of cotton to *Verticillium albo-atrum*. *Phytopathology* 57, 1264.

Bugbee, W.M. and Sappenfield, W.P. (1970) Effect of *Verticillium* wilt on cotton yield, fiber properties, and seed quality. *Crop Science* 10, 649–652.

Bugbee, W.M. and Sappenfield, W.P. (1972) Greenhouse evaluation of *Verticillium*, *Fusarium*, and root knot nematode on cotton. *Crop Science* 12, 112–114.

Buĭmistru, L.D. (1976) [*Trichoderma* in the control of *Verticillium* wilt of eggplant]. *Vertitsillezn. Vilt kul'turn. rast. Moldavii*, 17–22.

Buĭmistru, L.D. (1977a) [Effect of antibiotics on *Verticillium dahliae* Kleb. and their use in the control of wilt of eggplant]. *Gribn. i virus. bolezni s.-kh. kul'tur Moldavii*, 59–64.

Buĭmistru, L.D. (1977b) [Eradicant spraying in the control of *Verticillium* wilt of eggplant]. *Izvestiya Akademii Nauk Moldavskoi SSR, Biologicheskikh i Khimicheskikh Nauk* 1, 89.

Buisman, C. (1932) [Report of the investigations on the elm disease conducted in the Phytopathological Laboratory 'Willie Commelin Scholten' at Baarn during 1932]. *Tijdschrift over Plantenziekten* 39, 77–94.

Bukar, B.I. and Salinskaya, I.A. (1986) [Principal fungal diseases of dahlias and their control]. *Sadovodstvo i Vinogradarstvo Moldavii* 3, 42–43.

Burden, O.J. (1968) A wilt of pawpaw caused by *Verticillium dahliae* Kleb. *Queensland Journal of Agricultural and Animal Science* 25, 153–156.

Burpee, L.L. and Bloom, J.R. (1974) Interaction of *Verticillium albo-atrum* and *Pratylenchus penetrans* on potato. *Phytopathology* 64, 579.

Burpee, L.L. and Bloom, J.R. (1978) The influence of *Pratylenchus penetrans* on the incidence and severity of *Verticillium* wilt of potato. *Journal of Nematology* 10, 95–99.

Busch, H. (1991) Variety breeding for disease resistance in double-low winter oilseed rape in Germany – situation and results. *Bulletin SCROP* 14, 212–218.

Busch, L.V. (1967) Distribution in Ontario of *Verticillium* strains causing wilt of potatoes. *Canadian Plant Disease Survey* 47, 76–78.

Busch, L.V. and Edgington, L.V. (1967) Correlation of photoperiod with tuberization and susceptibility of potato to *Verticillium albo-atrum*. *Canadian Journal of Botany* 45, 691–693.

Busch, L.V. and Hall, R. (1971) *Verticillium* wilt of chrysanthemum: suppression of fungal colonization of leaves and prevention of wilt symptom development by foliar applications of benomyl. *Canadian Journal of Botany* 49, 1987–1991.

Busch, L.V. and Schooley, H.D. (1970) Environmental influence on symptom expression in *Verticillium* wilt of *Chrysanthemum*. *Canadian Journal of Botany* 48, 1939–1941.

Busch, L.V. and Smith, E.A. (1981) Susceptibility of Ontario-grown alfalfa cultivars and certain *Medicago* species to *Verticillium albo-atrum*. *Canadian Journal of Plant Pathology* 3, 169–172.

Busch, L.V. and Smith, E.A. (1982) Reaction of a number of cultivated plants and weed species to an alfalfa isolate of *Verticillium albo-atrum*. *Canadian Journal of Plant Pathology* 4, 266–268.

Busch, L.V., Smith, E.A. and Njoh-Elango, F. (1978) The effect of weeds on the value of rotation as a practical control for *Verticillium* wilt of potato. *Canadian Plant Disease Survey* 58, 61–64.

Butcher, H. (1975) Iwa. *New Zealand Potato Bulletin* 58, 7.

Butler, E.J. (1933) Cotton diseases. *Empire Cotton Growing Review* 10, 91–99.

Butler, M.J., Lazarovits, G., Higgins, V.J. and Lachance, M.A. (1988) Partial purification and characterization of a dehydratase associated with the pentaketide melanogenesis pathway of *Phaeococcomyces* sp. and other fungi. *Experimental Mycology* 12, 367–376.

Butterfield, E.J. (1975) Effects of cultural practices on the ecology of *Verticillium dahliae* and the epidemiology of *Verticillium* wilt of cotton. PhD dissertation. University of California, Davis, 71pp.

Butterfield, E.J. and DeVay, J.E. (1975a) Association of a bacterium with microsclerotia of *Verticillium dahliae* inhibited by soil fungistasis. *Proceedings of the American Phytopathological Society* 2, 41.

Butterfield, E.J. and DeVay, J.E. (1975b) An improved soil assay for *Verticillium dahliae*. *Proceedings of the American Phytopathological Society* 2, 111.

Butterfield, E.J. and DeVay, J.E. (1977) Reassessment of soil assays for *Verticillium dahliae*. *Phytopathology* 67, 1073–1078.

Butterfield, E.J., DeVay, J.E. and Garber, R.H. (1978) The influence of several crop sequences on the incidence of *Verticillium* wilt of cotton and on the population of *Verticillium dahliae* in field soil. *Phytopathology* 68, 1217–1220.

Buxton, E.W. and Hastie, A.C. (1962) Spontaneous and ultraviolet irradiation-induced mutants of *Verticillium albo-atrum*. *Journal of General Microbiology* 28, 625–632.

Byrne, T.C., McCain, A.H. and Kretchun, T.M. (1980) Testing chrysanthemums for disease resistance. *California Agriculture* 34(11–12), 14–15.

Caballero, J.M., Perez-Hernandes, J., Blanco-Lopez, M.A. and Jimenez-Diaz, R.M. (1981) Olive, a new host for *Verticillium dahliae* Kleb. in Spain. *Olea* 1981(December), 7–10.

Cai, Y.-F., Tan, Y.-J., He, H.-H. and Liu, B.-Y. (1994) [The characteristics and genetic effects of cotton germplasm resistant to various strains of *Verticillium* spp.]. *China Cottons* 21(12), 10–11.

Callens, D., de Rooster, L. and de Reycke, L. (2000) Rassenproef zomerbloemkool: Fremont blijft interessant, behalve bij vaatziekte. *Proeftuinnieuws* 10(2), 16–17.

Campbell, C.L. and Madden, L.V. (1990) *Introduction to Plant Disease Epidemiology*. John Wiley & Sons, New York.

Campbell, W.A. and Presley, J.T. (1946) Diseases of cultivated guayule and their control. *US Department of Agriculture Circular* 749, 1–42.

Campbell, W.A., Leach, L.D., Presley, J.T. and Snyder, W.C. (1943) Some diseases of guayule in California. *Plant Disease Reporter* 27, 63–66.

Campbell, W.P. and Griffiths, D.A. (1974) Morphological variation in isolates of *Verticillium albo-atrum*. *Canadian Journal of Microbiology* 20, 163–166.

Camporota, P. and Rouxel, F. (1977) Intérêt du broyage et du tamisage à sec dans les techniques d'isolement du *Verticillium dahliae* à partir du sol. *Annals of Phytopathology* 9, 77–81.

Cancino, E.L., Latorre, G.B. and Aronowosky, C. (1971) *Verticillium albo-atrum*, un grave problema en las nuevas plantaciones de durazneros de la Zona Central. *Agricultura Técnica* 31, 161–164.

Canevascini, V. and Caccia, R. (1976) Prove di lotta contro la verticilliosi della melanzana mediante innesto su pomodoro. *Informatore di Ortoflorofrutticoltura* 17(7), 15–20.

Cannon, O.S. and Waddoups, V. (1952) Loran Blood and V R Moscow. Two new *Verticillium* wilt resistant tomatoes for Utah. *Utah Farm and Home Science* 13(4), 74.

Canter-Visscher, T.W. (1970) *Verticillium* wilt of grapevine, a new record in New Zealand. *New Zealand Journal of Agricultural Research* 13, 359–361.

Cantrell, R.G., Roberts, C.L. and Barnes, C.E. (1995) Registration of 'Acala 1517-95' cotton. *Crop Science* 35(4), 1227–1228.

Cao, Y.Q., Wang, K.R., Ju, L.H. and Lu, J.Y. (1988) [Determination of virulence of the cotton wilt pathogen *Verticillium dahliae* Kleb.]. *Journal of Nanjing Agricultural University* 11, 40–42.

Cappaert, M.R., Powelson, M.L., Christensen, N.W., Stevenson, W.R. and Rouse, D.I. (1994) Assessment of irrigation as a method of managing potato early dying. *Phytopathology* 84(8), 792–800.

Cappelli, C., Winter, W. and Paul, V.H. (1998) Detection of seed-transmitted pathogens of rape (*Brassica napus* var. *oleifera* DC. *Bulletin OILB/SROP* 21(5), 1–13.

Carder, J.H. (1989) Distinctions between cellulase isoenzyme patterns of five plant-pathogenic *Verticillium* species. *Mycological Research* 92(3), 297–301.

Carder, J.H. and Barbara, D.J. (1990) Molecular variation in six species of *Verticillium* examined using restriction fragment length polymorphism (RFLP) analysis. *Proceedings of the Fifth International* Verticillium *Symposium*. Leningrad. p. 95.

Carder, J.H. and Barbara, D.J. (1991) Molecular variation and restriction fragment length polymorphisms (RFLPs) within and between six species of *Verticillium*. *Mycological Research* 95(8), 935–942.

Carder, J.H. and Barbara, D.J. (1999) Taxonomic status of putative *Verticillium albo-atrum* isolates. *FEMS Microbiology Letters* 170, 211–219.

Carder, J.H., Hignett, R.C. and Swinburne, T.R. (1987) Relationship between the virulence of hop isolates of *Verticillium albo-atrum* and their *in vitro* secretion of cell-wall degrading enzymes. *Physiological and Molecular Plant Pathology* 31, 441–452.

Carder, J.H., Morton, A., Tabrett, A.M. and Barbara, D.J. (1994) Detection and differentiation by PCR of subspecific groups within two *Verticillium* species causing vascular wilts in herbaceous hosts. In: Schots, A., Dewey, F.M. and Oliver, R.P. (eds) *Modern Assays for Plant Pathogenic Fungi: Identification, Detection and Quantification*. CAB International, Wallingford, UK, pp. 91–97.

Carns, H.R., Presley, J.T., Schnathorst, W.C., Harrison, G.K. and Taylor, E.E. (1964) *Proceedings of the Beltwide Cotton Production Research Conference*. National Cotton Council, Memphis, Tennessee, pp. 83–89.

Caroselli, N.E. (1955) Investigations of toxins produced *in vitro* by the maple with fungus *Verticillium* sp. *Phytopathology* 45, 183–184.

Caroselli, N.E. (1957) *Verticillium* wilt of maples. *Rhode Island Agricultural Experiment Station Bulletin* 355, 1–84.

Caroselli, N.E., Mahadevan, A. and Mozumder, B.G. (1964) The effect of light quality on the growth and microsclerotial production of *Verticillium albo-atrum*. *Plant Disease Reporter* 48, 484–486.

Carpenter, C.W. (1914) The *Verticillium* wilt problem. *Phytopathology* 4, 393.

Carpenter, C.W. (1918) Wilt diseases of okra and the *Verticillium* wilt problem. *Journal of Agricultural Research* 12, 529–546.

Carr, A.J.H. (1960) *Verticillium* wilt of lucerne. *Report of the Welsh Plant Breeding Station 1959*, pp. 64–66.

Carr, A.J.H. and Potter, L.R. (1975) *Verticillium* wilt of lucerne. *Report of the Welsh Plant Breeding Station 1974*, p. 88.

Carr, A.J.H., Wilkins, P.W., Beeraj, D.R., Simmonds, P.W., Catherall, P.L. and A'brook, J. (1972) Report for 1971. *Report of the Welsh Plant Breeding Station 1971*.

Carter, J.C. (1938) *Verticillium* wilt of woody plants in Illinois. *Plant Disease Reporter* 22, 253–254.

Carter, J.C. (1940) Some isolations of *Verticillium* in Illinois during 1938 and 1939. *Plant Disease Reporter* 24, 133–134.

Carter, J.C. (1945) Isolations of *Verticillium* from trees and shrubs in Illinois, 1940–1944. *Plant Disease Reporter* 29, 95–96.

Carter, J.C. (1975) Diseases of midwest trees. *University of Illinois Special Publication* 35, 168.

Cartia, G. (1989) La solarizzione del terreno: esperienze maturate in Sicilia. *Informatore Fitopatologico* 39(5), 49–52.

Cartia, G. and Lio, G.M. Di S. (1981) Infezioni di *Verticillium dahliae* su carciofo in Sicilia. *Tecnica Agricola, Italy* 33, 59–62.

Cartia, G., Greco, N. and Cirvilleri, G. (1991) Soil solarization in a plastic house. *FAO Plant Production and Protection Paper* 109, 266–275.

Cartia, G., Causerano, R. and Di Primo, P. (1996) La fumagazione del terreno in ambiente protetto: impiego di dosi ridotte di bromuro di metile e di coperture a diversa permeabilita. Terzo contributo. *Difesa delle Piante* 19, 49–54.

Casarini, B. (1970) Attivita di miglioramento della coltivazione del pomodoro con particcolare riguardo ai problemi dell'Italia settentrionale. *Annali Acc. Nazionali Agricoltura* 90, 277–301.

Casarini, B. and Antonellini, E. (1971) Ulteriori osservazioni su nuove stirpi di pomodoro della serie 'Bt.' (Battipaglia). *Annali Acc. Nazionali Agricoltura*, p. 91.

Cash, S.D., Knipe, W.J. and McCaslin, M.H. (1998a) Registration of 'Tahoe' alfalfa. *Crop Science* 38(2), 536.

Cash, S.D., Knipe, W.J. and McCaslin, M.H. (1998b) Registration of 'Parade' alfalfa. *Crop Science* 38(2), 536.

Cassini, R., Cassini, R., Alabouvette, C., Rouxel, F., Bonnel, L., Bremeersch, P. and Casenave, J.G. (1971) Desinfection du sol au bromure de methyle. Action sur les desperissements du persil, du celeri, du concombre et du melon. *Phytiatrie Phytopharmacie* 20, 215–231.

Castejon-Munoz, M. and Bollen, G.J. (1993) Induction of heat resistance in *Fusarium oxysporum* and *Verticillium dahliae* caused by exposure to sublethal heat treatments. *Netherlands Journal of Plant Pathology* 99(2), 77–84.

Castoria, R., Fanelli, C., Zoina, A. and Scala, F. (1995) Analysis of fatty acids in lipids of *Verticillium dahliae* and induction of lubimin accumulation in eggplant. *Plant Pathology* 44(5), 791–795.

Castrejón Sanguino, A. (1978) Especies de *Verticillium* casuantes del marchitamiento del algodonero en la Comarca Lagunera. *Agricultura Técnica en México* 4, 127–135.

Castrejon Sanguino, A. (1994) Deteccion, *in vitro*, de hongos antagonicos a *Verticillium dahliae* Kleb. raza T-9. *ITEA Produccion Vegetal* 90(2), 129–131.

Catani, C.S. and Peterson, J.L. (1966) Antagonistic relationships between *Verticillium dahliae* and fungi isolated from the rhizosphere of *Acer platanoides*. *Phytopathology* 57, 363–366.

Catesson, A.M., Czaninski, Y., Péresse, M. and Moreau, M. (1976) Sécretions intravasculaires de substances 'gommeuses' par les cellules associées aux vaisseaux en réaction à une attaque parasitaire. *Bulletin de la Société Botanique de France* 123, 93–107.

Cauquil, J. (1973) Les microorganismes pathogenes du cotonnier dans le sud-oest de Madagascar. *Coton et Fibres Tropicales* 28, 337.

Cavanni, P., Fantuz, F. and Ponti, I. (1987) Le malattie crittogamiche del legno della vite. *Informatore Fitopatologico* 37(1), 27–34.

Cebolla, V., Martinez, P.F., del Busto, A., Gomez De Barreda, D. and Tuset, J.J. (1995) Dosage reduction of methyl bromide fumigation in the Spanish Mediterranean coast. In: Vanachter, A. (ed) *Acta Horticulturae. Fourth International Symposium on Soil and Substrate Infestation and Disinfestation*. Leuven, Belgium, 6–12 September 1993. pp. 156–163.

Cech, T.L. (1998) Absterben von Gotterbaumen (*Ailanthus altissima*) in der Sudsteiermark. *Forstschutz Aktuell* 22, 16–18.

Celetti, M.J. and Platt, H.W. (1987a) Occurrence, inoculation technique, varietal response, and chemical control of *Verticillium* wilt. *Canadian Journal of Plant Pathology* 9, 78–79.

Celetti, M.J. and Platt, H.W. (1987b) A new cause for an old disease: *Verticillium dahliae* found on Prince Edward Island. *American Potato Journal* 64, 209–212.

Cenis, J.L., Martinez, P.F., Gonzalez-Benavente, A. and Aragon, R. (1988) Ensayo de control de *Verticillium dahliae* y *Rhizoctonia solani* mediante desinfeccion solar, en el campo de Cartagena. In: Llobet, L.G. (ed.) *Proceedings of the Third National Congress of Phytopathology*. Tenerife, Canaries, Spain, 1984. Centro de Investigacion y Tecnologia Agrarias, La Laguna, Tenerife, p. 107.

Cerenak, A., Dolinar, M. and Rak, M. (1999) Izolacija in identifikacija gliv *Verticillium albo-atrum* Reinke & Berthold in *Verticillium dahliae* Klebahn iz hmelja (*Humulus lupulus* L.). *Zbornik Predavanj in Referatov 4 Slovenskega Posvetovanja o Varstvu Rastlin v Portorozu*. 3 – 4 Marca 1999. pp. 489–492.

Chambers, A.Y. and Bush, T.D. (1998) Utilizing a disease nursery to evaluate cotton cultivar reaction to *Verticillium* wilt. In: Dugger, P. and Richter, D.J. (eds) *Proceedings of the Beltwide Cotton Conferences*. San Diego, California, 5–9 January 1998. Vol. 1, pp. 164–165.

Chambers, A.Y. and McCutchen, T.G. (1978) Reaction of eight upland cotton varieties to *Verticillium* wilt. In: Brown, J.M. (ed.) *Proceedings of the Beltwide Cotton Production Research Conference*. Dallas, Texas. National Cotton Council, Memphis, Tennessee, pp. 26–28.

Chambers, D.A. and Darby, P. (1995) Field assessment of wilt resistance in hop. 6th International *Verticillium* Symposium, Dead Sea, Israel. *Phytoparasitica* 23, 63.

Chambers, D.A. and Harris, D.C. (1997) Methods of screening *Acer platanoides* L. seedlings for resistance to wilt (*Verticillium dahliae* Kleb.). *Journal of Horticultural Science* 72(4), 601–608.

Chambers, D.A., Davies, M.K. and Talboys, P.W. (1983) *Verticillium* wilt (*Verticillium albo-atrum*) of hops. *Report of the East Malling Research Station* 1982, p. 86.

Chambers, D.A., Hignett, R.C., Carder, J.H. and Marks, M.J. (1986) Diseases of hops. *Report of the East Malling Research Station* 1985, pp. 137–140.

Chambers, D.A., Darby, P. and Harris, D.C. (1997) Developing a method for assessing wilt resistance in hop seedlings. *Proceedings of the Seventh International* Verticillium *Symposium*. Cape Sounion, Athens, Greece. p. 57.

Chan, B.G. and Wilhelm, S. (1984) A chemical investigation of a die-back cotton disease. *Proceedings of the Beltwide Cotton Production Research Conference*. National Cotton Council, Memphis, Tennessee. pp. 22–31.

Chang, R.J. and Eastburn, D.M. (1994) Host range of *Verticillium dahliae* from horseradish and pathogenicity of strains. *Plant Disease* 78(5), 503–506.

Charles, J.P., Chalet, C., Lehmann, J. and Briner, H.U. (1989) Essais 1986–1988 de varietés de lucerne (*Medicago sativa* L.) et liste des varietés recommandées. *Revue Suisse d'Agriculture* 21, 245–247.

Charudattan, R. and DeVay, J.E. (1972) Common antigens among varieties of *Gossypium hirsutum* and isolates of *Fusarium* and *Verticillium* species. *Phytopathology* 62, 230–234.

Chattaway, M.M. (1949) The development of tyloses and secretion of gum in heartwood formation. *Australian Journal of Scientific Research Series B, Biological Science* 2, 227–240.

Chaudhuri, H. (1923) A study of the growth in culture of *Verticillium albo-atrum* R. & B. *Annals of Botany* 37, 519–539.

Chen, J., Bird, G.W. and Mather, R.L. (1995) Impact of multi-year cropping regimes on *Solanum tuberosum* tuber yields in the presence of *Pratylenchus penetrans* and *Verticillium dahliae*. *Journal of Nematology* 27 (4 Suppl.), 654–660.

Chen, Weidong (1994) Vegetative compatibility groups of *Verticillium dahliae* from ornamental woody plants. *Phytopathology* 84(2), 214–219.

Chen, X., Wang, M., Chen, Y., Davisson, V.J. and Heinstein, P. (1996) Cloning and heterologous expression of a second (+)-δ-cadinene synthase from *Gossypium arboreum*. *Journal of Natural Products* 59, 944–951.

Chen, X., Chen, Y. and Huang, J. (1998) [Biochemical property of the toxic protein secreted by the VD8 strain of *Verticillium dahliae* from cotton]. *Jiangsu Journal of Agricultural Sciences* 14(2), 126–128.

Chen, Z.Y., Chen, Y., Heinstein, P. and Davisson, V.J. (1995) Cloning, expression, and characterization of (+)-δ-cadinene synthase: a catalyst for cotton phytoalexin biosynthesis. *Archives of Biochemistry and Biophysics* 324(2), 255–266.

Cheng, Z.S. and Zhang, J.G. (1985) [Analysis of fibre quality and productivity of the cotton cultivar Jinmian 7]. *Shanxi Nongye Kexue Shanxi Agricultural Science* 8, 10–11.

Cheo, P.C. (1982) Effects of tannic acid on rhizomorph production by *Armillaria mellea*. *Phytopathology* 72, 676–679.

Chepenko, L.I., Runov, V.I. and Borodin, G.I. (1969) [On the nature of electrophoretically mobile toxic substances of the *Verticillium* wilt pathogen of cotton]. *Uzbekskii Biologicheskii Zhurnal* 13, 13–15.

Chernyaeva, I.I., Ortikov, T.N., Globus, G.A., Khashinov, F.K. and Tuev, N.A. (1989) Use of calcium carbide in cotton growing. *Soviet Agricultural Science* 12, 21–25.

Chernyayeva, I.I., Hasanova, F.S. and Muromtsev, G.S. (1984) Effect of various nitrogen forms upon *Verticillium dahliae* and the metabolism of some soil bacteria. In: Szeji, J. (ed.) *Soil Biology and Conservation of the Biosphere*. Academiai Kiado, Budapest, pp. 703–710.

Cherrab, M., Serrhini, M.N. and Charest, P.M. (2000) Characterization of Moroccan isolates of *Verticillium dahliae* Kleb. using RAPD markers. *Journal of Phytopathology* 148(4), 243–249.

Chet, I., Havkin, D. and Katan, J. (1978) The role of catechol in inhibition of *Fusarium* wilt. *Phytopathologische Zeitschrift* 91, 60–66.

Chi, C.C. (1963) Fungistatic and fungicidal effects of *Streptomyces rimosus* on some soil-inhabiting pathogenic fungi *in vitro*. *Phytopathology* 53, 872.

Chikrizova, V.L. (1987) [*Verticillium* wilt of *Hippophae rhamnoides* and *Rosa* in Moldavia]. *Lesnoe Khozyaïstvo* 1, 57–58.

Chilton, J.E. (1956) *Verticillium* wilt report. *Proceedings of the 17th Annual Meeting of the Cotton Disease Council*. pp. 9–10.

Chilton, J.E. (1957) Sasame (*Sesamum indicum* L.) a host for *Verticillium albo-atrum* Reinke & Berth. *Plant Disease Reporter* 41, 803.

Chitzanidis, A. (1995) Pistachio diseases in Greece. In: Kaska, N., Kuden, A.B., Ferguson, L. and Michailides, T. (eds) *Acta Horticulturae. First International Symposium on Pistachio Nut*. Adana, Turkey, 20–24 September 1994. pp. 345–348.

Chloupek, O. (1996) Vojteska seta Niva. *Genetika a Slechteni* 32(2), 159–160.

Chloupek, O., Babinec, J. and Holubar, J. (1996) Development and evaluation of new synthetic varieties of lucerne Jitka and Niva. *Acta Universitatis Agriculturae et Silviculturae Mendelianae Brunensis* 44(1/4), 19–23.

Chlumská, J. and Krátká, J. (1984) [Changes in pH in lucerne plants after inoculation with *Verticillium albo-atrum* and *Corynebacterium michiganense* pv. *insidiosum*]. *Sbornik-UVTIZ, Ochrana Rostlin 1984* 20, 189–194.

Choy, Y.M. and Unrau, A.M. (1971) Metabolites of *Verticillium dahliae*: the isolation and structure elucidation of some disaccharides and a trisaccharide. *Canadian Journal of Biochemistry* 49, 894–899.

Christen, A.A. (1982a) A selective medium for isolating *Verticillium albo-atrum* from soil. *Phytopathology* 72, 47–49.

Christen, A.A. (1982b) Demonstration of *Verticillium albo-atrum* within alfalfa seed. *Phytopathology* 72, 412–414.

Christen, A.A. (1983) Incidence of external seedborne *Verticillium albo-atrum* in commercial seed lots of alfalfa. *Plant Disease* 67, 17–18.

Christen, A.A. and French, R.C. (1982) Growth and pathogenicity of alfalfa strains of *Verticillium albo-atrum*. *Plant Disease* 66, 416–418.

Christen, A.A. and Peaden, R.N. (1981) *Verticillium* wilt in alfalfa. *Plant Disease* 65, 319–321.

Christen, A.A. and Peaden, R.N. (1982) Relative importance of *Verticillium* wilt infestation in alfalfa. *Phytopathology* 72, 960.

Christen, A.A., Peaden, R.N., Harris, G.P. and Heale, J.B. (1983) Virulence of North American and European isolates of *Verticillium albo-atrum* on alfalfa cultivars. *Phytopathology* 73, 1051–1054.

Christensen, P.D., Stith, L.S. and Lyerly, P.J. (1954) The occurrence of *Verticillium* wilt in cotton as influenced by the level of salt in the soil. *Plant Disease Reporter* 38, 309–310.

Christie, B.R., Papadopoulos, Y.A. and Busch, L.V. (1985) Genetics and breeding for resistance to *Verticillium* wilt in alfalfa. *Canadian Journal of Plant Pathology* 7, 206–210.

Christie, T. (1956) A note on an outbreak of *Verticillium* wilt in a Nelson hop garden. *New Zealand Journal of Science and Technology* 38, 15–19.

Church, V.J. and McCartney, H.A. (1995) Occurrence of *Verticillium dahliae* on sunflower (*Helianthus annus*) in the UK. *Annals of Applied Biology* 127(1), 49–56.

Cia, E. (1977) Ocorrência e conhecimento das doenças de algodoeiro anual *Gossypium hirsutum* L. no Brazil. *Summa Phytopathologica* 3, 167–193.

Cia, E., Ferraz, C.A.M., Machado da Silva, N. and Fuzatto, M.G. (1970) Comportamento de variedades paulistas de algodeiro em solo infestado pelo fungo causador de murcho verticilar. *Bragantia* 29, 73–79.

Cia, E., Fuzatto, M.G., Kondo, J.I., Gridi Papp, I.L. and Chiavegato, E.J. (1999) Evolução da resistência multipla a doènças em linhagens avançadas de algodoeiro. *Anais II Congresso Brasileiro de Algodão: o Algodão no Seculo XX, Perspectivas para o Seculo XXI.* Ribeirão Preto, SP, Brasil, 5–10 September 1999. pp. 431–434.

Ciccarese, F., Cirulli, M. and Frisullo, S. (1985) Prove di lotta chimica contro la verticilliosi del carciofo. *Informatore Fitopatologico* 35(5), 39–42.

Ciccarese, F., Frisullo, S. and Cirulli, M. (1987) Severe outbreaks of *Verticillium* wilt on chicory (*Cichorium intybus* L.) and broccoli rab (*Brassica rapa* L.) and pathogenic variations among isolates of the fungus. *Plant Disease* 71, 1144–1145.

Ciccarese, F., Frisullo, S. and Cirulli, M. (1990) Natural disease recovery from *Verticillium* wilt in peach. *Proceedings of the Fifth International* Verticillium *Symposium*. Leningrad, p. 6.

Ciccarese, F., Amenduni, M., Schiavone, D. and Cirulli, M. (1994) Influenza della verticilliosi sulla produzione di melanzane suscettibili e resistenti ('slow wilting') in campo. *Phytopathologia Mediterranea* 33(3), 212–216.

Ciccarese, F., Cirulli, M. and Amenduni, M. (1995) Progress in the search for *Verticillium* wilt resistant artichoke. 6th International *Verticillium* Symposium, Dead Sea, Israel. *Phytoparasitica* 23, 62.

Ciccarone, A. (1974) A survey of olive diseases in Mediterranean countries. *Options-Mediterraneenes* 24, 71–80.

Ciddi, V., Srinivasan, V. and Shuler, M.L. (1995) Elicitation of *Taxus* sp. cell cultures for production of Taxol. *Biotechnology Letters* 17(12), 1343–1346.

Cirulli, M. (1969) Un isolato di *Verticillium dahliae* Kleb. virulento verso varieta resistenti di pomodoro. *Phytopathologia Mediterranea* 8, 132–136.

Cirulli, M. (1974) Infezioni locali da *Verticillium dahliae* Kleb. su foglie di pomodoro. *Phytopathologia Mediterranea* 13, 23–26.

Cirulli, M. (1975) Il deperimento dell'olivo da *Verticillium dahliae* Kleb. *Italia Agricola* 112(6), 120–124.

Cirulli, M. (1981) Attuali cognizioni sulla verticilliosi dell'olivo. *Informatore Fitopatologico* 31(1/2), 101–105.

Cirulli, M. (1984) Pathobiology of *Verticillium* species. Proceedings of the Third International Symposium on *Verticillium*, Bari, Italy. *Phytopathologia Mediterranea* 23, 95–221.

Cirulli, M. and Ciccarese, F. (1984) Progress in the search for *Verticillium* resistant eggplant. Pathobiology of *Verticillium* species. Third International *Verticillium* Symposium, Bari, Italy. *Phytopathologia Mediterranea* 23, 109.

Cirulli, M. and Montemurro, G. (1976) A comparison of pathogenic isolates of *Verticillium dahliae* and sources of resistance in olive. *Poljoprivredna Znanstvena Smotra* 39, 469–476.

Cirulli, M., Frisullo, S. and Ciccarese, F. (1985) Estese epidemie di verticilliosi sul pomodoro in Puglia. *Informatore Fitopatologico* 35(7/8), 29–32.

Cirulli, M., Ciccarese, F. and Frisullo, S. (1987) *Verticillium* wilt of artichoke: aetiological studies and chemical control. *Canadian Journal of Plant Pathology* 9, 79.

Cirulli, M., Ciccarese, F. and Amenduni, M. (1990) Progress in the search for *Verticillium* wilt-resistant eggplant. *Phytopathologia Mediterranea* 29, 184–190.

Cirulli, M., Ciccarese, F. and Amenduni, M. (1994) Evaluation of Italian clones of artichoke for resistance to *Verticillium dahliae*. *Plant Disease* 78(7), 680–682.

Cirulli, M., Amenduni, M. and Ciccarese, F. (1997) Resistance to *Verticillium* wilt in eggplant. *Proceedings of the Seventh International* Verticillium *Symposium*. Cape Sounion, Athens, Greece. p. 42.

Clancy, K.J. (1986) New or unusual diseases recorded in 1984/5. *Research Report 1984–1985, Faculty of General Agriculture, University College, Dublin*. pp.165–166.

Clarkson, J.M. and Heale, J.B. (1985a) Pathogenicity and colonization studies on wild-type and auxotrophic isolates of *Verticillium albo-atrum* from hop. *Plant Pathology* 34, 119–128.

Clarkson, J.M. and Heale, J.B. (1985b) Heterokaryon compatibility and genetic recombination within a host plant between hop isolates of *Verticillium albo-atrum*. *Plant Pathology* 34, 129–138.

Clarkson, J.M. and Heale, J.B. (1985c) A preliminary investigation of the genetics of pathogenicity in hop wilt isolates of *Verticillium albo-atrum*. *Transactions of the British Mycological Society* 85, 345–350.

Cline, S.D. and Coscia, C.J. (1988) Stimulation of sanguinarine production by combined fungal elicitation and hormonal deprivation in cell suspension cultures of *Papaver bracteatum*. *Plant Physiology* 86, 161–165.

Cline, S.D., McHale, R.J. and Coscia, C.J. (1993) Differential enhancement of benzophenanthridine alkaloid content in cell suspension cultures of *Sanguinaria canadensis* under conditions of combined hormonal deprivation and fungal elicitation. *Journal of Natural Products* 56(8), 1219–1228.

Cole, H., Mills, W.R. and Massie, L.B. (1972) Influence of chemical seed and soil treatments on *Verticillium*-induced yield reduction and tuber defects. *American Potato Journal* 49, 79–92.

Collingwood, E.F. (1965) Report of the mycologist, 1964. *Report of the States Experimental Station, Jersey 1964*. pp. 47–50.

Colson, E. (1998) Systemic induced resistance helps natural cotton plant defences. *Australian Cottongrower* 19(4), 30–32, 34.

Concibido, V.C., Secor, G.A. and Jansky, S.H. (1994) Evaluation of resistance to *Verticillium* wilt in diploid, wild potato interspecific hybrids. *Euphytica* 76(1/2), 145–152.

Congly, H. and Hall, R. (1976) Effects of osmotic potential on germination of microsclerotia and growth of colonies of *Verticillium dahliae*. *Canadian Journal of Botany* 54, 1214–1220.

Conn, K.L. and Lazarovits, G. (1999) Impact of animal manures on *Verticillium* wilt, potato scab, and soil microbial populations. *Canadian Journal of Plant Pathology* 21(1), 81–92.

Connel, S. and Heale, J.B. (1985) Development of an *in vitro* selection system for novel sources of resistance to *Verticillium* wilt in hops. In: Withers, L. and Alderson, P.G. (eds) *Tissue Culture and Agriculture. Proceedings of Nottingham Conference*, September 1984.

Conroy, J.J. and Green, R.J. (1974) Interaction of the root-knot nematode *Meloidogyne incognita* and the stubby root nematode *Trichodorus christiei* with *Verticillium albo-atrum* on tomato at controlled inoculum densities. *Phytopathology* 64, 1118–1124.

Conroy, J.J., Green, R.J. and Ferris, J.M. (1972) Interaction of *Verticillium albo-atrum* and the root lesion nematode, *Pratylenchus penetrans*, in tomato roots at controlled inoculum densities. *Phytopathology* 62, 362–366.

Conway, K.E. and Martin, M.J. (1983) The potential of soil solarization to control *Verticillium dahliae* in Oklahoma. *Proceedings of the Oklahoma Academy of Science* 63, 25–27.

Cooke, C.J., Smith, C.J., Walton, T.J. and Newton, R.P. (1994) Evidence that cyclic AMP is involved in the hypersensitive response of *Medicago sativa* to a fungal elicitor. *Phytochemistry* 35(4), 889–895.

Cooper, H.B. (1963) Breeding for blight and *Verticillium* resistant high quality cotton. *Proceedings, 15th Cotton Improvement Conference*. pp. 115–123.

Cooper, H.B., Dobbs, J., Lehman, M., Wilton, A.C., Turner, J.H. and Schaller, C.W. (1967) Breeding and evaluation of cotton for *Verticillium* wilt tolerance in Tulare county. *Cotton Research in the San Joachin Valley, California: Research Report for 1966–67*. pp. 106–153.

Cooper, R.M. (1975) Cell wall degrading enzymes of vascular wilt fungi. PhD thesis. University of London.

Cooper, R.M. and Durrands, P.K. (1989) Selection characterisation, pathogenicity and virulence of pectinase-deficient mutants of *Verticillium albo-atrum*. In: Tjamos, E.C.

and Beckman, C.H. (eds) *Vascular Wilt Diseases of Plants.* Springer-Verlag, Berlin, pp. 325–335.

Cooper, R.M. and Wood, R.K.S. (1973) Induction of synthesis of extracellular cell-wall degrading enzymes in vascular wilt fungi. *Nature (London)* 246, 309–311.

Cooper, R.M. and Wood, R.K.S. (1974) Scanning electron microscopy of *Verticillium albo-atrum* in xylem vessels of tomato plants. *Physiological Plant Pathology* 4, 443–446.

Cooper, R.M. and Wood, R.K.S. (1975) Regulation of synthesis of cell-wall degrading enzymes by *Verticillium albo-atrum* and *Fusarium oxysporum* f. sp *lycopersici. Physiological Plant Pathology* 5, 135–156.

Cooper, R.M. and Wood, R.K.S. (1980) Cell wall degrading enzymes of vascular wilt fungi. III. Possible involvement of endo-pectin lyase in *Verticillium* wilt of tomato. *Physiological Plant Pathology* 16, 285–300.

Cooper, R.M., Rankin, B. and Wood, R.K.S. (1978) Cell-wall degrading enzymes of vascular wilt fungi. II. Properties and modes of action of polysaccharidases of *Verticillium albo-atrum* and *Fusarium oxysporum* f. sp *lycopersici. Physiological Plant Pathology* 13, 101–134.

Cooper, R.M., Flood, J., Rowan, M.G., Resende, M.L.V. and Beale, M.H. (1995) The role of tannins and phytoalexins in the resistance of cocoa (*Theobroma cacao* L.) to *Verticillium dahliae. Aspects of Applied Biology. Physiological Responses of Plants to Pathogens.* 11–13 September 1995, University of Dundee, UK. pp. 315–322.

Cooper, R.M., Resende, M.L.V., Flood, J., Rowan, M.G., Beale, M.H. and Potter, U. (1996) Detection and cellular localization of elemental sulphur in disease-resistant genotypes of *Theobroma cacao. Nature (London)* 379(6561), 159–162.

Coosemans, J. (1977) Interaction and population dynamics of *Pratylenchus penetrans* (Cobb) and *Verticillium* spp. on flax. *Parasitica* 33, 53–58.

Coosemans, J. (1979) Partial control by systemic nematicides and/or fungicides of complex diseases caused by *Pratylenchus penetrans* (Cobb) Filipjev & Stekhoren, *Verticillium albo-atrum* Reinke & Berthold, and *Verticillium dahliae* Klebahn. *Parasitica* 35, 63–72.

Corazza, L. and Chilosi, G. (1986) Il comportamento di alcuni portinnesti di drupacee nei confronti di *Verticillium dahliae* e *Fusarium oxysporum. Annali dell'Istituto Sperimentale per la Patologia Vegetale* 11, 113–116.

Corbett, D.C.M. and Hide, G.A. (1971) Interactions between *Heterodera rostochiensis* Woll. and *Verticillium dahliae* Kleb. on potatoes and the effect of CCC on both. *Annals of Applied Biology* 68, 71–80.

Correll, J.C., Gordon, T.R. and McCain, A.H. (1988) Vegetative compatibility and pathogenicity of *Verticillium albo-atrum. Phytopathology* 78, 1017–1021.

Corsini, D.L., Davis, J.R. and Pavek, J.J. (1985) Stability of resistance of potato to strains of *Verticillium dahliae* from different vegetative compatibility groups. *Plant Disease* 69, 980–982.

Corsini, D.L. and Pavek, J.J. (1996) Agronomic performance of potato germplasm selected for high resistance to *Verticillium* wilt. *American Potato Journal* 73(6), 249–260.

Corsini, D.L., Pavek, J. and Davis, J.R. (1986) *Verticillium* wilt resistance in non-cultivated tuber-bearing solanums. *Proceedings of the Fourth International* Verticillium *Symposium.* Guelph, Canada.

Corsini, D.L., Pavek, J. and Davis, J.R. (1987) *Verticillium* wilt resistance in non-cultivated tuber bearing solanums. *Canadian Journal of Plant Pathology* 9, 79.

Corsini, D.L., Thompson, C. and Pavek, J.J. (1989) The effect of plant growth regulators on *Verticillium* wilt of potato. *American Potato Journal* 66, 125–136.

Corsini, D.L., Pavek, J., Brown, C., Inglis, D., Martin, M., Powelson, M., Dorrance, A. and Lozoya Saldana, H. (1999) Late blight resistant potato germplasm release AWN86514-2. *American Journal of Potato Research* 76(1), 45–49.

Costache, M. (1974) [Vascular diseases caused by fungi in vegetable crops in the glasshouse]. *Anale Institutul de Cercetari pentru Legumicultura si Floricultura, Vidra* 3, 603–619.

Costache, M. (1977) Results obtained in the control of vascular diseases in vegetables by means of systematic fungicides. *Acta Horticulturae* 58, 461–468.

Costache, M. and Raicu, C. (1976) [Eggplant wilt caused by *Verticillium* spp. in the glasshouse]. *Analele Institutului de Cercetări Legumicult şi Floricult* 3, 343–348.

Costache, M. and Stoenescu, A.T. (1978) [Prevention and control of diseases of sweet pepper in the glasshouse]. *Producţia Vegetală, Horticultura* 27, 3–5.

Costache, M. and Tomescu, A. (1980) [Control of cucumber wilts in the glasshouse. III Preventive and curative effects of treatments with systemic fungicides]. *Probleme de Proteçtia Plantelor* 8, 75–84.

Costache, M. and Tomescu, A. (1987) Research on the cucumber wilt produced by *Fusarium oxysporum* (Schlecht) Snyder et Hansen and *Verticillium dahliae* Kleb. II. Specialization of the pathogens. *Bulletin de l'Academie des Sciènces Agricoles et Forestières* 15, 199–204.

Costache, M., Lăcătus, V., Tomescu, A. and Mândricel, M. (1978a) [Control of cucumber wilt in the glasshouse, l. Factors predisposing cucumber plants to infection by *Fusarium oxysporum* f. sp. *cucumerinum* Owen and *Verticillium dahliae* Kleb.]. *Producţia Vegetală, Horticultura* 27(2), 34–38.

Costache, M., Mihăilescu, S., Mândricel, M. and Tomescu, A. (1978b) [Control of cucumber wilt in the glasshouse II. Factors favouring infection by *Fusarium oxysporum* f. sp. *cucumerinum* Owen and *Verticillium dahliae* Kleb.]. *Producţia Vegetală, Horticultura* 27(11), 3–6.

Costache, M., Lăcătus, V., Tomescu, A. and Mândricel, M. (1979) [Studies on the factors predisposing cucumber plants to infection by *Fusarium oxysporum* f. sp. *cucumerinum* Owen and *Verticillium dahliae* Kleb.]. *Probleme de Proteçtia Plantelor* 7, 51–60.

Costache, M., Lăcătus, V., Lăcătus, F. and Donoiu, E. (1980) Recherches concernant l'influence de la nutrition à azote sur le jaunissement des tomates produit par *Verticillium dahliae* Kleb. *Bulletin de l'Académie des Sciences Agricoles et Forestières* 10, 129–131.

Costache, M., Tomescu, A. and Dragomir, N. (1981) [Studies on the effect of nitrogen sources on the growth and sporulation of *Fusarium oxysporum* f. sp. *cucumerinum* Owen and *Verticillium dahliae* Kleb.]. *Probleme de Proteçtia Plantelor* 9(9), 1–8.

Cotton, J.R. (1965) *Breeding Cotton for Tolerance to* Verticillium *Wilt*. Publication ARS 34–80. United States Department of Agriculture, Washington, DC.

Coxon, D.T. (1982) Phytoalexins from other plant families. In: Bailey, J.A. and Mansfield, J.W. (eds) *Phytoalexins*. Blackie & Sons, Glasgow.

Craig, D.L., Aalders, L.E. and Bishop, G.W. (1978) Micmac strawberry. *Canadian Journal of Plant Science* 58, 903–904.

Crill, P., Stroebel, J.W., Burgis, D.S., Bryan, H.H., John, C.A., Everett, P.H., Bartz, J.A., Hayslip, N.C. and Deen, W.W. (1971) Florida MH-1, Florida's first machine harvest fresh market tomato. *Circular, Agricultural Experiment Station, University of Florida* S212, 1–12.

Crill, P., Bryan, H.H., Everett, P.H., Bartz, J.A., Jones, J.P. and Matthews, R.F. (1977) Floramerica. *Circular, Agricultural Experiment Station, University of Florida* S248, 1–10.

Cristinzio, G. and Scala, F. (1994) Electrolyte leakage as a potential method for measuring of eggplant resistance and *Verticillium dahliae* virulence. *Capsicum and Eggplant Newsletter*, 110–113.

Cristinzio, G., Rotino, G.L. and Scala, F. (1995) Investigation of *in-vitro* methods for measuring eggplant resistance to *Verticillium dahliae*. *Plant Pathology* 44(4), 704–709.

Cronshaw, D.K. and Pegg, G.F. (1976) Ethylene as a toxin synergist in *Verticillium* wilt of tomato. *Physiological Plant Pathology* 9, 33–44.

Crooks, M. (1975) Degenerative diseases of lucerne. *New Zealand Journal of Agriculture* 131, 19–20.

Crosse, J.E., Butt, D.J., Talboys, P.W., Sewell, G.W.F., Wilson, J.F. and Blake, C.M. (1977) Fungal diseases. *Report of the East Malling Research Station 1976*, pp. 131–145.

Crowe, F.J., Debons, J., Farris, N. and Yang, C. (1997) Peppermint performance and changes in inoculum density of *Verticillium dahliae* associated with management practices. *Proceedings of the Seventh International* Verticillium *Symposium*. Cape Sounion, Athens, Greece. p. 75.

Cuartero, J. (1994) Breeding new tomato cultivars for protected cultivation in mild winter climates. In: Cockshull, K.E., Tuzel, Y. and Gul, A. (eds) *Acta Horticulturae. Second Symposium on Protected Cultivation of Solanacea in Mild Winter Climates*. Adana, Turkey, 13–16 April 1993. pp. 209–220.

Cui, Y.-X., Bell, A.A., Joost, O. and Magill, C. (2000) Expression of potential defense response genes in cotton. *Physiological and Molecular Plant Pathology* 56(1), 25–31.

Cun, M. and Guiliang, J. (1995) Evaluation on resistance of a new cotton strain 86-6 to *Verticillium dahliae*. *Plant Protection* 21(6), 31.

Cunningham, G.H. (1949) *Twenty-third Annual Report of the Department of Scientific and Industrial Research, New Zealand, 1949*.

Curl, E.A. and Gudauskas, R.T. (1985) Effects of soil insects on populations and germination of fungal populations. In: Parker, C.A., Rovira, A.D., Moore, K.G., Wong, P.T. and Kollmorgen, J.F. (eds) *Ecology and Management of Soil-borne Plant Pathogens*. American Phytopathological Society, St Paul, Minnesota, pp. 20–23.

Curzi, M. (1925) Intorno alla causa dell'avizzimento del peperone (*Capsicum annuum*). *Congresso delle Scienze di Pavia (1925) Rivista di Patologia Vegetale* 15, 88.

Curzi, M. (1930) Intorno alle tracheomicosi e a nuovi gravi casi di verticillosi. *Bollettino della Roma Stazione di Patologia Vegetale NS* 1, 44–62.

Czaplinska, S. (1973) [Studies on the diseases of lucerne caused by the fungi *Verticillium albo-atrum* and *Ascochyta imperfecta*]. *Acta Mycologica* 9, 23–52.

Czaplinska, S. and Cebrat, J. (1975) [The influence of *Verticillium albo-atrum* Rke et Berth. on the course of mitosis in meristem cells of the root tips of lucerne (*Medicago media* Pers.)]. *Hadowla Roslin, Aklimatyzacja i Nasiennictwo* 19, 343–348.

Czarnecki, H. (1923) Studies on the so-called black heart disease of the apricot. *Phytopathology* 13, 216–224.

Daayf, F., Nicole, M. and Geiger, J.P. (1995) Differentiation of *Verticillium dahliae* populations on the basis of vegetative compatibility and pathogenicity on cotton. *European Journal of Plant Pathology* 101(1), 69–79.

Daayf, F., Nicole, M., Boher, B., Pando, A. and Geiger, J.P. (1997) Early vascular defense reactions of cotton roots infected with a defoliating mutant strain of *Verticillium dahliae*. *European Journal of Plant Pathology* 103(2), 125–136.

Daayf, F., Nicole, M., Belanger, R.R. and Geiger, J.P. (1998) Hyaline mutants from *Verticillium dahliae*, an example of selection and characterization of strains for host–parasite interaction studies. *Plant Pathology* 47, 523–529.

Daebeler, F., Amelung, D. and Engel, K.H. (1985) Zur Verwechlungsmoglichkeit der durch *Phoma lingam* (Tode ex Fr.) Desm. verursachten Wurzelhals-faule mit *Rhizoctonia solani* K. und *Verticillium dahliae* Kleb. *Nachrichtenblatt für den Pflanzenschutzdienst in der DDR* 39, 180–181.

Daebeler, F., Seidel, D. and Makowski, N. (1987) Phytosanitare Gesichtspunkte bei der Gestaltung von Rapsfruchtfolgen. *Nachrichtenblatt für den Pflanzenschutzdienst in der DDR* 41(2), 30–32.

Daebeler, F., Amelung, D. and Zeise, K. (1988) *Verticillium* Welke an Winterraps – Auftreten und Bedeutung. *Nachrichtenblatt für den Pflanzenschutzdienst in der DDR* 42, 71–73.

Dale, A., Gray, V.P. and Ricketson, C.L. (1986) Governor Simcoe and Secord strawberries. *Canadian Journal of Plant Science* 66, 1031–1032.

Danti, S. and Broggio, M. (1997) Impact of salinity stress on the biometric responses of an isolate of *Verticillium dahliae* Kleb. from potato. *Proceedings of the Seventh International* Verticillium *Symposium*. Cape Sounion, Athens, Greece. p. 53.

D'armini, M., Zazzarini, A. and Nava, M. (1976) Biologia e controllo del *Verticillium dahliae* Kleb. su peperone. Azione di alcuni fitofarmaci sulla microflora deterreno in relazione al loro impiego nella lotta contro la verticilliosi (1969–1971). *Notiziario sulle Malattie delle Piante* 94/95, 301–330.

da Silva, G.S. and Tokeshi, H. (1979) Reação de algunas ervas daninhas a *Verticillium albo-atrum*. *Summa Phytopathologica* 5, 85–89.

Daugaard, H., Todsen, T. and Polesny, F. (1997) Preliminary work on integrated production of strawberries in Denmark. Integrated Plant Protection in Orchards, Soft Fruits. Workshop held in Vienna, Austria, 7–10 October 1997. *Bulletin OILB SROP* 21(10), 77–80.

D'aulerio, A.Z. and De Polzer, S. (1982) Il crisantemo e le sue principal malattie crittogamiche. *Informatore di Ortoflorofrutticoltura* 23(7), 7–12.

Davidse, L.C. (1973) Antimitotic activity of methyl benzimidazol-2-yl carbamate (MBC) in *Aspergillus nidulans*. *Pesticide Biochemistry and Physiology* 3, 317–325.

Davidse, L.C. (1975) *Systemfungicide*. Longmans, Green, New York, pp. 92–114.

Davidse, L.C. and Flach, W. (1977) Differential binding of methyl benzimidazol-2-yl carbamate to fungal tubulin as a mechanism of resistance to this antimitotic agent in mutant strains of *Aspergillus nidulans*. *Journal of Cell Biology* 72, 174–193.

Davies, H.T., Young, D.A. and Johnston, G.R. (1977) Tobique, a new chipping variety. *American Potato Journal* 54, 419–424.

Davies, R.R. and Isaac, I. (1958) Dissemination of *Verticillium albo-atrum* through the atmosphere. *Nature (London)* 181, 649.

Davies, W.E. and Young, N.R. (1975) The performance of Sabilt lucerne, a *Verticillium* wilt resistant variety. *Report of the Welsh Plant Breeding Station* 1974, p. 25.

Davis, D.A., Legendre, L., Heinstein, P. and Low, P.S. (1990) Elicitation of plant defense responses by components from *Verticillium dahliae*. *Plant Physiology*. 93, S-53, Abstract 304.

Davis, D.A., Merida, J., Legendre, L., Low, P.S. and Heinstein, P. (1993) Independent elicitation of the oxidative burst and phytoalexin formation in cultured plant cells. *Phytochemistry* 32(3), 607–611.

Davis, D.A., Low, P.S. and Heinstein, P. (1998) Purification of a glycoprotein elicitor of phytoalexin formation from *Verticillium dahliae*. *Physiological and Molecular Plant Pathology S2* 4, 259–273.

Davis, D.D. (1993) Registration of five upland-type parental A- and B-lines for hybrid cotton. *Crop Science* 33(6), 1426.

Davis, D.D., Malm, N.R., Roberts, C.R., Chew, C.F., Barnes, C.E., Staten, G. and Wood, R.L. (1978) Registration of Acala 1517-70 cotton (Reg.No. 66). *Crop Science* 18, 164.

Davis, D.D., Malm, N.R., Roberts, C.L. and Barnes, C.E. (1980) Registration of Acala 1517E-2 Upland cotton. *Crop Science* 20, 113.

Davis, J.R. (1984) Approaches for the control of potato early dying caused by *Verticillium dahliae*. *American Potato Journal* 61(8, II), 519–520.

Davis, J.R. (1985) Approaches to control of potato early dying caused by *Verticillium dahliae*. *American Potato Journal* 62, 177–185.

Davis, J.R. and Everson, D.O. (1986) Relation of *Verticillium dahliae* in soil and potato tissue, irrigation method, and N-fertility to *Verticillium* wilt of potato. *Phytopathology* 76, 730–736.

Davis, J.R. and McDole, R.E. (1979) Influence of cropping sequences on soil-borne populations of *Verticillium dahliae* and *Rhizoctonia solani*. In: Schippers, B. and Gams, W. (eds) *Soil-borne Plant Pathogens. Proceedings of the 4th International Symposium on Factors Determining the Behaviour of Plant Pathogens in the Soil.* Munich, Germany, August 1978. Academic Press, New York, pp. 399–405.

Davis, J.R. and Sorensen, L.H. (1986) Influence of soil solarization at moderate temperatures on potato genotypes with differing resistance to *Verticillium dahliae*. *Phytopathology* 76, 1021–1026.

Davis, J.R., Jones, W.B. and Santos, G.S. (1981) Evaluation of aldicarb for *Verticillium* wilt suppression of potato: *Verticillium dahliae* stem colonization and relationship to nematode populations. *Phytopathology* 71, 104.

Davis, J.R., Pavek, J.J. and Corsini, D.L. (1983a) A sensitive method for quantifying *Verticillium dahliae* colonization in plant tissue and evaluating resistance among potato genotypes. *Phytopathology* 73, 1009–1014.

Davis, J.R., Sorensen, L.H., Corsini, D.L. and Hafez, S.L. (1983b) Effects of continuous cropping with potato genotypes on *Verticillium dahliae* and *Pratylenchus neglectus*. *Phytopathology* 73, 958.

Davis, J.R., Sorensen, L.H. and Corsini, G.S. (1983c) Interaction of *Erwinia* spp. and *Fusarium roseum* 'Sambucinum' on the Russet Burbank potato. *American Potato Journal* 60, 409–421.

Davis, J.R., Hafez, S.L. and Sorensen, L.H. (1986) The Butte potato cultivar: a tool for interpreting field interactions between *Verticillium dahliae* and *Pratylenchus* spp. *Phytopathology* 76, 842.

Davis, J.R., Azad, H.R., Hafez, S.L., Sorensen, L.H., Pavek, J.J. and Corsini, D.L. (1987) Interactions between potato genotypes and rhizosphere organisms and the potential for the biological control of *Verticillium*. *Canadian Journal of Plant Pathology* 9, 79–80.

Davis, J.R., Stark, J.C. and Sorrensen, L.H. (1988) Reduced *Verticillium dahliae* colonizations in potato stems, wilt suppression, and reductions of *Verticillium dahliae* in soil with optimal N and P. *Biological and Cultural Tests. American Phytopathological Society* 3, 22.

Davis, J.R., Sorensen, L.H. and Schneider, A.T. (1990) Ecological effects of cropping practices upon soilborne populations of *Verticillium dahliae* and *V. tricorpus*. *Proceedings of the Fifth International* Verticillium *Symposium*. Leningrad, p. 7.

Davis, J.R., Pavek, J.J., Corsini, D.L., Sorensen, L.H., Schneider, A.T., Everson, D.O., Westermann, D.T. and Huisman, O.C. (1994a) Influence of continuous cropping of several potato clones on the epidemiology of *Verticillium* wilt of potato. *Phytopathology* 84, 207–214.

Davis, J.R., Stark, J.C., Sorensen, L.H. and Schneider, A.T. (1994b) Interactive effects of nitrogen and phosphorus on *Verticillium* wilt of Russet Burbank potato. *American Potato Journal* 71(7), 467–481.

Davis, J.R., Huisman, O.C., Westermann, D.T., Hafez, S.L., Everson, D.O., Sorensen, L.H. and Schneider, A.T. (1996) Effects of green manures on *Verticillium* wilt of potato. *Phytopathology* 86(5), 444–453.

Davis, J.R., Huisman, O.C., Sorensen, L.H. and Schneider, A.T. (1997a) Field studies comparing the susceptibility between crops for the colonization of *Verticillium dahliae* in plant tissue. *Proceedings of the Seventh International* Verticillium *Symposium*. Cape Sounion, Athens, Greece. p. 25.

Davis, J.R., Huisman, O.C., Westermann, D.T., Everson, D.O., Sorensen, L.H. and Schneider, A.T. (1997b) Alternative approaches for control of *Verticillium* wilt of potato with sudangrass and correlations with related factors. *Proceedings of the Seventh International* Verticillium *Symposium*. Cape Sounion, Athens, Greece. p. 74.

Davis, R.M., Menge, J.A. and Erwin, D.C. (1979) Influence of *Glomus fasciculatus* and soil phosphorus on *Verticillium* wilt of cotton. *Phytopathology* 69, 453–456.

Davis, S.H. and Peterson, J.L. (1973) A tree wound dressing to prevent spread of the *Ceratocystis* causing canker stain disease of the plane tree. *Plant Disease Reporter* 57, 28–30.

de Badilla, H.S. (1974) Estudoi preliminar de resistancia de lineas y/o variadades de tomate a *Verticillium albo-atrum* Reinke & Berth. *Agricultura Técnica* 34, 39–40.

De Bary, A. (1863) Récherches sur le développement de quelques champignons parasitaires. *Annales des Sciences Naturelles Botanique, Série 4* 20, 5–148.

Dedio, W. and Hoes, J.A. (1983) Registration of sunflower restorer parental lines CM 586 and CM 587 (Reg.No. PL 34 and PL 35). *Crop Science* 23, 196.

Dedio, W. and Hoes, J.A. (1988) Registration of sunflower parental lines CM 593, CM 594, CM 596 and CM 597. *Crop Science* 28, 1042.

Dedio, W., Hoes, J.A. and Huang, H.C. (1983) Registration of sunflower parental lines CM 338 and CM 361 (Reg.No. PL 32 and PL 33). *Crop Science* 23, 195.

Deese, D.C. and Stahmann, M.A. (1962) Formation of pectic enzymes by *Verticillium albo-atrum* on susceptible and resistant tomato stem tissues and on wheat bran. *Phytopathologische Zeitschrift* 46, 53–70.

de Leeuw, G.T.N., Reerink, J.A. and Elgersma, D.M. (1986) Histopathology of leaves of resistant and susceptible tomato plants infected with *Verticillium albo-atrum*. *Mededelingen van de Faculteit Landbouwwetenschappen Rijksuniversiteit Gent* 51, 563–567.

Delwiche, P.A., Grau, C.R. and Maxwell, D.P. (1981) Reaction of alfalfa cultivars and other crops to Wisconsin isolates of *Verticillium albo-atrum* (alfalfa strain). *Phytopathology* 71, 870.

de Melo, I.S. and Brar, J.K. (1995a) Use of isolated cells to detect phytotoxic activity of *Verticillium dahliae* culture extracts on *Solanum melongena*. 6th International *Verticillium* Symposium, Dead Sea, Israel. *Phytoparasitica* 23, 48.

de Melo, I.S. and Brar, J.K. (1995b) Use of isolated cells to detect phytotoxic activity of *Verticillium dahliae* culture extracts on *Solanum melongena*. *Fitopatologia Brasileira* 20(2), 160–163.

de Melo, I.S. and da Costa, C.P. (1985) Reaçao de resistência de progênies de berinjela a *Verticillium albo-atrum* Reinke & Berth. *Summa Phytopathologica* 11, 180–185.

de Melo, I.S., da Costa, C.P. and Silveira, A.P.D. (1985) Influência de micorrizas vesículo-arbusculares sobre a murcha da berinjela causada por *Verticillium albo-atrum* Reinke & Berth. *Summa Phytopathologica* 11, 173–179.

Demetriades, S.D. and Emmanouil, V.N. (1971) Sur l'hydrolse de l'arbutine par quelques champignons. *Annales de l'Institut Phytopathologique Benaki* 10, 106–209.

Demetriades, S.D., Zachos, D.G., Constantinou, P.T., Panagopoulos, C.G. and Holevas, C.D. (1958) Rapport sommaire sur les principales maladies des plantes cultivées, observées en Grèce au cours de l'année 1958. *Annales de l'Institut Phytopathologique Benaki NS* 1, 323–329.

Denby, L.G. and Woolliams, G.E. (1962) The development of *Verticillium* resistant strains of established tomato varieties. *Canadian Journal of Plant Sciences* 42, 681–685.

Denner, F.D.N. and Millard, C.P. (1997) Occurrence of *Verticillium* wilt on potatoes in South Africa. *Proceedings of the Seventh International* Verticillium *Symposium*. Cape Sounion, Athens, Greece. p. 91.

de Oliveira, M.L. (1983) Sensibilidade de *Verticillium dahliae* Kleb., agente causal da murcha de *Verticillium* do cacaueiro (*Theobroma cacao* L.) a fungicidas *in vitro*. *Revista Theobroma* 13, 35–39.

D'Ercole, N. (1975) La verticilliosi della fragola. *Informatore Fitopatologico* 25(4), 5–10.

D'Ercole, N. (1976) La verticilliosi del cetriolo. *Informatore Fitopatologico* 26(11/12), 37–39.

D'Ercole, N. and Gavina, F. (1975) Verticilliosi della fragola. II. Richerche sulla diffusione dell'agente patogeno nei tessuti dello stolene e nelle piante figlie. *Informatore Fitopatologico* 25(7), 5–10.

D'Ercole, N. and Nipoti, P. (1980) La verticilliosi della fragola: suscettibilita di piante micropropagate e frigoconservate. *Informatore Agrario* 36, 11465–11466.

D'Ercole, N. and Nipoti, P. (1986) Lotta biologica contro fusariosi e verticilliosi del pomodoro in coltura protetta. *Colture Protette* 15, 55–59.

D'Ercole, N., Nipoti, P., Gavina, F. and Vitale, C. (1980) Miglioramento genético–sanitario della fragola nei confronti di *Verticillium dahliae*, di *Rhizoctonia fragariae* e di *Rhizoctonia solani*. *Informatore Agrario* 36, 12591–12593.

D'Ercole, N., Sportelli, M. and Nipoti, P. (1984) Differenti tipi di antagonismo di *Trichoderma* sp. nei confronti di miceti fitopatogeni del terreno. *Informatore Fitopatologico* 34(11), 43–47.

D'Ercole, N., Manzali, D., Gennari, S. and Nipoti, P. (1989) La verticilliosi dell' *Amoracia rusticana* Gaert., Mey. et Sch. *Difesa delle Piante* 12, 195–203.

D'Ercole, N., Nipoti, P., Di Pillo, L. and Gavina, F. (1997) *In vitro* and *in vivo* tests of *Trichoderma* spp. as biocontrol agents of *Verticillium dahliae* in egg plants. *Proceedings of the Seventh International* Verticillium *Symposium*. Cape Sounion, Athens, Greece. p. 68.

de Ramallo, N.E.V. (1972) Marchitamiento en paltos producido por *Verticillium* sp. *Revista Industrial y Agricola de Tucumán* 49, 16–67.

Dermovskaya, L.I. (1972) [*Verticillium* wilt of plum in the Alma-Ata region]. *Povyshenie Produktivnosti Plod. yagod. Nasazhdenii* 2, 238–241.

Desai, D.B. and Wiles, A.B. (1976) Reaction to foliar applications of microelements. *Proceedings of the Beltwide Cotton Production Research Conference*. National Cotton Council, Memphis, Tennessee. p. 21.

de Samblanx, G.W., Fernandez Del Carmen, A., Sijtsma, L., Plasman, H.H., Schaaper, W.M.M., Posthuma, G.A., Fant, F., Meloen, R.H., Broekaert, W.F. and van Amerongen, A. (1996) Antifungal activity of synthetic 15-mer peptides based on the Rs-AFP2 (*Raphanus sativus* antifungal protein 2) sequence. *Peptide Research* 9(6), 262–268.

Deshkova, A.D., Popushoi, I.S., Shatrova, D.L., Davidovich, R.E. and Buĭmistru, L.D. (1977) [Effect of nitrogen nutrition sources on the growth and morphological–cultural features of *Verticillium* species]. Микология и Фитопатология [*Mycology and Phytopathology*] 11, 122–127.

Desjobert, A. and Herisset, A. (1972) L'aconit napel (*Aconitum napellus* L.). II. Essais de culture realises en Anjou. *Plantes Medicinates et Phytotherapie* 6, 239–246.

Devaux, A.L. and Sackston, W.B. (1966) Taxonomy of *Verticillium* species causing wilt of horticultural crops in Quebec. *Canadian Journal of Botany* 44, 803–811.

DeVay, J.E. (1989) Physiological and biochemical mechanisms in host resistance and susceptibility to wilt pathogens. In: Tjamos, E.C. and Beckman, C.H. (eds) *Vascular Wilt Diseases of Plants*. Springer Verlag, Berlin, pp. 197–217.

DeVay, J.E. (1990) Physiological responses of cotton (*Gossypium hirsutum*) cultivars to different pathotypes of *Verticillium dahliae*. *Proceedings of the Fifth International* Verticillium *Symposium*. Leningrad, p. 62.

DeVay, J.E. and Pullman, G.S. (1984) Epidemiology and ecology of diseases caused by *Verticillium* species, with emphasis on *Verticillium* wilt of cotton. *Phytopathologia Mediterranea* 23, 95–108.

DeVay, J.E., Forrester, L.L., Garber, R.H. and Butterfield, E.J. (1974) Characteristics and concentration of propagules of *Verticillium dahliae* in air-dried field soils in relation to prevalence of *Verticillium* wilt of cotton. *Phytopathology* 64, 22–29.

DeVay, J.E., Tzeng, D.D. and Kapulnik, E. (1987) Effect of methionine–riboflavin mixtures on ethylene production and on the viability of propagules of *Verticillium dahliae* in field soils and liquid culture media. *Canadian Journal of Plant Pathology* 9, 80.

DeVay, J.E., El Zik, K.M., Bourland, F.M., Garber, R.H., Kappleman, A.M., Lyda, S.D., Minton, E.B., Roberts, P.A. and Wallace, T.P. (1989) Strategies and tactics for managing plant pathogens and nematodes. In: Frisbie, R.E., El Zik, K.M. and Wilson, L.T. (eds) *Integrated Pest Management Systems and Cotton Production.* John Wiley & Sons, New York, pp. 225–266.

DeVay, J.E., Weir, B.L., Wakeman, R.J. and Stapleton, J.J. (1997) Effects of *Verticillium dahliae* infection of cotton plants (*Gossypium hirsutum*) on potassium levels in leaf petioles. *Plant Disease* 81(9), 1089–1092.

Devey, M.E. and Roose, M.L. (1987) Genetic analysis of *Verticillium* wilt tolerance in cotton using pedigree data from three crosses. *Theoretical and Applied Genetics* 74, 162–167.

Devey, M.E. and Rosielle, A.A. (1986) Relationship between field and greenhouse ratings for tolerance to *Verticillium* wilt on cotton. *Crop Science* 26, 1–4.

Di Corato, U., Frisullo, S., Lops, F. and Sisto, D. (1996) Studi sulla verticilliosi del kenaf in Basilicata. *Difesa delle Piante* 19(2–3), 65–69.

Dikiĭ, S.P. and Neklyudova, E.T. (1975) [Study on a collection of eggplants for resistance to wilt]. *Byulleten' Vsesoyuznogo Nauchno-Issledovatel'skogo Instituta Zashchity Rasteniĭ* 50, 65–69.

Dikiĭ, S.P. and Neklyudova, E.T. (1977) [Pests and diseases of eggplant in the Krasnodar region]. *Byulleten' Vsesoyuznogo Ordena Lenina i Ordena Drushby Narodov Instituta Rastenievodstva Imeni N.I. Vavilova* 74, 18–24.

Dikii, S.P. and Neklyudova, E.T. (1979) [Field resistance of egg plants from the world collection to *Verticillium* and big bud]. *Trudy po Prikladnoi Botanike, Genetike i Selektsii* 64, 89–92.

Dimock, A.W. (1943) A method of establishing *Verticillium*-free clones of perennial plants. *Phytopathology* 33, 3.

Dimond, A.E. (1955) Pathogenesis in the wilt diseases. *Annual Review of Plant Physiology* 6, 329–350.

Dimond, A.E. (1962) Wilt pathogens and oxygen levels in tracheal fluid of stems. *Phytopathology* 52, 7.

Dimond, A.E. (1966) Pressure and flow relations in vascular bundles of the tomato plant. *Plant Physiology* 41, 119–131.

Dimond, A.E. (1970) Biophysics and biochemistry in the vascular wilt syndrome. *Annual Review of Phytopathology* 8, 301–322.

Dimond, A.E. (1972) The nature and control of epidemics. *Proceedings of the Indian National Science Academy B* 37, 267–279.

Dimond, A.E. and Horsfall, J.G. (1965) The theory of inoculum. In: Baker, K.F. and Snyder, W.C. (eds) *Ecology of Soil-borne Pathogens.* John Murray, London, pp. 404–419.

Dimond, A.E. and Waggoner, P.E. (1953a) On the nature and role of vivotoxins in plant disease. *Phytopathology* 43, 229–235.

Dimond, A.E. and Waggoner, P.E. (1953b) The water economy of *Fusarium* wilted tomato plants. *Phytopathology* 43, 619–623.

Diwan, N., Fluhr, R., Eshed, Y., Zamir, D. and Tanksley, S.D. (1999) Mapping of Ve in tomato: a gene conferring resistance to the broad-spectrum pathogen, *Verticillium dahliae* race 1. *Theoretical and Applied Genetics* 98, 315–319.

Dixon, G.R. (1972) Effects of thiabendazole on *Verticillium* wilt of lucerne. *Plant Pathology* 21, 129–132.

Dixon, G.R. and Doodson, J.K. (1971) Assessment keys for some diseases of vegetable, fodder and herbage crops. *Journal of The National Institute of Agricultural Botany* 12, 299–307.

Dixon, G.R. and Pegg, G.F. (1969) Hyphal lysis and tylose formation in tomato cultivars infected by *Verticillium albo-atrum*. *Transactions of the British Mycological Society* 53, 109–118.

Dixon, G.R. and Pegg, G.F. (1972) Changes in aminoacid content of tomato xylem sap following infection with strains of *Verticillium albo-atrum*. *Annals of Botany* 36, 147–154.

Dixon, G.R., Kershaw, C.D. and Hunter, E.A. (1989) Crop yields from lucerne (*Medicago sativa*) cultivars displaying gradations in resistance to wilt (*Verticillium albo-atrum*). *Journal of Agricultural Science* 112, 387–394.

Dobinson, K.F. (1995) Genetic transformation of the vascular wilt fungus *Verticillium dahliae*. *Canadian Journal of Botany* 73(5), 710–715.

Dobinson, K.F., Tenuta, G.K. and Lazarovits, G. (1996) Occurrence of race 2 of *Verticillium dahliae* in processing tomato fields in southwestern Ontario. *Canadian Journal of Plant Pathology* 18(1), 55–58.

Dobinson, K.F., Lecomte, N. and Lazarovits, G. (1997) Production of an extracellular trypsin-like protease by the fungal plant pathogen *Verticillium dahliae*. *Canadian Journal of Microbiology* 43(3), 227–233.

Dobinson, K.F., Patterson, N.A., White, G.J. and Grant, S. (1998) DNA fingerprinting and vegetative compatibility analysis indicate multiple origins for *Verticillium dahliae* race 2 tomato isolates. *Mycological Research* 102, 1089–1095.

Dochinger, L.S. (1956) Maple wilt progress report. *Shade Tree* 29, 1–3.

Dogo, M., Toyoda, H., Tampo, H., Matsuda, Y., Nonomura, T. and Ouchi, S. (1999) Antimicrobial activities of indole derivatives against soil-borne pathogens of tomato. *Bulletin of the Institute for Comprehensive Agricultural Sciences Kinki University* 7, 91–95.

Dolar, M.S. (1984) [Studies on the susceptibility of some cotton varieties to wilt (*Verticillium dahliae* Kleb.) in the Mediterranean region]. *Bitki Koruma Bülteni* 24, 148–158.

Domsch, K.H., Gams, W. and Anderson, T.-H. (1980a, publ. 1981) *Compendium of Soil Fungi*, Vol. 1. Academic Press, London.

Domsch, K.H., Gams, W. and Anderson, T.-H. (1980b, publ. 1981) *Compendium of Soil Fungi*, Vol. 2. Academic Press, London.

Donald, Z. and Czeslaw, S. (1998) Effect of temperature on infestation and development of *Verticillium dahliae* Kleb. on winter oilseed rape. *Bulletin OILB/SROP* 21(5), 41–47.

Donant, S. (1932) Untersuchungen über die Pathogenität des Wirtelpilzes *Verticillium albo-atrum* R. u B. *Zeitschrift für Parasitenkunde* 4, 653–711.

Donegan, K.K., Schaller, D.L., Stone, J.K., Ganio, L.M., Reed, G., Hamm, P.B. and Seidler, R.J. (1996) Microbial populations, fungal species diversity and plant pathogen levels in field plots of potato plants expressing the *Bacillus thuringiensis* var. *tenebrionis* endotoxin. *Transgenic Research* 5(1), 25–35.

Dong, Z.-Q., He, Z.-P. and Zhai, X.-J. (2000) [Studies of DPC (mepiquat) inhibition on *Verticillium dahliae* Kleb. infected cotton and the mechanism of its action]. *Acta Gossypii Sinica* 12(2), 77–80.

Dotsenko, A.S. (1971) [The distribution of *Verticillium* wilt of cotton in the Kirghiz SSR]. *Byulleten' Vsesoyuznogo Nauchno-Issledovatel'skogo Instituta Zashchity Rasteniĭ* 22, 18–21.

Douglas, S.M. and MacHardy, W.E. (1981) The relationship between vascular alterations and symptom development in *Verticillium* wilt of chrysanthemum. *Physiological Plant Pathology* 19, 31–39.

Dowson, W.J. (1922) On the symptoms of wilting of michaelmas daisies produced by a toxin secreted by a *Cephalosporium*. *Transactions of the British Mycological Society* 7, 383–386.

Dowson, W.J. (1923) The wilt disease of michaelmas daisies. *Journal of the Royal Horticultural Society* 48, 38–57.

Draper, A.D., Galletta, G.J. and Swartz, H.J. (1981) 'Tribute' and 'Tristar' everbearing strawberries. *HortScience* 16(6, I), 794–795.

Drayton, F.L. (1924) Survey of the prevalence of plant diseases in the Dominion of Canada, 1923. *Canada Department of Agriculture: Experimental Farms Branch Fourth Annual Report.*

Drew, M.C. and Lynch, J.M. (1980) Soil anaerobiosis, micro-organisms and root function. *Annual Review of Phytopathology* 18, 37–66.

Dube, H.C. (1971) Studies on permeability changes in cotton (*Gossypium hirsutum*) caused by *Verticillium albo-atrum*. *Proceedings of the Indian Academy of Sciences Section B* 74, 185–193.

Dube, H.C. and Mathur, S. (1975, published 1976) Hydrolytic and trans-eliminative degradation of pectic substances by extra-cellular enzymes of *Verticillium dahliae*. *Indian Phytopathology* 28, 346–351.

Dube, H.C. and Mathur, S. (1977) Extracellular pectic and cellulytic enzymes of *Verticillium albo-atrum* and *V. dahliae*. In: Bilgrami, K.S. (ed.) *Physiology of Micro-organisms*. Today and Tomorrow's Printers and Publishers, New Delhi, India, pp. 235–247.

Dube, H.C. and Mathur, S. (1982) Effect of calcium ions on pectolytic enzyme activity of *Verticillium albo-atrum* and *V. dahliae*. *Proceedings of the National Academy of Science, India* 13, 59–64.

Dubery, I.A. and Meyer, R. (1996) Specific binding of a *Verticillium dahliae* phytotoxin to protoplasts of cotton, *Gossypium hirsutum*. *Plant Cell Reports* 15(10), 777–780.

Dubery, I.A. and Slater, V. (1997) Induced defence responses in cotton leaf disks by elicitors from *Verticillium dahliae*. *Phytochemistry* 44(8), 1429–1434.

Dufrenoy, J. (1927a) Dépérissements des arbres dans le Massif Central. *Bulletin d'Office Agri. Régional Massif Central* 8, 101–119.

Dufrenoy, J. (1927b) *Dépérissement des Arbres Fruitiers dans le Massif Central*. Mont-Louis, Clermont-Ferrand, France.

Dufrenoy, J. (1927c) Étude expérimental des reations des *Verticillium* et leur hôtes. *Revue de Pathologie Végétale et d'Entomologie Agricole de France* 14, 207–214.

Dufrenoy, J. (1928) Recherches relatives aux dépérissements des arbres fruitiers. *Bulletin Trimestriel de l'Office Régional Agricole du Midi* 26, 94–96.

Dufrenoy, J. (1929) Les hadromycoses des états de l'est Etats-Unis. *Annales Cryptogamique Exotique* 2, 158–167.

Dufrenoy, J. and Dufrenoy, M.L. (1927) Hadromycoses. *Annales des Épiphyties* 13, 195–212.

Dunbier, M.W., Close, R.C. and Ellis, T.J. (1976) Disease and pest-resistant lucerne cultivars for New Zealand. *Proceedings of the Twenty-ninth New Zealand Weed and Pest Control Conference*. pp. 46–49.

Duncan, D.R. and Himelick, E.B. (1986) Inhibition of conidial production of *Verticillium dahliae* with ammonium suphate. *Phytopathology* 76, 788–792.

Duncan, D.R. and Himelick, E.B. (1987) The effects of amino acids on the growth and sporulation of *Verticillium dahliae*. *Canadian Journal of Botany* 65, 1299–1302.

Durrands, P.K. and Cooper, R.M. (1988a) Development and analysis of pectic screening media for use in the detection of pectinase mutants. *Applied Microbiology and Biotechnology* 28, 463–467.

Durrands, P.K. and Cooper, R.M. (1988b) Selection and characterization of pectinase-deficient mutants of the vascular wilt pathogen *Verticillium albo-atrum*. *Physiological and Molecular Plant Pathology* 32, 343–362.

Durrands, P.K. and Cooper, R.M. (1988c) The role of pectinases in vascular wilt disease as determined by defined mutants of *Verticillium albo-atrum*. *Physiological and Molecular Plant Pathology* 32, 363–371.

Durrands, P.K., Keene, R.A., Cooper, R.M., O'Garro, L.W. and Clarkson, J.M. (1988) Polygalacturonase isoenzyme profiles of *Verticillium dahliae* isolates races 1 and 2 from different geographical origins. *Transactions of the British Mycological Society* 91, 533–536.

Dutta, B.K. (1980) Studies on wilt disease caused by *Verticillium albo-atrum* R & B in relation to some systemic fungicides. *Journal of Indian Botanical Science* 59, 81–89.

Dutta, B.K. (1981) Growth substances in the control of *Verticillium* wilt of tomato. *Indian Phytopathology* 34, 421–425.

Dutta, B.K. and Bremner, E. (1981) Trace elements as plant chemotherapeutants to control *Verticillium* wilt. *Zeitschrift für Pflanzenkrankheiten und Pflanzenschutz* 88, 405–412.

Dutta, B.K. and Isaac, I. (1979a) Effects of inorganic amendments (N, P and K) to soil on the rhizosphere microflora of antirrhinum plants infected with *Verticillium dahliae* Kleb. *Plant and Soil* 52, 561–569.

Dutta, B.K. and Isaac, I. (1979b) Effects of organic amendments to soil on the rhizosphere microflora of antirrhinum infected with *Verticillium dahliae* Kleb. *Plant and Soil* 53, 99–103.

Dutta, B.K. and Isaac, I. (1979c) Seasonal variation of fungistasis in some soils. *Transactions of the British Mycological Society* 73, 157–159.

Dutta, B.K. and Isaac, I. (1981) Control of *Verticillium* wilt of tomato by antibiotic chemotherapy. *Zeitschrift für Pflanzenkrankheiten und Pflanzenschutz* 88, 276–285.

Dwinell, L.D. and Sinclair, W.A. (1967) Effects of N, P, K, and inoculum density of *Verticillium dahliae* on populations of *Pratylenchus penetrans* in roots of American elm and sugar maple. *Phytopathology* 57, 810.

Eastburn, D.M. and Butler, E.E. (1988) Microhabitat characterization of *Trichoderma harzianum* in natural soil: evaluation of factors affecting population density. *Soil Biology and Biochemistry* 20, 541–545.

Eastburn, D.M. and Chang, R.J. (1994) *Verticillium dahliae*: a causal agent of root discoloration of horseradish in Illinois. *Plant Disease* 78(5), 496–498.

Easton, G.D. (1976) Effect of soil temperature in fumigation control of *Verticillium* wilt of potato. *Proceedings of the Second International* Verticillium *Symposium*. University of California, Berkeley, p. 15.

Easton, G.D. (1990) Effect of foliage fungicides on natural inhibitions to *Verticillium dahliae* and *Sclerotinia sclerotiorum* on potato. *Proceedings of the Fifth International* Verticillium *Symposium*. Leningrad, p. 80.

Easton, G.D. and Nagle, M.E. (1986) *Verticillium* wilt control and enhanced potato production following cropping to a green pea–sudangrass rotation. *Proceedings of the Fourth International* Verticillium *Symposium*, Guelph, Canada.

Easton, G.D. and Nagle, M.E. (1987) Giant hill selection for control of *Verticillium dahliae* in potato. *Plant Disease* 71, 937–940.

Easton, G.D., Nagle, M.E. and Bailey, D.L. (1969) A method of estimating *Verticillium albo-atrum* propagules in field soil and irrigation waste water. *Phytopathology* 59, 1171–1172.

Easton, G.D., Nagle, M.E. and Bailey, D.L. (1972a) Effect of annual soil fumigation and preharvest vine burning on *Verticillium* wilt of potato. *Phytopathology* 62, 520–524.

Easton, G.D., Nagle, M.E. and Bailey, D.L. (1972b) *Verticillium albo-atrum* carried by certified seed potatoes into Washington and control by chemicals. *American Potato Journal* 49, 397–402.

Easton, G.D., Nagle, M.E. and Bailey, D.L. (1975) Residual effect of soil fumigation with vine burning on control of *Verticillium* wilt of potato. *Phytopathology* 65, 1419–1422.

Ebbels, D.L. (1979) A historic review of certification schemes for vegetatively-propagated crops in England and Wales. *ADAS Quarterly Review* 32, 21–58.

Ebbels, D.L. and Dickens, J.S.W. (1981) *Verticillium* wilt of hops. *MAFF, ADAS leaflet* (revised) 413, 1–8.

Ebbels, D.L. and King, J.E. (1979) *Plant Health. The Scientific Basis for Administrative Control of Plant Diseases and Pests*. Blackwell Scientific Publications, London.

Ebihara, Y., Nagao, H., Koike, M. and Shiraish, T. (1997) Assessment of vegetative compatibility of Japanese isolates of *Verticillium dahliae* with the standardised VCGs. *Proceedings of the Seventh International* Verticillium *Symposium*. Cape Sounion, Athens, Greece. p. 30.

Ebihara, Y., Nagao, H., Koike, M., Shiraishi, T. and Iijima, T. (1999a) Vegetative compatibility relationships among weakly pathogenic isolates (pathotype E) of *Verticillium dahliae*. *Mycoscience* 40(1), 41–49.

Ebihara, Y., Nagao, H., Koike, M., Shiraishi, T. and Iijima, T. (1999b) How do Japanese isolates of *Verticillium dahliae* correspond with standardized VCG testers? *Mycoscience* 40(4), 333–343.

Ecvil, F. and Yalcin, O. (1977) [Preliminary studies on fungi causing melon wilt in the Aegean region]. *Zirai Mücadele Araştirma Yilliği* 11, 78.

Edgington, L.V. (1981) Structural requirements of systemic fungicides. *Annual Review of Phytopathology* 19, 107–124.

Edgington, L.Y. and Walker, J.C. (1957) Influence of soil air temperature on *Verticillium* wilt on tomato. *Phytopathology* 47, 594–598.

Edson, H.A. and Shapovalov, M. (1920) Temperature relations of certain potato-rot and wilt producing fungi. *Journal of Agricultural Research* 18, 511–524.

Edson, H.A. and Wood, J.I. (1936) Diseases of plants in the United States in 1935. *Plant Disease Reporter Supplement* 96, 115–289.

Edwards, M. and Taylor, C. (1998) Dieback and canker in Australian pistachios. Proceedings of the Second International Symposium on Pistachios and Almonds. Davis, California, USA, 24–29 August 1997. *Acta Horticulturae* 470, 596–603.

Efimova, G. (1978) [Selection of specimens with complex resistance to wilt]. *Khlopkovodstvo* 1, 30–31.

Egamberdiyev, A.E. (1990) [Amphidiploids of cotton as a new source of wilt resistance]. *Proceedings of the Fifth International* Verticillium *Symposium*. Leningrad, p. 69.

Egamov, I. (1976) [Rye and sorghum for wilt control]. *Khlopkovodstvo* 10, 18.

Egamov, I. (1979) [Effects of lucerne in crop rotation on the viability of microsclerotia of *Verticillium dahliae* Kleb.]. Микология и Фитопатология [*Mycology and Phytopathology*] 13, 80–82.

Egger, E. (1973) Considerazioni su una sconosciuta moria di viti durante la scorsa primavera. *Rivista di Viticoltura e di Enologia* 26, 414–415.

Ehteshamul-Haque, S. and Ghaffar, A. (1994) New records of root infecting fungi from Pakistan. *Pakistan Journal of Phytopathology* 6(1), 50–57.

El-Abyad, M.S., El-Sayad, M.A., El-Shanshoury, A.R. and El-Sabbagh, S.M. (1993) Towards the biological control of fungal and bacterial diseases of tomato using antagonistic *Streptomyces* species. *Plant and Soil* 149, 185–195.

El-Abyad, M.S., El-Sayad, M.A., El-Shanshoury, A.R. and El-Sabbagh, S.M. (1996) Antimicrobial activities of *Streptomyces pulcher*, *S. canescens* and *S. citreofluorescens* against fungal and bacterial pathogens of tomato *in vitro*. *Folia Microbiologica* 41(4), 321–328.

Elad, Y., Katan, J. and Chet, I. (1980) Physical, biological, and chemical control integrated for soil-borne diseases of potatoes. *Phytopathology* 70, 418–422.

El Aissami, A., Brhada, F. and Lahlou, H. (1998) Effet de l'adaptation progressive à la luzerne d'un isolat de *Verticillium dahliae* (d'origine tomate) sur la synthese *in vitro* de ses enzymes pectocellulolytiques. *Cryptogamie, Mycologie* 19, 227–234.

El Aissami, A., El Amri, H., Mrabet, N., Chekti, R. and Lahlou, H. (1999) Contribution to the study of the chemical composition of *Verticillium albo-atrum* secretions in liquid media. *Mycopathologia* 144(2), 93–95.

Elango, D., Robb, J., Newcombe, G. and Busch, L.V. (1986) Growth pouch technique for the observation of cellular interactions between alfalfa seedling roots and *Verticillium albo-atrum*. *Canadian Journal of Plant Pathology* 8, 78–84.

El-Deeb, A.A., Hilal, A.A., Radwan, I.A., Ali, A.A. and Mohamed, H.A. (1985) Varietal reaction and fungicidal effect to root rot and wilt diseases of sesame. *Annals of Agricultural Science Moshtohor* 23, 713–721.

Elena, K. (1997) Genetic relatedness of *Verticillium dahliae* isolates from cotton, tomato and watermelon plants. *Proceedings of the Seventh International* Verticillium *Symposium*. Cape Sounion, Athens, Greece. p. 27.

Elena, K. (1999) Genetic relationships among *Verticillium dahliae* isolates from cotton in Greece based on vegetative compatibility. *European Journal of Plant Pathology* 105(6), 609–616.

Elena, K. (2000) Vegetative compatibility among *Verticillium dahliae* isolates from watermelon in Greece. *Phytoparasitica* 28(2), 115–120.

Elena, K. and Paplomatas, E.J. (1998) Vegetative compatibility groups within *Verticillium dahliae* isolates from different hosts in Greece. *Plant Pathology* 47, 635–640.

Elgersma, D.M. (1980) Accumulation of rishitin in susceptible and resistant tomato plants after inoculation with *Verticillium albo-atrum*. *Physiological Plant Pathology* 16, 149–153.

Elgersma, D.M. and Liem, J.I. (1989) Accumulation of phytoalexins in susceptible and resistant near-isogenic lines of tomato infected with *Verticillium albo-atrum* or *Fusarium oxysporum* f.sp. *lycopersici*. *Physiological and Molecular Plant Pathology* 34, 545–555.

Elgersma, D.M. and Overeem, J.C. (1981) Identification of a polyacetylenic phytoalexin in tomato plants after inoculation with *Verticillium albo-atrum*. *Netherlands Journal of Plant Pathology* 87, 69–70.

Elgersma, D.M., Weijman, A.C.M., Roeymans, H.J. and van Eijk, G.W. (1984) Occurrence of falcarinol and falcarindiol in tomato plants after infection with *Verticillium albo-atrum* and characterization of four phytoalexins by capillary gas chromatography–mass spectrometry. *Phytopathologische Zeitschrift* 109, 237–240.

Elmer, W.H. and Ferrandino, F.J. (1991) Effect of black plastic mulch and nitrogen side-dressing on *Verticillium* wilt of eggplant. *Plant Disease* 75, 1164–1167.

Elmer, W.H. and Ferrandino, F.J. (1994) Comparison of ammonium sulfate and calcium nitrate fertilization effects on *Verticillium* wilt of eggplant. *Plant Disease* 78(8), 811–816.

El-Quakfaoui, S., Potvin, C., Brzezinski, R. and Asselin, A. (1995) A *Streptomyces* chitosanase is active in transgenic tobacco. *Plant Cell Reports* 15, 222–226.

El-Zik, K.M. (1984) Integrated control of *Verticillium* wilt of cotton. In: Brown, J.M. (ed.) *Proceedings of the Beltwide Cotton Producers Research Conference*. National Cotton Council, Memphis, Tennessee, pp. 53–54.

El-Zik, K.M. (1985) Integrated control of *Verticillium* wilt of cotton. *Plant Disease* 69, 1025–1032.

El-Zik, K.M. (1986) Half a century dynamics and control of cotton disease: dynamics of cotton diseases and their control. *Proceedings of the Beltwide Cotton Production Research Conference*. National Cotton Council, Memphis, Tennessee, pp. 29–33.

El-Zik, K.M. and Thaxton, P.M. (1990) Registration of Tamcot HQ95 cotton. *Crop Science* 30, 1359–1360.

El-Zik, K.M. and Thaxton, P.M. (1996) Registration of 'Tamcot Sphinx' cotton. *Crop Science* 36(4), 1074.

El-Zik, K.M. and Yamada, H. (1981) Effects of row spacing, irrigation scheduling and nitrogen rate on *Verticillium* wilt, yield and fibre quality of two Acala cotton (*Gossypium hirsutum* L.) cultivars. *Proceedings of the Beltwide Cotton Production Research Conference*. National Cotton Council, Memphis, Tennessee, pp. 31–32.

Emechebe, A.M. (1975) Some host factors affecting inoculation of cacao seedlings with *Verticillium dahliae*. *East African Agricultural and Forestry Journal* 40, 271–277.

Emechebe, A.M. (1980) The effect of soil moisture and of N, P and K on incidence of infection on cacao seedlings inoculated with *Verticillium dahliae*. *Plant and Soil* 54, 143–147.

Emechebe, A.M., Leakey, C.L.A. and Banage, W.B. (1971) *Verticillium* wilt of cacao in Uganda: symptoms and establishment of pathogenicity. *Annals of Applied Biology* 69, 223–227.

Emmanouil, V. and Wood, R.K.S. (1982) Behaviour *in vitro* and *in vivo* of a benomyl insensitive strain of *Verticillium dahliae*. *Phytopathologische Zeitschrift* 103, 13–24.

Emmanouil, V. and Wood, R.K.S. (1983) Fungicides as inducers of the accumulation of fungitoxic substances in plants. *Physiological Plant Pathology* 22, 51–54.

Emmatty, D.A. and Green, R.J. (1969) Fungistasis and the behaviour of the microsclerotia of *Verticillium albo-atrum* in soil. *Phytopathology* 59, 1590–1595.

Emmond, G.S. and Ledingham, R.J. (1972) Effects of crop rotation on some soil-borne pathogens of potato. *Canadian Journal of Plant Science* 52, 605–611.

Engelhard, A.W. (1957) Host index of *Verticillium albo-atrum* (including *Verticillium dahliae* Kleb.). *Plant Disease Reporter Supplement* 244, 23–49.

Engelhard, A.W. (1989) *Soilborne Plant Pathogens: Management of Diseases with Macro- and Microelements*. American Phytopathological Society, St Paul, Minnesota.

Engelhard, A.W. and Carter, J.C. (1956) Isolations of *Verticillium albo-atrum* from woody hosts in Illinois, 1945–1955. *Plant Disease Reporter* 40, 459–462.

Engelkes, C.A. and Fravel, D.R. (1998) Procedure for electrophoretic analysis of proteins from metham sodium-treated microsclerotia of *Verticillium dahliae*. *Soil Biology and Biochemistry* 30, 699–672.

Engelkes, C.A., Nuclo, R.L. and Fravel, D.R. (1997) Effect of carbon, nitrogen, and C:N ratio on growth, sporulation, and biocontrol efficacy of *Talaromyces flavus*. *Phytopathology* 87(5), 500–505.

Erikson, P.I., Melouk, H.A. and Ketring, D.L. (1986) Apparent sap flow velocity in peanut genotypes susceptible and resistant to *Verticillium dahliae*. *Phytopathology* 76, 500–502.

Ershad, D. (1972) [Tracheomycosis of *Cercis canadensis* caused by *Verticillium dahliae* Klebahn]. *Iranian Journal of Plant Pathology* 8, 52–59.

Erturk, H., Ozkut, S., Borazanci, N., Hekimoglu, G. and Arinc, Y. (1975) [The relationship between plant parasitic nematodes and cotton wilt disease and their control methods]. *Bitki Koruma Bülteni* 15, 69–96.

Erwin, D.C. (1958) *Verticillium* wilt of *Cicer arietinum* in Southern California. *Plant Disease Reporter* 42, 1111.

Erwin, D.C. (1969) Methods of determination of the systemic and fungitoxic properties of chemicals applied to plants with emphasis on control of *Verticillium* wilt with thiabendazole and benlate. *World Review of Pest Control* 8, 6–22.

Erwin, D.C. (1970) Progress in the development of systemic fungitoxic chemicals for control of plant diseases. *FAO Plant Protection Bulletin* 18, 73–82.

Erwin, D.C. (1972) *Proceedings of the Western Cotton Production Conference, Bakersfield.* pp. 17–20.

Erwin, D.C. (1973) Systemic fungicides: disease control, translocation and mode of action. *Annual Review of Phytopathology* 11, 389–422.

Erwin, D.C. (1977a) *Antifungal Compounds,* Vol. 1. Dekker, New York, pp. 163–224.

Erwin, D.C. (1977b) *Crop Protection Agents: Their Biological Evaluation,* Vol. 1. Academic Press, New York, pp. 491–504.

Erwin, D.C. (1981) Chemical control. In: Mace, M.E., Bell, A.A. and Beckman, C.H. (eds) *Fungal Wilt Diseases of Plants.* Academic Press, New York, pp. 563–600.

Erwin, D.C. and DeWölfe, T.A. (1968) Lack of control of *Verticillium* wilt of cotton by three systemic insecticides. *Plant Disease Reporter* 52, 281–283.

Erwin, D.C. and Howell, A.B. (1998) *Verticillium* survives heat in Mojave desert alfalfa. *California Agriculture* 52(4), 24–27.

Erwin, D.C. and Khan, R.A. (1993) Registration of six nondormant alfalfa germplasms resistant to *Verticillium* wilt. *Crop Science* 33(6), 1425–1426.

Erwin, D.C., Moje, W. and Malca, I. (1965) An assay of the severity of *Verticillium* wilt on cotton plants inoculated by stem puncture. *Phytopathology* 55, 663–665.

Erwin, D.C., Malca, I. and Harris, R. (1966) Delay in symptoms of *Verticillium* wilt associated with soil treatment by certain growth regulating chemicals. *Proceedings of the 26th Cotton Disease Council,* pp. 132–133.

Erwin, D.C., Mee, H. and Sims, J.J. (1968a) The systemic effect of 1-(butyl carbamoyl)-2-benzimidazole carbamic acid methyl ester on *Verticillium* wilt of cotton. *Phytopathology* 58, 528–529.

Erwin, D.C., Munneche, D.E. and Sims, J.J. (1968b) Thiabendazole and DuPont 1991 for systemic chemical control of *Verticillium* wilt of cotton. *Proceedings of the Beltwide Cotton Production Research Conference,* Hot Springs. p. 49.

Erwin, D.C., Sims, J.J. and Partridge, J. (1968c) Evidence for the systemic fungitoxic activity of 2-(4′-thiazolyl) benzimidazole in the control of *Verticillium* wilt of cotton. *Phytopathology* 58, 860–865.

Erwin, D.C., Garber, R.H., Carter, L. and DeWölfe, T.A. (1969) Studies on thiabendazole and benlate as systemic fungicides against *Verticillium* wilt of cotton in the field. *Proceedings of the Beltwide Cotton Production Research Conference.* New Orleans, Louisiana. p. 29.

Erwin, D.C., Sims, J.J., Borum, D.E. and Childers, J.R. (1971) Detection of the systemic fungicide thiabendazole in cotton plants and soil by chemical analysis and bioassay. *Phytopathology* 61, 964–967.

Erwin, D.C., Khan, R.A. and Buchenauer, H. (1974) Effect of oil emulsions on the uptake of benomyl and thiabendazole in relation to control of *Verticillium* wilt of cotton. *Phytopathology* 64, 485–489.

Erwin, D.C., Tsai, S.D. and Khan, R.A. (1976) Reduction of the severity of *Verticillium* wilt of cotton by the growth retardant tributyl [(5 chloro-2-thienyl) methyl] phosphonium chloride. *Phytopathology* 66, 106–110.

Erwin, D.C., Tsai, S.D. and Khan, R.A. (1978) Reduced number of microsclerotia formed by *Verticillium dahliae* in cotton tissue exposed to systemic benzimidazole fungicides and desiccation. *Phytopathology* 68, 1488–1494.

Erwin, D.C., Tsai, S.D. and Khan, R.A. (1979) Growth retardants mitigate *Verticillium* wilt. *Phytopathology* 69, 283–287.

Erwin, D.C., Khan, R.A. and Baameur, A. (1988) *Verticillium* wilt of alfalfa in southern California caused by *Verticillium albo-atrum*. *Plant Disease* 72, 453.

Erwin, D.C., Khan, R.A., Howell, A., Baameur, A. and Orloff, S.B. (1989) *Verticillium* wilt found in southern California alfalfa. *California Agriculture* 43(5), 12–14.

Esentepe, M. (1974) Investigations on determination of the cotton wilt disease agent and its distribution, severity, loss degree and ecology in Adana and Antalya provinces. *Journal of Turkish Phytopathology* 3, 29–38.

Esentepe, M., Karcilioglu, A. and Sezgin, E. (1972) The first report of *Verticillium* wilt on sesame and okra. *Journal of Turkish Phytopathology* 1, 127–129.

Esentepe, M., Karcilioglu, A., Sezgin, E. and Onan, E. (1986) Investigations on relation between the severity of cotton wilt disease (*Verticillium dahliae* Kleb.) and yield loss in Aegean region. *Journal of Turkish Phytopathology* 15, 27–36.

Eshita, S.M., Roberto, N.H., Beale, J.M., Mamiya, B.M. and Workman, R.F. (1995) Bacillomycin L_c, a new antibiotic of the iturin group: isolations, structures, and antifungal activities of the congeners. *Journal of Antibiotics* 48(11), 1240–1247.

Eshtiaghi, H. (1975) Effects of the northern root-knot nematode (*Meloidogyne hapla* Chitwood, 1949) on Mitcham peppermint (*Mentha piperita* L.) and Scotch spearmint (*Mentha cardiaca* Baker). *Dissertation Abstracts International* 35B, 5734–5735.

Evans, G. (1968) Infection of *Xanthium pungens* by seed-borne *Verticillium dahliae*. *Plant Disease Reporter* 52, 976–978.

Evans, G. (1971a) Influence of weed hosts on the ecology of *Verticillium dahliae* in newly cultivated areas of the Namoi Valley, New South Wales. *Annals of Applied Biology* 67, 169–175.

Evans, G. (1971b) The pre-infection phase of diseases caused by *Verticillium dahliae* Kleb. *Proceedings of the First International* Verticillium *Symposium*. Wye College, University of London.

Evans, G. and Gleeson, A.C. (1973) Observations on the origin and nature of *Verticillium dahliae* colonizing plant roots. *Australian Journal of Biological Sciences* 26, 151–161.

Evans, G. and Gleeson, A.C. (1980) An evaluation of the sampling variation when estimating the population of *Verticillium dahliae* in field soil. *Annals of Applied Biology* 95, 177–184.

Evans, G. and McKeen, C.D. (1975a) Influence of crops on numbers of microsclerotia of *Verticillium dahliae* in soils and development of wilt in South-western Ontario. *Canadian Journal of Plant Science* 55, 827–834.

Evans, G. and McKeen, C.D. (1975b) A strain of *Verticillium dahliae* pathogenic to sweet pepper in south-western Ontario. *Canadian Journal of Plant Science* 55, 857–859.

Evans, G., Snyder, W.C. and Wilhelm, S. (1966a) Inoculum increase of the *Verticillium* wilt fungus in cotton. *Phytopathology* 56, 590–594.

Evans, G., Wilhelm, S. and Snyder, W.C. (1966b) Dissemination of the *Verticillium* wilt fungus with cotton seed. *Phytopathology* 56, 460–466.

Evans, G., Wilhelm, S. and Snyder, W.C. (1967) Quantitative studies by plate counts of propagules of the *Verticillium* wilt fungus in cotton fields. *Phytopathology* 57, 1250–1255.

Evans, G., McKeen, C.D., Moore, K.J. and Gleeson, A.C. (1972) A quantitative bioassay for determining low numbers of microsclerotia of *Verticillium dahliae* in naturally infested field soils. *Proceedings of the Thirty-eighth Session of the Canadian Phytopathological Society*. University of Western Ontario, London, Ontario, June 1972. 39, 30.

Evans, G., Keen, C.D. and Gleeson, A.C. (1974) A quantitative bioassay for determining low numbers of microsclerotia of *Verticillium dahliae* in soil. *Canadian Journal of Microbiology* 20, 119–124.

Evans, K. (1982) Interactions between *Globodera pallida, G. rostochiensis* and *Verticillium dahliae* on four potato cultivars. *Nematologica* 28, 143–144.

Evans, K. (1987) The interactions of potato cyst nematodes and *Verticillium dahliae* on early and maincrop potato cultivars. *Annals of Applied Biology* 110, 329–339.

Evans, W.D. (1975) Centennial everbearing strawberry. *Canadian Journal of Plant Science* 55, 647.

Ezruch, Ė.N. and Babushkina, I.N. (1973) [Some physiological features of pathogenic species of the genus *Verticillium*]. Микология и Фитопатология [*Mycology and Phytopathology*] 7, 536–539.

Ezrukh, Ė.N. (1978) [Interaction between actinomycetes and fungi of the genus *Verticillium* Wallr.]. Микология и Фитопатология [*Mycology and Phytopathology*] 12, 443–445.

Ezrukh, E.N. and Strelkova, L.A. (1979) [Interactions between bacteria and pathogenic fungi of the genus *Verticillium* Wallr.]. Микология и Фитопатология [*Mycology and Phytopathology*] 13, 283–286.

Faedi, W. (1983) Fragola: orietamento varietale per gli ambienti di montagna. *Rivista di Frutticoltura e di Ortofloricoltura* 45(12), 43–46.

Faedi, W., Rosati, P. and D'ercole, N. (1986) Tre nuove cultivar di fragola: Etna, Ferrara e Gea. *Informatore Agrario* 42(6), 41–43.

Fahima, T. and Henis, Y. (1995) Quantitative assessment of the interaction between the antagonistic fungus *Talaromyces flavus* and the wilt pathogen *Verticillium dahliae* on eggplant roots. *Plant and Soil* 176(1), 129–137.

Farley, J.D., Wilhelm, S. and Snyder, W.C. (1971) Repeated germination and sporulation of microsclerotia of *Verticillium albo-atrum* in soil. *Phytopathology* 61, 260–264.

Farrar, J.J., Davis, R.M., Canevari, W.M. and Fouche, C.F. (2000) First report of *Verticillium dahliae* on stevia (*Stevia rebaudiana*) in North America. *Plant Disease* 84(8), 922.

Fassuliotis, G. and Dukes, P.D. (1972) Disease reactions of *Solanum melongena* and *S. sisymbrifolium* to *Meloidogyne incognita* and *Verticillium albo-atrum*. *Journal of Nematology* 4, 222–223.

Faulkner, L.R. and Skotland, C.B. (1963) Control of *Verticillium* wilt of mint and plant parasitic nematodes associated with mint by soil fumigation. *Plant Disease Reporter* 47, 662–665.

Faulkner, L.R. and Skotland, C.B. (1965) Interactions of *Verticillium dahliae* and *Pratylenchus minyus* in *Verticillium* wilt of peppermint. *Phytopathology* 55, 583–586.

Faulkner, L.R., Bolander, W.J. and Skotland, C.B. (1970) Interaction of *Verticillium dahliae* and *Pratylenchus minyus* in *Verticillium* wilt of peppermint: influence of the nematode as determined by a double root technique. *Phytopathology* 60, 100–103.

Fedorinchik, N.S. (1964) [Biological methods of controlling plant diseases]. *Trudy Vsesoyuznogo Nauchno-Isseldovatel'skogo Instituta Zashchity Rastenii* 23, 201–210.

Fedotova, T.I. and Guseva, N.N. (1970) [Induced resistance of cotton to wilt]. Микология и Фитопатология [*Mycology and Phytopathology*] 4, 512–516.

Felix, E.L. (1955) Some tree diseases in Tennessee. *Plant Disease Reporter* 39, 882.

Fellows, H. (1928) Some chemical and morphological phenomena attending infection of the wheat plant by *Ophiobolus graminis*. *Journal of Agricultural Research* 37, 647–661.

Fernandes, N.G., Tokeshi, H. and Vencovsky, R. (1975) *Verticillium albo-atrum* e *Pseudomonas solanacearum* provocando proteção cruzada em tomateiro. *Cientifica* 3, 55–68.

Fernandez, M.C. and Tobar, C.G. (1989) Identifacion de *Verticillium dahliae* Kleb en alcachofas *Cynara scolymus*. *Agricultura Técnica Santiago de Chile* 49, 161–163.

Ferrari, V., Acciarri, N. and Ficcadenti, N. (1990) Polluce 88–083: a new tomato hybrid with indeterminate plant development, medium size and firmness of fruit for fresh market. In: Cuatero, J., Gomez-Guillamon, M.L. and Fernandez-Munoz, R.J. (eds) *Proceedings of the XIth Eucarpia Meeting on Tomato Genetics and Breeding*. Malaga, Spain, pp. 149–152.

Ferreira, J.F., van Der Merwe, P.C. and Naude, S.P. (1990) First report of race 2 of *Verticillium dahliae* on tomatoes in South Africa. *Plant Disease* 74, 530.

Fick, G.N., Kiman, M.L. and Zimmer, D.E. (1979a) Registration of HA 290 sunflower parental line (Reg.No. PL 10). *Crop Science* 19, 422.

Fick, G.N., Zimmer, D.E. and Zimmerman, D.C. (1979b) Registrtion of nine sunflower parental lines (Reg.No. PL 17 to PL 25). *Crop Science* 19, 423.

Filatova, O.F., Denisov, Yu.M., Khodzhibaeva, S.M. and Stepanichenko, N.N. (1987) [Study of plant host/fungal pathogen interactions by means of a mathematical model]. *Uzbekskii Biologicheskii Zhurnal* 5, 33–37.

Fiola, J.A. and Swartz, H.J. (1994) Inheritance of tolerance to *Verticillium albo-atrum* in raspberry. *HortScience* 29(9), 1071–1073.

Fiori, M. and Carta, C. (1982) L'avvizimento' del cartamo (*Carthamus tinctorius* L.) causato da *Verticillium dahliae* Kleb. *Studi Sassarese III* 29, 435–439.

Fisher, W.D. (1968) Breeding of cotton for tolerance to *Verticillium* wilt. *Proceedings of the Beltwide Cotton Production Research Conference*. Hot Springs. p. 230.

Fitt, B.D.L., Bauers, F., Burhenne, S. and Paul, V.H. (1992) Occurrence of *Verticillium dahliae* on linseed (*Linum usitatissimum*) in the UK and Germany. *Plant Pathology* 41, 86–90.

Fitt, B.D.L., Cook, A.J. and Harold, J.F.S. (1998) Effects of *Verticillium dahliae* on linseed (*Linum usitatissimum*) crops. *Bulletin OILB/SROP* 21(5), 57–63.

Fitzell, R., Evans, G. and Fahy, P.C. (1980a) Studies on the colonization of plant roots by *Verticillium dahliae* Klebahn with use of immunofluorescent staining. *Australian Journal of Botany* 28, 357–368.

Fitzell, R., Fahy, P.C. and Evans, G. (1980b) Serological studies on some Australian isolates of *Verticillium* spp. *Australian Journal of Biological Sciences* 33, 115–124.

Fletcher, J.T. and Griffin, M.J. (1972) *Verticillium* wilt of pelargoniums. *Plant Pathology* 21, 158–162.

Flocco, C.G., Alvarez, M.A. and Giulietti, A.M. (1998) Peroxidase production *in vitro* by *Armoracia lapathifolia* (horseradish)-transformed root cultures: effect of elicitation on level and profile of isoenzymes. *Biotechnology and Applied Biochemistry* 28, 33–38.

Flood, J. and Milton, J.M. (1982) Studies on the incompatible and compatible interactions between lucerne leaves and *Verticillium albo-atrum*. *Physiological Plant Pathology* 21, 97–104.

Flood, J., Isaac, I. and Milton, J.M. (1978a) Reactions of some cultivars of lucerne to various isolates of *Verticillium albo-atrum*. *Plant Pathology* 27, 166–169.

Flood, J., Khan, F.Z. and Milton, J.M. (1978b) The role of phytoalexins in *Verticillium* wilt of lucerne. *Annals of Applied Biology* 89, 329–332.

Fodale, A.S. and Mule, R. (1999) Prova di lotta chimica al deperimento da *Verticillium dahliae* Kleb. in olivo con diversi fungicidi somministrati per via xilematica. *Informatore Fitopatologico* 49(11), 19–24.

Fodale, A.S., Mule, R. and Briccoli Bati, C. (1999) The antifungal activity of olive oil waste water on isolates of *Verticillium dahliae* Kleb. *in vitro*. In: Metzidakis, I.T. and Voyiatzis, D.G. (eds) *Acta Horticulturae. Proceedings of the Third International Symposium on Olive Growing.* Chania, Crete, Greece, 22–26 September 1997, Vol. 2. 474, 753–756.

Foëx, E. (1923) Une pouriture verticillienne du tubercule de pomme de terre. *Annales des Épiphyties* 9, 121–133.

Foller, I., Henneken, M. and Paul, V.H. (1998) Occurrence of false flax diseases (*Camelina sativa* (L.) CRTZ.) in field trials in Germany in 1995 and 1996. In: Paul, V.H., Kohr, K., Thomas, J. and Dupprich, P.D. (eds) Integrated Control in Oilseed Crops. Proceedings of a Meeting held at Poznan, Poland, 10–12 April 1997. *Bulletin OILB/SROP* 21(5), 65–76.

Fordyce, C. and Green, R.J. (1960) Studies in the host-specificity of *Verticillium albo-atrum* var. *menthae*. *Phytopathology* 50, 635.

Fordyce, C. and Green, R.J. (1964) Mechanisms of variation in *Verticillium albo-atrum*. *Phytopathology* 54, 795–798.

Formigoni, A., Guidi, G. and Tumino, S. (1973) Ulteriori esperienze con funghicide endoterapici impiegati da soli od associati con nematocidi contro la verticilliosi del pomodoro coltivato in serra (*Verticillium dahliae* Kleb.) ed i nematodi galligeni (*Meloidogyne* spp.). *Notiziario sulle Malattie delle Piante* 88/89, 241–246.

Fortnagel, B. and Schlosser, E. (1995) Interference of a fungal lectin with ELISA. *Zeitschrift für Pflanzenkrankheiten und Pflanzenschutz* 102(3), 331–332.

Fowler, M.F. (1937) *Verticillium* wilt of the smoke tree. *Plant Disease Reporter* 21, 10.

Fox, C.C. and Knipe, W.J. (1987) Registration of 'Commandor' alfalfa. *Crop Science* 27, 1305.

Foy, C.D., Webb, H.W. and Jones, J.E. (1981) Adaptation of cotton genotypes to an acid manganese toxic soil. *Agronomy Journal* 73, 107–110.

Francl, L.J., Madden, L.V., Rowe, R.C. and Riedel, R.M. (1987) Potato yield loss prediction and discrimination using preplant population densities of *Verticillium dahliae* and *Pratylenchus penetrans*. *Phytopathology* 77, 579–589.

Francl, L.J., Rowe, R.C., Riedel, R.M. and Madden, L.V. (1988) Effects of three soil types on potato early dying disease and associated yield reduction. *Phytopathology* 78, 159–166.

Francl, L.J., Madden, L.V., Rowe, R.C. and Riedel, R.M. (1990) Correlation of growing season environmental variables and the effect of early dying on potato yield. *Phytopathology* 80, 425–432.

Franco, J. and Bendezú, E. (1985) Estudio del complejo *Verticillium dahliae* Kleb. y *Globodera pallida* Stone y su efecto en el comportamiento de algunos cultivares pervanos de papa. *Fitopatologia* 20, 21–27.

Frank, J.A., Webb, R.E. and Wilson, D.R. (1973) The relationship between *Verticillium* wilt and the pinkeye disease on potatoes. *American Potato Journal* 50, 431–438.

Frank, Z.R. and Krikun, J. (1968) *Verticillium* wilt of groundnut in Israel: screening for varietal resistance. *Israel Journal of Agricultural Research* 18, 83–85.

Frank, Z.R. and Krikun, J. (1969) Evaluation of peanut (*Arachis hypogaea*) varieties for *Verticillium* wilt resistance. *Plant Disease Reporter* 53, 744–746.

Frankowski, J., Berg, G. and Bahl, H. (1998) Mechanisms involved in the antifungal activity of the rhizobacterium *Serratia plymuthica*. In: Duffy, B., Rosenberger, U. and Defago, G. (eds) *Bulletin OILB/SROP. Molecular Approaches in Biological Control.* Delemont, Switzerland, 15–18 September 1997. *IOBC/WPRS*, 21(9), 45–50.

Fravel, D.R. (1989) Biocontrol of *Verticillium* wilt of eggplant and potato. In: Tjamos, E.C. and Beckman, C.H. (eds) *Vascular Wilt Diseases of Plants, Basic Studies and Control. NATO ASI Series H: Cell Biology*, Vol. 28. Springer-Verlag, Berlin, pp. 487–492.

Fravel, D.R. (1996) Interaction of biocontrol fungi with sublethal rates of metham sodium for control of *Verticillium dahliae*. *Crop Protection* 15(2), 115–119.

Fravel, D.R. (1997) Effect of sublethal stresses on microsclerotia of *Verticillium dahliae*. *Proceedings of the Seventh International* Verticillium *Symposium*, Cape Sounion, Athens, Greece. p. 76.

Fravel, D.R. and Adams, P.B. (1986) Estimation of United States and world distribution of *Talaromyces flavus*. *Mycologia* 78, 684–686.

Fravel, D.R. and Marois, J.J. (1986) Edaphic parameters associated with establishment of the biocontrol agent *Talaromyces flavus*. *Phytopathology* 76, 643–646.

Fravel, D.R. and Tjamos, E.C. (1995) Augmentation of performance of *Talaromyces flavus* by combination with sublethal metham sodium or heat. 6th International *Verticillium* Symposium, Dead Sea, Israel. *Phytoparasitica* 23, 67.

Fravel, D.R., Davis, J.R. and Sorensen, L.H. (1986) Effect of *Talaromyces flavus* and metham on *Verticillium* wilt incidence and potato yield 1984–1985. *Biological Culture Tests* 1, 17.

Fravel, D.R., Kim, K.K. and Papavizas, G.C. (1987) Viability of microsclerotia of *Verticillium dahliae* reduced by a metabolite produced by *Talaromyces flavus*. *Phytopathology* 77, 616–619.

Fravel, D.R., Lewis, J.A. and Chittams, J.L. (1995) Alginate prill formulations of *Talaromyces flavus* with organic carriers for biocontrol of *Verticillium dahliae*. *Phytopathology* 85(2), 165–168.

Friebertshauser, G.E. and DeVay, J.E. (1982) Differential effects of the defoliating and nondefoliating pathotypes of *Verticillium dahliae* upon the growth and development of *Gossypium hirsutum*. *Phytopathology* 72, 872–877.

Frisullo, S., Ciccarese, F. and Cirulli, M. (1987) La viola matronale (*Hesperis matronalis* L.) nuovo ospite di *Verticillium dahliae* Kleb. *Informatore Fitopatologico* 37(12), 48.

Frisullo, S., de Corato, U., Lops, F. and Trombetta, N.M. (1995) Le malattie del Kenaf in Basilicata. *Informatore Fitopatologico* 45(1), 37–41.

Frusciante, L., Carputo, D., Biagetti, M. and Ceoloni, C. (1996) Advanced cytogenetic techniques for the transfer of genes for biotic stress resistance in *Solanum* and *Triticum* species. Proceedings of the workshop: Plant breeding for resistance, 6 (Supplemento 1), 219–232.

Fu, Z.-Q., Xia, Z.-J., Wu, A.-M., Yang, Y.-B., Zheng, Q. and Gu, B.-K. (1999a) [Inhibition of mycelial growth and toxin production of *Verticillium dahliae* and growth-promotion of cotton by endophytic bacteria]. *Acta Phytopathologica Sinica* 29(4), 374–375.

Fu, Z.-Q., Xia, Z.-J., Wu, A.-M., Yang, Y.-B., Zheng, Q. and Gu, B.-K. (1999b) [The mechanism for controlling cotton wilt (*Verticillium dahliae*) by endophytic bacteria]. *Jiangsu Journal of Agricultural Sciences* 15(4), 211–215.

Fuchs, A. (1977) Use of systemic fungicides in practice (1). Internal therapy of plants. In: Fuchs, A. (ed.) *Netherlands Journal of Plant Pathology* 287–352.

Fuchs, A. and Drandarevski, C.A. (1973) Wirkungsbreite und Wirkungsgrad von Triforine *in vitro* und *in vivo*. *Zeitschrift für Pflanzenkrankheiten und Pflanzenschutz* 80, 403–417.

Fuentes-Granados, R.G. and Widrlechner, M.P. (1995) Evaluation of *Agastache* and other Lamiaceae species for reaction to *Verticillium dahliae*. *Journal of Herbs, Spices and Medicinal Plants* 3(3), 3–11.

Fujinaga, M., Takehana, Y., Sekiguchi, H., Wada, T. and Arai, Y. (1999) [Control of *Verticillium* yellows in Chinese cabbage by metham-sodium injection into mulched ridge and spraying followed by cultivation]. *Annual Report of the Kanto Tosan Plant Protection Society* 46, 23–26.

Furgał-Wegrzycka, H. (1997) Vegetative compatibility of *Verticillium albo-atrum* Reinke et. Berth. *Acta Microbiologica Polonica* 46(1), 95–103.

Gafoor, A. and Heale, J.B. (1971a) Melanin formation and peroxidase activity in *Verticillium*. *Microbios* 3(10–11), 87–95.

Gafoor, A. and Heale, J.B. (1971b) Near-UV radiation and morphogenesis in *Verticillium*. *Microbios* 3, 131–141.

Gafurov, I.E., Lutsetti, V.S. and Asorov, K.A. (1990) New technique of lessening cotton wilt virulence. *Proceedings of the Fifth International* Verticillium *Symposium*, Leningrad, p. 89.

Gagné, S. and Richard, C. (1982) La verticilliose de la luzerne en Amerique du Nord. *Canadian Journal of Plant Pathology* 4, 47–53.

Gaibullaev, I., Avazkhodzhaev, M.K. and Egamberdiev, A.E. (1975) [Response of chemical mutants of cotton to infection with the causative agent of *Verticillium* wilt]. *Genetika* 11, 47–50.

Gaibullaev, N. (1972) [On the possibility of decreasing wilt infection by grading seeds]. *Sbornik Rabot po Genetike, Selektsii i Semenovodstvy Khlopchatnika* 133–137.

Galanopoulos, N. and Tribe, H.T. (1974) Conidial survival in *Verticillium dahliae*. *Transactions of the British Mycological Society* 63, 85–91.

Galletta, G.J., Draper, A.D., Stiles, H.D. and Swartz, H.J. (1980) 'Scott' strawberry. *HortScience* 15, 541–542.

Galletta, G.J., Draper, A.D. and Swartz, H.J. (1981) 'Allstar' strawberry. *HortScience* 16(6, I), 792–794.

Galperina, M.I. (1990) Genetic effects of some pesticides on *Verticillium* spp. *Proceedings of the Fifth International* Verticillium *Symposium, Leningrad*, p. 101.

Gamliel, A. and Stapleton, J.J. (1995) Improved soil disinfestation by biotoxic volatile compounds generated from solarized, organic-amended soil. *Acta Horticulturae* 129–135.

Gamliel, A., Grinstein, A., Klein, L., Peretz, I. and Katan, J. (1995) Control of *Verticillium* wilt with methyl bromide at reduced dosage using virtually impermeable films. 6th International *Verticillium* Symposium, Dead Sea, Israel, 1994. *Phytoparasitica* 23, 65.

Gamliel, A., Grinstein, A., Peretz, Y., Klein, L., Nachmias, A., Tsror, L., Livescu, L. and Katan, J. (1997) Reduced dosage of methyl bromide for controlling *Verticillium* wilt of potato in experimental and commercial plots. *Plant Disease* 81(5), 469–474.

Gams, W. and Van Zaayen, A. (1982) Contribution to the taxonomy and pathogenicity of fungicolous *Verticillium* species. I. Taxonomy. *Netherlands Journal of Plant Pathology* 88, 57–58.

Gan, L., Lu, J. and Wang, P. (1995) [The relationship between glycoprotein toxin secreted from *Verticillium dahliae* of cotton and its pathogenicity]. *Scientia Agricultura Sinica* 28(2), 58–65.

Gangopadhyay, S. and Kapoor, K.S. (1975) Control of damping off of solanaceous vegetables in Kulu Valley. *Pesticides* 9(5), 40–41.

Gao, Y.-R. and Gu, C.-Y. (2000) [Identification of cotton varieties (lines) resistant to defoliating strain of *Verticillium dahliae* and their utilization]. *Jiangsu Agricultural Sciences* 3, 25–26.

Garas, N.A. and Waiss, A.C. (1986) Differential accumulation and distribution of antifungal sesquiterpenoids in cotton stems inoculated with *Verticillium dahliae*. *Phytopathology* 76, 1011–1017.
Garas, N.A., Wilhelm, S. and Sagen, J.E. (1986) Relationship of cultivar resistance to distribution of *Verticillium dahliae* in inoculated cotton plants and to growth of single conidia on excised stem segments. *Phytopathology* 76, 1005–1010.
Garber, R.H. (1973) *United States Department of Agriculture Publication 1*. ARS-S-19. pp. 69–77.
Garber, R.H. and Houston, B.R. (1966) Penetration and development of *Verticillium albo-atrum* in the cotton plant. *Phytopathology* 56, 1121–1126.
Garber, R.H. and Houston, B.R. (1967) Nature of *Verticillium* wilt resistance in cotton. *Phytopathology* 57, 885–888.
Garber, R.H. and Presley, J.T. (1971) Relation of air temperature to development of *Verticillium* wilt on cotton in the field. *Phytopathology* 61, 204–207.
Garber, R.H., Carter, L., DeVay, J.E. and Miller, J. (1976) The effects of certain cultural practices on the incidence of *Verticillium* wilt of cotton. *Proceedings of the Second International* Verticillium *Symposium*, University of California, Berkeley. p. 18.
Garber, R.H., Hyer, A.H. and Jorgenson, E.C. (1983) Differential disease reaction of cotton cultivars to the *Fusarium* wilt-root-knot nematode complex. *Phytopathology* 73, 796.
García, E., Garcia, R.S. and Téliz Ortiz, D. (1984) Problemas fitopatológicos del aguacatero (*Persea americana* Mill.) en Tetela del Volcán, Morelos. *Agrociencia* 56, 129–149.
Gardner, R.G. (1982) NC50–7 breeding line, 'Cherokee', and 'Mountain Pride' tomato. *HortScience* 17, 92–93.
Gardner, R.G. (1993a) 'Mountain Gold' tomato. *HortScience* 28(4), 348–349.
Gardner, R.G. (1993b) 'Mountain Belle' cherry tomato; NC 1C and NC 2C cherry tomato breeding lines. *HortScience* 28(4), 349–350.
Garg, V.K., Ram Partap, Relan, P.S. and Girdhar, K.K. (1995) Toxicity of transition metal complexes to the growth of plant pathogenic fungi. *Haryana Agricultural University Journal of Research* 25(3), 87–89.
Garibaldi, A. (1976) Sull' impiego per via fogliare dei benzimidazolici contro la verticilliosii del pomodoro. *Colture Protette* 5(6/7), 41–43.
Garibaldi, A. and Gullino, G. (1973) Malattie nuove o poco note delle piante da fiore e ornamentali in Italia. *Notiziario sulle Malattie delle Piante* 88/89, 53–71.
Garibaldi, A. and Gullino, G. (1976) Malattie delle piante da fiore ed ornamentali nuove o poco note in Italia. III. Malattie radicali e vascolari. *Agricoltura Italiana* 105, 273–284.
Garibaldi, A. and Lamonarca, F. (1974) Ulteriori studi sull'impiego di fungicidi sistemici contro la verticilliosi del pomodoro e del peperone coltivati in serra. Proceedings of the First National Conference on Methods of Glasshouse Cultivation of Market Garden Plants. Ragusa, 22–24 November, 1971. *Tecnica Agricola, Italy* 26, 757–765.
Garibaldi, A., Rapetti, S., Aloi, C., Gullino, M.L. and Bozzano, G. (1990) Malattie delle piante de fiore e ornamentali recentemente osservate in Italia. *Informatore Fitopatologico* 40(5), 9–15.
Garofalo, F. (1970) Una tracheomicosi dell'assenzio (*Artemisia absinthium* L.) in Piemonte e prove orientative di lotta. *Phytopathologia Mediterranea* 9, 179–181.

Garrett, S.D. (1956) *Biology of Root-infecting Fungi*. Cambridge University Press, Cambridge.

Garrett, S.D. (1960) Inoculum potential. In: Horsfall, J.G. and Dimond, A.E. (eds) *Plant Pathology. An Advanced Treatise,* Vol. 3. Academic Press, New York, pp. 23–57.

Gaudreault, S.M., Powelson, M.L., Christensen, N.W. and Crowe, F.J. (1995) Soil water pressure and *Verticillium dahliae* interactions on potato. *Phytopathology* 85(12), 1542–1546.

Gavrilenko, L.A., Goantsa, I.K. and Kropis, E.P. (1977) [*Verticillium* wilt of plum]. *Sadovodstvo, Vinogradarstvo i Vinodelie Moldavii* 1, 42–46.

Gazikhodzhaeva, M.A. and Becker, Z.E. (1968) [Adaptation of antagonistic fungi in the rhizosphere of cotton and their influence on yield and *Verticillium* wilt]. *Uzbekskii Biologicheskii Zhurnal* 12(3), 24–28.

Gazizov, Kh.G., Magrupova, D.A., Khakimova, V.K. and Mukhamedzhanov, M.A. (1975) [Application of copper-containing ammophos to cotton]. *Khimiya v Sel'skom Khozvaĭstve* 13(9), 29–30.

Gazquez, P., Sierra, J.M. and González Ruíz, R.G. (1998) Aplicación experimental de *Verticillium dahliae* WCS850, en el control preventivo de la grafiosis del olmo, *Ophiostoma novo-ulmi*. *Ecologia Madrid* 12, 319–331.

Gennari, M., Gentile, I.A. and Cugudda, L. (1987) Activity of a sesquiterpene-γ-lactone against phytopathogenic fungi. *Zeitschrift für Pflanzenkrankheiten und Pflanzenschutz* 94, 68–73.

Gennari, S., D'ercole, N., Mirotti, A., Fantino, M.G. and Baldazzi, D. (1997) Vegetative compatibility groupings among strains of *Verticillium dahliae* Kleb. isolated from vegetable crops. *Proceedings of the Seventh International* Verticillium *Symposium*. Cape Sounion, Athens, Greece. p. 29.

Gent, M.P., Ferrandino, F.J. and Elmer, W.H. (1995) Effect of *Verticillium* wilt on gas exchange of entire eggplants. *Canadian Journal of Botany* 73(4), 557–565.

Gent, M.P.N., Lamondia, J.A., Ferrandino, F.J., Elmer, W.H. and Stoner, K.A. (1999) The influence of compost amendment or straw mulch on the reduction of gas exchange in potato by *Verticillium dahliae* and *Pratylenchus penetrans*. *Plant Disease* 83, 371–376.

Georgescu, C.C., Orenschi, S. and Petrescu, M. (1959) [Investigations of a verticilliosis of *Quercus cerris*]. *Omagiu lui T. Săvulescu* 39, 743–744.

Gerik, J.S. and Huisman, O.C. (1985) Mode of colonization of roots by *Verticillium* and *Fusarium*. In: Parker, C.A., Rovira, A.D., Moore, K.G., Wong, P.T. and Kollmorgen, J.F. (eds) *Ecology and Management of Soil-borne Plant Pathogens*. American Phytopathological Society, St Paul, Minnesota, pp. 80–83.

Gerik, J.S. and Huisman, O.C. (1988) Study of field grown cotton roots infected with *Verticillium dahliae* using an immunoenzymatic staining technique. *Phytopathology* 78, 1174–1178.

Gerik, J.S., Lommel, S.A. and Huisman, O.C. (1987) A specific serological staining procedure for *Verticillium dahliae* in cotton root tissue. *Phytopathology* 77, 261–265.

Gerlach, W. and Franz, W. (1973) *Verticillium*-Welke und *Thielaviopsis*-Wurzelfaule, zwei bisher unbekannte Krankheiten des Arznei-Baldrians (*Valeriana officinalis* L.). *Phytopathologische Zeitschrift* 76, 172–178.

Gerstel, D.V. (1950) Is resistance to *Verticillium* wilt of guayule related to chromosome number? *Agronomy Journal* 42, 310–311.

Gersti, Z., Mingelgrin, N., Krikun, J. and Yaron, B. (1977) Behaviour and effectiveness of Vapam applied to soil in irrigation water. Proceedings of the Israel France Symposium on Behaviour of Pesticides in Soil. *Special Publication Agricultural Research Organ* 82, 42–50.

Gilad, Z. and Borochov, A. (1993) Hot-water treatment of *Liatris* tubers. *Scientia Horticulturae* 56(1), 61–69.

Gilbert, R.G. and Griebel, G.E. (1969) The influence of volatile substances from alfalfa on *Verticillium dahliae* in soil. *Phytopathology* 59, 1400–1403.

Gilbert, R.G. and Peaden, R.N. (1988) Dissemination of *Verticillium albo-atrum* in alfalfa by internal seed inoculum. *Canadian Journal of Plant Pathology* 10, 73–77.

Gilles, G.L. (1971) Control of *Verticillium* wilt on strawberry. *Proceedings of the First International* Verticillium *Symposium*. Wye College, University of London.

Gil-Ortega, R., Luis-Arteaga, M. and Palazon-Espanol, C. (1986) 'Luesia' (INIA 225), a selected pepper cultivar for processing. *Capsicum Newsletter* 5, 22.

Gindrat, D., Vallotton, R. and Neury, G. (1973) Essais de desinfection chimique du sol en culture maraichère. *Revue Suisse de Viticulture, Arboriculture, Horticulture* 5, 149–156.

Gindrat, D., Ducrot, V. and Caccia, R. (1976) Lutte contre la verticilliose de l'aubergine par greffage sur tomates resistantes. *Revue Suisse de Viticulture, Arboriculture, Horticulture* 8, 71–76.

Ginoux, G. (1974) Bilan de quatre années d'experimentation sur le greffage de solanacées dans le Sud-Est. In: Beyries, A. (ed.) *Pepinieristes Horticulteurs Maraichers* 152, 35–54.

Ginoux, G. and Dauple, P. (1982) Un moyen de lutte contre la verticilliose de l'aubergine. *Revue Horticole* 230, 17–40.

Gkilpathi, D. and Fitt, B.D.L. (1999) Pathogenicity of *Verticillium dahliae* to spring linseed. In: Barrow, A., Biddle, A., Cook, S., Knott, C.M., Lainsbury, M.A. and Walker, K. (eds) *Aspects of Applied Biology*. Protection and Production of Combinable Break Crops, Royal Agricultural College, Cirencester, UK, 14–15 December 1999. 56, 173–176.

Gladkikh, T.A., Davydova, M.A., Mddosanova, G.A., Vasil'eva, K.V. and Metlitskii, L.V. (1979) Effect of pectin-trans-eliminase preparations of *Verticillium dahliae* Klebahn on cotton tissues and plants. *Prikl. Biokhimiya i Mikrobiol.* 15(2), 194–206.

Glaser, T. (1977) [The disease complex (CVFN) on greenhouse tomatoes and present-day possibilities of its control]. *Ochrona Roslin* 21(6), 13–16.

Globus, G.A. and Muromtsev, G.S. (1990) The use of *Gliocladium roseum* as antagonist for defence of cotton from phytopathgenic fungi. *Proceedings of the Fifth International* Verticillium *Symposium*. Leningrad. p. 90.

Godoy Avila, S., Palomo Gil, A., Hernandez Hernandez, V. and Garcia Castaneda, E.A. (1994) Comportamiento de tres nuevas variedades de algodonero, *Gossypium hirsutum* L., en suelos infestados por *Verticillium dahliae* K. *Agricultura Técnica en México* 20(2), 133–146.

Godoy Avila, S., Palomo Gil, A. and Garcia, C.E.A. (1995) Performance of new cotton cultivars on *Verticillium dahliae* K. infested soils at the Comarca Lagunera, Mexico. *Proceedings of the Beltwide Cotton Conferences*. San Antonio, Texas, 4–7 January 1995. Vol. 1, 498–500.

Goethal, M. (1971) *Verticillium dahliae* Kleb. agent d'une trachcomycose du carthame au Maroc. *Awamia* 39, 39–54.

Goetz, G. (1935) Einiges über die Johannesbeer-Welkekrankheit. *Obst-und Gemüseb.* 81(11), 172.

Goheen, R.C. (1949) *Verticillium* wilt on apricots. *Plant Disease Reporter* 33, 99.

Goidanich, G. (1934a) La moria degli Aceri. *Italia Agricultura* 71, 1043–1055.

Goidanich, G. (1934b) La verticilliosi dell' *Acer platanoides* L., dell *A. pseudoplatanus* L. e della *Maclura aurantiaca* L. *Bollettino della Roma Stazione di Patologia Vegetale* NS 14, 268–272.

Goidanich, G. (1935) Nuovi casi di tracheomicosi da *Verticillium* in Italia. Osservazioni su una specie nuova di *Verticillium* tracheicolo. *Bollettino della Roma Stazione di Patologia Vegetale* NS 15, 548–554.

Gold, J. and Robb, J. (1995a) Possible resistance to *Verticillium dahliae* Race 2 in tomato. 6th International *Verticillium* Symposium, Dead Sea, Israel. *Phytoparasitica* 23, 61.

Gold, J. and Robb, J. (1995b) The role of the coating response in Craigella tomatoes infected with *Verticillium dahliae*, races 1 and 2. *Physiological and Molecular Plant Pathology* 47(3), 141–157.

Gomaa, E.A.A. and Matolcsy, G. (1974) Leaching and decomposition of sazol in Hungarian soil. *Indian Journal of Agricultural Sciences* 44, 181–183.

Gondran, J. (1976) Répartition en France de la verticilliose de la luzerne. *Annals of Phytopathology* 8, 203–212.

Gondran, J. (1977) Facteurs du développement de *Verticillium albo-atrum* Rke. et Berth. dans les luzernières. *Annales de Phytopathologie* 9, 473–485.

Gondran, J. (1984) Stability of the *Verticillium* wilt resistance of lucerne cultivars. *Phytopathologia Mediterranea* 23, 204.

Gonzalez-Salan, M.M. and Bosland, P.W. (1993) Enhanced *Capsicum* germplasm resistant to *Verticillium* wilt. *Capsicum & Eggplant Newsletter* 85–87.

Goodell, J.J., Powelson, M.L. and Allen, T.C. (1982) Interrelations between potato virus X, *Verticillium dahliae* and *Colletotrichum atramentarium* in potato. *Phytopathology* 72, 631–634.

Gordee, R.S. and Porter, C.L. (1961) Structure, germination and physiology of microsclerotia of *Verticillium albo-atrum*. *Mycologia* 53, 171–182.

Gordon, S.E. and Paleg, L.G. (1961) Formation of auxin from tryptophan through action of polyphenols. *Plant Physiology Lancaster* 36, 838–845.

Gordon, T.R., Correll, J.C., Gilchrist, D.G. and Martensen, A.N. (1989) *Verticillium* wilt of alfalfa in California. *Plant Disease* 73, 18–20.

Gordon, T.R., Duniway, J.M., Paulus, A.O., Vilchez, M. and Westerlund, F. (1997) The effect of chemical fumigants and cover crops on soilborne pathogens of strawberries in California. *Proceedings of the Seventh International* Verticillium *Symposium*. Cape Sounion, Athens, Greece, p. 82.

Gorkovtseva, E. and Azizova, Z. (1997) [Interaction of callus tissue with *Verticillium dahliae*]. *Proceedings of the Seventh International* Verticillium *Symposium*. Cape Sounion, Athens, Greece, p. 97.

Gossen, B.D. and Jesperson, G.D. (1990) *Verticillium* wilt of irrigated alfalfa in Saskatchewan, 1987–1989. *Canadian Plant Disease Survey for 1990* 70(2), 129–131.

Goth, R.W. and Haynes, K.G. (2000) Evaluation of potato clones for severity of *Verticillium* wilt, yield and specific gravity in Maine. *American Journal of Potato Research* 77(2), 89–94.

Goth, R.W. and Webb, R.E. (1981) Sources and genetics of host resistance in vegetable crops. In: Mace, M.E., Bell, A.A. and Beckman, C.H. (eds) *Fungal Wilt Diseases of Plants*. Academic Press, New York, pp. 377–411.

Goth, R.W., Haynes, K.G. and Wilson, D.R. (1993) Relationship of *Verticillium* wilt with pink-eye of potato in Maine. *Plant Disease* 77(4), 402–405.

Goth, R.W., Haynes, K.G. and Wilson, D.R. (1994a) Independent segregation in potato for resistance to *Verticillium* wilt and pink-eye. *Plant Disease* 78(6), 562–564.

Goth, R.W., Haynes, K.G. and Wilson, D.R. (1994b) *Verticillium* wilt resistant germplasm: release of Russet clone B0169-56. *American Potato Journal* 71(11), 735–742.

Götz, J., Pett, B. and Bulnheim, U. (1982) Zur Diagnose der wichtigsten Knollenfäulen der Kartoffel. *Nachrichtenblatt für den Pflanzenschutzdienst in der DDR* 36, 184–187.

Goud, J.C. and Termorshuizen, A.J. (1997) Morphological characteristics of *Verticillium dahliae* and *V. tricorpus* on selective media for the detection of *Verticillium dahliae* in soil. *Proceedings of the Seventh International* Verticillium *Symposium*. Cape Sounion, Athens, Greece, p. 36.

Gould, M.D. (1974) The root lesion nematode – *Verticillium* wilt disease complex of potatoes and pathogenicity studies of the lesion nematode on selected ornamentals and cover crops. *Disease Abstracts International* 35, 783.

Gour, H.N. (1987) Isolation and partial purification of toxins produced by *Verticillium dahliae*. *Canadian Journal of Plant Pathology* 9, 80–81.

Gour, H.N. and Dube, H.C. (1982) Utilization of carbohydrates by *Verticillium albo-atrum* and *V. dahliae*. *Indian Journal of Mycology and Plant Pathology* 12, 55–56.

Gour, H.N. and Dube, H.C. (1985) Effects of ouabain and phytotoxic metabolites from *Verticillium dahliae* on the cell membranes of cotton plants. *Physiological Plant Pathology* 27, 109–118.

Gourley, C.O. (1979) *Verticillium dahliae* from stunted plants of summer savory. *Canadian Plant Disease Survey* 59, 1, 18.

Govorova, G.F. (1979) [Results of an immunity evaluation of the world tomato collection]. *Trudy po Prikladnoi Botanike, Genetike i Selektsii* 64, 66–69.

Govorova, G.F. (1983) [Sources of resistance in strawberry breeding]. Микология и Фитопатология [*Mycology and Phytopathology*] 17, 213–218.

Govorova, G.F. (1985) [New strawberry varieties with combined resistance to fungal pathogens]. *Sbornik Nauchnykh Trudov po Prikladnoĭ Botanike, Genetike i Selektsii* 97, 96–100.

Govorova, G.F. (1989) [*Verticillium* resistance of tomato varieties from the collection and commercial varieties]. *Sbornik Nauchnykh Trudov po Prikladnoi Botanike, Genetike i Selektsii* 123, 106–111.

Govorova, G.F. and Govorov, D.N. (1997a) Species and race specialization of *Verticillium* fungi in the former Soviet Union: report of a long-term study. *Proceedings of the Seventh International* Verticillium *Symposium*. Cape Sounion, Athens, Greece, p. 55.

Govorova, G.F. and Govorov, D.N. (1997b) Methodological bases of strawberry breeding for resistance to *Verticillium* wilt. *Proceedings of the Seventh International* Verticillium *Symposium*. Cape Sounion, Athens, Greece, p. 58.

Govorova, G.F., Khrustaleva, V.V. and Shcherbakov, V.K. (1987) Mutant lines of currant tomato, valuable germplasm with multiple disease resistance. *Mutation Breeding Newsletter* 30, 8–10.

Graham, J.H., Peaden, R.N. and Evans, D.W. (1977) *Verticillium* wilt of alfalfa found in the United States. *Plant Disease Reporter* 61, 337–340.

Graham, J.H., Stuteville, D.L., Frosheiser, F.I. and Erwin, D.C. (1979) *A Compendium of Alfalfa Diseases*. American Phytopathological Society, St Paul, Minnesota.

Gram, E. and Rostrup, S. (1925) Oversigt over Sygdomme hos Landbrugets ov Havebrugets Kulturplanter i 1924. *Tidsskrift for Planteavl* 31, 353–417.

Granges, A. (1981) Résultats de quatre années d'essais de variétés de tomates pour leur résistance à la verticilliose et au 'corky-root'. *Revue Suisse de Viticulture, Arboriculture, Horticulture* 13, 211–216.

Grasso, S. and La Rosa, R. (1982) *Verticillium* wilt of *Peperomia* in Italy. *Plant Disease* 66, 960.

Grasso, S. and Pacetto, M. (1969) Una nuova tracheoverticlliosi (*Verticillium dahliae* Kleb.) della poinsettia (*Euphorbia pulcherrima* Willd.) in Sicilia. *Tecnica Agricola, Catania* 31(4), 3–6.

Grasso, S. and Tirro, A. (1982) Primi riseltati sull'effecto della preinoculazione di *Verticillium dahliae* in piante di arancio amaro inoculate con *Phoma tracheiphila. Tecnica Agricola* 34, 179–186.

Grau, C.R., Delwiche, P.A., Norgren, R.L., O'connell, T.E. and Maxwell, D.P. (1981) *Verticillium* wilt of alfalfa in Wisconsin. *Plant Disease* 65, 843–844.

Gravatt, G.F. (1926) Maple wilt. *US Department of Agriculture Circular* 382, 1–13.

Gray, F.A. and Page, M.S. (1988) Assessing loss in alfalfa from *Verticillium* wilt. In: Moutray, J.B. and Elgin, J.H. (eds) *Report of the Thirty-first North American Alfalfa Improvement Conference*. Beltsville, Maryland, 19–23 June 1988. p. 21.

Gray, F.A. and Roth, D.A. (1982) *Verticillium* wilt of alfalfa in Wyoming. *Plant Disease* 66, 1080.

Green, R.J. (1954) A preliminary investigation of toxins produced *in vitro* by *Verticillium albo-atrum. Phytopathology* 44, 319–326.

Green, R.J. (1958) Deep ploughing for controlling *Verticillium* wilt of muck soils. *Phytopathology* 48, 575–577.

Green, R.J. (1964) Control of *Verticillium* wilt of peppermint by soil fumigation in muck soils. *Plant Disease Reporter* 48, 960–963.

Green, R.J. (1967) Control of *Verticillium* wilt of peppermint by crop rotation sequence. *Plant Disease Reporter* 51, 449–453.

Green, R.J. (1969) Survival and inoculum potential of conidia and microsclerotia of *Verticillium albo-atrum* in soil. *Phytopathology* 59, 874–876.

Green, R.J. (1971) Factors affecting survival of microsclerotia of *Verticillium dahliae* in soil. *Proceedings of the First International* Verticillium *Symposium*. Wye College, University of London.

Green, R.J. (1976) Effect of various physical factors of the soil environment on survival of microsclerotia of *Verticillium dahliae*. *Proceedings of the Second International* Verticillium *Symposium*. University of California, Berkeley.

Green, R.J. (1977) Alteration of pathogenicity of *Verticillium dahliae* from *Mentha* sp. under field conditions. *Plant Disease Reporter* 61, 373–374.

Green, R.J. (1980) Soil factors affecting survival of *Verticillium dahliae. Phytopathology* 70, 353–355.

Green, R.J. (1981) An overview. In: Mace, M.E., Bell, A.A. and Beckman, C.H. (eds) *Fungal Wilt Diseases of Plants*. Academic Press, New York, pp. 1–24.

Green, R.J. and Papavizas, G.C. (1968) The effect of carbon source, carbon to nitrogen ratios and organic amendments on survival of propagules of *Verticillium albo-atrum* in soil. *Phytopathology* 58, 567–570.

Green, R.J. and Wiley, G.L. (1987) *Verticillium dahliae* as a biocontrol agent of velvet leaf *Abutilon theophrasti. Canadian Journal of Plant Pathology* 9, 81.

Greenberger, A., Yogev, A. and Katan, J. (1987) Induced suppressiveness in solarised soils. *Phytopathology* 77, 1663–1667.

Greig, B.J.W., Lonsdale, D., Mitchell, A.G. and Gibbs, J.N. (1982) *Report on Forest Research for the Year Ended March 1982.* Forestry Commission, HMSO, London.

Griffen, A.M., Heale, J.B. and Bainbridge, B.W. (1994) Cloning and characterization of the ribosomal RNA gene complex from the plant pahogen *Verticillium albo-atrum. FEMS Microbiology Letters* 118, 291–296.

Griffen, A.M., Bainbridge, B.W. and Heale, J.B. (1997) Ribosomal, mitochondrial and amplified DNA polymorphisms in *Verticillium albo-atrum* pathogenic to hops, lucerne and other plants. *Mycological Research* 101(9), 1085–1091.

Griffin, G.J. and Roth, D.A. (1979) Nutritional aspects of soil mycostasis. In: Schippers, B. and Gams, W. (eds) *Soil-borne Plant Pathogens. Proceedings of the 4th International Symposium on Factors determining the Behaviour of Plant Pathogens in the Soil.* Munich, Germany, August 1978. Academic Press, New York, pp. 79–96.

Griffin, G.J., Hora, T.S. and Baker, R. (1975) Soil fungistasis: evaluation of the exogenous carbon and nitrogen requirements for spore germination by fungistatic volatiles in soils. *Canadian Journal of Microbiology* 21, 1468–1475.

Griffiths, D.A. (1970) The fine structure of developing microsclerotia of *Verticillium dahliae* Kleb. *Archiv für Mikrobiologie* 74, 207–212.

Griffiths, D.A. (1971a) The fine structure of *Verticillium dahliae* Kleb. colonizing cellophane. *Canadian Journal of Microbiology* 17, 79–81.

Griffiths, D.A. (1971b) The development of lignitubers in roots after infection by *Verticillium dahliae* Kleb. *Canadian Journal of Microbiology* 17, 441–444.

Griffiths, D.A. (1982) Structure of *Verticillium nigrescens* and *V. nubilum* chlamydospores. *Transactions of the British Mycological Society* 78, 141–145.

Griffiths, D.A. and Campbell, W.P. (1971) The fine structure of resting mycelium of *Verticillium albo-atrum* R. & B. *Canadian Journal of Microbiology* 17, 1533.

Griffiths, D.A. and Isaac, I. (1963) Wilt of lupin and sunflower caused by species of *Verticillium. Horticultural Research* 2, 104–114.

Griffiths, D.A. and Isaac, I. (1966a) Colonization of cellulose membranes by species of *Verticillium. Annals of Applied Biology* 57, 59–65.

Griffiths, D.A. and Isaac, I. (1966b) Host/parasite relationships between tomato and pathogenic isolates of *Verticillium. Annals of Applied Biology* 58, 259–272.

Griffiths, D.A. and Lim, W.C. (1964) Mechanical resistance in root hairs to penetration by species of vascular wilt fungi. *Mycopathologia et Mycologia Applicata* 24, 103–112.

Griffiths, D.A. and Lim, W.C. (1966) Intercellular colonization of *Vigna sesquipedalis* and *Hibiscus esculentus* and the production of pectic enzymes by Malayan isolates of *Fusarium. Plant Disease Reporter* 50, 116–118.

Grimes, D.W. and Yamada, H. (1982) Relation of cotton growth and yield to minimum leaf water potential. *Crop Science* 22, 134–139.

Grinstein, A., Katan, J. and Eshel, Y. (1976) Effect of dinitroaniline herbicides on plant resistance to soil-borne pathogens. *Phytopathology* 66, 517–522.

Grinstein, A., Orion, D., Greenberger, A. and Katan, J. (1978) Solar heating of the soil for the control of *Verticillium dahliae* and *Pratylenchus thornei* in potatoes. *Third International Congress of Plant Pathology.* Münich, Germany, 16–23 August 1978. p. 373.

Grinstein, A., Orion, D., Greenberger, A. and Katan, J. (1979a) Solar heating of the soil for the control of *Verticillium dahliae* and *Pratylenchus thornei* in potatoes. In:

Schippers, B. and Gams, W. (eds) *Soil-borne Plant Pathogens. Proceedings of the 4th International Symposium on Factors Determining the Behaviour of Plant Pathogens in the Soil.* Munich, Germany, August 1978. Academic Press, New York, pp. 431–438.

Grinstein, A., Orion, D. and Katan, J. (1979b) Solar heating for control of *Verticillium dahliae–Pratylenchus thornei* complex disease in potatoes. *Nematropica* 9, 100.

Grinstein, A., Lisker, N., Katan, J. and Eshel, Y. (1981) Trifluralnin – a 'sensitizer' for *Fusarium* resistance in tomatoes. *Phytoparasitica* 9, 235.

Grishechkina, L.D. (1988) [Role of different crops and their post-harvest remnants in preserving *Verticillium* infection in the soil]. Микология и Фитопатология [*Mycology and Phytopathology*] 22, 543–548.

Grishechkina, L.D. (1990) The effect of some agrotechnical methods on the quantity of causative agent of *Verticillium* wilt of cotton. *Proceedings of the Fifth International* Verticillium *Symposium*, Leningrad, p. 26.

Grishechkina, L.D. and Sidorova, S.F. (1984a) [Effect of mineral compounds on leaf infection by cotton wilt pathogen]. Микология и Фитопатология [*Mycology and Phytopathology*] 18, 66–70.

Grishechkina, L.D. and Sidorova, S.F. (1984b) [Comparative studies of different methods for quantitative assessments of *Verticillium dahliae* in the soil]. Микология и Фитопатология [*Mycology and Phytopathology*] 18(2), 153–156.

Grogan, R.G., Ioannou, N., Schneider, R.W., Sall, M.A. and Kimble, K.A. (1979) *Verticillium* wilt on resistant tomato cultivars in California; virulence of isolates from peanuts and soil and the relationship of inoculum density to disease incidence. *Phytopathology* 69, 1176–1180.

Gruenhagen, R.H. and Fordyce, C. (1963) *Osmanthus ilicifolius*, a new host of *Verticillium albo-atrum*. *Plant Disease Reporter* 47, 688.

Grybauskas, A.P. and Dutky, E.M. (1987) First report of *Verticillium* wilt of alfalfa in Maryland. *Plant Disease* 71, 376.

Grzelinska, A. (1980) Isolation and identification of tomato (*Lycopersicon esculentum*) phytoalexins: 2. Isolation of lupeol. *Bulletin Academie Polonaise des Sciences Sciences Biologiques* 28, 293–298.

Grzelinska, A. and Sierakowska, J. (1978) Isolation of rishitin from tomato plants. *Phytopathologische Zeitschrift* 91, 320–321.

Grzybowska, T. (1993) Podatnosc linii i wybranych rodow rzepaku ozimego na porazenie przez Phoma lingam i *Verticillium dahliae*. *Postepy Nauk Rolniczych* 40–45(5), 115 125.

Grzybowska, T. (1996) Poszukiwanie genotypow tolerancyjnych na porazenie przez *Phoma lingam* i *Verticillium dahliae* w rodach hodowlanych rzepaku ozimego. *Rosliny Oleiste* 17(1), 149–160.

Grzybowska, T. and Olechnowicz, D. (1996) Poszukiwanie form odpornych na porazenie przez *Verticillium dahliae* Kleb. w mieszancach podwojnie ulepszonego rzepaku ozimego. *Rosliny Oleiste* 17(1), 137–148.

Gu, K.-Y., Ji, R.-L., Guo, C.-Z., Huang, Z.-Y. and Fei, Y. (1998) [Su Mian 9, a high yielding high quality cotton cultivar]. *Crop Genetic Resources* 1, 54.

Gu, Z., Zhi, Y., Hu, Y. and Zhu, Z. (1995) A study of *Verticillium dahliae* Kleb. on eggplant and biological determination of its toxin. *Journal of Shanghai Agricultural College* 13, 218–221.

Gubanov, G.Ya. (1962) [On the conversion of phenolic substances in cotton infected by wilt]. *Fiziologiya Rastenii* 9, 614–619.

Gubanov, G.Ya., Bredikhina, A.L., Sobirov, B.G. and Shadmanova, M. (1985) [Growth and development of *Verticillium dahliae* Kleb. in the presence of various nitrogen sources in the medium]. Микология и Фитопатология [*Mycology and Phytopathology*] 19, 58–61.
Gubis, V. (1979) [Combined inoculation of Palava alfalfa with some pathogens]. *Pol'nohospodárstvo* 25, 718–726.
Gubis, V. (1982) [Methods of testing lucerne for resistance to bacterial and *Verticillium* wilt]. *Pol'nohospodárstvo* 28, 680–687.
Gubis, V., Zakova, M. and Drbal, J. (1990) [Resistance of lucerne cultivars to vascular wilt and longevity of lucerne]. *Pol'nohospodarstvo* 36, 520–526.
Gubler, W.D., Grogan, R.G. and Greathead, A.S. (1978) Wilt of cucumber caused by *Verticillium albo-atrum* in California. *Plant Disease Reporter* 62, 786–788.
Guéguen, M.F. (1906) *Acrostalagmus vilmorinii* n.sp. mucèdinèe preduisant une maladie à sclèrotes du collet des reines-marguerites. *Bulletin Trimestriel de la Société Mycologique de France* 22, 254–265.
Guidi, L., Nali, C., Ciompi, S., Lorenzini, G. and Soldatini, G.F. (1997) Effects of two wilt fungi on leaf gas exchange of asymptomatic leaves of tomato plants. *Advances in Horticultural Science* 11(3), 123–127.
Gullino, G. (1983) Segnalazione di tre nuove malattie di arbusti ornamentale. *Informatore Fitopatologico* 33, 51–53.
Gullino, M.L. and Garibaldi, A. (1996) Diseases of roses: evolution of problems and new approaches for their control. In: Morisot, A. and Ricci, P. (eds) *Acta Horticulturae Second International Symposium on Roses*. Antibes, France, 20–24 February 1995, pp. 195–201.
Gullino, M.L. and Garibaldi, A. (1998) Le malattie della margherita. *Colture Protette* 27(6), 53–55.
Gullino, M.L., Concaro, G. and Garibaldi, A. (1987) Gravi epidemie di verticilliosi su patata in Piedmonte nel 1987. *Informatore Agrario* 43(45), 79–81.
Guo, H., Dong, Z., Lin, Y., Li, Z., Li, J., Huang, G., Cui, S. and Pan, X. (1995) Effect of infection of *Verticillium* wilt on the SOD, POD activities and photosynthetic character in cotton leaves. *Scientia Agricultura Sinica* 28(6), 40–46.
Guo, Y. and Yan, Z.-W. (1999) [Study on the spatial distribution pattern and sampling method of *Verticillium* wilt on eggplant]. *Plant Protection* 25(6), 1–4.
Guozhong, Z., Hengwen, F., Aiguo, L., Zengshu, L., Shuxiang, S. and Zhenglan, L. (1994) Improvement effect of the Upland cotton with eight wild species and two diploid cultivated species. *Acta Agriculturae Boreali-Sinica* 9(4), 44–48.
Gupta, D.P. (1973) Endopolygalacturonase stimulation in *Verticillium albo-atrum*. *Indian Phytopathology* 26, 90–106.
Gupta, D.P. (1974, published 1975) A disc electrophoretic study of the protein of susceptible, tolerant and resistant lucerne with reference to *Verticillium* wilt. *Indian Phytopathology* 27, 542–548.
Gupta, D.P. (1977, published 1978) Purification of *in vitro* pectolytic enzymes secreted by *Verticillium albo-atrum*. *Indian Phytopathology* 30, 315–319.
Gusakova, S.D., Khomova, T.V., Mezhlum'yan, L.G. and Tuldashev, P.Kh. (1994, published 1995) Comparative investigation of the composition of the protease inhibitors of wilt-resistant and wilt-susceptible varieties of cotton. *Chemistry of Natural Compounds* 30(4), 434–438.
Guseinov, V.A. and Runov, V.I. (1971) [A study of the nucleic-acids of certain fungi of the genus *Verticillium*]. *Mikrobiologiya* 40, 112–115.

Guseinov, V.A. and Runov, V.I. (1974) Strain differences in the DNA structure of the fungi *Verticillium dahliae*. *Doklady Akademii Nauk SSSR Ser. Biokhim* 212, 450–452.

Guseinov, V.A., Achilova, G.S., Otroshchenko, O.S. and Sadykov, A.S. (1974) Properties of nucleic acids of *Verticillium dahliae* isolated from varieties of cotton differing in resistance to wilt. *Uzbekskii Biologicheskii Zhurnal* 18, 52–54.

Guseinov, A.M., Babaev, F.A. and Eminbeili, A.G. (1979) [Wilt-resistant cultivars from the Azerbaijan Institute of Cotton Growing]. *Khlopkovodstvo* 9, 35–36.

Guseva, N.N. (1972) [Composition of water-soluble proteins in *Verticillium* mycelium as a taxonomic indicator]. *Byulleten' Vsesoyuznogo Nauchno-Issledovatel'skogo Instituta Zashchity Rastenii* 23, 25–30.

Guseva, N.N., Akhmedzhanov, A.N. and Khalkhodzhaev, T. (1974) [Focal microinfection of wilt-resistant varieties of Tashkent group]. *Byulleten' Vsesoyuznogo Nauchno Issledovatel'skogo Instituta Zashshity Rastenii* 30, 58–61.

Guthrie, J.W. (1960) Early dying (*Verticillium* wilt) of potatoes in Idaho. *Research Bulletin, University of Idaho Agricultural Experiment Station* 45.

Gutierrez, A.P., DeVay, J.E., Pullman, G.S. and Friebertshauser, G.E. (1983) A model of *Verticillium* wilt in relation to cotton growth and development. *Phytopathology* 73, 89–95.

Gutierrez, J.C., Lopez, M. and El-Zik, K.M. (1994) AMMI (additive main effects and multiplicative interactions analysis): a tool to determine adaptability of upland cotton genotypes in Spain. *1994 Proceedings Beltwide Cotton Conferences*. 5–8 January, San Diego, California, pp. 688–689.

Guzhova, N.V., Voronkov, L.A. and Zhivopistseva, I.V. (1976) [Oxidative apparatus of the pathogens of *Fusarium* and *Verticillium* wilt of cotton]. *Patol. Fiziol. i Immunitet Rast.*. 157–164.

Haass, J. and Märtin, B. (1977) Untersuchungen zur Übertragung des Luzernewelkeerregers *Verticillium albo-atrum* Rke. & Berth. innerhalb des Pflanzenbestandes durch Befahren der Luzerne mit Ernteaggregaten und Transportfarhzeugen. *Nachrichtenblatt für den Pflanzenschutzdienst in der DDR* 31, 240–243.

Haensler, C.M. (1922) A new peach disease. *Phytopathology* 12, 56–57.

Haensler, C.M. (1924) Okra wilt (*Mimeo*). *New Jersey Agricultural Experiment Station, Department of Plant Pathology, Plant Disease Notes* 2(7), 4.

Haensler, C.M. (1925) Wilt of *Chrysanthemum*. *Plant Disease Reporter Supplement* 37, 390.

Hafez, A.A.R., Stout, P.R. and DeVay, J.E. (1975) Potassium uptake by cotton in relation to *Verticillium* wilt. *Agronomy Journal* 67, 359–361.

Halila, H.M. and Harrabi, M.M. (1987) Wilt of chickpea in Tunisia caused by *Verticillium albo-atrum*. *Plant Disease* 71, 101.

Halisky, P.M., Garber, R.H. and Schnathorst, W.C. (1959) Influence of soil temperature on *Verticillium* hadromycosis of cotton in California. *Plant Disease Reporter* 43, 584–587.

Hall, B.H., Jones, M.K., Wicks, T.J., Walker, G. and Barth, G. (1996) First report of diseases of *Ixodia achillaeoides* in South Australia. *Australasian Plant Pathology* 25(3), 215.

Hall, R. (1969) *Verticillium albo-atrum* and *Verticillium dahliae* distinguished by acrylamide gel electrophoresis of proteins. *Canadian Journal of Botany* 47, 2110–2111.

Hall, R. (1975) Differential sensitivity of *Verticillium dahliae*, *Verticillium albo-atrum* and *Verticillium nigrescens* to benomyl. *Canadian Journal of Botany* 53, 452–455.

Hall, R. (1986) Effects of root pathogens on plant water relations. *Water, Fungi and Plants* 241–265.

Hall, R. and Busch, L.V. (1971) *Verticillium* wilt of chrysanthemum: colonization of leaves in relation to symptom development. *Canadian Journal of Botany* 49, 181–185.
Hall, R. and Ly, H. (1972a) Development and quantitive measurement of microsclerotia of *Verticillium dahliae*. *Canadian Journal of Botany* 50, 2097–2102.
Hall, R. and Ly, H. (1972b) Nutritional regulation of development of microsclerotia in *Verticillium dahliae*. *Proceedings of the Thirty-eighth Session of the Canadian Phytopathological Society*. University of Western Ontario, 26–28 June, 1972. 39, 31–32.
Hall, R. and MacHardy, W.E. (1981) Water relations. In: Mace, M.E., Bell, A.A. and Beckman, C.H. (eds) *Fungal Wilt Diseases of Plants*. Academic Press, New York, pp. 255–298.
Hall, T.J., Schreiber, L.R. and Leben, C. (1986) Effects of xylem-colonizing *Bacillus* on *Verticillium* wilt of maples. *Plant Disease* 70, 521–524.
Hamdollah-Zadeh, A. (1993) Properties of defoliant and non-defoliant strains of *Verticillium dahliae*, the causal agent of cotton wilt in northern Iran. *Iranian Journal of Plant Pathology* 29(3–4), 53–54 (en), 125–131 (Pe).
Hammerschlag, R.S. and Sisler, H.D. (1973) Benomyl and methyl-2-benzimidazole-carbamate (MBC): biochemical, cytological and chemical aspects of toxicity to *Ustilago maydis* and *Saccharomyces cerevisiae*. *Pesticide Biochemistry and Physiology* 3, 42–54.
Han, S.-J., Han, G.-Q., Cui, S.-Y., Ge, Z.-N. and Leng, S.-F. (1996) A brief introduction to Nantongmian 13. *China Cottons* 23(6), 20.
Hanna, M.R. and Huang, H.C. (1987) Barrier alfalfa. *Canadian Journal of Plant Science* 67, 827–830.
Hansen, H.N. (1938) The dual phenomenon in imperfect fungi. *Mycologia* 30, 442–455.
Hansen, H.R. and Weber, A. (1948) Plant diseases and pests in Denmark in 1945. *Tidsskrift for Planteavl* 51, 434.
Hansford, C.G. (1939) Report of the Senior Plant Pathologist. *Report of the Department of Agriculture, Uganda, for 1937–1938*, pp. 20–25.
Hanson, A.A., Kugler, J.L., Huset, D.E., Peterson, M.A., Ostazeski, S.A. and Bedard, F.L. (1987) Registration of 'WL 320' alfalfa. *Crop Science* 27, 1084.
Happstadius, I. and Melander, M. (1995) Foradling for sjukdomsresistens i oljevaxter. *Nordisk Jordbruksforskning* 77(2), 125.
Harada, Y., Furueda, T. and Murata, K. (1997) *Verticillium* wilt of *Tilia japonica* and *Acer palmatum*, the first report on the occurrence of *Verticillium dahliae* on trees in Japan. *Annals of the Phytopathological Society of Japan* 63(4), 345–350.
Hardy, G.E.St-J. and Sivasithamparam, S. (1985) Soil solarization: effects on *Fusarium* wilt of carnation and *Verticillium* wilt of eggplant. In: Parker, C.A., Rovira, A.D., Moore, K.G., Wong, P.T. and Kollmorgen, J.F. (eds) *Ecology and Management of Soil-borne Plant Pathogens*. American Phytopathological Society, St Paul, Minnesota, pp. 279–281.
Harig, R. (1976) Untersuchungen über Wechselbeziehungen zwischen *Heterodera rostochiensis* (Wollenweber 1923) und bodengebundenen phytopathogenen Pilzen unter besonderer Berucksichtigung von *Fusarium oxysporum* (Schlecht.) f.sp. *lycopersici* (Sacc.) Snyd. & Hans. Dissertation, Fakultät für Gartenbau und Landeskultur der Technischen Universität Hannover.
Harland, S.C. (1957) Breeding for *Verticillium* resistance in Peruvian cotton. *Empire Cotton Growing Review* 34, 161–168.
Harley, J.L. and Waid, J.S. (1955) A method of studying active mycelia on roots and other surfaces in the soil. *Transactions of the British Mycological Society* 38, 104–118.

Harling, R., Scheffer, R.J. and Elgersma, D.M. (1986) Protein-lipopolysaccharide complexes produced by virulent and non-virulent isolates of *Verticillium albo-atrum* and *Verticillium dahliae* are non-specific toxins to tomato and potato. *Journal of Phytopathology* 117, 198–215.

Harper, A.M. and Huang, H.C. (1984) Contamination of insects by the plant pathogen *Verticillium albo-atrum* in an alfalfa field. *Environmental Entomology* 13, 117–120.

Harper, A.M., Huang, H.C. and Kozub, G.C. (1988) Survival of *Verticillium albo-atrum* on insect bodies and in insect faeces at various temperatures. *Journal of Economic Botany* 81, 1799–1802.

Harrar, J.G. (1937) Some unusual diseases of ornamentals in Virginia. *Plant Disease Reporter* 21, 217.

Harrington, T.C. and Cobb, F.W. (1984) *Verticillium albo-atrum* on *Ceanothus* in a Californian forest. *Plant Disease* 68, 1012.

Harris, D.C. (1989) Control of *Verticillium* wilt of strawberry in Britain by chemical soil disinfestation. *Journal of Horticultural Science* 64, 683–686.

Harris, D.C. (1990a) Control of *Verticillium* wilt and other soil-borne diseases of strawberry in Britain by chemical soil disinfestation. *Journal of Horticultural Science* 65, 401–408.

Harris, D.C. (1990b) Predicting strawberry wilt risk through pre-planting soil analysis for *Verticillium dahliae*. *Proceedings of the Fifth International* Verticillium *Symposium.* Leningrad. p. 8.

Harris, D.C. (1991) A comparison of dazomet, chloropicrin and methyl bromide as soil disinfestants for strawberries. *Journal of Horticultural Science* 66, 51–58.

Harris, D.C. and Fitt, B.D.L. (1995) The threat posed to wilt-susceptible crops in the United Kingdom by linseed cultivation. 6th International *Verticillium* Symposium, Dead Sea, Israel. *Phytoparasitica* 23, 47.

Harris, D.C. and Yang, J. (1990) Pre-planting prediction of strawberry wilt (*Verticillium dahliae*) risk as an aid in disease management. *Brighton Crop Protection Conference, Pests and Diseases* 1, 117–122.

Harris, D.C. and Yang, J.R. (1995) Characteristics of English isolates of *Verticillium dahliae*. *Phytoparasitica* 23(1), 42.

Harris, D.C. and Yang, J.R. (1996) The relationship between the amount of *Verticillium dahliae* in soil and the incidence of strawberry wilt as a basis for disease risk prediction. *Plant Pathology* 45(1), 106–114.

Harris, D.C., Funnell, P.S., Davies, M.K., Talboys, P.W., Cardon, J.A., Simpson, D.W., Adams, A.N. and Blake, V.S.A. (1986) Diseases of strawberries. *Report, East Malling Research Station for 1985*, p. 121.

Harris, D.C., Yang, J.R. and Ridout, M.S. (1993) The detection and estimation of *Verticillium dahliae* in naturally infested soil. *Plant Pathology* 42, 238–250.

Harris, J.R. (1985) Use of aerial photography for assessing soil-borne diseases. In: Parker, C.A., Rovira, A.D., Moore, K.G., Wong, P.T. and Kollmorgen, J.F. (eds) *Ecology and Management of Soil-borne Plant Pathogens.* American Phytopathological Society, St Paul, Minnesota, pp. 27–30.

Harris, R.V. (1925a) A wilt disease of hops. *Report of the East Malling Research Station 1925, Supplement* 2, pp. 92–93.

Harris, R.V. (1925b) The blue stripe wilt of the raspberry. *Annual Report, East Malling Research Station for 1924*, pp. 126–133.

Harris, R.V. (1936) The *Verticillium* wilt of hops. Some facts and recommendations. *Report of the East Malling Research Station 1935*, pp. 158–162.

Harris, R.V. (1953) Wilt tolerant hops. *Report of the East Malling Research Station 1953*, pp. 208–209.
Harris, R.V. (1958) Progress in research on *Verticillium* wilt and virus diseases in hops in 1957. *Report of the East Malling Research Station 1957*, pp. 22–27.
Harris, R.V. and Furneaux, B.S. (1938) *Verticillium* wilt of hops. A summary of information for growers. *Report of the East Malling Research Station 1937*, pp. 257–258.
Harrison, A.F. and Clare, B.G. (1970) Host reactions involved in the recovery of apricot trees from *Verticillium* rot. *Australian Journal of Biological Sciences* 23, 1027–1032.
Harrison, G.J. (1955) A History of Cotton in California. Manuscript, Agriculture Library, University of California, Berkeley.
Harrison, G.J. and Brinkerhoff, L.A. (1947) *Verticillium* wilt of cotton in California. *Plant Disease Reporter* 30, 20–21.
Harrison, J.A.C. (1971) Association between the potato cyst nematode *Heterodera rostochiensis* Woll. and *Verticillium dahliae* Kleb. in early dying disease of potatoes. *Annals of Applied Biology* 57, 185–193.
Harrison, J.A.C. and Isaac, I. (1968) Leaf area development in King Edward potato plants inoculated with *Verticillim albo-atrum* and *V. dahliae*. *Annals of Applied Biology* 61, 217–230.
Harrison, J.A.C. and Isaac, I. (1969) Survival of the causal agents of early dying disease (*Verticillium* wilt) of potatoes. *Annals of Applied Biology* 63, 277–288.
Harrison, J.G., Lowe, R., Fox, R.A., Dashwood, E.P., Wilson, H.M., Nicholson, T.H. and Spencer, D. (1985) Fungal diseases of the growing potato plant. *Report of the Scottish Crop Research Institute 1984*, pp. 107–110.
Harrison, M.D. (1976) The effect of barley straw on the survival of *Verticillium albo-atrum* in naturally infested field soil. *American Potato Journal* 53, 385–394.
Harrison, M.D. and Livingston, C.H. (1966) A method for isolating *Verticillium* from field soil. *Plant Disease Reporter* 50, 897–899.
Harrison, N.A. and Beckman, C.H. (1982) Time/space relationships of colonization and host response in wilt-resistant and wilt susceptible cotton (*Gossypium*) cultivars inoculated with *Verticillium dahliae* and *Fusarium oxysporum* f. sp. *vasinfectum*. *Physiological Plant Pathology* 21, 193–207.
Hartill, W.F.T. (1971) Control of *Verticillium* wilt of tobacco with benomyl. *Plant Disease Reporter* 55, 889–893.
Hartill, W.F.T. and Campbell, J.M. (1976) Some host–parasite relationships of *Verticillium* wilt in flue-cured tobacco. *New Zealand Journal of Agricultural Research* 19, 377–381.
Hartman, R.E. and Keen, N.T. (1973) Enzymes catalyzing anaplerotic carbon dioxide fixation in *Verticillium albo-atrum*. *Phytopathology* 63, 947–953.
Hartman, R.E. and Keen, N.T. (1974a) The pyruvate carboxylase of *Verticillium albo-atrum*. *Journal of General Microbiology* 81, 15–19.
Hartman, R.E. and Keen, N.T. (1974b) The phosphoenolpyruvate carboxykinase of *Verticillium albo-atrum*. *Journal of General Microbiology* 81, 21–26.
Hartmann, H.T. and Whisler, J.E. (1970) Some root stock and inter stock influences in the olive-D (*Olea europaea*) D cv Sevillano. *Journal of the American Society for Horticultural Science* 95, 562–565.
Hartmann, H.T., Schnathorst, W.C. and Whisler, J.E. (1971) Oblonga, a clonal olive rootstock resistant to *Verticillium* wilt. *California Agriculture* 25(6), 12–15.
Hartman, R.E., Keen, N.T. and Long, M. (1972) Carbon dioxide fixation by *Verticillium albo-atrum*. *Journal of General Microbiology* 73, 29–34.

Hartz, T.K., DeVay, J.E. and Elmore, C.L. (1993) Solarization is an effective soil disinfestation technique for strawberry production. *HortScience* 28(2), 104–106.

Hartzler, R.G. (1996) Velvetleaf (*Abutilon theophrasti*) population dynamics following a single year's seed rain. *Weed Technology* 10(3), 581–586.

Harvey, I.C. (1982) Assessment keys for some crown and root diseases of lucerne (*Medicago sativa* L.). *New Zealand Journal of Experimental Agriculture* 10, 317–322.

Harvey, I.C., Seyb, A.M., Warren, A.F.J. and Van Den Ende, H. (1996) The biological control of Russell lupin in riverbeds with endemic plant pathogens. In: O'Callaghan, M.J. (ed.) *Proceedings of the 49th New Zealand Plant Protection Conference*, pp. 119–125.

Hasan, A. and Khan, M.N. (1985) The effect of *Rhizoctonia solani*, *Sclerotium rolfsii*, and *Verticillium dahliae* on the resistance of tomato to *Meloidogyne incognita*. *Nematologia Mediterranea* 13, 133–136.

Hashimoto, K. (1989) [Studies on *Verticillium* wilt of eggplant]. *Saitama Horticultural Experiment Station* 2(3), 1–110.

Hashmi, M.H. (1988) Seedborne mycoflora of *Trigonella foenum-graecum* L. *Pakistan Journal of Botany* 20, 233–237.

Hastie, A.C. (1962) Genetic recombination in the hop-wilt fungus, *Verticillium albo-atrum*. *Journal of General Microbiology* 27, 373–382.

Hastie, A.C. (1964) The parasexual cycle in *Verticillium albo-atrum*. *Genetical Research* 5, 305–315.

Hastie, A.C. (1967) Mitotic recombination in conidiophores of *Verticillium albo-atrum*. *Nature* 214, 249–252.

Hastie, A.C. (1968) Phialide analysis of mitotic recombination in *Verticillium*. *Molecular and General Genetics* 102, 232–240.

Hastie, A.C. (1970) The genetics of asexual phytopathogenic fungi with special reference to *Verticillium*. In: Tousson, T.A., Bega, R.V. and Nelson, P.E. (eds) *Root Diseases and Soil-borne Pathogens. Second International Symposium on Factors Determining the Behaviour of Plant Pathogens in Soil.* Imperial College, London. University of California Press, California, Berkeley, pp. 55–62.

Hastie, A.C. (1971) Genetics and *Verticillium*. *Proceedings of the First International* Verticillium *Symposium.* Wye College, University of London, p. 16.

Hastie, A.C. (1973) Hybridization of *Verticillium albo-atrum* and *Verticillium dahliae*. *Transactions of the British Mycological Society* 60, 511–523.

Hastie, A.C. (1978) Genetic analysis of species limits in *Verticillium*. In: Subramanian, C.V. (ed.) *Taxonomy of Fungi.* University of Madras, India, pp. 155–163.

Hastie, A.C. (1981) The genetics of conidial fungi. In: Cole, G.T. and Kendrick, W.B. (eds) *Biology of Conidial Fungi*, Vol. 2. Academic Press, New York, pp. 511–543.

Hastie, A.C. and Gadd, G.M. (1981) Mutants of *Verticillium albo-atrum* specifically affecting conidium germination. *FEMS Microbiology Letters* 12, 139–141.

Hastie, A.C. and Heale, J.B. (1984) Genetics of *Verticillium*. *Phytopathologia Mediterranea* 23, 130–162.

Haunold, A., Likens, S.T., Nickerson, G.B., Horner, C.E. and Zimmermann, C.E. (1978) Registration of USDA 21055 hop germplasm (Reg.No. GP 5). *Crop Science* 18, 919.

Haunold, A., Nickerson, G.B., Gampert, U., Kling, D.S. and Kenny, S.T. (1995) Registration of 'Crystal' hop. *Crop Science* 35(1), 279–280.

Hawke, M.A. and Lazarovits, G. (1994) Production and manipulation of individual microsclerotia of *Verticillium dahliae* for use in studies of survival. *Phytopathology* 84(9), 883–890.

Hawke, M.A. and Lazarovits, G. (1995a) The role of melanin in the survival of microsclerotia of *Verticillium dahliae*. 6th International *Verticillium* Symposium, Dead Sea, Israel. *Phytoparasitica* 23, 54.
Hawke, M.A. and Lazarovits, G. (1995b) The production of microsclerotia of *Verticillium dahliae* for use in studies of survival. 6th International *Verticillium* Symposium, Dead Sea, Israel. *Phytoparasitica* 23, 53.
Hawksworth, D.L. and Talboys, P.W. (1970) *CMI Descriptions of Pathogenic Fungi and Bacteria*. International Mycological Institute, Kew, London, pp. 255–259.
Hawksworth, D.L., Sutton, B.C. and Ainsworth, G.C. (1983) *Ainsworth & Bisby's Dictionary of the Fungi*, 7th edn. Commonwealth Mycological Institute, Kew, London.
Hawthorne, B.T. (1983) Variation in pathogenicity among isolates of *Verticillium albo-atrum* from lucerne. *New Zealand Journal of Agricultural Research* 26, 405–408.
Hawthorne, B.T. (1987) Qualitative and quantitive assessments of the reaction of lucerne plants to *Verticillium* wilt. *New Zealand Journal of Agricultural Research* 30, 349–359.
Hawthorne, B.T. and Triggs, C.M. (1986) Resistance of lucerne cultivars to *Verticillium albo-atrum*. *Agronomy Society of New Zealand Special Publication* 5, 213–218.
Head, M. and Finch, R. (1975) Lucerne variety trial in a *Verticillium* wilt situation. *UK Agricultural Development and Advisory Service: Experiments and Development in the Eastern Region – 1974*, pp. 215–217.
Heale, J.B. (1966) Heterokaryon synthesis and morphogenesis in *Verticillium*. *Journal of General Microbiology* 45, 419–427.
Heale, J.B. (1985) *Verticillium* wilt of alfalfa, background and current research. *Canadian Journal of Plant Pathology* 7, 191–198.
Heale, J.B. (1988) *Verticillium* spp., the cause of vascular wilts in many species. *Advances in Plant Pathology* 6, 291–312.
Heale, J.B. (1989) Implications of genetic/molecular evidence with respect to virulence/avirulence of fungal wilt pathogens. In: Tjamos, E.C. and Beckman, C.H. (eds) *Vascular Wilt Diseases of Plants. Basic Studies and Control. NATO ASI Series H: Cell Biology*, Vol. 28. Springer-Verlag, Berlin, pp. 259–275.
Heale, J.B. (1997) Diversification and speciation in *Verticillium* – an overview. *Proceedings of the Seventh International* Verticillium *Symposium*. Cape Sounion, Athens, Greece. p. 9.
Heale, J.B. and Connel, S. (1987) Use of tissue culture to produce novel sources of resistance to *Verticillium albo-atrum* attacking hop. *Canadian Journal of Plant Pathology* 9, 81.
Heale, J.B. and Gupta, D.P. (1970) The utilization of cellobiose by *Verticillium albo-atrum*. *Journal of General Microbiology* 63, 175–181.
Heale, J.B. and Gupta, D.P. (1972) Mechanism of vascular wilting induced by *Verticillium albo-atrum*. *Transactions of the British Mycological Society* 58, 19–28.
Heale, J.B. and Isaac, I. (1963) Wilt of lucerne caused by species of *Verticillium* IV. Pathogenicity of *V. albo-atrum* and *V. dahliae* to lucerne and other crops; spread and survival of *V. albo-atrum* in soil and in weeds, effect upon lucerne production. *Annals of Applied Biology* 52, 439–451.
Heale, J.B. and Isaac, I. (1964) Dark pigment formation in *Verticillium albo-atrum*. *Nature* 202, 412–413.
Heale, J.B. and Isaac, I. (1965) Environmental factors in the production of dark resting structures in *Verticillium albo-atrum*, *V. dahliae* and *V. tricorpus*. *Transactions of the British Mycological Society* 48, 39–50.

Heale, J.B., Gafoor, A. and Rajasingham, K.C. (1968) Nuclear division in conidia and hyphae of *Verticillium albo-atrum*. *Canadian Journal of Genetics and Cytology* 10, 321–340.

Heale, J.B., Isaac, I. and Milton, J.M. (1979) The administrative control of *Verticillium* wilt of lucerne. In: Ebbels, D.L. and King, J.E. (eds) *Plant Health. The Scientific Basis for Administrative Control of Plant Diseases and Pests.* Blackwell Scientific Publications, London, pp. 71–78.

Hedlund, T. (1923) Om några sjukdomar och skador på våra lantbruksväxter, 1922. *Allmän Jordbrukstidskrift* 5, 166–168.

Heesung, P. and Yongse, L. (1996) Increase in 3-hydroxy-3-methylglutaryl coenzyme A reductase mRNA level in tomato by fungal elicitors and mechanical wounding. *Korean Journal of Plant Pathology* 12, 285–290.

Heffer, V.J. and Regan, R.P. (1996) First report of *Verticillium* wilt caused by *Verticillium dahliae* of ash trees in Pacific Northwest nurseries. *Plant Disease* 80(3), 342.

Hefner, J.J. (1969) Effects of a fungicide and method of application on *Verticillium* wilt incidence in Acala cotton – a preliminary report. *Proceedings of the Beltwide Cotton Production Research Conference.* New Orleans, Louisiana. pp. 28–29.

Heinstein, P., Lin, H., Chen, X.Y. and Davisson, V.J. (1996) Differential expression of (+)δ-cadinene synthases in cotton cells. *1996 Proceedings Beltwide Cotton Conferences.* Nashville, Tennessee, 9–12 January, 1996, Vol. 2, pp. 1182–1183.

Heinz, R.A. and Platt, H.W. (2000a) A competitive PCR-based assay to quantify *Verticillium tricorpus* propagules in soil. *Canadian Journal of Plant Pathology* 22(2), 122–130.

Heinz, R., Lee, S.W., Saparno, A., Nazar, R.N. and Robb, J. (1998) Cyclical systemic colonization in *Verticillium*-infected tomato. *Physiological and Molecular Plant Pathology* 52(6), 385–396.

Heinz, R.A. and Platt, H.W. (2000b) Improved DNA extraction method for *Verticillium* detection and quantification in large-scale studies using PCR-based techniques. *Canadian Journal of Plant Pathology* 22(2), 117–121.

Hellmers, E. (1975) [Healthy pelargoniums 3. Fungus diseases]. *Gartner-Tidende* 19, 325–328.

Henis, Y. (1971) Factors affecting the germination of *Verticillium dahliae*. *Proceedings of the First International* Verticillium *Symposium.* Wye College, University of London, p. 17.

Henis, Y. and Bar, J. (1973) Sensitivity to chloropicrin of microsclerotia of *Verticillium dahliae* as related to their germination on growth media and in soil. *Phytoparasitica* 1, 95–99.

Henis, Y. and Fahima, T. (1990) Biological control of *Verticillium* wilt of eggplant by *Talaromyces flavus*. *Proceedings of the Fifth International* Verticillium *Symposium.* Leningrad, p. 81.

Henneken, M., Foller, I. and Paul, V.H. (1998) First laboratory investigations on the reaction of cultivars and breeding lines of *Camelina sativa* (L.) Crtz to downy mildew, clubroot and *Verticillium* wilt (*Verticillium dahliae*). *Bulletin OILB/SROP* 21(5), 77–84.

Henni, J.E. (1987a) Evaluation de l'efficacité de certains champignons antagonistes vis-à-vis de *Verticillium dahliae* Klebahn. *Cryptogamie, Mycologie* 8, 203–207.

Henni, J.E. (1987b) Utilisation des ultra-sons pour purifier les microsclérotes de *Verticillium dahliae* Klebahn. *Agronomie* 7, 743–744.

Heppner, C. and Heitefuss, R. (1995) Untersuchungen zum Auftreten von *Verticillium dahliae* Kleb. und anderen pilzliche Erregern am Erntegut von Winterraps (*Brassica napus* L. var. *oleifera* Metzger). *Nachrichtenblatt des Deutschen Pflanzenschutzdienstes* 47(3), 57–61.
Hepting, G.H. and Toole, E.R. (1950) Some southeastern tree diseases – 1948 and 1949. *Plant Disease Reporter* 34, 135–137.
Herrington, M.E. and Saranah, S. (1985) 'Redlands Summertaste' tomato. *HortScience* 20, 956–959.
Hibben, C.R. (1959) A new host for *Verticillium albo-atrum* Reinke & Berth. *Plant Disease Reporter* 43, 1137.
Hide, G.A. (1981) Fungus diseases on potato seed tubers planted in England and Wales, 1963–76. *Annals of Applied Biology* 98, 377–393.
Hide, G.A. and Corbett, D.C.M. (1971) Disease of potatoes caused by *Verticillium dahliae* and *Heterodera rostocheinsis* and its control. *Proceedings of the First International* Verticillium *Symposium*. Wye College, University of London, p. 18.
Hide, G.A. and Corbett, D.C.M. (1973) Controlling early death of potatoes caused by *Heterodera rostochiensis* and *Verticillium dahliae*. *Annals of Applied Biology* 75, 461–462.
Hide, G.A. and Corbett, D.C.M. (1974) Field experiments in the control of *Verticillium dahliae* and *Heterodera rostochiensis* on potatoes. *Annals of Applied Biology* 78, 295–307.
Hide, G.A., Corbett, D.C.M. and Evans, K. (1984) Effects of soil treatments and cultivars on 'early dying' disease of potatoes caused by *Globodera rostochiensis* and *Verticillium dahliae*. *Annals of Applied Biology* 104, 277–289.
Hiemstra, J.A. (1995) Recovery of *Verticillium*-infected ash trees. 6th International *Verticillium* Symposium, Dead Sea, Israel. *Phytoparasitica* 23, 64.
Hiemstra, J.A. (1997) Petioles from infected trees spread *Verticillium dahliae*. *Proceedings of the Seventh International* Verticillium *Symposium*. Cape Sounion, Athens, Greece, p. 47.
Hiemstra, J.A. (1997) Screening for *Verticillium* resistance in Norway maple. *Proceedings of the Seventh International* Verticillium *Symposium*. Cape Sounion, Athens, Greece, p. 54.
Hiemstra, J.A. and Rataj-Guranowska, M. (1997) Vegetative compatibility groups in *Verticillium dahliae* in the Netherlands. *Proceedings of the Seventh International* Verticillium *Symposium*. Cape Sounion, Athens, Greece, p. 24.
Hijwegen, T. and Termorshuizen, A.J. (2000) Retarded development of *Verticillium* wilt in eggplant after application of Bion R or Oryzalin R. *Zeitschrift für Pflanzenkrankheiten und Pflanzenschutz* 107(3), 267–273.
Hill, C.F. (1982) New plant disease records in New Zealand. *New Zealand Journal of Botany* 20, 355–359.
Hill, M.K., Lyon, K.J. and Lyon, B.R. (1999) Identification of disease response genes expressed in *Gossypium hirsutum* upon infection with the wilt pathogen *Verticillium dahliae*. *Plant Molecular Biology* 40(2), 289–296.
Hillocks, R.J. (1981) Cotton disease research in Tanzania. *Tropical Pest Management* 27, 1–12.
Hillocks, R.J. (1991) Screening for resistance to *Verticillium* wilt in Zimbabwe. *Tropical Agriculture Trinidad* 68, 144–148.
Himelick, E.B. (1969) Tree and shrub hosts of *Verticillium albo-atrum*. *Biological Notes, Illinois Natural History Survey* 66, 3–6.

Hine, R.B. and McCain, A.H. (1984) *Verticillium* wilt of *Tagetes* spp. *Plant Disease* 68, 1098.

Hine, R.B. and Simbwa-Bunnya, M. (1969) Studies on uptake and translocation of benlate in cotton. *Proceedings of the Beltwide Cotton Production Research Conference*. New Orleans, Louisiana, p. 27.

Hine, R.B., Johnson, D.L. and Wenger, C.J. (1969) The persistency of two benzimidazole fungicides in soil and their fungistatic activity against *Phymatotrichum omnivorum*. *Phytopathology* 59, 798–801.

Hine, R.B., Ashworth, L.J. and McMeans, J.L. (1970) Absorption and movement of benomyl into cotton bolls. *Phytopathology* 60, 1295–1296.

Hinkle, D.A. and Fulton, N.D. (1963) Yield and *Verticillium* wilt incidence in cotton as affected by crop rotations. *Bulletin, Arkansas Experimental Station* 165, 1–15.

Hinkle, D.A. and Staten, G. (1941) Fertilizer experiments with Acala cotton on irrigated soils. *Bulletin of the New Mexico Agricultural Experimental Station* 280, 1–15.

Hirota, K. and Miyagawa, T. (1988) Measures for the control of Chinese cabbage *Verticillium* yellow. *Research Bulletin of the Aichi-ken Agricultural Research Centre* 20, 238–244.

Hockey, J.F. (1923) Blue stem of the black raspberry. *Phytopathology* 13, 293.

Hoes, J.A. (1972) Losses due to *Verticillium dahliae* wilt in sunflower. *Phytopathology* 62, 764.

Hoes, J.A. and Dedio, W. (1988) Registration of eleven disease resistant oilseed sunflower germplasm lines. *Crop Science* 28, 875.

Hoffman, G.M. and Zinkernagel, V. (1972) Zur Abtotung von Dauerorganen pathogener Bodenpilze durch Bodenentseuchung mit Terabol. *Gartenwelt* 72, 291–293.

Hoitink, H.A.J. and Krause, M.S. (1999) Control of nuisance and detrimental molds (fungi) in mulches and composts. *Special Circular – Ohio Agricultural Research and Development Center* 165, 66–69.

Hoitink, H.A.J., Sydnor, T.D. and Wilson, C.L. (1979) Resistance of maple cultivars and species to *Verticillium* wilt – a preliminary report. *Ohio Agricultural Research and Development Center Research Circular* 246, 46–47.

Holliday, P. (1980) *Fungus Diseases of Tropical Crops*. Cambridge University Press, Cambridge.

Holliday, P. (1989) *Dictionary of Plant Pathology*. Cambridge University Press, Cambridge.

Holmes, F.W. (1957) *Verticillium* wilt of dogwood. *Phytopathology* 47, 17.

Holmes, F.W. (1967) Resistance of certain elm clones to *Ceratocystis ulmi* and *Verticillium albo-atrum*. *Phytopathology* 57, 1247–1249.

Holmes, F.W. (1976) *Verticillium* wilt of salt-injured sugar maple – preliminary study. *Proceedings of the American Phytopathological Society* 3, 305–306.

Honda, N., Hirai, M., Ano, T. and Shoda, M. (1999) Control of tomato damping-off caused by *Rhizoctonia solani* by the heterotrophic nitrifier *Alcaligenes faecalis* and its product, hydroxylamine. *Annals of the Phytopathological Society of Japan* 65, 153–162.

Hora, T.S., Baker, R. and Griffin, G.J. (1977) Experimental evaluation of hypotheses explaining the nature of soil fungistasis. *Phytopathology* 67, 373–379.

Horiuchi, S., Hagiwara, H. and Takeuchi, S. (1990) *Biological Control of Soil-borne Plant Pathogens*. CAB International, Wallingford, UK, pp. 285–298.

Horner, C.E. (1954) Pathogenicity of *Verticillium* isolates to peppermint. *Phytopathology* 44, 239–242.

Horner, C.E. (1965) *Verticillium* wilt of hops in Oregon. *Plant Disease Reporter* 49, 538–540.
Horner, C.E. and Dooley, H.L. (1965) Propane flaming kills *Verticillium dahliae* in peppermint stubble. *Plant Disease Reporter* 49, 581–582.
Horner, C.E. and Melouk, H.A. (1976) Irradiation-induced resistance to *Verticillium* wilt in mints. *Proceedings of the Second International* Verticillium *Symposium*, University of California, Berkeley, p. 20.
Horst, R.K. (1983) *Compendium of Rose Diseases.* American Phytopathological Society, St Paul, Minnesota,
Hoskinson, P.E. and Hancock, N.I. (1972) Registration of Hancock cotton. *Crop Science* 12, 714.
Hougas, R.W., Peloquin, S.J. and Ross, R.W. (1958) Haploids of the common potato. *Journal of Heredity* 49, 103.
Howard, B.H. (1993) Investigations into inconsistent and low bud-grafting success in *Acer platanoides* 'Crimson King'. *Journal of Horticultural Science* 68(3), 455–462.
Howard, C.M. and Albregts, E.E. (1982) Outbreak of *Verticillium* wilt of strawberries in central Florida. *Plant Disease* 66, 856–857.
Howard, R.J. (1985) Local and long-distance spread of *Verticillium* species causing wilt of alfalfa. *Canadian Journal of Plant Pathology* 7, 199–202.
Howell, A.B. and Erwin, D.C. (1990) *Verticillium* wilt of alfalfa in southern California. *Proceedings of the Fifth International* Verticillium *Symposium.* Leningrad, p. 11.
Howell, A.B. and Erwin, D.C. (1995) Characterization and persistence of *Verticillium albo-atrum* isolated from alfalfa growing in high temperature regions of southern California. *Plant Pathology* 44(4), 734–748.
Howell, A.B., Francois, L. and Erwin, D.C. (1994) Interactive effect of salinity and *Verticillium albo-atrum* on *Verticillium* wilt disease severity and yield of two alfalfa cultivars. *Field Crops Research* 37(3), 247–251.
Howell, C.R. (1970) Differential enzyme synthesis by haploid and diploid forms of *Verticillium albo-atrum. Phytopathology* 60, 488–490.
Howell, C.R. (1976) Use of enzyme-deficient mutants of *Verticillium dahliae* to assess the importance of pectolytic enzymes in symptom expression of *Verticillium* wilt of cotton. *Physiological Plant Pathology* 9, 279–283.
Howell, C.R., Bell, A.A. and Stipanovic, R.D. (1973) Virulence to cotton and tolerance to sanguinarine among *Verticillium* species. *Canadian Journal of Microbiology* 19, 1367–1371.
Howell, C.R., Bell, A.A. and Stipanovic, R.D. (1976) Effect of ageing on flavonoid content and resistance of cotton leaves to *Verticillium* wilt. *Physiological Plant Pathology* 8, 181–188.
Hoyman, W.G. (1974a) Consequence of planting Norgold Russet seed infected with *Verticillium albo-atrum. American Potato Journal* 51, 22–25.
Hoyman, W.G. (1974b) Searching for resistance to *Meloidogyne hapla. American Potato Journal* 51, 278.
Hoyos, G.P., Lauer, F.I. and Anderson, N.A. (1993) Early detection of *Verticillium* wilt resistance in a potato breeding program. *American Potato Journal* 70(7), 535–541.
Hu, X., Nazar, R.N. and Robb, J. (1993) Quantification of *Verticillium* biomass in wilt disease development. *Physiological and Molecular Plant Pathology* 42(1), 23–36.
Hu, X., Nazar, R.N. and Robb, J. (1995) *Verticillium* colonisation in selected resistant, tolerant and susceptible alfalfa plants as determined by PCR-based assay. 6th International *Verticillium* Symposium, Dead Sea. *Phytoparasitica* 23, 59.

Huang, H.C. (1989) Distribution of *Verticillium albo-atrum* in symptomed and symptomless leaflets of alfalfa. *Canadian Journal of Plant Pathology* 11, 235–241.

Huang, H.C. and Harper, A.M. (1985) Survival of *Verticillium albo-atrum* from alfalfa in feces of leaf-chewing insects. *Phytopathology* 75, 206–208.

Huang, H.C. and Kokko, E.G. (1983) A simple technique for observation and photography of verticillate conidiophores *in situ* using a stereomicroscope. *Canadian Journal of Plant Pathology* 5, 36–37.

Huang, H.C. and Kokko, E.G. (1985) Infection of alfalfa pollen by *Verticillium albo-atrum*. *Phytopathology* 75, 859–865.

Huang, H.C. and Richards, K.W. (1983) *Verticillium albo-atrum* contamination on leaf pieces forming cells for the alfalfa leafcutting bee. *Canadian Journal of Plant Pathology* 5, 248–250.

Huang, H.C. and Richards, K.W. (1987) Alfalfa leafcutter bee: a potential vector for *Verticillium* wilt. *Research Highlights, Lethbridge Research Station, Agriculture Canada*. Alberta, Canada, p. 74.

Huang, H.C., Harper, A.M., Kokko, E.G. and Howard, R.J. (1983) Aphid transmission of *Verticillium albo-atrum* to alfalfa. *Canadian Journal of Plant Pathology* 5, 141–147.

Huang, H.C., Hanna, M.R. and Kokko, E.G. (1985) Mechanisms of seed contamination by *Verticillium albo-atrum* in alfalfa. *Phytopathology* 75, 482–488.

Huang, H.C., Harper, A.M. and Kozub, G.C. (1986a) Effect of temperature on survival of *Verticillium albo-atrum* on insect bodies or in insect faeces. *Proceedings of the Fourth International* Verticillium *Symposium*. Guelph, Canada.

Huang, H.C., Hironaka, R. and Howard, R.J. (1986b) Survival of *Verticillium albo-atrum* in alfalfa tissue buried in manure or fed to sheep. *Plant Disease* 70, 218–221.

Huang, H.C., Richards, K.W. and Kokko, E.G. (1986c) Role of the leafcutter bee in dissemination of *Verticillium albo-atrum* in alfalfa. *Phytopathology* 76, 75–79.

Huang, H.C., Harper, A.M. and Kozub, G.C. (1987) Effect of temperature on survival of *Verticillium albo-atrum* on insect bodies or in insect feces. *Canadian Journal of Plant Pathology* 9, 81.

Huang, H.C., Kozub, G.C. and Kokko, E.G. (1994) Survival of *Verticillium albo-atrum* in alfalfa seeds. *Canadian Journal of Botany* 72, 1121–1125.

Huang, H.C., Kozub, G.C. and Kokko, E.G. (1995) Survival of *Verticillium albo-atrum* in lucerne seeds. 6th International *Verticillium* Symposium, Dead Sea, Israel. *Phytoparasitica* 23, 48.

Huang, H.C., Acharya, S.N., Hou, T.J., Erickson, R.S., Dalton, R.E. and Mueller, C.A. (1999) Susceptibility of Chinese alfalfa cultivars to *Verticillium* wilt. *Plant Pathology Bulletin* 8(2), 67–72.

Huang, L.K. and Mahoney, R.R. (1999) Purification and characterization of an endo-polygalacturonase from *Verticillium albo-atrum*. *Journal of Applied Microbiology* 86(1), 145–156.

Hubbeling, N. and Basu Chaudhary, K.C. (1969) Influence of pH and calcium nutrition on the resistence of tomato seedlings to *Verticillium* wilt. *Mededelingen van de Faculteit Landbouwwetenschappen Rijksuniversiteit Gent* 34, 937–943.

Hubbeling, N. and Basu Chaudhary, K.C. (1970) Mutagenic effects of a herbicide on *Verticillium dahliae*. *Mededelingen van de Faculteit Landbouwwetenschappen Rijksuniversiteit Gent* 35, 625–635.

Hubbeling, N., Alexander, L.J. and Cirulli, M. (1971) Resistance to *Fusarium* and *Verticillium* wilts in tomato. *Mededelingen van de Faculteit Landbouwwetenschappen Rijksuniversiteit Gent* 36, 1006–1016.

Huber, D.M. (1980) The role of mineral nutrition in defence. In: Horsfall, J. and Cowling, E. (eds) *Plant Disease. An Advanced Treatise.* Academic Press, New York, pp. 381–406.

Huber, D.M. and Watson, R.D. (1970) Effect of organic amendment on soil-borne plant pathogens. *Phytopathology* 60, 22–26.

Huber, G.A. and Jones, L.K. (1934) The *Verticillium* disease of chrysanthemum. *Bulletin of Washington Agricultural Experiment Station* 290, 1–14.

Huettel, R.N., Francl, L.J., Reise, R.W., Meyer, S.L.F. and Henn, R.A. (1991) Plant parasitic nematodes in the potato growing areas of Maine. *American Potato Journal* 68, 345–354.

Hughes, S.J. (1951) Studies on micro-fungi. XI. Some hyphomycetes which produce phialides. *Mycological Papers* 45, 1–36.

Huisman, O.C. (1988a) Seasonal colonization of roots of field-grown cotton by *Verticillium dahliae* and *Verticillium tricorpus. Phytopathology* 78, 708–716.

Huisman, O.C. (1988b) Colonization of field-grown cotton roots by pathogenic and saprophytic soilborne fungi. *Phytopathology* 78, 716–722.

Huisman, O.C. and Ashworth, L.J. (1974a) Quantitative assessment of *Verticillium albo-atrum* in field soils: procedural and substrate improvements. *Phytopathology* 64, 1043–1044.

Huisman, O.C. and Ashworth, L.J. (1974b) *Verticillium albo-atrum*: quantitive isolation of microsclerotia from field soils. *Phytopathology* 64, 1159–1163.

Huisman, O.C. and Ashworth, L.J. (1976) Influence of crop rotation on survival of *Verticillium albo-atrum* in soils. *Phytopathology* 66, 978–981.

Huisman, O.C. and Davis, J.R. (1997) Root colonization by *Verticillium dahliae*: influence of organic soil amendments. *Proceedings of the Seventh International* Verticillium *Symposium*. Cape Sounion, Athens, Greece, p. 89.

Huisman, O.C. and Grimes, W.D. (1989) Cultural practices: the effect of planting density and irrigation regimes on *Verticillium* wilt of cotton. In: Tjamos, E.J. and Beckman, C.H. (eds) *Vascular Wilt Diseases of Plants, Basic Studies and Control. NATO ASI Series H: Cell Biology*, Vol. 28. Springer-Verlag, Berlin, pp. 537–541.

Huisman, O.C., Davis, J.R., Sorensen, L.H. and Schneider, A.T. (1995) Soil supressiveness of *Verticillium dahliae* infection on potato with cropping practices. 6th International *Verticillium* Symposium, Dead Sea, Israel. *Phytoparasitica* 23, 71.

Hulea, A., Iliescu, H. and Bunescu, S. (1973) [Detection of an unusual fungal complex pathogenic on sunflower in Romania]. *Probleme de Proteçtia Plantelor* 1, 73–85.

Hung, K.S. and Whitney, N.J. (1976) Leaf inoculation as a rapid means of screening potatoes for resistance to *Verticillium* wilt. *Proceedings of the Second International* Verticillium *Symposium*. University of California, Berkeley, p. 22.

Husz, B. (1936) [Some wilt diseases of cultivated plants]. *Botanikai Közl.* 32, 38–51.

Husz, B. (1947) [Mycological investigations on apoplectic apricot trees]. *Acta Mycologica Hungaricae* 4, 6–7.

Hutchinson, C.M., McGiffen, M.E. Jr, Ohr, H.D., Sims, J.J. and Becker, J.O. (2000) Efficacy of methyl iodide and synergy with chloropicrin for control of fungi. *Pest Management Science* 56(5), 413–418.

Hutson, R.A. and Smith, I.M. (1980) Phytoalexins and tyloses in tomato cultivars infected with *Fusarium oxysporum* f. sp. *lycopersici* or *Verticillium albo-atrum. Physiological Plant Pathology* 17, 245–257.

Hutson, R.A. and Smith, I.M. (1983) The response of tomato seedling roots to infection by *Verticillium albo-atrum* or *Fusarium oxysporum* f. sp. *lycopersici. Annals of Applied Biology* 102, 89–97.

Hyer, A.H., Garber, R.H. and Jorgenson, E.C. (1983) *Reaction of Nematode or* Fusarium *Wilt Resistant and Susceptible Cottons to* Fusarium *Wilt-root-knot Nematode Disease Complex.*

Ibragimov, A.A. and Khodzhibayeva, S.M. (1990) Extrachromasome genetic elements of the fungus *Verticillium dahliae. Proceedings of the Fifth International* Verticillium *Symposium, Leningrad,* p. 103.

Ibragimov, A.P. (1990) Molecular mechanisms of interactions between *Verticillium dahliae* and its cotton plant host. *Proceedings of the Fifth International* Verticillium *Symposium.* Leningrad, p. 54.

Ibragimov, A.P. and Ismailov, E.M. (1976) Comparative study of some properties of tRNA and aminoacyl-tRNA synthesis (ARS) in cotton and the fungus *Verticillium dahliae. Doklady Vsesoyuznoi Ordena Lenina Akadamii Sel'skokhozyaistvennykh Nauk Imeni V.I. Lenina* 12, 13–15.

Ibragimov, Sh., Straumal, B. and Tashabaev, K. (1977) [Cotton in dense stands (preliminary information)]. *Kklopkovodstvo* 2, 39.

Ibrahimllari, L. (1987) [Some data on wilt organisms of pepper in the district of Tiranë]. *Buletini i Shkencare Bujgësore* 26, 94–100.

Iglesias-Olivas, J., Bosland, P.W. and Lindsey, D.L. (1987) Resistance of *Capsicum* species to *Verticillium dahliae* K. *Capsicum Newsletter* 6, 73–74.

Ikramov, Yu. (1979) [*Growth, Development and Variability of Cotton under the Action of Temperature and Light Factors*]. FAN UzbekSSR, Tashkent, Uzbek SSR.

Ikramov, Yu. and Miradaev, M. (1974) [Effect of different air temperature conditions on infection of cotton by *Verticillium* wilt]. *Voprosy Genetiki, Selektsii i Semenovodstva Khlopchatnika i Lyutserny* 261–264.

Ikramov, Yu.I., Musaev, D.A. and Berdiev, O.M. (1976) [Effect of temperature of the air on infection of cotton by *Verticillium* wilt]. *Trudy Uzb. Opyt. St. Lub. Kul'tur (Vopr. Genet. Selektsii i Semenovodstva Khlopchatnika i Lyutserny)* 13, 123–130.

Ikramov, Yu., Berdiev, S.M., Usmanov, S.A. and Inoyatov, M.Kh. (1979) [Effect of air temperature on susceptibility of the cotton *Gossypium barbadense* L. to *Verticillium* wilt]. *Uzbekistan Biologija Zurnali* 4, 21–22.

Illidge, L.N. (1987) Marchitez del algodonero en El Tolima. *ASCOLFI Informa* 13, 2–3.

Imaizumi, M. and Tairako, K. (1996) Control of strawberry *Verticillium* wilt disease [*V. dahliae*] by solar heating sterilization. *Annual Report of the Society of Plant Protection of North Japan,* pp. 69–71.

Infantino, A. and Porta-Puglia, A. (1990) *Verticillium dahliae* Kleb. on chickpea in Italy. *Phytopathologia Mediterranea* 29, 120–121.

Ingle, M.R. and Hastie, A.C. (1971) Factors affecting the production and recovery of heterozygous diploids of *Verticillium albo-atrum. Proceedings of the First International* Verticillium *Symposium.* Wye College, University of London, p. 20.

Ingle, M.R. and Hastie, A.C. (1974) Environmental factors affecting the formation of diploids in *Verticillium albo-atrum. Transactions of the British Mycological Society* 62, 313–321.

Ingram, R. (1968) *Verticillium dahliae* var. *longisporum,* a stable diploid. *Transactions of the British Mycological Society* 51, 339–341.

Insunza, V. and Eriksson, B. (1981) [Beet cyst nematode (*Verticillium* wilt) oil seed crops – soil pathogens interacting in plant disease aetiology]. *Nordisk Vaxtskyddskonferens 1981.* Skokloster, 17–19 Mars 1981. Del II Zoologisk och Nematologisk Sektion, pp. 90–95.

Ioannou, N. (1999) Management of soil-borne pathogens of tomato with soil solarization. *Technical Bulletin Cyprus Agricultural Research Institute,* Vol. 205.

Ioannou, N., Schneider, R.W. and Grogan, R.G. (1976) The influence of soil flooding on the production and survival of microsclerotia of *Verticillium dahliae. Proceedings of the Second International* Verticillium *Symposium.* University of California, Berkeley, p. 23.

Ioannou, N., Schneider, R.W. and Grogan, R.G. (1977a) Effect of oxygen, carbon dioxide and ethylene on growth, sporulation and production of microsclerotia by *Verticillium dahliae. Phytopathology* 67, 645–650.

Ioannou, N., Schneider, R.W. and Grogan, R.G. (1977b) Effect of flooding on the soil gas composition and the production of microsclerotia by *Verticillium dahliae. Phytopathology* 67, 651–656.

Ioannou, N., Schneider, R.W., Grogan, R.G. and Duniway, J.M. (1977c) Effect of water potential and temperature on growth, sporulation and production of microsclerotia by *Verticillium dahliae. Phytopathology* 67, 637–644.

Iordanov, M. and Yordanov, M. (1978) [Promising new tomato varieties for early field production]. *Gradinarstvo* 59(3), 41.

Iordanov, M. and Yordanov, M. (1979) [Once more on some problems of early tomato production]. *Gradinarstvo* 60(1), 4–7.

Ĭotov, L. (1962) New hop diseases in Bulgaria. *Rastitelna Zashchita* 10, 36–47.

Ireland, K.F. and Leath, K.T. (1987) Potential of using culture filtrates from *Verticillium albo-atrum* to evaluate germ plasm for resistance to *Verticillium* wilt. *Plant Disease* 71, 900–903.

Isaac, I. (1946) *Verticillium* wilt of sainfoin. *Annals of Applied Biology* 33, 28–34.

Isaac, I. (1949) A comparative study of pathogenic isolates of *Verticillium. Transactions of the British Mycological Society* 32, 137–157.

Isaac, I. (1953a) Studies on the interactions between species of *Verticillium. Annals of Applied Biology* 40, 623–629.

Isaac, I. (1953b) The spread of diseases caused by a species of *Verticillium. Annals of Applied Biology* 40, 630–638.

Isaac, I. (1953c) A further comparative study of pathogenic isolates of *Verticillium: V. nubilum* Pethybr. and *V. tricorpus* sp. nov. *Transactions of the British Mycological Society* 36, 180–185.

Isaac, I. (1956) Some factors affecting *Verticillium* wilt of *Antirrhinum. Annals of Applied Biology* 44, 105–112.

Isaac, I. (1957a) *Verticillium* wilt of brussels sprout. *Annals of Applied Biology* 45, 276–283.

Isaac, I. (1957b) The effects of nitrogen supply upon the *Verticillium* wilt of *Antirrhinum. Annals of Applied Biology* 45, 512–515.

Isaac, I. (1957c) Wilt of lucerne caused by species of *Verticillium. Annals of Applied Biology* 45, 550–558.

Isaac, I. (1959) *Verticillium* wilt of lucerne. *NAAS Quarterly Review* 46, 75–79.

Isaac, I. (1967) Speciation in *Verticillium. Annual Review of Phytopathology* 5, 201–222.

Isaac, I. and Davics, R.R. (1955) A new hyaline species of *Verticillium: V. intertextum* sp.nov. *Transactions of the British Mycological Society* 38(2), 143–156.

Isaac, I. and Harrison, J.A.C. (1968) The symptoms and casual agents of early-dying disease (*Verticillium* wilt) of potatoes. *Annals of Applied Biology* 61, 231–244.

Isaac, I. and Heale, J.B. (1961) Wilt of lucerne caused by species of *Verticillium*. III. Viability of *V. albo-atrum* carried with lucerne seed; effects of seed dressings and fumigants. *Annals of Applied Biology* 49, 675–691.

Isaac, I. and Keyworth, G.W. (1948) *Verticillium* wilt of the hop (*Humulus lupulus*) III. A study of the pathogenicity of the isolates from fluctuating and from progressive outbreaks. *Annals of Applied Biology* 35, 243–249.

Isaac, I. and Levy, J.E. (1971) The importance of the non-host to the survival of *Verticillium dahliae*. *Proceedings of the First International* Verticillium *Symposium*. Wye College, University of London.

Isaac, I. and Lloyd, A.T.E. (1959) Wilt of lucerne caused by species of *Verticillium*. II. Seasonal cycle of disease; range of pathogenicity; host–parasite relations; effects of seed dressings. *Annals of Applied Biology* 47, 673–684.

Isaac, I. and MacGarvie, Q.D. (1962) Germination of resting bodies in *Verticillium* species. *Nature* 195, 826–827.

Isaac, I. and MacGarvie, Q.D. (1966) Dormancy and germination of resting structures of *Verticillium* spp. *Transactions of the British Mycological Society* 49, 669–678.

Isaac, I. and Milton, J.M. (1967) Response of *Verticillium* species to griseofulvin. *Journal of General Microbiology* 46, 273–283.

Isaac, I. and Rogers, W.G. (1974) *Verticillium* wilt of pea (*Pisum sativum*). *Annals of Applied Biology* 76, 27–35.

Isaac, I., Fletcher, P. and Harrison, J.A.C. (1971) Quantitative isolation of *Verticillium* spp. from soil and moribund potato haulm. *Annals of Applied Biology* 67, 177–183.

Isaac, I., Panadian, T.T., Saraswathi-Devi, L. and Dube, H.C. (1972) *Verticillium* wilt of cotton in Tamil Nadu. *Transactions of the British Mycological Society* 59, 313–314.

Isaev, B.M. (1963) [Microelements and wilt]. *Khlopkovodstvo* 13, 44–45.

Isaev, B.M. (1966) [Important microelements in the control of *Verticillium* wilt of cotton]. *Trudy Vsesoyuznogo Nauchno-issledovatel'skogo Instituta Khlopskovodstva* 9, 86–90.

Ismailov, E.M. and Rysbayeva, A.S. (1990) Molecular genetic analysis of cotton wilt pathogen variability and its adaptability to different host cultivars. *Proceedings of the Fifth International* Verticillium *Symposium*. Leningrad, p. 72.

Itoh, S., Komada, H., Momma, T. and Amano, T. (1989) [Development of field diagnosis system (FDS) for preventing continuous cropping injury of crop. 12. Study of factors related to the development of a prediction model of *Verticillium* yellows in Chinese cabbage]. *Bulletin of the National Agricultural Research Center Japan* 16, 33–53.

Ivanova, L.V. and Kasyanenko, A.G. (1990) Vegetative hybridization between *Verticillium dahliae* and *Verticillium tricorpus*. *Proceedings of the Fifth International* Verticillium *Symposium*. Leningrad, p. 104.

Jackson, C.W. and Heale, J.B. (1985) Relationship between DNA content and spore volume in sixteen isolates of *Verticillium lecanii* and two new diploids of *Verticillium dahliae* equals *Verticillium dahliae* var. *longisporum*. *Journal of General Microbiology* 131, 3229–3236.

Jacobs, K.A. and Berg, L.C. (2000) Inhibition of fungal pathogens of woody plants by the plant growth regulator paclobutrazol. *Pest Management Science* 56(5), 407–412.

Jacobs, K.A., Bentz, S.E. and Johnson, G.R. (1994) Screening three *Cercis* species for susceptibility to *Verticillium dahliae*. *Plant Disease* 78(9), 925.

Jacobsen, B.J. (1974) Interaction between *Meloidogyne hapla* and *Verticillium* wilt disease of potato. *Dissertation Abstracts International* 34B, 5279.

Jacobsen, B.J., MacDonald, D.H. and Bissonette, H.L. (1979) Interaction between *Meloidogyne hapla* and *Verticillium albo-atrum* in the *Verticillium* wilt disease of potato. *Phytopathology* 69, 288–292.

Jagger, I.C. and Stewart, V.B. (1918) Some *Verticillium* diseases. *Phytopathology* 8, 15–19.

Janes, L.D., Ooosterhuis, D.M., Bourland, F.M. and Rothrock, C.S. (1993) Foliar potassium fertilization of different cotton cultivars. *Special Report – Agricultural Experiment Station, Division of Agriculture, University of Arkansas*, pp. 188–191.

Janowicz, K. and Mazurkiewicz, K. (1982) [Involvement of nematodes in tomato wilt]. *Ochrona Roslin* 26(4), 14.

Jarl, C.I., Rietveld, E.M. and de Haas, J.M. (1999) Transfer of fungal tolerance through interspecific somatic hybridisation between *Solanum melongena* and *S. torvum*. *Plant Cell Reports* 18(9), 791–796.

Jenkins, J.N., McCarty, J.C., Creech, R.G. and Parrott, W.L. (1979) Registration of 77 cotton germplasm (Reg.No. GP 64 to GP 140). *Crop Science* 19, 933–934.

Jensen, H.J. (1978) Interrelations of nematodes and other organisms in disease complexes. *International Potato Center. Report of the 2nd Planning Conference on the Developments in the Control of Nematode Pests of Potatoes.* Lima, Peru, 13–17 November, 1978, pp. 154–160.

Ji, H., Guo, X. and Pan, J. (1995) Physiological and biochemical indexes of resistance to *Verticillium dahliae* in cotton. *Acta Agriculturae Boreali-Sinica* 10(3), 73–75.

Jian, G.-L. and Ma, C. (1999) [Effects of Pix on *Verticillium* wilt of cotton]. *Acta Gossypii Sinica* 11(1), 23, 45–47.

Jiménez-Díaz, R.M. and Millar, R.L. (1986) Lack of systemic colonization of alfalfa plants after inoculation of uninjured leaves with conidia of *Verticillium albo-atrum*. *Plant Disease* 70, 509–515.

Jiménez-Díaz, R.M. and Millar, R.L. (1987) Fatores que afectan la esporulatión de *Verticillium albo-atrum* sobre tejidos de alfalfa afectados de marchitez. *Investigación Agraria, Producción y Protección Vegetales* 2, 295–303.

Jiménez-Díaz, R.M. and Millar, R.L. (1988) Sporulation on infected tissues, and presence of airborne *Verticillium albo-atrum* in alfalfa fields in New York. *Plant Pathology* 37, 64–70.

Jiménez-Díaz, R.M. and Montes-Agusti, F. (1974) Notas sobre *Verticillium dahliae* Kleb., causante de marchítez en diversas huéspedes. *Anales del Instituto Nacional de Investigaciones Agrarias Protección Vegetal* 4, 11–18.

Jiménez-Díaz, R.M., Blanco-López, M.A., Melero-Vara, J.M. and García Baudin, C. (1980) *Verticillium dahliae* Kleb. patogeno del girasol en España. *Comunicaciones Instituto Nacional de Investigaciones Agrarias* 10, 5–33.

Jiménez-Díaz, R.M., Blanco-Lopez, M.A. and Melero-Vara, J.M. (1985) [*Verticillium* wilt of safflower caused by *Verticillium dahliae* in Andalusia]. *Anales del Instituto Nacional de Investigaciones Agrarias Protección Vegetal* 28(Suppl.), 157–170.

Jiménez-Díaz, R.M., Bejarano-Alcazar, J., Melero Vara, J.M. and Blanco Lopez, M.A. (1990) The influence of the inoculum density of defoliating and non-defoliating pathotypes of *Verticillium dahliae* on epidemics of *Verticillium* wilt and yield components of cotton. *Proceedings of the Fifth International* Verticillium *Symposium*. Leningrad, p. 12.

Jiménez-Díaz, R.M., Bejarano, J., Blanco, M.A., Gomez, J., Gonzalez, R. and Melero, J.M. (1991) Control of *Verticillium* wilt and *Fusarium* wilt diseases by soil solarization in southern Spain. *FAO Plant Production and Protection Paper* 109, 94–107.

Jing, Y.-L., Liu, Y.-B., Fan, W.-F., Xiao, B.-C. and Jiang, K.-M. (1999) [Breeding of a new cotton variety – Shan 2234]. *Journal of Southwest Agricultural University* 21(6), 510–513.

Joaquim, T.R. and Rowe, R.C. (1990) Reassessment of vegetative compatibility relationships among strains of *Verticillium dahliae* using nitrate-nonutilizing mutants. *Phytopathology* 80, 1160–1166.

Joaquim, T.R. and Rowe, R.C. (1991) Vegetative compatibility and virulence of strains of *Verticillium dahliae* from soil and potato plants. *Phytopathology* 81, 552–558.

Joaquim, T.R., Smith, V.L. and Rowe, R.C. (1988) Seasonal variation and effects of wheat rotation on populations of *Verticillium dahliae* Kleb. in Ohio potato field soils. *American Potato Journal* 65, 439–447.

Joggi, D., Lehmann, J. and Briner, H.U. (1982a) Considérations sur la culture de la luzerne (*Medicago sativa* L.) et essais varietaux 1978–80. *Revue Suisse d'Agriculture* 14, 197–201.

Joggi, D., Lehmann, J. and Meister, E. (1982b) Sortenversuche mit Luzerne (1978 bis 1980). *Mitteilungen für die Schweizerische Landwurtschaft* 30, 109–117.

Joham, H.E. (1971) Some aspects of the zinc nutrition of cotton. *Proceedings of the Beltwide Cotton Production Research Conference.* National Cotton Council, Memphis, Tennessee. pp. 44–45.

Johansen, R.H., Farnsworth, B., Secor, C.A., Nelson, D.C., Orr, P. and Lana, E.P. (1981) Crystal, a new processing potato variety. *North Dakota Farm Research* 38(5), 7–8.

Johansen, R.H., Farnsworth, B.L., Secor, G.A., Gudmestad, N.C. and Thompson-Johns, A. (1993) Goldrush – a new russet potato variety. *North Dakota Farm Research* 50(1), 41–43.

Johnson, D.A. and Cummings, T.F. (2000) Evaluation of mint mutants, hybrids, and fertile clones for resistance to *Verticillium dahliae*. *Plant Disease* 84(3), 235–238.

Johnson, K.B. (1988) Modeling the influences of plant infection rate and temperature on potato foliage and yield losses caused by *Verticillium dahliae*. *Phytopathology* 78, 1198–1205.

Johnson, K.B., Radcliffe, E.B. and Teng, P.S. (1986) Effects of interacting populations of *Alternaria solani, Verticillium dahliae* and the potato leafhopper (*Empoasca fabae*) on potato yield. *Phytopathology* 76, 1046–1052.

Johnson, K.B., Teng, P.S. and Radcliffe, E.B. (1987) Analysis of potato foliage losses caused by interacting infestations of early blight, *Verticillium* wilt, and potato leafhopper, and the relationship to yield. *Zeitschrift für Pflanzenkrankheiten und Pflanzenschutz* 94, 22–33.

Johnson, K.B., Apple, J.D. and Powelson, M.L. (1988) Spatial patterns of *Verticillium dahliae* propagules in potato field soils of Oregon's Columbia basin. *Plant Disease* 72, 484–488.

Johnson, R.S. and Bosland, P.W. (1987) Development of a new disease screening technique for *Verticillium* wilt of pepper, *Capsicum annuum*. *Capsicum Newsletter* 6, 75–76.

Johnson, W.M., Johnson, E.K. and Brinkerhoff, L.A. (1980) Symptomatology and formation of microsclerotia in weeds inoculated with *Verticillium dahliae* from cotton. *Phytopathology* 70, 31–35.

Johnston, G.R. and Rowberry, R.G. (1980) Rideau: a new red-skinned maincrop table variety with high resistance to *Verticillium* wilt strains. *American Potato Journal* 57, 47–51.

Johnston, H.W., Celetti, M.J., Kimpinski, J. and Platt, H.W. (1994) Fungal pathogens and *Pratylenchus penetrans* associated with preceding crops of clovers, winter wheat, and annual ryegrass and their influence on succeeding potato crops on Prince Edward Island. *American Potato Journal* 71(12), 797–808.

Jones, H.G. (1986) Movement of liquid water and the effects of fungal infection. *Water, Fungi and Plants* 223–239.

Jones, J.P. and Crill, P. (1973) The effect of *Verticillium* wilt on resistant, tolerant and susceptible tomato varieties. *Plant Disease Reporter* 57, 122–124.

Jones, J.P. and Overman, A.J. (1976a) Tomato wilts, nematodes, and yields as affected by soil reaction and a persistent contact nematicide. *Plant Disease Reporter* 60, 913–917.

Jones, J.P. and Overman, A.J. (1976b) Failure of the *Ve* gene to control *Verticillium* wilt of tomato at high temperatures and short days. *Proceedings of the American Phytopathological Society* 3, 273.

Jones, J.P. and Overman, A.J. (1978) Evaluation of chemicals for the control of *Verticillium* and *Fusarium* wilt of tomato. *Plant Disease Reporter* 62, 451–455.

Jones, J.P. and Overman, A.J. (1986) Management of *Fusarium* wilt, *Fusarium* crown rot, *Verticillium* wilt (race 2), southern blight, and root-knot of tomato on fine sandy soils. *Proceedings of the Florida State Horticultural Society* 98, 229–231.

Jones, J.P. and Woltz, S.S. (1972) Effect of soil pH and micronutrient amendments on *Verticillium* and *Fusarium* wilt of tomato. *Plant Disease Reporter* 56, 151–153.

Jones, J.P., Overman, A.J. and Geraldson, C.M. (1971) Fumigants for the control of *Verticillium* wilt of tomato. *Plant Disease Reporter* 55, 26–30.

Jones, J.P., Crill, P. and Geraldson, C.M. (1973, published 1974) Effect of *Verticillium albo-atrum* wilt on tomato plants transplanted to fumigated and nonfumigated soil. *Proceedings of the Florida State Horticulture Society* 86, 162–165.

Jones, J.P., Crill, P. and Volin, R.B. (1975) Effect of light duration on *Verticillium* wilt of tomato. *Phytopathology* 65, 647–648.

Jones, J.P., Overman, A.J. and Crill, P. (1977, published 1978) Effect of temperature and short day on development of *Verticillium* wilt of susceptible, tolerant and resistant tomato cultivars. *Proceedings of the Florida State Horticultural Society* 90, 397–399.

Jones, R.A. and Millet, A.H. (1984) 'Sierra Sweet' fresh market tomato. *HortScience* 19, 133.

Jones, R.A., Smith, P.G., Millett, A.H. and Kimble, K.A. (1980) 'Royal Red Cherry' and 'Short Red Cherry' tomatoes. *HortScience* 15, 98.

Joost, O., Bianchini, G., Bell, A.A., Benedict, C.R. and Magill, C.W. (1995) Differential induction of 3-hydroxy-3-methylglutaryl CoA reductase in two cotton species following inoculation with *Verticillium*. *Molecular Plant–Microbe Interactions* 8(6), 880–885.

Jordan, E.G. and Barnett, H.L. (1978) Nutrition and parasitism of *Melanospora zamiae*. *Mycologia* 70, 300–312.

Jordan, V.W.L. (1972) Evaluation of fungicides for the control of *Verticillium* wilt (*V. dahliae*) of strawberry. *Annals of Applied Biology* 70, 163–168.

Jordan, V.W.L. (1973) The control of *Verticillium* wilt disease of strawberry. *Prace Instytutu Sadownictwa* E 2.

Jordan, V.W.L. (1974) Treatments to control strawberry wilt in fruiting plantations. *Grower* 82, 466.

Jordan, V.W.L. (1975) Control of *Verticillium* wilt transmission in strawberry runners with systemic benzimidazole-type fungicides. *Annals of Applied Biology* 81, 43–50.

Jordan, V.W.L. and Tarr, H.S. (1978) Inoculum suppression of *Verticillium dahliae*. *Annals of Applied Biology* 89, 139–141.

Jordan, V.W.L., Sneh, B. and Eddy, B.P. (1972) Influence of organic soil amendments on *Verticillium dahliae* and on the microbial composition of the strawberry rhizosphere. *Annals of Applied Biology* 70, 139–148.

Jordan, V.W.L., Hunter, T., Fielding, E.C., Stinchcombe, G.R., Hutcheon, J.A., Rooney, J.M., Huxley, M.W. and Clay, D.V. (1987) Controlling diseases. *Report, Long Ashton Research Station 1986*, pp. 27–29.

Julén, G. (1980) Development of high yielding varieties of lucerne resistant to *Verticillium albo-atrum*. *EPPO Bulletin* 10, 323–327.

June, R.I. (1950) *Verticillium* wilt of stone fruit. *New Zealand Journal of Agriculture* 81, 307–308.

Kaiser, W.J. (1962) Influences of light on the production of microsclerotia by *Verticillium albo-atrum*. *Phytopathology* 52, 362.

Kaiser, W.J. (1964a) Biotin requirement of the *Verticillium* wilt fungus. *Phytopathology* 54, 481–487.

Kaiser, W.J. (1964b) Effects of light on growth and sporulation of the *Verticillium* wilt fungus. *Phytopathology* 54, 765–770.

Kaiser, W.J. (1964c) Influence of nutrition and environment on the growth and sporulation of the plant pathogen *Verticillium albo-atrum*. *Dissertation Abstracts* 24, 3920–3921.

Kalandarov, S. (1990) New sources of cotton resistance to *Verticillium*. *Proceedings of the Fifth International* Verticillium *Symposium*. Leningrad, p. 105.

Kalb, D.W. and Millar, R.L. (1984) Dispersal of *V. albo-atrum* by fungus gnats (*Bradysia*). *Phytopathology* 74, 853.

Kalb, D.W. and Millar, R.L. (1986) Dispersal of *Verticillium albo-atrum* by the fungus gnat (*Bradysia impatiens*). *Plant Disease* 70, 752–753.

Kamal, M. and Saydam, C. (1970) [Studies on *Verticillium* wilt observed on cotton in the Aegean region]. *Bitki Koruma Bülteni* 10, 171–179.

Kamal, M. and Wood, R.K.S. (1956) Pectic enzymes secreted by *Verticillium dahliae* and their role in the development of wilt disease in cotton. *Annals of Applied Biology* 44, 322.

Kamran, R. (1985) [Reports on two new diseases caused by *Verticillium dahliae* in Fars Province]. *Iranian Journal of Plant Pathology* 21, 23, 71.

Kandov, A.I. and Bakhramov, K.B. (1976) [Earliness, wilt resistance and yield in new cotton varieties and lines]. *Uzbekistan Biologija Zurnali* 4, 67–69.

Kandrov, K.V., Vasil'ev, A.O., Kapkov, D.V., Fuentes, A.M. and Stepanov, A.S. (1990) Comparative analysis of the chromatographic, kinetic and antigenic properties of type 2 casein kinases from bovine liver and the phytopathogenic fungus *Verticillium dahliae*. *Biochemistry (New York)* 55, 1573–1578.

Kannan, A. and Srinivasan, K.V. (1984) A new race of *Verticillium dahliae* affecting cotton. *Current Science* 53, 104–105.

Kanno, H. and Horiuchi, S. (1999) *Verticillium* wilt of Perilla (*Perilla ocymoides* L.), Nalta jute (*Corchorus olitorius* L.) and Malabar nightshade (*Basella rubra* L.) caused by *Verticillium dahliae*. *Annual Report of the Society of Plant Protection of North Japan* 50, 120–124.

Kapkin, A. and Ari, M. (1982) A new host of *Verticillium dahliae* Kleb. in Turkey. *Journal of Turkish Phytopathology* 11, 77.

Karaca, I., Karcihoğlu, A. and Ceylân, S. (1971) Wilt disease of cotton in the Ege region of Turkey. *Journal of Turkish Phytopathology* 1, 4–11.

Karaca, I., Ceylan, S. and Karciloglu, A. (1973) The importance of cotton seed in the dissemination of *Verticillium* wilt. *Journal of Turkish Phytopathology* 2, 30–33.

Karadimos, D.A., Karaoglanidis, G.S. and Klonari, K. (2000) First report of *Verticillium* wilt of sugarbeet. *Plant Disease* 84(5), 593.

Karajeh, M. and Al-Raddad, A. (1999) Effect of VA mycorrhizal fungus (*Glomus mosseae* Gerd. & Trappe) on *Verticillium dahliae* Kleb. of olive. *Dirasat. Agricultural Sciences* 26(3), 338–341.

Karapapa, V.K., Bainbridge, B.W. and Heale, J.B. (1994) Diploid isolates of *Verticillium dahliae*, viz *V. dahliae* var. *longisporum* Stark, causing vascular disease in oilseed rape. *Proceedings of the Sixth International* Verticillium *Symposium*. Dead Sea, Israel. p. 42.

Karapapa, V.K., Bainbridge, B.W. and Heale, J.B. (1995) Diploid isolates of *Verticillium dahliae*, viz. *Verticillium dahliae* var. *longisporum* Stark causing vascular wilt disease in oilseed rape. *Phytoparasitica* 23(1), 42–43.

Karapapa, V.K., Baig, M.A., Heale, J.B. and Rossiter, J.T. (1997a) Glucosinolate response in winter oilseed rape *Brassica napus* ssp. *oleifera* to *Verticillium dahliae* (non-pathogenic) and *V. longisporum* comb. nov. [Karapapa, Bainbridge & Heale, 1997] (pathogenic). *Proceedings of the Seventh International* Verticillium *Symposium*. Cape Sounion, Athens, Greece, p. 50.

Karapapa, V.K., Bainbridge, B.W. and Heale, J.B. (1997b) Morphological and molecular characterization of *Verticillium longisporum* comb. nov., pathogenic to oilseed rape. *Mycological Research* 101(11), 1281–1294.

Karapapa, V.K., Bainbridge, B.W. and Heale, J.B. (1997c) Classical and molecular characterisation of the new species *Verticillium longisporum* pathogenic to oilseed rape. *Proceedings of the Seventh International* Verticillium *Symposium*. Cape Sounion, Athens, Greece, p. 12.

Karban, R., Adamchak, R. and Schnathorst, W.C. (1987) Induced resistance and interspecific competition between spider mites and a vascular wilt fungus. *Science* 235, 678–680.

Karcilioglu, A., Esentepe, M. and Sezgin, E. (1982) Investigations on the determination of susceptibility of some cotton varieties against cotton wilt disease caused by *Verticillium dahliae* Kleb. *Journal of Turkish Phytopathology* 11, 55–59.

Karoleva, N. and Kas'yanenko, A. (1978) [Resistance of the pathogen of *Verticillium* wilt to benlate]. *Khlopkovodstvo* 10, 31.

Karthaus, J.P. (1927) *Het Afsterven van Stengels en Knoppen bij de Roode Framboos.* Hollandia Drukkeril, Baarn, The Netherlands, pp. 40–44.

Kas'yanenko, A.G. (1980) [The structure of intraspecific variability of the pathogen of wilt and methods of plant breeding for resistance]. Микология и Фитопатология [*Mycology and Phytopathology*] 14, 247–253.

Kas'yanenko, A. (1987) [Genetics of the virulence of the wilt pathogen and breeding for immunity]. *Khlopkovodstvo* 8, 28–32.

Kas'yanenko, A.G. (1990) [*Verticillium* microevolution and cotton breeding for wilt resistance]. *Proceedings of the Fifth International* Verticillium *Symposium*. Leningrad, p. 108.

Kas'yanenko, A.G. and Portenko, L.G. (1978a) [Somatic recombination as a source of new races of cotton *Verticillium* wilt casual agent]. In: Maksumov, A. (ed.) [*Genetic Variability of the Wilt-causing Agent and Prospects for Increasing Cotton Wilt Resistance*]. Donish Publishing House, Dushanbe, USSR, pp. 61–67.

Kas'yanenko, A.G. and Portenko, L.G. (1978b) [Mutational variability of *Verticillium dahliae*]. In: Maksumov, A. (ed.) [*Genetic Variability of the Wilt-causing Agent and Prospects for Increasing Cotton Wilt Resistance*]. Donish Publishing House, Dushanbe, USSR, pp. 107–109.

Kas'yanenko, A.G. and Ryabova, I.M. (1986) [The infection conditions in accelerating breeding for resistance to soil phytopathogens]. *Selektsiya i Semenovodstvo USSR* 5, 24–25.

Kas'yanenko, A.G. and Shevtsova, V.M. (1981) [Genetic structure of physiologic races of the pathogen of *Verticillium* wilt of cotton]. Микология и Фитопатология [*Mycology and Phytopathology*] 15, 409–414.

Kas'yanenko, A.G., Alyamov, A.A. and Gorkovtseva, E.A. (1976) *Gossypium arboreum* and *G. herbaceum* as possible sources of wilt-resistant genes. *Proceedings of the Second International* Verticillium *Symposium*. University of California, Berkeley, p. 26.

Kasymova, T.D. and Yuldashev, P.Kh. (1995) Isolation and characterization of protease inhibitors from cotton seeds. *Chemistry of Natural Compounds* 31(3), 369–371.

Katan, J. (1981) Solar heating (solarization) of soil for control of soil-borne pests. *Annual Review of Phytopathology* 19, 211–236.

Katan, J., Greenberger, A., Alon, H. and Grinstein, A. (1975) Increasing soil temperatures by mulching for the control of soil-borne diseases. *Phytoparasitica* 3, 69.

Katan, J., Greenberger, A., Alon, H. and Grinstein, A. (1976a) Solar heating by polyethylene mulching for the control of diseases caused by soil-borne pathogens. *Phytopathology* 66, 683–688.

Katan, J., Greenberger, A., Grinstein, A. and Alon, H. (1976b) Additional studies on the control of *Verticillium dahliae* by polyethylene soil mulching. *Proceedings of the Second International* Verticillium *Symposium*. University of California, Berkeley, p. 27.

Katan, J., Grinstein, A., Greenberger, A., Yarden, O. and DeVay, J.E. (1987) The first decade (1976–1986) of soil solarization (solar heating): a chronological bibliography. *Phytoparasitica* 15, 229–255.

Katan, J., DeVay, J.E. and Greenberger, A. (1988) The biological control induced by soil solarization. In: Tjamos, E.C. and Beckman, C.H. (eds) *Vascular Wilt Diseases of Plants, Basic Studies and Control*. NATO ASI Series H: Cell Biology, Vol. 28. Springer-Verlag, Berlin, pp. 493–499.

Katsumata, Y., Koike, M., Amemiya, Y. and Shimada, T. (1993) Alfalfa–*Verticillium albo-atrum* interactions VII. Selection and characterization of high peroxidase cell lines in response to fungal cell wall products. *Research Bulletin of Obihiro University, Natural Science* 18(3), 151–156.

Kaufman, D.D. and Williams, L.E. (1965) Influence of soil reaction and mineral fertilization on numbers and types of fungi antagonistic to four soil-borne plant pathogens. *Phytopathology* 55, 570–574.

Kaufman, Z., Nachmias, A., Livescu, L., Meiri, A. and Tibor, M. (1990) [*Verticillium* wilt of potatoes under irrigation with saline water]. *Hassadeh* 70, 898–901.

Kawaradani, M., Kusakari, S., Kimura, M., Takizawa, H. and Nishihashi, H. (1998) [A new method for measuring β-1,3-glucanase activity using *p*-nitrophenyl-β-D-laminaritetraoside as a substrate to diagnose *Verticillium* wilt of eggplant]. *Annals of the Phytopathological Society of Japan* 64, 489–493.

Kawchuk, L.M., Lynch, D.R., Hachey, J., Bains, P.S. and Kulcsar, F. (1994) Identification of a codominant amplified polymorphic DNA marker linked to the verticillium wilt resistance gene in tomato. *Theoretical and Applied Genetics* 89(6), 661–664.

Kawchuk, L.M., Hachey, J. and Lynch, D.R. (1998) Development of sequence characterized DNA markers linked to a dominant *Verticillium* wilt resistance gene in tomato. *Genome* 41, 91–95.

Keen, N.T. and Erwin, D.C. (1971) Endopolygalacturonase – evidence against involvement in *Verticillium* wilt of cotton. *Phytopathology* 61, 198–203.

Keen, N.T. and Long, M. (1972) Isolation of a phytotoxic protein–lipopolysaccharide complex from *Verticillium albo-atrum*. *Physiological Plant Pathology* 2, 307–315.

Keen, N.T., Long, M. and Malca, I. (1970) Induction and repression of β-galactosidase synthesis by *Verticillium albo-atrum*. *Physiologia Plantarum* 23, 691–696.

Keen, N.T., Wang, M.C., Long, M. and Erwin, D.C. (1971) Dimorphism in *Verticillium albo-atrum* as affected by initial spore concentration and antisporulant chemicals. *Phytopathology* 61, 1266–1269.

Keen, N.T., Long, M. and Erwin, D.C. (1972) Possible involvement of a pathogen-produced protein–lipopolysaccharide complex in *Verticillium* wilt of cotton. *Physiological Plant Pathology* 2, 317–331.

Keinath, A.P. and Millar, R.L. (1986) Persistence of an alfalfa strain of *Verticillium albo-atrum* in soil. *Phytopathology* 76, 576–581.

Keinath, A.P., Fravel, D.R. and Papavizas, G.C. (1990) Evaluation of potential antagonists for biocontrol of *Verticillium dahliae*. *Proceedings of the Fifth International* Verticillium *Symposium*. Leningrad, p. 82.

Kelsheimer, E.G. and May, C. (1940) *Verticillium* wilt of elm. *Plant Disease Reporter* 24, 282–284.

Kendrick, J.B. and Middleton, J.T. (1959) Influence of soil temperature and strains of the pathogen on severity of *Verticillium* wilt of pepper. *Phytopathology* 49, 23–28.

Kendrick, J.B. and Schroeder, F.R. (1934) Inoculation tests with *Verticillium* wilt of muskmelons. *Phytopathology* 24, 1250–1252.

Kening, L., Rouse, D.I. and German, T.L. (1995) *Verticillium* species-specific molecular probes and their application for biological assay and phylogenetic studies. *Phytoparasitica* 23(1), 39–40.

Kennedy, D.M. (1985) The components of pathogen virulence and host resistance in cane fruit. *Annual Report, Scottish Crop Research Institute 1984*, p. 124.

Kennedy, D.M. (1987) *Verticillium* wilt of sea buckthorn (*Hippophae rhamnoides*). *Plant Pathology* 36, 420–422.

Kernasyuk, V.G. and Radzievskii, L.L. (1981) [Increase in yield]. *Zashchita Rastenĭĭ* 4, 24.

Kerr, A. (1961) A study of tomato root surface organisms antagonistic to *Verticillium albo-atrum*. *Transactions of the British Mycological Society* 44, 365–371.

Kerr, A., Bloss, H.E., Cartwright, D., Ellis, J. and Murphy, P.J. (1978) *Biennial Report, Waite Agricultural Research Institute 1976–1977*. pp. 120–121.

Kerr, E.A. and Cook, F.I. (1979) Crimsonvee tomato. *Canadian Journal of Plant Science* 59, 895–896.

Kerr, E.A. and Cook, F.I. (1982) Bonnyvee and Harvestvee tomatoes. *Canadian Journal of Plant Science* 62, 529–531.

Kerr, E.A. and Cook, F.I. (1983a) Wondervee tomato. *Canadian Journal of Plant Science* 63, 1103–1105.

Kerr, E.A. and Cook, F.I. (1983b) Ontario 7710 – a tomato breeding line with resistance to bacterial speck, *Pseudomonas syringae* pv *tomato* (Okabe). *Canadian Journal of Plant Science* 63, 1107–1109.

Kessler, K.J. (1966) Xylem sap as a growth medium for four tree wilt fungi. *Phytopathology* 56, 1165–1169.

Ketudat, V. (1969) The effects of some soil-borne fungi on the sex ratio of *Heterodia rostochiensis* on tomato. *Nematologica* 15, 229–233.

Keyworth, W.G. (1939) *Verticillium* wilt of hops. A summary of observations during 1938. *Report of the East Malling Research Station 1938*, pp. 244–249.

Keyworth, W.G. (1942) *Verticillium* wilt of the hop (*Humulus lupulus*). *Annals of Applied Biology* 29, 346–357.

Keyworth, W.G. (1944a) Diseases of hops. *Report of the East Malling Research Station 1943*, p. 28.

Keyworth, W.G. (1944b) *Verticillium* wilt of plum. *Report of the East Malling Research Station 1943*, pp. 52–54.

Keyworth, W.G. (1945) Three important hop diseases. *Report of the East Malling Research Station 1944*, pp. 130–134.

Keyworth, W.G. (1946) Release of two new varieties OR55 and OJ47, both with some resistance to *Verticillium* wilt. *Annual Report, East Malling Research Station 1945*, p. 43.

Keyworth, W.G. (1947) Notes on varieties of hop resistant to *Verticillium* wilt. *Report of the East Malling Research Station 1946*, pp. 157–159.

Keyworth, W.G. (1951) The progressive wilt of hops order. *Annals of Applied Biology* 38, 537–538.

Keyworth, W.G. (1953a) *Verticillium* wilt of the hop. VI. The relative roles of root and stem in the determination of wilt severity. *Annals of Applied Biology* 40, 344–361.

Keyworth, W.G. (1953b) Resistance of hop stems to invasion by *Verticillium albo-atrum*. *Nature (London)* 171, 656–657.

Keyworth, W.G. (1964) Hypersensitivity of monogenic resistant tomato scions to toxins produced in Bonny Best rootstocks invaded by *Fusarium oxysporum* f. *lycopersici*. *Annals of Applied Biology* 54, 99–105.

Keyworth, W.G. and Hewitt, E.J. (1948) *Verticillium* wilt of hops, V. The influence of nutrition on the reaction of the plant to infection with *Verticillium albo-atrum*. *Journal of Horticultural Science* 24, 219–227.

Keyworth, W.G., Hitchcock, M.M. and Goode, P.M. (1953) *Verticillium* wilt of the hop. VII. Further studies on wilt-resistant varieties. *Report of the East Malling Research Station 1952*, pp. 112–119.

Khaidarov, D.Kh. (1990) Wilt incidence in cotton cultivars of separate and mixed backgrounds. *Proceedings of the Fifth International* Verticillium *Symposium*. Leningrad, p. 73.

Khakimov, A. and Ynusov, M. (1980) [Lignin in combination with *Trichoderma*]. *Khlopkovodstvo* 5, 22–23.

Khakimov, A.Kh., Takhirov, M.T. and Usmanov, I.A. (1981) [Trichodermin and cotton wilt]. *Zashchita Rasteniĭ* 3, 36–37.

Khalida, A.-Ahmed, Michail, S.H. and Tarabeth, A.M. (1983) Testing certain cotton cultivars for resistance against *Verticillium dahliae* by using the soil inoculation method. *Iraqi Journal of Agricultural Science 'ZANCO'* 1, 149–156.

Khalitov, A.Kh., Sikorskii, I.A., Nemachenko, V.V. and Galaktionov, K.V. (1977) [The preparation tur as a means for plant protection]. *Zashchita Rasteniĭ* 10, 48.

Khalkhodzhaev, T. and Akhmedzhanov, A. (1971) [Use of the fluorescence effect in selecting individual plants]. *Khlopkovodstvo* 21, 34.

Khamidov, M. and Umarov, A. (1980) [Effect of fungicides on fibre and seed quality of cotton]. *Khlopkovodstvo* 9, 18–19.

Khan, A., Atibalentja, N. and Eastburn, D.M. (2000) Influence of inoculum density of *Verticillium dahliae* on root discoloration of horseradish. *Plant Disease* 84(3), 309–315.

Khan, A.J. and Minocha, S.C. (1989) Biosynthetic arginine decarboxylase in phytopathogenic fungi. *Life Sciences, USA* 44, 1215–1222.

Khan, F.Z. and Milton, J.M. (1975) Phytoalexin production by lucerne (*Medicago sativa* L.) in response to infection by *Verticillium*. *Physiological Plant Pathology* 7, 179–187.

Khan, F.Z. and Milton, J.M. (1978) Phytoalexin production and the resistance of lucerne (*Medicago sativa* L.) to *Verticillium albo-atrum*. *Physiological Plant Pathology* 13, 215–221.

Khan, F.Z. and Milton, J.M. (1979) Some factors affecting the production of medicarpin and sativan by lucerne leaflets in response to *Verticillium albo-atrum*. *Physiological Plant Pathology* 14, 1–17.

Khandrov, K.V., Vasil'ev, A.D., Kapkov, D.V., Fuéntes, A.M. and Stepanov, A.S. (1990) Comparative analysis of the chromatographic, kinetic, and antigenic properties of type 2 caseinkinases from bovine liver and the phytopathogenic fungus *Verticillium dahliae*. *Biochemistry New York* 55, 1573–1578.

Khar'kova, A. (1975) [The resistance of different varieties of eggplant to *Verticillium* disease]. *Selektsiya i Genetika Ovoshchynykh Kul'tur, Kishinev* 3, 204–206.

Khar'kova, A.P. (1990) [Results of sweet pepper and egg plant breeding for resistance to *Verticillium* wilt]. *Proceedings of the Fifth International* Verticillium *Symposium*. Leningrad. p. 74.

Khar'kova, A.P. and Okhova, E.P. (1972) [Premature wilting of tomato leaves]. Микология и Фитопатология [*Mycology and Phytopathology*] 6, 360–363.

Khasanov, T. (1975a) [Benomyl against *Verticillium* infection of cotton]. *Sintez i Primenenie Novykh Khimicheskikh Preparatov Protiv Vilta Khlopchatnika*. 72–74.

Khasanov, T. (1975b) [Results of trials with Uzgen-1 applied against wilt infection on slightly saline soils]. *Sintez i Primenenie Novykh Khimicheskikh Preparatov Protiv Vilta Khlopchatnika* 17–18.

Khasanov, T., Davlyalov, A. and Uranov, I. (1986) [A growth bioregulator for cotton]. *Khlopkovodstvo* 6, 23–24.

Khatib, A.El., Aramouni, A., Hassan, A. and Ravise, A. (1974) Accumulation des composés phénoliques et de phytoalexines par des varietés de tomate cultivées sur les sols infectés au Liban. *Phytopathologische Zeitschrift* 81, 23–27.

Khokhlacheva, V.E. (1990) [Infection potential of *Verticillium dahliae* Kleb. in soil and agroecological factors]. *Proceedings of the Fifth International* Verticillium *Symposium*. Leningrad. p. 27.

Khokhryakov, M.K. (1976) [The means of variability of *Verticillium dahliae* Kleb. and related fungi]. *Al'goflora i Mikroflora Sredn. Azii*. 263–270.

Khoury, F.Y. (1970) The influence of *Rhizoctonia solani* Kuhn and of *Meloidogyne incognita* Chitwood on the infection of cotton plants by *Verticillium albo-atrum* Reinke & Berthold. *Dissertation Abstracts International* 31B, 999B.

Khoury, F.Y. and Alcorn, S.M. (1973) Effect of *Meloidogyne incognita* on the susceptibility of cotton plants to *Verticillium albo-atrum*. *Phytopathology* 63, 485–490.

Khripunova, E.F. (1976) [The pathogens of fungus diseases of essential oil rose in Moldavia]. *Griby na Kul't. Rast. Moldavii* 45–54.

Khristov, A. (1977) [*Verticillium* wilt of sunflower in Bulgaria]. *Rastitelna Zashchita* 25, 40.

Khristova, E. (1958) [New diseases on decorative plants in Bulgaria and their control]. *Ovoshtarst and Gradinarst* 1, 38–41.

Kibishauri, V.D. and Tsiklauri, M.S. (1979) [The effect of chemical nutrition on the resistance of apricot to *Verticillium* wilt]. *Trudy Nauchno-Issledovatel'skogo Instituta Rasteniĭ Gruzinskoĭ SSR* 30, 10–14.

Kichanova, I.M., Salikhbayeva, M.T. and Dzhurayeva, Sh. (1990) Mineral nutrition of cotton and *Verticillium* with rabe. In: *Proceedings of the Fifth International* Verticillium *Symposium*. Leningrad, p. 93.

Kiessig, R. and Haller-Kiessig, R. (1957) Beitrag zur Kenntnis einen infecktiosen Welkekrankheit der Luzerne (*Verticillium albo-atrum* R. et B.). *Phytopathologische Zeitschrift* 31, 185–222.

Kim, K.K., Fravel, D.R. and Papavizas, G.C. (1988) Identification of a metabolite produced by *Talaromyces flavus* as glucose oxidase and its role in the biocontrol of *Verticillium dahliae*. *Phytopathology* 78, 488–492.

Kimpinski, J., Platt, H.W., Perley, S. and Walsh, J.R. (1998) *Pratylenchus* spp. and *Verticillium* spp. in New Brunswick potato fields. *American Journal of Potato Research* 75(2), 87–91.

Kingsley, P., Scriber, J.M., Grau, C.R. and Delwiche, P.A. (1983) Feeding and growth performance of *Spodoptere eridania* (Noctuidae: Lepidoptera) on 'vernal' alfalfa, as influenced by *Verticillium* wilt. *Protection Ecology* 5, 127–134.

Kirkpatrick, B.L. and Harrison, H.F. (1979) Effect of *Verticillium* on velvetleaf (*Abutilon theophrasti* Medic.) survivorship and its theoretical implications. *Proceedings of 1979 Meeting of the Weed Science Society of America. Abstracts*. pp. 76–77.

Kitazawa, K. and Sato, R. (1981) [Wilt of alfalfa caused by *Verticillium albo-atrum* Reinke & Berthold]. *Annals of the Phytopathological Society of Japan* 47, 272–274.

Kitazawa, K. and Suzui, T. (1980) [*Verticillium* black-spot disease of radish]. *Annals of the Phytopathological Society of Japan* 46, 271–273.

Klebahn, H. (1913) Beiträge zur Kenntnis der Fungi Imperfecti I. Eine *Verticillium-Krankheit* auf Dahlien. *Mykologisches Zentralblatt* 3, 49–66.

Klebahn, H. (1937) Untersuchungen über die Krankheiten des Meerrettichs. II Bericht. *Phytopathologische Zeitschrift* 10, 121–167.

Klingner, A.E., Hildebrand, D.C. and Wilhelms, S. (1971) Occurrence of *Erwinia carotovora* in the rhizosphere of cotton plants which escape *Verticillium* wilt. *Plant and Soil* 34, 215–218.

Klisiewicz, J.M. (1974a) Role of infected safflower seed in survival and dissemination of *Verticillium dahliae*. *Annual Proceedings of the American Phytopathological Society for 1974* 1, 37.

Klisiewicz, J.M. (1974b) Assay of *Verticillium* in safflower seed. *Plant Disease Reporter* 58, 926–927.

Klisiewicz, J.M. (1975) Survival and dissemination of *Verticillium* in infected safflower seed. *Phytopathology* 65, 696–699.

Klisiewicz, J.M. (1981) Reaction of sunflower and safflower germ plasm to *Verticillium dahliae*. *Plant Disease* 65, 237–239.

Klisiewicz, J.M. (1986) Susceptibility of yellow starthistle to selected plant pathogens. *Plant Disease* 70, 295–297.

Knoll, F.A. (1972) Untersuchungen zur Ausbreitung gefässbesiedelnder *Verticillium* Arten in Luzernepflanzen. *Zentralblatt für Bakteriologie, Parasitenkunde, Infektionskrankheiten und Hygiene* 127, 332–345.

Knoll, F.A. (1978) Beitrag zum systematisch-chemotherapeutischen Verhalten von Carbendazim und dessen präkursorischen Verbindungen. *Archiv für Phytopathologie und Pflanzenschutz* 14, 313–325.

Kocaturk, S. and Karcilioglu, A. (1976a) [Research on identification of hosts and species of *Verticillium* in the Aegean region]. *Zirai Mücadele Araştirma Yilligĭ* 10, 19–20, 50.

Kocaturk, S. and Karcilioglu, A. (1976b) [Investigations on the determination of hosts and species of *Verticillium* in the Aegean region]. *Bitki Koruma Bülteni* 19, 237–242.

Kocaturk, S. and Karcilioglu, A. (1980) [Investigations to determine the host range and species of *Verticillium* in the Aegean region]. *Bitki Koruma Bülteni* 19, 237–242.

Kohlmann, J., Kremheller, H.T. and Liebl, H. (1974) Schädlinge und Krankheiten des Hopfens und ihre Bekämpfung. *Hopfen-Rundschau* 25, 439–441.

Koike, M., Saito, J. and Shimada, T. (1992) Alfalfa-*Verticillium albo-atrum* interactions: V. Phytoalexin accumulation in calli in response to the cytotoxic components of fungal culture filtrates. *Research Bulletin of Obihiro University Natural Science* 18, 35–39.

Koike, M., Shimada, T. and Amemiya, Y. (1993) Alfalfa-*Verticillium albo-atrum* interactions: IV. Reactions of alfalfa protoplasts to fungal culture filtrates and cell wall components. *Nippon Sochi Gakkaishi [Journal of Japanese Society of Grassland Science]* 39(1), 101–107.

Koike, M., Katsumata, Y., Amemiya, Y. and Shimada, T. (1995a) Alfalfa–*Verticillium albo-atrum* interactions. 6. Selection and characterization of high PAL cell lines in response to fungal cell wall components. *Grassland Science* 41(2), 128–132.

Koike, M., Watanabe, M., Nagao, H. and Ohshima, S. (1995) Molecular analysis of Japanese isolates of *Verticillium dahliae* and *V. albo-atrum*. *Letters in Applied Microbiology* 21(2), 75–78.

Koike, M., Watanabe, M. and Nagao, H. (1995c) RAPD analysis of *Verticillium dahliae* in five different pathogenicity groups in Japan. *Phytoparasitica* 23(1), 40–41.

Koike, M., Fujita, M., Nagao, H. and Oshima, S. (1996a) Random amplified polymorphic DNA analysis of Japanese isolates of *Verticillium dahliae* and *Verticillium albo-atrum*. *Plant Disease* 80, 1224–1227.

Koike, M., Kozawa, T., Okino, K., Nagao, H. and Shimada, T. (1996b) *In vitro* evaluation of potato resistance to *Verticillium albo-atrum*. *Research Bulletin of Obihiro University, Natural Science* 20(1), 11–16.

Koike, M., Nagao, H., Robb, J. and Dobinson, K. (1997) RAPD and PCR-RFLP analyses of Japanese and Canadian isolates of plant pathogenic *Verticillium* spp. *Proceedings of the Seventh International* Verticillium *Symposium*. Cape Sounion, Athens, Greece, p. 11.

Koike, S.T. (1988) First report of *Verticillium* wilt (*V. albo-atrum*) of *Vinca major* in California. *Plant Disease* 72, 1077.

Koike, S.T. (1991) *Verticillium* wilt of dusty-miller (*Senecio cineraria*) caused by *Verticillium dahliae*. *Plant Disease* 75, 319.

Koike, S.T. and Subbarao, K.V. (1995) *Verticillium* wilt of cauliflower: a new disease in California. 6th International *Verticillium* Symposium, Dead Sea, Israel. *Phytoparasitica* 23, 52.

Koike, S.T., Nelson, J.H. and Perry, D.K. (1991) *Verticillium* wilt of protea (*Leucospermum cordifolium*) caused by *Verticillium dahliae*. *Plant Disease* 75, 1074.

Koike, S.T., Subbarao, K.V., Davis, R.M., Gordon, T.R. and Hubbard, J.C. (1994) *Verticillium* wilt of cauliflower in California. *Plant Disease* 78(11), 1116–1121.

Koike, S.T., Xiao, C.L., Hubbard, J.C., Schulbach, K.F. and Subbarao, K.V. (1997) Effects of broccoli residue on the cauliflower–*Verticillium dahliae* host–pathosystem. *Proceedings of the Seventh International* Verticillium *Symposium*. Cape Sounion, Athens, Greece, p. 80.

Kondo, N., Kodama, F., Ozaki, M. and Akai, J. (1987) *Bulletin of Hokkaido Prefectural Agricultural Experiment Station* 56, 59–64.

Konnova, E.V., Sattarova, R.S. and Khalmuradov, A.G. (1995) Polysaccharides of the cell walls of fungi pathogenic for the cotton plant. *Chemistry of Natural Compounds* 31(2), 163–165.

Korolev, N. and Katan, T. (1997) Improved medium for selecting nitrate-nonutilizing (*nit*) mutants of *Verticillium dahliae*. *Phytopathology* 87(10), 1067–1070.

Korolev, N. and Katan, T. (1999) Vegetative compatibility grouping in *Verticillium nigrescens* and *V. tricorpus*. *Mycological Research* 103, 65–76.

Korolev, N., Jimenez-Dias, R.M., Katan, J., Perez-Artes, E., Garcia-Pfdrajas, M., Bejarano-Alcazar, J., Rodriguez-Jurado, D. and Katan, T. (1997) Comparative study of vegetative compatibility among cotton isolates of *Verticillium dahliae* from Spain and Israel. *Proceedings of the Seventh International* Verticillium *Symposium*. Cape Sounion, Athens, Greece, p. 28.

Korolev, N., Katan, J. and Katan, T. (2000) Vegetative compatibility groups of *Verticillium dahliae* and their regional distribution in Israel. *Phytopathology* 90, 529–536.

Koroleva, N.S. (1985) [Methods of determining the content of *Verticillium dahliae* Kleb. diaspores in soil]. Микология и Фитопатология [*Mycology and Phytopathology*] 19(1), 73–78.

Koroleva, N.S. and Kas'yanenko, A.G. (1977) [Isolation of benlate-resistant mutants of *Verticillium dahliae* and the study of their virulence]. *Doklady Akademii Nauk Tadzh-SSR* 20(5), 55–57.

Koroleva, N.S. and Kas'yanenko, A.G. (1978) [Resistance of the pathogen of *Verticillium* wilt to benlate]. *Khlopkovodstvo* 10, 31.

Koroleva, N.S. and Kas'yanenko, A.G. (1987) [Studies on the ecology, population and evolution of microfungi of the genus *Verticillium* II. Dynamics of the racial composition of a population of *Verticillium dahliae*]. Микология и Фитопатология [*Mycology and Phytopathology*] 21, 164–168.

Koroleva, N.S., Kas'yanenko, A.G. and Ryabova, I.M. (1978) [Resistance of *Verticillium dahliae* Kleb. to benomyl]. Микология и Фитопатология [*Mycology and Phytopathology*] 12, 204–208.

Koroleva, N.S., Kas'yanenko, A.G. and Miller, V.R. (1986) [Studies on the ecology, populations and evolution of *Verticillium* species. I. Population density dynamics of *Verticillium dahliae* Kleb.]. Микология и Фитопатология [*Mycology and Phytopathology*] 20(6), 509–512.

Kort, J. and van Rheenen, H.A. (1959) [Some observations on the occurrence of lucerne wilt disease in Zeeland in 1957]. *Tijdschrift over Plantenziekten* 65, 147–157.

Kostenko, I.R. and Nesterov, Yu.B. (1971) [Protection of cotton from pests and diseases in the Uzbek SSR]. *Trudy Vsesoynznogo Nauchno-issledovatel'skogo Instituta Zashchity Rastenii* 32, 24–32.

Kostroma, E.Yu. (1973) [Activity of pectolytic enzymes of strains of *Verticillium dahliae* differing in virulence and of other species of pathogens of *Verticillium* wilt of plum and apricot]. *Byulleten' Vsesoyuznogo Nauchno-Issledovatel'skogo Instituta Zashchity Rasteniĭ* 26, 53–57.

Kostroma, E.Yu. and Kropis, E.P. (1972) [Toxicity of extracellular exudates of *Verticillium dahliae* Kleb. the causal agent of wilt of stone fruit trees]. Микология и Фитопатология [*Mycology and Phytopathology*] 6, 252–255.

Kostyuk, N.A. (1973) [Association of nematodes with *Verticillium* wilt in cotton]. *Trudy Gel'mintologicheskoi Laboratorii Ekologiya i Taksonomiya Gel'mintov* 23, 77–82.

Kosuge, E., Iijima, T. and Ida, S. (1977) [Experimental studies on the breeding of tomato resistant to *Verticillium* wilt breeding the NFVR and VR lines]. *Bulletin of the Tokyo-to Agricultural Experiment Station* 10, 3–29.

Kotcon, J.B. and Rouse, D.I. (1984) Root deterioration in the potato early dying syndrome: causes and effects of root biomass reductions associated with colonization by *Verticillium dahliae*. *American Potato Journal* 61, 557–568.

Kotcon, J.B., Rouse, D.I. and Mitchell, J.E. (1984) Dynamics of root growth in potato fields affected by the early dying syndrome. *Phytopathology* 74, 462–467.
Kotcon, J.B., Rouse, D.I. and Mitchell, J.E. (1985) Interactions of *Verticillium dahliae, Colletotrichum coccoides, Rhizoctonia solani* and *Pratylenchus penetrans* in the early dying syndrome of Russet Burbank potatoes. *Phytopathology* 75, 68–74.
Koury, F.Y. (1970) The influence of *Rhizoctonia solani* Kuhn and of *Meloidogyne incognita acida* Chitwood on the infection of cotton plants by *Verticillium albo-atrum* Reinke & Berthold. *Dissertation Abstracts International* 31B, 999.
Kovacs, G. (1976) [A comparison of soil disinfectants for effectiveness against *Verticillium dahliae*]. *Gyumolcstermesztes* 3, 137–141.
Kowalik, M. (1994) Grzyby zasiedlajace srodowisko systemu korzeniowego pomidora w uprawie hydroponicznej. *Zeszyty Naukowe Akademii Rolniczej im. Hugona Kołłataja w Krakowie, Ogrodnictwo* 87–95.
Krangauz, R.A. (1958) [New disease of oak seedlings in the south-eastern European part of the USSR]. *Sbornik Rabot po Lesnomu Khozyaistvu* 35, 258–260.
Krasil'nikov, N.A., Khodzhibaeva, S.M. and Mirchink, T.G. (1965) [Formation of toxins by *Verticillium dahliae*, the casual agent of cotton wilt]. *Agrokhimiya* 1965(10), 128–134.
Krasil'nikov, N.A., Khodzhibaeva, S.M. and Mirchink, T.G. (1969) Properties of toxins of *Verticillium dahliae*, the causative agent of cotton wilt disease. *Journal of General and Applied Microbiology* 15, 1–9.
Krasnoshchekova, O.F. and Runov, V.I. (1962) [The effect of infection by *Verticillium dahliae* and *Fusarium vasinfectum* on the tannin substance content of cotton varieties differing in resistance]. *Uzbekskii Biologicheskii Zhurnal* 6(5), 13–15.
Krasnyanski, S., Ball, T.M. and Sink, K.C. (1998) Somatic hybridization in mint: identification and characterization of *Mentha piperita* × *M. spicata* hybrid plants. *Theoretical and Applied Genetics* 96, 683–687.
Kratirov, O. (1965) [*Verticillium* fungus and resistance of cotton varieties to wilt]. *Sel'sko Khoziaistvo Uzbek.* 2, 24–25.
Kratka, J. and Kudela, V. (1981) Biochemical changes in alfalfa plants inoculated with *Verticillium albo-atrum*. *Phytopathologische Zeitschrift* 100, 289–299.
Kratka, J. and Kudela, V. (1982) [The effect of the application of IAA and kinetin on the degree of infection of lucerne by bacterial and *Verticillium* wilt]. *Sbornik-UVTIZ, Ochrana Rostlin* 18, 107–113.
Krátka, J. and Kůdela, V. (1987) [Evaluation of plant resistance to vascular pathogens by conventional and biochemical methods]. *Ochrana Rostlin* 23, 247–254.
Krause-Heuther, J.P., Sotiropoulos, P. and Adams, L.E. (1975) Systemic insecticide control of aphids and potato *Verticillium* wilt. *Plant Disease Reporter* 59, 159–163.
Kravtsova, T.I. (1990) The area of distribution of *Verticillium*. *Proceedings of the Fifth International* Verticillium *Symposium*, Leningrad, p. 29.
Kreitlow, K.W. (1962) *Verticillium* wilt of alfalfa, a destructive disease in Britain and Europe not yet observed in the United States. *US Department of Agriculture, Agricultural Research Service* 34–20, 1–15.
Krikun, J. and Bernier, C.C. (1987) Infection of several crop species by two isolates of *Verticillium dahliae*. *Canadian Journal of Plant Pathology* 9, 241–245.
Krikun, J. and Bernier, C.C. (1990) Morphology of microsclerotia of *Verticillium dahliae* in roots of gramineous plants. *Canadian Journal of Plant Pathology* 12(4), 439–441.
Krikun, J. and Chorin, M. (1966) *Verticillium* – a new soil pathogen in the Negev region of Israel. *Israel Journal of Agricultural Research* 16, 177–178.

Krikun, J. and Frank, Z.R. (1975) Effect of sequestrene on the reaction to *Verticillium dahliae* of peanuts growing in calcareous loess soil. *Phytoparasitica* 3, 77–78.

Krikun, J. and Frank, Z.R. (1982) Metham sodium applied by sprinkler irrigation to control pod rot and *Verticillium* wilt of peanut. *Plant Disease* 66, 128–130.

Krikun, J. and Orion, D. (1977) Studies of the interaction of *Verticillium dahliae* and *Pratylenchus thornei* on potato. *Phytoparasitica* 5, 67.

Krikun, J. and Orion, D. (1979) *Verticillium* wilt of potato: importance and control. *Phytoparasitica* 7, 107–116.

Krikun, J. and Susnoski, M. (1971) Introduction and establishment of *Verticillium dahliae* in a newly developed arid zone. *Proceedings of the First International* Verticillium *Symposium*. Wye College, University of London, p. 21.

Krikun, J., Chorin, M. and Vaadia, Y. (1971) Kinetics of the hormonal imbalance of tomato plants infected with *Verticillium dahliae*. Proceedings of the Third Meeting of the Phytopathological Society of Israel. *Israel Journal of Agricultural Research* 21, 149.

Krikun, J., Orion, D., Netzer, D. and Schlevin, E. (1976) [An improved method of applying Edigan (metham-sodium) to control soil pathogens]. *Hassadeh* 57, 369–371.

Krikun, J., Orion, D. and Nachmias, A. (1984) Integrated control of *Verticillium dahliae* in potato. Pathobiology of *Verticillium* species. Third International *Verticillium* Symposium, Bari, Italy. *Phytopathologia Mediterranea* 23, 114.

Krikun, J., Barash, Y. and Nachmias, A. (1990) Mineral imbalance in *Verticillium dahliae* infected hosts and its correction by nutrient application. *Proceedings of the Fifth International* Verticillium *Symposium*, Leningrad, p. 83.

Kristyanne, E.S., Kim, K.S. and Stewart, J.Mcd. (1997) Magainin 2 effects on the ultrastructure of five plant pathogens. *Mycologia* 89(3), 353–360.

Kritzman, G., Nachmias, A. and Ben Yephet, Y. (1987) Interaction of *Verticillium dahliae* and *Erwinia* spp. pathogens of potato. *Canadian Journal of Plant Pathology* 9, 82.

Kroeker, G. (1976) *Verticillium dahliae* on brassica oilseed crops in Sweden. *Proceedings of the Second International* Verticillium *Symposium*. University of California, Berkeley. pp. 28–29.

Kr'stev, K. and Savov, S.G. (1977) [Physiological and biochemical changes in cotton plants attacked by *Verticillium* wilt]. *Rasteniev'dni Nauki* 14(3), 134–139.

Krüger, W. (1987) Schwierige Diagnose bei Befall von Raps mit *Verticillium dahliae* and *Phoma lingam*. *Bulletin SROP* 10(4), 35–37.

Krüger, W. (1989) Untersuchungen zur Verbreitung von *Verticillium dahliae* Kleb. und anderen Krankheits – und Schaderregern bei Raps in der Bundesrepublik Deutschland. *Nachrichtenblatt des Deutschen Pflanzenschutzdienstes, Stuttgart* 41, 49–56.

Kruger, W. (1991) Resistance testing in oilseed rape. *Bulletin SCROP* 14, 205–211.

Ksaymova, T.D., Mezhumyan, L.G. and Yuldasheva, P.Kh. (1990) The investigation of cotton seed protease inhibitors. *Proceedings of the Fifth International* Verticillium *Symposium*. Leningrad, p. 55.

Kůdela, V. (1970) [Method for assessing the resistance of lucerne varieties to vascular wilt]. *Rostlinná Vyroba* 16(43), 1041–1050.

Kudela, V. (1973) [Brown lucerne seeds, their origin and distribution in the plant]. *Rostlinna Vyroba* 19, 417–423.

Kůdela, V. (1974) [Comparison of vascular wilt ratings on lucerne after artificial and natural infection]. *Ochrana Rostlin* 10, 103–111.

Kudela, V. (1975) [Reaction of lucerne to *Verticillium albo-atrum* at different photoperiod lengths]. *Ochrana Rostlin* 11, 275–282.

Kudela, V. and Pirkl, J. (1978) Influence of soil pH and calcium nutrition on resistance of alfalfa to bacterial and *Verticillium* wilt. *Zentralblatt für Bakteriologie, Parasitenkunde, Infektionskrankeiten und Hygiene Abt.* 2 133, 503–511.

Kůdela, V. and Řezáč, A. (1972) [Testing European lucerne varieties for resistance to bacterial and *Verticillium* wilt]. *Ochrana Rostlin* 8, 269–276.

Kugler, J.L., Nygaard, S.L., Huset, D.E. and Peterson, M.A. (1995) Registration of 'WL 323' alfalfa. *Crop Science* 35(5), 1501.

Kuhn, G.B. (1981) Morte de plantas de videira (*Vitis* spp.) devido à ocorrência de fungos causadores de podridões radiculares e doenças vasculares. *Circular Tecnica EMBRAPA* 6, 30.

Kuhne, H. (1980) Pilzkrankheiten an Himbeeren. *Erwerbsobstbau* 22, 161–163.

Kulibaba, Yu.F. (1972) [Diseases of *Gerbera* under glass]. *Byulleten' Glavnogo Botanicheskogo Sada* 83, 115–119.

Kunkel, R. and Weller, M. (1965) Fumigation of potato soils in Washington. *American Potato Journal* 42, 57–69.

Kurahashi, Y., Araki, Y., Kinbara, T., Pontzen, R. and Yamaguchi, I. (1998) Intermediates accumulation and inhibition sites of carpropamid in the melanin biosynthesis pathway of *Pyricularia oryzae*. *Journal of Pesticide Science* 23(1), 22–28.

Kurbanov, M. and Nasridinov, M. (1980) [Chemical preparations and wilt-infection of cotton]. *Khlopkovodstvo* 9, 19.

Kuter, G.A. (1984) Hyphal interactions between *Rhizoctonia solani* and some *Verticillium* species. *Mycologia* 76, 936–940.

Kuznetsov, L.V. (1979) [Taxonomy of the pathogens of *Verticillium* wilt of cotton]. Микология и Фитопатология [*Mycology and Phytopathology*] 13, 247–253.

Kuznetsov, L.V., Kornilova, V. and Ageeva, I. (1977) [Lipid composition in strains of *Verticillium dahliae* differing in pathogenicity]. *Prikladnaya Biokhimiya Mikrobiologiya* 13, 527–529.

La Mondia, J.A., Gent, M.P.N., Ferrandino, F.J., Elmer, W.H. and Stoner, K.A. (1999) Effect of compost amendment or straw mulch on potato early dying desease. *Plant Disease* 83, 361–366.

Lacey, C.N.D., Jellis, G.I., Currell, S.B. and Thomson, A.J. (1984) Changes in the apparent tolerance of potato cultivars to *Globodera* spp. in the presence of *Verticillium dahliae*. In: Winiger, F.A. and Stockli, A. (eds) *Proceedings of the 9th Triennial Conference of the European Association for Potato Research.* Interlaken, July 1984. pp. 359–360.

Lacy, M.L. and Horner, C.E. (1966) Behaviour of *Verticillium dahliae* in the rhizosphere and on roots of plants susceptible, resistant and immune to wilt. *Phytopathology* 56, 427–430.

Lambert, F. and Pujarniscle, S. (1984) Purification and properties of the proteinase produced *in vitro* by *Verticillium dahliae*. *Canadian Journal of Microbiology* 30, 1488–1493.

Lambeth, V.N. (1982) Origin and release of tomato line 31-St-15 and hybrid 'Conquest' VFN. *Research Bulletin, Agricultural Experiment Station, College of Agriculture, University of Missouri* 1044, 1–4.

Lamers, J.G. (1995) *Verticillium dahliae* incidence on potatoes grown after a monoculture of onions, forage maize, flax, kidney beans, peas or field beans. 6th International *Verticillium* Symposium, Dead Sea, Israel. *Phytoparasitica* 23, 50.

Lamers, J.G. and Termorshuizen, A.J. (1998) *Verticillium*-besmetting omlaag? Aardappelloof behandelen om *Verticillium*-besmetting van de grond te verlagen. *PAV Bulletin Akkerbouw* 6–9.

Langcake, P. (1981) Alternative chemical agents for controlling plant disease. *Philosophical Transactions of the Royal Society of London B* 295, 83–101.

Larkin, R.P. and Fravel, D.R. (1998) Biological control of wilt pathogens with fungal antagonists. *Proceedings of the Beltwide Cotton Conferences*. San Diego, California, 5–9 January 1998, pp. 125–127.

Laskin, A.I. and Lechevalier, H.A. (1973) *Handbook of Microbiology – Microbial Composition*. CRC Press, Cleveland, Ohio.

Laterrout, H. and Pécaut, P. (1966) *Proceedings, First Congress of the Mediterranean Phytopathological Union*. pp. 431–433.

Latorre, B.A. and Allende, P.T. (1983) Occurrence and incidence of *Verticillium* wilt of Chilean avocado groves. *Plant Disease* 67, 445–447.

Latorre, B.A., Flores, V. and Marholz, G. (1987) Wilt of tobacco in Chile caused by *Verticillium dahliae*. *Plant Disease* 71, 101.

Latorre, B.A., Lolas, M. and Marholz, G.M. (1989) *Verticillium* wilt, a limiting factor for tobacco production in Chile. *Plant Disease* 73, 664–666.

Latunde-Dada, A.O. and Lucas, J.A. (1983) Somaclonal variation and reaction of *Verticillium* wilt in *Medicago sativa* L. plants regenerated from protoplasts. *Plant Science Letters* 32, 205–211.

Latunde-Dada, A.O. and Lucas, J.A. (1985) Involvement of the phytoalexin medicarpin in the differential response of callus lines in lucerne (*Medicago sativa*) to infection by *Verticillium albo-atrum*. *Physiological Plant Pathology* 26, 31–42.

Latunde-Dada, A.O. and Lucas, J.A. (1986) Influence of temperature on host resistance and fungal sensitivity to medicarpin in lucerne callus lines infected with *Verticillium albo-atrum*. *Physiological and Molecular Plant Pathology* 28, 89–97.

Latunde-Dada, A.O. and Lucas, J.A. (1988) Somaclonal variation and resistance to *Verticillium* wilt in lucerne, *Medicago sativa* L., plants regenerated from callus. *Plant Science Irish Republic* 58, 111–119.

Latunde-Dada, A.O., Dixon, R.A. and Lucas, J.A. (1987) Induction of phytoalexin biosynthetic enzymes in resistant and susceptible lucerne callus lines infected with *Verticillium albo-atrum*. *Physiological and Molecular Plant Pathology* 31, 15–23.

Lawrence, W.H. (1912) Bluestem of the black raspberry. *Bulletin of the Washington Agricultural Experiment Station* 108, 1–30.

Lazarovits, G. (1997) Organic soil amendments for the control of soilborne plant pathogens. *Proceedings of the Seventh International* Verticillium *Symposium*. Cape Sounion, Athens, Greece, p. 73.

Lazarovits, G. and Stoessl, A. (1988) Tricyclazole induces melanin shunt products and inhibits altersolanol A accumulation by *Alternaria solani*. *Pesticide Biochemistry and Physiology* 31, 36–45.

Lazarovits, G., Bar-Joseph, M. and Nachmias, A. (1987) Detection of *Verticillium* antigens by dot-ELISA on nitrocellulose membranes. *Canadian Journal of Plant Pathology* 9, 82.

Lazarovits, G., Hawke, M.A., Tomlin, A.D., Olthof, T.H.A. and Squire, S. (1991) Soil solarization to control *Verticillium dahliae* and *Pratylenchus penetrans* on potatoes in Ontario. *Canadian Journal of Plant Pathology* 13, 116–123.

Lazarovits, G., Conn, K.L. and Potter, J. (1999) Reduction of potato scab, *Verticillium* wilt, and nematodes by soymeal and meat and bone meal in two Ontario potato fields. *Canadian Journal of Plant Pathology* 21(4), 345–353.

Leach, C.M. (1962) Sporulation of diverse species of fungi under near-ultraviolet radiation. *Canadian Journal of Botany* 40, 151–161.

Leach, L.D., Huffman, E.M. and Reische, W.C. (1969) Some factors influencing protection of cotton from *Verticillium* wilt in benomyl-treated soil on bioassay and performance. *Proceedings of the Beltwide Cotton Production Research Conference.* New Orleans, Louisiana, pp. 36–40.

Leakey, C.L.A. (1965) Sudden death disease of cocoa in Uganda associated with *Verticillium dahliae* Kleb. *East African Agricultural and Forestry Journal* 31, 21.

Leal, J.A. and Villanueva, J.R. (1962) Lack of pectic enzyme production by non-pathogenic species of *Verticillium. Nature (London)* 195, 1328–1329.

Leath, K.T. (1984) North and central counties hit by new wilt on alfalfa. *Science in Agriculture* 31, 6–7.

Leath, K.T. and Hower, A.A. (1988) Role of *Sitona hispudulus* on the development of *Fusarium* wilt. In: Moutray, J.B. and Elgin, J.H. (eds) *Report of the Thirty-first North American Alfalfa Improvement Conference.* Beltsville, Maryland, June 1988. p. 24.

Leben, S.D., Wadi, J.A. and Easton, G.D. (1987) Effects of *Pseudomonas fluorescens* on potato plant growth and control of *Verticillium dahliae. Phytopathology* 77, 1592–1595.

Lee, S., Heinz, R., Robb, J. and Nazar, R.N. (1994) Differential utilization of alternate initiation sites in a plant defense gene responding to environmental stimuli. *European Journal of Biochemistry* 226, 109–114.

Lee, S.-F. and Le Tourneau, D. (1958) Chlorogenic acid content and *Verticillium* wilt resistance of potatoes. *Phytopathology* 48, 268–274.

Lehman, M., Miravalle, R.J. and Turner, J.H. (1963) Wilt tolerance in cotton varieties. *California Agriculture* 17(4), 10–11.

Leibelt, W. and Senser, F. (1972) Die Welke des Hopfens und ihre Bekämpfung. *Hopfen-Rundschau* 23(2–3), 19–25, 40–43.

Lemeshev, N.K. and Shakhmedova, G.S. (1990) [The use of the world cotton germplasm for creation of donors of wilt resistance on a new genetic basis]. *Proceedings of the Fifth International* Verticillium *Symposium.* Leningrad, p. 75.

Lemeshev, N.K., Alimov, D.A. and Bolbekov, M.E. (1987) [Problem of producing donors of resistance to wilt in cotton]. *Sbornik Nauchnykh Trudov po Prikladnoi Botanike, Genetike i Selektsii* 113, 4–9.

Leski, B. (1974) [Investigations on *Verticillium* wilt of strawberry (*Verticillium dahliae* and *Verticillium albo-atrum*)]. *Roczniki Nauk Rolniczych* 4, 253–269.

Lester, D.T. (1975) Variance in tolerance of American elm to the *Verticillium* wilt fungus. *Forest Science* 21, 227.

Le Tourneau, D. (1958) Component sugars of mycelium and of polysaccharide material produced in liquid culture by *Verticillium albo-atrum. Physiologia Plantarum* 11, 23–28.

Le Tourneau, D. (1961) Oligosaccharide production from sucrose by *Verticillium albo-atrum. Plant Physiology* 36, 102–104.

Le Tourneau, D. (1974) Alcohol-soluble sugars and sugar alcohols produced from glucose by *Verticillium* spp. *Transactions of the British Mycological Society* 62, 619–622.

Le Tourneau, D., Keith, R.S. and Hammond, D.L. (1976) Degradation of some phenols in liquid media by *Verticillium dahliae*. *Transactions of the British Mycological Society* 66, 327–328.

Leyendecker, P.J. (1949) *Verticillium* wilt on cotton in New Mexico. *Bulletin of the New Mexico Agricultural Experimental Station* 1032, 1–8.

Leyendecker, P.J. (1950) Effects of certain cultural practices on *Verticillium* wilt of cotton in New Mexico. *Bulletin of the New Mexico Agricultural Experimental Station*, p. 356.

Leyendecker, P.J. (1951) 1950 plant disease survey for New Mexico. *Plant Disease Reporter* 35, 168.

Leyendecker, P.J., Blank, L.M. and Nakayama, R.M. (1952) *Bulletin of the New Mexico Agricultural Experimental Station*, p. 1062.

Li, A. and Zhang, Y. (1995) Control of cotton *Fusarium* wilt and *Verticillium* wilt by using systemic beneficial microorganisms. *Chinese Journal of Biological Control* 11, 185.

Li, G.X. and Yang, F.X. (1989) [Regional *Fusarium* and *Verticillium* wilt resistant cotton variety tests over 16 years in China]. *China Cottons* 5, 26–27.

Li, H.-L., Yuan, H.-X., Wang, Y., Cai, J.-X., Huang, J.-L. and Wang, S.-Z. (1998) [Study of the relationship between diversity of microbes in rhizosphere and resistance of cotton cultivars to *Verticillium dahliae*]. *Acta Phytopathologica Sinica* 28, 341–345.

Li, H.-L., Yuan, H.-X., Wang, Y., Huang, J.-L., Cai, J.-X. and Wang, S.-Z. (1999) [Studies on the relationship between diversity of microbes in rhizosphere and resistance of cotton cultivars to *Verticillium dahliae*. II. The fungi flora in the rhizosphere of different cotton cultivars and its inhibition to *V. dahliae*]. *Acta Phytopathologica Sinica* 29(3), 242–246.

Li, J., Li, M., Zhai, X., Li, Z. and Wang, G. (1995) Analyses of biochemical characteristics in cotton leaves infected by *Verticillium dahliae* Kleb. *Acta Agriculturae Boreali-Sinica* 10(Supplement), 134–138.

Li, K.-N., German, T.L. and Rouse, D.I. (1993) A unique RAPD fragment for *Verticillium dahliae* and its application to the specific detection of the pathogen. *Phytopathology* 83, 1370.

Li, K.-N., Rouse, D.I. and German, T.L. (1994) PCR primers that allow intergeneric differentiation of ascomycetes and their application to *Verticillium* spp. *Applied and Environmental Microbiology* 60(12), 4324–4331.

Li, K.-N., Rouse, D.I., Eyestone, E.J. and German, T.L. (1999) The generation of specific DNA primers using random amplified polymorphic DNA and its application to *Verticillium dahliae*. *Mycological Research* 103, 1361–1368.

Li, Q.-F., Cai, Y.-F., Tan, Y.-J. and He, H.-H. (1996) Research on new cotton sources resistant to many biological strains of *Verticillium* wilt. *Scientia Agricultura Sinica* 29(4), 47–52.

Li, T., Kerong, W. and Jiayun, L. (1997) Heredity of microsclerotia forming ability in *Verticillium dahliae*. *Mycosystema* 16(3), 197–201.

Li, T., Wang, K. and Lu, J. (1998) [Morphology and virulence variance of heterocaryon of *Verticillium dahliae* and its progenies]. *Mycosystema* 17, 226–232.

Li, T.-Q., Tian, L. and Jia, X.-F. (1999) [Control of *Verticillium* wilt in eggplant by grafting on tomato rootstock]. *China Vegetables* 6, 30–31.

Liese, J. (1934) Anzucht gesunder Pappeln- und Aspenpflanzen II Teil. *Forstarchiv* 9, 111–115.

Ligoxigakis, E.K. and Vakalounakis, D.J. (1994) The incidence and distribution of races of *Verticillium dahliae* in Crete. *Plant Pathology* 43(4), 755–758.

Ligoxigakis, E.K. and Vakalounakis, D.J. (1997) Hosts of *Verticillium dahliae* race 2 in Greece. *Proceedings of the Seventh International* Verticillium *Symposium*. Cape Sounion, Athens, Greece, p. 49.

Lin, B.Q. and Xiao, Y.H. (1995) Sources of resistance to *Verticillium* wilt in *Solanum melongena* and its affinities identified by improved root dip method. *Capsicum and Eggplant Newsletter* 81–84.

Linardakis, D.C., Psilakis, N.E., Micros, L.G. and Malathrakis, N.E. (1980) Susceptibility of different olive varieties to *Verticillium* wilt (*Verticillium dahliae* Klebahn). *Phytopathologia Mediterranea* 19, 70–71.

Lindemann, J., Arny, D.C. and Delwiche, P.A. (1982) Detection of *Verticillium albo-atrum* in the air over infected alfalfa fields in Wisconsin. *Phytopathology* 72, 1382.

Lindfors, T. (1917) Om vissnesjuka hos gurkor fororsaked av *Verticillium albo-atrum* Rke. & Berth. *Centralanstalt Försöksv. Jordbruksomr. Bot. Avdel.* Med. n. 159 13, 3–14.

Litvinov, M.A. and Babushkina, I.N. (1978) [Cultural and morphological variability of species of *Verticillium* Wallr. under the influence of antibiotics]. Микология и Фитопатология [*Mycology and Phytopathology*] 12, 378–385.

Liu, M. and Kolattukudy, P.E. (1996) Oxidative burst in tomato cell culture induced by the elicitor from *Verticillium albo-atrum*. *Acta Phytophysiologica Sinica* 22, 137–144.

Liu, C.-J., Heinstein, P. and Chen, X.-Y. (1999) Expression pattern of genes encoding farnesyl diphosphate synthase and sesquiterpene cyclase in cotton suspension-cultured cells treated with fungal elicitors. *Molecular Plant–Microbe Interactions* 12(12), 1095–1104.

Liu, J.Y. (1985) The identification of cotton resistant to *Verticillium dahliae* in seedling stage. *Journal of Nanjing Agricultural University* 2, 59–65.

Liu, J.-G., Benedict, C.R., Stipanovic, R.D. and Bell, A.A. (1999) Purification and characterization of *S*-adenosyl-L-methionine: desoxyhemigossypol-6-*O*-methyltransferase from cotton plants. An enzyme capable of methylating the defense terpenoids of cotton. *Plant Physiology* 121(3), 1017–1024.

Liu, R.J., Shen, C.Y., Li, H.F. and Chiu, W.F. (1993) Induction of pathogenesis-related proteins in cotton seedlings by VAM fungi [*Glomus hoi*, *G. mosseae*, *G. versiforme*] and *Verticillium dahliae*. *Acta Phytopathologica Sinica* 23(2), 162.

Liu, R.J., Li, H.F., Shen, C.Y. and Chiu, W.F. (1995) Detection of pathogenesis-related proteins in cotton plants. *Physiological and Molecular Plant Pathology* 47(6), 357–363.

Liu, X.-T., Guo, J.-C., Zhang, Y.-E., Song, X.-X. and Sun, J.-L. (1999a) [RAPD analysis of cotton *Verticillium* wilt pathogens in main cotton production area in China]. *Acta Agriculturae Boreali Sinica* 14(1), 107–114.

Liu, X.-T., Guo, J.-C., Zhang, Y.-E., Song, X.-X., Sun, J.-L. and Li, W. (1999b) [Studies on the gDNA extraction method and RAPD analysis of cotton *Verticillium* wilt pathogen]. *Acta Agriculturae Universitatis Henanensis* 33(3), 274–278.

Livescu, L., Kaufman, Z. and Nachmias, A. (1990) The effect of salinity on *Verticillium* wilt of potato. *Proceedings of the Fifth International* Verticillium *Symposium*. Leningrad, p. 9.

Locke, J.C. and Green, R.J. (1971) Systemic activity of benomyl fungicide against *Cladosporium* leaf mold and *Verticillium* wilt of greenhouse tomatoes. *Phytopathology* 61, 1024.

Locke, T. and Buck, S. (1997) The effect of linseed cropping on *Verticillium dahliae* soil infestation levels. *Proceedings of the Seventh International* Verticillium *Symposium*. Cape Sounion, Athens, Greece, p. 88.

Locke, T. and Thorpe, I.G. (1976) Benomyl tolerance in *Verticillium dahliae* Kleb. *Plant Pathology* 25, 59.

Locke, T. and Thorpe, I.G. (1977) Author's correction. Benomyl tolerance in *Verticillium dahliae* Kleb. *Plant Pathology* 26, 152.

Lockhart, C.L., McNab, A.A. and Bolwyn, B. (1969) A systemic fungicide for control of *Verticillium* wilt in strawberries. *Canadian Plant Disease Survey* 49, 46–48.

Lomelí Osuna, C., Leyva Mir, G. and Castro Acero, L. (1991) Enfermedades radicales de la alfalfa (*Medicago sativa* L.) en Chapingo, México y su transmisión por semilla. *Revista Mexicana de Fitopatología* 9(1), 8–10.

Longenecker, D.E. and Hefner, J.J. (1961) Effect of soil fertility level, soil moisture and trace elements on the incidence of *Verticillium* wilt in Upland irrigated cotton. *Progress Report, Texas Agricultural Experiment Station* 2175, 1–5.

Lopez-Escudero, F.J. and Blanco-Lopez, M.A. (1997) Control of *Verticillium* wilt by soil solarization in established olive orchards in Andalucia (Southern Spain). *Proceedings of the Seventh International* Verticillium *Symposium*. Cape Sounion, Athens, Greece, p. 84.

López-Escudero, F.J. and Blanco-López, M.A. (1999) First report of transmission of *Verticillium dahliae* by infested manure in olive orchards in Andalucia (southern Spain). *Plant Disease* 83(12), 1178.

Lorenzini, G., Triolo, E. and Scaramuzzi, G. (1983) Ulteriore contributo alla cognoscenza delle malattie nuove o poco note delle piante ornamentali in Toscana. *Revista della Ortoflorofrutticoltura Italiana* 67, 23–34.

Lottmann, J. and Berg, G. (1998) Interactions between *Stenotrophomonas maltophilia* and the soilborne pathogen *Verticillium dahliae* var. *longisporum*. In: Duffy, B., Rosenberger, U. and Defago, G. (eds) *Molecular Approaches in Biological Control*, Delemont, Switzerland, 15–18 September 1997. *Bulletin OILB/SROP* 21(9), 57–60.

Lottmann, J., Heuer, H., Smaller, K. and Berg, G. (1997) Beneficial bacteria in microenvironments of underground organs of potato. *Proceedings of the Seventh International* Verticillium *Symposium*. Cape Sounion, Athens, Greece, p. 69.

Low, P.S. and Heinstein, P.F. (1986) Elicitor stimulation of the defense response in cultured plant cells monitored by fluorescent dyes. *Archives of Biochemistry and Biophysics* 249, 472–479.

Low, P.S. and Merida, J.R. (1996) The oxidative burst in plant defense: function and signal transduction. *Physiologia Plantarum* 96(3), 533–542.

Low, P.S., Legendre, L., Heinstein, P.F. and Horn, M.A. (1993) Comparison of elicitor and vitamin receptor-mediated endocytosis in culture of soybean cells. *Journal of Experimental Botany* 44(Supplement), 269–274.

Lu, H.S., Hu, N.C. and Wu, L.Y. (1985) [Effects of the time of *Verticillium* wilt incidence on cotton yield]. *China Cottons* 4, 43–45.

Lu, J., Yang, J. and Ji, R. (1990) Studies on technique of soil assay for *Verticillium dahliae*. *Proceedings of the Fifth International* Verticillium Symposium. Leningrad, p. 13.

Lu, J.Y., Yu, C.F. and Ju, L.H. (1984) [Studies on differentiation of the virulence of *Verticillium dahliae* on pepper]. *Acta Phytopathologica Sinica* 14, 213–218.

Lucas, M.T. and Borges, M. de L.V. (1972) *Verticillium dahliae* agent of pepper and tomato wilt in Portugal. *Boletim da Sociedade Broteriana* 46, 529–534.

Luck, J.V. (1954) Studies on the *Verticillium* wilt of *Mentha piperita* L. with special emphasis on the causal organism, *Verticillium albo-atrum* R. & B. *Dissertation Abstracts* 14, 916–917.

Luckwill, L.C., Woodcock, D., Morgan, N.G. and Byrde, R.J.W. (1973) *Report, Long Ashton Research Station, University of Bristol.*

Ludbrook, W.V. (1933) Pathogenicity and environmental studies on *Verticillium* hadromycosis. *Phytopathology* 23, 117–154.

Luisi, N. and Sicoli, G. (1993) Una grave moria del mandorlo da *Verticillium dahliae* Kleb. in Italia. *Informatore Fitopatologico* 43(2), 45–48.

Luisi, N., Ciccarese, F., Sicoli, G. and Amenduni, M. (1994) Outbreaks of *Verticillium* wilt on almond and pathogenic variations among isolates of *V. dahliae*. In: Barbera, G. (ed.) *Acta Horticulturae. First International Congress on Almond.* Agrigento, Italy, 17–19 May 1993. pp. 287–292.

Lukomski, S. (1962) *Verticillium albo-atrum* moze bye grozne. *Las Polski* 2, 10–11.

Lumsden, R.D. (1981) A nylon fabric technique for studying the ecology of *Pythium aphanidermatum* and other fungi in soil. *Phytopathology* 71, 282–285.

Lundin, P. and Jonsson, A.A. (1975) Weibull's Vertus, a lucerne variety with high resistance to stem nematodes and *Verticillium* wilt. *Agricultura Hortique Genetica* 33, 17–32.

Luth, P. and Pfeffer, H. (1989) Gehört Rotklee in einer Rapsfruchtfolge? Anmerkungen zür Fruchtfolgegestaltung unter dem Gesichtpunkt des Auftretens von *Verticillium dahliae* Kleb. *Nachrichtenblatt für den Pflanzenschutzdienst in der DDR* 43, 246–247.

Lynch, D.R., Kawchuk, L.M., Schaupmeyer, C.A., Holley, J., Waterer, D., Rex, B.L., Pritchard, M.K. and Panford, J. (1995) Amisk: a clonal variant of Ranger Russett. *American Potato Journal* 72(3), 185–189.

Lynch, D.R., Kawchuk, L.M., Hachey, J., Bains, P.S. and Howard, R.J. (1997) Identification of a gene conferring high levels of resistance to *Verticillium* wilt in *Solanum chacoense*. *Plant Disease* 81(9), 1011–1014.

Lynov, S.I., Miroshnikov, V.G. and Bol'shakova, V.V. (1985) [The possibility of applying the method of delayed fluorescence to diagnosis of the functional state of seabuckthorn in relation to wilting]. *Fiziol., Ekol. i Agrotekhn. Sad. Kul'tur* 45–48.

Ma, C. and Jian, G.-L. (1995) Evaluation on resistance of a new cotton strain 86-6 to *Verticillium dahliae*. *Plant Protection* 21(6), 31.

Ma, P. and Shezeng, L. (1997) Incidence and damage by *Verticillium* wilt of cotton in China. *Proceedings of the Seventh International* Verticillium *Symposium.* Cape Sounion, Athens, Greece, p. 92.

Ma, Z.-Y., Wang, S.-F., Zhang, G.-Y. and Liu, Z.-G. (1998) [Distribution of the isolates of different virulence groups of *Verticillium dahliae* in Hebei cotton-growing areas]. *Journal of Hebei Agricultural University* 21(3), 1–6.

Maas, J.L., Draper, A.D. and Galletta, G.I. (1985) Inoculation methods for evaluating *Verticillium* wilt resistance in strawberry germ-plasm. *HortScience* 20, 739–741.

Maas, P.W.T. and Klaare, J.T. (1973) *Rotylenchulus reniformis* and *Verticillium albo-atrum* on *Codiaeum variegatum*. Proceedings of the Twenty-fifth International Symposium on Phytopharmacy and Phytiatry, 8 May 1973 I, II. *Mededelingen Faculteit Landbowwetenschappen, Gent* 38, 1291–1296.

MacDonald, D.H. (1976) Interaction of *Verticillium albo-atrum* and *Pratylenchus* sp. in potato. *Proceedings of the American Phytopathological Society* 3, 243.

MacDonald, J.D., Hartman, J.R. and Shapiro, J.D. (1984) Pathogens of ice plant in California. *Plant Disease* 68, 965–967.

Mace, M.E. (1978) Contributions of tyloses and terpenoid aldehyde phytoalexins to *Verticillium* wilt resistance in cotton. *Physiological Plant Pathology* 12, 1–11.

Mace, M.E. (1983) Elicitation of accumulation terpenoid aldehyde phytoalexins in *Verticillium dahliae* infected cotton. *New Phytologist* 95, 115–119.

Mace, M.E. and Howell, C.R. (1974) Histochemistry and identification of condensed tannin precursors in roots of cotton seedlings. *Canadian Journal of Botany* 52, 2423–2426.

Mace, M.E. and Stipanovic, R.D. (1995) Mode of action of the phytoalexin desoxyhemigossypol against the wilt pathogen, *Verticillium dahliae. Pesticide Biochemistry and Physiology* 53(3), 205–209.

Mace, M.E., Bell, A.A. and Stipanovic, R.D. (1974) Histochemistry and isolation of gossypol and related terpenoids in roots of cotton seedlings. *Phytopathology* 64, 1297–1302.

Mace, M.E., Bell, A.A. and Beckman, C.H. (1976) Histochemistry and identification of disease-induced terpenoid aldehydes in *Verticillium* wilt resistant and susceptible cottons. *Canadian Journal of Botany* 54, 2095–2099.

Mace, M.E., Bell, A.A. and Stipanovic, R.D. (1978) Histochemistry and identification of flavenols in *Verticillium* wilt-resistant and susceptible cottons. *Physiological Plant Pathology* 13, 143–149.

Mace, M.E., Bell, A.A. and Beckman, C.H. (eds) (1981) *Fungal Wilt Diseases of Plants.* Academic Press, New York.

Mace, M.E., Stipanovic, R.D. and Bell, A.A. (1985) Toxicity and role of terpenoid phytoalexins in *Verticillium* wilt resistance in cotton. *Physiological Plant Pathology* 26, 209–218.

Mace, M.E., Stipanovic, R.D. and Bell, A.A. (1989) Histochemical localization of desoxyhemigossypol, a phytoalexin in *Verticillium dahliae* infected cotton stems. *New Phytologist* 111, 229–232.

Mace, M.E., Elissalde, M.H., Stipanovic, R.D. and Bell, A.A. (1990a) A rapid, tetrazolium-based assay for toxicity of the phytoalexin desoxyhemigossypol to *Verticillium dahliae. Pesticide Biochemistry and Physiology* 38, 57–59.

Mace, M.E., Stipanovic, R.D. and Bell, A.A. (1990b) *Verticillium* wilt of cotton. *Proceedings of the Fifth International* Verticillium *Symposium*, Leningrad, p. 44.

Mace, M.E., Stipanovic, R.D. and Bell, A.A. (1993) Toxicity of cotton phytoalexins to zoopathogenic fungi. *Natural Toxins* 1(5), 294–295.

Macedo, E. de C. and Chiba, S. (1985) Influência de herbicidas residuais no crescimento *in vitro* de *Verticillium albo-atrum* (Reinke & Berth.) causador da murcha do algodoeiro. *Biológico* 51, 285–290.

Macfayden, A. (1967) Thermal energy as a factor in the biology of soils. In: Rose, E.H. (ed.) *Thermobiology.* Academic Press, New York and London, pp. 535–553.

MacGarvie, Q.D. and Hide, G.A.R. (1966) *Verticillium* species from potato seed stocks in Britain in 1965. *Plant Pathology* 15, 72–75.

MacGarvie, Q.D. and Isaac, I. (1966) Structure and behaviour of the nuclei of *Verticillium* spp. *Transactions of the British Mycological Society* 49, 687–693.

MacGuidwin, A.E. and Rouse, D.I. (1990a) Effect of *Meloidogyne hapla*, alone and in combination with subthreshold populations of *Verticillium dahliae*, on disease symptomology and yield of potato. *Phytopathology* 80, 482–486.

MacGuidwin, A.E. and Rouse, D.I. (1990b) Role of *Pratylenchus penetrans* in potato early dying disease of Russet Burbank potato. *Phytopathology* 80, 1077–1082.

MacHardy, W.E., Hall, R. and Busch, L.V. (1972) Relative water content and chlorophyll, protein and RNA levels in *Verticillium*-infected chrysanthemum leaves. *Proceedings of the Canadian Phytopathological Society, Thirty-eighth Session.* London, Ontario, June 1972, 39, 34.

Macit, F. and Saydam, C. (1970) [Preliminary test of the resistance of egg plant to *Verticillium dahliae* Kleb.]. *Ege Universitesi Ziraat Facultesi Dergisi A* 7, 135–141.

MacMillan, P.C. and Brandt, W.H. (1960) Possible role of a peroxidative system in melanin synthesis in *Verticillium. Antonie van Leeuwenhoek* 32, 202–211.

MacNeill, M.M. (1972) Temperature control of plant containers under glass. *Horticultural Research* 12(3), 173–176.

Madar, Z., Gottlieb, H.E., Cojocaru, M., Riov, J., Solel, Z. and Sztejnberg, A. (1995) Antifungal terpenoids produced by cypress after infection by *Diplodia pinea* f.sp. *cupressi. Phytochemistry* 38(2), 351–354.

Maden, S. (1987) Seed-borne fungal diseases of chick-pea in Turkey. *Journal of Turkish Phytopathology* 16, 1–8.

Madhosingh, C. (1996) A rapid tomato seedling assay for *Verticillium* wilt. *Journal of Phytopathology* 144(3), 131–134.

Madi, L., Katan, T., Katan, J. and Henis, Y. (1997) Biological control of *Sclerotium rolfsii* and *Verticillium dahliae* by *Talaromyces flavus* is mediated by different mechanisms. *Phytopathology* 87(10), 1054–1060.

Maharshak, G., Heiman, O., Sadan, D., Varshavski, S., Tsror, L. and Nachmias, A. (1995a) Formaldehyde as a soil disinfestant for control of *Verticillium dahliae* in crop rotation. 6th International *Verticillium* Symposium, Dead Sea, Israel. *Phytoparasitica* 23, 66.

Maharshak, G., Klein, L., Sadan, D., Tsror, L. and Nachmias, A. (1995b) Methyl bromide as a soil disinfestant for control of *Verticillium dahliae* in crop rotation. 6th International *Verticillium* Symposium, Dead Sea, Israel. *Phytoparasitica* 23, 65.

Mahuku, G.S., Platt, H.W. and Maxwell, P. (1999) Comparison of polymerase chain reaction based methods with plating on media to detect and identify *Verticillium* wilt pathogens of potato. *Canadian Journal of Plant Pathology* 21(2), 125–131.

Makhbubov, M., Voitenok, F. and Kas'yanenko, A. (1985) [A promising wilt-resistant cotton line]. *Khlopkovodstvo* 9, 27–28.

Makhmadaliev, K.M. (1984) [Inheritance of wilt resistance in interspecific cotton hybrids]. *Uzbekskii Biologicheskii Zhurnal* 1, 51–52.

Makhmudov, A.M., Siderova, S.F. and Gubaĭdullina, M.G. (1978) [Effect of nitrogen fertilization on cotton wilt infection]. *Khimiya v Sel'skom Khozvaĭstve* 16, 24–27.

Makhubov, M.B. (1990) [Inheritance of resistance in intraspecific hybridisation (*Gossypium* L.)]. *Proceedings of the Fifth International* Verticillium *Symposium.* Leningrad, p. 106.

Maksudov, N.Kh., Avezov, K. and Maksumova, Kh.G. (1975) [Study of the eradicating effect of chemical preparations on microsclerotia of *Verticillium dahliae*]. *Uzbekskii Biologicheskii Zhurnal* 19, 13–14.

Maksumov, A.N. (ed.) (1978) *Genetic Variability of the Pathogen of Wilt and the Ways for Increasing Wilt-resistance in Cotton.* Dushanbe, USSR, 209 pp.

Malaguti, G., Valásquez, F. and Subero, L. (1973) Marchitez por *Verticillium* en algodón. *Agronomía Tropical* 23, 65–70.

Malamud, O.S. and O'Keefe, R.B. (1976) Evaluation and inheritance of the reactions of *Verticillium dahliae* in tuber-forming *Solanum* species hybrids. *Proceedings of the Second International* Verticillium *Symposium.* University of California, Berkeley, p. 31.

Malatin, K.M. and Manninger, S. (1980) [Field test of resistance to infectious wilt in lucerne at Kompolt]. *Növénytermelés* 29, 497–504.

Malca, I., Erwin, D.C., Moje, W. and Jones, B. (1966) Effects of pH and carbon and nitrogen sources on the growth of *Verticillium albo-atrum*. *Phytopathology* 56, 401–406.

Malca, I., Erwin, D.C., Sims, J.J., Long, M. and Jones, B. (1968) Respiratory and enzymatic studies with *Verticillium albo-atrum*. *Phytopathology* 58, 348–353.

Malek, K., Stefan, K. and Malinowska, E. (1991) [Vascular gangrene of potato – previously unrecorded tuber infection in Poland]. *Ochrona Roslin* 35(2–3), 14–15.

Malia, M.E. and Tattar, T.A. (1975) Presymptomatic detection of *Verticillium* wilt in red maple by use of electrical resistance of roots. *Proceedings of the American Phytopathological Society* 2, 101.

Malia, M.E. and Tattar, T.A. (1978) Electrical resistance, physical characteristics, and cation concentrations in xylem of sugar maple infected with *Verticillium dahliae*. *Canadian Journal of Forest Research* 8, 322–327.

Malik, N.K. and Milton, J.M. (1980) Survival of *Verticillium* in monocotyledonous plants. *Transactions of the British Mycological Society* 75, 496–498.

Malinin, V.M. (1964) [On methods of increasing the wilt resistance of cotton]. *Khlopkovodstvo* 14, 43–44.

Malinin, V.M. (1968) [Wilt-resistant cotton varieties]. *Khlopkovodstvo* 18, 32.

Malm, N.R., Davis, D.D., Roberts, C.R., Barnes, C.E., Wood, R.L. and Staten, G. (1978) Registration of Acala 1517-75 Upland cotton. *Crop Science* 18, 164.

Malm, N.R., Davis, D.D., Roberts, C.R., Barnes, C.E., Wood, R.L. and Staten, G. (1978) Registration of Acala 1517V Upland cotton (Reg.No. 65). *Crop Science* 18, 163–164.

Malysheva, K.M. and Zel'tser, S.S. (1968) [Protein–lipid–polysaccharide complex of culture liquid and mycelium of *Verticillium dahliae* Kleb., the causative agent of *Verticillium* wilt in cotton]. *Doklady Akademii Nauk SSSR* 179, 231–234.

Mamadaliev, A. and Altunina, N. (1976) [Cobalt in the composition of ammonium polyphosphate]. *Khlopkovodstvo* 7, 21–22.

Mamadaliev, A., Altunina, N. and Akhmedov, A. (1976) [Effectiveness of cobalt applied to cotton on meadow-bog soils]. *Nauchnye-Trudy, Tashkentskii-Sel'skokhozyaistvennyi-Institut* 66, 54–61.

Mamadaliyev, A.K. and Biorov, A.A. (1990) The influence of lignin and mineral fertilizers on wilt incidence in cotton. *Proceedings of the Fifth International* Verticillium *Symposium.* Leningrad, p. 91.

Mamluk, O.F. (1974) First report of *Verticillium* wilt of cotton in Iraq. *Plant Disease Reporter* 58, 996–997.

Mamluk, O.F. (1975) Beitrage zur Taxonomie der mittleren Orient auf Baumwolle auftretenden Verticilliose. *Phytopathologische Zeitschrift* 84, 307–315.

Mamluk, O.F. and Skaria, M. (1979) *Verticillium* wilt on vegetables in Jordan. *FAO Plant Protection Bulletin* 27, 125–129.

Manandhar, J.B. and Bruehl, G.W. (1973) *In vitro* interactions of *Fusarium* and *Verticillium* wilt fungi with water, pH and temperature. *Phytopathology* 63, 413–419.

Mancini, G., Rameri, G. and Villani, R. (1973) Sull' attivita *in vitro* di alcuni fungicide contro *Verticillium albo-atrum* Reinke et Berth. *Annali della Facoltà di Scienze Agrarie della Università di Napoli in Portici IV* 7, 1–14.

Manici, L.M., Leoni, D., Lazzeri, L., Galletti, S. and Palmieri, S. (1999) Fungitoxic activity of four thio-functionalised glucosinolate enzyme-derived products on ten soil-borne pathogens. *Pesticide Science (9th International Congress of Pesticide Chemistry. London,1998)* 55, 486–488.

Mankau, S.K. and Mankau, R. (1962) Multiplication of *Aphelenchus avenae* on phytopathogenic soil fungi. *Phytopathology* 52, 741.

Mannanov, N.M., Yarovenko, G.I. and Émikh, B.A. (1971) [Measures to improve the health of cotton and increase its yield]. *Vestnik Sel'skokhozyaĭstvennoĭ Nauki, Moscow, USSR* 16, 16–22.

Mannapova, M. (1976) [Catch crops and wilt infection]. *Khlopkovodstvo* 7, 19–20.

Manninger, K., Buglos, J. and Manninger, S. (1978) [Relation between saponins and pathogens of lucerne wilt]. *Novenytermeles* 27, 395–401.

Maraite, H. and Meyer, J.A. (1976) La verticilliose du manioc en Afrique Centrale. *Mededelingen der Zittingen. Koninklijke Academie voor Overzeese Wetenschappen* 3, 443–446.

Maraite, H. and Meyer, J.A. (1977) Cassava wilt due to *Verticillium dahliae. Plant Protection Bulletin FAO* 25, 129.

Marchal, E. (1939) Observations et récherches effectuées à la Station de Phytopathologie de l'état pendant l'année 1939–40. *Bulletin de l'Institut Agronomique de Gembloux* 8, 77–85.

Marchoux, G., Conti, M., Boccardo, G., Lisa, V., Lovisolo, O., Zimmerman-Gries, S., Pilowsky, M., Marrou, J., Migliori, *et al.* (1973) Tomato, pepper, eggplant. *Proceedings of the Third Conference on Phytiatry and Phytopharmacy in the Mediterranean Region, 1973*, pp. 11–85.

Marinkovic, N., Miladinovic, Z. and Aleksic, Z. (1984) [Resistance of progenies of some interspecies hybrids of *Capsicum annuum* L. to *Verticillium albo-atrum* Reinke & Berth.]. *Zaštita Bilja* 35, 363–372.

Markocic, M., Roudeillac, P. and Pommier, J.J. (2000) Trois nouvelles variétés de fraisiers rémontants du Ciref. *Arboriculture Fruitière* 537, 38–41.

Marocke, R., Trouvelot, A. and Trendel, R. (1977) Effets compares de trois formes azoteés sur l'expression de la verticilliose chêz le houblon (*Humulus lupulus* L.). *Compte Rendu des Séances de l'Academie d'Agriculture de France* 63, 1082–1094.

Marois, J.J., Johnston, S.A., Dunn, M.T. and Papavizas, G.C. (1982) Biological control of *Verticillium* wilt of eggplant in the field. *Plant Disease* 66, 1166–1168.

Marsh, P.B., Taylor, E.E. and Bassler, L.M. (1959) A guide to the literature on certain effects of light on fungi: reproduction, morphology, pigmentation and phototrophic phenomena. *Plant Disease Reporter Supplement* 261, 249–312.

Marshunova, G.N. and Fedorova, N.N. (1980) [The effect of green manure catch crops on grey soil fungistasis]. Микология и Фитопатология [*Mycology and Phytopathology*] 14, 233–237.

Marshunova, G.N. and Muromtsev, G.S. (1975) [On the effects of intermediate crops on the behaviour of *Verticillium dahliae* Kleb. in the soil]. Микология и Фитопатология [*Mycology and Phytopathology*] 9, 34–37.

Martelli, G.P. and Prota, U. (1997) Malattie e patogeni dell'olivo trasmessi col materiale di propagazione. *Georgofili* 44(Suppl. 2), 73–96.

Martelli, G.P., Savino, V., Catalano, L., Di Terlizzi, B., Sabanadzovic, S. and Greco, N. (1995) Viruses and certification of olive in Apulia (Southern Italy). In: Barber, M. and Hadidi, A. (eds) *Acta Horticulturae. International Symposium on Fruit Tree Virus Diseases.* Rome, Italy, pp. 569–573.

Martin, C. (1985) *Verticillium* wilt of potato in central Peru. *American Potato Journal* 62, 195–199.

Martin, G.H. (1926) Diseases of forest and shade trees, ornamental and miscellaneous plants in the United States in 1925. *Plant Disease Reporter Supplement* 50, 413–478.

Martin, G.H. (1929) Diseases of forest trees, ornamental and miscellaneous plants in the United States in 1928. *Plant Disease Reporter Suppl.* 73, 372.

Martin, J.M., Bruce, M.R., Ditterline, R.L. and Mathre, D.E. (1993) Influence of plant age, disease incubation time, and host plant resistance on *Verticillium* wilt symptom expression in lucerne (*Medicago sativa*). *Euphytica* 69(1/2), 123–128.

Martin, M.J., Riedel, R.M. and Rowe, R.C. (1982) *Verticillium dahliae* and *Pratylenchus penetrans*: interactions in the early dying complex of potato in Ohio. *Phytopathology* 72, 640–644.

Martins-Corder, M.P. and de Melo, I.S. (1998) Antagonismo *in vitro* de *Trichoderma* spp. a *Verticillium dahliae* Kleb. *Scientia Agricola* 55, 1–7.

Martinson, C.A. (1964) Active survival of *Verticillium dahliae* in soil. *Dissertation Abstracts* 25, 18.

Martinson, C.A. and Horner, C.E. (1962) Importance of nonhosts in maintaining the inoculum potential of *Verticillium*. *Phytopathology* 52, 742.

Marupov, A. (1976) [*Trichoderma* suppresses wilt]. *Khlopkovodstvo* 10, 30.

Marupov, A. (1990) Effect of predecessors and green manure cultures on manifestation of *Verticillium* wilt infection level and biogenousness of soil. *Proceedings of the Fifth International* Verticillium *Symposium*. Leningrad, p. 94.

Massee, G. (1896) The sleeping disease of tomatoes. *Journal of the Royal Horticultural Society* 12, 20–25.

Mathre, D.E. (1968) Photosynthetic activities of cotton plants infected with *Verticillium albo-atrum*. *Phytopathology* 58, 137–141.

Mathre, D.E. (1986) Occurrence of *Verticillium dahliae* on barley. *Plant Disease* 70, 981.

Mathre, D.E. (1989) Pathogenicity of an isolate of *Verticillium dahliae* from barley. *Plant Disease* 73, 164–167.

Mathre, D.E. and Otta, J.D. (1967) Sources of resistance in the genus *Gossypium* to several soil-borne pathogens. *Plant Disease Reporter* 51, 864–866.

Mathre, D.E., Garber, R.H. and Ravenscroft, A.V. (1967) Effect of *Thielaviopsis basicola* root rot of cotton on incidence and severity of *Verticillium* wilt. *Phytopathology* 57, 90–92.

Mathur, S. and Dube, H.C. (1978) Cellulose degrading potential of two pathogenic species of *Verticillium*. *Indian Journal of Experimental Biology* 16, 135–136.

Mathur, S. and Dube, H.C. (1982) Effect of growth hormones on cellulolytic enzyme (CMCase) production by *Verticillium albo-atrum* and *V. dahliae*. *Geobios* 9, 190–191.

Mathur, S.B. and Kabeere, F. (1975) Seed-borne fungi of sesame, in Uganda. *Seed Science and Technology* 3, 655–660.

Matsubara, Y.I., Tamura, H. and Harada, T. (1995) Growth enhancement and *Verticillium* wilt control by vesicular–arbuscular mycorrhizal fungus inoculation in eggplant. *Journal of the Japanese Society for Horticultural Science* 64(3), 555–561.

Matta, A. (1976) Dannosita della *Pyrenochaeta lycopersici* nelle colture di pomodoro in serra della riviera ligure. *Colture Protette* 5(5), 31–33.

Matta, A. (1989) Induced resistance to *Fusarium* wilt diseases. In: Tjamos, E.C. and Beckman, C.H. (eds) *Vascular Wilt Diseases of Plants, Basic Studies and Control*. NATO ASI Series H: Cell Biology, Vol. 28. Springer-Verlag, Berlin, pp. 173–196.

Matta, A. and Dimond, A.E. (1963) Symptoms of *Fusarium* wilt in relation to quantity of fungus and enzyme activity in tomato stems. *Phytopathology* 53, 574–578.

Matta, A. and Garibaldi, A. (1970) Attività di fungicidi sistemici contro le verticillosi del pomodoro e della melanzana. *Agricoltura Italiana* 70, 331–340.

Matta, A. and Garibaldi, A. (1974) Linee di pomodoro 'Marmande' resistanti a *Verticillium dahliae* Kleb. adatte per la coltura in serra in Liguria. *Tecnica Agricola, Italy* 26, 767–775.

Matta, A. and Garibaldi, A. (1977) Control of *Verticillium* wilt of tomato by preinoculation with avirulent fungi. *Netherlands Journal of Plant Pathology Supplement* 83(1), 457–462.

Matta, A. and Garibaldi, A. (1984) Strains of *Verticillium dahliae* virulent on tomato hybrids with the *Ve* gene for resistance to race 1 in Italy. *Proceedings of the 6th Congress of Mediterranean Phytopathological Union.* Cairo, Egypt, pp. 283–285.

Matta, A., Cirulli, M., D'Ercole, N. and Ciccarese, F. (1980) Indagini sulla specializzazione fisiologica di *Verticillium dahliae* Kleb. in Italia. *Informatore Fitopatologico* 30(11–12), 5–11.

Matveev, G.G. and Tribunskii, A.N. (1979) [The rotation of cultivars in the control of cotton wilt]. Микология и Фитопатология [*Mycology and Phytopathology*] 13, 407–414.

McCain, A.H. (1963) New or unusual hosts of *Verticillium albo-atrum*. *Plant Disease Reporter* 47, 233.

McCain, A.H. (1975) Chrysanthemum cultivars resistant to *Verticillium* wilt. *Flower and Nursery Report* 1975(Mar–April), 1–2.

McCain, A.H. and Farnham, D.S. (1974) Control of *Verticillium* wilt of *Chrysanthemum* – progress report. *Flower and Nursery Report* 1974(March), 4–5.

McCain, A.H. and Welch, K.E. (1976) *Chrysanthemum* cultivars resistant to *Verticillium* wilt. *Flower and Nursery Report* 1976(May–June), 4.

McClean, J.G. (1955) Selecting and breeding potatoes for field resistance to *Verticillium* wilt in Idaho. *Research Bulletin of the Idaho Agricultural Experiment Station* 30, 1–19.

McClellan, W.D., Borthwick, H.A., Bjornsson, I. and Barton, H.M. (1955a) Some responses of fungi to light. *Phytopathology* 45, 465.

McClellan, W.D., Wilhelm, S. and George, A. (1955b) Incidence of *Verticillium* wilt in cotton not affected by root-knot nematodes. *Plant Disease Reporter* 39, 226–227.

McConnell, J.S., Baker, W.H. and Kirst, R.C., Jr (1998) Irrigation methods and nitrogen fertilization rates in cotton production. *Special Report – Arkansas Agricultural Experiment Station* 188, 131–133.

McConnell, J.S., Baker, W.H. and Kirst, R.C., Jr (1999) Long-term irrigation methods and nitrogen fertilization rates in cotton production: the last five years. *Research Series Arkansas Agricultural Experiment Station* 463, 34–37.

McFadden, H., De Feyter, R. and Murray, F. (1997) Genetic engineering approaches to the improvement of cotton's tolerance to *Verticillium* wilt. *Proceedings of the Seventh International* Verticillium *Symposium.* Cape Sounion, Athens, Greece, p. 46.

McFerran, J., Goode, M.J. and Montgomery, W. (1982) VF Pink – a *Verticillium* wilt resistant pink fruit, fresh market tomato. *Arkansas Farm Research* 31, 6–7.

McGeary, F.M. (1980) Studies on the genetics and pathogenicity of *Verticillium albo-atrum*. PhD thesis, University of Dundee, UK.

McGeary, F.M. and Hastie, A.C. (1982) Hybridization of *Verticillium albo-atrum* strains from tomato and lucerne. *Physiological Plant Pathology* 21, 437–444.

McHugh, J.B. and Schreiber, L.R. (1984) Tolerance of *Verticillium dahliae* to benzimidazoles. *Plant Disease* 68, 424–427.
McIntyre, G.A. (1965) Absence of pectic plugs in tomato cuttings treated with polygalacturonases of *Verticillium albo-atrum*. Fifty-seventh Annual Meeting of the American Phytopathological Society. Miami, Florida, 3–7 October 1965. *Phytopathology*, 55, 1051–1085.
McKay, M.B. (1921) Transmission of some wilt disease in seed potatoes. *Journal of Agricultural Research* 21, 821–848.
McKay, M.B. (1926) Further studies of potato wilt caused by *Verticillium albo-atrum*. *Journal of Agricultural Research* 32, 437–470.
McKeen, C.D. (1943) A study of some factors affecting the pathogenicity of *Verticillium albo-atrum* R. & B. *Canadian Journal of Research* 21, 95–117.
McKeen, C.D. (1969) *Proceedings of the Forty-first Congress, ANZAAS*, p. 10.
McKeen, C.D. (1970) The occurrence of *Verticillium* species in horticultural crop plants and weeds in south-western Ontario [abstract in Proceedings of the 36th Meeting of the Canadian Phytopathological Society, 3–5 June 1970]. *Proceedings of the Canadian Phytopathological Society* 36, 25.
McKeen, C.D. (1972a) A measure of control of *Verticillium* wilt in eggplant and tomato with soil applications of benomyl. *Proceedings of the Canadian Phytopathological Society* 39, 21.
McKeen, C.D. (1972b) Observations on the invasion and subsequent development of *Verticillium albo-atrum* and *Verticillium dahliae* in eggplant, tomato and potato. *Proceedings of the Canadian Phytopathological Society* 39, 35.
McKeen, C.D. (1976) *Verticillium* wilt of potato in southwestern Ontario and survival of *Verticillium albo-atrum* and *Verticillium dahliae* in field soil. *Proceedings of the Second International* Verticillium *Symposium*. University of California, Berkeley, p. 31.
McKeen, C.D. and Mountain, W.B. (1960) Synergism between *Pratylenchus penetrans* (Cobb) and *Verticillium albo-atrum* R. & B. in eggplant wilt. *Canadian Journal of Botany* 38, 789–794.
McKeen, C.D. and Mountain, W.B. (1967) Soil fumigation for control of *Verticillium* wilt of eggplant. *Canadian Journal of Plant Science* 47, 1–10.
McKeen, C.D. and Sayre, R.M. (1964) Control of *Verticillium dahliae* and *Meloidogyne incognita* in greenhouse tomatoes by soil fumigation. *Canadian Journal of Plant Science* 44, 466–470.
McKeen, C.D. and Thorpe, H.J. (1971) An adaptation of a moist-chamber method for isolating and identifying *Verticillium* spp. *Canadian Journal of Microbiology* 17, 1139–1141.
McKeen, C.D. and Thorpe, H.J. (1973) Pathogenic species of *Verticillium* in horticultural crops and weeds in south-western Ontario. *Canadian Journal of Plant Science* 53, 615–622.
McKeen, C.D. and Thorpe, H.J. (1981) *Verticillium* wilt of potato in south-western Ontario and survival of *Verticillium albo-atrum* and *V. dahliae* in field soil. *Canadian Journal of Plant Pathology* 3, 40–46.
McKinley, R.T. and Talboys, P.W. (1979) Effects of *Pratylenchus penetrans* on development of strawberry wilt caused by *Verticillium dahliae*. *Annals of Applied Biology* 92, 347–357.
McLean, J.G., Letourneau, D.J. and Guthrie, J.W. (1961) Relation of histochemical tests for phenols to *Verticillium* wilt resistance of potatoes. *Phytopathology* 51, 84–89.

McLeod, A.G. and Smith, H.C. (1960) *Verticillium* wilt of tobacco III. The effect of culture filtrates of *Verticillium dahliae* Kleb. on tobacco. *New Zealand Journal of Agricultural Research* 4, 123–128.

McLeod, A.G. and Thomson, R. (1959) *Verticillium* wilt of tobacco. II. Field and glasshouse trials to determine resistance of varieties and species to *Verticillium dahliae* Kleb. *New Zealand Journal of Agricultural Research* 2, 792–796.

McSorley, R., McMillan, R.T. and Parrado, J.L. (1985) Soil fumigants for tomato production on Rockdale soils. *Proceedings of the Florida State Horticultural Society* 98, 232–237.

Mekhmanov, M. (1967) [Effect of zinc and molybdenum on cotton resistance to *Verticillium* wilt]. *Uzbekskii Biologicheskii Zhurnal* 11, 47–50.

Melero, J.M., Gil, A. and Risueno, J. (1988) Enfermedades del cartamo en Andalucia; importancia y etiologia. In: Llobet, L.G.I. (ed.) *Comunicaciones del III Congresso Nacional de Fitopatologia.* Puerto de la Cruz (Tenerife Islas Canarias) 1984, pp. 11–14.

Melero-Vara, J.M., Blanco-Lopez, M.A., Bejanaro-Alcazar, J. and Jimenez-Diaz, R.M. (1990) Use of soil solarization to control cotton wilt induced by defoliation pathotypes of *Verticillium dahliae* in southern Spain. *Proceedings of the Fifth International* Verticillium *Symposium*. Leningrad, p. 14.

Melero-Vara, J.M., Blanco-Lopez, M.A., Bejarano-Alcazar, J. and Jimenez-Diaz, R.M. (1995b) Control of *Verticillium* wilt of cotton by means of soil solarization and tolerant cultivars in southern Spain. *Plant Pathology* 44(2), 250–260.

Melouk, H.A. and Horner, C.E. (1975) Cross protection in mints by *Verticillium nigrescens* against *V. dahliae*. *Phytopathology* 65, 767–769.

Melouk, H.A. and Horner, C.E. (1976) Recovery of *Verticillium dahliae* pathogenic to mints from castings of earthworms. *Proceedings of the American Phytopathological Society* 3, 265.

Melouk, H.A. and Wadsworth, D.F. (1990) Infection of peanut seed with *Verticillium dahliae*. *Proceedings of the Fifth International* Verticillium *Symposium*. Leningrad, p. 15.

Melouk, H.A., Wadsworth, D.F. and Sherwood, J.L. (1983) Effect of *Verticillium* wilt on root and top weight of peanut cultivar Tamnut 74. *Plant Disease* 67, 1349–1350.

Melouk, H.A., Li, X., Damincone, J.P. and Jackson, K.E. (1995) Inhibitory effects of volatile compounds from rapeseed meal to *Verticillium dahliae*. Proceeding of the 6th International *Verticillium* Symposium, Dead Sea, Israel. *Phytoparasitica* 23.

Menzies, J.D. (1962) Effect of anaerobic fermentation in soil on survival of sclerotia of *Verticillium dahliae*. *Phytopathology* 52, 743.

Menzies, J.D. and Griebel, G.E. (1967) Survival and saprophytic growth of *Verticillium dahliae* in uncropped soil. *Phytopathology* 57, 703–709.

Merlo, P.A., Rollan, M.C. and Ronco, B.L. (1985/1986, published 1985) Marchitamiento del conejito (*Antirrhinum majus* L.) ocasionado por *Verticillium dahliae* Klebahn en la Republica Argentina. *Revista de la Facultad de Agronomía, Universidad Nacional de la Plata* 61/62, 181–185.

Messner, R., Schweigkofler, W., Ibl, M., Berg, G. and Prillinger, H. (1996) Molecular characterization of the plant pathogen *Verticillium dahliae* Kleb. using RAPD-PCR and sequencing of the 18S rRNA-gene. *Journal of Phytopathology* 144(7/8), 347–354.

Metcalf, J.G., Reyes, A.A. and Mohr, W.P. (1984) Bellestar, a *Verticillium* wilt resistant processing tomato. *HortScience* 19(1), 895–896.

Metcalf, J.G., Reyes, A.A. and Mohr, W.P. (1986) Earlibright, an early processing tomato. *Canadian Journal of Plant Science* 66, 417–419.

Metlitskii, L.V., Lyubimova, N.V. and Mukhamedova, R.A. (1971) [The phytoalexin activity and wilt resistance of the cotton plant]. *Izvestiya Akademii Nauk SSSR Ser. Biol.* 268–275.

Meyer, R. and Dubery, I.A. (1993) High-affinity binding of a protein–lipopolysaccharide phytotoxin from *Verticillium dahliae* to cotton membranes. *FEBS Letters* 335(2), 203–206.

Meyer, R., Slater, V. and Dubery, I.A. (1994) A phytotoxic protein–lipopolysaccharide complex produced by *Verticillium dahliae*. *Phytochemistry* 35(6), 1449–1453.

Mezhlum'yan, L.G., Redina, E.F., Kasymova, G.A., Yuldashev, P.Kh. and Khodzhibaeva, S.M. (1994, published 1995) Functional properties of a serine protease inhibitor from cotton seeds. *Chemistry of Natural Compounds* 30(4), 492–495.

Michail, S.H. (1989) *Fusarium* and *Verticillium* wilts of cotton. In: Agrihotri, V.P. (ed.) *Perspectives in Phytopathology*. Today and Tomorrow's Printers & Publishers, New Delhi, India, pp. 199–217.

Michail, S.H. and Carr, A.J.H. (1966) Use of culture filtrates as a rapid technique for screening lucerne for resistance to *Verticillium albo-atrum*. *Transactions of the British Mycological Society* 49, 133–138.

Middleton, J.T. and Bohn, G.W. (1953) Cucumber, melons, squash. In: Stefferand, A. (ed.) *Plant Diseases. The Yearbook of Agriculture*. USDA, Washington, DC, pp. 498–493.

Mielke, J.L. (1935) *Verticillium* wilt of maple in California. *Plant Disease Reporter* 19, 303.

Miles, L.E. (1934) *Verticillium* wilt of cotton in Greece. *Phytopathology* 24, 558–559.

Miles, L.E. and Persons, T.D. (1932) *Verticillium* wilt of cotton in Mississippi. *Phytopathology* 22, 767–773.

Milev, V. and Nechev, G. (1973) [Elimination of *Verticillium dahliae* from vegetable crops in rotation with rice]. *Proceedings of the Third Conference on Phytiatry and Phytopharmacy in the Mediterranean Region*. Sassari, Italy, September, 1971, pp. 195–197.

Milkova, L., Vitanov, M. and Daskalov, S. (1988) Phytostop – a new cultivar resistant to *Phytophthora capsici*. *Capsicum Newsletter* 7, 62.

Milkova, L., Daskalov, S. and Vitanov, M. (1989) [Fitostop – a new red pepper variety resistant to *Phytophthora capsici*]. *Ovoshcharstvo, Gradinarstvo i Konservna Promishlenost* 70, 19.

Miller, J.F. (1986) Registration of six *Verticillium* wilt resistant nonoilseed restorer sunflower germplasm lines. *Crop Science* 26, 1265–1266.

Miller, J.F. (1993) Registration of two reduced-height (HA 378 and HA 379) and two early-maturity (HA 380 and RHA 381) sunflower germplasm lines. *Crop Science* 33(1), 221.

Miller, J.F. and Gulya, T.J. (1985) Registration of four *Verticillium* wilt resistant sunflower germplasm lines. *Crop Science* 25, 718.

Miller, J.F. and Gulya, T.J. (1995) Registration of four maintainer (HA 382 to HA 385) and four restorer (RHA 386 to RHA 389) sunflower germplasm lines. *Crop Science* 35(1), 286.

Miller, P.M. (1975) Effect of the tobacco cyst nematode *Heterodera tabacum* on severity of *Verticillium* and *Fusarium* wilts of tomato. *Phytopathology* 65, 81–82.

Miller, R.W., Nigh, E.L. and Alcorn, S.M. (1967) Nematodes associated with *Verticillium albo-atrum* attacking cotton in the southwest. *Proceedings of the Beltwide Cotton Production Research Conference.* National Cotton Council, Memphis, Tennessee. p. 52.

Miller, V.R. and Bekker, Z.E. (1983) [Role of trace elements in the resistance of cotton to *Verticillium* wilt]. *Sel'skokhozyaĭstvennaya Biologiya* 11, 54–56.

Milton, J.M. and Isaac, I. (1967) Studies on a biotin requiring strain of *Verticillium dahliae. Transactions of the British Mycological Society* 50, 539–547.

Milton, J.M. and Isaac, I. (1976) *Verticillium* wilt of clover. *Plant Pathology* 25, 119–121.

Milton, J.M., Rogers, W.G. and Isaac, I. (1971) Application of acrylamide gel electrophoresis of soluble fungal proteins to taxonomy of *Verticillium* spp. *Transactions of the British Mycological Society* 56, 61–65.

Min, L.-F., Zhu, X.-F., Tang, C.-M., Zhou, Z.-H. and Zhang, T.-Z. (1999) [Studies on recurrent selection for the development of cotton cultivars resistant to *Verticillium dahliae* – construction of basic pools]. *Acta Gossypii Sinica* 11(4), 182–184.

Mingochi, D.S. and Clarkson, J.M. (1994) Pathogenicity of race-1 and race-2 tomato wilt isolates of *Verticillium dahliae* to four solanaceous and two non-solanaceous hosts. *Journal of Phytopathology* 141(3), 225–232.

Minton, E.B. (1972) Effects of weed control in grain sorghum on subsequent incidence of *Verticillium* wilt in cotton. *Phytopathology* 62, 582–583.

Minton, E.B. (1973) Chemical control. Verticillium *Wilt of Cotton. Proceedings of a Work Conference of the National Cotton Pathology Research Laboratory.* College Station, Texas 1971, ARS-S19, pp. 105–134.

Minton, E.B. (1978) Neutralizers and pesticides and their sequence of application to acid-delucted cottonseed: effects on germination, stand, and *Verticillium* wilt of cotton. *Crop Science* 18, 831–835.

Minton, E.B. (1980) Effects of row spacing and cotton cultivars on seedling diseases, *Verticillium* wilt, and yield. *Crop Science* 20, 347–350.

Minton, E.B. and Ebelhar, M.W. (1991) Potassium and aldicarb-disulfoton effects on *Verticillium* wilt, yield and quality of cotton. *Crop Science* 31, 209–212.

Minton, E.B. and Gipson, J.R. (1978) Effects of controlled night temperatures on *Verticillium* wilt of field grown cotton. In: Brown, J.M. (ed.) *Proceedings of the Beltwide Cotton Production Research Conference.* Dallas, Texas. National Cotton Council, Memphis, Tennessee.

Minton, E.B., Bradshears, A.D., Kirk, I.W. and Hudspeth, E.B. (1972) Effects of row and plant spacings on *Verticillium* wilt of cotton. *Crop Science* 12, 764–767.

Minton, E.B., Gipson, J.R. and Quisenberry, J.E. (1979) Effects of controlled night temperatures on incidence of *Verticillium* wilt in field-grown cotton. *Phytopathology* 69, 977–979.

Mirakhmedov, S.M. and Adylkhodzhaev, I.N. (1964) [On the resistance of the originally wild form of cotton species, *mexicanum* to *Verticillium* disease]. *Akad. Nauk. SSR Dok.* 169, 1216–1219.

Mirakhmedov, S.M., Rakhimova, M., Khidyaev, Kh. and Gafurova, R. (1969) [Initial material in the section of wilt-resistant varieties]. *Khlopkovodstvo* 19(9), 40–43.

Mirpulatova, N.S. (1961) [Some results from research on wilt]. *Khlopkovodstvo* 11, 30–33.

Mirpulatova, N.S., Shadmanova, N.A. and Salikhbaeva, T. (1977) [Ammonium sulphate and wilt infection of cotton]. *Khlopkovodstvo* 10(10), 30–31.

Mirpulatova, N.S., Akbarov, S. and Shadmanova, N.A. (1978) [The development of *Verticillium dahliae* Kleb. in the soil and infection of cotton by wilt with different sources of nitrogen nutrition]. Микология и Фитопатология [*Mycology and Phytopathology*] 12, 341–343.

Miryakobova, M. (1977) [Dehydrogenase activity of *Verticillium dahliae* Kleb. and *Fusarium oxysporum* f. *vasinfectum*]. *Trudy Sredneaz. Nii Zashchity Rast.* 11, 53–56.

Mirzaazimova, Kh.T., Gusakova, S.D. and Glushenkova, A.I. (1994a) Effect of infection by *Verticillium dahliae* on the lipid complex of cotton seeds from varieties with contrasting resistance. *Chemistry of Natural Compounds* 30(5), 555–558.

Mirzaazimova, Kh.T., Gusakova, S.D. and Glushenkova, A.I. (1994b) Effect of infection with *Verticillium dahliae* on the bound and strongly bound lipids of cotton seeds of varieties with contrasting resistance. *Chemistry of Natural Compounds* 30(5), 559–562.

Mirzaazimova, Kh.T., Gusakova, S.D. and Glushenkova, A.I. (1995, published 1996) Glycolipids of the leaves of a cotton plant of the 175-F variety. *Chemistry of Natural Compounds* 31(6), 674–676.

Mirzabaev, M.A. (1977) [The effectiveness of *Trichoderma* sp. with glauconite sand against *Verticillium* wilt of cotton]. *Trudy Srednaez. NII Zashchity Rast.* 11, 52–53.

Mirzabaev, M.M., Kuchkarov, G.K., Khakimov, A.Kh. and Yuldashev, A.Yu (1979) [Measures against wilt]. *Khlopkovodstvo* 7, 33–34.

Misaghi, I., DeVay, J.E. and Ravenscroft, A. (1969) Cytokinin activity in cotton plants infected by *Verticillium albo-atrum*. *Phytopathology* 59, 1041.

Misaghi, I., DeVay, J.E. and Kosuge, T. (1972) Changes in cytokinin activity associated with the development of *Verticillium* wilt and water stress in cotton plants. *Physiological Plant Pathology* 2, 187–196.

Mishina, G.N. (1974) [Some characteristics of the cultural development of *Verticillium dahliae* Kleb. on artificial nutrient media]. Микология и Фитопатология [*Mycology and Phytopathology*] 8, 147–149.

Misra, A.K. and Shedha, M.D. (1990) Remote sensing for detection of wilt of *Erythrina* species. *Indian Phytopathology* 43, 210–212.

Mistinova, A. (1999) Zdroje lucerny s roznym stupnom odolnosti voci korenovym chorobam. *Vedecke Prace Vyskumneho Ustavu Rastlinnej Vyroby v Piest'any* 29, 75–81.

Moens, M. and Ben Aicha, B. (1986) Control of *Verticillium dahliae* on tomato in Tunisia by resistant cultivars and soil disinfestation. *EPPO Bulletin* 16, 317–320.

Mohamed, H.A., Abdel-Sattar, M.A. and Hilal, A.A. (1987) Factors affecting *Pelargonium* growth and oil yield. *Agricultural Research Review* 65, 297–308.

Mohamed, M.K. (1982) Nebenwirkungen von Herbiziden auf den Befall von Weizen durch *Puccinia recondita* f.sp. *tritici* und von Baumwolle durch *Verticillium albo-atrum*. Thesis, Universität Hohenheim, Germany.

Mohan, R., Vijayan, P. and Kolattukudy, P.E. (1993) Developmental and tissue-specific expression of a tomato anionic peroxidase (tap1) gene by a minimal promoter, with wound and pathogen induction by an additional 5′-flanking region. *Plant Molecular Biology* 22(3), 475–490.

Mohan, S.B. and Ride, J.P. (1982) An immunoelectrophoretic approach to the identification of progressive and fluctuating isolates of the hop wilt fungus *Verticillium albo-atrum*. *Journal of General Microbiology* 128, 255–265.

Mohan, S.B. and Ride, J.P. (1983) Fluorescent antibody staining of conidia of isolates of *Verticillium albo-atrum* in relation to their virulence for hop. *Journal of General Microbiology* 129, 3581–3589.

Mohan, S.B. and Ride, J.P. (1984) Some biochemical and morphological characteristics of serotypes of *Verticillium albo-atrum* from hop. *Journal of General Microbiology* 130, 3203–3218.

Mohan, S.K., Davis, J.R., Corsini, D.L., Sorensen, L.H. and Pavek, J.J. (1987) Evaluation of response of potato genotypes to *Verticillium dahliae*. In: Foldo, N.E., Hansen, S.E., Nielsen, N.K. and Rasmussen, R.I. (eds) *Abstracts of Conference Papers and Posters of the Tenth Triennial Conference of the European Association for Potato Research*. Aalborg, Denmark, 26–31 July 1987, pp. 36–37.

Mol, L. (1995a) The influence of plant roots on the germination behaviour of microsclerotia of *Verticillium dahliae* in soil. 6th International *Verticillium* Symposium, Dead Sea, Israel. *Phytoparasitica* 23, 54.

Mol, L. (1995b) Formation of microsclerotia of *Verticillium dahliae* on various crops. *Netherlands Journal of Agricultural Science* 43(2), 205–215.

Mol, L. (1995c) Effect of plant roots on the germination of microsclerotia of *Verticillium dahliae*. II. Quantitative analysis of the luring effect of crops. *European Journal of Plant Pathology* 101(6), 679–685.

Mol, L. and Meijer, E.M.J. (1995) Quantification of microsclerotia of *Verticillium dahliae* in plant material by image analysis. *European Journal of Plant Pathology* 101(6), 705–709.

Mol, L. and Scholte, K. (1995a) Formation of microsclerotia of *Verticillium dahliae* Kleb. on various plant parts of two potato cultivars. *Potato Research* 38(2), 143–150.

Mol, L. and Scholte, K. (1995b) Effect of haulm treatments on the formation of microsclerotia of *Verticillium dahliae* Kleb. on potato. *Potato Research* 38(2), 151–157.

Mol, L. and Scholte, K. (1996) A method to observe antagonism to soil-borne pathogens of potato. *Potato Research* 39(1), 11–14.

Mol, L. and Van Riessen, H.W. (1995) Effect of plant roots on the germination of microsclerotia of *Verticillium dahliae* 1. Use of root observation boxes to assess differences among crops. *European Journal of Plant Pathology* 101, 673–678.

Mol, L., Scholte, K. and Vos, J. (1995) Effects of crop rotation and removal of crop debris on the soil population of two isolates of *Verticillium dahliae*. *Plant Pathology* 44(6), 1070–1074.

Mol, L., Huisman, O.C., Scholte, K. and Struik, P.C. (1996a) Theoretical approach to the dynamics of the inoculum density of *Verticillium dahliae* in the soil: first test of a simple model. *Plant Pathology* 45(2), 192–204.

Mol, L., Van Halteren, J.M., Scholte, K. and Struik, P.C. (1996b) Effects of crop species, crop cultivars and isolates of *Verticillium dahliae* on the population of microsclerotia in the soil, and consequences for crop yield. *Plant Pathology* 45, 205–214.

Molchanova, R.P. (1990) [The study on soil inoculum density of *Verticillium* wilt in the rhizosphere of hybrid cotton lines]. *Proceedings of the Fifth International* Verticillium *Symposium*. Leningrad, p. 31.

Molchanova, R.P. and Kas'yanenko, A.G. (1978) [Induced mutants of *Verticillium tricorpus*]. In: Maksumov, A. (ed.) [*Genetic Variability of the Wilt-causing Agent and Prospects for Increasing Cotton Wilt Resistance*]. Donish Publishing House, Dushanbe, USSR, pp. 122–129.

Molchanova, R.P. and Portenko, L.G. (1990) The effect of long term storage under different conditions on the collection of *Verticillium* spp. strains. *Proceedings of the Fifth International* Verticillium *Symposium*. Leningrad, p. 32.

Molchanova, R.P., Kas'yanenko, A.G. and Azizova, Z.S. (1978) [Polyploidisation and haploidisation as one of the possible mechanisms of genetic variability of *Verticillium tricorpus*]. In: Maksumov, A. (ed.) [*Genetic Variability of the Wilt-causing Agent and Prospects for Increasing Cotton Wilt Resistance*]. Donish Publishing House, Dushanbe, USSR, pp. 130–138.

Møller, N.H. and Andreasen, B. (1971) *Verticillium albo-atrum* in lucerne: I. The effect of different methods of inoculation. *Aarsskrift. K. Veterinaer-og Landbohojskole 1971*, 35–49.

Monnington, S.J. (1996) Infection of potato by *Verticillium dahliae* under controlled environment conditions. *Brighton Crop Protection Conference: Pests & Diseases – 1996. Volume 1. Proceedings of an International Conference*. Brighton, UK, 18–21 November, 1996, pp. 417–418.

Montgomerie, I.G. and Kennedy, D.M. (1982) Analysis of and screening for resistance to diseases of soft fruit. *Report of the Scottish Crop Research Institute* 1981, pp. 90–91.

Montgomerie, I.G., Kennedy, D.M. and Campbell, A.M. (1983) Analysis and screening for resistance to diseases of soft fruit. *Report of the Scottish Crop Research Institute* 1982, pp. 126–127.

Moore, W.C. (1959) *British Parasitic Fungi*. Cambridge University Press, Cambridge.

Moorman, G.W. (1981) *Verticillium* wilt infection rates in eggplant and the influence of black plastic mulching. *Phytopathology* 71, 896.

Moorman, G.W. (1982) The influence of black plastic mulching on infection rates of *Verticillium* wilt and yield of eggplant. *Phytopathology* 72, 1412–1414.

Morales-Garcia, J.L. (1989) Otras enfermedades radicales en el Estado de Michoacan. *Revista Mexicana de Fitopatología* 7, 243–245.

Moreau, M. and Catesson, A.M. (1982) Nouvelle approche des maladies vasculaires et réflexions sur la spécificité parasitaire. *Cryptogamie, Mycologie* 3, 11–13.

Moreau, M. and Catesson, A.M. (1984) Modulation de la résponse du xylème selon la nature de l'agression vasculaire. *Phytopathologische Zeitschrift* 111, 133–154.

Moreau, M. and Catesson, A.M. (1985) Responses of carnation xylem parenchyma contact cells to incompatible and compatible fungal vascular parasites introduced by wounding. *Annals of Botany* 55, 225–236.

Moreau, M., Catesson, A.M., Peresse, M. and Czaninski, Y. (1978) Dynamique comparée des réactions cytologiques du xylème de l'oeillet en présence de parasites vasculaires. *Phytopathologische Zeitschrift* 91, 289–306.

Morehart, A.L. and Melchior, G.L. (1982) Influence of water stress on *Verticillium* wilt of yellow poplar. *Canadian Journal of Botany* 60, 201–209.

Morehart, A.L., Blits, K.C. and Melchior, G.L. (1985) Pathogenic potential of *Verticillium albo-atrum* protoplasts. *Mycologia* 77, 784–790.

Morgan, D.P., Liebman, J.A., Epstein, L. and Jimenez, M.J. (1991) Solarizing soil planted with cherry tomatoes vs. solarizing fallow ground for control of *Verticillium* wilt. *Plant Disease* 75, 148–151.

Morocke, R., Trouvelot, A. and Trendel, R. (1977) Effets comparés de trois formes azotées sur l'expression de la verticilliose chez le houblon. *Compte Rendu des Séances de l'Academie d'Agriculture de France* 63, 1082–1092.

Morsink, F. and Rich, A.E. (1968) Interaction between *Verticillium albo-atrum* and *Pratylenchus penetrans* in *Verticillium* wilt of potatoes. *Phytopathology* 58, 401.

Morton, A., Tabrett, A.M., Carder, J.H. and Barbara, D.J. (1995) Sub-repeat sequences in the ribosomal RNA intergenic regions of *Verticillium alboatrum* and *V. dahliae*. *Mycological Research* 99(3), 257–266.

Moser, P.E. and Sackston, W.E. (1973) Effect of concentration of inoculum and method of inoculation on development of *Verticillium* wilt of sunflowers. *Phytopathology* 63, 1521–1523.

Moshirabadi, H. (1978) *Verticillium dahliae* in Iran. *Third International Congress of Plant Pathology*. Münich, Germany, 16–23 August 1978, p. 303.

Mosley, A.R., James, S.R., Hane, D.C., Rykbost, K.A., Shock, C.C., Charlton, B.A., Pavek, J.J., Love, S.L., Corsini, D.L. and Thornton, R.E. (2000a) Umatilla Russet: a full season long Russet for processing and fresh market use. *American Journal of Potato Research* 77(2), 83–87.

Mosley, A.R., James, S.R., Rykbost, K.A., Hane, D.C., Stanger, C.E., Shock, C.C., Pavek, J.J., Corsini, D.L., Miller, J.C. Jr, Love, S.L., Thornton, R.E., Holm, D.G. and Voss, R.E. (2000b) Century Russet: a high-yielding fresh market cultivar with *Verticillium* resistance. *American Journal of Potato Research* 77(3), 161–165.

Mosley, A.R., James, S.R., Shock, C.C., Hane, D.C., Rykbost, K.A., Charlton, B.A., Holm, D.G., Love, S.L., Corsini, D.L., Pavek, J.J. and Thornton, R.E. (2000c) Russet Legend: a full season long Russet for processing and fresh market use. *American Journal of Potato Research* 77(2), 77–81.

Mostafa, A.K.M. (1967) Studies on the control of vascular wilt diseases of cotton by fungi associated with the cotton plant. *Dissertation Abstracts International* 27B, 3746.

Moukhamedov, R., Hu, X., Nazar, R.N. and Robb, J. (1994) Use of polymerase chain reaction-amplified ribosomal intergenic sequences for the diagnosis of *Verticillium tricorpus*. *Phytopathology* 84(3), 256–259.

Moulton, J.E. and Andersen, R.L. (1978) Scarlet – a new strawberry variety. *Research Report, Agricultural Experiment Station, Michigan State University No. 356, Farm Science*, pp. 1–4.

Mountain, W.B. and McKeen, C.D. (1960) Increase in the incidence of *Verticillium* wilt of eggplant in the presence of *Pratylenchus penetrans*. *Phytopathology* 50, 647.

Mountain, W.B. and McKeen, C.D. (1962a) Interaction of *Verticillium dahliae* and *Pratylenchus penetrans* in tomato wilt. *Phytopathology* 52, 744.

Mountain, W.B. and McKeen, C.D. (1962b) Effect of *Verticillium dahliae* on the population of *Pratylenchus penetrans*. *Nematologica* 7, 261–266.

Mozumder, B.K.G. and Caroselli, N.E. (1966) The influence of substrate moisture on the growth of *Verticillium albo-atrum* R. & B. *Advancing Frontiers of Plant Science* 16, 77–83.

Mozumder, B.K.G. and Caroselli, N.E. (1974) Influence of microelements in a sporulation medium on subsequent water requirements for germination of *Verticillium albo-atrum* conidia. *Indian Phytopathology* 27, 537–541.

Mozumder, B.K.G., Caroselli, N.E. and Albert, L.S. (1970) Influence of water activity, temperature and their interaction germination of *Verticillium albo-atrum* conidia. *Plant Physiology* 46, 347–349.

Mshana, D.E. (1988) *Verticillium* wilt of cotton in eastern cotton growing areas. *TARO Newsletter* 3(2), 5–6.

Mueller, J.P., Percich, J.A. and Mitchell, J.E. (1982) Root deterioration associated with *Verticillium* wilt of horseradish. *Plant Disease* 66, 410–414.

Mueller, W.C. and Morgham, A.T. (1993) Ultrastructure of the vascular responses of cotton to *Verticillium dahliae*. *Canadian Journal of Botany* 71(1), 32–36.

Mujica, F. (1957) Determinaciones micológícas IX. *Agricultura Técnica Santiago de Chile* 17, 106–107.

Mukhamedalieva, S.Z. (1972) [Resistance of melon varieties to *Verticillium* wilt]. *Nauchno Tr. NII Ovoshche Bakhch. Kul'tur i Kartofelya* 9, 152–155.

Mukhamedalieva, S.Z. (1973) [Susceptability of melons to *Verticillium*]. *Kartofel'i Ovoshchi* 8, 42.

Mukhamedalieva, S.Z. (1977) [Virulence of *Fusarium* under mixed conditions of infection]. *Materialy 9-oi Konf. Molodykh Uchenykh Uzbekistana po s. kh. 1977* 161–163.

Mukhamedov, R.S., Irisbayev, B.K. and Khanazarova, D.K. (1990) Restriction fragment length polymorphism of *Verticillium dahliae* Kleb. fungi. *Proceedings of the Fifth International* Verticillium *Symposium*. Leningrad, p. 107.

Mukhamedova, R.A., Lyubinova, N.V., Avazkhodzhaev, M.K. and Metlitskii, L.V. (1975) [Phytoalexin activity of cotton as a factor of wilt resistance]. Микология и Фитопатология [*Mycology and Phytopathology*] 9, 505–510.

Mukhamedzhanov, M.V. (1966) [Present and prospective scientific research on cotton wilt and its control]. In: Mukhamedzhanov, M.V. (ed.) *Cotton Wilt*. 'Fan', Tashkent, Uzbek SSR, pp. 1–22.

Mukhamedzhanov, M.V. and Zubenko, T.F. (1990) The role of field soil saprophyte microflora in experimental cotton–alfalfa crop rotation for the reduction of stock infection of *Verticillium dahliae* Kleb. *Proceedings of the Fifth International* Verticillium *Symposium*. Leningrad, p. 23.

Mukhina, R.V., Agarkova, I.V. and Lebedev, V.B. (1990) The plant nematodes–*Verticillium* wilt of cotton relationships. *Proceedings of the Fifth International* Verticillium *Symposium*. Leningrad, p. 342.

Müller, H.L. (1969) Ergänzende Untersuchungen über die Luzernevertizilliose (*Verticillium albo-atrum*) inbesondere die Möglichkeit einer systematischen Auslese resistenter Luzernepflanzen. *Phytopathologische Zeitschrift* 65, 69–97.

Muller, J. (1973a) Über den *Verticillium–Pratylenchus*–Krankheitskom plex an Erdbeeren. *Gartenbauwissenschaft* 38, 43–55.

Müller, J. (1973b) Zum Einfluss von *Pratylenchus penetrans* auf die *Verticillium* Welke von *Impatiens balsamina*. *Zeitschrift für Pflanzenkrankheiten und Pflanzenschutz* 80, 295–311.

Müller, J. (1973c) Vergleichende Versuche über den *Verticillium–Pratylenchus*–Krankheitskom plex unter sterilen und unsterilen Bedingungen. *Nematologica* 19, 249–258.

Müller, J. (1977) Wechselwirkungen zwischen fünf *Pratylenchus*–Arten und *Verticillium albo-atrum*. *Zeitschrift für Pflanzenkrankheiten und Pflanzenschutz* 84, 215–220.

Muller, M., Kruse, L., Tabrett, A.M. and Barbara, D.J. (1997) Detection of a single base exchange in PCR-amplified DNA fragments using agarose gel electrophoresis containing bisbenzimide-PEG. *Nucleic Acids Research* 25(24), 5125–5126.

Muller, R. (1990) Johannisbeertriebschwellung (*Verticillium dahliae* Kleb.). *Nachrichtenblatt für den Pflanzenschutzdienst in der DDR* 44, 1, inside back cover.

Munnecke, D.E., Bricker, J.L. and Kolbezen, M.J. (1978) Comparative toxicity of gaseous methyl bromide to ten soil borne phytopathogenic fungi. *Phytopathology* 68, 1210–1216.

Murakami, T., Koike, M. and Shimada, T. (1993) Evaluation of tomato resistance to *Verticillium dahliae* culture filtrate in tissue culture. *Research Bulletin of Obihiro University, Natural Science* 18(3), 157–162.

Murakami, T., Koike, M. and Shimada, T. (1996) Effects of phytotoxic metabolites from *Verticillium dahliae* and ouabain on tissues and callus cultures from susceptible and

resistant cultivars of tomato. *Research Bulletin of Obihiro University, Natural Science* 20(1), 17–23.
Muromtsev, G.S. (1980) [Microbiological aspects of wilt control of cotton]. *Zashchita Rastenii* 10, 35–36.
Muromtsev, G.S. and Chernyaeva, I.I. (1979) [The effect of various forms of nitrogen fertilizer on the viability of *Verticillium dahliae*, the causative organism of cotton wilt]. *Doklady Vsesoyuznoi Ordena Lenina Akademii Sel'skokhozyaistvennykh Nauk Imeni V. I. Lenina* 12, 7–9.
Muromtsev, G.S. and Strunnikova, O.K. (1981) [Study of the phytopathogenic fungus *Verticillium dahliae* using immunofluorescent staining]. *Doklady Vsesoyvznogo Ordena Lenina Ordena Trudovogo Krasnogo Znameni Akademii Sel'skokhozyaistvennyen Nauk Imeni V.I. Lenina* 2, 24–25, 39–41.
Muromtsev, G.S., Lagutina, T.M. and Chernayeva, I.I. (1977) [The membrane chamber method for studying *Verticillium dahliae* Kleb. in the soil]. Микология и Фитопатология [*Mycology and Phytopathology*] 11, 358–363.
Muromtsev, G.S., Marshunova, G.N., Chernikova, N.K., Astanov, R.A. and Mannapova, M. (1979) [Role of green manure crops in forming the antibiotic potential of soils under cotton]. *Sel'skokhozyaĭstvennaya Biologiya* 14, 45–49.
Murphy, R.P. and Lowe, C.C. (1989) Registration of 'Oneida' alfalfa. *Crop Science* 29, 487.
Murphy, R.P., Lowe, C.C. and Viands, D.R. (1989) Registration of 'Mohawk' alfalfa. *Crop Science* 29, 487.
Murray, F., Llewellyn, D., McFadden, H., Last, D., Dennis, E.S. and Peacock, W.J. (1999) Expression of the *Talaromyces flavus* glucose oxidase gene in cotton and tobacco reduces fungal infection, but is also phytotoxic. *Molecular Breeding* 5(3), 219–232.
Murray, M.J. (1969) *Induced Mutations in Plants.* International Atomic Energy Authority, Vienna, Austria, pp. 345–371.
Murray, M.J. (1971) Additional observations on mutation breeding to obtain *Verticillium* resistant strains of peppermint. *Mutation Breeding for Disease Resistance* 171–195.
Muslimov, Z.M. (1974) Immunization of cotton by metabolites of *Verticillium dahliae.* Микология и Фитопатология [*Mycology and Phytopathology*] 8, 119–122.
Mussell, H.W. (1972) Toxic proteins secreted by cotton isolates of *Verticillium albo-atrum.* In: Wood, R.K.S., Ballio, A. and Graniti, A. (eds) *Phytotoxins in Plant Disease*, pp. 443–445.
Mussell, H.W. (1973) Endopolygalacturonase – evidence for involvement in *Verticillium* wilt of cotton. *Phytopathology* 63, 62–70.
Mussell, H.W. and Green, R.J. (1968) Production of polygalacturonase by *Verticillium albo-atrum* and *Fusarium oxysporum* f. sp. *lycopersici in vitro* and in vascular tissue of susceptible and resistant hosts. *Phytopathology* 58, 1061.
Mussell, H.W. and Green, R.J. (1970) Host colonization and polygalacturonase production by two tracheomycotic fungi. *Phytopathology* 60, 192–195.
Mussell, H.W. and Strand, L.L. (1974) Symptoms generated in intact cotton plants by purified endo-polygalacturonase. *Proceedings of the American Phytopathological Society* 1, 78.
Mussell, H.W. and Strand, L.L. (1977) Pectic enzymes: involvement in pathogenesis and possible reference to tolerance and specificity. In: Solheim, B. and Raa, R. (eds) *Cell Wall Biochemistry Related to Specificity in Host–Plant Pathogen Interactions.* Oslo Universitetsforlaget, Oslo, Norway, pp. 31–70.
Mussell, H.W. and Strouse, B. (1971) Proteolytic enzyme production by *Verticillium albo-atrum. Phytopathology* 61, 904.

Mussell, H.W. and Strouse, B. (1972) Characterization of two polygalacturonases produced by *Verticillium albo-atrum*. *Canadian Journal of Biochemistry* 50, 625–632.

Mussell, H.W., Stillwell, P. and Peck, S. (1982) The possible origin of ethylene in *Verticillium* wilt of tomato. *Phytopathology* 72, 968.

Mussel, H., Osmeloski, J., Wettlaufer, S.H. and Weinstein, L. (1987) Suppression of *Verticillium* wilt of tomato by difluoromethylornithine, a suicidal inhibitor of polyamine biosynthesis. *Plant Disease* 71, 313–316.

Mustakimova, F. and Muratmukhamedov, K.H. (1982) [Increasing cotton wilt resistance due to the effect of zinc sulphate and hairy vetch]. *Izvestiya Akademii Nauk Turkmenskoĭ SSR Seriya Biologicheskikh Nauk* 2, 67–70.

Nachmias, A. and Krikun, J. (1984a) Transmission of *Verticillium dahliae* in potato tubers. *Phytopathology* 74, 535–537.

Nachmias, A. and Krikun, J. (1984b) Diagnosis of *Verticillium dahliae* in potato seed tubers with immunoassay techniques. *Potato Research* 27, 423–426.

Nachmias, A. and Krikun, J. (1985) *Verticillium* wilt of potato in Israel. *American Potato Journal* 62, 201–205.

Nachmias, A. and Orenstein, J. (1986) Selection of potato germplasm tolerant to *Verticillium* wilt using a phytotoxin produced by *Verticillium dahliae*. *Proceedings of the Fourth International* Verticillium *Symposium*. Guelph, Canada.

Nachmias, A. and Orenstein, J. (1987) Selection of potato germplasm tolerant to *Verticillium* wilt using a phytotoxin produced by *Verticillium dahliae*. *Canadian Journal of Plant Pathology* 9, 83.

Nachmias, A., Buchner, V. and Krikun, J. (1982a) Differential diagnosis of *Verticillium dahliae* in potato with antisera to partially purified pathogen-produced extracellular antigens. *Potato Research* 25, 321–328.

Nachmias, A., Buchner, V. and Krikun, J. (1982) Comparison of protein–lipopolylsaccharide complexes produced by pathogenic and non-pathogenic strains of *Verticillium dahliae* Kleb. from potato. *Physiological Plant Pathology* 20, 213–221.

Nachmias, A., Buchner, V. and Burstein, Y. (1985) Biological and immunochemical characterization of a low molecular weight phytotoxin isolated from a protein-lipopolysaccharide complex produced by a potato isolate of *Verticillium dahliae* Kleb. *Physiological Plant Pathology* 26, 43–55.

Nachmias, A., Buchner, V., Tsror, L., Burstein, Y. and Keen, N. (1987) Differential phytotoxicity of peptides from culture fluids of *Verticillium dahliae* races 1 and 2 and their relationship to pathogenicity of the fungi on tomato. *Phytopathology* 77, 506–510.

Nachmias, A., Kaufman, Z., Livescu, L., Tsror, L., Meiri, A. and Caligari, P.D.S. (1993) Effects of salinity and its interactions with disease incidence on potatoes grown in hot climates. *Phytoparasitica* 21(3), 245–255.

Nachmias, A., Livescu, L., Tsror, L., Ben Hador, G., Orenstein, J., Sabach, S. and Orion, D. (1995a) Enhancement of *Verticillium* infection in potato under stress. 6th International *Verticillium* Symposium, Dead Sea, Israel. *Phytoparasitica* 23, 51.

Nachmias, A., Tsror, L., Livescu, L. and Krikun, J. (1995b) The use of methyl bromide as a soil disinfestant in potato cultivation in Israel. In: Vanachter, A. (ed.) *Acta Horticulturae Fourth International Symposium on Soil and Substrate Infestation and Disinfestation*. Leuven, Belgium, 6–12 September 1993, pp. 145–151.

Nachmias, A., Livescu, L. and Tsror (lahkim), L. (1997a) *Verticillium* tolerance in potato. *Proceedings of the Seventh International* Verticillium *Symposium*. Cape Sounion, Athens, Greece, p. 44.

Nachmias, A., Livescu, L., Tsror, L., Holczer, O. and Krikun, J. (1997b) *Verticillium* wilt on potato in Israel – an overview. *Proceedings of the Seventh International* Verticillium *Symposium*. Cape Sounion, Athens, Greece, p. 79.

Nadakavukaren, M.J. (1960) The effect of soil moisture and temperature on survival of *Verticillium* microsclerotia. *Dissertation Abstracts* 21, 419.

Nadakavukaren, M.J. (1963) Fine structure of microsclerotia of *Verticillium albo-atrum* Reinke & Berth. *Canadian Journal of Microbiology* 9, 411–413.

Nadakavukaren, M.J. and Horner, C.E. (1959) An alcohol agar medium selective for determining *Verticillium* microsclerotia in soil. *Phytopathology* 49, 527–528.

Nagao, H., Arai, H., Oshima, S., Koike, M. and Iijima, T. (1998) Vegetative compatibility of an isolate of *Verticillium dahliae* pathogenic to both tomato and pepper. *Mycoscience* 39, 37–42.

Nagao, H., Wakatabe, D. and Iijima, T. (1994) Difficulty to establish vegetative compatibility of Japanese isolates of *Verticillium dahliae* Kleb. using melanin-synthesis deficient mutants. *Journal of General and Applied Microbiology* 40(4), 277–285.

Nagao, H., Wakatabe, D., Shiraishi, T. and Iijima, T. (1995) Vegetative compatibility groups in Japanese isolates of *Verticillium dahliae* using nitrate non-utilizing mutants. 6th International *Verticillium* Symposium, Dead Sea, Israel, 1994. *Phytoparasitica* 23(1), 44.

Nagtzaam, M.P.M. (1995) *Talaromyces flavus* as a potential biocontrol agent for controlling *Verticillium dahliae* in potatoes. 6th International *Verticillium* Symposium, Dead Sea, Israel. *Phytoparasitica* 23, 69.

Nagtzaam, M.P.M. (1998) Biological control of *Verticillium dahliae* by *Talaromyces flavus*. Thesis, Landbouwuniversiteit Wageningen, Wageningen, The Netherlands.

Nagtzaam, M.P.M. and Bollen, G.J. (1997) Colonization of roots of eggplant and potato by *Talaromyces flavus* from coated seed. *Soil Biology and Biochemistry* 29(9/10), 1499–1507.

Nagtzaam, M.P.M., Termorshuizen, A.J. and Bollen, G.J. (1997) The relationship between soil inoculum density and plant infection as a basis for a quantitative bioassay of *Verticillium dahliae*. *European Journal of Plant Pathology* 103(7), 597–605.

Nagtzaam, M.P.M., Bollen, G.J. and Termorshuizen, A.J. (1998) Efficiency of *Talaromyces flavus* alone or in combination with other antagonists in controlling *Verticillium dahliae* in growth chamber experiments. *Journal of Phytopathology* 146, 165–173.

Nan, Z.B., Skipp, R.A. and Long, P.G. (1991) Fungal invasion of red clover roots in a soil naturally infested with a complex of pathogens: effects of soil temperature and moisture content. *Soil Biology and Biochemistry* 23, 415–421.

Nance, N.W. (1939) Diseases of plants in the United States in 1938. *Plant Disease Reporter Supplement* 119, 119–289.

Napier, E.J., Turner, D.I. and Rhodes, A. (1956) The *in vitro* action of griseofulvin against pathogenic fungi of plants. *Annals of Botany* 20, 461.

Narisawa, K., Ohki, K.T. and Hashiba, T. (2000) Suppression of clubroot and *Verticillium* yellows in Chinese cabbage in the field by the root endophytic fungus, *Heteroconium chaetospira*. *Plant Pathology* 49(1), 141–146.

Naser, Z.W. and Al-Raddad Al-Momany, A. (1998) Dissemination factors of *Verticillium* wilt of olive in Jordan. *Dirasat. Agricultural Sciences* 25, 16–21.

Nazar, R.N., Hu, X., Schmidt, J., Culham, D. and Robb, J. (1991) Potential use of PCR-amplified ribosomal intergenic sequences in the detection and differentiation of *Verticillium* wilt pathogens. *Physiological and Molecular Plant Pathology* 39, 1–13.

Nazirov, N.N. (1962) [The reaction of different varieties and hybrids of cotton to wilt infection]. *Uzbekskii Akad. Nauk SSR Dok.* 19, 43–46.

Ndubizu, T.O.C. (1977) Effect of earthworms, nematodes, cultivations and host plants on *Verticillium* wilt of peach and cherry. *Annals of Applied Biology* 86, 153–161.

Neal, D.C. and Sinclair, J.B. (1960) Nitrogen supplement in the Louisiana–Mississippi river delta as a possible control of *Verticillium* wilt of cotton. *Plant Disease Reporter* 44, 478.

Nees von Esenbeck, C.G.D. (1816, published 1817) *Das System der Pilze und Schwämme.* Stahelschen Buchhandlung, Würtzburg, Germany.

Negmatov, M.N., Rybova, I.M. and Jivonov, S.O. (1990) [Investigation of reaction of *Gossypium barbadense* dwarf line to inocula of physiologic races 1 2 3 of *Verticillium dahliae*]. *Proceedings of the Fifth International* Verticillium *Symposium*. Leningrad, p. 76.

Nelson, P.E. and Wilhelm, S. (1958) Thermal death range of *Verticillium albo-atrum*. *Phytopathology* 48, 613–616.

Nelson, R. (1926) *Verticillium* wilt of peppermint. *US Department of Agriculture Disease Reporter Supplement* 50, 474.

Nelson, R. (1950) *Verticillium* wilt of peppermint. *Michigan Agricultural Station Technical Bulletin* 221, 1–260.

Nervo, G., Morra, L., Bilotto, M. and Mennella, G. (1999) Valutazione di portainnesti per il peperone in coltura protetta. *Colture Protette* 28(2), 79–84.

Neve, R.A. (1972) Wye Northdown, Wye Challenger and Wye Target. A progress report. *Annual Report, 1971, Wye College Department of Hop Research, Ashford, Kent*. pp. 46–52.

Neve, R.A. (1976) The breeding system and selection of the hop. *Proceedings of the Scientific Commission of the International Hop Growers Convention at Wye College*. pp. 73–80.

Neve, R.A. (1979) Hop diseases: the risks and consequences of spread in a vegetatively propagated crop. In: Ebbels, D.L. and King, J.E. (eds) *Plant Health. The Scientific Basis for Administrative Control of Plant Diseases and Pests.* Blackwell Scientific Publications, London, pp. 155–161.

Newbanks, D. and Tattar, T.A. (1982) Electrical resistance and stem water relations in *Verticillium* infected red maple. *European Journal of Forest Pathology* 12, 113–123.

Newcombe, A.G. and Robb, J. (1988) The function and relative importance of the vascular coating response in highly resistant, moderately resistant and susceptible alfalfa infected by *Verticillium albo-atrum*. *Physiological and Molecular Plant Pathology* 33, 47–58.

Newcombe, A.G. and Robb, J. (1989a) The chronological development of a lipid-to-suberin response at *Verticillium* trapping sites in alfalfa. *Physiological and Molecular Plant Pathology* 34, 55–73.

Newcombe, A.G. and Robb, J. (1989b) An improved method of determining the colonization ratio in *Verticillium*-infected alfalfa plants. *Canadian Journal of Plant Pathology* 11, 60–64.

Newcombe, A.G., Papadopoulos, Y.A., Robb, J. and Christie, B.R. (1989) The colonization ratio: a measure of pathogen invasiveness and host resistance in *Verticillium* wilt of alfalfa. *Canadian Journal of Botany* 67, 365–370.

Newton, W. and Van Adrichem, A.C.W. (1958) Resistance to *Verticillium* wilt in F_1 generations of self-fertilized species of *Fragaria*. *Canadian Journal of Botany* 36, 297–299.

Nguyen, V.Q. (1994) 'Juliette' fresh-market tomato. *HortScience* 29(4), 332.

Nicholls, H., Richards, C. and Martin, J.G. (1987) *Verticillium* wilt of alfalfa in Quebec. *Canadian Plant Disease Survey* 67, 17–19.

Nicholson, J.F., Sinclair, J.B., White, E.R. and Kirkpatrick, B.L. (1972) Upward and lateral translocation of benomyl in strawberry. *Phytopathology* 62, 1183–1185.

Nicoletti, R. and Lahoz, E. (1997) Recovery of *Verticillium biguttatum* from *Rhizoctonia solani* pathogenic on tobacco in southern Italy. *Proceedings of the Seventh International* Verticillium *Symposium*. Cape Sounion, Athens, Greece, p. 71.

Nicot, R.C. and Rouse, D.I. (1983) Evaluation of methods of assessing inoculum density of *Verticillium dahliae* in field soil. *Phytopathology* 73, 813.

Nicot, R.C. and Rouse, D.I. (1987a) Precision and bias of three quantitive soil assays for *Verticillium dahliae*. *Phytopathology* 77, 875–881.

Nicot, R.C. and Rouse, D.I. (1987b) Relationship between soil inoculum density of *Verticillium dahliae* and systemic colonization of potato stems in commercial fields over time. *Phytopathology* 77, 1346–1355.

Nieder, G. (1980) Holzzerstörende Pilze der Rebe – Ursachen für Kümmerwuchs und Schlagtreffen (Apoplexie). *Pflanzenarzt* 33, 7–9.

Nieto, P.L.E. (1989) Comportamiento de algunas variedades comerciales y materiales de la Colección Central Colombiana de Papa, al ataque de *Verticillium albo-atrum* y *Verticillium dahliae*. *Fitopatología Colombiana* 13, 78–84.

Nilsson, H.E. (1976) Influence of herbicides on infection by *Verticillium dahliae* on oil seed rape, *Brassica napus* var. *oleifera* Metzger. *Proceedings of the Second International* Verticillium *Symposium*. University of California, Berkeley, p. 3.

Nilsson, H.E. (1977) Influence of herbicides on infection by *Verticillium dahliae* on oil seed rape, *Brassica napus* var. *oleifera*. *Phytopathologische Zeitschrift* 90, 361–363.

Nilsson, H.E. (1985) Remote sensing of oil seed rape infected by *Sclerotinia* stem rot and *Verticillium* wilt. *Vaxtskyddsrapporter Jordbruk* 33, 33.

Nipoti, P. (1982a) Comportamento di pianti di melanzana 'Bella di Romagna' inoculate con diverse concentrazione di *Verticillium dahliae* Kleb. *Informatore Fitopatologico* 32(2), 31–34.

Nipoti, P. (1982b) Comportamento i quattro 'morfotipi' di melanzana 'Bella di Romagna' nei confronti della Verticilliosi. *Informatore Fitopatologico* 32(9/10), 47–52.

Nipoti, P. and D'Ercole, N. (1982) La verticilliosi della vite: II. Comportamento di diversi portainnesti. *Vignevini* 9(3), 47–50.

Niu, S.-G., Jiang, S.-R. and Tang, W.-H. (1999) [Positive regulations of *Pseudomonas fluorescens* by carbendazim and its application in controlling cotton *Verticillium* wilt]. *Acta Phytophylacica Sinica* 26(2), 171–176.

Noble, M., Robertson, N.F. and Dowson, W.J. (1953) *Verticillium* wilt of lucerne in Britain. *Plant Pathology* 2, 31–33.

Noling, J.W. (1987) Multiple pest problems and control on tomato. *Nematology Circular, Division of Plant Industry, Florida Department of Agriculture and Consumer Service*, Vol. 139.

Nothman, J. and Ben-Yephet, Y. (1979) Screening eggplant and other *Solanum* species for resistance to *Verticillium dahliae*. *Plant Disease Reporter* 63, 70–73.

Novruzov, M. (1976) [Importance of potassium for controlling wilt infection of cotton]. *Khlopkovodstvu* 121–126.

O'Brien, M.J. (1983) Evaluation of eggplant accessions and cultivars for resistance to *Verticillium* wilt. *Plant Disease* 67, 763–764.

O'Brien, R.G. and Hutton, D.G. (1981) Identification of race 2 of *Verticillium* wilt in tomatoes in south-east Queensland. *Australian Plant Pathology* 10, 56–58.

O'Garro, L.W. and Clarkson, J.M. (1988a) Heterokaryon incompatibility and heterozygous diploid production in tomato wilt isolates of *Verticillium dahliae*. *Journal of General Microbiology* 134, 2977–2984.

O'Garro, L.W. and Clarkson, J.M. (1988b) Pathogenicity of race 1 and race 2 tomato wilt isolates of *Verticillium dahliae* from different geographical origins. *Journal of Phytopathology* 123, 297–303.

Ohlsson, E. (1988) [Oilseed plants as catch crops]. *Växtskyddsnotiser* 52, 106–111.

Ohto, Y., Nakajima, T., Naito, S., Akasaka, Y. and Nakatani, F. (1993) Occurrence of *Verticillium* wilt disease of soyabean in Iwate prefecture. *Annual Report of the Society of Plant Protection of North Japan*, pp. 39–40.

Oingii, L. and Chiyi, S. (1990) Species, physiological form and vegetative compatibility group of cotton *Verticillium* wilt pathogen in China. In: *Proceedings of the Fifth International* Verticilium *Symposium*. Leningrad, USSR, p.17.

Okamoto, K., Shimada, A., Shirai, R., Sakamoto, H., Yoshida, S., Ojima, F., Ishiguro, Y., Sakai, T. and Kawagishi, H. (1993) Antimicrobial chlorinated orcinol derivatives from mycelia of *Hericium erinaceum*. *Phytochemistry* 34(5), 1445–1446.

Okasha, K.A., Ryugo, K. and Bringhurst, R.S. (1968a) Relationships of tannins, polyphenolics and reducing sugars to *Verticillium* wilt resistance. *Phytopathology* 58, 1118–1122.

Okasha, K.A., Ryugo, K., Wilhelm, S. and Bringhurst, R.S. (1968b) Inhibition of growth of *Verticillium alboatrum* sporelings by tannins and polyphenols from infected crowns of *Verticillium*-resistant and -susceptible strawberry cultivars. *Phytopathology* 58, 1114–1117.

Okie, W.R. and Gardner, R.G. (1982a) Screening tomato seedlings for resistance to *Verticillium dahliae* races 1 and 2. *Plant Disease* 66, 34–37.

Okie, W.R. and Gardner, R.G. (1982b) Breeding for resistance to *Verticillium dahliae* race 2 of tomato in North Carolina. *Journal of the American Society for Horticultural Science* 107, 552–555.

Okoli, C.A.N., Carder, J.H. and Barbara, D.J. (1993) Molecular variation and sub-specific groupings within *Verticillium dahliae*. *Mycological Research* 97(2), 233–239.

Okoli, C.A.N., Carder, J.H. and Barbara, D.J. (1994) Restriction fragment length polymorphisms (RFLPs) and the relationships of some host-adapted isolates of *Verticillium dahliae*. *Plant Pathology* 43(1), 33–40.

Oliva, A., Lahoz, E., Contillo, R. and Aliotta, G. (1999) Fungistatic activity of *Ruta graveolens* extract and its allelochemicals. *Journal of Chemical Ecology* 25, 519–526.

Olivier, C., Vaughn, S.F., Mizubuti, E.S.G. and Loria, R. (1999) Variation in allyl isothiocyanate production within *Brassica* species and correlation with fungicidal activity. *Journal of Chemical Ecology* 25(12), 2687–2701.

Olsson, S., Bååth, E. and Söderström, B. (1987) Growth of *Verticillium dahliae* Kleb. hyphae and of bacteria along the roots of rape (*Brassica napus* L.) seedlings. *Canadian Journal of Microbiology* 33, 916–919.

Onan, E. and Karcilioglu, A. (1998) Pathotypes of *Verticillium dahliae* from cotton in Aegean region and review of *Verticillium* wilt tolerance in Nazilli 84 cotton. *Journal of Turkish Phytopathology* 27(2–3), 113–120.

Onuorah, O.M.O. (1987) Isolation of pathogen-synthesized fraction that elicits phytoalexins in lucerne (*Medicago sativa* L.). *Acta Biologica Hungarica* 38, 247–256.

Orellana, R.G., Foy, C.D. and Fleming, A.L. (1975) Effect of soluble aluminium on growth and pathogenicity of *Verticillium albo-atrum* and *Whetzelinia sclerotiorum* from sunflower. *Phytopathology* 65, 202–205.

Orenstein, J., Nachmias, A., Colon, L.T. and Hoogendoorn, J. (1994) The effect of a *Verticillium dahliae* phytotoxin on germination of potato pollen. *Israel Journal of Plant Sciences* 42(1), 29–36.

Orenstein, J., Sabbah, S. and Nachmias, A. (1995) Enhancement of *Verticillium dahliae* infection of potato plants under stress conditions. 6th International *Verticillium* Symposium, Dead Sea, Israel. *Phytoparasitica* 23, 51.

Orijkowski, L.B. (1995) Wertycylioza czy fytoftoroza pelargonii? *Ochrona Roslin* 39(11), 3–4.

Orion, D. and Krikun, J. (1976) Response of *Verticillium*-resistant and *Verticillium*-susceptible tomato varieties to inoculation with the nematode *Meloidogyne javanica* and with *Verticillium dahliae*. *Phytoparasitica* 4, 41–44.

Orlowski, J.I., Powelson, M.L. and Ingham, R.E. (1988) Relationship of *Verticillium dahliae*, *Meloidogyne chitwoodi*, and irrigation level to potato early dying. *American Potato Journal* 65, 494.

Orton, W.A. (1914) Potato wilt, leaf roll and related diseases. *USDA Bulletin*, Vol. 64.

Orum, T.V., Alcorn, S.M., Herrera-Perez, T. and Miller, G. (1981) Susceptibility of jojoba (*Simmondsia chinensis*) to *Verticillium dahliae* and to *Phymatotrichum omnivorum*. *Plant Disease* 65, 243–245.

Oshima, N., Livingston, C.H. and Harrison, M.D. (1963) Weeds as carriers of two potato pathogens in Colorado. *Plant Disease Reporter* 47, 466–469.

Osmolovskaya, M.B. and Zlotina, G.D. (1969) [*Verticillium* species on fruit trees]. Микология и Фитопатология [*Mycology and Phytopathology*] 3(3), 266–268.

O'Sullivan, J. and Reyes, A.A. (1980) Effects of soil fumigation, rotation, and nitrogen on yield, petiole NO_3-N, and *Verticillium* wilt of potatoes. *Journal of the American Society for Horticultural Science* 105, 809–812.

Otazu, V., Gudmestad, N.C. and Zink, R.T. (1978) The role of *Colletotrichum atramentarium* in the potato wilt complex in North Dakota. *Plant Disease Reporter* 62, 847–851.

Otazu, V., Rodriguez, M., Arbizu, C. and Lizarraga, C. (1998) Arrosetamiento del ulluco (*Ullucus tuberosus*) Caldas, un sintoma inusual causado por *Verticillium dahliae*. *Fitopatología* 33, 153–159.

Ouf, S.A., Hady, F.K.A., Elgamal, M.H. and Shaker, K.H. (1994) Isolation of antifungal compounds from some *Zygophyllum* species and their bioassay against two soil-borne plant pathogens. *Folia Microbiologica* 39, 215–221.

Ourecky, D.K. (1977) 'Brandywine' purple raspberry. *HortScience* 12, 268–269.

Ouyang, B.-L. and Huang, D.-Y. (1996) A very early quality cotton variety – Xinluzao 5. *China Cottons* 23(12), 19.

Overman, A.J. (1982) Soil fumigation via drip irrigation under full-bed mulch cover for row crops. *Proceedings of the Soil and Crop Science Society of Florida*, pp. 153–155.

Overman, A.J. and Jones, J.P. (1977, published 1978) Efficacy of one stream versus three of a soil fumigsant for production of tomato. *Proceedings of the Florida State Horticultural Society* 90, 407–409.

Overman, A.J. and Jones, J.P. (1980, published 1981) Efficacy of methyl bromide–chloropicrin and ethylene dibromide–chloropicrin mixtures for control of nematodes and *Verticillium* wilt of tomato. *Proceedings of the Florida State Horticultural Society* 93, 248–250.

Overman, A.J. and Jones, J.P. (1986) Soil solarization, reaction and fumigation effects on double-cropped tomato under full-bed mulch. *Proceedings of the Florida State Horticultural Society* 99, 315–318.

Oxenham, B.L. (1963) Safflower wilt. *Report of the Department of Agriculture for Queensland* 1962–1963, pp. 23–24.

Page, M.S. and Gray, F.A. (1988) Interaction of *Phytophthora megasperma* f.sp. *medicaginis* and *Verticillium albo-atrum* in resistant and susceptible alfalfa. In: Moutray, J.B. and Elgin, J.H. (eds) *Report of the Thirty-first North American Alfalfa Improvement Conference.* Beltsfield, Maryland, USA, 19–23 June 1988, p. 18.

Palatinus, L. (1983) [Vascular wilt of lucerne under irrigation]. *Pol'nohospodarstvo* 29, 127–134.

Palazon, C. and Palazon, I. (1989) The epidemiological studies of 'tristesa' of pepper in the area of Valle Medio del Ebro. *Boletín de Sanidad Vegetal Plagas* 15, 233–262.

Palomo Gil, A. and Godoy Avila, S. (1994) Laguna 89 and CIAN 95: *Verticillium* wilt tolerant and high fiber quality new cotton varieties. *Proceedings of the Beltwide Cotton Conferences.* San Diego, California, 5–8 January 1994, p. 678.

Palomo Gil, A. and Quirarte, R.H. (1976) Effects of high plant populations and the number of auxiliary irrigation cycles on the yield and quality of fibres of two cotton varieties in soils infested with *Verticillium dahliae* Kleb. *Proceedings of the Second International* Verticillium *Symposium.* University of California, Berkeley, pp. 36–37.

Palomo Gil, A., Godoy Avila, S. and Garcia Castaneda, E.A. (1993a) Laguna 89: cultivar de algodon resistente a secadera tardia (*Verticillium dahliae* K.). *Revista Fitotecnia Méxicana* 16(1), 88.

Palomo Gil, A., Godoy Avila, S. and Garcia Castaneda, E.A. (1993b) CIAN 95: nuevo cultivar de algodon con alta calidad de fibra. *Revista Fitotecnia Méxicana* 16(1), 90.

Palomo Gil, A., Godoy Avila, S. and Santamaria Cesar, J. (1996) Estabilidad del rendimiento de nueve genotipos des algodon en la Comarca Lagunera, Mexico. *ITEA Producción Vegetal* 92, 238–243.

Pan, J.J., Zhang, T.Z., Kuai, B., Guo, X.P. and Wang, M. (1994) Studies on the inheritance of resistance to *Verticillium dahliae* in cotton. *Journal of Nanjing Agricultural University* 17(3), 8–18.

Pandian, T.T. and Isaac, I. (1971) *Verticillium dahliae* wilt of cotton. *Current Science* 40, 416.

Panella, A., Ribaldi, M. and Lorenzetti, F. (1969) Screening of alfalfa lines for resistance to *Verticillium* wilt. *Phytopathologia Mediterranea* 8, 116–123.

Panteleev, A.A. and Bagdasaryan, A.Z. (1972) [Antifungin against wilt]. *Khlopkovodstvo* 22(7), 27–28.

Panton, C.A. (1964) A review of some aspects of the wilt pathogen *Verticillium albo-atrum* Rke. et Berth. *Acta Agriculturae Scandinavica* 14, 97–112.

Panton, C.A. (1965) The breeding of lucerne, *Medicago sativa* L. for resistance to *Verticillium albo-atrum* Rke. et Berth. I. Preliminary studies on the effectiveness of selection and investigations on methods for inducing symptom development and facilitating selection in early breeding stage. *Acta Agriculturae Scandinavica* 15, 85–100.

Panton, C.A. (1966) Genetic control of resistance of lucerne, *Medicago sativa* L. to *Verticillium albo-atrum* Rke. et Berth. *Proceedings of the 10th International Grass Congress*, pp. 741–745.

Panton, C.A. (1967a) Observations on the nature of inheritance of resistance of lucerne to *Verticillium albo-atrum*. *Hereditas* 57, 333–338.

Panton, C.A. (1967b) The breeding of lucerne, *Medicago sativa* L. for resistance to *Verticillium albo-atrum* Rke. & Berth. III. Transgressive segregation in F3 and the effect on resistance of certain systems of inbreeding. *Hereditas* 57, 115–126.

Panton, C.A. (1967c) The breeding of lucerne, *Medicago sativa* L. for resistance to *Verticillium albo-atrum* Rke. et Berth. II. The quantitative nature of the genetic mechanism controlling resistance in inbred and hybrid generations. *Acta Agriculturae Scandinavica* 17, 43–52.

Panton, C.A. (1967d) The breeding of lucerne, *Medicago sativa* L. for resistance to *Verticillium albo-atrum* IV. Evidence for the presence of a toxic substance in culture filtrate of the causal organism and its effect on plants. *Acta Agriculturae Scandinavica* 17, 59–77.

Papaioannou, A.J. (1956) Notes phytopathologiques. *Annales de l'Institut Phytopathologique Benaki* 8, 96–102.

Papavizas, G.C. (1985) *Trichoderma* and *Gliocladium*: biology, ecology, and potential for biocontrol. *Annual Review of Phytopathology* 23, 23–54.

Paplomatas, E.E. and DeVay, J.E. (1990) Biological control activity of *Verticillium tricorpus* against seedling diseases and *Verticillium* wilt of cotton. *Proceedings of the Fifth International* Verticillium *Symposium*. Leningrad, p. 86.

Paplomatas, E.J. and Elena, K. (1995) Use of DNA hybridisation and vegetative compatibility groups for the detection and differentiation of *Verticillium* species. 6th International *Verticillium* Symposium, Dead Sea, Israel, 1994. *Phytoparasitica* 23, 45.

Paplomatas, E.J. and Lampropoulos, C.J. (1997) Molecular characterization of *Verticillium* species by random amplified polymorphic DNA analysis. *Proceedings of the Seventh International* Verticillium *Symposium*. Cape Sounion, Athens, Greece, p. 17.

Paplomatas, E.J., Tzalavaras, S. and DeVay, J.E. (1997) Use of *Verticillium tricorpus* as a biocontrol of *Rhizoctonia solani* on cotton seedlings. *Proceedings of the Seventh International* Verticillium *Symposium*. Cape Sounion, Athens, Greece, p. 66.

Pardossi, A. and Tognoni, F. (1985) Possibilità di controllo dei patogeni terricoli nella coltivazione del melone. *Colture Protette* 14, 27–31.

Parent, J.G., Hamelin, C. and Laberge, S. (1995) Physiological, pathological and molecular characterisation of *Verticillium albo-atrum* strains isolated from potato and alfalfa. 6th International *Verticillium* Symposium, Dead Sea, Israel, 1994. *Phytoparasitica* 23, 41.

Park, C.I. and Kim, W.G. (1986) [Fungi detected in the seeds of vegetable crops imported from Japan]. *Korean Journal of Mycology* 14, 89–91.

Park, H.-S. and Lee, Y.-S. (1996) Increase in 3-hydroxy-3-methylglutaryl coenzyme A reductase mRNA level in tomato by fungal elicitors and mechanical wounding. *Korean Journal of Plant Pathology* 12(3), 285–290.

Park, Y., Kim, J., Kwak, Y., Cho, M. and Kim, W. (1995) Tomato wilt caused by *Verticillium dahliae*. *RDA Journal of Agricultural Science, Crop Protection* 37, 303–307.

Parker, C.A., Rovira, A.D., Moore, K.J., Wong, P.T.W. and Kollmorgen, J.F. (eds) (1985) *Ecology and Management of Soil-borne Plant Pathogens*. American Phytopathological Society, St Paul, Minnesota.

Parker, K.G. (1959) *Verticillium* hadromycosis of deciduous tree fruits. *Plant Disease Reporter Supplement* 255, 39–61.

Parks, R.L. and Powelson, M.L. (1997) *Verticillium* wilt of potato in two soil types: effect of a sudangrass green manure on disease and pathogen soil populations. *Proceedings of the Seventh International* Verticillium *Symposium*. Cape Sounion, Athens, Greece, p. 77.

Parnis, E.M. and Sackston, W.E. (1979) Invasion of lupin seed by *Verticillium albo-atrum*. *Canadian Journal of Botany* 57, 597–601.

Paternotte, S.J. and Van Kesteren, H.A. (1993) A new aggressive strain of *Verticillium albo-atrum* in *Verticillium* resistant cultivars of tomato in the Netherlands. *Netherlands Journal of Plant Pathology* 99(3), 169–172.

Patil, S.S. and Dimond, A.E. (1967) Induction and repression of polygalacturonase synthesis in *Verticillium albo-atrum*. *Phytopathology* 57, 825.

Patil, S.S., Powelson, R.L. and Young, R.A. (1964) Relation of chlorogenic acid and free phenols in potato roots to infection by *Verticillium albo-atrum*. *Phytopathology* 54, 531–535.

Patrick, T.W., Hall, R. and Fletcher, R.A. (1977) Cytokinin levels in healthy and *Verticillium*-infected tomato plants. *Canadian Journal of Botany* 55, 377–382.

Pavek, J.J., Douglas, D.R., McKay, H.C. and Ohms, R.E. (1973) Targhee: an oblong russet potato variety with attractive tubers and high resistance to common scab. *American Potato Journal* 50, 293–296.

Peaden, R.N., Gilbert, R.G. and Christen, A.A. (1985) Control of *Verticillium albo-atrum* on alfalfa. *Canadian Journal of Plant Pathology* 7, 211–214.

Peaden, R.N., Evans, D.W. and Elgin, J.H., Jr (1995) Registration of $W12SR_2W_1Fu_1$ alfalfa germplasm. *Crop Science* 35(1), 287–288.

Pearson, R.C. and Goheen, A.C. (eds) (1988) *Compendium of Grape Diseases*. American Phytopathological Society, St Paul, Minnesota.

Pegg, G.F. (1957) A hyaline variant of *Verticillium albo-atrum* pathogenic to tomato plants. *Phytopathology* 47, 57–58.

Pegg, G.F. (1959) Rôle of indole-3-acetic acid in the development of disease symptoms in *Verticillium* wilt of tomato. *Proceedings of the Ninth International Botanical Congress*. Montreal, Canada, p. 297.

Pegg, G.F. (1965) *Nature (London)* 208, 1228–1229.

Pegg, G.F. (1972) The first International *Verticillium* Symposium, Wye College, London University, 19–24 September 1971. *International Newsletter of Plant Pathology* 2, 3.

Pegg, G.F. (1974) *Verticillium* diseases. *Review of Plant Pathology* 53, 157–182.

Pegg, G.F. (1976a) The response of ethylene-treated tomato plants to infection by *Verticillium albo-atrum*. *Physiological Plant Pathology* 9, 215–226.

Pegg, G.F. (1976b) The occurrence of 1, 3-Ì-glucanase in healthy and *Verticillium* infected resistant and susceptible tomato plants. *Journal of Experimental Botany* 27, 1093–1101.

Pegg, G.F. (1977) Glucanohydrolases of higher plants: a possible defence mechanism against parasitic fungi. In: Solheim, B. and Raa, J. (eds) *Cell Wall Biochemistry Related to Specificity in Host–Plant Pathogen Interactions*. Universitetsforlaget, Oslo, Norway, pp. 305–342.

Pegg, G.F. (1978) Effect of host substrate on germination and growth of *Verticillium albo-atrum* and *V. dahliae* conidia and mycelia. *Transactions of the British Mycological Society* 71, 483–489.

Pegg, G.F. (1981a) Biochemistry and physiology of pathogenesis. In: Mace, M.E., Bell, A.A. and Beckman, C.H. (eds) *Fungal Wilt Diseases of Plants*. Academic Press, New York, pp. 193–253.

Pegg, G.F. (1981b) The impact of *Verticillium* diseases on agriculture. *Abstracts of the 3rd International* Verticillium *Symposium*. University of Bari, Italy, p. 50.

Pegg, G.F. (1984) The impact of *Verticillium* diseases in agriculture. *Phytopathologia Mediterranea* 23, 176–192.

Pegg, G.F. (1985) Life in a black hole – the micro-environment of the vascular pathogen. *Transactions of the British Mycological Society* 85, 1–20.
Pegg, G.F. (1987) Metabolism of IAA and its precursors by *Verticillium albo-atrum* and *Verticillium*-infected tomatoes. *Canadian Journal of Plant Pathology* 9, 84.
Pegg, G.F. (1989) Pathogenesis in vascular diseases of plants. In: Tjamos, E.C. and Beckman, C.H. (eds) *Vascular Wilt Diseases of Plants, Basic Studies and Control. NATO ASI Series H: Cell Biology*, Vol. 28, pp. 51–95.
Pegg, G.F. and Cronshaw, D.K. (1976) Ethylene production in tomato plants infected with *Verticillium albo-atrum*. *Physiological Plant Pathology* 8, 279–295.
Pegg, G.F. and Dixon, G.R. (1969) The reactions of susceptible and resistant tomato cultivars to strains of *Verticillium albo-atrum*. *Annals of Applied Biology* 63, 389–400.
Pegg, G.F. and Jonglaekha, N. (1981) Assessment of colonization in chrysanthemum grown under different photoperiods and infected with *Verticillium dahliae*. *Transactions of the British Mycological Society* 76, 353–360.
Pegg, G.F. and Selman, I.W. (1959) An analysis of the growth response of young tomato plants to infection by *Verticillium albo-atrum*. II. The production of growth substances. *Annals of Applied Biology* 47, 222–231.
Pegg, G.F. and Street, P.F.S. (1984) Measurement of *Verticillium albo-atrum* in high and low resistance hop cultivars. *Transactions of the British Mycological Society* 82, 99–106.
Pegg, G.F. and Vessey, J.C. (1973) Chitinase activity in *Lycopersicon esculentum* and its relation to the *in vivo* lysis of *Verticillium albo-atrum* mycelium. *Physiological Plant Pathology* 3, 207–222.
Pegg, G.F. and Woodward, S. (1986) Synthesis and metabolism of alpha-tomatine in tomato isolines in relation to *Verticillium albo-atrum*. *Physiological and Molecular Plant Pathology* 28, 187–201.
Pegg, G.F. and Young, D.H. (1981) Changes in glucosidase activity and their relationship to fungal colonisation during infection of tomato by *Verticillium albo-atrum*. *Physiological Plant Pathology* 19, 371–382.
Pegg, G.F. and Young, D.H. (1982) Purification and characterization of chitinase enzymes from healthy and *Verticillium albo-atrum*-infected tomato plants, and from *V. albo-atrum*. *Physiological Plant Pathology* 21, 389–409.
Pegg, G.F., Gull, K. and Newsam, R.J. (1976) Transmission electron microscopy of *Verticillium albo-atrum* hyphae in xylem vessels of tomato plants. *Physiological Plant Pathology* 8, 221–224.
Pelletier, G. and Aubé, C. (1970) Conidial size and contents in *Verticillium* as affected by environmental factors. *Canadian Journal of Microbiology* 16, 231–236.
Pelletier, G. and Hall, R. (1971) Relationships among species of *Verticillium*: protein composition of spores and mycelium. *Canadian Journal of Botany* 49, 1293–1297.
Pennypacker, B.W. (1989) The role of mineral nutrition in the control of *Verticillium* wilt. In: Engelhard, A.W. (ed.) *Management of Diseases with Macro- and Microelements*. American Phytopathological Society, St Paul, Minnesota, pp. 33–45.
Pennypacker, B.W. and Leath, K.T. (1983) Dispersal of *Verticillium albo-atrum* in the xylem of alfalfa. *Plant Disease* 67, 1226–1229.
Pennypacker, B.W. and Leath, K.T. (1986) Anatomical response of a susceptible alfalfa clone infected with *Verticillium albo-atrum*. *Phytopathology* 76, 522–527.
Pennypacker, B.W. and Leath, K.T. (1993) Anatomical response of resistant alfalfa infected with *Verticillium albo-atrum*. *Phytopathology* 83(1), 80–85.

Pennypacker, B.W., Leath, K.T. and Hill, R.R. (1985) Resistant alfalfa plants as symptomless carriers of *Verticillium albo-atrum*. *Plant Disease* 69, 510–511.
Pennypacker, B.W., Leath, K.T. and Hill, R.R. (1988) Growth and flowering of resistant alfalfa infected by *Verticillium albo-atrum*. *Plant Disease* 72, 397–400.
Pennypacker, B.W., Knievel, D.P., Risius, M.L. and Leath, K.T. (1994) Photosynthetic photon flux density × pathogen interaction in growth of alfalfa infected with *Verticillium albo-atrum*. *Phytopathology* 84(11), 1350–1358.
Pennypacker, B.W., Knieval, D.P., Risius, M.L. and Leath, K.T. (1995a) Regulation of resistance of *Verticillium albo-atrum* in alfalfa. 6th International *Verticillium* Symposium, Dead Sea, Israel. *Phytoparasitica* 23, 60.
Pennypacker, B.W., Knievel, D.P., Risius, M.L. and Leath, K.T. (1995b) Impact of *Verticillium albo-atrum* and photosynthetic photon flux density on ribulose-1,5-biphosphate carboxylase/oxygenase in resistant alfalfa. *Phytopathology* 85, 132–138.
Perches, E.S., Prado, M.R., Aguilar, S.H. and Rodrigues, A.E. (1971) La marchitez por *Verticillium* del algodonero en la comarco Lagunera. *Folleto Tecnico Instituto Nacional Investigaciones Agricolas, México* 56, 30 pp.
Peresse, M., Moreau, M. and El Mahjoub, M. (1971) Aptitude parasitaire du *Verticillium dahliae* Kleb. vis-à-vis de l'oeillet. *Compte Rendu des Séances de la Société Biologie et ses Filiales* 165, 1783–1786.
Perez-Artes, E., Garcia-Petrajas, M., Bejarno-Alcazar, J., Korolev, N., Rodriguez-Jurado, Katan, T., Katan, Y. and Jimenez-Dias, R.M. (1997) Comparative RAPD analysis of cotton isolates of *Verticillium dahliae* from Israel and Spain. *Proceedings of the Seventh International* Verticillium *Symposium*. Cape Sounion, Athens, Greece, p. 14.
Perez-Lara, C., Perez-Artes, E. and Jiménez-Diaz, R.M. (1995) Use of RAPD to characterise cotton defoliating and non-defoliating isolates of *Verticillium dahliae*. 6th International *Verticillium* Symposium, Dead Sea, Israel. *Phytoparasitica* 23, 41.
Perry, J.W. and Evert, R.F. (1983) Histopathology of *Verticillium dahliae* within mature roots of Russet Burbank potatoes. *Canadian Journal of Botany* 61, 3405–3421.
Perry, J.W. and Evert, R.F. (1984) Structure of microsclerotia of *Verticillium dahliae* in roots of Russet Burbank potatoes. *Canadian Journal of Botany* 62, 396–401.
Pessoa, H.B.S.V., Miranda, J.E.C., Maluf, W.R. and Huang, S.P. (1988) Nemadoro, a processing tomato resistant to gall nematode. *Horticultura Brasileira* 6, 41–42.
Pethybridge, G.H. (1916) The *Verticillium* disease of the potato. *Scientific Proceedings of the Royal Dublin Society* 15, 63–92.
Pethybridge, G.H. (1919) Notes on some saprophytic species of fungi associated with diseased potato plants and tubers. *Transactions of the British Mycological Society* 6, 104–120.
Petrova, O.P. (1982) [Fungal wilt pathogen of *Hippophae rhamnoides*]. *Byulleten' Glavnogo Botanicheskogo Sada* 124, 96–97.
Petrovskaya, N.N. (1985) [*Verticillium* wilt resistance in tomatoes]. *Sbornik Nauchnykh Trudov po Prikladnoĭ Botanike, Genetiki i Selektsii* 92, 76–81.
Petsikos-Panayotarov, N. (1980, published 1981) Comportement d'un fongicide systémique après injection dans le tronc de l'olivier afin de lutter contre la verticilliose. *Annales de l'Institut Phytopathologique Benaki* 12, 227–235.
Pfyffer, G.E., Boraschi-Gaia, C., Weber, B., Hoesch, L., Orpin, C.G. and Rast, D.M. (1990) A further report on the occurrence of acyclic sugar alcohols in fungi. *Mycological Research* 94(2), 219–222.

Phillips, D.H. and Burdekin, D.A. (1983) *Diseases of Forest and Ornamental Trees*. Macmillan Press, London.

Phillips, D.J. and Wilhelm, S. (1971) Root distribution as a factor influencing symptom expression of *Verticillium* wilt of cotton. *Phytopathology* 61, 1312–1313.

Phillips, E.H. (1916) Observations on sour sap disease of apricots. *Phytopathology* 6, 309.

Phillips, J.C. (1979) Description and evaluation of chickpea germplasm in coastal California. *Agronomy Abstracts* 72.

Piazek, A. (1996a) Laboratory evaluation of alfalfa susceptibility to *Verticillium albo-atrum* Rke et Berth. Part II. Comparison of the laboratory and greenhouse methods. *Plant Breeding and Seed Science* 40, 51–56.

Piazek, A. (1996b) Laboratoryjna ocena wrazliwosei roslin lucerny mieszancowej odmiany Kometa na *Verticillium albo-atrum* Rke. *et* Berth. Czesc 1. Przydatnosc poszczegolnych metod we wstepnj selekeji materialu hodowlanego. *Biuletyn Instytutu Hadlowi i Aklimatyzacji Roslin*. pp. 179–185.

Picado, C. (1923) Sur l'action à distance des champignons phytopathogènes. *Procèdures du Congrès de Pathologie Végétale Strasbourg*. pp. 28–34.

Picman, A.K. and Schneider, E.F. (1993) Inhibition of fungal growth by selected sesquiterpene lactones. *Biochemical Systematics and Ecology* 21(3), 307–314.

Picman, A.K., Schneider, E.F. and Gershenzon, J. (1990) Antifungal activities of sunflower terpenoids. *Biochemical Systematics and Ecology* 18, 325–328.

Picman, A.K., Schneider, E.F. and Picman, J. (1995) Effect of flavonoids on mycelial growth of *Verticillium albo-atrum*. *Biochemical Systematics and Ecology* 23(7/8), 683–693.

Piearce, G.D. and Gibbs, J.N. (1981) *Verticillium* wilt of trees and shrubs. *Arboricultural Leaflet* 9, ii + 9 pp.

Piir, R. (1996) Astelpaju viljapuuna. *Agraarteadus* 7(2), 162–176.

Pilkington, S. and Heale, J.B. (1969) Respiration and morphogenesis in *Verticillium*. *Transactions of the British Mycological Society* 53, 467–470.

Pimentel, C.P.V., Pitta, G.P.B. and Figueiredo, M.B. (1980) Preservaçâo da patogenicidade de alguns fungos conservados em água destilada. *Biológico* 46, 281–287.

Pineau, R. (1977) [Reservoirs of infection of *Verticillium dahliae* Kleb. in tomato crops in Morocco]. *Bulletin de l'école Nationale Supérieure d'Agronomie et des Industries Alimentaires* 19, 91–100.

Pineau, R. and Besri, M. (1991) Inoculation simultanèe de *Fusarium oxysporum* f.sp. *lycopersici* et *Verticillium dahliae* à la tomate. *Phytopathologia Mediterranea* 30, 103–111.

Pinillos, O., Martín, C. and Galindez, W. (1987) Occurrence and distribution of *Verticillium* wilt of potatoes in farmer's fields in the Mantaro Valley, Perú. *Fitopatología* 22(2), 91–94.

Pinkas, J. and Chorin, M. (1970) *Verticillium* wilt of mango trees. *Hassadeh* 50, 1124–1125.

Pinkas, Y., Kariv, A. and Katan, J. (1984) Soil solarization for the control of *Phytophthora cinnamomi*: thermal and biological effects. *Phytopathology* 74, 796.

Piotrowski, J. and Milczek, M. (1982) [Biochemical indices of resistance of hop to *Verticillium albo-atrum* and *Fusarium sambucinum*]. *Acta Agrobotanica* 34, 277–284.

Pirone, P.P. (1959) Why shade trees die along city streets. *Proceedings of the National Shade Tree Conference* 35, 6–15.

Pirozhenko, T.K. and Shevtsova, V.M. (1988) [Study of the genetic control of adenine biosynthesis in the cotton wilt pathogen *Verticillium dahliae* Kleb.]. Микология и Фитопатология [*Mycology and Phytopathology*] 22, 568–570.

Pitt, J. (1980) *The Genus* Penicillium, p. 470.

Plaisted, R.L., Halseth, D.E., Brodie, B.B., Slack, S.A., Thurston, H.D., Loria, R., Sieczka, J.B., Porter, G.A. and Plissey, E.S. (1990) Allegany: a late season golden nematode resistant variety. *American Potato Journal* 67, 227–231.

Plasencia, J. and Banttari, E.E. (1997) Comparison between a culture plate method and an immunoassay to evaluate vascular colonization of potato by *Verticillium dahliae. Plant Disease* 81(1), 53–56.

Plasencia, J., Jemmerson, R. and Banttari, E.E. (1996) Production and characterization of monoclonal antibodies to *Verticillium dahliae* and development of a quantitative immunoassay for fungal biomass. *Phytopathology* 86(2), 170–176.

Platt, H.W. (1986) Varietal response and crop loss due to *Verticillium* wilt of potato caused by *V. albo-atrum. Phytoprotection* 67, 123–127.

Platt, H.W. and Bollen, G.J. (1993) Use of tissue culture potato plantlets for investigations of diseases of subterranean plant parts. *Plant Disease* 77(11), 1112–1113.

Platt, H.W. and Bollen, G.J. (1995) The influence of isolation procedure on recovery of *Verticillium* species and *Colletotrichum coccodes* from colonized potato stems. *Mycological Research* 99(8), 942–944.

Platt, H.W., Robb, J., Nazar, R. and Hu, X. (1995) *Verticillium* wilt of potatoes: causal agents and pathogen detection. 6th International *Verticillium* Symposium, Dead Sea, Israel. *Phytoparasitica* 23, 47.

Platt, H.W., Maclean, V., Mahuku, G. and Maxwell, P. (1997a) *Verticillium* wilt of potatoes caused by three *Verticillium* species. *Proceedings of the Seventh International* Verticillium *Symposium*. Cape Sounion, Athens, Greece, p. 19.

Platt, H.W., Maxwell, P., Maclean, V. and Mahuku, G. (1997b) Detection techniques for research on *Verticillium* species in potato soils. *Proceedings of the Seventh International* Verticillium *Symposium*. Cape Sounion, Athens, Greece, p. 48.

Poletaeva, V.F. and Bekker, Z.Ė. (1986) [Balance of dehydrogenases and oxidases and the susceptibility of two cotton species to different wilt pathogen species]. *Türkmenistan SSR Ylymlar Akademijasynyn, Habarlary, Biologik Ylymlaryn* 4, 3–8.

Polizzi, G. (1996) L'avvizzimento della *Lagunaria patersonii* da *Verticillium dahliae. Informatore Fitopatologico* 46(5), 46–49.

Pollock, C.J. and Drysdale, R.B. (1976) The role of phenolic compounds in the resistance of tomato cultivars to *Verticillium albo-atrum. Phytopathologische Zeitschrift* 86, 56–66.

Polyanichko, O.F. and Runov, V.I. (1964) [Reaction of cotton varieties of different resistance to infection by *Verticillium dahliae* and *Fusarium vasinfectum*]. *Mater. Vses. Simp po Bor'be s Viltom Khlopchatnika, Tashkent*, pp. 61–63.

Pontecorvo, G. (1954) Mitotic recombination in the genetic systems of filamentous fungi. *Caryologia* 6(Suppl.), 192–200.

Pontecorvo, G., Roper, J.A. and Forbes, E. (1953) Genetic recombination without sexual reproduction in *Aspergillus niger. Journal of General Microbiology* 8, 198–210.

Popov, V.I. (1979) [Scientific bases of the use of fungicides for the protection of cotton against soil infection]. *Problemy Obshcheĭ i Chastnoĭ Fitotoksikologii, Leningrad*, pp. 128–132.

Popov, V.I. and Stepanova, M.Yu. (1973) [Determination of the amount of infection units of *Verticillium dahliae* Kleb. in naturally infested soil]. Микология и Фитопатология [*Mycology and Phytopathology*] 7(2), 116–119.

Popov, V.I., Stepanova, M.Yu. and Kumacheva, E.M. (1976) [Effect of urea on the pathogen of cotton *Verticillium* wilt in the soil]. Микология и Фитопатология [*Mycology and Phytopathology*] 10, 324–325.
Popov, V.J., Tarunina, T.A. and Usmanov, Z.V. (1972) [On the physiologic races of *Verticillium dahliae* Kleb., a causal agent of cotton wilt]. Микология и Фитопатология [*Mycology and Phytopathology*] 6, 500–502.
Popova, I.V., Konstantinova, A.F., Kapustin, N.I., Mishin, V.V., Orekhova, O.V., Drobysheva, N. and Pichuzhkina, L.I. (1975) [A study of foreign strawberry varieties]. *Sadovodstvo* 8, 31–32.
Popushoĭ, I.S. (1977) *Verticillium* wilt of apricots. *EPPO Bulletin* 7, 95–104.
Popushoĭ, I.S. and Kulik, M.F. (1976) [Apricot wilt]. *Zashchita Rasteniĭ* 7, 46–47.
Porcelli, S., Fiume, F. and Saponaro, A. (1987) La resistenza genetica come metodo di lotta efficace al *Verticillium dahliae* Kleb. nel peperone. *Sementi Elette* 33, 41–44.
Porta-Puglia, A. and Montorsi, F. (1982) Osservazioni sulla micoflora di semi di melanzana. *Informatore Fitopatologico* 32, 37–41.
Portenko, L.G. (1990) Mutations in virulence and their role in the process of race formation in *Verticillium dahliae*, causal agent of cotton wilt. *Proceedings of the Fifth International* Verticillium *Symposium*. Leningrad, p. 109.
Portenko, L.G. (1997) Study of *Verticillium dahliae* strains attacking rapeseed. *Proceedings of the Seventh International* Verticillium *Symposium*. Cape Sounion, Athens, Greece, p. 94.
Portenko, L.G. and Akimov, G.I. (1997) Vegetative compatibility among *Verticillium dahliae* populations of cotton growing regions of Middle Asia. *Proceedings of the Seventh International* Verticillium *Symposium*. Cape Sounion, Athens, Greece, p. 93.
Portenko, L.G. and Kas'yanenko, A.G. (1978) [Virulence and aggressiveness of induced mutants of *Verticillium dahliae*]. In: Maksumov, A. (ed.) [*Genetic Variability of the Wilt-causing Agent and Prospects for Increasing Cotton Wilt Resistance*]. Donish Publishing House, Dushanbe, USSR, pp. 74–83.
Portenko, L.G. and Kas'yanenko, A.G. (1987) [Genetics of the cotton wilt pathogen *Verticillium dahliae* Kleb. for virulence and aggressiveness]. *Genetika USSR* 23, 1859–1865.
Portenko, L.G., Ryabova, I.M. and Azizova, Z.S. (1985) [Induction of phytoalexins in the course of cross-reaction defence in the cotton/*Verticillium* system]. *Ahboroti Akademijai Fanhoi RSS Točikiston, Šū'bai Fanhoi Biologija* 3, 30–33.
Portenko, L.G., Akimov, G.I. and Poshnagova, V.G. (1995) Studies of vegetative compatibility in *Verticillium dahliae*. 6th International *Verticillium* Symposium, Dead Sea, Israel, 1994. *Phytoparasitica* 23, 46.
Porter, C.L. and Green, R.J. (1952) Production of exotoxin in the genus *Verticillium*. *Phytopathology* 42, 472.
Porter, I.J. and Merriman, P.R. (1985) Evaluation of soil solarization for control of root diseases of row crops in Victoria. *Plant Pathology* 34, 108–118.
Post, T.B. (1935) Some diseases of ornamental and miscellaneous plants. *Plant Disease Reporter* 19, 213–214.
Pötschke, A. (1923) Über das Schwartzwerden des Meerrettichs. *Biologische Reichsanstalt für Land-und Forstwirtschaft* 11, 337–338.
Powell, C.C. (1977) *Nursery Notes* 10, 4–5.
Powell, D., Robb, J. and Ellis, B. (1987) Induction of phenylalanine ammonia-lyase activity in *Verticillium*-infested tomato plants. *Canadian Journal of Plant Pathology* 9, 84.

Powelson, M.L. (1985) Potato early dying disease in the Pacific northwest caused by *Erwinia carotovora* pv. *carotovora* and *E. carotovora* pv. *atroseptica. American Potato Journal* 62, 173–176.

Powelson, M.L. and Rowe, R.C. (1993) Biology and management of early dying of potatoes. *Annual Review of Phytopathology* 31, 111–126.

Powelson, M.L., Cappaert, M.R. and Christensen, N.W. (1995) Prospects for managing irrigation water to suppress potato early dying. 6th International *Verticillium* Symposium, Dead Sea, Israel, 1994. *Phytoparasitica* 23, 70.

Powelson, R.L. (1970) Significance of population level of *Verticillium* in soil. In: Tousson, T.A., Bega, R.V. and Nelson, P.E. (eds) *Root Diseases and Soil-borne Pathogens. Second International Symposium on Factors Determining the Behaviour of Plant Pathogens in Soil.* Imperial College, London. University of California Press, Berkeley, California, pp. 31–33.

Powelson, R.L. and Carter, G.E. (1973) Efficacy of soil fumigants for control of *Verticillium* wilt of potatoes. *American Potato Journal* 50, 162–167.

Powelson, R.L. and Gross, A.E. (1962) Thermal inactivation of *Verticillium albo-atrum* in diseased potato vines. *Phytopathology* 52, 364.

Pramateftaki, P. and Typas, M.A. (1997) The complete DNA sequence of the nuclear ribosomal RNA gene complex of *Verticillium dahliae. Proceedings of the Seventh International* Verticillium *Symposium.* Cape Sounion, Athens, Greece, p. 18.

Prashar, P. and Enevoldsen, M. (1993) 'Super Chief' tomato hybrid. *HortScience* 28(9), 959–960.

Prato, J.D. (ed.) (1971) California agricultural extension service agronomy notes. Certified seed notes. June 1971.

Presley, J.T. (1941) Saltants from a monosporic culture of *Verticillium albo-atrum. Phytopathology* 31, 1135–1139.

Presley, J.T. (1950) *Verticillium* wilt of cotton with particular emphasis on variation of the causal organism. *Phytopathology* 40, 497–511.

Presley, J.T. (1966) Current status of breeding for disease resistance in cotton in the United States. *Proceedings of a Seminar on Cotton Production Research.* Lima, Peru, pp. 1–6.

Presley, J.T. and Dick, J.B. (1951) Fertilizer and weather affect *Verticillium* wilt. *Mississippi Farm Research* 14, 1–6.

Presley, J.T., Carns, H.R., Taylor, E.E. and Schnathorst, W.C. (1966) Movement of conidia of *Verticillium albo-atrum* in cotton plants. *Phytopathology* 56, 375.

Pressey, R. (1997) Two isoforms of NP24: a thaumatin-like protein in tomato fruit. *Phytochemistry* 44(7), 1241–1245.

Price, D.W. (1976) Passage of *Verticillium albo-atrum* propagules through the alimentary canal of the bulb mite. *Phytopathology* 66, 46–50.

Price, T.V., McLeod, R.W. and Sumeghy, J.B. (1980) Studies on the interactions between *Fusarium oxysporum* f.sp. *lycopersici*, *Verticillium dahliae* and *Meloidogyne* spp. in resistant and susceptible tomatoes. *Australian Journal of Agricultural Research* 31, 1119–1127.

Profic-Alwasiak, H. (1982) [The occurrence of *Verticillium* disease on strawberry grown after different crops]. *Prace Instytutu Sadownictwa i Kwiaciarstwa w Skierniewicach A* 23, 81–90.

Profic-Alwasiak, H., Szczygiel, A. and Niezborala, E. (1977/78, published 1977) [The effectiveness of the preparations Benlate and Topsin M in the control of *Verticillium* wilt of strawberry]. *Prace Instytutu Sadownictwa i Skierniewicach, A* 20, 177–184.

Proska, B., Sturdikova, M., Pronayova, N. and Liptaj, T. (1996) (−) usnic acid and its derivatives. Their inhibition of fungal growth and enzyme activity. *Pharmazie* 51, 195–196.

Puckhaber, L.S., Stipanovic, R.D. and Bell, A.A. (1998) Kenaf phytoalexin: toxicity of *o*-hibiscanone and its hydroquinone to the plant pathogens *Verticillium dahliae* and *Fusarium oxysporum* f.sp. *vasinfectum*. *Journal of Agricultural and Food Chemistry* 46, 4744–4747.

Pudovkina, Z.M. and Alikhodzhzeva, S.S. (1963) [Wilt resistance in foreign cotton varieties]. *Khlopkovodstvo* 13, 34–35.

Pugacheva, T.I. (1981) [*Fusarium* and *Verticillium* wilt in tomato]. *Byulleten' Vsesoyuznogo Ordena Lenina i Ordena Druzhby Narodov Instuta Rastenievodstva Imeni N.I. Vavilova* 114, 63–65.

Pugacheva, T.I. (1982) [Resistance of tomato to *Fusarium* and *Verticillium* wilt]. *Trudy po Prikladnoi Botanike, Genetike i Selektsii* 71, 117–119.

Puhalla, J.E. (1973a) Differences in sensitivity of *Verticillium* species to ultraviolet irradiation. *Phytopathology* 63, 1488–1492.

Puhalla, J.E. (1973b) Genetics of the fungus. Verticillium *Wilt of Cotton. Proceedings of a Work Conference of the National Cotton Pathology Research Laboratory, College Station, Texas 1971.* ARS-S19, pp. 39–41.

Puhalla, J.E. (1975) Induction, isolation and characterisation of mutants of *Verticillium dahliae* with deficiencies in melanin synthesis. *Proceedings of the Beltwide Cotton Production Research Conference.* National Cotton Council, Memphis, Tennessee, p. 26.

Puhalla, J.E. (1976) Glycerol as a selective agent for auxotrophs of *Verticillium dahliae. Journal of General Microbiology* 94, 409–412.

Puhalla, J.E. (1977) A highly mutable locus in the fungus *Verticillium dahliae. Genetics* 86, s49–s50.

Puhalla, J.E. (1979) Classification of isolates of *Verticillium dahliae* based on heterokaryon incompatibility. *Phytopathology* 69, 1186–1189.

Puhalla, J.E. and Bell, A.A. (1981) Genetics and biochemistry of wilt pathogens. In: Mace, M.E., Bell, A.A. and Beckman, C.H. (eds) *Fungal Wilt Diseases of Plants.* Academic Press, New York, pp. 145–192.

Puhalla, J.E. and Howell, C.R. (1975) Significance of endo-polygalacturonase activity to symptom expression of verticillium wilt in cotton, assesed by the use of mutants of *Verticillium dahliae* Kleb. *Physiological Plant Pathology* 7, 147–152.

Puhalla, J.E. and Hummel, M. (1983) Vegetative compatibility groups within *Verticillium dahliae. Phytopathology* 73, 1305–1308.

Puhalla, J.E. and Hummel, M. (1981) Further studies on heterokaryon-compatible populations within *Verticillium dahliae. Phytopathologia Mediterranea* 23, 211.

Puhalla, J.E. and Mayfield, J.E. (1974) The mechanisms of heterokaryotic growth in *Verticillium dahliae. Genetics* 76, 411–422.

Pulido, M. (1969) 2-(thiocyanomethylthio) benzothiazol: un tratomiento efectivo para la semilla de algodón. *Fitopatología* 4, 19–25.

Pullman, G.S. and DeVay, J.E. (1981) Effect of soil flooding and paddy rice culture on the survival of *Verticillium dahliae* and incidence of *Verticillium* wilt in cotton. *Phytopathology* 71, 1285–1289.

Pullman, G.S. and DeVay, J.E. (1982a) Epidemiology of *Verticillium* wilt of cotton: a relationship between inoculum density and disease progression. *Phytopathology* 72, 549–554.

Pullman, G.S. and DeVay, J.E. (1982b) Epidemiology of *Verticillium* wilt of cotton: effects of disease development on plant phenology and lint yield. *Phytopathology* 72, 554–559.

Pullman, G.S., DeVay, J.E., Garber, R.H. and Weinhold, A.R. (1979) Control of soil-borne fungal pathogens by plastic tarping of soil. In: Schippers, B. and Gams, W. (eds) *Soil-borne Plant Pathogens. Proceedings of the 4th International Symposium on Factors determining the Behaviour of Plant Pathogens in the Soil.* Munich, Germany, August 1978. Academic Press, New York, pp. 439–446.

Pullman, G.S., DeVay, J.E. and Garber, R.H. (1980) Thermal sensitivity and death of *Verticillium dahliae* in relation to soil solarization and cotton lint yields. In: Brown, J.M. (ed.) *Proceedings of the Beltwide Cotton Production Research Conference.* January 6–8, 1980. National Cotton Council, Memphis, Tennessee, p. 16.

Pullman, G.S., DeVay, J.E. and Garber, R.H. (1981a) Soil solarization and thermal death: a logarithmic relationship between time and temperature for four soil-borne plant pathogens. *Phytopathology* 71, 959–964.

Pullman, G.S., DeVay, J.E., Garber, R.H. and Weinhold, A.R. (1981b) Soil solarization: effects on *Verticillium* wilt of cotton and soil-borne populations of *Verticillium dahliae, Pythium* spp., *Rhizoctonia solani* and *Thielaviopsis basicola. Phytopathology* 71, 954–959.

Purss, G.S. (1961) Wilt of peanut (*Arachis hypogaea* L.) in Queensland, with particular reference to *Verticillium* wilt. *Queensland Journal of Agricultural Science* 18, 453–462.

Putnam, M.L. and Crowe, F. (1999) *Verticillium dahliae*, a cause of wilt in *Echinacea purpurea. Plant Disease* 83(9), 877.

Qasem, J.R. and Abu-Blan, H.A. (1995) Antifungal activity of aqueous extracts from some common weed species. *Annals of Applied Biology* 127(1), 215–219.

Qian, Q.H., Gao, Y.R., Li, J.Y. and Gu, B.K. (1987) [Host range of *Verticillium dahliae* infecting cotton]. *Journal of Agricultural Science Jiangsu Academy of Agricultural Science* 3, 39–44.

Qiongfang, L., Yingfan, C., Yongjiu, T. and Honghua, H. (1996) Research on new cotton sources resistant to many biological strains of *Verticillium* wilt. *Scientia Agricultura Sinica* 29(4), 47–52.

Quak, F. (1957) Meristeemcultuur, gecombineerd met warmtebehandeling, voor het verkrijgen van virusvrije anjerplanten. *Tijdschrift over Plantenziekten* 63, 13–14.

Quanjer, H.M. (1916) Aard, verspreidingswijze en bestrijding van phloeem-necrose (bladrol) en verwante ziekten. *Mededelingen van de Rijks Hoogere Land-, Tuinen Boschbouwschool* 10, 1–138.

Qui, X., Wu, Y., Jaynes, J., Goodwin, P. and Erickson, L.R. (1995) Effect of a membrane interactive peptide on plant cells of canola (*Brassica napus*) and two fungal pathogens. *Plant Cell Reports* 15(1–2), 115–118.

Raabe, R.D. and Wilhelm, S. (1978) Susceptibility of several *Pistacia* species to *Verticillium albo-atrum. Plant Disease Reporter* 62, 672–673.

Ragazzi, A. and Fedi, I.D. (1984) Effetto di alcune sorgenti di azoto sulla crescita di *Verticillium dahliae* Kleb. isolato da olivo. *Rivista di Patologia Vegetale* 20, 21–24.

Ragazzi, A. and Parrini, C. (1982) Caratteristiche colturali di *Verticillium dahliae* Kleb., isolato da olivo. *Rivista di Patologia Vegetale* 18, 129–141.

Ragazzi, A., Parrini, C., Mesturino, L. and Dellavalle-Fedi, I. (1987) Epidemiologia di *Verticillium dahliae* Kleb. su olivo. 1 contributo. *Rivista di Patologia Vegetale IV* 23(3), 132–139.

Ragoni Galluci, M.M. (1959) Una verticilliosi in hibridi de *Gerbera jamesonii* Bolus. *Bollettino Laboratorio Sperimentale Fitopatologica Torino NS* 21, 57–61.

Rahimian, M.K. and Mitchell, J.E. (1984) Relationship of *Verticillium dahliae* and *Erwinia carotovora* pv. *carotovora* in the early dying disease of potato. *Phytopathology* 74, 327–332.

Raj, S.A. (1974) *Annamalai University Agricultural Research Annual* 4–5, 198–200.

Rakhimzhanov, A.G. (1977) [Restriction of varietal populations of cotton to *Verticillium dahliae*]. *Mirov. Rastitel'n Resursy v Srednei Azii* 3, 128–131.

Rakhimzhanov, A.G. (1980) [Resistance of Indian forms of *Gossypium hirsutum* L. to *Verticillium* wilt]. *Mirov. Rastitel'n Resursy v Srednei Azii* 7, 100–104.

Rakhimzhanov, A.G. (1987) [Inheritance of wilt resistance in distant intraspecific cotton hybrids]. *Sbornik Nauchnykh Trudov po Prikladnoi Botanike, Genetike i Selektsii* 113, 15–18.

Rakhmankulova, M.E., Tukhtaeva, G.M., Rodimtseva, N.E. and Khodzhaev, A.S. (1973) [Effect of *Verticillium dahliae* on the structure and function of cotton chloroplasts]. *Uzbekskii Biologicheskii Zhurnal* 17(3), 21–23.

Ramsay, J.R., Multani, D.S. and Lyon, B.R. (1996) RAPD-PCR identification of *Verticillium dahliae* isolates with differential pathogenicity on cotton. *Australian Journal of Agricultural Research* 47(5), 681–693.

Ranella, A.V., Ribaldi, M. and Lorenzetti, F. (1969) Screening of alfalfa lines for resistance to *Verticillium* wilt. *Phytopathologia Mediterranea* 8, 116–123.

Ranganathiah, K.G. and Swarmy, B.C.N. (1969) *Verticillium* wilt of *Pelargonium* in Mysore. *Mysore Journal of Agricultural Science* 3, 354–355.

Rankin, W.H. (1914) Thrombotic disease of maple. *Phytopathology* 4, 395.

Ranney, C.D. (1959) The effect of growth-regulators and nutrition on the development of *Verticillium* wilt of cotton. PhD dissertation, Texas A. & M. University, College Station.

Ranney, C.D. (1962) Effects of nitrogen source and rates on the development of *Verticillium* wilt of cotton. *Phytopathology* 52, 38–41.

Ranney, C.D. (1971) Studies with benomyl and thiabendazole on control of cotton disease. *Phytopathology* 61, 783–786.

Ranney, C.D. (1973) The effect of fallowing obtained with skip-row culture on *Verticillium* wilt of cotton. *Journal of Environmental Quality* 2, 303–306.

Ranney, C.D. (1995) Control of cotton diseases with host plant resistance: a success story with opportunities. *1995 Proceedings Beltwide Cotton Conferences*. San Antonio, Texas, January 4–7, 1995, Vol. 1. pp. 532–535.

Rao, A.S. (1959) A comparative study of competitive saprophytic ability in twelve root-infecting fungi by an agar plate method. *Transactions of the British Mycological Society* 42, 97–111.

Rataj-Guranowska, M. and Hiemstra, J.A. (1997) Comparison of European and American testers for vegetative compatibility groups in *Verticillium dahliae*. *Proceedings of the Seventh International* Verticillium *Symposium*. Cape Sounion, Athens, Greece, p. 22.

Rataj-Guranowska, M., Walkowiak, I., Carder, J.H. and Barbara, D.J. (1995) Vegetative compatibility and pathogenicity groupings within *Verticillium albo-atrum*. 6th International *Verticillium* Symposium, Dead Sea, Israel. *Phytoparasitica* 23, 46.

Rattigan, A. and Ayres, P.G. (1975) Growth of five phytopathogenic fungi in liquid media containing a uronic acid as the sole carbohydrate. *Transactions of the British Mycological Society* 65, 315–317.

Rauscher, K.J., Lester, D.T. and Smalley, E.B. (1974) Response of elm species and clones to inoculation with *Verticillium albo-atrum*. *Phytopathology* 64, 702–705.

Rava, C.A. (1970) Marchitamiento del garasol incitado por *Verticillium albo-atrum* Reinke et Berth., 1879. Variabilidad del huesped y del parasito. *Boletín Tecnico, Centro de Investigaciones Agricolas 'Alberto Boerger'* No. 11.

Rava, C.A. (1974) Selección de la variedad de girasol 'Estanzuela 60' por resistancia a *Verticillium albo-atrum*. *Boletín Técnico Centro de Investigaciones Agricolas 'Alberto Boerger'* 21, 11.

Ravisé, A. and Chopin, J. (1981) Influence de la structure decomposés phénoliques sur l'inhibition du *Phytophthora parasitica* et d'enzymes participant aux processus parasitaires V. Flavones, O- et C-glycosides. *Phytoparasitische Zeitschrift* 100, 257–269.

Rawlins, T.E. and Booth, J.A. (1968) Tween 20 as an adjuvant for systemic soil fungicides for *Verticillium* in cotton. *Plant Disease Reporter* 52, 944–945.

Ray, D.T., Orum, T.V., Bigelow, D.M. and Alcorn, S.M. (1995) Selection of diploid and screening of polyploid guayule lines for *Verticillium* tolerance. *Industrial Crops and Products* 4(4), 303–310.

Ray, L.L. (1976) Short-season cotton research. *Public-supported Cotton Research* ARS-S70, pp. 13–15.

Rayle, D.L. (1973) Auxin-induced hydrogen-ion secretion in *Avena* coleoptiles and its implications. *Planta* 114, 63–73.

Razakov, R.R., Kasimov, A.K., Aslanov, K.A. and Sadykov, A.S. (1985) [Study of the complex mixtures of natural compounds by defocusing and direct analysis of daughter ions: 5. Steroid metabolites of the phytopathogenic fungus *Verticillium dahliae*]. *Khimiya Prirodnykh Soedinenii Tashkent* 1985, pp. 227–232.

Rebandel, Z. (1981) [Effectiveness of the fungicide Benlate in controlling *Verticillium* on strawberries]. *Informator o Badaniach Prowadzonych w Zakladzie Sadownictwa Akademii Rolniczej w Poznaniu*, pp. 135–138.

Redondo Juárez, E. (1979) La secadura del jitomate causada por el hongo *Verticillium* sp. en Bajío. *Agricultura Técnica en México* 5, 3–9.

Reed, R.M. and Verhalen, L.M. (1970) Altus irrigated cotton-d nitrogen fertilizer *Verticillium* wilt interaction study. *Oklahoma Agricultural Experiment Station Progress Report* 639, 5–6.

Reeves, A.F., Cunningham, C.E., Nickson, R.L., Akeley, R.V., Murphy, H.J., Manzer, F.E. and True, R.H. (1985) Campbell 14: a *Verticillium* resistant round white potato variety. *American Potato Journal* 62, 215–220.

Reeves, A.F., Porter, G.A., Manzer, F.E., Work, T.M., Davis, A.A. and Plissey, E.S. (1994) MaineChip: a new chipping potato variety for cool storage processing. *American Potato Journal* 71(4), 237–247.

Reeves, A.F., Porter, G.A., Manzer, F.E., Work, T.M., Davis, A.A., Hensel, D.R. and Shumaker, J.R. (1996) St Johns: a round white potato variety for fresh market. *American Potato Journal* 73(2), 89–98.

Reeves, A.F., Porter, G.A., Work, T.M., Lambert, D.H., Davis, A.A. and Plissey, E.S. (1997) Mainestay: a high-yielding, round white potato variety for fresh market. *American Potato Journal* 74(4), 255–263.

Regnani, C. and Matta, A. (1986) Colonizzazione ed espressione dei sintomi nella verticilliosi del peperone. *Phytopathologia Mediterranea* 25, 1–9.

Regulski, F.J. and Peterson, J.L. (1983) Growth of *Verticillium dahliae* on sap from five maple species. *Phytopathology* 73, 115–118.

Reichenbach, G.L. (1845) *Icones Florae Germanicae et Helveticae*, Vol. 7.

Reimers, F., Smolka, S.E., Werres, S., Plank-Schumacher, K. and Wagner, G. (1993) Effect of ajoene, a compound derived from *Allium sativum*, on phytopathogenic and epiphytic micro-organisms. *Zeitschrift für Pflanzenkrankheiten und Pflanzenschutz* 100(6), 622–633.

Reinke, J. and Berthold, G. (1879) Die Zersetzung der Kartoffel durch Pilze. *Untersuchungen aus dem Botanischen Laboratorium der Universität Göttingen* 1, 67–96.

Resende, M.L.V., Flood, J. and Cooper, R.M. (1994) Host specialization of *Verticillium dahliae*, with emphasis on isolates from cocoa (*Theobroma cacao*). *Plant Pathology* 43(1), 104–111.

Resende, M.L.V., Cooper, R.M. and Flood, J. (1995a) Studies on the resistance of cocoa (*Theobroma cacao* L.) to vascular wilt caused by *Verticillium dahliae* Kleb. 6th International *Verticillium* Symposium, Dead Sea, Israel, 1994. *Phytoparasitica* 23, 58.

Resende, M.L.V., Cooper, R.M., Flood, J. and Beale, M.H. (1995b) The role of tannins and a multiple phytoalexin response in the resistance of cocoa (*Theobroma cacao* L.) to *Verticillium dahliae* Kleb. 6th International *Verticillium* Symposium, Dead Sea, Israel. *Phytoparasitica* 23, 59.

Resende, M.L.V., Flood, J. and Cooper, R.M. (1995c) Effect of method of inoculation, inoculum density and seedling age at inoculation on the expression of resistance of cocoa (*Theobroma cacao* L.) to *Verticillium dahliae* Kleb. *Plant Pathology* 44(2), 374–383.

Resende, M.L.V., Flood, J., Ramsden, J.D., Rowan, M.G., Beale, M.H. and Cooper, R.M. (1996a) Novel phytoalexins including elemental sulphur in the resistance of cocoa (*Theobroma cacao* L.) to *Verticillium* wilt (*Verticillium dahliae* Kleb.). *Physiological and Molecular Plant Pathology* 48(5), 347–359.

Resende, M.L.V., Mepsted, R., Flood, J. and Cooper, R.M. (1996b) Water relations and ethylene production as related to symptom expression in cocoa seedlings infected with defoliating and non-defoliating isolates of *Verticillium dahliae*. *Plant Pathology* 45(5), 964–972.

Restaino, F. (1986) 'Picentia', 'Macla', 'Floralba': new eggplant cultivars obtainment through mutagenesis and cross. *Genetica Agraria* 40, 469–470.

Reuveni, R. and Ferreira, J.F. (1985) The relationship between peroxidase activity and the resistance of tomatoes (*Lycopersicon esculentum*) to *Verticillium dahliae*. *Phytopathologische Zeitschrift* 112, 193–197.

Richard, C. (1987) Outbreak of *Verticillium* wilt of alfalfa in Quebec. *Plant Disease* 71, 761.

Richter, H. and Klinkowski, M. (1938) Wirtelpilz-Welkekrankheit an Luzerne und Esparsette (Erreger: *Verticillium albo-atrum* Rke et Berth.). *Nachrichtenblatt für den Deutschen Pflanzenschutzdienstes Berlin* 18, 57–58.

Rick, C.M. (1978) The tomato. *Scientific American* 239, 76–87.

Ricketson, C.L., Gray, V.P., Cook, F.I., Vandenberg, A.A. and Dale, A. (1986) Vantage strawberry. *Canadian Journal of Plant Science* 66, 1025–1026.

Ride, J.P. and Drysdale, R.B. (1971) A chemical method of estimating *Fusarium oxysporum* f.sp. *lycopersici* in infected tomato plants. *Physiological Plant Pathology* 1, 409–420.

Ride, J.P. and Drysdale, R.B. (1972) A rapid method for the chemical estimation of filamentous fungi in plant tissue. *Physiological Plant Pathology* 2, 7–15.

Riedel, R.M. and Rowe, R.C. (1985) Lesion nematode involvement in potato early dying disease. *American Potato Journal* 62, 163–171.

Riedel, R.M., Rowe, R.C. and Martin, M.J. (1985) Differential interactions of *Pratylenchus crenatus, Pratylenchus penetrans* and *Pratylenchus scribneri* with *Verticillium dahliae* in potato early dying disease. *Phytopathology* 75, 419–422.

Riggs, J.L. and Graham, C.J. (1995) A screening of New Mexico *Verticillium dahliae* isolates for cross-infectivity to cotton and chilli. *1995 Proceedings Beltwide Cotton Conferences*. San Antonio, Texas, January 4–7, 1995, Vol. 1, pp. 218–221.

Rijkers, A.J.M., Hiemstra, J.A. and Bollen, G.J. (1992) Formation of microsclerotia of *Verticillium dahliae* in petioles of infected ash trees. *Netherlands Journal of Plant Pathology* 98, 261–264.

Riley, M.K. and Bosland, P.W. (1995) Additional sources of resistance to *Verticillium* wilt and Phytophthora root rot of Capsicum. *Capsicum and Eggplant Newsletter* 65–67.

Rintelen, J. (1974) Die *Verticillium*-Welke des Hopfens. *Zeitschrift für Pflanzenkrankheiten und Pflanzenschutz* 81, 304–315.

Roath, W.W., Miller, J.F. and Gulya, T.J. (1981) Registration of RHA 801 sunflower germplasm (Reg.No. GP 5). *Crop Science* 21, 479.

Robb, J. and Busch, L.V. (1979) Nuclear invaginations in plants infected by *Verticillium dahliae*. *Canadian Journal of Plant Pathology* 1, 23–26.

Robb, J. and Busch, L.V. (1982) Ultrastructural changes in drought-induced wilt: a comparison with pathogen-induced flaccidity. *Canadian Journal of Plant Pathology* 4, 97–105.

Robb, J. and Busch, L.V. (1983) Structure and significance of 'blocked veins' in *Verticillium*-infected chrysanthemums. *Physiological Plant Pathology* 23, 35–53.

Robb, J. and Nazar, R.N. (1996) PCR diagnostics for potato and other crops. *Agri-food Research in Ontario* 19(2), 21–22.

Robb, J., Busch, L.V., Brisson, J.D. and Lu, B.C. (1975a) Ultrastructure of wilt syndrome caused by *Verticillium dahliae* II. In sunflower leaves. *Canadian Journal of Botany* 53, 2725–2739.

Robb, J., Busch, L.V. and Lu, B.C. (1975b) Ultrastructure of wilt syndrome caused by *Verticillium dahliae* I. In chrysanthemum leaves. *Canadian Journal of Botany* 53, 901–913.

Robb, J., Busch, L.V., Brisson, J.D. and Lu, B.C. (1977) Ultrastructure of wilt syndrome caused by *Verticillium dahliae* III. Chronological symptom development in sunflower leaves. *Canadian Journal of Botany* 55, 139–152.

Robb, J., Brisson, J.D., Busch, L.V. and Lu, B.C. (1978) Ultrastructure of wilt syndrome caused by *Verticillium dahliae* V. Attempted localization of phenolic compounds in the vascular region. *Canadian Journal of Botany* 56, 2594–2612.

Robb, J., Brisson, J.D., Busch, L.V. and Lu, B.C. (1979a) Ultrastructure of wilt syndrome caused by *Verticillium dahliae* VII. Correlated light and transmission electron microscope identification of vessel coatings and tyloses. *Canadian Journal of Botany* 57, 822–834.

Robb, J., Smith, A., Brisson, J.D. and Busch, L.V. (1979b) Ultrastructure of wilt syndrome caused by *Verticillium dahliae* VI. Interpretive problems in the study of vessel coatings and tyloses. *Canadian Journal of Botany* 57, 795–821.

Robb, J., Smith, A. and Busch, L.V. (1982) Wilts caused by *Verticillium* species. A cytological survey of vascular alterations in leaves. *Canadian Journal of Botany* 60, 825–837.

Robb, J., Street, P.F.S. and Busch, L.V. (1983) Basic fuchsin: a vascular dye in studies of *Verticillium*-infected chrysanthemum and tomato. *Canadian Journal of Botany* 61, 3355–3365.

Robb, J., Powell, D.A. and Street, P.F.S. (1989) Vascular coating: a barrier to colonization by the pathogen in *Verticillium* wilt of tomato. *Canadian Journal of Botany* 67, 600–607.

Robb, J., Hu, X., Schmidt, J. and Nazar, R. (1990) DNA hybridization probes for the identification and quantification of *Verticillium dahliae* and *Verticillium albo-atrum*. *Proceedings of the Fifth International* Verticillium *Symposium*. Leningrad, p. 97.

Robb, J., Lee, S.W., Mohan, R. and Kolattukudy, P.E. (1991) Chemical characterization of stress-induced vascular coating in tomato. *Plant Physiology* 97, 528–536.

Robb, J., Moukhamedov, R., Hu, X., Platt, H. and Nazar, R.N. (1993) Putative subgroups of *Verticillium albo-atrum* distinguishable by PCR-based assays. *Physiological and Molecular Plant Pathology* 43(6), 423–436.

Robb, J., Hu, X., Platt, H. and Nazar, R. (1994) PCR-based assays for the detection and quantification of *Verticillium* species in potato. In: Schots, A., Dewey, F.M. and Oliver, R.P. (eds) *Modern Assays for Plant Pathogenic Fungi: Identification, Detection and Quantification.* CAB International, Wallingford, UK, pp. 83–90.

Roberts, D.M., Bainbridge, B.W., Evans, H.C. and Heale, J.B. (1995) Genome analysis in isolates of *Verticillium, V. albo-atrum, V. dahliae* and *V. lecanii. Phytoparasitica* 23(1), 40.

Roberts, E.T. and Large, E.C. (1963) Surveys of *Verticillium* wilt in lucerne, England and Wales, 1958–60. NAAS Survey. *Plant Pathology* 12, 47–58.

Roberts, F.M. (1943) Factors influencing infection of the tomato by *Verticillium albo-atrum. Annals of Applied Biology* 30, 327–331.

Roberts, F.M. (1944) Factors influencing infection of the tomato by *Verticillium albo-atrum* II. *Annals of Applied Biology* 31, 191.

Robinson, D.B. and Ayers, G.W. (1961) *Verticillium* wilt of potato in relation to vascular infection of the tuber. *Canadian Journal of Plant Science* 41, 703–708.

Robinson, D.B., Larson, R.H. and Walker, J.C. (1957) *Verticillium* wilt of potato in relation to symptoms, epidemiology and variability of the pathogen. *Bulletin of the Wisconsin Agricultural Experimental Station* 202, 1–49.

Robinson, P. and Smith, C.J. (1995) Signal transduction in phytoalexin elicitation in *Medicago sativa. Aspects of Applied Biology. Physiological Responses of Plants to Pathogens.* 11–13 September 1995, University of Dundee, UK, pp. 209–216.

Rodriguez Jurado, D., Blanco Lopez, M.A., Rapoport, H.F. and Jimenez-Diaz, R.M. (1993) Present status of *Verticillium* wilt of olive in Andalucia (southern Spain). *Bulletin OEPP* 23(3), 513–516.

Rogers, H.H. (1976) Forage legumes (with particular reference to lucerne and red clover). *Report of the Plant Breeding Institute* 1975, pp. 22–57.

Rogerson, C.T. (1957) *Verticillium* wilt in Kansas. *Plant Disease Reporter* 41, 1053–1054.

Romanko, R.R. and Jaeger, J. (1987) Evidence for competitive antagonism in strains of pathogenic verticillia. *Canadian Journal of Plant Pathology* 9, 85.

Romanko, R.R., Jaeger, J., Nickerson, G.B. and Zimmermann, C.E. (1979) Registration of Galena hop. *Crop Science* 19, 563.

Romanko, R.R., Jaeger, J., Nickerson, G.B. and Zimmermann, C.E. (1982) Registration of Eroica Hop (Reg.No. 8). *Crop Science* 22, 1261–1262.

Romanko, R.R., Nickerson, G.B., Jaeger, J., Kenny, S.T. and Skotland, C.B. (1996) Registration of 'Banner' hop. *Crop Science* 36(5), 1417.

Rosati, P., Faedi, W. and D'Ercole, N. (1982) Schede pomologiche e agronomich-sanitarie di quattro nuove cultivar di fragola per la Valle Padana, 'Addie', 'Brio', 'Cesena', 'Dana'. *Frutticoltura* 44(5), 47–50.

Ross, R.W. and Rowe, P.R. (1965) *Bulletin of the US Department of Agriculture* 533, 1–73.

Roth, J.N. and Brandt, W.H. (1964a) Some effects of vitamins on growth and differentiation in *Verticillium*. *Plant Disease Reporter* 48, 649–652.

Roth, J.N. and Brandt, W.H. (1964b) Nuclei in spores and mycelium of *Verticillium*. *Phytopathology* 54, 363–364.

Roth, J.N. and Brandt, W.H. (1964c) Influence of some environmental factors on hereditary variation in monospore cultures of *Verticillium albo-atrum*. *Phytopathology* 54, 1454–1458.

Rotino, G.L., Falavigna, A. and Restaino, F. (1987) *In vitro* selection of eggplant cells resistant to culture filtrate of *Verticillium dahliae* Kleb. and regeneration of plants. *Capsicum Newsletter* 6, 94–95.

Rouse, D.I. (1985) Some approaches to prediction of potato early dying disease severity. *American Potato Journal* 62, 187–193.

Rowe, R.C. (1985) Potato early dying – a serious threat to the potato industry. *American Potato Journal* 62, 157–161.

Rowe, R.C. (1995) Recent progress in understanding relationships between *Verticillium* species and subspecific groups. *Phytoparasitica* 23(1), 31–38.

Rowe, R.C. and Botseas, D.D. (1995) Variability in virulence among pathotypes of *Verticillium dahliae* and possible implications in IPM programs for potato early dying. 6th International *Verticillium* Symposium, Dead Sea, Israel, 1994. *Phytoparasitica* 23, 43.

Rowe, R.C. and Riedel, R.M. (1984) Synergistic interactions between *Verticillium dahliae* and *Pratylenchus* species in potato early dying disease. *Phytopathology* 74, 845.

Rowe, R.C. and Riedel, R.M. (1990) Modification of the resistance response to *Verticillium dahliae* in potato as a result of co-infection by the root-lesion nematode *Pratylenchus penetrans*. *Proceedings of the Fifth International* Verticillium *Symposium*, Leningrad, p. 66.

Rowe, R.C., Riedel, R.M. and Martin, M.J. (1985) Synergistic interactions between *Verticillium dahliae* and *Pratylenchus penetrans* in potato early dying disease. *Phytopathology* 75, 412–418.

Rowe, R.C., Davis, J.R., Powelson, M.L. and Rouse, D.I. (1987) Potato early dying: causal agents and management. *Plant Disease* 71, 482–489.

Rowe, R.C., Johnson, D.A., Beery, W.R. and Omer, M.A. (1997) Vegetative compatibility analysis of strains of *Verticillium dahliae* from potato seed tubers and plants from the western and eastern United States. *Proceedings of the Seventh International* Verticillium *Symposium*. Cape Sounion, Athens, Greece, p. 23.

Rubin, B.A. and Perevyazkina, L.M. (1951) [The role of tannic substances in the phenomena of resistance of the cotton plant to wilt]. *Doklady Akademii Nauk SSSR* 79, 303–306.

Rubin, B.A. and Tukeeva, M.I. (1978) [The lysosomal-vacuolar system of the pathogen of *Verticillium* wilt of cotton and its role in protective reactions of the organism]. *Sel'skokhozyaĭstvennaya Biologiya* 13, 583–592.

Rubin, B.A., Shevyreva, V.V., Merzlyak, M.N. and Veronkov, L.A. (1977) [Higher fatty acids of cotton chloroplasts in infection by *Verticillium* wilt]. *Fiziologiya Rastenii* 24, 1060–1066.

Rudolph, B.A. (1926a) *Verticillium* wilt of tomatoes in California. *Phytopathology* 16, 234.
Rudolph, B.A. (1926b) Now they know – it's verticilliosis. *Pacific Rural Press* 111, 763,766.
Rudolph, B.A. (1931) *Verticillium* hadromycosis. *Hilgardia* 5, 201–361.
Rudolph, B.A. and Snyder, W.C. (1936) *Verticillium* from snapdragon, watermelon, celery and cowpea. *Plant Disease Reporter* 20, 125–126.
Rudolph, B.A. and Snyder, W.C. (1937) *Verticillium* wilt of pepper. *Plant Disease Reporter* 21, 404.
Rudolph, B.A. and Snyder, W.C. (1938) *Verticillium* wilt of Persian and honeydew melons. *Plant Disease Reporter* 22, 447.
Rudolph, E. (1968) Untersuchungen zur Welkekrankheiten des Hopfens (*Humulus lupulus* L.). *Zeitschrift für Pflanzenkrankheiten und Pflanzenschutz* 75, 401–412.
Ruggieri, G. (1946) Nuova malattia dell'olivo. *Italia Agricultura* 83, 369–372.
Ruggieri, G. (1948) Richerche ed esperienze su una tracheovertilliosi dell'olivio. *Annali della Sperimentazione Agraria Roma Nuova Serie* 2, 1–8.
Rumbos, I.C. and Karayiannis, I. (1999) Apricot decline and integrated plant protection. *Acta Horticulturae. Proceedings of the XIth International Symposium on Apricot Culture.* Veria, Makedonia, Greece, 25–30 May 1997, Vol. 2 488, 635–642.
Runia, W.T., van Os, E.A. and Bollen, G.J. (1988) Disinfection of drainwater from soilless cultures by heat treatment. *Netherlands Journal of Agricultural Science* 36, 231–238.
Russel, S. (1975a) The role of cellulase produced by *Verticillium albo-atrum* in *Verticillium* wilt of tomatoes. *Phytopathologische Zeitschrift* 82, 35–48.
Russel, S. (1975b) Characteristics of *Verticillium albo-atrum* cellulase. *Phytopathologische Zeitschrift* 84, 222–232.
Russell, E.J. (1920) The partial sterilization of soils. *Journal of the Royal Horticultural Society* 45, 237–246.
Russomanno, O.M.R., Coutinho, L.N., Pimentel, C.P.V. and Figueiredo, M.B. (1995) Preservacao da patogenicidade de culturas de *Verticillium dahliae* conservadas em agua destilada. *Summa Phytopathologica* 21(2), 178–181.
Ryabova, I.M. (1990) The cotton genotype effect on the ratio of marking races in the populations of *Verticillium dahliae*. *Proceedings of the Fifth International* Verticillium *Symposium*, Leningrad, p. 110.
Rysbayeva, A.S. and Urunov, I.S. (1990) Fruit setting stage of cotton as a stage of resistance deficiency. *Proceedings of the Fifth International* Verticillium *Symposium*, Leningrad, p. 77.
Saccardo, F. and Sree Ramulu, K. (1977) Mutagenesis and cross breeding in *Capsicum* for disease resistance against *Verticillium dahliae*. In: Poichard, E. (ed.) *Proceedings of the 3rd Congress of EUCARPIA on Genetics and Selection of Pepper*, Avignon-Montfavet, 5–8 July 1977, pp. 161–169.
Sackston, W.E. (1949) Sunflower: observations on sunflower diseases in Manitoba. *Canadian Plant Disease Survey for 1948* 28, 31–33.
Sackston, W.E. (1957) Diseases of sunflower in Uruguay. *Plant Disease Reporter* 41, 885–889.
Sackston, W.E. (1959) *Verticillium albo-atrum* on red clover (*Trifolium pratense*). *Report of the Quebec Society for the Protection of Plants from Insects and Fungus Diseases* 41, 116–120.
Sackston, W.E. (1971) Factors affecting *Verticillium* infection. *Proceedings of the First International* Verticillium *Symposium*. Wye College, University of London.

Sackston, W.E. (1980) Some factors influencing infection of sunflower seed by *Verticillium dahliae*. *Canadian Journal of Plant Pathology* 2, 209–212.

Sackston, W.E. (1983a) Sunflower diseases, with special reference to tropical and subtropical areas. *Garcia de Orta, Estudos Agronómicos* 10, 105–112.

Sackston, W.E. (1983b) Epidemiology and control of seed-borne *Verticillium* spp. causing vascular wilt. *Seed Science and Technology* 11, 731–747.

Sackston, W.E. and Martens, J.W. (1959) Dissemination of *Verticillium albo-atrum* on seed of sunflower (*Helianthus annuus*). *Canadian Journal of Botany* 37, 759–768.

Sackston, W.E. and Sheppard, J.W. (1973) Effect of day length on reaction of sunflowers to *Verticillium* wilt. *Canadian Journal of Botany* 51, 23–26.

Sackston, W.E., McDonald, W.C. and Martens, J. (1957) Leaf mottle or *Verticillium* wilt of sunflower. *Plant Disease Reporter* 41, 337–343.

Sadasivan, T.S. (1961) Physiology of wilt disease. *Annual Review of Plant Physiology* 12, 449–468.

Sadasivan, T.S., Subramanian, C.V., Kalyanasundaram, R. and Saraswathi-Devi, L. (1978) *Pathological Wilting of Plants*. University of Madras, India.

Sadras, V.O., Trapani, N., Pereyra, V.R., López Pereira, M., Quiroz, F. and Mortarini, M. (2000) Intraspecific competition and fungal diseases as sources of variation in sunflower yield. *Field Crops Research* 67(1), 51–58.

Sadykhov, A.S., Metlitskii, L.V., Karimdzhanov, A.K., Ismailov, A.I., Mukhamedova, R.A., Avazkhodzhaev, M.Kh. and Kamaev, F.G. (1974) [Isohemigossypol as a phytoalexin of the cotton plant]. *Doklady Akademii Nauk SSSR* 2181, 506–509 [1472–1475].

Saeed, I.A.M., MacGuidwin, A.E. and Rouse, D.I. (1997a) Synergism of *Pratylenchus penetrans* and *Verticillium dahliae* manifested by reduced gas exchange in potato. *Phytopathology* 87(4), 435–439.

Saeed, I.A.M., MacGuidwin, A.E. and Rouse, D.I. (1997b) Disease progress based on effects of *Verticillium dahliae* and *Pratylenchus penetrans* on gas exchange in russet burbank potato. *Phytopathology* 87(4), 440–445.

Saeed, I.A.M., Rouse, D.I., Harkin, J.M. and Smith, K.P. (1997c) Effects of soil water content and soil temperature on efficacy of metham-sodium against *Verticillium dahliae*. *Plant Disease* 81(7), 773–776.

Saeed, I.A.M., MacGuidwin, A.E. and Rouse, D.I. (1998) Effect of initial nematode population density on the interaction of *Pratylenchus penetrans* and *Verticillium dahliae* on 'Russet Burbank' potato. *Journal of Nematology* 30(1), 100–107.

Saeed, I.A.M., MacGuidwin, A.E., Rouse, D.I. and Sharkey, T.D. (1999) Limitation to photosynthesis in *Pratylenchus penetrans*- and *Verticillium dahliae*-infected potato. *Crop Science* 39(5), 1340–1346.

Safiyazov, J.S., Mannanov, R.N. and Sattarova, R.K. (1995) The use of bacterial antagonists for the control of cotton diseases. *Field Crops Research* 43(1), 51–54.

Safiyazov, Zh. (1977) [*Cytological and Morphological Features of the Pathogen Causing Wilt in Cotton*]. Fan, Tashkent, Uzbek SSR.

Safiyazov, Zh. (1978) [Effect of *Verticillium dahliae* Kleb. on the ultrastructure of cotton chloroplasts]. *Uzbekskii Biologicheskii Zhurnal* 6, 9–11.

Safiyazov, Zh. and Cherkasova, G.V. (1979) Ultrastructural features of mutant strains of *Verticillium dahliae*. *UzSSR Fanlar Akadamii Doklady* 11, 72–74.

Safiyazov, Zh. and Shaabdurakhmanova, G.A. (1974) [Cytofluorometric study of the content of DNA-proteids and total protein in cells of the causal agent of cotton wilt]. *Uzbekskii Biologicheskii Zhurnal* 18(2), 10–13.

Safiyazov, Zh., Rodimitsev, A.I. and Shoabdurakhmanova, G. (1972) [The cytology of the causal agent of *Verticillium* wilt of cotton]. Микология и Фитопатология [*Mycology and Phytopathology*] 6, 127–130.
Safiyazov, Zh., Mukhamedov, R.S. and Abdukarimov, A.A. (1990) [The extraction and purification of the fungus *Verticillium dahliae* virus particles]. *Proceedings of the Fifth International* Verticillium *Symposium, Leningrad*, p. 111.
Sagdieva, M.G. (1974) [Effects of spore concentration and culture conditions of *Verticillium dahliae* Kleb. on the production of pectin-trans-eliminase in connexion with the appearance of dimorphism]. Микология и Фитопатология [*Mycology and Phytopathology*] 8, 102–105.
Sagdieva, M.G., Vasilyeva, K.V. and Metlitskii, L.V. (1974) [Effects of gossypol on the activity and production of pectin-trans-eliminase of *Verticillium dahliae* Kleb.]. Микология и Фитопатология [*Mycology and Phytopathology*] 8, 201–207.
Saito, A.B., Melo, I.S., Azevedo, J.L. and Faull, J.L. (1995) Antagonistic properties of wild-types and mutants of *Talaromyces flavus* against *Verticillium dahliae*. 6th International *Verticillium* Symposium, Dead Sea, Israel. *Phytoparasitica* 23, 69.
Sakhacheva, N.V., Zainutdinova, K.M. and Ignatenko, M.A. (1988) [Fungicidal and nematicidal ability of substituted pyrimidines and dipyridyls]. *Uzbekskii Biologicheskii Zhurnal* 57–59.
Salikhova, B.S., Borodin, G.I., Runov, V.I. and Chepenko, L.I. (1970) [Effect of toxic compounds of the mycelium of *V. dahliae* on the gas-exchange of cotton leaves]. *Uzbekskii Biologicheskii Zhurnal* 14, 28–31.
Sampson, P.J. (1977) Contamination with *Erwinia carotovora* and *Verticillium albo-atrum* during multiplication of pathogen-tested seed potato crops, cultivar Kennebec. *American Potato Journal* 54(1), 1–9.
Sampson, P.J. (1980) Infection of Kennebec potato with *Verticillium albo-atrum*. *Australian Journal of Agricultural Research* 31, 525–532.
Sanderdson, F.R. (1976) Lucerne diseases in New Zealand as recognised in the national lucerne survey 1975–76. *Proceedings of the 29th New Zealand Weed and Pest Control Conference*, pp. 38–41.
Sandrock, R.W. and Van Etten, H.D. (1998) Fungal sensitivity to and enzymatic degradation of the phytoanticipin α-tomatine. *Phytopathology* 88(2), 137–143.
Santo, G.B. (1974) Influence of *Pratylenchus vulnus*, *Meloidogyne hapla* and *Verticillium dahliae* on the growth of Manetti rose root stock. *Dissertation Abstracts International* 35B, 4, 1475.
Sappenfield, W.P. (1980) Registration of Delcot 311 cotton (Reg.No. 79). *Crop Science* 20, 669.
Sappenfield, W.P. (1981) Registration of HYC76-59 cotton germplasm (Reg.No. GP 163). *Crop Science* 21, 991–992.
Sappenfield, W.P. (1985) Registration of Delcot 390 cotton. *Crop Science* 25, 198.
Saribay, A., Delen, N. and Ercivan, S. (1973) *Verticillium* wilt of apricot in the Ege region of Turkey. *Journal of Turkish Phytopathology* 2, 90–92.
Saribay, A. and Demir, T. (1984) *Verticillium* wilt of plum in Turkey. *Journal of Turkish Phytopathology* 13, 105–107.
Satlarov, Kh. (1976) [Proportion of nutrients in relation to soil fertility]. *Khlopkovodstvo* 5, 18–19.
Sattarov, B.Kh. (1981a) [Susceptibility of cotton varieties of *Gossypium barbadense* L. to *Verticillium* wilt]. *Mirov. Rastitel'n Resursy v Srednei Azii* 8, 49–53.

Sattarov, B.Kh. (1981b) [Forms of *Gossypium barbadense* L. resistant to *Verticillium dahliae* Kleb.]. *Byulleten' Vsesoyuznogo Ordena Lenina i Ordena Druzhby Naroda Instituta Rastenievodstva Imeni N.I. Vavilova* 115, 35–37.

Savov, S.G. (1978a) [The spread of *Verticillium dahliae* Kleb. in different organs of cotton plants]. *Rasteniev'dni Nauki* 15(7), 121–125.

Savov, S.G. (1978b) [Effect of some agrotechnical measures on *Verticillium* wilt of cotton]. *Rasteniev'dni Nauki* 15(4), 122–127.

Savov, S.G. (1979) [*Verticillium* wilt of cotton in Bulgaria and species classification of its pathogen]. *Rasteniev'dni Nauki* 16, 95–104.

Savov, S.G. (1986a) [Effectiveness of some trace elements in the control of cotton *Verticillium* wilt]. *Rasteniev'dni Nauki* 23, 68–71.

Savov, S.G. (1986) [Comparative resistance of Bulgarian and foreign cotton varieties to *Verticillium* wilt]. *Rasteniev'dni-Nauki* 23(5), 63–67.

Sawamura, K. and Soma, A. (1976) *Verticillium* wilt of udo (*Aralia cordata* Thunb.) caused by *Verticillium dahliae* Klebahn. *Bulletin of the Faculty of Agriculture, Hirosaki University* 26, 10–17.

Saydam, C. (1976) *Verticillium* wilt on different host plants in Turkey. *Probleme de Proteçtia Plantelor* 4, 213–218.

Saydam, C. and Copcu, M. (1972) *Verticillium* wilt of olives in Turkey. *Journal of Turkish Phytopathology* 1, 45–49.

Saydam, C. and Copcu, M. (1973) *Verticillium* wilt in Ege region and its importance on some vegetables. *Journal of Turkish Phytopathology* 2, 34–40.

Saydam, C. and Kamel, M. (1970) Outbreaks and new records. *Plant Protection Bulletin FAO* 18, 43–46.

Saydam, C., Saribay, A. and Ogut, M. (1971) Occurrence of *Verticillium* wilt of peach in Turkey. *Journal of Turkish Phytopathology* 1, 12–13.

Saydam, C., Copcu, M. and Sezgin, E. (1973) Studies on the inoculation techniques of cotton wilt caused by *Verticillium dahliae* Kleb. I. Investigations on the laboratory inoculation techniques. *Journal of Turkish Phytopathology* 2(2), 69–75.

Sboeva, Zh.N., Nematov, G.N. and Yunusov, M.R. (1977) [Anti-wilt effectiveness of benzimidazole derivatives and their behaviour in plants and soil]. *Byulleten' Vsesoyuznogo Nauchno-Issledovatel'skogo Instituta Zashchity Rasteniĭ* 40, 23–25.

Sboeva, Zh.N., Siderova, S.F., Borovkov, A.V. and Tellyaev, R. (1980) [The effect of systemic fungicides benomyl and uzgen on leaf infection of cotton by *Verticillium* wilt]. *Byulleten' Vsesoyuznogo Nauchno-Issledovatel'skogo Instituta Zashchity Rasteniĭ* 49, 57–60.

Schaible, L., Cannon, O.S. and Waddoups, V. (1951) Inheritence of resistance to *Verticillium* wilt in a tomato cross. *Phytopathology* 41, 986–990.

Scheffer, R.J. and Elgersma, D.M. (1990) *Verticillium dahliae* as a control agent of Dutch elm disease. *Proceedings of the Fifth International* Verticillium *Symposium*, Leningrad, p. 85.

Scheffer, R.P., Gothoskar, S.S., Pierson, C.F. and Collins, R.P. (1956) Physiological aspects of *Verticillium* wilt. *Phytopathology* 46, 83–87.

Scheller, H. (1971) Luzerne. *Bayerisches Landwirtschaftliches Jahrbuch* 48, 191–199.

Schiefer, C., Hülsbergen, H., Hempel, K. and Kohls, D. (1987) Untersuchungen zur Einschätzung der Resistenz gegen *Verticillium albo-atrum* R. et B. sowie zum Entragsverhalten von Sorten und Neuzuchtmaterial der Luzerne. *Archiv fur Zuchtungsforschung* 17, 371–379.

Schippers, B. and Gams, W. (1979) *Soil-borne Plant Pathogens. Proceedings of the 4th International Symposium on Factors Determining the Behaviour of Plant Pathogens in the Soil.* Munich, Germany, August 1978. Academic Press, London.

Schippers, B. and Schermer, A.K.F. (1966) Effect of anti-fungal properties of soil on dissemination of the pathogen and seedling infection originating from *Verticillium*-infected achenes of *Senecio. Phytopathology* 56, 549–552.

Schleier, S., Voigt, K. and Worstemeyer, J. (1997) RAPD-based molecular diagnosis of mixed fungal infections on oilseed rape (*Brassica napus*): evidence for genus and species-specific sequences in the fungal genome. *Journal of Phytopathology* 145, 81–87.

Schmid, G. (1969) Beziehungen zwischen Bodeneigenschaften und Hopfenwelke (*Verticillium albo-atrum* Reinke et Berth.). *Zeitschrift für Pflanzenkrankheiten und Pflanzenschutz* 76, 79–83.

Schmidt, H.E. and Skadow, K. (1967) *Verticillium albo-atrum* Reinke et Berth. als Erreger der Welkekrankheit des Hopfens im Gebiet der Deutschen Demokratischen Republik. *Nachrichtenblatt für den Pflanzenschutzdienst in der DDR* 21, 206–209.

Schnathorst, W.C. (1963) Theoretical relationships between inoculum potential and disease severity based on a study of the variation in virulence among isolates of *V. albo-atrum. Phytopathology* 53, 888.

Schnathorst, W.C. (1964) A fungal complex associated with the sudden wilt syndrome in California cotton. *Plant Disease Reporter* 48, 90–92.

Schnathorst, W.C. (1965) Origin of new growth in dormant microsclerotial masses of *Verticillium albo-atrum. Mycologia* 57, 343–351.

Schnathorst, W.C. (1969) Serological relationships among several *Verticillium* spp. and their virulence in cotton plants. *Proceedings of the Beltwide Cotton Production Research Conference.* New Orleans, Louisiana, pp. 31–32.

Schnathorst, W.C. (1971) Some biological aspects of *Verticillium albo-atrum. Cotton Disease Research in the San Joachin Valley, California.* Division of Agricultural Science, University of California, Research Report 1968–1969, pp. 37–41.

Schnathorst, W.C. (1973) Nomenclature and physiology of *Verticillium* species with emphasis on the *V. albo-atrum* versus *V. dahliae* controversy. *Verticillium* wilt of cotton. *Proceedings of a Work Conference of the National Cotton Pathology Research Laboratory, College Station, Texas.* ARS-S-1–19.

Schnathorst, W.C. (1975) New approaches for obtaining field tolerance in *Gossypium hirsutum* to severe strains of *Verticillium dahliae*: a pathologist's perspective. *Proceedings of the Beltwide Cotton Production Research Conference, National Cotton Council.* Memphis, Tennessee, p. 148.

Schnathorst, W.C. (1981) Life cycle and epidemiology of *Verticillium.* In: Mace, M.E., Bell, A.A. and Beckman, C.H. (eds) *Fungal Wilt Diseases of Plants.* Academic Press, New York, pp. 81–111.

Schnathorst, W.C. and Davis, J.R. (1978) *Phytopathology News* 12, 205.

Schnathorst, W.C. and Evans, G. (1971) Comparative virulence of American and Australian isolates of *Verticillium albo-atrum* in *Gossypium hirsutum. Plant Disease Reporter* 55, 977–980.

Schnathorst, W.C. and Fogle, D. (1973) Survival structures of *Verticillium dahliae* in naturally infested soils. *Proceedings of the Beltwide Cotton Production Research Conference.* Phoenix, Arizona, p. 24.

Schnathorst, W.C. and Goheen, A.C. (1977) A wilt disease of grapevines (*Vitis vinifera*) in California caused by *Verticillium dahliae. Plant Disease Reporter* 61, 909–913.

Schnathorst, W.C. and Mathré, D.E. (1966a) Host range and differentiation of a severe form of *Verticillium albo-atrum* in cotton. *Phytopathology* 56, 1155–1161.

Schnathorst, W.C. and Mathré, O.E. (1966b) Cross protection in cotton by strains of *Verticillium albo-atrum*. *Phytopathology* 56, 1204–1209.

Schnathorst, W.C. and Sibbett, G.S. (1971) The relation of strains of *Verticillium albo-atrum* to severity of *Verticillium* wilt in *Gossypium hirsutum* and *Olea europaea* in California. *Plant Disease Reporter* 55, 780–782.

Schnathorst, W.C., Presley, J.T. and Carns, H.R. (1967) Determination of the internal inoculum potential of *Verticillium albo-atrum* in cotton plants. *Phytopathology* 57, 101–102.

Schnathorst, W.C., Presley, J.T. and Carns, H.R. (1968) Factors affecting the development of *Verticillium* wilt in cotton plants as revealed by studies on tracheal fluids. *Proceedings of the Beltwide Cotton Production Research Conference*. Hot Springs. pp. 227–228.

Schnathorst, W.C., Howell, C.R., Bell, A.A., Brinkerhoff, L.A. and Barrow, J.R. (1973) *Verticillium* wilt of cotton. *Proceedings of a Work Conference, National Cotton Pathology Reserch Laboratory, College Station, Texas*. August 30 – September 1, 1971.

Schnathorst, W.C., Reeve, T.A. and Fogle, D. (1975) *Verticillium dahliae* strains in cotton in the Pahrump valley, Nevada. *Plant Disease Reporter* 59, 863–865.

Schneider, H. (1945) Surveys and observations on *Verticillium* wilt of guayule in California from 1943 to 1945. *Plant Disease Reporter* 29, 615–617.

Schneider, H. (1948) Susceptibility of guayule to *Verticillium* wilt and influence of soil temperature and moisture on development of infection. *Journal of Agricultural Research* 76, 129–143.

Scholte, K. (1989a) The effect of netted scab (*Streptomyces* spp.) and *Verticillium dahliae* on growth and yield of potato. *Potato Research* 32, 65–73.

Scholte, K. (1989b) Effects of crop rotation and granular nematicides on the incidence of *Verticillium dahliae* Kleb. and *Colletotrichum coccodes* (Wallr.) Hughes, in potato. *Potato Research* 32, 377–385.

Scholte, K. and S'Jacob, J.J. (1989) Synergistic interactions between *Rhizoctonia solani* Kuhn, *Verticillium dahliae* Kleb., *Meloidogyne* spp. and *Pratylenchus neglectus* (Rensch) Chitwood & Oteifa, in potato. *Potato Research* 32, 387–395.

Scholte, K., Veenbaas-Rijks, J.W. and Labruyere, R.E. (1985) Potato growing in short rotations and the effect of *Streptomyces* spp., *Colletotrichum coccoides*, *Fusarium tabacinum* and *Verticillium dahliae* on plant growth and tuber yield. *Potato Research* 28, 331–348.

Schots, A., Dewey. F.M. and Oliver, R. (1994) *Modern Assays for Plant Pathogenic Fungi: Identification, Detection and Quantification*. CAB International, Wallingford, UK.

Schreiber, L.R. and Green, R.J. (1962) Comparative survival of mycelium, conidia and microsclerotia of *Verticillium albo-atrum* in mineral soil. *Phytopathology* 52, 288–289.

Schreiber, L.R. and Green, R.J. (1963) Effect of exudates on germination of conidia and microsclerotia of *Verticillium albo-atrum* inhibited by the soil fungistatic principle. *Phytopathology* 53, 260–264.

Schreiber, L.R. and Green, R.J. (1966) Anastomosis in *Verticillium albo-atrum* in soil. *Phytopathology* 56, 1110.

Schreiber, L.R. and Townsend, A.M. (1981) Sources and genetics of host resistance in shade trees. In: Mace, M.E., Bell, A.A. and Beckman, C.H. (eds) *Fungal Wilt Diseases of Plants*. Academic Press, New York, pp. 413–429.

Schreiber, L.R., Fordyce, C. and Green, R.J. (1961) *Verticillium* wilt of saucer magnolia, *Magnolia soulangeana*. *Plant Disease Reporter* 45, 108.

Schuring, W. and Van der Schaaf, K. (1999) [The assessment of different provenances of ash of the susceptibility to ash wilting disease]. *Nederlands Bosbouwtijdschrift* 71, 25–28.

Schuster, M.L. and Nuland, D.S. (1960) Seed transmission of safflower *Verticillium* wilt fungus. *Plant Disease Reporter* 44, 901–903.

Scott, J.W., Olson, S.M., Bryan, H.H., Howe, T.K., Stoffella, P.J. and Bartz, J.A. (1989) Solar Set. A heat tolerant, fresh market tomato hybrid. *Circular, Agricultural Experimemt Station, University of Florida* S-359, 1–10.

Scott, J.W., Jones, J.B., Somodi, G.C., Chellemi, D.O. and Olson, S.M. (1995) 'Neptune', a heat-tolerant, bacterial-wilt-tolerant tomato. *HortScience* 30(3), 641–642.

Scott, P.D. (1977) Hops in Britain and West Germany. *Outlook on Agriculture* 9, 174–179.

Sebastiani, L., Lenzi, A., Pugliesi, C. and Fambrini, M. (1994) Somaclonal variation for resistance to *Verticillium dahliae* in potato (*Solanum tuberosum* L.) plants regenerated from callus. *Euphytica* 80(1–2), 5–11.

Sedun, F. (1980) *Verticillium* wilt of alfalfa in British Colombia. *Phytoprotection* 61, 117.

Sedun, F.S. and Sackston, W.E. (1982) Effect of daylength, host line, and pathogen isolate on development of *Verticillium* wilt in sunflower. *Canadian Journal of Plant Pathology* 4, 109–114.

Segers, R., Butt, T.M., Carder, J.H., Keen, J.N., Kerry, B.R. and Peberdy, J.F. (1999) The subtilisins of fungal pathogens of insects, nematodes and plants: distribution and variation. *Mycological Research* 103, 395–402.

Seidel, D. and Zeise, K. (1990) *Verticillium* wilt in oilseed rape. *Proceedings of the Fifth International* Verticillium *Symposium*. Leningrad, p. 18.

Selman, I.W. and Buckley, W.R. (1959) Factors affecting the invasion of tomato roots by *Verticillium albo-atrum*. *Transactions of the British Mycological Society* 42, 227–234.

Selman, I.W. and Pegg, G.F. (1957) An analysis of the growth response of young tomato plants to infection by *Verticillium albo-atrum*. *Annals of Applied Biology* 45, 674–681.

Selvaraj, J.C. (1971) Morphological and physiological studies of isolates of *Verticillium dahliae*. *Indian Phytopathology* 24, 471–480.

Selvaraj, J.C. (1974a) Effect of the carbon nitrogen ratio on the growth and cultural characteristics of *Verticillium* spp. *Indian Phytopathology* 26, 746–748.

Selvaraj, J.C. (1974b) Repression of the production of *Verticillium* polygalacturonase and cellulase by glucose. *Indian Phytopathology* 27, 663–665.

Selvaraj, J.C. (1975) Differential growth response of the isolates of *Verticillium dahliae* and *V. albo-atrum* to the nutrition of certain amino acids. *Indian Phytopathology* 27, 117–119.

Selvaraj, J.C. and Meyer, J.A. (1974) Electrophoretic protein and enzyme patterns and antigenic structure in *Verticillium dahliae* and *V. albo-atrum*. *Mycopathologia et Mycologia Applicata* 54, 549–558.

Seman, É.O. (1968) On a new species of fungus, *Verticillium fumosum* Seman, isolated from cotton. Микология и Фитопатология [*Mycology and Phytopathology*] 2, 246–249.

Seman, É.O. (1970) [A variant of *Verticillium dahliae* Kleb. with orange-brown microsclerotia]. Микология и Фитопатология [*Mycology and Phytopathology*] 4, 479–481.

Sequeira, L. (1963) Growth regulators in plant disease. *Annual Review of Phytopathology* 1, 5–30.

Sequeira, L. (1973) Hormone metabolism in diseased plants. *Annual Review of Plant Physiology* 24, 353–380.

Sewell, G.W.F. (1959) Direct observation of *Verticillium albo-atrum* in soil. *Transactions of the British Mycological Society* 42, 312–321.

Sewell, G.W.F. and Glasscock, H.H. (1958) *Verticillium* wilt of apple. *Plant Pathology* 7, 76.

Sewell, G.W.F. and Wilson, J.F. (1958) Weed hosts of the 'progressive' hop strain of *Verticillium albo-atrum* Reinke & Berth. *Report of the East Malling Research Station* 1957, pp. 126–128.

Sewell, G.W.F. and Wilson, J.F. (1961) Machine picking in relation to progressive *Verticillium* wilt of the hop. I. A study of the infectivity of machine-picked hop waste. *Report of the East Malling Research Station* 1960. pp. 100–104.

Sewell, G.W.F. and Wilson, J.F. (1964a) The occurrence and dispersal of *Verticillium* conidia in xylem sap of the hop (*Humulus lupulus* L.). *Nature (London)* 204, 901.

Sewell, G.W.F. and Wilson, J.F. (1964b) *Verticillium* wilt of the hop: a summary of information and recommendations for control. *Report of the East Malling Research Station* 1963, pp. 193–198.

Sewell, G.W.F. and Wilson, J.F. (1966) *Verticillium* wilt of the hop: the survival of *Verticillium albo-atrum* in soil. *Annals of Applied Biology* 58, 241–249.

Sewell, G.W.F. and Wilson, J.F. (1967) *Verticillium* wilt of the hop: field studies on wilt in a resistant cultivar in relation to nitrogen fertilizer applications. *Annals of Applied Biology* 59, 265–273.

Sewell, G.W.F. and Wilson, J.F. (1974) The influence of normal tillage and of non-cultivation on *Verticillium* wilt of the hop. *Annals of Applied Biology* 76, 37–47.

Sewell, G.W.F. and Wilson, J.F. (1975) Virulent races of hop *Verticillium*. *Report of the East Malling Research Station* 1974, p. 155.

Sewell, G.W.F. and Wilson, J.F. (1976) Virulence of hop *Verticillium* isolates. *Report of the East Malling Research Station* 1975, pp. 120.

Sewell, G.W.F. and Wilson, J.F. (1977) Effect of soil temperature on wilt. *Report of the East Malling Research Station* 1977, p. 88.

Sewell, G.W.F. and Wilson, J.F. (1978) Virulence of hop *Verticillium* isolates. *Report of the East Malling Research Station* 1977, pp. 86–87.

Sewell, G.W.F. and Wilson, J.F. (1979) *Verticillium* wilt of hops (*V. albo-atrum*). *Report of the East Malling Research Station* 1978, pp. 95–97.

Sewell, G.W.F. and Wilson, J.F. (1984) The nature and distribution of *Verticillium albo-atrum* strains highly pathogenic to the hop. *Plant Pathology* 33, 39–51.

Sewell, G.W.F., Wilson, J.F. and Martin, G.D. (1962) Machine picking in relation to progressive *Verticillium* wilt of the hop II. The effect of composting on the infectivity of machine-picked hop waste. *Report of the East Malling Research Station* 1961, pp. 102–106.

Sezgin, E., Karcilioğlu, A. and Yemişçioğlu, Ü. (1982) Investigations on the effects of some cultural applications and antagonistic fungi on *Rhizotonia solani* Kühn. and *Verticillium dahliae* Kleb. in the Aegean region. II. Effects of herbicides and antagonistic fungi. *Journal of Turkish Phytopathology* 11, 79–91.

Sezgin, E., Onan, E., Karcilioglu, A. and Esentepe, M. (1985) A new host for *Verticillium dahliae*: *Cosmos*. *Journal of Turkish Phytopathology* 14, 43–44.

Shabi, E., Pinkas, Y. and Katan, J. (1986) Soil solarization for the combat of *Verticillium* wilt in live orchard trees. *Proceedings of the Fourth International* Verticillium *Symposium*. Guelph, Canada.

Shabi, E., Pinkas, Y. and Katan, J. (1987) Soil solarization for the control of *Verticillium* wilt in live orchard trees. *Canadian Journal of Plant Pathology* 9, 85–86.
Shabi, E., Pinkas, Y. and Katan, J. (1990) *Verticillium* wilt in orchard trees and its control by soil solarization. *Proceedings of the Fifth International* Verticillium *Symposium*. Leningrad, p. 84.
Shadmanov, R.K. (1995) Mechanism of plant tolerance to wilt. 6th International *Verticillium* Symposium, Dead Sea, Israel. *Phytoparasitica* 23, 64.
Shadmanov, R.K. and Saranskaya, L.B. (1990) [Improvement of wilt tolerance of cotton]. *Proceedings of the Fifth International* Verticillium *Symposium*. Leningrad, p. 78.
Shadmanova, N., Karimov, Kh. and Ishankulova, M. (1981) [Effectiveness of *Trichoderma* against complex infectious backgrounds]. *Khlopkovodstvo* 5, 27–28.
Shakhova, L.N. (1974) [Strawberry varieties for the nonchernozen area of the RSFSR]. *Sadovodstvo* 7, 42–43.
Shal'tyanis, B.V. and Saltenis, B. (1974) [*Verticillium* wilt of *Rosa rugosa*]. *Lietvos TSR Mokslu Akademijos Darbai C* 3, 67–70.
Shami, A. (1971) L'équation du 'wilt' comme indice de sélection de la tolérance à la verticilliose en Syrie. *Coton et Fibres Tropicales* 26, 201–210.
Shang, Hongsheng; Yang, Jiarong and Zhao, Xiaoming (1998) A comparative study on the pathogen of *Verticillium* wilt of alfalfa and other *Verticillium* species. *Acta Prataculturae Sinica* 7(4), 32–37.
Shanmugam, N., Rajendran, G., Vinayagamurthi, A., Muthukrishnan, T.S. and Kandaswamy, T.K. (1976) Studies on chemical control of cotton wilts. *Madras Agricultural Journal* 63, 386–388.
Shanmugam, N., Vinayagamurthy, A., Rajendran, G., Muthukrishnan, T.S. and Kandaswamy, T.K. (1977) Occurrence of nematode–fungus interactions on cotton around Coimbatore. *Science and Culture* 43, 346–348.
Shao, F.M. and Foy, C.D. (1982) Interaction of soil manganese and reaction of cotton to *Verticillium* wilt and *Rhizoctonia* root rot. *Commun. Soil Science Plant Analysis* 1, 21–38.
Shapovalov, M. and Lesley, J.W. (1940) Wilt resistance of the Riverside variety of tomato to both *Fusarium* and *Verticillium* wilts. *Phytopathology* 30, 760–768.
Shapovalov, M. and Rudolph, B.A. (1930) *Verticillium* hadromycosis (wilt) of cotton in California. *Plant Disease Reporter* 14, 9–10.
Sharma, V.K. and Novak, J. (1998) Enhancement of *Verticillium* wilt resistance in tomato transplants by *in vitro* co-culture of seedlings with a plant growth promoting rhizobacterium (*Pseudomonas* sp. strain PsJN). *Canadian Journal of Microbiology* 44, 528–536.
Shatrova, G.L. (1977) [Eggplant spacing and wilt]. *Izvestiya Akademii Nauk Moldavskoi SSR, Biologicheskikh i Khimicheskikh Nauk* 2, 87–88.
Shatrova, G.L. and Buimistru, L.D. (1976) [Importance of microelements, urea and potassium chloride in the control of *Verticillium* wilt of eggplant]. *Izvestiya Akademii Nauk Moldavskoi SSR, Biologicheskikh i Khimicheskikh Nauk* 6, 46–50.
Shaw, D.V., Gubler, W.D., Larson, K.D. and Hansen, J. (1996) Genetic variation for field resistance to *Verticillium dahliae* evaluated using genotypes and segregating progenies of California strawberries. *Journal of the American Society for Horticultural Science* 121(4), 625–628.
Shaw, D.V., Gubler, W.D. and Hansen, J. (1997a) Field resistance of California strawberries to *Verticillium dahliae* at three conidial inoculum concentrations. *HortScience* 32(4), 711–713.

Shaw, D.V., Gubler, W.D., Hansen, J. and Larsen, K.D. (1997b) Response to family selection for field resistance to *Verticillium dahliae* in California strawberries. *Journal of the American Society for Horticultural Science* 122(5), 653–655.

Shen, C.Y. (1985) Integrated management of *Fusarium* and *Verticillium* wilts of cotton in China. *Crop Protection* 4, 337–345.

Sheppard, J.W. (1979) *Verticillium* wilt, a potentially dangerous disease of alfalfa in Canada. *Canadian Plant Disease Survey* 59, 60.

Sheppard, J.W. and Needham, S.N. (1980) *Verticillium* wilt of alfalfa in Canada: occurrence of seed-borne inoculum. *Canadian Journal of Plant Pathology* 2, 159–162.

Sheppard, J.W. and Peterson, J.F. (1976) Chlorogenic acid and *Verticillium* wilt of tobacco. *Canadian Journal of Plant Science* 56, 157–160.

Sheppard, J.W. and Viswanathan, M.A. (1974) Survey for *Verticillium* wilt of tobacco in Quebec. *Canadian Plant Disease Survey* 54, 57–60.

Sherbakoff, C.D. (1928) Wilt caused by *Verticillium albo-atrum*. *Plant Disease Reporter Supplement* 61, 283–284.

Sherbakoff, C.D. (1929) *Verticillium* wilt of cotton. *Phytopathology* 19, 94.

Sherbakoff, C.D. (1934) Tobacco wilt caused by *Verticillium albo-atrum*. *Plant Disease Reporter* 18, 153.

Sherbakoff, C.D. (1949) Breeding for resistance to *Fusarium* and *Verticillium* wilts. *Botanical Review* 15, 377–422.

Sherengovyi, P.Z. (1968) [*Verticillium* disease of black currant]. *Zashchita Rasteniĭ* Moscow 13(7), 41.

Sherrod, L.L. and Elliot, A.M. (1967) Grain sorghum as a host for *Verticillium albo-atrum*. *Phytopathology* 57, 14–17.

Sherzhanov, K. (1987) [Resistance to the pathogen causing *Verticillium* wilt in varieties and ecologically distant hybrids]. *Aktual'nye Voprosy Genetiki, Selektsii i Semenovodstva Khlopchatnika*, 77–81.

Sherzhanov, K. (1989) [Inheritance of the wilt resistance of foreign varieties]. *Teoreticheskie i Metodicheskie Problemy Selekstii i Semenovodstva Khlopchatnika*, 119–122.

Shetty, K.G., Subbarao, K.V., Martin, F.N. and Koike, S.T. (1999) Management of *Verticillium* wilt in strawberry using vegetable crop rotation. *Annual International Research Conference on Methyl Bromide Alternatives and Emissions Reductions*, San Diego, California, 1–4 November 1999.

Shetty, K.G., Subbarao, K.V., Huisman, O.C. and Hubbard, J.C. (2000) Mechanism of broccoli-mediated *Verticillium* wilt reduction in cauliflower. *Phytopathology* 90(3), 305–310.

Shevtsova, V.M. (1978) [Complementation tests with pigment mutants of *Verticillium dahliae*]. In: Maksumov, A. (ed.) [*Genetic Variability of the Wilt-causing Agent and Prospects for Increasing Cotton Wilt Resistance.*] Donish Publishing House, Dushanbe, USSR, pp. 116–121.

Shevtsova, V.M. (1990) Developmental genetics of *Verticillium* genus. *Proceedings of the Fifth International* Verticillium *Symposium*. Leningrad, p. 112.

Shevtsova, V.M., Kas'yanenko, A.G., Ten, L.N., Stepanichenko, N.N. and Samoilov, G.G. (1982) [Genetics of *Verticillium dahliae*, the causing agent of *Verticillium* wilt of cotton. III. Use of the methods of biochemical genetics aimed at studying *Verticillium dahliae* melaninogenesis]. *Genetika USSR* 18, 1868–1877.

Shevtsova, V.M. and Zummer, N.A. (1988) [Study of the genetic control of mycelial-yeast dimorphism in the cotton wilt pathogen *Verticillium dahliae*]. Микология и Фитопатология [*Mycology and Phytopathology*] 22, 117–119.

Shevtsova, V.M., Stepanichenko, N.N., Ten, L.N. and Tyshchenko, A.A. (1983) [On the homology of the genetic systems controlling the biosynthesis of melanin in *Verticillium dahliae* Kleb. and *V. tricorpus* Isaac]. Микология и Фитопатология [*Mycology and Phytopathology*] 17, 499–503.

Shi, L.-Y., Jian, G.-L., Sun, W.-J. and Feng, J. (1998) [Discussion of the research status and direction on cotton *Verticillium* wilt in China]. *Acta Phytophylacica Sinica* 25(2), 103–107.

Shibkova, N.A. and Tkachenko, M.P. (1971) [Results of trials of the systemic fungicide benlate against *Verticillium* wilt of cotton]. *Byulleten' Vsesoyuznogo Nauchno-Issledovatel'skogo Instituta Zashchity Rasteniĭ* 22, 58–62.

Shijie, H., Guangqin, H., Shiyou, C., Zhinan, G. and Sufeng, L. (1996) A brief introduction to Nantongmian 13. *China Cottons* 23(6), 20.

Shimotsuma, M., Kuc, J. and Jones, C.M. (1972) The effects of prior inoculations with non-pathogenic fungi on *Fusarium* wilt of watermelon. *HortScience* 7, 72–73.

Shishkanu, G.V. and Istrati, L.N. (1970) [Content of pigments in plum leaves with *Verticillium* disease]. *Infektsionnye Zabolevaniya Kul'turnykh Rastenii Moldavii* 24–28.

Shivanandappa, N. and Govindu, H.C. (1965) *Verticillium* wilt of *Pyrethrum* in Mysore. *Current Science* 34, 513–514.

Shmotina, G.E., Kokurina, M.A. and Gorlenko, M.V. (1971) [DNA nucleotide composition of the causal agent of plant *Verticillium* wilt]. Микология и Фитопатология [*Mycology and Phytopathology*] 5, 311–313.

Shoemaker, N.P., Swartz, H.J. and Galletta, G.J. (1985) Cultivar dependent variation in pathogen resistance due to tissue culture-propagation of strawberries. *HortScience* 20, 253–254.

Shoemaker, P.B. (1985) Fumigants in bedded rows for tomato root knot and *Verticillium* wilt. *Fungicide and Nematicide Tests* 41, 82.

Shoemaker, P.B. and Barker, K.R. (1979) Influence of *Meloidogyne incognita* and *Verticillium dahliae* on tomato at different inoculum densities. *Phytopathology* 69, 1A9.

Shufelt, C. and Linderman, R.G. (1987) The influence of irrigation on the incidence and severity of *Verticillium* wilt of Norway Maple. *Canadian Journal of Plant Pathology* 9, 86.

Shuxiang, S. (1995) A preliminary research report on breeding of the *Verticillium* wilt resistant cotton variety. *China Cottons* 22(6), 12–13.

Sickinger, S.M., Grau, C.R. and Harvey, R.G. (1987) *Verticillium* wilt of velvet leaf (*Abutilon theophrasti*). *Plant Disease* 71, 415–418.

Siddiqui, M.R. and Deshprabhu, N.N. (1970) Screening of cotton germplasm for resistance to *Verticillium* wilt. *Indian Journal of Agricultural Sciences* 40, 1001–1003.

Sidhu, G.S. and Webster, J.M. (1979) Genetics of tomato resistance to the *Fusarium–Verticillium* complex. *Physiological Plant Pathology* 15, 93–98.

Sidorova, S.F. (1974) [Effects of preceding crops on the pathogens of cotton]. Микология и Фитопатология [*Mycology and Phytopathology*] 8, 412–414.

Sidorova, S.F. (1978) [Biological immunization of cotton plants by avirulent strains of the pathogen of *Verticillium* wilt]. Микология и Фитопатология [*Mycology and Phytopathology*] 12, 136–141.

Sidorova, S.F. (1981) [The role of cross protection in the pathogenesis of *Verticillium* and *Fusarium* wilt of cotton]. *Trudy Vsesoyuznogo Nauchno-Issledovatel'skogo Instituta Zashchity Rasteniĭ*, 107–112.

Sidorova, S.F. (1990) [Specialisation of *Verticillium dahliae* Kleb. and features of its survival in soil]. *Proceedings of the Fifth International* Verticillium *Symposium*. Leningrad, p. 37.

Siimisker, T. (1996) Astelpaju sordivordluse tulemusi Louna-Eesti sordikatsepunktis. *Agraarteadus* 7(2), 177–181.

Simpson, D. and Blanke, M. (1989) 'Pandora' – die neue Erdbeersorte aus East Malling. *Erwerbsobstbau* 31, 19–21.

Simpson, D. and Blanke, M. (1991) 'Pegasus' – die mittelspate Erdbeersorte aus East Malling mit *Verticillium*-Resistenz. *Erwerbsobstbau* 33, 18–20.

Simpson, D. and Blanke, M. (1994) Calypso und Tango – remontierende Erdbeersorten aus East Malling mit *Verticillium*-Resistenz. *Erwerbsobstbau* 36(5), 121–123.

Simpson, D.W., Bell, J.A. and Harris, D.C. (1994) Breeding for resistance to fungal diseases in strawberry. In: Schmidt, H. and Kellerhals, J.J. (eds) *Progress in Temperate Fruit Breeding. Proceedings of the Eucarpia Fruit Breeding Section Meeting.* Wadenswil/Einsiedeln, Switzerland, 30 August – 3 September, 1993, pp. 63–66.

Sims, J.J., Mee, H. and Erwin, D.C. (1969) Methyl-2-benzimidazole-carbamate a fungitoxic compound isolated from cotton plants treated with methyl-1-(butyl carbamoyl)-2-benzimidazole carbamate (benomyl). *Phytopathology* 59, 1775–1776.

Sinclair, W.A., Smith, K.L. and Larsen, A.O. (1981) *Verticillium* wilt of maples: symptoms related to movement of the pathogen in stems. *Phytopathology* 71, 340–345.

Singh, C.P., Mishra, M.M. and Kapoor, K.K. (1982) Solubilization of insoluble phosphates by mesophilic fungi. *Revu d'Ecologie et de Biologie du Sol* 19(1), 17–25.

Singh, D., Brinkerhof, L.A. and Gunn, G. (1971) Effect of alanine on development of *Verticillium* wilt in cotton cultivars with different levels of resistance. *Phytopathology* 61, 881–882.

Sinha, A.K. and Wood, R.K.S. (1967a) An analysis of responses of resistant and susceptible tomato plants to *Verticillium* infection. *Annals of Applied Biology* 59, 143–154.

Sinha, A.K. and Wood, R.K.S. (1967b) The effect of growth substances on *Verticillium* wilt of tomato plants. *Annals of Applied Biology* 60, 117–128.

Sinha, A.K. and Wood, R.K.S. (1968) Studies on the nature of resistance in tomato plants to *Verticillium albo-atrum*. *Annals of Applied Biology* 62, 319–327.

Sink, K.C. and Grey, W.E. (1999) A root-injection method to assess *Verticillium* wilt resistance of peppermint (*Mentha* × *piperita* L.) and its use in identifying resistant somaclones of cv Black Mitcham. *Euphytica* 106(3), 223–230.

Siti, E., Krikun, J., Orion, D. and Cohn, E. (1979) *Verticillium–Pratylenchus* interactions in potato. *Phytoparasitica* 7, 42.

Sivaprakasam, K. and Rajagopalan, C.K.S. (1971) Leaf infection and systemic invasion of brinjal by *Verticillium dahliae*. *Annamalai University Agricultural Research Annual* 3, 100–101.

Sivaprakasam, K. and Rajagopalan, C.K.S. (1974a) Inoculum concentration of *Verticillium dahliae* in relation to development and severity of wilt in brinjal. *Madras Agricultural Journal* 61, 49–50.

Sivaprakasam, K. and Rajagopalan, C.K.S. (1974b, published 1975) Studies on the control of *Verticillium* wilt disease of brinjal. *Indian Phytopathology* 27, 304–308.

Sivaprakasam, K. and Rajagopalan, C.K.S. (1975, published 1976) Varietal resistance of eggplant to *Verticillium dahliae*. *Indian Phytopathology* 28, 312–313.

Sivaprakasam, K. and Rajagopalan, C.K.S. and Vidhyasekaran, P. (1972–1973, published 1973) Role of amino acids on *Verticillium* disease resistance in brinjal. *Annamalai University Agricultural Research Annual* 4–5, 101–106.

Skadow, K. (1966) Feinde der Acker-Kratzdistel. *Biologische Schule* 15, 353–355.

Skadow, K. (1969a) Untersuchungen über die Welkeerreger *Verticillium albo-atrum* Rke. et Berth. und *V. dahliae* Kleb. I. Unkräuter als Wirtspflanzen. *Zentralblatt für Bakteriologie, Parasitenkunde, Infektionskrankheiten und Hygiene* 123, 715–735.

Skadow, K. (1969b) Ein Beitrag zur *Verticillium* Taxonomie. *Archiv für Pflanzenschutz* 5, 155–166.

Skarmoutsos, G. and Skarmoutsou, H. (1998) Occurrence of wilt disease caused by *Verticillium dahliae* on *Ailanthus glandulosa* in Greece. *Plant Disease* 82(1), 129.

Skipp, R.A. (1986) Fungal root pathogens in pasture soils and effects of fungicides. *Proceedings of the 39th New Zealand Weed and Pest Control Conference*. Palmerston North, New Zealand, pp. 60–64.

Skotland, C.B. (1964) The efficacy of various fumigants in controlling *Verticillium* wilt in south central Washington. *Plant Disease Reporter* 48, 502–505.

Skotland, C.B. (1971) Pathogenic and non-pathogenic *Verticillium* species from south central Washington. *Phytopathology* 61, 435–436.

Slattery, R.J. (1981) Inoculum potential of *Verticillium* infested potato cultivars. *American Potato Journal* 58, 135–142.

Slattery, R.J. (1983) Inoculum potential of *Verticillium* species in crop and weed residues overwintering in infested soil. *American Potato Journal* 60, 47–53.

Slattery, R.J. and Eide, C.J. (1980) Prevalence of *Verticillium* wilt in potatoes in the Red River Valley area of Minnesota. *American Potato Journal* 57, 293–299.

Slawson, D.D. (1987) Occurrence of *Verticillium dahliae* on various woody ornamental hosts. *Plant Pathology* 36, 606–607.

Smalley, E.B. and Lester, D.T. (1973) 'Sapporo Autumn Gold' elm. *HortScience* 8, 514–515.

Smalley, E.B. and Lester, D.T. (1983) 'Regal' elm. *HortScience* 18, 960–961.

Smit, F. and Dubery, I.A. (1997) Cell wall reinforcement in cotton hypocotyls in response to a *Verticillium dahliae* elicitor. *Phytochemistry* 44(5), 811–815.

Smith, A.L. and Dick, J.B. (1960) Inheritance of resistance to *Fusarium* wilt in Upland and Sea Island Cottons as complicated by nematodes under field conditions. *Phytopathology* 50, 44–48.

Smith, D. and Onions, A.H.S. (1983) *The Preservation and Maintenance of Living Fungi*. CAB International, Wallingford, UK.

Smith, D.G., McInnes, A.G., Higgins, V.J. and Millar, R.L. (1971) Nature of the phytoalexin produced by alfalfa in response to fungal infection. *Physiological Plant Pathology* 1, 41–44.

Smith, E.G., Acharya, S.N. and Huang, H.C. (1995) Economics of growing verticillium wilt-resistant and adapted alfalfa cultivars in western Canada. *Agronomy Journal* 87(6), 1206–1210.

Smith, H.C. (1965) The morphology of *Verticillium albo-atrum, V. dahliae* and *V. tricorpus*. *New Zealand Journal of Agricultural Research* 8, 450–478.

Smith, I.M., Dunez, J., Lelliott, R.A., Phillips, D.H. and Archer, S.A. (1988) *European Handbook of Plant Diseases*. Blackwell, Oxford, UK.

Smith, L.D. (1979) *Verticillium* wilt of landscape trees. *Journal of Arboriculture* 5, 193–197.

Smith, L.D. (1983) Major nutrient influence on *Verticillium dahliae* infections of *Acer saccharum*. *Journal of Arboriculture* 9, 277–281.

Smith, T.A. and Richards, F.J. (1962) The biosynthesis of putrescine in higher plants and its relation to potassium nutrition. *Biochemical Journal* 84, 292–294.

Smith, T.E. (1960) Occurrence of *Verticillium* wilt on peanut. *Plant Disease Reporter* 44, 435.

Smith, T.E. (1961) *Verticillium* wilt in peanuts (*Arachis hypogaea*). *Phytopathology* 51, 411–412.

Smith, V.L. and Rowe, R.C. (1984) Characteristics and distribution of propagules of *Verticillium dahliae* in Ohio potato field soils and assessment of two assay methods. *Phytopathology* 74, 553–556.

Snyder, W.C. and Rudolph, B.A. (1939) *Verticillium* wilt of pepper, *Capsicum annuum*. *Phytopathology* 29, 359–362.

Snyder, W.C. and Smith, S.N. (1981) Current Status. In: Mace, M.E., Bell, A.A. and Beckman, C.H. (eds) *Fungal Wilt Diseases of Plants*. Academic Press, New York, pp. 25–50.

Snyder, W.C. and Wilhelm, S. (1962) Seed transmission of *Verticillium* wilt of spinach. *Phytopathology* 52, 365.

Snyder, W.C., Hansen, H.N. and Wilhelm, S. (1950) New hosts of *Verticillium albo-atrum*. *Plant Disease Reporter* 34, 26–27.

Sobirov, K., Esamov, Yu. and Marupov, A. (1980) [Mustard for wilt control]. *Khlopkovodstvo* 8, 26.

Soesanto, L. (2000) *Ecology and Biological Control of* Verticillium dahliae. Landbouwuniversiteit Wageningen, The Netherlands.

Soesanto, L. and Termorshuizen, A.J. (1997) Formation of microsclerotia of *Verticillium dahliae*. *Proceedings of the Seventh International* Verticillium *Symposium*. Cape Sounion, Athens, Greece, p. 33.

Soileau, J.M., Englestad, O.P. and Martin, J.B. (1969) Cotton growth in an acid frangipan subsoil: II Effect of soluble calcium, magnesium and aluminium on roots and tops. *Proceedings of the Soil Science Society of America* 33, 919–924.

Solarska, E. (1994a) Niechemiczne metody ograniczania wystepowania werticiliozy chmielu. *Ochrona Roslin* 38(8), 11–13.

Solarska, E. (1994b) Investigations on the effect of rhizosphere microorganisms on the sanitary status of hops under different cultivation conditions. *Phytopathologia Polonica* 115–121.

Solarska, E. (1995) Effect of rye intercropping-enhanced fluorescent pseudomonads on the control of *Verticillium* wilt in hops. 6th International *Verticillium* Symposium, Dead Sea, Israel. *Phytoparasitica* 23, 70.

Solarska, E. (1997) Antagonistic action of *Talaromyces flavus* and *Trichoderma viride* against *Verticillium albo-atrum* on hops. *Proceedings of the Seventh International* Verticillium *Symposium*. Cape Sounion, Athens, Greece, p. 62.

Solarska, E., Migdal, J. and Dwornikiewicz, J. (1996) Oddziaływanie wybranych roslin uprawianych w miedzyrzedziach na plonowanie chmielu i wystepowanie werticiliozy (*Verticillium albo-atrum*). *Pamietnik Puławski* 107, 77–83.

Solel, Z., Sandler, D. and Dinoor, A. (1979) Mobility and persistence of carbendazim and thiabendazole applied to soil via drip irrigation. *Phytopathology* 69, 1273–1277.

Solovyeva, A.I. (1959) [Results of testing samples of the world's cotton collection for resistance to *Verticillium* wilt]. In: Mirpulatov, K.I. (ed.) *Protection of Cotton, Lucerne and Maize from Diseases, Pests and Weeds*. Publication 5, 141–161.

Solovyeva, A.I. and Polyarkova, L.V. (1940) *Wilt of Cotton*. Tashkent Publishing Department, Uzbekistan.

Solyanova, E. (1978) [The perennial use of herbicides and the incidence of cotton wilt]. *Khlopkovodstvo* 1, 26.

Somers, T.C. and Harrison, A.F. (1967) Wood tannins. Isolation and significance in host resistance to *Verticillium* wilt disease. *Australian Journal of Biological Sciences* 20, 475–479.

Song, M., Zhu, H., Mao, S. and Xing, J. (1995) [Potash fertilizer for treatment of withered cotton, research on control of *Verticillium* wilt]. *Henan Nongye Kexue* 15–16.

Song, X.-X., Zhu, H.-Q. and Guo, J.-C. (1997) Study on pathogenicity differentiation for *Verticillium dahliae* Kleb. Anyang strains on cotton. *Scientia Agricultura Sinica* 30(1), 13–18.

Sorensen, E.L., Stuteville, D.L., Horber, E.K. and Peaden, R.N. (1989) Registration of K81-7Ve2 alfalfa germplasm with resistance to five diseases and four insects. *Crop Science* 29, 492.

Spink, D.S. and Rowe, R.C. (1989) Evaluation of *Talaromyces flavus* as a biological control agent against *Verticillium dahliae* in potato. *Plant Disease* 73, 230–236.

Sportelli, M., Nipoti, P. and D'Ercole, N. (1983) Prove di 'lotta biologica' contro alcune micopatie del pomodoro in coltura protetta. *Informatore Fitopatologico* 33(6), 35–38.

Springer, J.K. (1967) Free amino acid content of tracheal fluids of *Verticillium* susceptible and resistant strawberry in relation to disease development. *Dissertation Abstracts* 27(6B), 1693–1694.

Sreeganga, C., Heinstein, P.F. and Low, P.S. (1996) Activation of phospholipase A by plant defense elicitors. *Plant Physiology* 110(3), 979–986.

Staffeldt, E.E. (1959) Elimination of *Verticillium albo-atrum* by composting cotton gin wastes. *Plant Disease Reporter* 43, 1150–1152.

Stapleton, J.J. and DeVay, J.E. (1982) Effect of soil solarization on populations of selected soilborne microorganisms and growth of deciduous fruit tree seedlings. *Phytopathology* 72, 323–326.

Stapleton, J.J. and DeVay, J.E. (1986) Soil solarization: a non-chemical approach for management of plant pathogens and pests. *Crop Protection* 5, 190–198.

Stapleton, J.J. and Duncan, R.A. (1999) Fumigation and/or solarization soil treatment for *Verticillium* wilt of apricot. *Annual International Research Conference on Methyl Bromide Alternatives and Emissions Reductions*. San Diego, California, 1–4 November 1999.

Stapleton, J.J., DeVay, J.E. and Lear, B. (1987) Effect of combining ammonium-based fertilizers with solarization on pathogen control of plant growth. *Phytopathology* 77, 1744.

Stapleton, J.J., Asai, W.K. and DeVay, J.E. (1989) Use of polymer mulches in integrated pest management programs for establishment of perennial fruit crops. *Acta Horticulturae* 255, 161–168.

Stapleton, J.J., DeVay, J.E. and Lear, B. (1991) Soil solarization and field effects of ammonia-based fertilizers and soil solarization on pathogen survival, soil fertility, and crop growth. *FAO Plant Production and Protection Paper* 109, 331–342.

Stapleton, J.J., Paplomatas, E.J., Wakeman, R.J. and DeVay, J.E. (1993) Establishment of apricot and almond trees using soil mulching with transparent (solarization) and black polyethylene film: effects on *Verticillium* wilt and tree health. *Plant Pathology* 42(3), 333–338.

Stapp, C. (1928) Über die Ursache des Ulmensterbens. *Mitteilungen der Deutschen Dendrologischen Gesellschaft* 40, 139–146.

Stark, C. (1961) Das Auftreten der *Verticillium-Tracheomykosen* in Hamburger Gartenbaukulturen. Ein Beitrag zur Kenntnis ihrer Erreger. *Gartenbauwissenschaft* 26, 493–528.

Steiner, G.W. and Lockwood, J.L. (1970) Soil fungistasis: mechanism in sterilized, reinoculated soil. *Phytopathology* 60, 89–91.

Stepanichenko, N.N., Tyshchenko, A.A., Ten, L.N., Agzamova, V.A., Khotyanovich, A.V., Kas'yanenko, A.G. and Zhdanova, N.N. (1990) Lipids and polyketides of phytopathogenic fungi of *Verticillium* genus. *Proceedings of the Fifth International* Verticillium *Symposium*. Leningrad, p. 57.

Stepanova, M.Yu. (1975) [Modified method for quantitative determinations of infectious units of *Verticillium dahliae* Kleb. in field soil]. Микология и Фитопатология [*Mycology and Phytopathology*] 9(3), 256–258.

Stepanova, M.Yu., Sidorova, S.F., Grishechkina, L.D. and Zhdarkina, Z.P. (1977) [Dynamics of soil populations of the pathogen of wilt in cultivation of different agricultural crops]. Микология и Фитопатология [*Mycology and Phytopathology*] 11(4), 341–344.

Sterne, R.E., McCarver, T.H. and Courtney, M.L. (1979) Survival of plant pathogens in composted soil gin trash. *Arkansas Farm Research* 28, 9.

Steuckardt, R. and Hempel, K. (1976) Untersuchungen über die Verbesserung und Stabilisierung der Resistenz gegen *Verticillium albo-atrum* R. et B. an Züchmaterial von Luzerne. *Archiv für Züchtungsforschung* 6, 201–204.

Stevenson, F.J. and Jones, H.A. (1953) Some sources of resistance in crop plants. *Plant Diseases. USDA Handbook of Agriculture*, pp. 192–216.

Stipanovic, R.D. and Bell, A.A. (1976) Pentaketide metabolites of *Verticillim dahliae* 3. Identification of 1-3, 4-dihydro-3, 8-dihydroxy-1 (2H)-naphthalenone (−) vermelone as a precursor to melanin. *Journal of Organic Chemistry* 41, 2468–2469.

Stipanovic, R.D. and Bell, A.A. (1977) Pentaketide metabolites of *Verticillium dahliae*. II. Accumulation of naphthol derivatives by the aberrant-melanin mutant *BRM-2*. *Mycologia* 69, 164–172.

Stipanovic, R.D., Bell, A.A. and Howell, C.R. (1975) Naphthofuran precursors of sesquiterpenoid aldehydes in diseased *Gossypium*. *Phytochemistry* 14, 1809–1811.

Stipanovic, R.D., Puckhaber, L.S. and Bell, A.A. (1997) Biotransformation of the phytoalexin hibiscanone by *Verticillium dahliae* and toxicity of the biotransformed product. *Proceedings of the Beltwide Cotton Conferences*. New Orleans, Louisiana, 6–10 January 1997. Vol. 1, pp. 104–105.

Stoddart, J.L. and Carr, A.J.H. (1966) Properties of wilt toxins produced by *Verticillium albo-atrum* Reinke & Berth. *Annals of Applied Biology* 58, 81–92.

Stokes, L.G. and Sappenfield, W.P. (1981) Registration of BW76-31 cotton germplasm (Reg.No. GP 162). *Crop Science* 21, 991.

Stoner, A.K. (1977) 'Wakefield' tomato. *HortScience* 12, 507.

Storey, G.W. and Evans, K. (1987) Interactions between *Globodera pallida* juveniles, *Verticillium dahliae* and three potato cultivars, with descriptions of associated histopathologies. *Plant Pathology* 36, 192–200.

Stosz, S.K., Fravel, D.R. and Roberts, D.P. (1996) *In vitro* analysis of the role of glucose oxidase from *Talaromyces flavus* in biocontrol of the plant pathogen *Verticillium dahliae*. *Applied and Environmental Microbiology* 62(9), 3183–3186.

Stosz, S.K., Roy, S., Murphy, C., Wergin, W. and Fravel, D.R. (1998) Localization of glucose oxidase with immunocytochemistry in the biocontrol fungus *Talaromyces flavus*. *Phytopathology* 88(6), 576–581.

Stout, D.G., Acharya, S.N., Hall, J.W. and Broersma, K. (1998) Yield and persistence from mixtures of a winter hardy and a *Verticillium* wilt resistant lucerne cultivar. *Journal of Agronomy and Crop Science* 180, 27–31.

Strand, L.L. and Mussell, H. (1975) Solubilization of peroxidase activity from cotton cell walls by endopolygalacturonases. *Phytopathology* 65, 830–831.
Stratton, S.D. (1994) Registration of 'Resistar' alfalfa. *Crop Science* 34(1), 305–306.
Stratton, S.D. (1997) Registration of 'Multistar' alfalfa. *Crop Science* 37(1), 294.
Straumal, B.P. (1964) [Principles for the selection of wilt-resistant cotton]. *Mater. Vses. Simp po Bor'be s Viltom Khlopchatnika, Tashkent*. pp. 79–85.
Straumal, B.P. and Tishin, A.I. (1976) [On the question of infection of cotton by different geographical populations of *Verticillium dahliae* Kleb.]. *Trudy Uzb. Opyt. St. Lub. Kul'tur (Vopr. Genet. Selektsii i Semenovodstva Khlopchatnika i Lyutserny)* 13, 149–155.
Strausbaugh, C.A. (1993) Assessment of vegetative compatibility and virulence of *Verticillium dahliae* isolates from Idaho potatoes and tester strains. *Phytopathology* 83(11), 1253–1258.
Strausbaugh, C.A., Schroth, M.N., Weinhold, A.R. and Hancock, J.G. (1990) Reassessment of vegetative compatibility groups of *Verticillium dahliae* and the compatbility of isolates from California potatoes. *Phytopathology* 80, 965–966.
Strausbaugh, C.A., Schroth, M.N. and Hancock, J.G. (1992) Assessment of vegetative compatibility of *Verticillium dahliae* tester strains and isolates from California potatoes. *Phytopathology* 82, 61–68.
Street, P.F.S. and Cooper, R.M. (1984) Quantitative measurement of vascular flow in petioles of healthy and *Verticillium*-infected tomato. *Plant Pathology* 33, 483–492.
Street, P.F.S., Robb, J. and Ellis, B.E. (1986) Secretion of vascular coating components by xylem parenchyma cells of tomatoes infected with *Verticillium albo-atrum*. *Protoplasma* 132, 1–11.
Strobel, J.W. (1963) *Verticillium* wilt of strawberry in Dade county, Florida. *Proceedings of the Florida State Horticultural Society* 76, 111–114.
Strunnikova, O.K. and Muromtsev, G.S. (1984) [Immunochemical characteristics of some *Verticillium* species]. Микология и Фитопатология [*Mycology and Phytopathology*] 18, 245–249.
Strunnikova, O.K. and Muromtsev, G.S. (1987) [Study of strains of *Verticillium dahliae* and *V. albo-atrum* by the method of immunoelectrophoresis]. Микология и Фитопатология [*Mycology and Phytopathology*] 21, 155–160.
Strunnikova, O.K. and Vishnevskaya, N.A. (1995) Development of the plant pathogenic fungus *Verticillium dahliae* Kleb. in soil. Микология и Фитопатология [*Mycology and Phytopathology*] 29(2), 59–61.
Strunnikova, O.K., Vishnevskya, N.A., Labutova, N.M. and Muromtsev, G.S. (1990) The use of immunofluorescence method to study ecology of *Verticillium dahliae*, causal organism of cotton wilt. *Proceedings of the Fifth International* Verticillium *Symposium*. Leningrad, p. 38.
Strunnikova, O.K., Vishnevskaya, N.A., Batyrov, A.A. and Muromtsev, G.S. (1994) Probability of *Verticillium* wilt appearance on alfalfa in cotton-growing regions of the CIS. *Russian Agricultural Sciences* 20–22.
Strunnikova, O.K., Vishnevskaya, N.A. and Muromtsev, G.S. (1997) Abundance of mycelium and microsclerotia as indices of the soil status of *Verticillium dahliae*. *Proceedings of the Seventh International* Verticillium *Symposium*. Cape Sounion, Athens, Greece, p. 32.
Stuteville, D.L., Willis, W.G., Houfek Christensen, J.A. and Sim, T. (1986) First report of *Verticillium* wilt of alfalfa in Kansas. *Plant Disease* 70, 475.

Subbarao, N.S. and Bailey, D.L. (1961) Rhizosphere studies in relation to varietal resistance to susceptibility of tomato to *Verticillium* wilt. *Canadian Journal of Botany* 39, 1747–1758.

Subbarao, K.V. and Hubbard, J.C. (1996) Interactive effects of broccoli residue and temperature on *Verticillium dahliae* microsclerotia in soil and on wilt in cauliflower. *Phytopathology* 86(12), 1303–1310.

Subbarao, K.V., Chassot, A., Gordon, T.R., Hubbard, J.C., Bonello, P., Mullin, R., Okamoto, D., Davis, R.M. and Koike, S.T. (1995) Genetic relationships and cross pathogenicities of *Verticillium dahliae* isolates from cauliflower and other crops. *Phytopathology* 85(10), 1105–1112.

Subbarao, K.V., Hubbard, J.C. and Koike, S.T. (1999) Evaluation of broccoli residue incorporation into field soil for *Verticillium* wilt control in cauliflower. *Plant Disease* 83, 124–129.

Suchorukov, K.T. (1957) Physiology of immunity of some agricultural plants. *Proceedings of the Second International Plant Protection Conference.* Fernhurst, 1957. pp. 42–52.

Sudakova, I.M. (1978) [The culturing of nematodes and the study of their role in cotton wilt]. *Printsipy i Metody Izucheniya Vzaimootnoshenii Mezhdu Paraziticheskimi Nematodami i Rasteniyami* 137–143.

Sui, S.-X. (1995) A preliminary research report on breeding of the *Verticillium* wilt resistant cotton variety. *China Cottons* 22(6), 12–13.

Sukhanberdina, E.Kh., Bolgov, A.N. and Frolova, A.G. (1986) [Resistance of eggplant varieties to wilt]. *Nauchno-Tekhnicheskii Byulleten' Vesesoyuznogo Ordena Lenina i Ordena Druzhby Narodov Nauchno Issledovatel'skogo Instituta Rastenievodstva imeni N.I. Vavilova* 166, 72–75.

Sultanov, B., Uronov, I.S. and Bicherov, E.V. (1977) [Complex fertilisers with microelements and cotton wilt]. *Trudy Srednaez. NII Zashchity Rast.* 11, 74–76.

Sumeghy, J.B. (1979) New hybrid tomatoes resist diseases. *Agricultural Gazette of New South Wales* 90(5), 22–23.

Sumino, A. and Abe, H. (1999) *Verticillium* wilt of sunflower caused by *Verticillium dahliae*. *Annals of the Phytopathological Society of Japan* 65(4), 498–500.

Summers, W.L. (1996a) 'Prairie Schooner' inbred cherry tomato. *HortScience* 31(2), 291.

Summers, W.L. (1996b) 'Prairie Dawn' fresh-market tomato inbred. *HortScience* 31(2), 292.

Summers, W.L. (1996c) 'Prairie Pastemaker' processing tomato inbred. *HortScience* 31(2), 293.

Sun, J.-L., Song, X.-X. and Zhu, H.-Q. (1998a) [A discussion on the amount of *Fusarium* and *Verticillium* pathogens borne on the cotton seed]. *China Cottons* 25(6), 11–12.

Sun, J.-L., Liu, X.-T., Song, X.-X., Zhu, H.-Q. and Wang, X.-L. (1998b) [Studies on the regularity of occurrence of *Verticillium* wilt in Bt transgenic cotton]. *Acta Gossypii Sinica* 10(6), 316–321.

Sun, X.-L., Yang, H.-L. and Song, W. (1999) [Identification, nitrogen-fixing and phytopathogen-antagonising activity of *Alcaligenes paradoxus* 1AG4 isolated from rice rhizoplane]. *Acta Phytopathologica Sinica* 29(4), 294–298.

Sussman, A.S. and Halvoresen, H.O. (1966) *Spores: Their Dormancy and Germination.* Harper and Row, New York.

Suta, E. and Stoenescu, A. (1986) [New varieties of sweet and hot peppers]. *Producţia Vegetală, Horticultura* 35(9), 25–27.

Suta, E., Tomescu, A., Adochitel, A. and Manolache, G. (1996) Contributii la imbogatirea structurii sotimentale a ardeiului gras cultivat in Romania. *Anale Institutul de Cercetari pentru Legumicultura si Floricultura, Vidra* 14, 117–135.

Sutherland, M.L., Mittempergher, L. and Brasier, C.M. (1995) Control of Dutch elm disease by induced host resistance. *European Journal of Forest Pathology* 25(6/7), 307–318.

Šutic, D., Aleksič, Ž. and Spasojevič, V. (1981) Résistance de certains hybrides interspécifiques du piment au *Verticillium albo-atrum* Reinke et Berth. *Compte Rendu des Séances de l'Academie d'Agriculture de France* 67, 471–475.

Svampa, G. (1974) Prove di lotta in serra control *Fusarium oxysporum* f.sp. *lycopersici* (Sacc.) Snyder & Hansen and *Verticillium dahliae* Kleb. con trattamenti fogliari alle colture. *Tecnica Agricola Italy* 26, 731–738.

Svensson, C. and Lerenius, C. (1987) An investigation on the effect of *Verticillium* wilt (*Verticillium dahliae* Kleb.) on oilseed rape. *Bulletin SROP* 10(4), 30–34.

Swanepoel, A. and De Kock, H. (1996) Progress with *Verticillium* wilt of cotton in the Republic of South Africa. *Proceedings of the Beltwide Cotton Conference.* Nashville, Tennessee 1, 273–275.

Swinburne, T.R., Carder, J.H., Hignett, R.C., Pritchard, S. and Welsh, M.P. (1985) Detection, estimation and virulence-assessment of *Verticillium* spp. *Report of the East Malling Research Station* 1984, 179–180.

Szczygiel, A. (1983) On the tolerance limit of strawberry plants to root lesion nematode *Pratylenchus penetrans* (Cobb). *Fruit Science Reports* 10, 135–142.

Szczygiel, A. and Rebandel, Z. (1981) [Effect of soil fumigation on the growth, health and cropping of strawberries]. *Informator o Badaniach Prowadzonych w Zakladzie Sadownictwa Akademii Rolniczej w Poznaniu* 139–141.

Szczygiel, A. and Rebandel, Z. (1982) [Control of root nematode and *Verticillium* on strawberries using Di-Trapex and Benlate]. *Roczniki Nauk Rolniczych E Ochrana-Roslin* 12, 163–174.

Tabaeizadeh, Z., Agharbaoui, Z., Harrak, H. and Poysa, V. (1999) Transgenic tomato plants expressing a *Lycopersicon chilense* chitinase gene demonstrate improved resistance to *Verticillium dahliae* race 2. *Plant Cell Reports* 19(2), 197–202.

Tabrett, A.M., Carder, J.H., Croomber, S. and Barbara, D.J. (1995) *Arabidopsis thaliana* as an experimental host of haploid isolates of *Verticillium dahliae*. 6th International *Verticillium* Symposium, Dead Sea, Israel. *Phytoparasitica* 23, 49.

Tai, F.L. (1979) *Sylloge Fungorum Sinicorum.* Science Press, Beijing, China,

Takacs, D.J. and Khoury, F.Y. (1970) Plants of greenhouse-grown cotton varieties Pima and Deltapine smooth leaf susceptibility to *Verticillium albo-atrum. Proceedings of the Beltwide Cotton Production Research Conference*, Houston, p. 6.

Takano, T., Takayama, M. and Hagiwara, J. (1985) [The disinfecting effectiveness of dry heat on 3 species of seed-borne pathogens of quarantine significance]. *Research Bulletin of the Plant Protection Service, Japan* 21, 1–9.

Takeuchi, J. and Horie, H. (1996) Occurrence of *Verticillium* wilt of balloonflower and safflower caused by *Verticillium dahliae* Klebahn. *Proceedings of the Kanto-Tosan Plant Protection Society*, pp. 125–127.

Takeuchi, J. and Horie, H. (1998) [First occurrence of *Verticillium* wilt of dahlia, small globe thistle and gay feather caused by *Verticillium dahliae* Klebahn in Japan]. *Proceedings of the Kanto-Tosan Plant Protection Society* 45, 123–125.

Talboys, P.W. (1955) Detection of vascular tissues available for water transport in the hop by colourless derivatives of basic dyes. *Nature (London)* 175, 510.

Talboys, P.W. (1957) The possible significance of toxic metabolites of *Verticillium albo-atrum* in the development of hop wilt symptoms. *Transactions of the British Mycological Society* 40, 415–427.

Talboys, P.W. (1958a) Degradation of cellulose by *Verticillium albo-atrum*. *Transactions of the British Mycological Society* 41, 242–248.

Talboys, P.W. (1958b) Association of tylosis and hyperplasia of the xylem with vascular invasion of the hop by *Verticillium albo-atrum*. *Transactions of the British Mycological Society* 41, 249–260.

Talboys, P.W. (1958c) Some mechanisms contributing to *Verticillium* resistance in the hop root. *Transactions of the British Mycological Society* 41, 227–241.

Talboys, P.W. (1960) A culture-medium aiding the identification of *Verticillium albo-atrum* and *V. dahliae*. *Plant Pathology* 9, 57–58.

Talboys, P.W. (1962) Systemic movement of some vascular pathogens. *Transactions of the British Mycological Society* 45, 280–281.

Talboys, P.W. (1964) A concept of the host–parasite relationship in *Verticillium* wilt diseases. *Nature (London)* 202, 361–364.

Talboys, P.W. (1968) Water deficits in vascular disease. In: Kozlowski, T.T. (ed.) *Water Deficits and Plant Growth,* Vol. 2. Academic Press Inc, New York, pp. 255–311.

Talboys, P.W. (1970) Variation in the virulence of *Verticillium dahliae* from strawberry. *Annals of Applied Biology* 66, 43–49.

Talboys, P.W. (1972) Resistance to vascular wilt fungi. *Proceedings of the Royal Society of London B* 181, 319–332.

Talboys, P.W. (1975) Simulation of the hop *Verticillium* wilt syndrome by application of ethrel. *Annals of Applied Biology* 81, 126.

Talboys, P.W. (1978) Simulation of the symptoms of hop *Verticillium* wilt by 'model' toxins. In: Sadasivan, T.S., Subramanian, C.V., Kalyanasundaram, R. and Saraswathi-Dev, L. (eds) *Pathological Wilting of Plants.* University of Madras, India, pp. 231–246.

Talboys, P.W. (1984) Chemical control of *Verticillium* wilts. *Phytopathologia Mediterranea* 23, 163–175.

Talboys, P.W. (1987) *Verticillium* wilt in English hops: retrospect and prospect. *Canadian Journal of Plant Pathology* 9, 68–77.

Talboys, P.W. and Busch, L.V. (1969) Potato wilt concerns strawberry growers. *Report of the East Malling Research Station* 1968. p. 171.

Talboys, P.W. and Busch, L.V. (1970) Pectic enzymes produced by *Verticillium* wilt species. *Transactions of the British Mycological Society* 55, 367–381.

Talboys, P.W. and Davies, M.K. (1976a) Benomyl tolerance in *Verticillium dahliae*. *Annals of Applied Biology* 82, 41–50.

Talboys, P.W. and Davies, M.K. (1976b) Pathogenicity of benomyl-tolerant variants of *Verticillium dahliae* to strawberry, cv. Cambridge Vigour. *Annals of Applied Biology* 83, 489–493.

Talboys, P.W. and Frick, E.L. (1972) *Report of the East Malling Research Station* 1971, p. 121.

Talboys, P.W. and Frick, E.L. (1973) *Report of the East Malling Research Station* 1972, pp. 159–160.

Talboys, P.W. and Frick, E.L. (1974) Benomyl-tolerance in *V. dahliae*. *Report of the East Malling Research Station* 1973, pp. 147–149.

Talboys, P.W. and Frick, E.L. (1975) Intergration of methods for controlling strawberry *Verticillium* wilt in fruiting beds. *Report of the East Malling Research Station* 1974, pp. 151–153.

Talboys, P.W. and Wilson, J.F. (1954) A method for determining the pathogenicity of strains of *Verticillium albo-atrum* isolates from hop. *Report of the East Malling Research Station* 1953, pp. 158–161.

Talboys, P.W. and Wilson, J.F. (1970) Effects of temperature and rainfall on the incidence of wilt (*Verticillium albo-atrum*) in hops. *Annals of Applied Biology* 66, 51–58.

Talboys, P.W., Bennett, M. and Wilson, J.F. (1961) Tolerance to *Verticillium* wilt diseases in strawberry. *Report of the East Malling Research Station* 1960, pp. 94–99.

Talboys, P.W., Frick, E.L. and Davies, M.K. (1975) Duration of effects of benomyl on growth and wilt (*Verticillium dahliae*) development in the strawberry, cv. Cambridge Vigour, in relation to time of inoculation. *Annals of Applied Biology* 81, 305–315.

Talboys, P.W., Frick, E.L. and Davies, M.K. (1976) Systemic fungicides in the control of strawberry wilt (*Verticillium dahliae*). *Annals of Applied Biology* 82, 537–555.

Talboys, P.W., Davies, M.K. and Twomlow, A.K. (1977) Host–parasite relations in strawberry *Verticillium* wilt. *Report of the East Malling Research Station* 1976. p. 140.

Tan, L.-W. (1998) [Wilt resistant cotton breeding viewed from the selection of CRI 12]. *China Cottons* 25(12), 10–12.

Tanasescu, M., Tanasescu, C., Tomescu, A., Botez, C., Lupascu, A., Necula, A. and Ambarus, S. (1996) Genotipuri noi obtinute la ardeiul gogosar (*Capsicum annuum* L. var. *grossum*). *Anale Institutul de Cercetari pentru Legumicultura si Floricultura, Vidra* 14, 175–188.

Tanev, I.I., Kazakova, K.L. and Tsalbukov, P.R. (1979) [Main diseases, pests and weeds of damask rose, lavender and mint and their control]. *Dostizheniya efirnomaslich, pr-ve NR Bolgarii i Moldavii* 232–249.

Tang, H.-Q., Hu, J., Yang, L. and Tan, R.-X. (2000) [Terpenoids and flavonoids from *Artemisia* species]. *Planta Medica* 66(4), 391–393.

Tarn, T.R., DeJong, H., Murphy, A.M., Tai, G.C.C., Arsenault, W.J., Thorpe, J.H.E., Bagnall, R.H., Platt, H.W., Young, D.A. and Davies, H.T. (1995) AC Belmont: a new early-maturing potato cultivar with short dormancy. *American Potato Journal* 72, 409–415.

Tartier, L.M. and Devaux, A. (1977) Enquête sur la flétrissure verticillienne de la pomme de terre de la region du nord de Montréal. *Phytoprotection* 58, 115–120.

Tarunina, T.A. (1980) [Criteria in the selection of strains of *Trichoderma* Pers. ex Fr. for the production of biopreparations]. *Sbornik Nauchnykh Trudov Vsesoyvznogo Nauchno-Issledovatel'skogo Instituta Zashchity Rasteniĭ, Leningrad, USSR* 89–96.

Tate, K.G. and Cheah, L.H. (1983) Ridge fumigation for control of *Verticillium* wilt in strawberries. *Proceedings of the 36th New Zealand Weed and Pest Control Conference.* Palmerston North, New Zealand, pp. 79–82.

Tattar, T.A. (1976) Use of electrical resistance to detect *Verticillium* wilt in Norway and sugar maple. *Canadian Journal of Forest Research* 6, 499–503.

Taubenhaus, J.J. (1936) *Verticillium* wilt of cotton. *49th Annual Report, Texas Agricultural Experimental Station*, p. 111.

Taylor, J.B. (1963) The inactivation of *Verticillium albo-atrum* in apricot trees. *Proceedings of the Forty-fifth Annual Meeting of the Pacific Division of the American Phytopathological Society at Stanford University*, California, June 19–21, 1963.

Taylor, J.B. (1968) Host range of *Verticillium tricorpus* Isaac. *New Zealand Journal of Agricultural Research* 11, 521–523.

Taylor, J.B. (1969) Fluctuations in natural soil populations of *Verticillium tricorpus. Canadian Journal of Botany* 47, 737–740.

Taylor, J.B. and Canter-Visscher, T.W. (1970) Prolonged control of *Verticillium* wilt and root-lesion nematodes by soil fumigation. *New Zealand Journal of Science* 13, 591–602.

Taylor, J.B. and Flentje, N.T. (1968) Infection, recovery from infection and resistance of apricot trees to *Verticillium albo-atrum*. *New Zealand Journal of Botany* 6, 417–426.

Taylor, N.J. (1993) First report of *Verticillium* wilt of *Impatiens walleriana* caused by *Verticillium dahliae* in the United States. *Plant Disease* 77(4), 429.

Tchatchoua, J. and Sikora, R.A. (1983) Veränderungen in der Anfälligkeit welke-resistenten Baumwollpflanzen gegenüber *Verticillium dahliae*, verursacht durch *Rotylenchulus reniformis*. *Zeitschrift für Pflanzenkrankheiten und Pflanzenschutz* 90, 232–237.

Teixeira, R. and Teixeira, A. (1953a) A verticiliose do algodoeiro. O caso concreto do vale do rio Limpopo. *Memórias e Trabalhos do Centro de Investigação Científica Algodoeira No 8. Revista da Faculdade de Ciéncia de Lisbõa* 2, 333–388.

Teixeira, R. and Teixeira, A. (1953b) A verticiliose do algodoeiro no vale do rio Limpopo 2. *Memórias e Trabalhos do Centro de Investigação Científica Algodoeira No 8. Revista da Faculdade de Ciéncia de Lisbõa* 3, 111–128.

Tellyaev, R. and Ligai, A. (1981) [Use of anti-wilt preparations jointly with defoliation]. *Khlopkovodstvo* 8, 31.

Tellyaev, R.Sh. (1978) [Effect of crop rotation on the effectiveness of systemic fungicides against wilt]. *Trudy Vsesoyuznogo Nauchno-Issledovatel'skogo Instituta Khlopskovodstva* 39, 35–39.

Ten, L.N., Otroshchenko, O.S., Stepanichenko, N.N. and Inoyatova, D.A. (1977a) Metabolites of the pathogenic fungus *Verticillium dahliae* VII. The phytotoxic pigment PKZh-1 from the culture liquid. *Khimiya Prirodnykh Soedinenii Tashkent* 5, 690.

Ten, L.N., Tyshchenko, A.A., Stepanichenko, N.N., Gusakova, S.D., Mukhamedzhanov, S.Z., Otroshchenko, O.S. and Kas'yanenko, A.G. (1977b) Metabolites of the pathogenic fungus *Verticillium dahliae* VI. Pentaketide metabolites and neutral lipids of virulent and avirulent strains. *Khimiya Prirodnykh Soedinenii Tashkent* 5, 632–635.

Ten, L.N., Mukhimedshanov, S.Z. and Otroshchenko, O.S. (1981) Structure of the peptide moiety of verticillin. *Khimiya Prirodnykh Soedinenii Tashkent* 4, 476–479.

Ten, L.N., Stepanichenko, N.N. and Aslanov, Kh.A. (1990) Chemical structure and biological activity of some metabolites of phytopathogenic fungi of the *Verticillium* genus. *Proceedings of the Fifth International* Verticillium *Symposium*. Leningrad, p. 58.

Ten, L.N., Faizlev, M.M. and Romanova, L.A. (1995) Lectins – role in *Verticillium* wilt of cotton. 6th International *Verticillium* Symposium, Dead Sea, Israel. *Phytoparasitica* 23, 57.

Teranisihi, J., Figueiredo, M.B., Cardoso, R.M.G. and Namekata, T. (1973) The value of serological techniques for the diffentiation between *Verticillium albo-atrum* Reink. and Berth. and *V. dahliae* Kleb. *Archivos do Instituto Biológico* São Paulo 40, 45–51.

Ter-Avanesyan, D.V. (1964) [Attraction of new forms for breeding wilt-resistant varieties of cotton]. *Khlopkovodstvo* 14, 32–34.

Ter-Avanesyan, D.V. (1976) Initial material in breeding of cotton varieties resistant to *Verticillium dahliae*. *Proceedings of the Second International* Verticillium *Symposium*. University of California, Berkeley, p. 40.

Ter-Avanesyan, D.V., Babaev, F.A. and Abdinov, F.A. (1977) [Wilt resistance of cotton from the USA in Azerbaijan]. *Vestnik Sel'skokhozyaistvennoi Nauki* 2, 39–41.

Termorshuizen, A.J. (1995) The search for the recovery in detection methods of microsclerotia of *Verticillium dahliae*. 6th International *Verticillium* Symposium, Dead Sea, Israel. *Phytoparasitica* 23, 55.

Termorshuizen, A.J. (1997a) Improvement of assays to detect microsclerotia of *Verticillium dahliae* in soil. Cooperation is necessary. *Proceedings of the Seventh International* Verticillium *Symposium*. Cape Sounion, Athens, Greece, p. 31.

Termorshuizen, A.J. (1997b) Sporulation of *Verticillium dahliae*. *Proceedings of the Seventh International* Verticillium *Symposium*. Cape Sounion, Athens, Greece, p. 34.

Termorshuizen, A.J. and Rouse, D.I. (1993) Towards a mechanistic model for the *Verticillium dahliae*–potato system. *Netherlands Journal of Plant Pathology* 99(Supplement 3), 201–218.

Termorshuizen, A.J., Blok, W.J., Coenen, G.C.M., Lamers, T.G., Slomp, P., Leffelaat, P.A. and Bollen, G.J. (1997) Biophysical soil disinfestation to control *Verticillium dahliae* by incorporation of fresh organic material followed by plastic mulching. *Proceedings of the Seventh International* Verticillium *Symposium*. Cape Sounion, Athens, Greece, p. 83.

Termorshuizen, A.J., Davis, J.R., Gort, G., Harris, D.C., Huisman, O.C., Lazarovits, G., Locke, T., Vara, J.M.M., Mol, L., Paplomatas, E.J., Platt, H.W., Powelson, M., Rouse, D.I., Rowe, R.C. and Tsror, L. (1998) Interlaboratory comparison of methods to quantify microsclerotia of *Verticillium dahliae* in soil. *Applied and Environmental Microbiology* 64(10), 3846–3853.

Terras, F.R.G., Schoofs, H.M.E., Thevissen, K., Osborn, R.W., Vanderleyden, J., Cammue, B.P.A. and Broekaert, W.F. (1993a) Synergistic enhancement of the antifungal activity of wheat and barley thionins by radish and oilseed rape 2S albumins and by barley trypsin inhibitors. *Plant Physiology* 103(4), 1311–1319.

Terras, F.R.G., Torrekens, S., van Leuven, F., Osborn, R.W., Vanderleyden, J., Cammue, B.P.A. and Broekaert, W.F. (1993b) A new family of basic cysteine-rich plant antifungal proteins from Brassicaceae species. *FEBS Letters* 316(3), 233–240.

Tervet, I.W. (1944) Diseases of plants in 1943. *Plant Disease Reporter Supplement* 149, 381.

Teshabaev, K.A. and Teshabaeva, R.I. (1974) [Affection of cotton varieties after artificial infection with black and white forms of the fungus *Verticillium dahliae* Kleb.]. *Voprosy Genetiki, Selektsii i Semenovodstva Khlopchatnika i Lyutserny* 2, 187–189.

Teshabaeva, R.I., Teshabaev, K.A., Rakhimov, Kh.R. and Rakhmankulov, S.A. (1973) [Susceptibility to wilt of different cotton varieties]. *Khlopkovodstvo* 2, 23–24.

Teviotdale, B.L., Epstein, L., Ferguson, L. and Reil, W. (1995) Susceptibility of pistachio rootstocks to *Verticillium dahliae* and *Armillaria mellea* – a progress report. In: Kaska, N., Kuden, A.B., Ferguson, L. and Michailides, T. (eds) *Acta Horticulturae*. First *international Sym*posium on Pistachio Nut. Adana, Turkey, 20–24 September 1994, pp. 353–357.

Thanassoulopoulos, C.C. (1993) Spread of *Verticillium* wilt by nursery plants in olive groves in the Halkidiki area (Greece). *Bulletin OEPP* 23(3), 517–520.

Thanassoulopoulos, C.C. and Kitsos, G.T. (1972) *Verticillium* wilt in Greece. *Plant Disease Reporter* 56, 264–267.

Thanassoulopoulos, C.C. and Kitsos, G.T. (1973) Notes on the control of soil-borne diseases of tomato under plastic. *Geoponika* 210, 5–9.

Thanassoulopoulos, C.C. and Kitsos, G.T. (1974) Stem length of tomato for evaluating the severity of attack by *Verticillium* wilt. *Phytopathologische Zeitschrift* 81, 258–261.

Thanassoulopoulos, C.C., Biris, D.A. and Tjamos, E.C. (1979) Survey of *Verticillium* wilt of olive trees in Greece. *Plant Disease Reporter* 63, 936–940.

Thanassoulopoulos, C.C., Biris, D.A. and Tjamos, E.C. (1981) Weed hosts as inoculum source of *Verticillium* in olive orchards. *Phytopathologia Mediterranea* 20, 164–168.

Thanassoulopoulos, C.C., Biris, D.A. and Tjamos, E.C. (1984) Occurrence of *Verticillium* wilt in weeds in olive orchards. Pathobiology of *Verticillium* species. Third International *Verticillium* Symposium, Bari, Italy. *Phytopathologia Mediterranea* 23, 215.

Thate, R. (1961) Die Apoplexie der Rebe: eine Vertilliose. *Mitteilungen aus der Biologischen Bundesanstalt für Land- und Forstwirtschaft Berlin-Dahlem* 104, 100–103.

Thaxton, P.M. and El Zik, K.M. (1998) Breeding for resistance to wilt diseases in Tamcot cultivars. In: Dugger, P. and Richter, D.J. (eds) *1998 Proceedings Beltwide Cotton Conferences*. San Diego, California, 5–9 January 1998. National Cotton Council, Memphis, Tennessee, Vol. 1, pp. 128–131.

Thomas, B.J. (1981) Some degeneration and dieback diseases of the rose. *Annual Report of the Glasshouse Crops Research Institute for 1979*, pp. 178–190.

Thomas, C.A. (1976) Resistance of VFR1 safflower to *Phytophthora* root rot and its inheritance. *Plant Disease Reporter* 60, 123–125.

Thomas, C.A. (1981) Susceptibility of safflower cultivars to cotton strains of *Verticillium dahliae*. *Phytopathology* 71, 771.

Thomas, C.A., Zimmerman, L.H. and Urie, A.L. (1978) Registration of LMVFP-1 safflower germplasm (Reg.GP No. 14). *Crop Science* 18, 1099.

Thomas, H.E. (1931) Verticilliosis of strawberries. *Phytopathology* 21, 996.

Thomas, W.D. (1950) New shade tree diseases in Colorado. *Plant Disease Reporter* 34, 83.

Thomas, W.D. and Boza, C.A. (1984) The oak decline complex. *Journal of Arboriculture* 10, 170–177.

Thompson, F.C. and Neve, R.A. (1972) The influence of various factors on the cost of production of hop alpha-acid. *Journal of the Institute of Brewing* 78, 156–161.

Thomson, R. and McLeod, A.G. (1959) *Verticillium* wilt of tobacco. I. A new disease of tobacco caused by *Verticillium dahliae* Kleb. *New Zealand Journal of Agricultural Research* 2, 785–791.

Thorat, S.C., More, B.B. and Konde, B.K. (1976) Chemical control of *Verticillium* wilt of eggplant (*Solanum melongena* L.) by soil and plant application. *Hindustan Antibiotics Bulletin* 18, 117–118.

Thornberry, G.O. (1973) Some physiological responses of *Verticillium albo-atrum* to zinc. *Canadian Journal of Botany* 51, 57–59.

Thorpe, H.J. and Jarvis, W.R. (1978) *Verticillium* wilt in Canadian redbud. *Canadian Plant Disease Survey* 58, 107.

Threlfall, R.J. (1959) Physiological studies on the *Verticillium* wilt disease of tomato. *Annals of Applied Biology* 47, 57–77.

Tian, L., Wang, K.-R. and Lu, J.-Y. (1997) Heredity of microsclerotia forming ability in *Verticillium dahliae*. *Mycosystema* 16(3), 197–201.

Tian, L., Wang, K.-R. and Lu, J.-Y. (1998a) [Genetic study on *nit* mutants and fungicide resistant mutants of *Verticillium dahliae*]. *Mycosystema* 17, 137–142.

Tian, L., Wang, K.-R. and Lu, J.-Y. (1998b) [Effect of carbendazim and tricyclazole on microsclerotia and melanin formation of *Verticillium dahliae*]. *Acta Phytopathologica Sinica* 28, 263–268.

Tian, L., Wang, K.-R. and Lu, J.-Y. (1998c) [Inhibition of *Stemphylium* sp. on *Verticillium dahliae* mycelial growth and microsclerotium formation]. *Chinese Journal of Biological Control* 14, 14–17.

Tigchelaar, E.C. and Dick, J.B. (1975) Induced resistance from simultaneous inoculation of tomato with *Fusarium oxysporum* Sacc. and *Verticillium albo-atrum* Reinke & Berth. *HortScience* 10, 623–624.

Tigchelaar, E.C., Tomes, M.L. and Nelson, P.E. (1974) 'Lafayette' tomato. *HortScience* 9, 80.

Tillaev, Kh. (1980) [Effectiveness of trichodermin against wilt]. *Khlopkovodstvo* 12, 19, 22.

Tjamos, E.C. (1979) Induction of resistance to *Verticillium* wilt in cucumber (*Cucumis sativus*). *Physiological Plant Pathology* 15, 223–227.

Tjamos, E.C. (1980, published 1981) Occurrence of race 2 of *Verticillium dahliae* in Greece. *Annales de l'Institut Phytopathologique Benaki* 12, 216–226.

Tjamos, E.C. (1981) Virulence of *Verticillium dahliae* and *V. albo-atrum* isolates in tomato seedlings in relation to their host of origin and the applied cropping system. *Phytopathology* 71, 98–100.

Tjamos, E.C. (1991) Post-plant application of soil solarization for tree crops. *FAO Plant Protection Paper* 109, 243–252.

Tjamos, E.C. (1993) Prospects and strategies in controlling *Verticillium* wilt of olive. *Bulletin OEPP* 23(3), 505–512.

Tjamos, E.C. and Beckman, C.H. (1989) *Vascular Wilt Diseases of Plants. Basic Studies and Control. NATO ASI Series H: Cell Biology*, Vol. 28. Springer-Verlag, Berlin.

Tjamos, E.C. and Despina, B. (1986) Occurrence of *Verticillium dahliae* in leaves of *Verticillium* wilted olive trees. *Proceedings of the Fourth International* Verticillium *Symposium*, Guelph, Canada.

Tjamos, E.C. and Faridis, A. (1981) Control of soil pathogens by solar heating in plastic houses. *Proceedings of the Fifth Congress of the Mediterranean Phytopathological Union.* Patras, Greece, September 1980, pp. 82–84.

Tjamos, E.C. and Fravel, D.R. (1995a) Distribution and establishment of *Talaromyces flavus* a biocontrol against *Verticillium dahliae* in soil and on roots of solanaceous crops. 6th International *Verticillium* Symposium, Dead Sea, Israel. *Phytoparasitica* 23, 68.

Tjamos, E.C. and Fravel, D.R. (1995b) Detrimental effects of sublethal heating and *Talaromyces flavus* on microsclerotia of *Verticillium dahliae*. *Phytopathology* 85(4), 388–392.

Tjamos, E.C. and Fravel, D.R. (1997) Distribution and establishment of the biocontrol fungus *Talaromyces flavus* in soil and on roots of solanaceous crops. *Crop Protection* 16(2), 135–139.

Tjamos, E.C. and Kornaros, E. (1978) Virulence of Greek *Verticillium dahliae* isolates on susceptible and tolerant cotton cultivars. *Plant Disease Reporter* 62, 456–458.

Tjamos, E.C. and Paplomatas, E.J. (1986) Long term effect of soil solarization on *Verticillium* wilt of artichokes in Greece. *Proceedings of the Fourth International* Verticillium *Symposium*, Guelph, Canada.

Tjamos, E.C. and Paplomatas, E.J. (1987a) Effect of soil solarization on the survival of fungal antagonists of *Verticillium dahliae*. *Bulletin OEPP* 17, 645–653.

Tjamos, E.C. and Paplomatas, E.J. (1987b) Long-term effect of soil solarization on *Verticillium* wilt of artichokes in Greece. *Canadian Journal of Plant Pathology* 9, 86.

Tjamos, E.C. and Paplomatas, E.J. (1988) Long-term effect of soil solarization in controlling *Verticillium* wilt of globe artichokes in Greece. *Plant Pathology* 37, 507–515.

Tjamos, E.C. and Skretis, L. (1990) Establishment of applied or increase of natural *Verticillium dahliae* antagonists in solarized soils. *Proceedings of the Fifth International* Verticillium *Symposium*, Leningrad, p. 87.

Tjamos, E.C. and Smith, I.M. (1974) The role of phytoalexins in resistance of tomato to *Verticillium* wilt. *Physiological Plant Pathology* 4, 249–259.

Tjamos, E.C. and Smith, I.M. (1975) The expression of resistance to *Verticillium albo-atrum* in monogenically resistant tomato varieties. *Physiological Plant Pathology* 6, 215–225.

Tjamos, E.C. and Tsougriani, H. (1990) Formation of *Verticillium dahliae* microsclerotia in partially disintegrated leaves of *Verticillium* affected olive trees. *Proceedings of the Fifth International* Verticillium *Symposium*. Leningrad, p. 20.

Tjamos, E.C. and Vellios, E. (1997) Studies on the effect of biocontrol agents on *Verticillium dahliae* microsclerotia formation, survival and germination in infected host tissues and soil. *Proceedings of the Seventh International* Verticillium *Symposium*. Cape Sounion, Athens, Greece, p. 37.

Tjamos, E.C., Paplomatas, E.J. and Biris, D.A. (1986) Recovery of *Verticillium* wilted olive trees after individual application of soil solarization. *Proceedings of the Fourth International* Verticillium *Symposium*. Guelph, Canada.

Tjamos, E.C., Karapapa, V. and Bardas, D. (1989) Low cost application of soil solarization in covered plastic houses for the control of *Verticillium* wilt of tomatoes in Greece. *Acta Horticulturae* 255, 139–149.

Tjamos, E.C., Biris, D.A. and Paplomatas, E.J. (1991) Recovery of olive trees with *Verticillium* wilt after individual application of soil solarization in established olive orchards. *Plant Disease* 75, 557–562.

Tjamos, E.C., Tjamos, S.E., Venieraki, A., Tsitsiyiannis, D. and Katinakis, P. (1997a) Ecology and effectiveness of two endorhizosphere isolates of *Bacillus* sp. as biocontrol agents against *Verticillium dahliae* of solanaceous hosts. *Proceedings of the Seventh International* Verticillium *Symposium*. Cape Sounion, Athens, Greece, p. 65.

Tjamos, E.C., Tsitsiyiannis, D., Tjamos, S.E. and Panagopoulos, C.G. (1997b) Selection and evaluation of rhizosphere bacteria as biocontrol agents against *Verticillium dahliae*. *Proceedings of the Seventh International* Verticillium *Symposium*. Cape Sounion, Athens, Greece, p. 63.

Tjamos, E.C., Rowe, R.C., Heale, J.B. and Fravel, D.R. (eds) (2000) *Advances in* Verticillium*: Research and Disease Management*. American Phytopathological Society, St Paul, Minnesota.

Todd, W.A., Green, R.J. and Horner, C.E. (1977) Registration of Murray Mitcham peppermint (Reg.No. 2). *Crop Science* 17, 188.

Todorov, I., Manuelyan, Kh. and Bakhariev, D. (1995) Results of intervarietal hybridization in red pepper. *Rasteniev'dni Nauki* 32(1–2), 59–60.

Todorova, T., Neykov, S. and Todorov, Y. (1997) Evaluation of local pepper accessions (*Capsicum annuum* L.) in Sadovo, Bulgaria. *Plant Genetic Resources Newsletter*. 75–76.

Tokousbalides, M.C. and Sisler, H.D. (1978) Effect of tricyclazole on growth and secondary metabolism in *Pyricularia oryzae*. *Pesticide Biochemistry and Physiology* 8, 26–32.

Tokunaga, J., Malca, I., Sims, J.J., Erwin, D.C. and Keen, N.T. (1969) Respiratory enzymes in the spores of *Verticillium albo-atrum*. *Phytopathology* 59, 1829–1832.

Tolkachev, N.Z. (1978) [The influence of the systematic use of herbicides in cotton crops on soil mycolytic activity against phytopathgenic fungi]. *Mikrobiologicheskie Protsessy v Pochvakh i Urozhainost' Sel'skokhozyaistvennykh Kul'tur. Materialy k Respublikanskoi Konferentskii, Vil'nyus 1978*. pp. 349–350.

Tolmsoff, W.J. (1960) Relation of age of potato plants to infection by *Verticillium albo-atrum*. *Phytopathology* 50, 86.

Tolmsoff, W.J. (1969) Ammonium-linked respiration in conidia of *Verticillium albo-atrum*. *Phytopathology* 59, 1053–1054.

Tolmsoff, W.J. (1972) Diploidization and heritable gene repression–depression as major sources for variability in morphology, metabolism, and pathogenicity of *Verticillium* species. *Phytopathology* 62, 407–413.

Tolmsoff, W.J. (1973) Life cycles of *Verticillium* species. Verticillium *Wilt of Cotton. Proceedings of a Work Conference of the National Cotton Pathology Research Laboratory*. College Station, Texas 1971. ARS-S19, pp. 20–38.

Tolmsoff, W.J. (1980) Report of the Disease and Pathogen Physiology Committee–Pathogen Genetics. In: Bell, A.A. (ed.) *Proceedings of the Beltwide Cotton Production Research Conference*. January 6–8, 1980. National Cotton Council, Memphis, Tennessee, pp. 7–8.

Tolmsoff, W.J. (1983) Heteroploidy as a mechanism of variability among fungi. *Annual Review of Phytopathology* 21, 317–340.

Tolmsoff, W.J. and Bell, A.A. (1971) Conversion of haploids to homozygous diploids in five *Verticillium* spp. *Proceedings of the First International* Verticillium *Symposium* Wye College, University of London, pp. 19–24.

Tolmsoff, W.T. and Wheeler, M.H. (1974) *Proceedings of the American Phytopathological Society* 1, 25.

Tolmsoff, W.J. and Wheeler, M.H. (1977) Report of the Disease and Pathogen Physiology Committee 1976. *Proceedings of the Beltwide Cotton Production Research Conference*. Atlanta, Georgia, p. 258.

Tomaszewski, Z. Furgal-Wegrzycka, H., Kubok, I. and Piszczek, W. (1983) [Assessment of the susceptibility of lucerne varieties (*Medicago sativa* L.) to infection by *Verticillium albo-atrum* (Rke. et Berth.) in laboratory and greenhouse.]. *Zeszyty Naukowe Akademii Rolniczo-Technicznej w Olsztynie* 37, 133–142.

Tomescu, A., Voican, V. and Racaru, V. (1997) Some physiological changes induced by the fungus *Verticillium dahliae* in pepper plants. *Proceedings of the Seventh International* Verticillium *Symposium*. Cape Sounion, Athens, Greece, p. 38.

Toppan, A., Esquerré-Tugayé, M.T. and Touzé, A. (1976) An improved approach for the accurate determination of fungal pathogens in diseased plants. *Physiological Plant Pathology* 9, 241–251.

Torgeson, D.C. (1951) Observations of *Verticillium* wilt of *Geranium* in Oregon. *Plant Disease Reporter* 36, 51.

Torres, H. and Gutierrez, H. (1981) *Verticillium dahliae* Kleb., identificación y sintomas que produce en seis cultivares comerciales de papas Peruanas. *Fitopatología* 16, 60–68.

Tosi, L. and Zazzerini, A. (1998) Investigations on the epidemiology of *Verticillium* wilt in olive in central Italy. *Olivæ* 71, 50–55.

Toussoun, T.A., Bega, R.V. and Nelson, P.E. (1970) *Root Diseases and Soil-borne Pathogens. Second International Symposium on Factors Determining the Behaviour of Plant Pathogens in Soil*. Imperial College, London. University of California Press, Berkeley, California.

Townsend, A.M. and Hock, W.K. (1973) Tolerance of half-sib families of red maple to *Verticillium* wilt. *Phytopathology* 63, 673–676.

Townsend, A.M., Schreiber, L.R., Hall, T.J. and Bentz, S.E. (1990) Variation in response of Norway maple cultivars to *Verticillium dahliae. Plant Disease* 74, 44–46.

Trentini, L., Ceroni, E. and Maioli, B. (1983) Lotta alla verticilliosi della melanzana. *Informatore di Ortoflorofrutticoltura* 24(12), 7–10.

Treschow, M. (1965) Response of some pathogenic fungi to sodium fluoride. *Mycologia* 57, 216–221.

Trevisan, M.T.S., Scheffer, J.J.C. and Verpoorte, R. (1997a) Effect of elicitation on the peroxidase activity in some cell suspension cultures of hop, *Humulus lupulus. Plant Cell, Tissue and Organ Culture* 48(2), 121–126.

Trevisan, M.T.S., Valdivia, A.C.R., Scheffer, J.J.C. and Verpoorte, R. (1997b) Enzyme activities in cell suspension cultures of two hop cultivars after elicitation by a fungal culture filtrate. *Biotechnology Letters* 19(3), 207–211.

Tribunskii, A.N., Mateev, G.G. and Siderova, S.F. (1976) [Dynamics of soil populations of *Verticillium dahliae* Kleb. in cultivation of different varieties of cotton]. Микология и Фитопатология [*Mycology and Phytopathology*] 10(1), 42–47.

Trocme, O. (1972) Contribution à l'étude d'une maladie du cacaoyer en Ougande: le dessechement eco-fongique des branches. *Cafe Cacao Thé* 16, 219–235.

Trotzenko, I.V. (1958) [*The Smoke Bush*]. Ukrainian Academy of Science, Kiev, USSR.

Trushkina, L.P., Bakhidov, G. and Nizamov, U. (1974) [Wilt infection of cotton sown at different inter-row spacings]. *Nauchnye-Trudy, Tashkentskii-Sel'skokhozyaistvennyi-Institut* 51, 52–57.

Truszkowska, W. and Legiec, B. (1973) [Observations on the effect of host residues on the occurrence of some fungus diseases of red clover and alfalfa]. *Acta Mycologica* 9, 53–66.

Tsai, S.D. and Erwin, D.C. (1975) A method of quantifying numbers of microsclerotia of *Verticillium albo-atrum* in cotton plant tissue and in pure culture. *Phytopathology* 65, 1027–1028.

Tsror, L. and Erlich, O. (1996) Jojoba diseases in Israel. In: Princen, L.H. and Rossi, C. (eds) *Proceedings of the Ninth International Conference on Jojoba and its Uses and of the Third International Conference on New Industrial Crops and Products*, pp. 26–29.

Tsror, L. and Nachmias, A. (1990) Field mapping for *Verticillium dahliae. Proceedings of the Fifth International* Verticillium *Symposium*. Leningrad, p. 22.

Tsror, L. and Nachmias, A. (1995) *Verticillium* wilt: comparison of the tolerance phenomenon in potato to *Ve* gene resistance in tomato. 6th International *Verticillium* Symposium, Dead Sea, Israel. *Phytoparasitica* 23, 61.

Tsror, L., Nachmias, A., Yaron, A. and Uner, M. (1990) [*Verticillium* wilt and *Alternaria* leaf spot in Pima cotton]. *Hassadeh* 70, 681–682.

Tsror, L., Nachmias, A., Barak, Z. and Keen, N.T. (1991) *Escherichia coli* carrying single pectate lyase genes from the phytopathogenic *Erwinia chrysanthemi* causes disease symptoms in potato. *Phytoparasitica* 19, 57–63.

Tsror, L., Erlich, O., Amitai, S. and Hazanovsky, M. (1998) *Verticillium* wilt of paprika caused by a highly virulent isolate of *Verticillium dahliae. Plant Disease* 82, 437–439.

Tsror, L., Aharon, M. and Erlich, O. (1999) Survey of bacterial and fungal seedborne diseases in imported and domestic potato seed tubers. *Phytoparasitica* 27(3), 215–226.

Tu, C.M. (1982) Effects of selected pesticides on *Rhizobium meliloti* and on seed germination and *Verticillium* wilt pathogens of alfalfa. *Chemosphere* 11, 1195–1202.

Tukeeva, M.I. and Tukeyeva, M.I. (1977) Activity of hydrolytic enzymes and the lysosomal system in cells of *Verticillium dahliae. Fiziologiya Rastenii* 24, 941–947.

Turakulov, Kh., Arbarov, S., Mirpulatova, N. and Shadmanova, Sh. (1977) [Effectiveness of ammonium sulphate on wilt-infected fields]. *Khlopkovodstvo* 1, 19.

Turconi, M. (1920) Sopra una nuova malattia del cacao (*Theobroma cacao* L.). *Atti, Laboratorio Crittogamico, Istituto Botanico della Università, Pavia Series II* 17(1–3), 1–8.

Turhan, G. and Grossmann, F. (1988) Antagonistic activity of *Neocosmospora vasinfecta* var. *africana* (von Arx) Cannon & Hawksworth against soil-borne fungi. *Journal of Phytopathology* 123, 199–206.

Turkensteen, L.J. (1988) Survey of bacterial and fungal diseases of potato in Baluchistan Province of Pakistan, 1987. *Pakistan–Swiss Potato Development Project, Pakistan Agricultural Research Council*, pp. 1–10.

Turkensteen, L.J. and Eduardo Nieto, L. (1987) Informe de un reconocimiento de enfermedades de la papa en Colombia. *ASCOLFI Informa* 13, 42–46.

Tursunov, Kh.Z. and Dariev, A.S. (1979) [Micromorphological study of the xylem in some cottons in connection with *Verticillium dahliae* Kleb. infection]. *Uzbekskii Biologicheskii Zhurnal*, 48–50.

Twomey, J.A., Hoyman, W.G. and Shaughnessy, J.J. (1977) Centenial Russet: an oblong russetted variety adapted to certain irrigated potato growing areas of the west. *American Potato Journal* 54, 603–605.

Typas, M.A. (1983) Heterokaryon incompatibility and interspecific hybridization between *Verticillium albo-atrum* and *Verticillium dahliae* following protoplast fusion and microinjection. *Journal of General Microbiology* 129, 3043–3056.

Typas, M.A. (1984) A cytoplasmic system of inheritance affecting the variability and differentiation of *Verticillium. Phytopathologia Mediterranea* 23, 218.

Typas, M.A. (1997) Molecular characterization of *Verticillium* species. *Proceedings of the Seventh International* Verticillium *Symposium*. Cape Sounion, Athens, Greece, p. 13.

Typas, M.A. and Heale, J.B. (1976a) Heterokaryosis and the role of cytoplasmic inheritance in dark resting structure formation by *Verticillium* spp. *Molecular and General Genetics* 146, 17–26.

Typas, M.A. and Heale, J.B. (1976) Acriflavine-induced hyaline variants of *Verticillium albo-atrum* and *V. dahliae. Transactions of the British Mycological Society* 66, 15–25.

Typas, M.A. and Heale, J.B. (1977) Analysis of ploidy levels in strains of *Verticillium* using a Coulter counter. *Journal of General Microbiology* 101, 177–180.

Typas, M.A. and Heale, J.B. (1978) Heterozygous diploid analyses via the parasexual cycle and a cytoplasmic pattern of inheritance in *Verticillium* species. *Genetical Research* 31, 131–144.

Typas, M.A. and Heale, J.B. (1979) Transfer of a cytoplasmic factor by microinjection in *Verticillium. Journal of General Microbiology* 3, 375–386.

Typas, M.A. and Heale, J.B. (1980) DNA content of germinating spores, individual hyphal cells and resting structure cells of *Verticillium* spp. measured by microdensimetry. *Journal of General Microbiology* 121, 231–242.

Typas, M.A., Griffen, A.M., Bainbridge, B.W. and Heale, J.B. (1992) Restriction fragment length polymorphisms in mitochondrial DNA and ribosomal RNA gene complexes as an aid to the characterization of species and sub-species populations in the genus *Verticillium. FEMS Microbiology Letters* 95(2–3), 157–162.

Tysdal, H.M., Estilai, A., Siddiqui, I.A. and Knowles, P.F. (1983) Registration of four guayule germplasms. *Crop Science* 23, 189.

Tyshchenko, A.A., Stepanichenko, N.N. and Ten, L.N. (1990) Stable biradicals in pigments of the fungus *Verticillium dahliae* Kleb. *Proceedings of the Fifth International* Verticillium *Symposium*. Leningrad, p. 59.

Tyuterev, S.L., Ksendova, E.N. and Plakunova, A.N. (1980) [Induction of phytoalexin reactions in wheat plants]. *Byulleten' Vsesoyuznogo Nauchno-Issledovatel'skogo Instituta Zashchity Rasteniĭ* 48, 66–70.

Tzeng, D.D. and DeVay, J.E. (1984) Ethylene production and toxicity of methionine and its derivatives with riboflavine in cultures of *Verticillium*, *Fusarium* and *Colletotrichum* species exposed to light. *Physiologia Plantarum* 62, 545–552.

Tzeng, D.D., Wakeman, R.J. and DeVay, J.E. (1985) Relationships among *Verticillium* wilt development, leaf water potential, phenology and lint yield in cotton. *Physiological Plant Pathology* 26, 73–82.

Ubaĭdullaev, Kh. (1980) [Use of *Trichoderma* with fertilizers]. *Khlopkovodstvo* 8, 27.

Urie, A.L. and Knowles, P.F. (1972) Safflower introductions resistant to *Verticillium* wilt. *Crop Science* 12, 545–546.

Urie, A.L., Zimmerman, L.H., Thomas, C.A., Peterson, W.F. and Knowles, P.F. (1976) Registration of VFSTP1 safflower germplasm (Reg.No. GP 12). *Crop Science* 16, 888.

Urie, A.L., Davia, D.J., Knowles, P.F. and Zimmerman, L.H. (1980) Registration of 14-5 safflower germplasm (Reg.No. CP 15). *Crop Science* 20, 115–116.

Uronov, I. and Sultanov, B. (1978) [Copper-containing ammophos reduces wilt infection of cotton]. *Khlopkovodstvo* 4, 30.

Uroševič, B. (1983, published 1987) Tracheiomycotic diseases in oak. *Communicationes Instituti Forestalis Čechosloveniae* 13, 85–100.

Urunov, N., Bicherov, E. and Tursunov, F. (1976) [Increasing wilt resistance in cotton]. *Khlopkovodstvo* 5, 20.

Ushiki, J., Hayakawa, Y. and Tadano, T. (1996) Medicinal plants for suppressing soil-borne plant diseases. I. Screening for medicinal plants with antimicrobial activity in roots. *Soil Science and Plant Nutrition* 42(2), 423–426.

Usmanov, Z.U. (1974) [Wilt resistance in conditions of infection]. *Khlopkovodstvo* 9, 30.

Vajna, L. (1999) [Tracheomycosis of pedunculate oak (*Quercus petraea*) caused by *Verticillium dahliae* Kleb.]. *Novenyvedelen* 35, 63–65.

Valadon, L.R.G. and Heale, J.B. (1964) Carotenoid pigments in a mutant strain of *Verticillium albo-atrum* Reinke & Berthold. *Nature (London)* 202, 512.

Valadon, L.R.G. and Heale, J.B. (1965) Carotenoids in mutants of *Verticillum albo-atrum*. *Journal of General Microbiology* 41, 341–348.

Van Beyma Thoe Kingma, F.H. (1939) Beschreibung einiger neuer Pilzarten aus dem 'Centraalbureau voor Schimmelcultures' (Holland). V Mitteilung. *Zentralblatt für Bakteriologie, Parasitenkunde, Infektionskrankeiten und Hygiene Abt. 2* 99, 381–394.

Van Beyma Thoe Kingma, F.H. (1940) Uber einige Formen von *Verticillium dahliae* Kleb. *Antonie van Leeuwenhoek* 6, 34–47.

Van de Koppel, M.M. and Schots, A. (1994) A double (monoclonal) antibody sandwich ELISA for the detection of *Verticillium* species in roses. In: Schots, A.,Dewey, F.M. and Oliver, R.P. (eds) *Modern Assays for Plant Pathogenic Fungi: Identification, Detection and Quantification.* CAB International, Wallingford, UK, pp. 99–104.

Van de Koppel, M.M. and Schots, A. (1995) Monoclonal antibody-based double-antibody sandwich-ELISA for detection of *Verticillium* spp. in ornamentals. *Phytopathology* 85, 608–612.

Van den Ende, G. (1958) Untersuchungen über den Pflazenparasiten *Verticillium albo-atrum* Reinke et Berth. *Acta Botanica Neerlandica* 7, 665–740.

Van Der Lek, H.A.A. (1918) Over de z. g. 'Verwelkingziekten' in het bijzonder die, welke door *Verticillium albo-atrum* veroorzaakt worden. *Tijdschrift over Plantenziekten* 24, 205–219.

Van der Meer, J.H.H. (1925) *Verticillium* wilt of herbaceous and woody plants. *Mededelingen Landbouwhogeschool Wageningen* 28(2), 1–82.

Van der Meer, J.H.H. (1926) *Verticillium* wilt of maple and elm seedlings in Holland. *Phytopathology* 16, 611–614.

Van Der Molen, G.E., Beckman, C.H. and Rodehorst, E. (1977) Vascular gelation: a general response phenomenon following infection. *Physiological Plant Pathology* 11, 95–100.

Van Der Molen, G.E., Labavitch, J.M. and DeVay, J.E. (1982) Carbohydrate composition and structure of vascular gels from wilt-diseased banana roots. *Plant Physiology* 69(Supplement), 143.

Van Der Molen, G.E., Labavitch, J.M., Strand, L.L. and DeVay, J.E. (1983) Pathogen-induced vascular gels: ethylene as a host intermediate. *Physiologia Plantarum* 59, 573–580.

Van Der Molen, G.E., Labavitch, J.M. and DeVay, J.E. (1986) *Fusarium*-induced vascular gels from banana roots – a partial chemical characterization. *Physiologia Plantarum* 66, 298–302.

Van Der Molen, G.E., Beckman, C.H. and Rodehorst, E. (1987) The ultrastructure of tylose formation in resistant banana following inoculation with *Fusarium oxysporum* f.sp. *cubense*. *Physiological and Molecular Plant Pathology* 31, 185–200.

Van Der Scheer, H.A.T., Wondergem, H.J. and Luimes, B.J. (1975) Evaluation of benomyl and thiophanate-methyl for the control of *Verticillium* wilt of strawberry in the Netherlands. *Netherlands Journal of Plant Pathology* 81, 94–101.

Van Der Spek, J. (1972) Internal carriage of *Verticillium dahliae* by seeds and its consequences. *24th International Symposium on Phytopharmacy and Phytiatry.* 9th May, 1972. Mededelingen Faculteit Landbouwwetenschappen Gent 37, 567–573.

Van Der Spek, J. (1973) Seed transmission of *Verticillium dahliae. 25th International Symposium on Phytopharmacy and Phytiatry.* 8th May, 1973. Mededelingen Faculteit Landbouwwetenschappen Gent 38, 1427–1434.

Vanderwalle, R. and Parmentier, G. (1951) Une maladie de la chicorée (*Cichorium intybus*) causée par *Verticillium dahliae* Kleb. *Parasitica* 7, 69–76.

Vanderweyen, A. (1978) La verticilliose de l'avocatier: maladie nouvelle au Maroc. *Awamia* 55, 41–55.

Van Wambeke, E., Deconinck, S., Descheemaeker, F. and Vanachter, A. (1984) Sodium azide for the control of soil-borne tomato pathogens. *Mededelingen van de Faculteit Landbouwwetenschappen Rijksuniversiteit Gent* 49, 373–381.

Van Wambeke, E., van Reet, J., Vanachter, A. and van Assche, C. (1985) Ultrasonic antifungal treatment of crop antifungal solutions. *Mededelingen van de Faculteit Landbouwwetenschappen Rijksuniversiteit Gent* 50, 1081–1086.

Van Wambeke, E., Ruelens, G., Vanachter, A. and van Assche, C. (1987) Microwave energy as a source for recycling inert crop growing substrates. *Mededelingen van de Faculteit Landbouwwetenschappen Rijksuniversiteit Gent* 3a, 851–858.

van Wyk, P.S. and Baard, S.W. (1971) Germination of conidia of *Verticillium dahliae* in soil. *Plant and Soil* 35, 601–611.

Vargas-Machuca, R., Martin, C. and Galindez, W. (1987) Recovery of *Verticillium dahliae* from weed plants in farmers' fields in Peru. *Plant Disease* 71, 756–758.

Vasil'eva, K.V. and Gladkikh, T.A. (1976) [Methods for studying the rôle of pectolytic enzymes of *Verticillium dahliae* Kleb. in the development of infectious wilt of cotton]. *Metody Issledovaniī Patologicheskikh Izmeneniī Rasteniī* 39–47.

Vasil'eva, K.V., Gladkikh, T.A., Ivanova, L.V., Davydova, M.A., Umrelina, A.B. and Shaposhnikov, G.L. (1984) [The use of ultrathin-layer isoelectric focusing in polyacrylamide gel for studying multiform polygalacturonases]. *Prikladnaya Biokhimiya i Mikrobiologiya* 20, 605–611.

Vasilieff, A.A. (1933) ['Wilt' of cultivated bast-yielding plants under Central Asian conditions]. [*Diseases and Pests of New Cultivated Textile Plants Moscow*] 22–24.

Vaughan, R.E. (1924) Diseases of field and vegetable crops in the United States in 1923. *Plant Disease Reporter Supplement* 34, 243.

Veech, J.A., Stipanovic, R.D. and Bell, A.A. (1976) Peroxidative conversion of hemigossypol to gossypol. A revised structure for isohemigossypol. *Journal of the Chemical Society Chemical Communications 1976*, 144–145.

Vega, R.R. and Le Tourneau, D. (1971) Trehalose and polyols as carbon sources for *Verticillium* spp. *Phytopathology* 61, 339–340.

Végh, I. (1987) *Champignons des Arbres et Arbustes d'Ornement. Premier Inventaire des Champignons Identifiés en France.* INRA, Paris, France.

Verhalen, L. (1967) Irrigated *Verticillium* wilt cotton-D nursery and strains test. *Oklahoma Agricultural Experiment Station Progress Report* 5, 622.

Verhalen, L.M., Brinkerhoff, L.A., Fun, K.C. and Morrison, W.G. (1971) A quantitative genetic study of *Verticillium* wilt resistance among selected lines of Upland Cotton. *Crop Science* 11, 407–412.

Verney, N.H., Moore, J.N. and Scott, D.H. (1959) Field resistance of various strawberry varieties and selections to *Verticillium*. *Plant Disease Reporter* 43, 567–569.

Vesper, S.J., Turner, J.T. and Phillips, D.V. (1983) Incidence of *Verticillium nigrescens* in soybeans. *Phytopathology* 73, 1338–1341.

Vessey, J.C. and Pegg, G.F. (1973) Autolysis and chitinase production in cultures of *Verticillium albo-atrum*. *Transactions of the British Mycological Society* 60, 133–143.

Veverka, K. and Kůdela, V. (1980) [The effectiveness of trimorphamid and benomyl against *Verticillium* wilt of lucerne]. *Ochrana Rostlin* 16, 207–212.

Veverka, K. and Kůdela, V. (1983) [The use of the mixture of trimorphamide and benomyl in the control of *Verticillium* wilt]. *Ochrana Rostlin* 19, 33–36.

Viands, D.R. (1985) Comparison of 'Maris Kabul' with 'Vertus' alfalfa for resistance to *Verticillium* wilt. *Crop Science* 25, 1096–1100.

Viands, D.R., Lowe, C.C., Vaughn, D.L. and Bergstrom, G.C. (1988) Benefits of alfalfa cultivars resistant to *Verticillium* wilt in New York. *Report of the Thirty-first North American Alfalfa Improvement Conference.* Beltsville, Maryland, June, 1988.

Viands, D.R., Lowe, C.C., Murphy, R.P. and Millar, R.L. (1990) Registration of 'Oneida VR' alfalfa. *Crop Science* 30, 955.

Viands, D.R., Hansen, J.L. and Lowe, C.C. (1995a) Registration of 'Victory' alfalfa. *Crop Science* 35(5), 1502.

Viands, D.R., Hansen, J.L. and Lowe, C.C. (1995b) Registration of 'Majestic' alfalfa. *Crop Science* 35(5), 1503.

Vigouroux, A. (1972) Étude, en conditions controllées, de efficacité de traitements par la chloropicrine et le bromure de methyle vis-a-vis de *Verticillium dahliae* Kleb. dans le sol. *Phytiatrie Phytopharmacie* 21, 195–202.

Vigouroux, A. (1975) *Verticillium dahliae*, agent d'un dépérissement de l'olivier en France. *Annals of Phytopathology* 7, 37–44.

Vigouroux, A. (1984) Verticilliose et bactériose, deux importants facteurs de dépérissement de l'abricotier. *Arboriculture Fruitière* 31, 31–35.

Vigouroux, A. and Castelani, C. (1969) La verticilliose de l'abricotier I. Premières obervations sur les symptômes et l'évolution de la maladie. Hypothèses sur quelques facteurs de variation. *Annales de Phytopathologie* 1, 427–448.

Villalobos, N. (1972) Estudio de la actividad amilasica en varias especies del genero *Verticillium*. *Revista Española de Fisiología* 28, 273–279.

Vishnevskaya, N.A., Strunnikova, O.K. and Mutomtsev, G.S. (1990) Life cyle of *Verticillium dahliae* in soil. In: *Proceedings of the Fifth International* Verticillum *Symposium*. Leningrad, p. 40.

Visser, S. (1980) *Verticillium albo-atrum* as the cause of a wilt disease of cucumber in the Republic of South Africa. *Phytophylactica* 12, 83–84.

Visser, S. and Hattingh, M.J. (1980) Criteria for quantal host response of tomato plants to infection by *Verticillium dahliae*. *Plant Disease* 64, 207–208.

Visser, S. and Kotze, J.M. (1974) *Verticillium* wilt of strawberries in the Transvaal. *Phytophylactica* 6, 63–64.

Visser, S. and Kotzé, J.M. (1975) Reactions of various crops to inoculation with *Verticillium dahliae* Kleb. isolated from wilting strawberry plants. *Phytophylactica* 7, 25–30.

Vitanov, M., Iliev, I. and Ilieva, E. (1972) [*Verticillium* wilt of almond trees in the Burgas region caused by *Verticillium dahliae*]. *Gradinarska i Lozarska Nuaka* 9(8), 33–39.

Vlasova, E.A. and Pugacheva, T.I. (1984) [Analysis of the gene bank of *Lycopersicon* Tourn. for wilt resistance]. *Sbornik Nauchnykh Trudov po Prikladnoi Botanike, Genetike i Selektsii* 88, 74–79.

Vlasova, E.A., Glushchenko, E.Ya. and Garan'ko, I.B. (1979) [Screening a tomato collection for disease resistance in the field and under cover]. *Trudy po Prikladnoi Botanike, Genetike i Selektsii* 64, 28–35.

Vlasova, T.A. (1986) [Pathological changes in isolated cotton root cells under inoculation with *Verticillium dahliae* Kleb.]. Микология и Фитопатология [*Mycology and Phytopathology*] 20, 127–131.

Vlasova, T.A. (1993) The effect of *Verticillium dahliae* Kleb. inoculation on peroxidase and phenol oxidase activity and total phenol content in isolated cotton roots. *Moscow University Biological Sciences Bulletin* 48(1), 44–50.

Vloutoglu, I., Paplomatas, E.J. and Lampopoulos, C.J. (1997) Differences in host reaction of *Verticillium dahliae* isolates from tomato, cotton and watermelon. *Proceedings of the Seventh International* Verticillium *Symposium*. Cape Sounion, Athens, Greece, p. 39.

Voitenok, F.V. (1958) [Variation in immunity of cotton to *Verticillium* wilt according to varieties]. *Khlopkovodstvo* 8, 32–34.

Voitenok, F.V. (1960) [The characteristics of breeding of cotton plant for wilt resistance]. *Khlopkovodstvo* 10, 35–37.

Volin, R.B., Augustine, J.J., Bryan, H.H., Burgis, D.S., Crill, P., Stroebel, J.W. and John, C.A. (1977) Florida 1011 tomato breeding line. *HortScience* 12, 508–509.

Volossiouk, T., Robb, E.J. and Nazar, R.N. (1995) Direct DNA extraction for PCR-mediated assays of soil organisms. *Applied and Environmental Microbiology* 61(11), 3972–3976.

von Mende, N. and McNamara, D.G. (1995) Biology of the hop cyst nematode *Heterodera humuli*. II. Host–parasite relationship of the nematode and its interaction with *Verticillium albo-atrum*. *Annals of Applied Biology* 126(3), 517–526.

von Tiedemann, A., Wolf, G. and Wilbert, S. (1983) Eine enfache Methode zur gezielten Isolierung von gefässbesiedelnden Mikroorganismen aus holzigen Pflanzen. *Phytopathologische Zeitschrift* 107, 87–91.

Voth, V., Munnecke, D.E., Paulus, A.O., Kolbergen, M.J. and Bringhurst, R.S. (1973) Effect of tarp thickness and dosage on response of California strawberries to fumigation. *California Agriculture* 27(12), 14.

Vrain, T.C. (1987) Effect of *Ditylenchus dipsaci* and *Pratylenchus penetrans* on *Verticillium* wilt of alfalfa. *Journal of Nematology* 19, 379–383.

Vulkova, Z.V. and Sotirova, V.G. (1993) Study of the three-genome hybrid *Lycopersicon esculentum* Mill.–*L. chilense* Dun.–*L. peruvianum* var. 'humifusum' Mill. and its use as a source for resistance. *Theoretical and Applied Genetics* 87(3), 337–342.

Waddle, B.A. and Fulton, N.D. (1956) Results of 1955 cotton *Verticillium* wilt studies at Oceola, Arkansas. *Proceedings, 16th Cotton Disease Council 1956*, p. 9.

Wadi, J.A. (1985) Control of *Verticillium dahliae* by coating potato seed pieces with antagonistic bacteria. In: Parker, C.A., Rovira, A.D., Moore, K.G., Wong, P.T. and Kollmorgen, J.F. (eds) *Ecology and Management of Soil-borne Plant Pathogens*. American Phytopathological Society, St Paul, Minnesota, pp. 134–136.

Wadi, J.A. (1999) Effect of soil solarization on some soil micro-organisms and tomato growth. *Egyptian Journal of Horticulture* 26(2), 167–176.

Waggoner, P.E. and Dimond, A.E. (1954) Reduction in water flow by mycelium in vessels. *American Journal of Botany* 41, 637–640.

Waggoner, P.E. and Dimond, A.E. (1956) Polyphenol oxidases and substrates in potato and tomato stems. *Phytopathology* 46, 495–497.

Wain, R.L. and Carter, G.A. (1977) Historical aspects. In: Marsh, R.W. (ed.) *Systemic Fungicides*. Wiley, New York, pp. 6–31.

Wakatabe, D., Nagao, H., Arai, H., Shiraishi, T., Koike, M. and Iijima, T. (1997) Vegetative compatibility groups of Japanese isolates of *Verticillium dahliae*. *Mycoscience* 38(1), 17–23.

Waker, H.J. and Onken, A.B. (1969) Fertilizing irrigated cotton, southern high plains of Texas. *Texas Agricultural Experimental Station Publication* MP 913.

Walker, J.C. (1971) Fusarium *Wilt of Tomato*. Monograph 6. American Phytopathological Society, St Paul, Minnesota.

Walker, R.F. and Thorneberry, G.O. (1971) Lipid content of *Verticillium albo-atrum*. *Phytochemistry* 10, 2979–2982.

Walter, J.M. (1967a) Breeding for combined resistance to diseases in tomato. *Annual Report, Florida Agricultural Experimental Station*, p. 145.

Walter, J.M. (1967b) Hereditary resistance to disease in tomato. *Annual Review of Phytopathology* 5, 131–162.

Wang, C.X., Wang, D. and Zhou, Q. (1997) Biological control of *Verticillium dahliae*. *Proceedings of the Seventh International* Verticillium *Symposium*. Cape Sounion, Athens, Greece, p. 95.

Wang, L.-M. and Shi, L.-Y. (1999) [Identification of defoliating strains of *Verticillium dahliae* from cotton in north China]. *Acta Phytopathologica Sinica* 29(2), 181–189.

Wang, M.C. and Bartnicki-Garcia, S. (1970) Structure and composition of walls of the yeast from of *Verticillium albo-atrum*. *Journal of General Microbiology* 64, 41–54.
Wang, M.C. and Keen, N.T. (1970) Purification and characterization of endopolygalacturonase from *Verticillium albo-atrum*. *Archives of Biochemistry and Biophysics* 141, 749–757.
Wang, M.C., Erwin, D.C., Sims, J.J., Keen, N.T. and Borum, D.E. (1971) Translocation of 14C-labelled thiabendazole in cotton plants. *Pesticide Biochemistry and Physiology* 1, 188–195.
Wang, X.-F., Zhang, G.-Y., Li, R.-Q. and Ma, Z.-Y. (1999) [Study on the inheritance of resistance of *Gossypium hirsutum* to *Verticillium* wilt]. *Journal of Hebei Agricultural University* 22(2), 10–14.
Wanjura, D.F. and Barker, G.L. (1987) Cotton yield decline analysis for the Southern Great Plains. In: Brown, J.M. and Nelson, T.C. (eds) *Proceedings of the Beltwide Cotton Production Research Conference*. National Cotton Council of America, Memphis, Tennessee, pp. 8–10.
Watanabe, T., Uzawa, M. and Sakai, R. (1973) A new disease of Chinese cabbage caused by *Verticillium albo-atrum* and some factors related to the incidence of the disease. *Annals of the Phytopathological Society of Japan* 39, 344–349.
Waterman, A.M. (1941) Diseases of shade and ornamental trees: annotated list of specimens received in 1940 at the New Haven Office, Division of Forest Pathology. *Plant Disease Reporter* 25, 181–182.
Waterman, A.M. (1956) *Verticillium* wilt of yellow poplar. *Plant Disease Reporter* 40, 349–350.
Webb, H.M., Gafoor, A. and Heale, J.B. (1972) Protein and enzyme patterns in strains of *Verticillium*. *Transactions of the British Mycological Society* 59, 393–402.
Webb, R.E. and Buck, R.W. (1955) *The National Potato Breeding Programme*, Vol. 216. USDA.
Webb, R.E. and Hougas, R.W. (1959) Preliminary evaluation of *Solanum* species and species hybrids for resistance to disease. *Plant Disease Reporter* 43, 144–151.
Wei, H. and Shang, H. (1998) An aetiological study on *Verticillium* wilt of *Astragalus adsurgens*. III Pathogenicity of isolates from different hosts to milk vetch and alfalfa. *Acta Prataculturae Sinica* 7(3), 41–45.
Wei, X.-T., Ma, B., Cui, L.-E. and Zhang, L.-B. (1999) [Chemical control of *Fusarium* and *Verticillium* wilts on cotton]. *Journal of Henan Agricultural Sciences* 4, 27–28.
Weir, B.L., DeVay, J.E., Stapleton, J.J., Wakeman, J. and Garber, D. (1987) Effect of solarisation on potassium disease complex of cotton. *Proceedings of the Beltwide Cotton Production Research Conference*. National Cotton Council, Memphis, Tennessee, pp. 51–53.
Weir, B.L., DeVay, J.E., Garber, R.H., Stapleton, J.J., Felix, R. and Wakeman, R.J. (1988) *Verticillium* wilt-bronzing leaf syndrome potassium deficiency complex in cotton. *California Agriculture*.
Weir, W.L., Garber, R.H., Stapleton, J.J., Felix-Gastelum, R., Wakeman, R.J. and DeVay, J.E. (1989) Control of potassium deficiency syndrome in cotton by soil solarization. *California Agriculture* 43(3), 26–28.
Welch, K.E. and McCain, A.H. (1977) *Chrysanthemum* varieties resistant to rust and *Verticillium* wilt. *Flower and Nursery Report* 1977 (Winter), 4–6.
Welvaert, W. and Poppe, J. (1986) Influence of plastic mulching and disinfection on the fungal flora of soils in Belgium. *EPPO Bulletin* 16, 311–316.

Westerdijk, J. (1917, published 1918) Een Verticilliumziekte der suikerbieten. *Mededelingen uit het Phytopathologisch Laboratorium 'Willie Commelin Scholten'* 8–10.

Westerdijk, J. and van Luijk, A. (1924) Über einige Gefässkrankheiten. *Mededelingen uit het Phytopathologisch Laboratorium 'Willie Commelin Scholten'* 8, 48–50.

Wheeler, M.H., Tolmsoff, W.J. and Meola, S. (1976) Ultrastructure of melanin formation in *Verticillium dahliae* with (+)-scytalone as a biosynthetic intermediate. *Canadian Journal of Microbiology* 22, 702–711.

Wheeler, M.H., Tolmsoff, W.J., Bell, A.A. and Mollenhauer, H.H. (1978) Ultrastructural and chemical distinction of melanins formed by *Verticillium dahliae* from (+)-scytalone; 1,8 dihydroxynaphthalene; catechol; and L-3,4-dihydroxyphenylalanine. *Canadian Journal of Microbiology* 24, 289–297.

Wheeler, T.A. and Gannaway, J.R. (1998) Effect of cotton pathogens on disease symptoms and yield of cotton varieties in large plot field trials. In: Dugger, P. and Richter, D.J. (eds) *Proceedings of the Beltwide Cotton Conferences*, San Diego, California, 5–9 January 1998, Vol. 1, pp. 165–168.

Wheeler, T.A. and Riedel, R.M. (1994) Interactions among *Pratylenchus penetrans, P. scribneri*, and *Verticillium dahliae* in the potato early dying disease complex. *Journal of Nematology* 26(2), 228–234.

Wheeler, T.A. and Rowe, R.C. (1995) Influence of soil characteristics and assay techniques on quantification of *Verticillium dahliae* in Ohio soils. *Plant Disease* 79(1), 29–34.

Wheeler, T.A., Madden, L.V., Riedel, R.M. and Rowe, R.C. (1994a) Distribution and yield-loss relations of *Verticillium dahliae, Pratylenchus penetrans, P. scribneri, P. crenatus* and *Meloidogyne hapla* in commercial potato fields. *Phytopathology* 84(8), 843–852.

Wheeler, T.A., Rowe, R.C., Riedel, R.M. and Madden, L.V. (1994b) Influence of cultivar resistance to *Verticillium* spp. on potato early dying. *American Potato Journal* 71(1), 39–57.

White, C. and Gadd, G.M. (1983) Effect of glucosamine on morphology of *Verticillium albo-atrum. Transactions of the British Mycological Society* 80, 533–536.

Whitney, P.J., Vaughan, J.G. and Heale, J.B. (1968) A disc electrophoretic study of the proteins of *Verticillium albo-atrum, V. dahliae* and *Fusarium oxysporum* with reference to their taxonomy. *Journal of Experimental Botany* 19, 415–426.

Whitney, P.J., Chapman, J.M. and Heale, J.B. (1969) Carboxymethylcellulase production by *Verticillium albo-atrum. Journal of General Microbiology* 56, 215–225.

Wickens, G.M. (1948) [Report from Barbeton, South Africa]. *Progress Reports from Experiment Stations, Season 1946–47.*

Wickens, G.M. (1949) *Progress Reports from Experiment Stations, Season 1947–48.*

Wiese, M.V. (1968) PhD thesis. University of California, Davis.

Wiese, M.V. and DeVay, J.E. (1969) Alterations in indolacetic acid, abscisic acid and ethylene in cotton infected with defoliating and nondefoliating isolates of *Verticillium albo-atrum. Proceedings of the Beltwide Cotton Production Research Conference*, New Orleans, Louisiana, p. 36.

Wiese, M.V. and DeVay, J.E. (1970) Growth regulator changes in cotton associated with defoliation caused by *Verticillium albo-atrum. Plant Physiology* 45, 304–309.

Wiese, M.V., DeVay, J.E. and Ravenscroft, A.V. (1970) Relationship between polygalacturonase activity and cultural characteristics of *Verticillium* isolates pathogenic in cotton. *Phytopathology* 60, 641–646.

Wilanowski-Guzdziol, B. and Sackston, W.E. (1972) Effect of fluctuating temperatures on *Verticillium dahliae* and *Sclerotium bataticola*. *Proceedings of the Thirty-eighth Session of the Canadian Phytopathological Society*, University of Western Ontario, London, Ontario, June 26–28, 1972, Vol. 39, p. 45.

Wilderspin, I.F. and Heale, J.B. (1984) *Verticillium* wilt (*V. albo-atrum*) of hops. *Report of the East Malling Research Station* 1983, p. 105.

Wilderspin, I.F. and Heale, J.B. (1985) Biological control. *Report of the East Malling Research Station* 1984, p. 180.

Wiles, A.B. (1960) Evaluation of cotton strains and progenies for resistance to *Verticillium* wilt. *Plant Disease Reporter* 44, 419–422.

Wiles, A.B. (1963) Comparative reactions of certain cottons to *Fusarium* and *Verticillium* wilts. *Phytopathology* 53, 586–588.

Wiles, A.B. (1967) Zinc deficiency in cotton induced by chloropicrin-methyl bromide soil fumigation to control *Verticillium* wilt. *Phytopathology* 57, 103.

Wiles, A.B. (1971) Inheritance of *Verticillium* wilt resistance in cotton. *Cotton Disease Research in the San Joaquin Valley, California Research Report 1968–1969*, pp. 65–96.

Wiles, A.B., Sagen, J.E. and Tietz, H. (1968) Sources and methods of identification of *Verticillium* wilt resistance in cotton. *Proceedings of the Beltwide Cotton Production Research Conference*. Hot Springs, pp. 165–167.

Wiley, G.L., Bauman, T.T. and Williams, J.L. (1985) The influence of disease on velvetleaf in competition with beans. *Proceedings, North Central Weed Control Conference, 1985*. St Louis, Missouri, Vol. 40, p. 9.

Wilhelm, S. (1948) The effect of temperature on the taxonomic characters of *Verticillium albo-atrum* Rke et Bert. *Phytopathology* 38, 919.

Wilhelm, S. (1950a) Vertical distribution of *Verticillium albo-atrum* in soils. *Phytopathology* 40, 368–376.

Wilhelm, S. (1950b) The inoculum potential of *Verticillium albo-atrum* as affected by soil amendments. *Phytopathology* 40, 970.

Wilhelm, S. (1951a) Effect of various soil amendments on the inoculum potential of *Verticillium* wilt fungus. *Phytopathology* 41, 684–690.

Wilhelm, S. (1951b) Is *Verticillium albo-atrum* a soil invader or a soil inhabitant? *Phytopathology* 41, 944–945.

Wilhelm, S. (1954) Aerial microsclerotia of *Verticillium albo-atrum* resulting from conidial anastomosis. *Phytopathology* 44, 609–610.

Wilhelm, S. (1955a) Longevity of the *Verticillium* wilt fungus in the laboratory and field. *Phytopathology* 45, 180–181.

Wilhelm, S. (1955b) *Verticillium* wilt of the strawberry with special reference to resistance. *Phytopathology* 45, 387–391.

Wilhelm, S. (1961) Diseases of strawberry, a guide for the commercial grower. *California Agricultural Experiment Station Circular* 494, 1–26.

Wilhelm, S. (1962) The control of wilt of strawberry and of weeds by preplant soil fumigation with chloropicrin–methyl bromide mixtures. *Proceedings of the 16th International Horticultural Congress*. Brussels, Belgium. Vol. I, Book of Summaries, pp. 223–225.

Wilhelm, S. (1966) Chemical treatments and inoculum potential of soil. *Annual Review of Phytopathology* 4, 53–78.

Wilhelm, S. (1971a) Principles and practice of *Verticillium* wilt control in strawberries in California by preplant soil fumigation with chloropicrin–methyl bromide mixtures.

Proceedings of the First International Verticillium *Symposium*, Wye College, University of London, p. 34.

Wilhelm, S. (1971b) Progress report for 1968–69. Division of Agricultural Sciences, University of California, Berkeley. *Cotton Disease Research in the San Joaquin Valley, California*, pp. 4–65.

Wilhelm, S. (1975) Sources and nature of *Verticillium* wilt resistance in some major crops. In: Bruehl, G.W. (ed.) *Biology and Control of Soil-borne Plant Pathogens. Third International Symposium on factors Determining the Behaviour of Plant Pathogens in Soil.* Minneapolis, Minnesota, pp. 166–171.

Wilhelm, S. (1976) Proceedings of the Second International Symposium on *Verticillium*, University of California, Berkeley, 7–10 September 1976. *Review of Plant Pathology* 59, 4889.

Wilhelm, S. (1981) Sources and genetics of host resistance in field and fruit crops. In: Mace, M.E., Bell, A.A. and Beckman, C.H. (eds) *Fungal Wilt Diseases of Plants.* Academic Press, New York, pp. 299–376.

Wilhelm, S. and Ferguson, J. (1953) Soil fumigation against *Verticillium albo-atrum*. *Phytopathology* 43, 593–596.

Wilhelm, S. and Koch, E.C. (1956) *Verticillium* wilt controlled. *California Agriculture* 10(6), 3 and 14.

Wilhelm, S. and Paulus, A.O. (1980) How soil fumigation benefits the California strawberry industry. *Plant Disease* 64, 264–270.

Wilhelm, S. and Sagen, J.E. (1974) *The History of the Strawberry.* Agricultural Publications, University of California, Berkeley.

Wilhelm, S. and Taylor, J.B. (1965) Control of *Verticillium* wilt of olive through natural recovery and resistance. *Phytopathology* 55, 310–316.

Wilhelm, S. and Thomas, H.E. (1950) *Verticillium* wilt of bramble fruits with special reference to *Rubus ursinus* derivatives. *Phytopathology* 40, 1103–1110.

Wilhelm, S., Nyland, G. and Charles, F. (1954) *Verticillium* wilt of sweet cherry in California. *Plant Disease Reporter* 38, 438.

Wilhelm, S., Raabe, R.D. and Smalley, E.B. (1955) Some previously unrecorded hosts of *Verticillium albo-atrum* in California. *Plant Disease Reporter* 39, 693.

Wilhelm, S., Storken, R.C. and Sagen, J.E. (1961) *Verticillium* wilt of strawberry controlled by fumigation of soil with chloropicrin and chloropicrin–methyl bromide mixtures. *Phytopathology* 51, 744–748.

Wilhelm, S., Kaiser, W.J., Georgopoulos, S.G. and Opitz, K.W. (1962) *Verticillium* wilt of olives in California. *Phytopathology* 52, 32.

Wilhelm, S., Bringhurst, R.S. and Voth, V. (1965) Origins of *Rubus* cultivars resistant to *Verticillium* wilt. *Phytopathology* 55, 731–733.

Wilhelm, S., Sagen, J.E., Snyder, W.C., Gedge, A., Mathre, D., Garber, R.H. and Hall, D. (1966) Cultural control of *Verticillium* in cotton – a three point approach. *California Agriculture* 20(4), 2–4.

Wilhelm, S., George, A. and Pendery, W. (1967) Zinc deficiency in cotton induced by chloropicrin–methyl bromide soil fumigation to control *Verticillium* wilt. *Phytopathology* 57, 103.

Wilhelm, S., Sagen, J.E. and Tietz, H. (1970) Seabrook (*Gossypium barbadense*) × Rex (*Gossypium hirsutum*) crosses give *Verticillium* wilt resistant, Upland-type, all fertile offspring. *Proceedings of the Beltwide Cotton Production Research Conference.* Houston. pp. 70–76.

Wilhelm, S., Sagen, J.E. and Tietz, H. (1971) *Cotton Disease Research in the San Joaquin Valley California Research Report for 1968–1969*. Division of Agricultural Sciences, University of California, Berkeley, pp. 56–96.

Wilhelm, S., Sagen, J.E. and Tietz, H. (1972a) Resistance to *Verticillium* wilt transferred from *Gossypium barbadense* to Upland cotton. *Phytopathology* 62, 798–799.

Wilhelm, S., Storkan, R.C., Sagen, J.E., George, A.G. and Tietz, H. (1972b) Cotton *Verticillium* wilt control with soil fumigation. *California Agriculture* 26(10), 4–6.

Wilhelm, S., Sagen, J.E. and Tietz, H. (1974a) Resistance to *Verticillium* wilt in cotton: sources, techniques of identification, inheritance trends and the resistance potential of multiline cultivars. *Phytopathology* 64, 924–931.

Wilhelm, S., Sagen, J.E. and Tietz, H. (1974b) *Gossypium hirsutum* subsp. *mexicanum* var. *nervosum*, Leningrad strain – a source of resistance to *Verticillium* wilt. *Phytopathology* 64, 931–939.

Wilhelm, S., Storkan, R.S. and Wilhelm, J.M. (1974c) Preplant soil fumigation with methyl bromide–chloropicrin mixtures for control of soil-borne diseases of strawberry – a summary of 15 years of development. *Agriculture and Environment* 1, 227–236.

Wilhelm, S., Sagen, J.E. and Tietz, H. (1976) Optimizing *Verticillium* wilt resistance in Upland Cotton with short-branching, arching phenotypes and in-row crowding. *Proceedings of the Second International* Verticillium *Symposium*. University of California, Berkeley, p. 43.

Wilhelm, S., Sagen, J.E. and Tietz, H. (1978) *Verticillium* wilt control in cotton through resistance breeding and dense plant cultivation. *Phytopathology News* 12, 163.

Wilhelm, S., Sagen, J.E. and Tietz, H. (1985) Phenotype modification in cotton for control of *Verticillium* wilt through dense plant population culture. *Plant Disease* 69, 283–288.

Williams, H. (1975) The effect on *Verticillium* wilt of strawberries of plant spacing, fertilizer application and defoliation. *Plant Pathology* 24, 45–47.

Williams, M.A.J., Brady, B.L., Bridge, P.D. and Paterson, R.R.M. (1992) Biochemical and physiological tests as aids to identification of *Verticillium* section *Nigrescentia*. *Mycopathologia* 119(2), 101–114.

Williams, P.H. (1962) Mycology. *Verticillium* wilt. *Report of the Glasshouse Research Institute* 1961, pp. 68–69.

Williams, P.H. (1979) Vegetable crop protection in the Peoples' Republic of China. *Annual Review of Phytopathology* 17, 311–324.

Wilson, K.S. and Porter, C.L. (1958) The pathogenicity of *Verticillium albo-atrum* as affected by muck soil antagonists. *Applied Microbiology* 6, 155–159.

Wiltshire, S.P. (1920) A disease of michaelmas daisies. *Annual Report, University of Bristol Agricultural and Horticultural Research Station*. pp. 84–85.

Wittmann, O. (1971) Boden und *Verticillium*-Welke des Hopfens. *Bayerisches Landwirtschaftliches Jahrbuch* 48, 206–236.

Wittmann, O. (1972) Welchen Einfluss hat der Boden auf den Befall mit Hopfenwelke? *Hopfen-Rundschau* 23(3), 43–45.

Wollenweber, H.W. (1913) Pilzparasitäre Welkekrankheiten der Kulturpflanzen. *Bericht der Deutschen Botanischen Gesellschaft* 31, 17–34.

Wollenweber, H.W. (1922) Tracheomykosen und andere Welkenkrankheiten nebst Aussichten ihren Abwehr. *Angewandte Botanik* 4, 1–4.

Wollenweber, H.W. (1929) Die Wirtelpilz-Welkekrankheit (Verticilliose) von Ulme, Ahorn und Linde usw. *Arbeiten aus der Biologischen Bundesanstalt für Land und Forstwirtschaft* 17, 273–299.

Wood, R.K.S. (1961) *Verticillium* wilt of tomatoes – the role of pectic and cellulolytic enzymes. *Annals of Applied Biology* 49, 120–139.

Wood, R.K.S. (1967) *Physiological Plant Pathology*. Blackwell Scientific Publications, Oxford.

Woodward, S. and Pegg, G.F. (1986) Rishitin accumulation elicited in resistant and susceptible isolines of tomato by mycelial extracts and filtrates from cultures of *Verticillium albo-atrum*. *Physiological and Molecular Plant Pathology* 29, 337–347.

Woodward, W.T.W., Miller, J.W., Eckman, L.D., Edmunds, L.K., Hoard, G.E., Hartman, B.J., Nash, L.M. and Poynor, E.F. (1987) Registration of '5432' alfalfa. *Crop Science* 27, 1305–1306.

Woodward, W.T.W., Hoard, G.E., Jessen, D.L., Miller, D.J., Satterlee, L.D. and Smith, M.A. (1990) Registration of '5364' alfalfa. *Crop Science* 30, 230.

Woodward, W.T.W., Satterlee, L.D., Hoard, G.E., Jessen, D.L., Poynor, E.F., Miller, D.J. and Smith, M.A. (1993) Registration of '5715' alfalfa. *Crop Science* 33(4), 875.

Woolliams, G.E. (1966) Host range and symptomatology of *Verticillium dahliae* in economic, weed and native plants in interior British Columbia. *Canadian Journal of Plant Science* 46, 661–669.

Worf, G.L., Spear, R.N. and Heimann, M.F. (1994) *Verticillium*-induced scorch and chlorosis in ash. *Journal of Environmental Horticulture* 12(3), 124–130.

Workman, M., Harrison, M.D. and Livingstone, C.H. (1977) Effects of soil fumigation with telone C on soil and plant nitrogen, *Verticillium* wilt and yield of Norland and Norgold potatoes. *American Potato Journal* 54, 103–113.

Wright, D.S.C. (1968) *Verticillium* wilt of tobacco. V. Varietal screening for resistance to *Verticillium dahliae* Kleb., and mode of inheritance of resistance. *New Zealand Journal of Agricultural Research* 11, 655–664.

Wright, D.S.C. (1969) *Verticillium* wilt of tobacco. IX. The relationship between leaf symptoms and levels of infection in the organs of mature tobacco plants. *New Zealand Journal of Agricultural Research* 12, 228–239.

Wright, D.S.C. (1973) Waimea, a new flue-cured tobacco variety. *New Zealand Journal of Agricultural Research* 16, 69.

Wright, D.S.C. and Abrahamson, J.P. (1970) Electron microscopic study of *Verticillium dahliae* Kleb. as a pathogen of tobacco (*Nicotiana tabacum* L.). *New Zealand Journal of Botany* 8, 326–343.

Wright, D.S.C. and Sackston, W.E. (1973) *Verticillium* wilt: a potential disease of tobacco in Canada. *Canadian Journal of Plant Science* 53, 391–393.

Wu, G., Shortt, B.J., Lawrence, E.B., Leon, J., Fitzsimmons, K.C., Levine, E.B., Raskin, I. and Shah, D.M. (1997) Activation of host defense mechanisms by elevated production of H_2O_2 in transgenic plants. *Plant Physiology* 115, 427–435.

Wu, L.-C. and Stahmann, M.A. (1975) Chromatographic estimation of fungal mass in plant materials. *Phytopathology* 65, 1032–1034.

Wu, Q.X. (1990) [A new low-phenol cotton variety, Xinluzao 3]. *Crop Genetic Resources* 2, 46.

Wu, X., Li, F., Zheng, C., Wang, Y. and Zhu, Y. (1995) Pathogenic differentiation of *Verticillium dahliae* in Shandong. *Plant Protection* 21(6), 2–5.

Wyllie, T.D. and DeVay, J.E. (1970a) Growth characteristics of several isolates of *Verticillium albo-atrum* and *Verticillium nigrescens* from cotton. *Phytopathology* 60, 907–910.

Wyllie, T.D. and DeVay, J.E. (1970b) Immunological comparison of isolates of *Verticillium albo-atrum* and *V. nigrescens* pathogenic to cotton. *Phytopathology* 60, 1682–1686.

Xia, Z.J. and Achar, P.N. (1999) Biochemical changes in cotton infected with *Verticillium dahliae*. *African Plant Protection* 5(1), 59–63.

Xia, Z.J., Gu, B.K., Wu, A.M., Li, J.Y., Chen, Y.L. and Chen, X.B. (1994) Studies on the biochemical components in cotton related to the resistance to *Verticillium dahliae*. *Acta Phytophylacica Sinica* 21(4), 305–310.

Xia, Z.J., Gu, B.K. and Wu, A.M. (1996) Studies on induced resistance of cotton plants against *Verticillium dahliae* by endophytic and rhizosphere bacteria. *Chinese Journal of Biological Control* 12, 7–10.

Xia, Z.J., Gu, B.K., Wu, A.M. and Fu, Z. (1997) Isoenzymatic activities involved in induced resistance of cotton to *Verticillium dahliae* by endophytic bacteria. *Jiangsu Journal of Agricultural Sciences* 13, 99–101.

Xia, Z.J., Achar, P.N. and Gu, B.K. (1998) Vegetative compatibility groupings of *Verticillium dahliae* from cotton in mainland China. *European Journal of Plant Pathology* 104, 871–876.

Xiao, C.L. and Subbarao, K.V. (1998) Relationships between *Verticillium dahliae* inoculum density and wilt incidence, severity, and growth of cauliflower. *Phytopathology* 88(10), 1108–1115.

Xiao, C.L., Hao, J.J. and Subbarao, K.V. (1997) Spatial patterns of microsclerotia of *Verticillium dahliae* in soil and *Verticillium* wilt of cauliflower. *Phytopathology* 87(3), 325–331.

Xiao, C.L., Subbarao, K.V., Schulbach, K.F. and Koike, S.T. (1998) Effects of crop rotation and irrigation on *Verticillium dahliae* microsclerotia in soil and wilt in cauliflower. *Phytopathology* 88(10), 1046–1055.

Xiao, Y.H. and Lin, B.Q. (1995) Identification of aubergine germplasm resources with resistance to *Verticillium* wilt. *China Vegetables* 32–33.

Xiaoxuan, S., Heqin, Z. and Jincheng, G. (1997) Study on pathogenicity differentiation for *Verticillium dahliae* Kleb. Anyang strains on cotton. *Scientia Agricultura Sinica* 30, 13–18.

Xing, J.-S. and Chin, C.-K. (2000) Modification of fatty acids in eggplant affects its resistance to *Verticillium dahliae*. *Physiological and Molecular Plant Pathology* 56(5), 217–225.

Xu, D.-Q. and Chen, Y.-X. (2000) Indirect ELISA for detection of *Verticillium* wilt of cotton and its application. *Journal of Nanjing Agricultural University* 23(1), 105–108.

Yakutkin, V.I. and Tavolzhanskri, N.P. (1999) Disease control in sunflower crops in Russia. *Research Progress in Plant Protection and Plant Nutrition* 214–218.

Yamada, M.M., Bartley, B.G.D., Lopes, U.V. and Pinto, L.R.M. (1996) Heranca do fator compatibilidade em *Theobroma cacao* L. II. Relacoes fenotipicas em genotipos adicionais do grupo Parinari (PA). *Agrotropica* 8(2), 51–52.

Yamaguchi, H., Uchiyama, K., Sawai, A., Gau, M., Ueda, S., Maki, Y., Matsu-Ura, M., Suginobu, K. and Sato, R. (1995) Breeding of 'Makiwakaba' alfalfa and its characteristics. *Research Bulletin of the Hokkaido National Agricultural Experiment Station*, pp. 1–15.

Yang, D., Plante, F., Bernier, L., Piche, Y., Dessureault, M., Laflamme, G. and Ouellete, G.B. (1993) Evaluation of a fungal antagonist, *Phaeotheca dimorphospora*, for biological control of tree diseases. *Canadian Journal of Botany* 71(3), 426–433.

Yang, J. and Harris, D.C. (1994) Inoculating methods of maple *Verticillium* wilt. *Journal of Northwest Forestry College* 9, 34–38.

Yang, S.M. and Wei, S.E. (1988) Diseases of cultivated sunflower in Liaoning Province, People's Republic of China. *Plant Disease* 72, 546.

Yarovenko, A., Mannanova, A. and Kadyrkhodzhaev, P. (1975) [Effect of rates and proportions of nitrogen to phosphorus on yield of cotton on wilt-infected fields]. *Sintez i Primenenie Novykh Khimicheskikh Preparatov Protiv Vilta Khlopchatnika* 81–82.

Yarovenko, G.I., Nuraliev, I. and Kir, I.N. (1976) [Effect of nitrogen and of combined use of fertilizers and pesticides on the increase of yield and wilt-resistance of cotton varieties]. *Trudy Uzb. Opyt. St. Lub. Kul'tur* 13, 135–141.

Yarris, L. (1981) New potato variety. *Agricultural Research, USA* 30(2), 15.

Yarris, L.C., Horner, C.E. and Melouk, H.A. (1977) Spearmint mutations resist wilt. *Agricultural Research, USA* 26(2), 5.

Yi, J.-J. and Chen, J.-H. (1998) [Progress in breeding research on resistance to *Verticillium* wilt in aubergine]. *China Vegetables* 6, 52–55.

Yingfan, C., Yongjiu, T., Honghua, H. and Biyu, L. (1994) The characteristics and genetic effects of cotton germplasm resistant to various strains of *Verticillium* spp. *China Cottons* 21(12), 10–11.

Yoshihara, T., Hagihara, Y., Nagaoka, T., Chiba, S. and Sakamura, S. (1988) Fungitoxic compounds from the roots of the eggplant stock. *Annals of the Phytopathological Society of Japan* 54, 453–459.

Young, D.A., Bagnall, R.H., de Jong, H., Murphy, A.M. and Tarn, T.R. (1988) Donna: a yellow-fleshed potato cultivar with potential in export seed markets. *American Potato Journal* 65, 509–515.

Young, D.H. and Pegg, G.F. (1981) Purification and characterization of 1,3 ß-glucan hydrolases from healthy and *Verticillium albo-atrum* infected tomato plants. *Physiological Plant Pathology* 19, 391–417.

Young, D.H. and Pegg, G.F. (1982) The action of tomato and *Verticillium albo-atrum* glycosidases on the hyphal wall of *Verticillium albo-atrum*. *Physiological Plant Pathology* 21, 411–423.

Young, R.A. (1956) Control of the early maturity disease of potatoes by soil treatment with VAPAM. *Plant Disease Reporter* 40, 781–784.

Young, R.A. and Tolmsoff, W.J. (1958) Current season and residual effects of vapam soil treatments for control of *Verticillium* wilt of potatoes. *Plant Disease Reporter* 42, 437–440.

Young, V.H., Foulton, N.D. and Waddle, B.A. (1959) Factors affecting the incidence and severity of *Verticillium* wilt disease of cotton. *Bulletin of the Arkansas Agricultural Experiment Station* 612, 1–26.

Yuan, H.-X., Li, H.-L., Li, W. and Wang, S.-Z. (1998) [Studies on the control of cotton *Verticillium* wilt by different chemical treatment]. *China Cottons* 25(9), 11–12.

Yuen, G.Y. and Raabe, R.D. (1984) Effects of small-scale aerobic composting on survival of some fungal plant pathogens. *Plant Disease* 68, 134–136.

Yui, S. (1986, published 1987) Breeding of *Verticillium* resistant Chinese cabbage. *The Breeding of Horticultural Crops. Proceedings of an International Symposium*, pp. 71–80.

Yui, S., Ashizawa, M. and Yamagishi, H. (1985) New soil born disease of Chinese cabbage caused by *Verticillium dahliae* Klebahn, and the resistance in *Brassica campestris* L. *Cruciferae Newsletter* 10, 106–107.

Yuldashev, A. (1978) [The use of *Trichoderma* sp. in combination with intermediate crops]. *Zashchita Rasteniĭ* 6, 35.

Yunusov, M. and Sultanov, A. (1978) [Systemic fungicides and resistance of cotton to wilt and root rot]. *Khlopkovodstvo* 7, 32.

Yunusov, M.R., Ibragimov, F.U., Popov, V.I. and Tkachenko, M.P. (1973) [Effectiveness of new preparations]. *Khlopkovodstvo* 4, 24–26.

Yunusov, M.R., Ibraginov, F.Yu., Utepova, R.U. and Teplyaev, R. (1979) [Effect of systemic fungicides on the technical properties of fibres and seeds]. *Khlopkovodstvo* 10, 30–31.
Yunusov, M., Tellyaev, R., Sultanov, A. and Utepova, R. (1980) [Effect of uzgen and benomyl on fibre and seed quality]. *Khlopkovodstvo* 8, 28.
Yunusov, S. *et al.* (1975) [Application of paraform against cotton wilt]. *Sintez i Primenenie Novykh Khimicheskikh Preparatov Protiv Vilta Khlopchatnika*, pp. 10–12.
Yuzbash'yan, O. (1976) [Benomyl against wilt and sucking pests]. *Khlopkovodstvo* 11, 24.
Zachos, D.G. (1963) La verticilliose de l'olivier en Grèce. *Annales de l'Institut Phytopathologique Benaki N.S.* 5, 105–107.
Zachos, D.G. and Panagopoulos, C.G. (1963) Une hadromycose de la vigne du au *Verticillium albo-atrum* Reinke et Berth. *Annales de l'Institut Phytopathologique Benaki N.S.* 5, 303–304.
Zaki, A.I., Keen, N.T. and Erwin, D.C. (1972a) Implication of vergosin and hemigossypol in the resistance of cotton to *Verticillium albo-atrum*. *Phytopathology* 62, 1402–1406.
Zaki, A.I., Keen, N.T., Sims, J.J. and Erwin, D.C. (1972b) Vergosin and hemigossypol, antifungal compounds produced in cotton plants inoculated with *Verticillium albo-atrum*. *Phytopathology* 62, 1398–1401.
Zaleski, A. (1957) Reactions of lucerne strains to *Verticillium* wilt. *Plant Pathology* 6, 137–142.
Zamani-Zadeh, H.R. and Zakii, Z. (1995) Etiology of apricot-tree decline in Shahrood. *Applied Entomology and Phytopathology* 62(1–2), 7 (En), 28–35 (Pe).
Zattler, F. (1960) Bericht über die Welke-, Nematoden- und Virusforschung im Hopfenbau in den Jahren 1958 und 1959. *Brauwissenschaft* 13, 159–161.
Zattler, F. and Chrometzka, P. (1960) Zur Biologie von *Verticillium albo-atrum* R. & B., dem Erreger der Welkekrankheit des Hopfens. *Praktische Blätter fur Pflanzenbau und Planzenschutz* 55, 17–23.
Zeise, K. (1995) Untersuchungen zur Virulenz von *Verticillium dahliae*-Isolaten unterschiedlicher Herkunft an *Brassica napus* L. *Nachrichtenblatt des Deutschen Pflanzenschutzdienstes* 47(3), 69–71.
Zeise, K. (1997) The potential of *Talaromyces flavus* (Kloekner) Stolk & Samson in controlling *Verticillium dahliae*. *Proceedings of the Seventh International* Verticillium *Symposium*. Cape Sounion, Athens, Greece, p. 61.
Zeise, K., Seidel, D. and Amelung, D. (1990) Zum Befallsverlauf von *Verticillium dahliae* Kleb. an Winterraps. *Archiv für Phytopathologie und Pflanzenschutz* 26, 205–207.
Zelle, M.A. (1932) [Sunflower diseases]. Publication 6 [*Pan-Soviet State Association for Control of Pests and Diseases in Agriculture and Sylviculture*], pp. 1–32.
Zeller, S.M. (1936) *Verticillium* wilt on cane fruits. *Bulletin of the Oregon Agricultural Experiment Station* 344, 1–25.
Zel'tser, S.S. and Malysheva, K.M. (1966) [Carbohydrate metabolites extracted from *Verticillium* in a culture medium]. *Vilt Khlopkovadstvs* 96–100.
Zentmeyer, G.A. (1949) *Verticillium* wilt of avocado. *Phytopathology* 39, 677–682.
Zhang, X., Zhang, S., Li, Q., Chen, B., Chen, J. and Yao, Y. (1981) [Identification of *Verticillium* wilt pathogen of cotton in China]. *Acta Phytopathologica Sinica* 11(3), 13–18.
Zhang, X., Mu, F., Huang, G., Mao, Z. and Shademanov, P.K. (1996) A preliminary research on biochemical breeding technique in cotton. *China Cottons* 23, 7–8.

Zhao, G.-Z., Feng, H.-W., Li, A.-G., Li, Z.-S., Sui, S.-X. and Liang, Z.-L. (1994) Improvement effect of the Upland cotton with eight wild species and two diploid cultivated species. *Acta Agriculturae Boreali-Sinica* 9(4), 44–48.

Zhao, X., Lu, J., Shang, H. and Yu, Z. (1997) Determination of esterase isoenzymes from isolates of *Verticillium dahliae*. *Journal of Hebei Agricultural University* 20, 37–45.

Zhdarkina, Z., Sidorova, S. and Stepanova, M. (1981) [Effect of crop rotation on cotton yield and wilt infection]. *Khlopkovodstvo* 7, 20–21.

Zhivopistseva, I.V. and Voronkov, L.A. (1974) [Content of polyphosphates in forms of *Verticillium dahliae* Kleb. differing in pathogenicity]. *Biologicheskie Nauki* 17(12), 70–73.

Zhou, B.-L., Gao, Y.-X., Lin, G.-R. and Fu, Y.-W. (1998a) [Relationship between disease resistance and electrolytic leakage, proline content and PAL activity in grafted eggplant]. *Acta Horticulturae Sinica* 25, 300–302.

Zhou, B.-L., Gao, Y.-X., Lin, G.-R. and Fu, Y.-W. (1998b) [The effectiveness of graftage in resisting to *Verticillium* wilt and its effect on physiological and biochemical characteristics in eggplant]. *Advances in Horticulture* 2, 429–433.

Zhou, Z.-H., Guo, W.-Z., Pan, J.-J. and Zhang, T.-Z. (1999) [Analyses on the genetic variation of cotton *Verticillium dahliae* isolate populations in China]. *Scientia Agricultura Sinica* 32(2), 60–65.

Zhou, Z.-H., Zhang, T.-Z., Pan, J.-J., Chen, Z.-X. and Jiang, W.-M. (2000) [Pathogenicity differentiation of *Verticillium dahliae* on cotton varieties]. *Scientia Agricultura Sinica* 33(2), 51–57.

Zhu, H., Song, X., Xing, J. and Guo, J. (1995) Relationship between anti-oxidative system of cotton and its resistance to *Verticillium* wilt stress. *Acta Agriculturae Boreali-Sinica* 10(Supplement), 130–133.

Zhu, Y.-Y., Wang, Y.-Y. and Lyon, B.R. (1999a) [PCR detection of *Verticillium dahliae* from diseased plants]. *Acta Phytopathologica Sinica* 29(3), 250–255.

Zhu, Y.-Y., Wang, Y.-Y., Multani, D.S. and Lyon, B.R. (1999b) The relationships between DNA polymorphism and geographical origin of Australian isolates of *Verticillium dahliae* from cotton plants. *Mycosystema* 18(4), 366–373.

Zielinski, D. and Sadowski, C. (1994) Porazenie wybranych odmian i rodow rzepaku ozimego przez *Verticillium dahliae* Kleb. w warunkach polowych. *Rosliny Oleiste* Oilseed crops. *XVIth Polish Research Conference.* 19–20 April 1994. 15(2), 93–96.

Zilbertsem, Y., Chet, I. and Henis, Y. (1983) Influence of microsclerotia source of *Verticillium dahliae* on inoculum quality. *Transactions of the British Mycological Society* 81, 613–617.

Zimm, L.A. (1918) A wilt disease of maples. *Phytopathology* 8, 80–81.

Zimmer, D.E. (1962) *Verticillium* wilt of safflower in the United States – a potential problem. *Plant Disease Reporter* 45, 665–666.

Zink, F.W. (1978) U.C. SR-91 Bush, U.C. Top Mark Bush, U.C. Perlita Bush muskmelon breeding lines. *HortScience* 13, 486.

Zink, F.W. (1979) U.C. Honeyloupe muskmelon. *HortScience* 14, 549.

Zink, F.W. (1980) U.C. Crensaw Bush breeding line. *HortScience* 15, 318–319.

Zink, F.W. and Gubler, W.D. (1986) U.C.PMR 45 *Fusarium* wilt-resistant muskmelon breeding line. *HortScience* 21, 333.

Zink, F.W. and Gubler, W.D. (1987) U.C. PMR 45 and U.C. Top Mark *Fusarium* wilt-resistant (Fom-3) muskmelon breeding lines. *HortScience* 22, 172.

Zink, R.T. and Secor, G.A. (1982) Interactions of fungal wilt pathogens and potato blackleg. *Plant Disease* 66, 1053–1056.
Zinkernagel, V. (1971) Control of *Verticillium dahliae* in strawberries with Di-Trapex. *Proceedings of the First International* Verticillium *Symposium.* Wye College, University of London, p. 37.
Zlotina, G.D. and Urakov, B. (1977) [*Verticillium* wilt of citrus trees]. *Sadovodstvo* 1977(3), 27.
Zunnunov, A.A. (1964) [Selection as a method for increasing cotton resistance to wilt]. *Mater. Vses. Simp po Bor'be s Viltom Khlopchatnika, Tashkent*, pp. 109–111.

Note Added in Proof

The Proceedings of the 8th International *Verticillium* Symposium held in November 2001, were published too late to be included in the monograph. Seventy-one abstracts are included in: 8th International *Verticillium* Symposium, Cordoba, Spain, 5–8 November, 2001, 125 pp.

Index